ARLIS

Alaska Resources
Library & Information Services
Anchorage Alaska

Second Edition

Ecological
Risk Assessment

Second Edition

Ecological Risk Assessment

Editor and Principal Author

Glenn W. Suter II

Contributing Authors

Lawrence W. Barnthouse

Steven M. Bartell

Susan M. Cormier

Donald Mackay

Neil Mackay

Susan B. Norton

 CRC Press
Taylor & Francis Group
Boca Raton London New York

CRC Press is an imprint of the
Taylor & Francis Group, an informa business

CRC Press
Taylor & Francis Group
6000 Broken Sound Parkway NW, Suite 300
Boca Raton, FL 33487-2742

International Standard Book Number-10: 1-56670-634-3 (Hardcover)
International Standard Book Number-13: 978-1-56670-634-6 (Hardcover)

Library of Congress Cataloging-in-Publication Data

Ecological risk assessment / edited by Glenn W. Suter II. -- 2nd ed.
    p. cm.
  Includes bibliographical references and index.
  ISBN 1-56670-634-3
  1. Ecological risk assessment. I. Suter, Glenn W.

QH541.15.R57E25 2006
333.95'14--dc22
                                               2006049394

Visit the Taylor & Francis Web site at
http://www.taylorandfrancis.com

and the CRC Press Web site at
http://www.crcpress.com

Dedication

To my parents,
Glenn W. Suter and Kathleen T. Suter

We are products of our heredity and environment,
and parents provide all of one and most of the other.

Preface to the Second Edition

The primary purpose of preparing this edition is to provide an update. In the 14 years since the first edition was published, ecological risk assessment has gone from being a marginal activity to being a relatively mature practice. There are now standard frameworks and guidance documents in the United States and several other countries. Ecological risk assessment is applied to the regulation of chemicals, the remediation of contaminated sites, the importation of exotic organisms, the management of watersheds, and other environmental management problems. Courses in ecological risk assessment have been taught at several universities. As a result, there is a much larger literature to draw on, including many case studies. This is reflected both in the citation of ecological risk assessments published in the open literature and in the use of more figures drawn from real assessments. Hence, the reader will notice a greater diversity in the graphical style, resulting from the many sources from which figures have been drawn so as to give a flavor of the diverse practice of ecological risk assessment.

The second edition also provides an opportunity for a new organization of the material that is more logically consistent. In particular, whereas the first edition had separate chapters for types of ecological risk assessments (i.e., predictive, retrospective, regional, surveillance, and exotic organisms), this edition presents a unitary process of ecological risk assessment that is applicable to various problems, scales, and mandates. All risk assessments are about the future consequences of decisions. Those that were described in the first edition as retrospective, following EPA terminology, are simply risk assessments that must begin with an analysis of the current consequences of past actions in order to predict future consequences (Chapter 1).

Since 1992, ecological risk assessment has become sufficiently important to acquire critics and opponents. Some criticisms deal with aspects of the technical practice. Ecological risk assessment is often criticized for being based on inadequate data and models, for not addressing large-scale spatial dynamics, and for using conservatism to compensate for those inadequacies (DeMott et al. 2004; Landis 2005; Tannenbaum 2005a). Other critics are opposed to ecological risk assessment per se (Pagel and O'Brien 1996; Lackey 1997; O'Brien 2000; Bella 2002). These criticisms arise from a misperception of the nature and purpose of risk assessment. In particular, risk assessment is technical support for decision making under uncertainty, but the critics hold risk assessment responsible for the decision itself. If decision makers listen to fishermen, loggers, chemical manufacturers, or utility companies more than to environmental advocates, critics say it is the fault of risk assessment. If risk assessments are limited by regulatory context to considering only one alternative, they say that also is the fault of risk assessment. If decisions are based on balancing of costs and benefits, it is again the fault of risk assessment. If the best available science does not address all of the important complexities of the system, they say that risk assessors who use that science are to blame. Similarly, risk assessors are blamed when holistic properties, endocrine disruptors, regional properties, or other favorite concerns are not addressed. Some of this criticism arises from an opposition to technology, science, and even rationality, but more generally it is based on anger that the environment is not being adequately protected. One partial solution is to avoid the phrase "risk-based decision making." Environmental decisions are, at best, "risk-informed." They are based on risk information plus economic considerations, technical feasibility, public pressures, political pressures, and the personal biases of the decision makers. Another partial solution is to be fastidious in quantifying, or at least describing, uncertainties and limitations of our assessments.

Some things have not changed since the first edition. The emphasis is still on providing clear, scientifically sound, and unbiased technical advice to environmental decision makers. Although other examples are included in this edition, the focus is still on risks from chemicals or chemical mixtures, indicating that most ecological risk assessments are concerned with these issues.

The text is still aimed at practitioners and advanced students with at least a basic knowledge of biology, chemistry, mathematics, and statistics. It does not assume any familiarity with ecological risk assessment or risk assessment in general. A glossary is provided, because terms from risk assessment, ecology, toxicology, and other disciplines are used.

As with the first edition, I have written most of the book myself in order to provide a common voice and a common vision of the topic. This is a service to the reader as well as an opportunity for me to share my particular vision of what ecological risk assessment is and what it could be. However, for some major topics, the readers would be ill-served by my meager expertise. Fortunately, Larry Barnthouse, Steve Bartell, and Don Mackay agreed to participate in this edition as they did in the first. I believe they are the preeminent experts in the application of population modeling, ecosystem modeling, and chemical transport and fate modeling, for the assessment of ecotoxicological effects. Fortunately, they have similar pragmatic approaches to mine.

The preface to the first edition described it as a manifesto. The program of that manifesto was that ecological assessors must become more rigorous in their methods and practices in order to be taken as seriously as human health and engineering risk assessors. That program is no longer needed. Ecological risk assessments are at least as rigorous as human health assessments and in some ways, particularly in the use of probabilistic analysis, ecological assessments are more advanced. As a result, ecological risks are more often the basis for environmental regulatory and management decisions. However, ecologically driven decisions are still far less common than health-driven decisions. To a certain extent, this is inevitable, because humans are making the decisions based on the concerns of other humans, the public. However, we can make progress in protecting the nonhuman environment by greater integration of ecological risk assessment with concerns for human health and welfare. Hence, the greatest challenge in the coming years is to estimate and communicate ecological risks in a way that makes people care.

Glenn Suter
Cincinnati, Ohio

Acknowledgments

I gratefully acknowledge the innumerable environmental scientists who contributed to this text. Those who are cited are thereby acknowledged, although you are probably not cited as much as you deserve. Many of you who are not cited at all deserve citation but must settle for this apologetic acknowledgment. I have heard your talks at meetings, exchanged ideas at your posters or in the halls, and even read your papers, but have forgotten that you were the source of those ideas. Even more sadly, many of you have done important work and produced important ideas that should appear in this text but do not, because I am unaware of them. There are forlorn piles of books, reports, and reprints on the table behind my back as I write this that I really wanted to read before completing this book, but could not. So, if you feel that I have not given your work the attention it deserves, you are probably right.

Parts of this book draw upon material in *Ecological Risk Assessment for Contaminated Sites*. Thanks to Rebecca Efroymson, Brad Sample, and Dan Jones who were coauthors of that book.

My 7 years with the US Environmental Protection Agency have improved this book by giving me a deeper understanding of the role of risk assessment in environmental regulation. Thanks to all of my agency colleagues. Particular thanks to Susan Cormier and Susan Norton who have been wonderful friends, inspiring collaborators, and guardians against sloppy thinking.

Finally, deep thanks to Linda who, after decades of marriage, has learned to tolerate my long hours in my study and even helped with the final rush to submit the manuscript.

Authors

Glenn W. Suter II is science advisor in the US Environmental Protection Agency's National Center for Environmental Assessment, Cincinnati, and was formerly a senior research staff member in the Environmental Sciences Division, Oak Ridge National Laboratory, United States. He has a PhD in ecology from the University of California, Davis, and 30 years of professional experience including 25 years in ecological risk assessment. He is the principal author of two texts in the field of ecological risk assessment, editor of two other books, and author of more than 100 open literature publications. He is associate editor for ecological risk of *Human and Ecological Risk Assessment*, and reviews editor for the Society for Environmental Toxicology and Chemistry (SETAC). He has served on the International Institute of Applied Systems Analysis Task Force on

Risk and Policy Analysis, the Board of Directors of SETAC, an expert panel for the Council on Environmental Quality, and the editorial boards of *Environmental Toxicology and Chemistry*, *Environmental Health Perspectives*, and *Ecological Indicators*. He is the recipient of numerous awards and honors; most notably, he is an elected fellow of the American Association for the Advancement of Science and he received SETAC's Global Founder's Award, its highest award for career achievement, and the EPA's Level 1 Scientific and Technical AchievementAward.

His research experience includes development and application of methods for ecological risk assessment and ecological epidemiology, development of soil microcosm and fish toxicity tests, and environmental monitoring. His workis currently focused on the development of methods for determining the causes of biological impairments.

Susan M. Cormier is a senior science advisor in the U.S. Environmental Protection Agency's National Risk Management Research Laboratory. Dr. Cormier received her BA in Zoology from the University of New Hampshire, her MA in biology from the University of South Florida, and her PhD in Biology from Clark University.

Donald Mackay (BSc, PhD (Glasgow)) is director of the Canadian Environmental Modelling Centre at Trent University, Peterborough, Ontario, Canada. He graduated in chemical engineering from the university of Glasgow. After working in the petrochemical industry he joined the University of Toronto, where he is now Professor Emeritus in the Department of Chemical Engineering and Applied Chemistry. He has been director of the Canadian Environmental Modelling Centre of Trent University Ontario since 1995. His primary research interest is the development, application, validation, and dissemination of mass-balance models describing the fate of chemicals in the environment in general, and in a variety of specific environments. These models include descriptions of bioaccumulation in a variety of organisms, water-quality models of contaminant fate in lakes, rivers, sewage-treatment plants, and in soils and vegetation. He has developed a series of multimedia mass-balance models employing the fugacity concept that are widely used for assessment of chemical fate in national regions in the global environment. A particular interest is the transport of persistent organic chemicals to cold climates such as the Canadian Arctic and their accumulation and migration in arctic ecosystems.

Susan B. Norton is a senior ecologist in the U.S. Environmental Protection Agency's National Center for Environmental Assessment. Since joining EPA in 1988, Dr. Norton has developed methods and guidance to better use ecological knowledge to inform environmental decisions. She was an author of many agency guidance documents including the *2000 Stressor Identification Guidance* document, the *1998 Guidelines for Ecological Risk Assessment*, the *1993 Wildlife Exposure Factors Handbook*, and the 1992 *Framework for Ecological Risk Assessment*. She has published numerous articles on ecological assessment and edited the book *Ecological Assessment of Aquatic Resources: Linking Science to Decision-Making*. She is currently enthusiastic about making methods and information for causal analysis more available via the World Wide Web at www.epa.gov/caddis. Dr. Norton received her BS in plant science from Penn State, her MS in natural resources from Cornell University, and her PhD in environmental biology from George Mason University.

Neil Mackay (BSc (Waterloo), DPhil (York)) is a senior research scientist for environmental modelling with DuPont (UK) Limited. As a member of the DuPont Crop Protection Global Modelling Team he is active in strategic development and regulatory exposure and risk assessment activities. Previous work experience includes employment as a consultant to both industry and government bodies, primarily in Europe. He was a participant in the European Commission Health and Consumer Protection Directorate General and the FOCUS Risk Assessment Working Group and is a member of the UK government expert advisory panel on veterinary medicines. Particular interests include aquatic risk assessment and use of spatial tools (GIS and remote sensing methods) to evaluate risks at various scales (field, catchment and regional scales) and assessment of long range transport potential for persistent organic pollutants (POPs).

Lawrence W. Barnthouse is the president of LWB Environmental Services, Inc. and adjunct associate professor of zoology at Miami University. He was formerly a senior research staff member and group leader in the Environmental Sciences Division at Oak Ridge National Laboratory. In 1981 he became co-principal investigator (with Glenn Suter) on EPA's first research project on ecological risk assessment. Since that time, he has been active in the development and application of ecological risk assessment methods for EPA, other federal agencies, state agencies, and private industry. He has chaired workshops on ecological risk assessment for the National Academy of Sciences and the Society of Environmental Toxicology and Chemistry, and served on the peer review panels for the EPA's Framework for Ecological Risk Assessment and the Guidelines for Ecological Risk Assessment. He continues to support the development of improved methods for ecological risk assessment as the Hazard/Risk Assessment Editor of *Environmental Toxicology and Chemistry* and a Founding Editorial Board Member of *Integrated Environmental Assessment and Management*.

Steven M. Bartell is a principal with E2 Consulting Engineers, Inc. He is also an adjunct faculty member in the Department of Ecology and Evolutionary Biology at the University of Tennessee, Knoxville. His education includes a PhD in oceanography and limnology (University of Wisconsin 1978), an MS in botany (University of Wisconsin 1973), and a BA in biology (Lawrence University 1971). Dr. Bartell's areas of expertise include systems ecology, ecological modeling, ecological risk analysis, risk-based decision analysis, vulnerability analysis, numerical sensitivity and uncertainty analysis, environmental chemistry, and environmental toxicology. He works with a variety of public and private sector clients in diverse projects in ecological risk assessment, environmental analysis, and more recently in ecological planning and restoration in the context of adaptive environmental management and ecological sustainability. Bartell has authored more than 100 peer-reviewed publications. He is a senior contributing author on several books including *Ecological Modeling in Risk Assessment* (2001), *Ecological Risk Assessment Decision-Support System: A Conceptual Design* (1998), *Risk Assessment and Management Handbook for Environmental, Health, and Safety Professionals* (1996), and *Ecological Risk Estimation* (1992). He currently serves on the editorial boards of *Aquatic Toxicology* and *Chemo-**sphere* having served previously on the editorial boards of *Human and Ecological Risk Assessment* and *Ecological Applications*. Bartell served for 11 years on the USEPA Science Advisory Board, mainly on the Environmental Processes and Effects Committee and review committees.

Contributors

Lawrence W. Barnthouse
LWB Environmental Services, Inc.
Hamilton, Ohio

Steven M. Bartell
E2 Consulting Engineers, Inc.
Maryville, Tennessee

Susan M. Cormier
US Environmental Protection Agency
Cincinnatti, Ohio

Donald Mackay
Canadian Environmental Modelling Centre
Trent University
Petersborough, Ontario, Canada

Neil Mackay
Cambridge Environmental Associates
Cambridge, United Kingdom

Susan B. Norton
US Environmental Protection Agency
Washington, DC

Glenn W. Suter II
US Environmental Protection Agency
Cincinnati, Ohio

Table of Contents

Part I

Introduction to Ecological Risk Assessment

Risk analysis is the closest thing science has given us to a method for analyzing our existential condition of uncertainty and doubt in the face of decisions.

Crawford-Brown (1999)

This book offers a number of processes for performing an ecological risk assessment. Part I provides a definition of the field, describes its relationship to other environmental assessment practices, and explains the framework that organizes it. It includes a chapter on ecological epidemiology, which has been treated as a type of ecological risk assessment, but is now recognized as a distinct practice. The last five chapters introduce important concepts that are relevant to many points in the ecological risk assessment process.

1 Defining the Field

Risk assessment is the product of a shotgun wedding between science and the law.

William Ruckelshaus

"Technical support for decision making under uncertainty" is the only definition of risk assessment that describes its many uses. As Bernstein (1996) plausibly argues, the use of rational methods for dealing with the uncertain future in place of prayers, prophecies, traditions, auguries, and hunches is the hallmark of modern culture. Risk assessment began with the need to calculate odds for gamblers, and subsequently, in seventeenth-century England and the Netherlands, with the need to determine premiums on annuities and the probability that a ship sent on a trading voyage would return successfully (Hacking 1975; Bernstein 1996). Most risk assessors are still involved in finance and insurance (Melnikov 2003). Risk assessment has since spread to many spheres of human endeavor including engineering, wildfire management, medicine, and environmental regulation. The general definition indicates that two features are common to all of these enterprises: a decision to be made and uncertainty concerning outcomes.

The conventional, objectivist definition of risk is: a combination of the severity (nature and magnitude) and the probability of effects from a proposed action. Severity may be variously described depending on the situation, e.g., the number of deaths, the reduction in abundance, and the reduction in areal extent. Probability may be derived from an estimate of the frequency of an effect among individuals in an exposed population or a hypothetical frequency of effects if the same decisions were made multiple times. For example, a risk might be a 0.3 annual frequency of mass mortalities in an exposed population, or a probability that an effluent will reduce the number of fish species in a lake by as much as 15%. Alternatively, risk may be defined subjectively as a state of mind of an individual making an uncertain decision or of those exposed to the consequences of a decision. This subjective risk is an important issue when assessing risks to humans, who are subject to anxiety and dread, but is less relevant to the topic of this book. Note that subjective risk is fundamentally different from the Bayesian, subjective interpretation of probability in estimates of objective risk (Chapter 5).

The terms "environmental risk" and "ecological risk" can cause confusion because of their similarity. In the United States, the term environmental risk has been used to describe risks to humans due to contaminants in the environment. Ecologists subsequently invented the term ecological risk to refer to risks to nonhuman organisms, populations, and ecosystems (Barnthouse and Suter 1986). However, the term environmental risk is commonly used in Europe in the way that ecological risk is used in the United States.

The decision to be supported is too often neglected in risk assessment. "It is hard to imagine risk analysis existing without the need for decisions, without the need for a systematic approach to aiding those who make decisions" (Crawford-Brown 1999). Yet, the most influential guidance for environmental risk assessment (ERA) has stressed the need to isolate risk assessors from the influence of decision makers in order to avoid bias (NRC 1983).

ERA practice has tended to emphasize the risk assessment process in the abstract without a grounding in a decision-making process. For example, due to the peculiarities of Superfund regulations, the guidance for baseline ERA at contaminated sites in the United States does not address the consequences of remedial decisions (Sprenger and Charters 1997). This situation is changing with the realization that the highest-quality assessment is worthless if it does not address the needs of the decision maker (National Research Council 1994; The Presidential/ Congressional Commission on Risk Assessment and Risk Management 1997). Risk-based environmental decisions generally fall into three categories: should we permit x (e.g., use of a new chemical, release of an effluent, or increased harvest of a resource); what should we do about x (e.g., remediate, treat, or restore); should we do x, y, or z (e.g., which pest management method poses the least risk)?

Probability, the other core concept in ERA, has also been surprisingly neglected. The probabilities that characterize risks may result from variability or uncertainty (Chapter 5). Although quantitative methods for analyzing uncertainty and variability in terms of probability have existed for centuries, most ERAs treat them qualitatively or in nonprobabilistic terms. This does not mean that uncertainty and variability are ignored or that, as some have contended, most current risk assessments are not truly assessments of risk. Rather, they are often dealt with by semiquantitative precautionary practices. That is, conservative assumptions and safety factors have been assumed to provide sufficient safety to avoid the need for a formal probabilistic analysis. However, formal probabilistic analysis of uncertainty is increasingly common. This is because the semiquantitative practices are subject to criticism that they are insufficiently precautionary, excessively precautionary, or precautionary to an undefined degree.

Risk assessment uses science, but is not science in the conventional sense, i.e., it does not seek to develop new theories or general knowledge. It rather uses scientific knowledge and tools to generate information that is useful for a specific purpose. In this sense, risk assessors are like engineers, and in fact much of the practice of ERA has been developed by engineers (see, e.g., Haimes 1998). However, contrary to some critics, risk assessment is based predominantly on factual information and scientific theory, and is not simply a scientific smoke screen for policy. Typically, risk assessments and their components are intensely and publicly reviewed and are often challenged in court. As a result, the use of bad science to justify a preordained decision is likely to be detected in contentious cases.

1.1 PREDICTIVE VS. RETROSPECTIVE RISK ASSESSMENT

The EPA's framework and guidelines for ecological risk assessment and the previous edition of this text distinguish retrospective from predictive risk assessment. This distinction has created some confusion, because it is nonsensical to speak of risks of events in the past. This text eliminates that distinction and focuses instead on the decision-supporting function of risk assessment. Hence, when assessing risks from spills or other past events, we are assessing risks associated with future consequences of those events. They include ongoing toxic effects, the spread of toxic levels of contaminants to other areas, loss of habitat due to failed restoration, and other sequela. Even when performing assessments to set monetary damages for past actions, we are not assessing risks of past events. For example, during the Exxon Valdez oil spill, a certain number of sea otters, bald eagles, and other wildlife were killed. The natural resource damage assessment (Section 1.3.8) did not assess risks to those organisms. Rather, to the extent that the uncertainties in the damage assessment could be interpreted as risks, they should be interpreted as risks that the level of monetary damages assessed will be either insufficient or excessive with respect to the cost of restoration of the populations of those species, making good lost services of nature, and otherwise remediating ecological and economic injuries.

Although all risk assessments are in some sense predictive, it does not mean that information concerning the past is irrelevant. Analyses of such data may be used to help formulate the assessment problem, elucidate trends that may extend into the future, identify causal relationships between agents and injuries, and define a baseline for remediation and restoration. By analogy to human health epidemiology, analyses of past ecological effects and their causes are termed ecological epidemiology (Chapter 4). Hence, what the US Environmental Protection Agency (US EPA) terms retrospective assessments should be thought of as predictive assessments of the future consequences of past actions.

1.2 RISKS, BENEFITS, AND COSTS

When assessing an action, it may be necessary to consider the risks associated with the action, the potential benefits, and the costs of carrying out the action. For example, when considering whether to apply a remedial technology to a contaminated site, it is important to consider the risks of ecological injury from the contaminants and from the remedial action itself (e.g., injuries to benthic communities due to dredging), the benefits of the action (e.g., reduced contaminant risks to epibenthic fish), and the cost of carrying out the remediation. Which of these dimensions is formally analyzed and how they are compared depends on the context. Some laws require consideration of costs while others do not allow it. Further, there is considerable variation in whose benefits, costs, and risks are considered. For example, the registration of pesticides or biocontrol agents in the United States may consider costs and benefits to farmers if there are no good alternatives to the pesticide in question, but not the costs to the manufacturer. Although the primary focus of regulatory agencies is on the risks from new chemicals, effluents, spills, and exotic organisms, the consideration of benefits and costs as well as ethical concerns and public preferences can provide a more complete basis for decision making (Chapter 36). Although costs to the regulated party can be readily identified and relatively easily estimated, the benefits to the environment are always incompletely identified and are difficult to quantify. Hence, cost–benefit analysis tends to be biased against environmental protection.

1.3 DECISIONS TO BE SUPPORTED

The form and content of an ecological risk assessment is determined by the decision to be supported. This is true not only because different decisions require different sorts of information, but also because of the different formal and informal traditions and constraints that have developed in the various decision-making cultures. For example, the assessments of new industrial chemicals in the United States must be performed within 90 days and normally cannot demand data generation. On the other hand, assessments of contaminated sites may require years of effort involving expensive field surveys, sample collection, analysis, and testing.

1.3.1 PRIORITIZATION OF HAZARDS

In the United States and many other countries, priorities for environmental management have been set in a highly inconsistent manner, based primarily on public pressures as translated by legislators into laws and budgets. Because resources for environmental management are limited, it would be desirable to devote them to the highest priority hazards rather than the ones that were given a strong mandate some decades ago. This concept of using risk assessment to prioritize hazards is appealing (Grothe et al. 1996), but controversial in practice (Finkel and Golding 1995).

The US EPA and its Science Advisory Board have performed comparative risk assessments for the purpose of prioritization (SAB 1990, 2000; MADEP 2002). The results have been controversial and have not greatly influenced the EPA's regulatory and management practices. However, they have led to the development of guidance for comparative assessment and to the performance of such assessments in most states of the United States (Bobek et al. 1995; EPA 1997a; Feldman et al. 1999). The performance of such assessments is difficult because of the paucity of information and the difficulty of comparing risks to disparate entities and processes over large ranges of spatial and temporal scales. As a result, expert judgment has been used in the absence of data analysis, as in the case of Harwell et al. (1992), and even that has been largely replaced as a means of prioritization by consensus of representatives of stakeholders and the public (EPA 1997a). Consensus-based assessments have some potential benefits beyond prioritization itself, such as better understanding of environmental issues and promotion of coordinated action by the participants, but they have not been followed and implemented by prioritized risk management programs (Feldman et al. 1999). Prioritization based on actual estimation of risks must await further development of assessment methods and a willingness to devote sufficient resources to the problem.

The need to replace expert judgment with technical approaches is illustrated by the consideration of oil spills in the prioritization of environmental hazards by the US EPA's Science Advisory Board. They gave a low rank to oil spills because of the perception that ecological effects of oil in the marine environment were short term (SAB 1990). However, that perception seems to be a result of a lack of high-quality long-term monitoring. It has been reported that some detectable effects of the Exxon Valdez spill persisted for at least a decade (Peterson et al. 2003).

Beyond the technical difficulties, risk-based prioritization has had little influence in regulatory agencies, in part because it may be considered illegal or immoral. The potential illegality arises because environmental laws in the United States and most other countries require protection independent of other laws. For example, the US EPA cannot decide to stop enforcing the Clean Air Act because resources would be more effectively spent enforcing the Clean Water Act. In addition, prioritization may assign high priority to hazards for which no legal authority for action exists. The accusations of immorality most commonly arise from the accusation that technical analysis is used to minimize the legitimate subjective concerns of citizens or to ignore risks to small groups with particular exposures (e.g., indigenous people consuming traditional foods). On the other hand, consensus-based prioritization may also provide unequal protection. When prioritization is based on a stakeholder process, there is a potential for higher ranking of the risks that concern the most articulate and influential segments of the population. Because of these issues, the potential benefits of a rational prioritization process must await a mandate from the highest levels of government to overcome the technical, social, and legal impediments (Davies 1996).

1.3.2 COMPARISON OF ALTERNATIVE ACTIONS

As discussed above, risk assessment is performed to inform decisions concerning alternative actions. Unfortunately, the range of alternatives is often small and the range of issues considered is often narrow. For example, registration of a pesticide often does not include consideration of the risks from the alternatives, existing pesticides that may be more persistent or toxic, and nonpesticide pest control techniques that have their own potentially severe ecological risks. Rather, the alternatives are typically restricted to registration, registration with restrictions, or rejection of the new pesticide. It is clear that the decision-making process can meet the legal mandate and be rational within its scope, but result in a less than optimum decision for the environment.

Comparative risk assessment raises two complications. First, it is not adequate to estimate risks; one must also estimate the benefits of the alternative actions. A relatively low-risk alternative action may be undesirable, because it produces small benefits or even net decrements. The comparison of risks and benefits (not just costs and benefits) is important in any case, but is essential when comparing a set of alternatives. Second, comparison of risks often involves the common units problem. That is, if the alternatives involve disparate risks and benefits, they may not be directly compared by simply quantifying future ecological conditions. The temporal integration of expected benefits and decrements of each action, expressed in common units, is termed net environmental benefit analysis (Efroymson et al. 2004). If the comparison must consider the costs of implementation, the net benefits must be monetized to yield a cost benefit analysis (Chapter 36).

Comparative risk assessment is a different way of looking at any of the decisions discussed in this chapter. Although all risk-based decisions involve the comparison of at least two alternatives (e.g., permit an action or not), a more comparative approach opens up the process to a range of potentially desirable alternatives. However, it complicates the assessment and decision-making process. Approaches to these issues are discussed in Chapter 34.

1.3.3 Permitting Releases

Ecological risk assessments have been concerned primarily with two activities: determining whether releases of chemicals or other agents should occur and determining how to deal with the releases that have already occurred. Clearly, we should do a better job of the former to reduce the need for the latter. Ecological risk assessments for permitting releases are distinguished by the type of agent released (i.e., chemicals, effluents and other wastes, and exotic organisms) and by whether the agents are novel or have been permitted before and are being reconsidered.

1.3.3.1 Chemicals

In the United States, new chemicals are regulated as pesticides under the Federal Insecticide, Fungicide, and Rodenticide Act (FIFRA), as industrial chemicals under the Toxic Substances Control Act (TOSCA) or under the Food, Drugs, and Cosmetics Act (under which ecological concerns have received little attention). The difference in ERA under FIFRA and TOSCA serves to illustrate the importance of legal constraints on assessment practices. Because pesticides are designed to be toxic, FIFRA allows the government to require relatively extensive characterization and testing of new chemicals by the manufacturer and allows the government time to complete the assessment. TOSCA does not allow characterization and testing requirements beyond basic descriptions of the compounds and allows only 90 days for assessment and decision making. As a result, assessment of pesticides has been based on a fairly elaborate tiered scheme and is moving to a system of probabilistic assessment (Urban and Cook 1986; Ecological Committee on FIFRA Risk Assessment Methods 1999a,b). The pesticide industry has responded with its own tiered and probabilistic ecological risk assessments of products such as atrazine (Section 32.4.4). In contrast, ERA under TOSCA relies on small data sets, quotient methods, and assessment factors of 10, 100, or 1000 (Zeeman 1995; Nabholz et al. 1997). Assessments of new chemicals in the European Union and elsewhere have their own testing requirements and schemes, but in general they rely on tiered testing approaches and simple methods using factors and quotients (RIVM 1996; Royal Commission on Environmental Pollution 2003). Similar assessment approaches have been developed by responsible chemical manufacturers to assure that their products "are safe for the environment" (Cowan et al. 1995). Methods for ecological assessment of chemicals are rapidly developing because of advances in science such as computational toxicology and because of renewed interest in existing chemicals, particularly the European Community's REACH regulations (Bradbury et al. 2004).

1.3.3.2   Effluents and Wastes

The release of aqueous and gaseous effluents and other waste streams is regulated in the United States and most other countries through a permitting process. The most important of these from an ecological perspective is the permitting of aqueous effluents, by a process known as National Pollutant Discharge Elimination System (NPDES) in the United States. This is accomplished primarily by specifying that the effluent will not violate water quality standards. Standards include concentrations, durations, and frequencies of exceedence that must not be violated (Section 2.2). In most states, standards are based on National Ambient Water Quality Criteria published by the EPA (1985). Equivalent criteria and standards are used in other nations (Roux et al. 1996; CCME 1999; ANZECC 2000). Alternatively, permits may specify that the toxicity of the effluent be tested using standard acute or subchronic tests (Section 24.2) or that the receiving community achieve biological criteria (EPA 1996a; Ohio EPA 1998).

1.3.3.3   New Organisms

Organisms may be deliberately imported for horticultural use, biological control, pets, or other purposes. Determining whether an importation should be permitted is conceptually difficult because organisms are complex and can display unexpected properties or may evolve new properties. Risk assessments in support of the regulation of importation of foreign organisms may be based on structured expert judgment as in the United States (Orr 2003) or more objective analyses. An example is the assessment of import of shrimp for aquaculture that may carry a virus, which is pathogenic to native shrimp (Fairbrother et al. 1999). Genetically engineered organisms are regulated in the United States as though they are chemicals. That is, novel biocontrol agents are regulated by the pesticides office of the EPA, and other novel organisms are regulated like industrial chemicals by the toxic substances office.

1.3.3.4   Items in International Trade

The World Trade Organization (WTO) 1995 Agreement on the Application of Sanitary and Phytosanitary Measures and some regional trade agreements require that a risk assessment be performed if a nation excludes an item from importation due to risks that it poses to human health, animals, or plants. The exclusion may be based on the determination that the item may be toxic, a pathogen, a pest, or otherwise pose an unacceptable risk, or there is a significant risk that an item may be a carrier of such a hazardous agent. Items that have been excluded from import range from controversial genetically modified crops to wood and plant products that may contain exotic pests. The WTO agreement is more demanding than most legal bases for risk assessment, in that risks must be expressed in terms of probabilities or likelihoods, not possibilities (Codex 1997; OIE 2001). New Zealand provides an excellent guide for conducting probabilistic risk assessments for imports of animals, animal products, and associated pathogens and pests (Murray 2002).

1.3.4   LIMITING LOADING

The regulation of the uses of chemicals and their disposal in effluents or solid wastes fails to be protective because of the effects of multiple releases from multiple sources. One solution is to define the rate at which an ecosystem may receive a pollutant from all sources without unacceptable effects. For atmospheric deposition, this is referred to as the critical load (Hettelingh and Downing 1991; Holdren et al. 1993; Hunsaker et al. 1993; Strickland et al. 1993). The same term is applied to aqueous pollution (Vollenweider 1976), but for water quality regulation in the United States, it is referred to as the total maximum daily load (TMDL) (Houck 2002). Setting limits on loading requires defining the resources to be protected (e.g., water quality, biotic communities, or human health) and the endpoints to be measured for each (e.g, water quality criteria, benthic invertebrate species richness, and soil pH). A mixture of field measurements and modeling is then used to determine whether a limit is exceeded, whether a new source will cause exceedence, the relative contributions of existing sources to an exceedence, or the likelihood that a remedial action will result in acceptable loading. In some cases, this is relatively straightforward. For example, if a persistent and soluble chemical is released at multiple points in a stream or watershed, simple transport and fate modeling can be used to determine contributions to exceedence of a water quality criterion at a downstream point. However, other cases are complex and difficult. For example, NO_x deposition in a watershed is difficult to trace back to point and nonpoint sources in the atmosphere, and effects on terrestrial and aquatic ecosystems are difficult to measure or predict because of the complexity of nitrogen cycling and acidification and eutrophication processes.

1.3.5 REMEDIATION AND RESTORATION

The predominant use of ecological risk assessment in the United States has been to support the remediation of contaminated sites under Superfund (Suter et al. 2000). A full ecological risk assessment for a contaminated site would consider the risks from the existing contamination (the no action alternative), from the remedial actions themselves, from residual contamination, and from subsequent land uses. In addition, if ecological restoration activities may be performed after remediation, risks associated with restoration must be considered.

In the United States, remedial assessments are performed in two stages: a baseline assessment that determines whether the unremediated contamination poses a significant risk and an assessment of remedial alternatives termed the feasibility study. Procedural and technical guidance are available for baseline assessment (Sprenger and Charters 1997) (see also the EPA's Environmental Response Team and Office of Solid Waste and Emergency Response web sites), and these assessments are often well performed and based on ample data. Because there is a contaminated site to be sampled, surveyed, and tested, the full range of assessment techniques is available to the assessor (Suter et al. 2000). In contrast, there is little guidance for the assessment of risks from dredging, soil removal, capping, construction of roads and other support facilities, chemical or thermal treatment of media, spills of treatment chemicals, and other remedial activities that pose obvious ecological hazards. Remedial decisions tend to focus on the efficacy of technologies in reducing the contaminant risks and on their costs rather than the risks from remediation. The ecological risks of remedial alternatives are usually given a serious assessment only when, as in the polychlorinated biphenyl (PCB) contamination of the Hudson River, New York, the remediation is particularly costly or controversial.

Restoration involves recreating to some extent the ecological structure and function of a site disturbed by a remedial action or any other action. Hazards associated with restoration include erosion and siltation, introduction of exotic species (e.g., as ground covers) with undesirable properties, use of pesticides and fertilizers, and conversion to parks with associated mowing and trampling. In addition to these risks from restoration, planted trees may die, instream structures may wash away, and for other reasons restoration activities may fail. Therefore, it may be appropriate to adapt engineering risk assessment techniques to restoration projects. As with other sorts of risk assessments, assessments of restoration should compare the risks of alternatives. For example, after the eruption of Mount St. Helens, parts of the area were seeded with exotic herbaceous plants. Although these plants reduced erosion of the ash, they apparently slowed the reestablishment of the native forest. Hence, the risks to stream and river ecosystems from eroding ash could be compared with the risks from delayed recovery of the native terrestrial community.

1.3.6 PERMITTING AND MANAGING LAND USES

The conversion of one land use to another, particularly the conversion of land supporting a natural community to agricultural, residential, or urban uses, is one of the most severe anthropogenic hazards that ecosystems face. However, land use conversion is little regulated in the United States, where land use permitting is usually performed at the local level of government. Hence, ecological risk assessment is seldom involved in land use decision. The exception is the major action of the United States federal government, which is subject to environmental impact assessment (Section 2.12). Nevertheless, the ecological risk assessment framework can be applied to land management decisions, even the complex decisions concerning land development and water use in South Florida (Harwell 1998).

Land use also involves decisions concerning the intensity of use: How many cattle should be allowed to graze on an area of rangeland? How often should a forest be logged? These decisions are the subject of their own well-developed assessment practices (Davis et al. 2000; Holchek et al. 2003). Although they make little use of risk assessment concepts and terminology, to the extent that they have clear quantifiable goals and analyze uncertainty to inform decision makers, they are ecological risk assessments.

1.3.7 SPECIES MANAGEMENT

Risks to species or species populations are estimated for the management of resource species and endangered species. The techniques developed for these purposes have been adapted for estimating risks to populations from pollutant chemicals and introduced organisms, and risk assessment concepts have been adapted to resource management (Chapter 27) (Francis and Shotton 1997). Management of game species, fisheries, timber trees, and other harvested plants requires assessments to determine harvest levels that do not pose unacceptable risks of extirpation. Probabilistic modeling has been particularly important in fisheries management. Other innovations in the assessment of resource management include modeling populations in a community or ecosystem and adaptive management (Walters 1986). Management of threatened or endangered species is often based on a form of population modeling termed population viability analysis that estimates the time to extinction or the probability of extinction within a prescribed time period such as the next 50 years (Sjogren-Gulve and Ebenhard 2000; Beissinger and McCollough 2002; Keedwell 2004). These two assessment practices overlap when harvesting is forcing a population to extinction (Musick 1999). Assessment of species management becomes particularly complicated when the valued species is potentially damaging to the ecosystem or to other valued species (Box 1.1). Risk assessment lends itself to species conservation, and ecological risk assessment for conservation and for pollution regulation have developed in parallel (Burgman et al. 1993; Burgman 2005).

1.3.8 SETTING DAMAGES

When ecological injuries occur due to negligence or criminal activities, the responsible parties may be required to pay monetary damages. These damages are used to restore the damaged ecosystems, or, when that is not possible, to acquire and protect other areas. In the United States, natural resource trustees are required to seek damages from polluters under the Clean

---

BOX 1.1
Elephant Management, Risks, and Environmental Ethics

The management of populations can have implications for co-occurring populations and communities that result in difficult decisions. This issue is particularly stark in the case of elephant management described by Whyte (2002). Elephants are potentially highly destructive of their habitats. They are capable of grazing, including uprooting entire tussocks of bunch grasses; of browsing, including ripping off branches or knocking down trees to feed on branches and roots; and of feeding on bark torn off with their tusks. In addition, they have no significant predators except humans. Hence, they are capable of destroying woodlands before succumbing to starvation, particularly during droughts, as occurred in Tsavo National Park, Kenya, in 1974 (Whyte 2002). On the other hand, elephants are quintessential charismatic megafauna, so they attract the concern of animal rights advocates. These conflicts became particularly acute in Kruger National Park, South Africa, which is a closed system where poaching is well controlled.

An ecological risk assessment of elephant management in Kruger National Park could be formulated in various ways. Elephants could be treated as agents that can cause risks of effects on woodlands and woodland-associated wildlife populations including extinction of some communities and populations within the park. This formulation could result in identification of an elephant population level that would maximize biodiversity by creating intermediate disturbance. Alternatively, an assessment could mix animal rights concerns with ecological concerns by comparing the risks to individual elephants with risks to populations and communities. Finally, it could take an organism rights approach to all species comparing risks to individuals of charismatic megafauna (elephants) with risks to individuals of other species, including charismatic megaflora (baobab trees).

If risks to individual elephants are admitted as an issue, it is not clear which management alternative is best for them. Conventional culling involves a shot in the brain, which is brutal but quick. Anesthesia followed by shooting is slower and potentially more cruel. Contraception requires frequent harassment of the females to administer the drug, can cause females to be harassed by males during pseudo estrus, and can upset the demographics of family units. In addition, it is tremendously expensive, since it requires repeated treatment of at least 70% of adult females that would require radio-collaring them so that they could be relocated. Finally, one could let nature take its course, which would result in slow and painful but natural deaths of elephants. Hence, the goals of minimum suffering and minimum interference could lead to different solutions.

The policy in Kruger was to maintain a population of 7000 elephants by culling, based on conventional analyses of habitat carrying capacity. This policy ended in 1994 due to animal rights protests. The compromise was to divide the park into six zones: two "high-impact" zones in which elephants would be undisturbed; two botanical reserves, which would preserve rare or ecologically important plants; and two "low-impact" zones. The various zones are monitored, and control implemented when damage is sufficient. The monitoring includes abundances of plants, frogs, birds, reptiles, and mammals, and some abiotic measures such as erosion rates. When changes exceed a threshold of potential concern in a low-impact area, the managers and stakeholders will consider management actions. For example, one threshold of potential concern is an 80% reduction in the number of mature trees. Hence, the abundance of mature trees is a measure of effects. Monitoring of elephants and the other biota should eventually yield empirical models of the biotic community's response to exposure to elephants. Although management decisions are currently made ad hoc, this experience could provide a basis for managing elephants based on clear biodiversity and elephant protection goals and endpoints. However, that will require reconciling the views of elephants as individuals with rights vs. elephants as components of an ecosystem with other species worthy of protection, even from elephants. The same ethical dilemma faces managers of other animals that are capable of detrimentally modifying ecosystems such as feral horses and donkeys in the western United States or whitetail deer in the eastern United States. Risk assessment cannot solve such dilemmas, but by clarifying the valued attributes of the system (endpoints), i.e., the causal relationships (conceptual models), the relationships between exposure and response, and the uncertainties involved, it can support decision making.

Water Act (1977 amendments), the Outer Continental Shelf Act (1989 amendments), the Oil Pollution Act of 1990, and the Comprehensive Environmental Response, Compensation, and Liability Act of 1980 (CERCLA). Natural resource trustees are public land managers such as the US Bureau of Land Management or managers of biological resources such as the US Fish and Wildlife Service. However, a forest owner or any other private owner of natural resources could sue for damages under civil laws. An assessment for setting damages must determine that a natural resource has been injured and that the injury is potentially associated with a responsible party. After this screening phase, the nature and extent of the injury must be quantified and the actions of the responsible parties must be demonstrated to be the cause. Conventionally the next step is to convert the injuries to the resource into monetary damages based on lost services of the resource, which are to be paid to the resource manager or owner by the responsible party. It is intended that the resource manager will use the damages to restore or replace the resource. Alternatively, the responsible party may be charged to restore the resource. This requires that an assessment of alternative restoration approaches be performed so that their costs and expected efficacies can be compared. Regulations and guidance for natural resource damage assessment for public resources in the United States are available at the Department of Interior (40 CFR 11 and http://www.doi.gov/oepc/frlist.html). Damage assessments in general are a subset of ecoepidemiological assessments (Chapter 4).

1.4   SOCIOPOLITICAL PURPOSES OF RISK ASSESSMENT

The primary purpose of risk assessment is to inform a decision-making process, but, because decision making is inevitably a sociopolitical process, risk assessment also serves sociopolitical purposes. First, a risk assessment provides a record of the technical basis for a decision. Second, it provides information on the legitimacy of stakeholder and public concerns. Third, it reduces controversy by providing a technical forum for resolving contentious issues. Finally, it provides a means for stakeholders to participate in framing and informing the decision-making process. Clearly, these functions are ideals that are not always achieved in practice. However, it is important for risk assessors to be aware that good decisions must be viewed as legitimate decisions, or the environmental management process will fail. Hence, they must be willing to participate in the ancillary activities such as public meetings and reviews that provide legitimacy. To that end, they must understand their role in the drama of environmental decision making.

1.5   CAST OF CHARACTERS

1.5.1   Assessors

Risk assessors are technical experts who perform assessments so as to support a decision. Ecological risk assessors typically work in teams that may include health risk assessors, ecologists, toxicologists, chemists, hydrologists, statisticians, system modelers, engineers, and other relevant technical experts. While most practitioners have learned risk assessment concepts and methods on the job, universities are increasingly providing training in risk assessment. Assessors may be employed by a regulatory agency, an applicant for a permit, a party responsible for a spill or dump, a local citizen's group, or an environmental advocacy organization. In any case, they are technical consultants, translating available science and practice into useful information.

1.5.2   Risk Managers

Risk managers are individuals or teams with the responsibility and authority for making a decision that involves risk. In some cases, the role is clearly assigned. For example, in the remediation of contaminated sites under Superfund, the Remedial Project Manager is the US

EPA official responsible for deciding what remedial actions should be taken. Risk assessors are clearly responsible for providing technical support to that individual. However, in other cases the role is less clear. For example, in both the United States and Europe, risk assessments of new chemicals are performed using a standard approach which leads to a conclusion that the chemical is acceptable or unacceptable in a proposed use. As a result, although individual risk assessors do not have decision-making authority, their analyses generate a decision, not simply a risk estimate. However, the authorized officials will step in when decisions are not routine or when new methods are proposed.

The relationship between risk assessors and risk managers is highly variable among nations and regulatory contexts. One reason is the relative concern in different risk assessment contexts for relevance and independence from biases. Clearly, if a risk assessment does not provide the information needed by a risk manager, it is largely a wasted effort. Therefore, the risk manager must provide the charge to the risk assessors, and should be available to inform the judgments that must be made on the basis of policy rather than fact in the course of the assessment. On the other hand, risk managers have biases that cause them to prefer certain outcomes to risk assessments a priori. Therefore, if a risk manager is too involved in the technical analyses, the results will appear biased and may in fact be biased. Therefore, the original guidance for risk assessment in the US federal government emphasized the need to isolate the risk manager who is politically accountable from the technical experts who must provide a credibly unbiased application of science (NRC 1983). Since then, the pendulum has swung to the other extreme so that the same august body has called for extensive input by risk managers and stakeholders (NRC 1994). As suggested in the previous paragraph, the relationship is also influenced by the extent to which the risk assessment is routine. Site-specific assessments and unconventional or high-profile assessments are more likely to receive attention from a risk manager.

The US EPA's framework for ERA, in keeping with the 1983 National Research Council guidance, shows the risk manager outside the risk assessment box (Figure 3.1). However, practice is highly variable, even within the US EPA. Risk assessors should be aware of who has the authority to make the risk management decision and how they prefer to receive technical support.

1.5.3 STAKEHOLDERS

Stakeholders are people or organizations that have a particular interest in the outcome of an environmental management decision. Examples include people who live on or near a contaminated site, parties responsible for contamination, environmental advocates, fishermen and other harvesters of biotic resources, manufacturers of a new chemical, and recreational users of a resource. Although the public as a whole has a stake in environmental management decisions, the stakeholders in a decision are a much smaller group with particular concerns. Therefore, the risk manager must consider public interests as distinct from stakeholder interests.

The role of stakeholders has been emphasized in recent risk assessment guidance (NRC 1993; The Presidential/Congressional Commission on Risk Assessment and Risk Management 1997). That emphasis is more appropriate for human health risk assessment both because stakeholders are typically focused on health or economic concerns, and because the stakeholder input is usually more relevant to those concerns.

Stakeholder involvement is important when remediation or treatment is driven by fears rather than risks or observed effects, and those fears differ among communities. For example, in most communities public fears could force remediation of radionuclides to levels "as low as reasonably achievable," even when those levels are far below background. However, in some communities such as Oak Ridge, Tennessee, or Los Alamos, New Mexico, risk assessments can be more influential because the public is well educated, knowledgeable, and familiar with radiation issues. Such differential fears do not occur among nonhuman organisms, but levels of concern for nonhuman organisms and ecosystems vary greatly among communities and interest groups.

Economic considerations associated with issues of fairness are also important stakeholder inputs. Some will bear the cost and others, the benefits of a decision. These issues are complicated by the problem of environmental justice, the concern that some racial or ethnic groups bear an unfair burden of environmental pollution. There are no such feelings of victimization or inequity among plants and nonhuman animals.

Although different levels of risk may be appropriate for different human communities because of differences in their risk aversion, stakeholder preferences do not necessarily provide a basis for differential protection of birds or plants in different communities. Agencies must enforce requirements of environmental laws, whether or not the local human community is concerned. For example, cranes are a national and world heritage resource, but local communities on the Platt River want to withdraw water that is needed for crane habitat. Those stakeholder preferences should not trump the legal protections or ethical obligations applied to those birds. However, if stakeholders are ignored or thwarted, they will, as in the Platt River case, use legal or political processes to achieve their aims. Alternatively, stakeholders may raise specific and legitimate community concerns for the nonhuman environment that would not be raised or given great significance by risk assessors or risk managers. For example, indigenous peoples along the Columbia River attribute religious and cultural as well as economic value to salmon. Similarly, fishermen are concerned about lesions, tumors, and deformities in fish that would not concern many ecologists.

In many cases, ecological risk assessors are in the position of educating stakeholders concerning the attributes of the environment that are at risk and their relationship to human welfare. However, assessors must also be open to learning from stakeholders who may have knowledge or concerns that are relevant to the assessment.

Risk assessors should realize that stakeholders might have agendas that are not consistent with increasing the understanding of risks to improve the rationality of decisions. Usually, some parties will have a stake in ignorance, because additional data and analyses are likely to weaken their position. Who they are depends in part on who bears the burden of proof. If regulators must prove that risks are excessive before taking action, the manufacturers of new products have an interest in limiting information. However, if manufacturers must prove the safety of a new product, the manufacturers of existing competitive products and environmentalists who are opposed to new technologies have an interest in limiting information. Similarly, if people feel that they have been injured by a product, and have attained public and political sympathy, they have an interest in rapid judgment rather than study and analysis. Some stakeholders will advocate additional data gathering and assessment as a delaying tactic rather than from a desire to improve the quality of the decision. For example, manufacturers of a hazardous product are likely to argue that there is not enough information to regulate it. An awareness of these unstated agendas is important when stakeholders participate in the planning of assessments or influence the information provided to risk managers.

Finally, stakeholders may be more than sources of goals or issues of concern. In some cases, stakeholders will generate data and even conduct their own risk assessment. Those risk assessments may be used as a basis for legally challenging a risk manager's decision or may be presented to the risk manager as an alternative to the assessment performed by his own assessors.

2 Other Types of Assessments

While risk assessment is broadly applicable to environmental decision making, other sorts of assessments and decision tools are more appropriate or more commonly employed in certain situations. Some apparent alternatives are actually complementary to risk assessment. The relationships of some of these alternatives to ecological risk assessment are discussed in this chapter. Assessments based on ecological epidemiology are discussed in Chapter 4.

2.1 MONITORING STATUS AND TRENDS

Concerns about the state of the environment have led to the development of programs to monitor status and trends in physical, chemical, and biological attributes. In particular, the expenditure of large sums of money for environmental regulation, treatment, remediation, and restoration has led to questions of efficacy: Are we achieving benefits that are commensurate with the expenditures? Is environmental degradation occurring that is not being addressed by existing programs? Hence, there is a need for environmental monitoring to quantify the state of the environment and generate environmental report cards. These programs may, like the US EPA's Environmental Monitoring and Assessment Program (EMAP) (Jones et al. 1997; Jackson et al. 2000) or the National Oceanic and Atmospheric Administration's National Status and Trends Program, generate their own data or, like the H. John Heinz III Center for Science Economics and the Environment (2002), analyze the existing data.

Monitoring to determine the environmental status and trends is conceptually distinct from monitoring for risk assessment. Such programs are concerned with broad patterns of common conditions, as indicated by results of random sampling, rather than the specific adverse effects that are of concern to risk assessors (Suter 2001). They are not concerned with the causal relationships that are the heart of risk assessment. Similarly, they monitor indicators, many of which are indices, rather than actual valued attributes of ecological entities (Jackson et al. 2000). Finally, they are not concerned with informing particular management decisions. The assessments generated by EMAP and similar programs are descriptive rather than predictive and serve as environmental report cards for regions, states, or nations (Jones et al. 1997; Office of Research and Development 1998; Jackson et al. 2000).

Although these progams are not designed for risk assessment, they can potentially provide relevant information. They may reveal conditions that should be addressed by follow-on assessments, which address the causes of conditions and the risks and benefits of corrective actions (Chapter 4). If they are continued over time, they may reveal trends that confirm the broad benefits of environmental regulation and management or reveal their failings. However, environmental monitoring programs in the United States have seldom achieved these goals (H. John Heinz III Center for Science Economics and the Environment 2002).

2.2 SETTING STANDARDS

Environmental standards are concentrations of contaminants in the environmental media that are legally enforced thresholds for acceptability. For the dynamic media, water and air, standards must also include an averaging period and a recurrence frequency. For example, US acute ambient water quality criteria are defined as 1 hour average concentrations that are not to be exceeded more than once in every 3 years (Stephan et al. 1985). The enforcement of standards is an alternative to estimate site-specific or case-specific risks. Standards are set by a variety of methods (Chapter 29).

Standards may be risk-based, but most often they are set so as to simply constitute reasonable and enforceable levels that do not correspond to defined ecological endpoints or defined levels of protection. In the latter case, risk assessments must treat a set of standards as effectively an endpoint in themselves, and risk assessments are limited to estimating the risk of exceeding a standard. However, standards that are set at levels that are estimated to constitute a defined risk to defined endpoints can be used in full risk assessments. For example, the maximum permissible concentrations in the Netherlands are intended to protect all species in the ecosystem from toxic effects (Sijm et al. 2002). In that case, once the term "toxic effects" is defined, one can estimate the risks and monitor for success.

Standards of enforcement and risk assessment may be combined in a common decision framework. For example, the Canadian Council of Ministers of the Environment integrates standards- and risk-based approaches into their framework for remediation of contaminated sites (CCME 1999). Remedial objectives for a site are normally set to the national guidelines, but the guidelines may be adapted to the site, or a site-specific risk assessment may be performed. Site-specific assessments are required if unusual or sensitive receptors, site conditions, or exposure conditions are identified, significant knowledge gaps exist, or guidelines are not applicable, because they do not exist or the contaminants constitute a complex mixture.

2.3 LIFE CYCLE ASSESSMENT

Life cycle assessment provides an integrative assessment of a product or industry by considering the potential environmental and health effects from the extraction of raw materials, through manufacturing, transport, use, and recycling to ultimate disposal. Like risk assessment, it has multiple uses. It can be used to determine the alternative products or processes to be used. Examples include paper vs. plastic bags, disposable vs. laundered diapers, or coal vs. nuclear power. Alternatively, it may be used by an industry as part of its due diligence to determine the environmental consequences of its activities and the potentially associated liabilities. Much of life cycle assessment has been devoted to inventorying materials and energy uses, and waste stream releases of manufacturing processes. Comparisons have been largely qualitative, employing ranking and scoring techniques, because of the difficulty of comparisons involved. For example, there is no obvious commonality to the environmental effects of extracting oil for a plastic bag vs. harvesting trees for a paper bag. However, life cycle assessments are becoming increasingly sophisticated, including analysis of uncertainty in parameters, scenarios, and models (Huijbegts et al. 2003). This trend could result in the convergence of life cycle assessment and comparative risk assessment (Sonnemann et al. 2004).

2.4 PROHIBITIONS

One alternative to risk-based decision making is the use of prohibitions against certain classes of agents or actions. For example, the 2001 Stockholm Convention on Persistent Organic Pollutants (POPs) commits signatories to eliminate or reduce production of, or trade in, chemicals that have the properties of being organic and relatively persistent. This commitment is based on the concern that, even if a chemical has little toxicity and extremely low concentrations, it may prove to have unexpected toxic properties or may accumulate in unexpected ways.

A clearly problematical example of this alternative to risk-based decision making is the prohibition on ocean disposal of sewage sludge (Weis 1996). Sewage sludge, the solid matter generated by the treatment of sewage, contains organic matter, nutrients, microbes, and household and industrial chemicals. Sludge may be dumped at sea, incinerated, buried, or incorporated into surface soil. The disposal of sewage sludge at an open ocean site (106 miles offshore) by the City of New York was permitted on the basis of studies, assessments, and reviews indicating that there were no significant ecological or health risks from the practice. However, sewage sludge and all other ocean dumping were banned by an act of Congress. The reason was public outrage over beach pollution in New Jersey and misinformation blaming New York sludge rather than the true sources (combined sewer overflow and a runoff from a landfill). This political decision has compelled the onshore disposal of sewage sludge without any consideration of the relative risks. Incineration and other land-based alternatives have also proven to be controversial, but have not led to reconsideration of the prohibition on ocean disposal.

Prohibitions of categories of chemicals or actions, in the absence of evidence of significant risks or of analysis of alternatives, may have unanticipated and undesired consequences. Persistent pesticides are replaced by more toxic pesticides, and sludge gets spread on land rather than deep ocean sediments. Because prohibitions are usually in the form of laws or treaties, they inhibit the choice of optimum decisions in individual cases. However, when formal analysis of individual decisions is difficult or unlikely, blanket prohibitions may result in net benefits to the environment.

2.5 TECHNOLOGY-BASED RULES

An alternative to risk-based regulatory decisions is to specify a particular technology or a set of approved technologies for treatment or remediation. This approach is less popular than it was formerly, because it tends to stifle technological innovation, which could reduce costs or improve performance. It also tends to preclude fitting the solution to the site-specific problem. However, when monitoring of performance is not feasible, requiring a particular technology is a relatively simple way to achieve a minimum level of protection.

2.6 BEST PRACTICES, RULES, OR GUIDANCE

A more flexible alternative to the use of a technology is to tell the engineers or managers to do the best they can. This is reflected in phrases such as "best management practices," "best available control technology," "as low as reasonably achievable," or "best practicable technology." These sorts of rules have been applied when a hazard has been identified, but could not be quantitatively assessed when the costs of achieving low risks were judged to be prohibitive relative to the benefits of the product or technology, or when risks were already low but public concern required regulatory action. Best practices and rules have become less popular as the utility of risk assessment has increased. In addition, ethical and practical considerations have intervened. Ethically, it is difficult to justify exposing one group to a higher risk than another simply because of the availability or cost of control technologies. This equity concern is reinforced by the observation that, when industries are required to achieve risk-based standards, new technologies are developed and costs of existing control technologies decline. On the other hand, strict interpretations of best practices and rules can result in treatment or remediation to levels of agents that pose minuscule risks or below-background levels of metals, radiation, or other naturally occurring agents. However, it has been argued that best available technology mandates are more effective than risk assessment and other science-based approaches, because they avoid conflicts and uncertainties concerning good science and are not delayed by data collection and analysis (Houck 2004).

Best practices and guidance are also a practical means of improving environmental quality when conventional regulatory approaches are impractical. For example, it is difficult to quantify and write emission permits for runoff of sediment and nutrients from tilled agricultural fields, but best practices such as buffer strips along streams can be encouraged or imposed (Cestti et al. 2003).

Trade or industry groups may define good practices in the form of codes of conduct. These codes allow members to take responsible actions without being disadvantaged by less responsible competitors and may forestall demands for regulation.

2.7   PRECAUTIONARY PRINCIPLE

The precautionary principle is sometimes presented as an alternative to risk assessment. However, like most environmental principles and slogans, it has many definitions and uses. There are at least 14 different versions of the principle (Foster et al. 2000). Some of them are alternatives to risk-based decision making and others are supplementary.

The most extreme versions require definitive proof of safety before allowing new technologies or actions (Foster et al. 2000). This would stop all new activities, including the introduction of greener technologies, because the criterion is impossible.

More commonly, as in the 1998 Wingspread Statement on the Precautionary Principle, actions should be regulated or existing contamination mitigated if they raise threats of harm to human health or the environment, even if the hypothesized causality is "not fully established scientifically" (Raffensperger and Tickner 1999). Similarly, the 1992 Rio Declaration on Environment and Development declares: "Where there are threats of serious or irreversible damage, lack of full scientific certainty shall not be used as a reason for postponing cost-effective measures to prevent environmental degradation." Hence, the emphasis is on precautionary actions in the absence of strong evidence. The trick is to distinguish credible threats from those that are incredible or even smoke-screens for protecting existing industries or practices from new industries or practices. This requires some sort of risk assessment.

Another definition of the precautionary principle is that the burden of proof should be placed on those who propose an action. This definition is usually vague, but in some cases it is a reaction to the use of conventional hypothesis testing in environmental assessments (Weis 1996). Such tests give priority to the null hypothesis (nothing is happening or will happen as a result of the action), and require that it be disproved with 95% confidence. A precautionary alternative would be to require proof of no significant effect, rather than of a significant effect. This would require specifying an effect, a magnitude of change, and a maximum acceptable probability of type II error. This formulation of the assessment process in terms of placing a burden of proof on any stakeholder is inconsistent with the idea of performing an unbiased assessment of the probability of effects. Ideally, risk assessments should estimate uncertainty rather than using it to force a particular decision.

Another class of definitions of the precautionary principle simply calls for a bias toward protection when making a decision. One particularly clear version states that the least hazardous alternative should be selected (Kreibel et al. 2001). This decision criterion implies the need for comparative risk assessment.

The case for a precautionary approach to environmental management is made in a European Environmental Agency review of 14 cases in which effects resulted when early warnings did not prompt precautionary actions (Harremoes et al. 2001). The lessons that they draw from the case studies amount to a call for more complete risk assessments that incorporate all uncertainties including gaps in scientific knowledge, consider alternatives, and carefully scrutinize claimed benefits as well as risks from products and technologies. This would result in better-informed decisions, which could reduce cases of both inadequate precaution and excessive precaution.

2.8 ADAPTIVE MANAGEMENT

Often, because of the complexity of ecosystems and their interactions with human activities, it is not possible to predict the consequences of an action with confidence. The precautionary approach in such situations would be to avoid action, but that is usually not possible. An alternative response is to try something and see if it works. That approach is problematical because the success or failure of a management action may result from the effects of uncontrolled variables such as the weather or population cycles. Adaptive management provides a more rigorous version of the "try it and see if it works" concept (Walters 1986). In adaptive management, the alternative actions are treated as experimental treatments. Ideally, the treatments are randomly assigned to replicate systems that may be replicated in space (e.g., lakes within a region) or in time (e.g., years for an annually breeding species). The results are then used to design or select an optimum management plan. The concept of adaptive management was developed for resource management where typical experiments have consisted of reducing harvesting fish or waterfowl in some years to determine the relationship between harvesting and recruitment. Other examples include applying alternative timber harvesting techniques to different forest watersheds or different farming practices to agricultural watersheds to determine the effects on water quality or stream habitat quality.

Adaptive management has obvious appeal in terms of realism of the assessment and defensibility of the resulting decision. However, it has seldom been applied. This is in part because of the difficulty of designing and carrying out the experiments. Like any other poorly designed experiment, an adaptive management experiment that does not have replication of treatments and systems or random assignment is subject to confounding, bias, or random effects that can cause misleading results. However, even unreplicated experiments may improve the quality of decisions, particularly if they are well monitored to minimize the potential for unrecognized confounding variables and if they monitor mechanistic parameters to confirm the plausibility of the assessment models. An example is the experimental release of water from the Glen Canyon Dam to the Grand Canyon reach of the Colorado River. The experiment was intended to determine whether simulated spring floods could restore habitat for rare species and improve recreation. While careful and extensive monitoring of the experiment gave apparently clear results, it has been recommended that the experiment be replicated (NRC 1999).

Another reason that adaptive management may fail is that the experiment is not based on an explicit model of exposure and response in the system. Such assessment models can (a) focus on the experimental design of critical system attributes, (b) suggest types and levels of treatments for the experiments, and (c) provide a means to extrapolate the experimental results so that managers know how to adjust management practices under current conditions or how to change them in future conditions (Holling 1978; Walters 1986).

Finally, adaptive management may not be used because risk managers and assessors are reluctant to admit that they do not know which alternative action is best (Walters 1986). Admission of uncertainty is particularly difficult when the costs are high or social disruption may occur. On the other hand, adaptive management experiments may be a means to defer a difficult decision.

Adaptive management can fit well with the ecological risk assessment framework. The emphasis of risk assessment on uncertainty and causal models provides a basis for determining when adaptive management is needed and what treatment levels are needed to distinguish among alternative assessment models. The clearly defined assessment endpoints provide the response parameters to be measured during the experiment. The alternative actions should be based on alternative risk models or parameterizations, so that the experiments could also be designed to elucidate the functional relationships and parameter values. The results could then be used to develop and refine assessment models for future applications. This approach has been applied to the assessment of risks to the largemouth bass population of Coralville Reservoir, Iowa, from agricultural use of dieldrin (Mauriello and Park 2002). The assessment model showed that recovery would occur if dieldrin were completely eliminated, and therefore it should be readily detected.

A trial-and-error approach may be useful, even when there is no explicit adaptive management. Many laws and regulations require that effluent permits be renewed or that products be reregistered. This provides an opportunity to examine the environmental results of the initial registration or permit, and to use the observed effects as a basis to revise the original decision. For example, a reassessment of the use of granular carbofuran considered bird kills as a basis for prohibiting previously permitted uses (OPP 1989). This process may be facilitated by monitoring of environmental effects of a permitted use.

2.9   ANALOGY

Prediction is inherently difficult, particularly when complex ecological systems are involved. Some ecologists have concluded that rather than applying conventional data and models to predict the effects of a new project or agent, one should simply study the effects of a similar project or agent and assume that the effects will be analogous (Goodman 1976). For example, if we wish to know the risk that a eutrophic system will develop rapidly in a proposed reservoir, we should look at other reservoirs in the region rather than modeling the proposed system. If the most similar reservoir is eutrophic, the new one will be so as well. This approach fails if the combination of project or agent and receiving ecosystem is not sufficiently similar. It converges on risk assessment if the analogous systems are used to develop empirical models of exposure and empirical exposure–response relationships.

2.10   ECOSYSTEM MANAGEMENT

Some have argued that real ecosystems, exposed to multiple agents, are too complex for rigorous analysis of clearly defined endpoints, as implied by the ecological risk assessment concept (Lackey 1994). One suggested alternative approach is variously termed ecosystem management or watershed approach (Christensen 1996). In practice, these assessments tend to have broadly defined goals like ecological integrity, sustainability, or health, and to emphasize stakeholder involvement in the process rather than analysis (Lackey 1998; Committee on Environment and Natural Resources 1999). This approach avoids the difficulties of analyzing complex situations by focusing on reconciling the perceptions and desires of the inhabitants or users of a watershed or regional ecosystem. The assessments may be simply descriptive, because their purpose is to achieve a common understanding of the state of the system rather than to predict the outcomes of choices (Berish et al. 1999). Ecosystem management may be implemented by land and resource managers, regulatory agencies, consortia of nongovernmental organizations (NGOs) (e.g., the Friends of the Potomac) and special interests (e.g., the Northwest Power Planning Council), and quasi-governmental commissions (e.g., the Interstate Commission on the Potomac River Basin), with various combinations of mandate, authority, and public legitimacy (Loucks 2003).

The inherently political character of many ecosystem management programs is problematical. While it is valuable to achieve consensus, or at least acquiescence, of the stakeholders, such processes are not necessarily protective of the ecosystem or its resources and may not even be socially just. Stakeholder processes may be dominated by those with the greatest financial stake or even by those with the most effective representatives. On the other hand, this approach may be effective when conventional management processes are stalemated.

2.11 HEALTH RISK ASSESSMENT

Human health risk assessment has been considered an alternative to ecological risk assessment based on the assumption that if you protect humans, nonhuman receptors will be protected as well. Although this proposition is heard less often than formerly, it is still believed by many risk managers. For a variety of reasons, some nonhuman species are likely to be more exposed to environmental contaminants than humans, or are more sensitive (Box 2.1). This is even more true for nonchemical agents. While the destruction of ecosystems for highways, reservoirs, etc. has implications for human health, it is devastating to the nonhuman species and ecosystem functions involved. Similarly, introduced species such as kudzu, nutria, Asian carp, and chestnut blight in North America, various deer in New Zealand, or mink in Europe have severe ecological effects with few, if any, health implications. The fact that nature is declining while the human species is expanding proves that protecting humans has not served to protect nonhuman species. Hence, rather than an alternative to ecological risk assessment, health risk assessment should be a complement to it (Chapter 37).

2.12 ENVIRONMENTAL IMPACT ASSESSMENT

Environmental impact assessment (EIA) is not really an alternative to ERA. It is distinguished by the nature of its mandate rather than by a different way of analyzing or managing hazards to the environment. In the United States, EIAs are legally mandated by the National Environmental Policy Act of 1969. Similar legal mandates exist in Canada, Europe, Australia, and elsewhere. In contrast, risk assessment is a practice that has been developed by regulatory agencies to provide scientific input for decision making. Hence, EIAs are imposed by law on often reluctant agencies, while ERAs are performed voluntarily by agencies to assist in performing their mandates. According to a prominent EIA practitioner, the test of EIA compliance is the enforcement of the assessment mandate, while the test of risk management is the achievement of desired risk levels (Lawrence 2003). Hence, EIAs are more concerned with acceptable process. In addition, they tend to be concerned with development projects such as dam construction or resource management, and not with typical ERA topics such as new chemicals, effluents, or contaminated sites. However, only the historical exigencies of laws and regulations are behind this distinction. Calls for more scientific rigor and analysis of uncertainty push EIA toward risk assessment practices, while calls for legal and policy reviews and for more stakeholder and public participation push risk assessment toward EIA practices.

2.13 SUMMARY

Ecological risk assessors must be aware that their assessment approach is not the only option. Some options such as prohibitions, best practices, technology-based rules, or standard setting are genuine alternatives that are useful when ecological risk assessment is not practical.

BOX 2.1
Why Human Health Risk Assessment Is Insufficient

Risk assessments have emphasized risks to human health and have largely neglected ecological effects. This bias results in part from anthropocentrism and in part from the common but mistaken belief that protection of human health automatically protects nonhuman organisms. The assumption that health risk assessments will be universally protective is justified by the protection of humans from very small risks (one in a million risks of cancer) and the use of conservative assumptions in most health risk assessments. However, there are obvious counter-examples, such as the fact that some chemicals that commonly cause severe effects to aquatic organisms like chlorine, ammonia, and aluminum, pose no risk or negligible risks to humans in drinking water. Nonhuman organisms, populations, or ecosystems may be more sensitive than humans for any of the following reasons:

1. Some routes of nonhuman exposure are not credible for humans, including respiring water, drinking from waste sumps, oral cleaning of pelt or plumage, and root uptake.

2. Chemicals are likely to be more toxic to some nonhuman species than to humans simply because there are far more nonhuman species, and some of them are likely to have properties that make them more susceptible than humans. In some cases this sensitivity is due to mechanisms of toxicity that do not occur in humans, such as eggshell thinning by DDE, stomatal closure in plants by sulfur dioxide, and imposition of male sex in snails by tributyl tin. In other cases, the cause is unknown, as in the greater sensitivity of birds and some nonhuman mammals to chlorinated dibenzodioxins.

3. There are mechanisms of action at the ecosystem level, such as eutrophication by nutrient chemicals, aquatic anaerobiosis due to degradable organic chemicals, and blockage of light by suspended solids, that have no human analogs.

4. Nonhuman organisms may be exposed more intensely to chemicals, even when the routes of exposure are the same. Any case of environmental pollution is likely to result in much higher exposure to some nonhuman organisms than to humans. Humans (at least in affluent cultures) inhabit closed dwellings, obtain a variety of food from a variety of locations, tend to move among a variety of locations, and in general are not immersed in a particular ambient environment. For example, most humans eat at most a few meals a week that contain fish as one component, and the fish come from a variety of sources, while a heron or river otter eats only fish for nearly every meal and eats the whole fish and not just the relatively uncontaminated muscle.

5. Most birds and mammals have higher metabolic rates than humans, so they receive a larger dose per unit body mass because proportionately they consume more contaminated food, drink more contaminated water, and breath more contaminated air. High metabolism may also result in more rapid metabolism of organic chemicals, but many nonhuman species have fewer metabolic enzymes than humans.

6. Some chemicals are designed to kill "pests" or "weeds" and are released to the environment at levels that are lethal to nonhuman organisms by design. In such cases, "nontarget" organisms that are physiologically and ecologically similar to the pest are inevitably affected.

7. Nonhuman organisms are highly coupled to their environments, so that even when they are resistant to a chemical they may experience secondary effects, such as loss of food or physical habitat. In contrast, humans in industrialized countries have alternate sources of food and materials for shelter if a portion of their environment is damaged by chemicals.

These arguments are mitigated somewhat by the fact that we are concerned about extremely low risks of human mortality that cannot be detected (10^{-6}) and have no significant consequences

Continued

BOX 2.1 (Continued)

for nonhuman populations. They also do not apply to mutagenic effects. Mutations are unacceptable in humans, but natural selection inconspicuously weeds them out of all but the smallest nonhuman populations. However, even for the polychlorinated dibenzo-*p*-dioxins (PCDDs), which are much-feared and regulated carcinogens, environmental exposures have been associated with significant effects on nonhuman organisms without clear and significant human effects, despite careful human monitoring. While the prompt effects of the PCDD release on humans at Seveso, Italy, were limited to some cases of chloracne, rabbits and other herbivorous animals were killed (Wipf and Schmidt 1981). No human effects were established at the Love Canal dump site that included PCDDs, but field mouse populations have been devastated by sterility and early mortality (Rowley et al. 1983; Christian 1983). The PCDD-contaminated oil at Times Beach, Missouri, killed horses, cats, dogs, chickens, and hundreds of sparrows, but apparently no humans (Sun 1983).

Others such as adaptive management can be supplementary, while those like the precautionary principle may provide guidance to risk management. Finally, some approaches such as life cycle assessment and EIA are equivalent practices developed in other contexts that may share methods and data with ecological risk assessment.

3 Ecological Risk Assessment Frameworks

Whilst a few persons, by extraordinary genius, or by the accidental acquisition of a good set of intellectual habits, may profitably work without pre-set principles, the bulk of mankind require either to understand the theory of what they are doing, or to have rules laid down for them by those who have understood the theory.

John Stewart Mill

One of the defining features of ecological risk assessment is that it follows a procedural framework that has evolved from the National Research Council framework for human health risk assessment (NRC 1983). Frameworks serve as guides for performing risk assessments, show the reader how the assessment is structured, and provide a basis for quality assurance by ensuring that necessary components are included. The health risk assessment framework was adapted to ecological risk assessment (Barnthouse and Suter 1986; EPA 1992a). The ecorisk framework has since been adapted for other nations including South Africa (Claassen et al. 2001), Australia and New Zealand (ANZ 1995; NEPC 1999), Canada (CCME 1996), the Netherlands (Gezondheidsraad 2003), and the United Kingdom (UK Department of the Environment, Food and Rural Affairs 2000). It is employed for various uses and legal contexts (Menzie and Freshman 1997; Power and McCarty 1998, 2002). These frameworks tend to be similar in the core processes of estimating ecological risk, but differ greatly in the extent to which they attempt to specify the nature of the decision-making process and the involvement of stakeholders. This chapter describes the standard US EPA framework and some alternatives that are potentially useful. It ends with a discussion of circumstances that lead to iteration of the assessment process and problem-specific frameworks.

3.1 BASIC US EPA FRAMEWORK

The most commonly used ecorisk framework is the US EPA framework, which is portrayed in Figure 3.1. It consists of planning, problem formulation, analysis, risk characterization, and risk management (Norton et al. 1992; EPA 1998a). The framework is outlined here, and developed in detail later.

Planning is a stage prior to risk assessment in which the risk manager, in consultation with the risk assessors and possibly with stakeholders, provides input to the assessment process. That input includes:

- Management goals—the assessors must know the desired condition of the environment.
- Management options—the assessors must know what actions might be taken so that they can be assessed and compared.

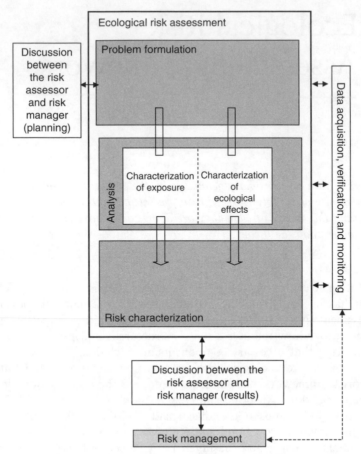

FIGURE 3.1 The US EPA framework for ecological risk assessment. (From US Environmental Protection Agency, *Framework for Ecological Risk Assessment*, EPA/630/R-92/001, Risk Assessment Forum, Washington, D.C., 1992; US Environmental Protection Agency, *Guidelines for Ecological Risk Assessment*, EPA/630/R-95/002F, Risk Assessment Forum, Washington, D.C., 1998. With permission.)

- Scope and complexity of the risk assessment—assessments are constrained by the nature of the decision (e.g., national or local), time, and resources to complete the assessment, and the risk manager's desire for completeness, accuracy, and detail.

Problem formulation is the phase in which the charge to the assessors from the risk manager is converted into a plan for performing the assessment. It includes:

- Integrating available information—assemble and summarize information concerning sources, contaminants or other agents, effects, and the receiving environment.
- Assessment endpoints—define in operational terms the environmental values that are to be protected.
- Conceptual model—develop a description of the hypothesized relationships between the sources and the endpoint receptors.
- Analysis plan—develop a plan for obtaining the needed data and performing the assessment.

Analysis is the phase in which a technical evaluation of the data concerning exposure and effects is performed.

Characterization of exposure component of the analysis consists of:

- Measures of exposure—results of measurements indicating the nature, distribution, and amount of the agent at points of potential contact with receptors
- Exposure analysis—a process of estimating the spatial and temporal distribution of exposure to the agent
- Exposure profile—a summary of the results of the exposure analysis

Characterization of effects component consists of:

- Measures of effect—results of measurements or observations indicating the responses of assessment endpoints to variation in exposure
- Ecological response analysis—a quantitative analysis of the effects data
- Stressor–response profile—the component of the ecological response analysis that specifically deals with defining a relationship between the magnitude and duration of exposure and the endpoint effects

Risk characterization is the phase in which the results of the analysis phase are integrated to estimate and describe risks. It consists of:

- Risk estimation—the process of using the results of the analysis of exposure to paramaterize and implement the exposure–response model and estimate risks, and of analyzing the associated uncertainty
- Risk description—the process of describing and interpreting the results of the risk estimation for communication to the risk manager

Risk management is the process of making a decision concerning the need for regulation, remediation, or restoration, and of determining the nature and extent of the action. Risk assessors may interact with the risk management process in two ways:

- At the end of the assessment, the results of the risk characterization may be simply communicated to the risk manager, who determines the course of action.
- The risk assessors may interface with other analysts who contribute to the decision such as cost–benefit analysts or decision analysts to provide integrated decision support.

Data acquisition is outside the ecorisk box. However, risk assessors may make calls for data during any of the three phases. In addition, the risk manager may demand that more data be collected and the assessment process be reiterated.

3.2 ALTERNATIVE FRAMEWORKS

The US EPA framework for ecological risk assessment is intended to be flexible, so in theory it can accommodate the features of all the following alternative frameworks. However, when a particular aspect of an assessment situation requires that an assessment be performed in a manner that is manifestly different from that portrayed in the standard framework, it is only fair to the user of the assessment results to present a framework that actually shows the way the assessment was done. Much of the criticism and confusion associated with the US EPA framework has to do with trying to fit all assessments into an agent-focused formalism

(Fairbrother et al. 1997; Harwell and Gentile 2000). The following alternative frameworks contain features that are useful in commonly encountered situations. These features may be selected and combined to generate a framework that represents an appropriate assessment process for the case at hand. However, any alternative to an ecorisk framework must contain the basic features of problem formulation, analysis of exposure and effects, and characterization of risks, although the names applied to the features might differ.

3.2.1 WHO-Integrated Framework

The US EPA framework for ecological risk assessment is different from its framework for human health risk assessment, which is the 1983 National Research Council framework. However, there are numerous advantages to integration of health and ecological risk assessment. They include providing a more consistent and coherent basis for decision making; greater efficiency and quality of assessments from the sharing of data, models, and insights; and the potential for assessing the effects of ecological injuries on human health and well-being. The International Programme on Chemical Safety of the World Health Organization (WHO) has developed a framework for integrated health and ecological risk assessment (Figure 3.2) (WHO 2001; Suter et al. 2003). The framework was based largely on the US EPA ecorisk framework, because it is more flexible and inclusive than any of the frameworks for health risk assessment.

The WHO framework has an advantage over the standard ecorisk framework in its treatment of interactions among risk assessors, risk managers, and stakeholders. The US EPA framework limits the contribution of risk managers to providing questions and goals prior to the problem formulation and limits stakeholders to providing input to the risk

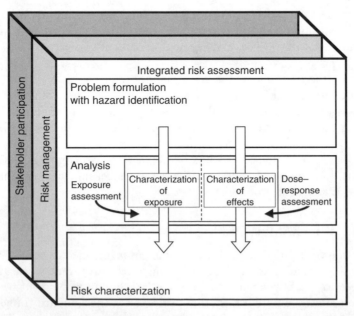

FIGURE 3.2 The framework of the World Health Organization (WHO) for integrated risk assessment. (WHO, *Report on Integrated Risk Assessment*, WHO/IPCS/IRA/01/12, World Health Organization, Geneva, Switzerland, 2001. With permission.)

manager's plans. This limited input is not consistent with all legal and regulatory contexts in the United States, and it is certainly not consistent with practices outside the United States. Hence, the WHO framework shows risk management and stakeholder processes as parallel with the risk assessment processes. Depending on the context, there may be a lot of interaction at numerous points in the process or none at all.

Even as a framework for ecological risk assessment in the US EPA, the WHO framework has the advantage of being consistent with advice from the Presidential/Congressional Commission on Risk Assessment and Management (1997) and National Research Council (1994). While the US EPA framework is designed to preserve the independence of the risk assessors from outside influences that might bias the assessment, these groups were concerned that the results of risk assessments have not been as useful as they should be because of insufficient input from decision makers and stakeholders, and that risk assessments performed in isolation have low credibility with the stakeholders and the public. The WHO framework accommodates those concerns by allowing for greater input and dialog among assessors, managers, and stakeholders. However, when implementing this framework in a particular context it is necessary to specify what interactions will actually occur. Otherwise, the assessment will not have sufficient procedural transparency to assure acceptance.

3.2.2 MULTIPLE ACTIVITIES

Most ecological risk assessments address a single activity, or a single chemical or other agent. However, some ecological risk assessments address risks from a set of activities, each of which may generate multiple agents at a site or in a region. Examples include military training exercises; management of a watershed with various land uses, effluents, and nonpoint sources; an energy technology; or the development of a transportation system. In such cases, the ecorisk process can become bogged down in its complexity if it is not broken down into modular components. Some examples are pasturage and confined feeding as activities within dairy farming, or mining, hauling, processing, combustion, condenser cooling, and ash disposal as activities within coal-fired generation of electricity. The framework shown in Figure 3.3 was developed for this purpose (Suter 1999a). An overall problem formulation is performed for the entire program which, in addition to the usual problem formulation components, divides the program into distinct activities. Each of these activities is then assessed with its own problem formulation, analysis, and characterization. Finally, the risk characterizations for the activities are combined to estimate the overall risk of the program. Clearly, this approach requires some insight in order to define activities that are either independent or dependent in such a way that their interactions can be modeled after their direct or principle effects have been estimated.

This framework also lends itself to ecosystem management. That is, rather than assessing risks from a particular chemical, waste, or even a complex program, one could assess risks to an ecosystem of all agents or activities that impinge upon it.

This framework also differs by including decision support systems and risks to humans and human activities resulting from ecological damage. Risks to human activities may be particularly important, because many activities that are the subject of risk assessment cause environmental effects that are detrimental to continuation of the activity. Obvious examples include overharvesting of fisheries, overgrazing, irrigation resulting in soil salinization, and tillage resulting in soil loss. Less obvious examples include overuse of military training sites resulting in the loss of a realistic setting, and overuse of parks resulting in the destruction of the natural resources and aesthetic qualities that attract users. This negative feedback between use of the environment and reduced utility of the environment for the users is hardly recognized in ecological risk assessments.

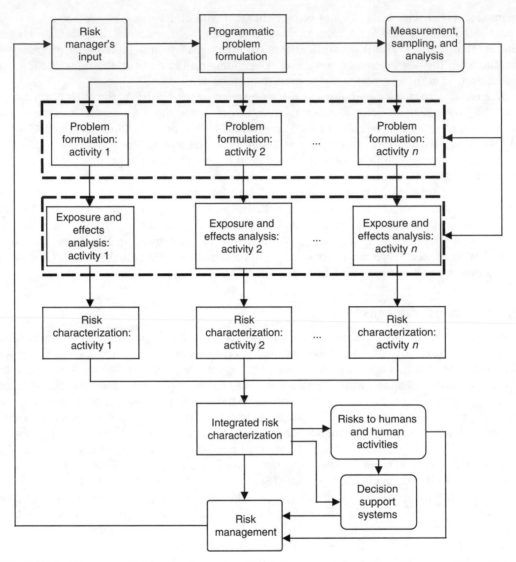

FIGURE 3.3 A framework for assessing a set of activities associated with a program. (From Suter, G.W., II., *Human and Ecological Risk Assessment*, 5, 397, 1999. With permission.)

3.2.3 ECOLOGICAL EPIDEMIOLOGY

The standard framework presumes that the agents of concern and their sources are identified and the assessors' task is to estimate the risks that they pose to the environment. However, in some cases, effects are observed but their causes are uncertain, or exposure is observed (e.g., mercury in fish flesh) but the source is unknown. In such cases, the analysis phase must address sources, exposure, and effects (Figure 3.4). Such assessments are termed ecological epidemiology or ecoepidemiology (Chapter 4). The goal of these assessments is to character-ize the causal chain from source to exposure and effects, so as to determine risks from allowing the current situation to continue or from alternative remedial or regulatory inter-ventions. This framework may be useful even in cases that begin with an identified source and agent. For example, in assessments of contaminated sites, biological surveys may identify

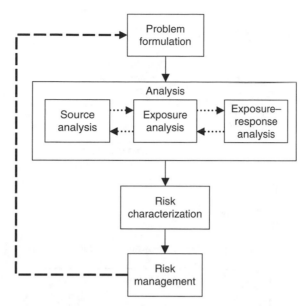

FIGURE 3.4 A framework for assessing ecoepidemiological risks in which the causal exposure or sources must be determined. (From Suter, G.W., II, Ed., *Ecological Risk Assessment*, Lewis Publishers, Boca Raton, FL, 1993. With permission.)

ecological impairments, but it may not be clear that the contaminants being assessed are the cause. Similarly, the observed toxicity of media may be due to contaminants or sources other than the subject of the assessment. For example, the waters of Poplar Creek on the Oak Ridge Tennessee Reservation, a Superfund site, were toxic, but so were the waters upstream of the reservation. The source of toxicity was an upstream municipal wastewater treatment plant. Hence, ecoepidemiological inference requires assessors to consider all credible causes of ecological effects, and not just the one that prompted the assessment.

While this ecoepidemiological framework provides considerable flexibility by simultaneously analyzing sources, exposures, and effects, it may not be the most efficient way to conduct such assessments. In many cases, an effect is observed but the cause is completely unknown, so it is not possible to reasonably formulate the problem. Rather, one should perform a causal analysis (Chapter 4) and then, once the cause is identified, perform a conventional risk assessment to determine the risks associated with remedial or regulatory alternatives.

3.2.4 CAUSAL CHAIN FRAMEWORK

The US EPA framework for ecological risk assessment, like its health risk predecessor, shows the analysis phase as consisting of an analysis of exposure and effects. This formulation is clearly appropriate to cases in which the endpoint entities are directly exposed to a chemical or other agent and in which the effect of concern is a direct result of that exposure. However, even in those cases, the division between exposure and effects is more pragmatic than inherent (Box 17.1). In ecological risk assessment, indirect effects commonly result in multistep causal chains. That is, toxic effects on one group of endpoint organisms such as forest trees must be estimated, but those effects in turn result in effects on other endpoint organisms due to loss of food or habitat structure or increased soil and nutrient export. Such indirect effects are identified during the development of the conceptual model for an assessment (Chapter 17). Indirect effects are recognized in the US EPA guidelines, but they are not explicitly incorporated in the framework.

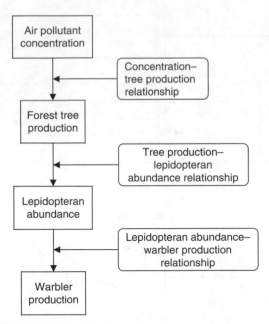

FIGURE 3.5 A hypothetical conceptual model of a risk assessment involving risks from air pollution on trees, lepidoptera, and birds.

An example of indirect effects is presented in Figure 3.5. Air pollutants damage forest trees, resulting in increased litter fall and tree mortality. This causes reduced abundance of forest lepidoptera, which in turn leads to reduced production of young birds that feed on lepidopteran larvae. Each of the system states (rectangles), except the pollutant concentration, is an endpoint effect, and each of the states except the last is an agent that may affect the succeeding states in the causal chain. Hence, the condition of the trees and the abundance of lepidoptera are both effects and causes. The probability that the effects (e.g., reduced bird production) will occur depends on the magnitude of exposure to the causal agent (e.g., the abundance of lepidopteran larvae) and an exposure–response relationship (e.g., productivity of birds as a function of larval abundance). Such chains of repeating units of cause and effect may be represented by a loop in which each effect may become a cause of additional effects (Figure 3.6) (Suter 1999a). The descending arm is the conventional exposure–response process. The ascending arm is the translation of an effect of interest (e.g., abundance of forest butterflies and moths) into properties relevant to exposure of the next receptor (e.g., abundance and biomass of larvae during the period of nestling and fledgling development). In the example, there are three loops: air pollution to trees, trees to lepidoptera, and lepidoptera to birds. If there are branches in the causal chains, the assessment process would loop through the causes and effects in each branch until the entire conceptual model is analyzed. When all loops are completed, the final risk characterization is performed to summarize all effects. This alternative version of the ecorisk framework is more explicitly ecological and ties the assessment more directly to the conceptual model.

This framework has the incidental advantage of clarifying the nature of the "characterization of effects." This step in the US EPA framework does not in fact characterize the effects; which is rather done in the risk characterization. It defines the functional relationship between exposure and effects. Hence, an exposure state is translated into an effects state and that translation is determined by a process (the triangle in Figure 3.6) that is estimated by an exposure–response function. Such state–process–state systems are drawn from the conceptual

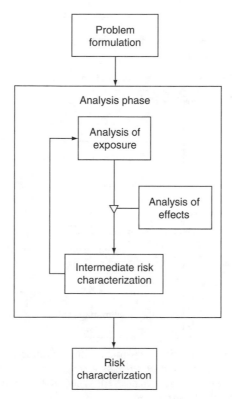

FIGURE 3.6 A framework for assessing ecological risks involving causal chains producing indirect effects. (From Suter, G.W., II, *Human and Ecological Risk Assessment*, 5, 397, 1999. With permission.)

model where they should be clearly represented (Chapter 17). Hence, this version of the ecorisk framework makes risk assessment a type of systems analysis, which has obvious potential benefits when dealing with complex systems.

3.3 EXTENDED FRAMEWORKS

Some frameworks include other types of assessments, in addition to risk assessments. These might include economic (e.g., benefit–cost; see Chapter 36), engineering feasibility, balancing of risks from contaminants at a site against risks from site remediation (Efroymson et al. 2004), environmental justice, and other assessments that contribute to decision making. An example is presented in Figure 3.7. Such frameworks have the advantage of showing the relationships among assessment activities contributing to a decision.

3.4 ITERATIVE ASSESSMENT

While a risk assessment can be performed by simply proceeding through the framework from planning to decision making, in many cases the assessments are iterated. That is, the process may be repeated one or more times until a sufficiently complete and defensible result is achieved. This may be done because more data or better models are needed to achieve sufficient confidence, the scope of the assessment must be expanded to include new issues, some issues must be analyzed in greater depth, a sequence of decisions requires a sequence of assessments, or for other reasons.

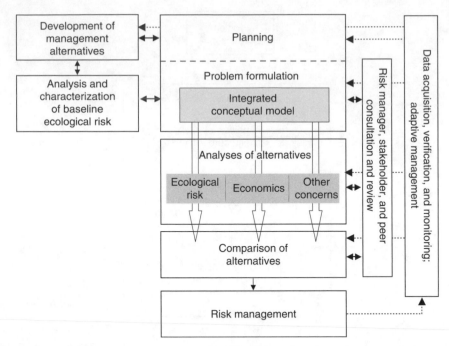

FIGURE 3.7 A framework integrating risk assessment with economic assessment. (Bruins, R.J.F. and Heberling, M.T., eds., *Integrating Ecological Risk Assessment and Economic Analysis in Watersheds: A Conceptual Approach and Three Case Studies*, EPA/600/R-03/140R, Environmental Protection Agency, Cincinnati, OH, 2004. With permission.)

One form of iterative assessment is the use of tiers of prescribed testing and measurement. The hazard assessment paradigm that preceded ecological risk assessment depended on tiering for its structure and decision logic (Figure 3.8) (Cairns et al. 1979). That is, bouts of testing and measurement were followed by simple assessments, which were followed by more testing and measurement until it was clear that a hazard did or did not exist. These tiered testing schemes were codified in the assessment methods for the regulation of pesticides and industrial chemicals (Urban and Cook 1986). Because of the potential cost-effectiveness of attempting to complete an assessment with a small and inexpensive data set before generating a more complete data set, tiered testing and assessment is still a common practice.

Recent tiered ecological risk assessment schemes are based on increasing complexity of modeling and quantitative analysis of a body of data, rather than increasing amounts of data to be analyzed. A prominent example in the United States is the aquatic and terrestrial ECOFRAM methodologies for assessing ecological risks of pesticides (ECOFRAM Aquatic Workgroup 1999; ECOFRAM Terrestrial Workgroup 1999). These methodologies define four tiers of assessment ranging from simple comparison of point estimates of exposure and effects to complex probabilistic analyses (Section 32.4). Similarly, a British framework calls for a screening assessment and two tiers of definitive assessment (UK Department of the Environment 2000). The two definitive tiers are generic quantitative risk assessments, which use standard models and assumptions, and tailored quantitative risk assessments, which use site-specific data, models, or assumptions.

Although iteration of risk assessments is often based simply on a desire for more or better information, it may also be based on a sequence of distinct types of assessments. The most common distinctions are between screening and definitive assessments and between baseline assessments and assessments of alternatives.

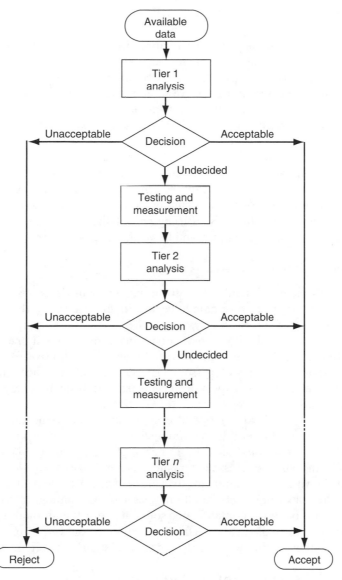

FIGURE 3.8 A framework for hazard assessment based on tiered testing and measurement.

3.4.1 SCREENING VS. DEFINITIVE ASSESSMENTS

Screening assessments are performed to narrow the scope of subsequent assessments by distinguishing issues that can clearly be ignored from those that require testing, measurement, and more complete assessment, or those that clearly require action without further defining the risk. They are analogous to the use of a screen to separate rocks from a soil sample. Screening assessments may be performed at the beginning of an assessment program to determine the objects of subsequent assessments. Examples include the screening of contaminated sites for inclusion in the US Superfund program (EPA 1990) or the screening of existing chemicals for regulatory assessments (Royal Commission on Environmental Pollution 2003). More often, they are performed at the beginning of individual site assessments. They may use existing data to quickly identify parts of a site, classes of receptors, or agents that need not be considered further (Chapter 31). Unless all issues are screened out, the results of a screening assessment will serve as input to the problem formulation for subsequent assessments. In some cases, there will be multiple iterations of screening assessment. For example, at contaminated sites a screening assessment with existing data is commonly followed by another screening assessment based on preliminary sampling and analysis, which in turn leads to focused and intensive sampling and analysis for the definitive assessment. In rare cases, screening assessments may indicate that risks are so large and manifest, that emergency actions must be taken without further measurement or assessment. Risk characterization in screening assessment is usually limited to the quotient method with conservative assumptions and safety factors. They are discussed in detail in Chapter 31.

Definitive assessments are those that are designed to define the risks and provide the basis for management decisions. Because of preceding screening assessments, they can be highly focused on the stressors, routes of exposure, and endpoints that are critical to a decision. Thus, it is possible to intensively assess those issues using probabilistic modeling, site-specific tests, or other labor- or cost-intensive techniques, and to weigh multiple types of evidence (Chapter 32).

3.4.2   BASELINE VS. ALTERNATIVES ASSESSMENTS

Although risk assessment is intended to inform management decisions that choose among alternative actions, it may be appropriate to begin with an assessment that simply determines whether any action is needed. Such assessments are termed baseline risk assessments; they determine the risks associated with current conditions, if no remedial or regulatory action is taken. Hence, they are risk assessments for the no-action alternative. Baseline assessments must consider temporal trends resulting from dispersal, degradation, accumulation, and other processes, and not simply current exposure and effects. In addition to determining the need for action, baseline assessments serve to guide the development of alternatives by defining the nature and sources of significant risks and even setting target remedial levels. Subsequent comparative assessments of alternatives estimate the risks for proposed actions and compare them to each other and to the baseline risks. For example, in an ecological risk assessment for a contaminated site, the baseline assessment might consider the risks from leaving contaminated soil in place and allowing the contaminants to degrade. The subsequent alternatives assessment would consider the risks from remedial alternatives such as capping, removal and burial, or removal and incineration. If the contaminated soil is in a forest, high-quality wetland, or other vulnerable ecosystem, the ecological risks from remediation could easily exceed those from the contaminants since remediation would destroy the ecosystem (Suter et al. 2000, Chapter 9).

3.4.3   ITERATIVE ASSESSMENT AS ADAPTIVE MANAGEMENT

Iterative assessment may be extended to include the management process. That is, a remedial action or restoration may be chosen and implemented based on a risk assessment of the alternative actions, recognizing that the goals may not be achieved. Hence, the management actions may be followed by monitoring of the results. If the goals are not achieved, the results of the monitoring can provide the basis for another iteration of risk assessment and another iteration of management actions. Hence, this type of iterative risk assessment may be an input to an adaptive management strategy (Section 2.8).

3.5   PROBLEM-SPECIFIC FRAMEWORKS

The EPA's ecological risk assessment framework is generic and applicable to any context; the assessment of particular classes of problems may benefit from a problem-specific framework. These frameworks show how to apply the generic ecorisk framework to a particular problem.

Examples include frameworks for the assessment of ecological risks from marine fish aquaculture (Nash et al. 2005), contaminated sites (Sprenger and Charters 1997), aircraft overflights (Efroymson and Suter 2001a,b), importation of animals (Murray 2002), nonnative fishes (Copp et al. 2005), and irrigation (Hart et al. 2005). These problem-specific frameworks show how the steps in the assessment process are performed for that problem. They include components such as lists of recommended assessment endpoints, generic conceptual models, and exposure–response models as well as pertinent examples. Although such frameworks do not require substantive changes in the generic assessment process, they may change the process or terminology to increase relevance to the problem or to more closely follow prior assessment practices.

3.6 CONCLUSIONS

The US EPA framework for ecological risk assessment was a real advance over prior risk assessment frameworks, particularly in terms of the problem formulation. However, ecological risk assessment in other nations, in specific regulatory contexts, and for specific classes of problems can benefit from modification or elaboration of that basic framework, so as to provide more relevant procedural guidance.

Advantages of using a single standard framework include familiarity and consistency, which reduce confusion and allow comparison and quality assurance of assessments. However, managers, stakeholders, and even assessors may reject the ecological risk assessment formalism if it does not appear to be applicable or if it requires too much stretching to fit their problem and context. Hence, the standard EPA framework is a preferred default for ecological risk assessment in the United States, but assessors should be willing to modify it as needed to produce a more useful and acceptable result.

Finally, problem-specific implementations of the ecological risk assessment framework should be developed for individual classes of assessments. These can provide specific guidance and useful information to facilitate state-of-the-science assessments.

4 Ecological Epidemiology and Causal Analysis

Glenn Suter, Susan Cormier, and Susan Norton

All reasonings concerning matters of fact seem to be founded on the relation of Cause and Effect. By means of that relation alone we can go beyond the evidence of our memory and senses.

David Hume

The assessment of the nature, causes, and consequences of observed ecological effects is termed ecological epidemiology or simply ecoepidemiology (Bro-Rasmussen and Lokke 1984; Suter 1990; Fox 1991). It is a process that uses risk assessment, but differs from conventional risk assessment in that it does not begin with the identification of a potentially hazardous agent. Rather, it begins with the observation of an undesirable biological effect. A general framework for this process of ecoepidemiology is presented in Figure 4.1. Ecoepidemiological studies may be prompted by incidental observations, assessments of contaminated sites, or studies to enforce environmental standards. They differ from the monitoring of environmental status and trends (Section 2.1) in their emphasis on causation and managerial actions.

The first impetus for ecoepidemiological assessments is observed incidents of mass mortality, deformities, or other apparent effects on organisms or ecosystems reported by ecologists performing field studies, resource managers, or members of the public. Familiar examples include observations of deformed frogs, snails with imposex, rafts of dead fish, and the decline of certain bird species. Some of these, such as fish kills, are so common that there are manuals for their investigation (Meyer and Barclay 1990).

Second, assessments of contaminated sites that have relied largely on conventional techniques based on laboratory toxicology increasingly employ observations of the biota on the site to determine the nature, magnitude, and extent of effects (Suter et al. 2000). Because the observed effects may be caused by agents other than the contaminants of concern, this requires that an epidemiological framework be employed (Section 3.2.3). In the United States, risk assessment and remediation of contaminated sites may be followed by Natural Resource Damage Assessments (NRDA) (DOI 1987) (Section 1.3.9). These assessments determine the nature, magnitude, and cause of environmental injuries for which compensatory damages may be demanded. Causal analysis is particularly important in NRDA, because of the requirement to provide legal proof that a particular pollutant caused the injury (Barnthouse and Stahl 2002).

Finally, ecoepidemiology is increasingly used in the enforcement of environmental criteria. In particular, the goal of achieving biological integrity found in the US Clean Water Act has been used as a basis for establishing biological criteria and standards based on designated uses of the water bodies (EPA 1991b). These biological criteria may be used in the enforcement of permits for point-source discharges or for achieving designated uses in the face of multiple-point and nonpoint sources. That is, the results of biological surveys may be used to help

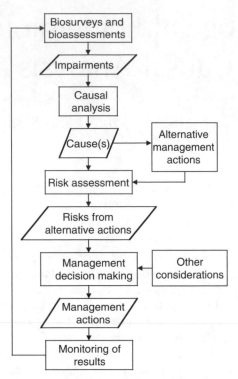

FIGURE 4.1 A logic diagram for the process of ecological epidemiology.

establish that an effluent or the pollution load from all sources is causing an unacceptable impairment. Guidance is available from the US Environmental Protection Agency (US EPA) for streams and small rivers (Gibson et al. 1996), estuaries and coastal marine waters (EPA 1997b), lakes and reservoirs (EPA 1998b), and wetlands (Danielson 1998). In addition, a semiquantitative version of the method for streams and small rivers, the Rapid Bioassessment Protocol (RBP), has been developed (Barbour et al. 1999). The concept may be further extended to watershed management programs that go beyond the limitations of laws and regulations to develop a plan to reduce or eliminate physical, chemical, and biological impairments (Serveiss 2002; EPA 2006c). These uses have inspired much of the work discussed in this chapter.

4.1   BIOLOGICAL SURVEYS

The biological surveys that prompt an ecoepidemiological assessment may be designed to support a regulatory or other management program or may be conducted for other purposes, particularly to define environmental status and trends. To the extent that the surveys are intended to support risk assessment and management, they should be based on a problem formulation (Part II): an understanding of the relevant management goals, the scope and nature of the environment, the types of effects that could prompt management actions (i.e., assessment endpoints), and the necessary data quality. The following are some of the approaches to biological surveys that have been commonly employed.

One strategy involves determining the presence or abundance of species that are presumed to be either sensitive or resistant to pollution. The original concept was to determine the presence of indicator species, which would indicate the presence or absence of a particular pollutant, usually oxygen-demanding organic material. This practice has largely fallen by the wayside and been replaced by multispecies or community properties that are indicative of a pollutant or class of pollutants. One approach is to identify sensitive and tolerant species and determine their relative abundance in a community. These approaches suffer from the fact that a sensitive or tolerant species will respond to many aspects of the environment other than the pollutant (Cairns and Pratt 1993). In addition, the classification of species as sensitive or tolerant has been subjective and inconsistent among investigators and agencies (Clements et al. 1992; Mathews et al. 2003). This problem may be remedied by basing the classification on controlled exposures, but that approach is more expensive and is rarely employed (Clements et al. 1992). Recently, some investigators have assumed that species can be defined as sensitive or resistant to anthropogenic disturbances in general, and not just to a class of pollutants. This assumption is clearly false since no species is sensitive or resistant to all chemical pollutants, much less the other agents that may impair an ecosystem (Cairns 1986). All of these problems are greater when higher-level taxa such as families are said to be sensitive or resistant, because sensitivity varies among species within taxa.

Another approach to biological surveys is to measure a large number of parameters and either combine them in a multimetric index or use multivariate statistical methods to group and compare samples, sites, or times. The multimetric index approach arithmetically combines the chosen metrics. The model for these indexes is the Index of Biotic Integrity (IBI), which combines 12 characteristics of fish communities based on presumed sensitivity, trophic status, and overt injuries (Karr et al. 1986). Similar indexes have been developed for benthic invertebrates (Kerans and Karr 1992) and peripiphyton (Hill 1997). Like other indexes, they have been criticized for their arbitrary mathematical structure, nonsense units, eclipsing of effects by other components of the metric, and other problems (Ott 1978; Suter 1993b, 2001; Taylor 1997). In addition, because they use the presumed sensitivity of species, these indexes are subject to all of the criticisms in the previous paragraph. Multivariate statistical approaches use classification, ordination, or similarity techniques to identify differences among sites, cluster similar sites, and quantify the degree of similarity (Norris and Georges 1993; Clements and Newman 2002). The most common are River Invertebrate Prediction and Classification System (RIVPACS) and Australian Rivers Assessment System (AUSRIVAS), which predict the community that should occur at a site and quantify the departure from that ideal (Wright et al. 1993; Simpson et al. 1996). These methods provide greater accuracy and precision in classifying communities than multimetric indexes do (Reynoldson et al. 1997). In addition, they are not subjective and their derivation and structure are transparent and defensible. They are criticized by the advocates of multimetric indexes for using complex statistics rather than simple arithmetic and for substituting statistics for expert judgment (Karr and Chu 1999). Both approaches tend to obscure the specific underlying biological responses. However, unlike multimetric indexes, which are claimed to estimate some immaterial ecosystem property such as integrity or health, multivariate methods make it clear that the results are simply statistical summaries of the biology.

Another approach is to survey species that are considered to be, in some sense, representative. These may be claimed to represent all species in an ecosystem or all species in a taxon. It may be assumed that other species respond similarly to the representative species, are less sensitive than the representative species, or are likely to be protected if the representative species is protected. For example, protection of spotted owls and grizzly bears is thought to protect other species in the old growth forests of the northwestern United States and the northern Rocky Mountains, respectively, because of their demanding habitat requirements. Depending on the assumed relationship between the representative species and other species to be protected, they may be referred to as flagship species, umbrella species, sentinel species, or focal species. The representative species approach has been criticized, like the sensitive species approach, for not ensuring protection and for being based on professional judgment rather than formal evidence (Simberloff 1998). However, a well-chosen flagship species may help to rally public support for conservation actions that might not be so popular otherwise. If condition measures are determined in addition to abundance, surveys of representative species are more likely to detect effects. For example, measures of age, growth, reproduction, and energy storage in white suckers are commonly used to detect effects of pulp mills and other pollutants and disturbances in Canada (Gibbons and Munkittrick 1994; Munkittrick et al. 2000; Borgmann et al. 2004).

Finally, biological surveys may measure aspects of the environment that are ecologically important and have sufficient societal or policy significance to influence decision making. When surveys are performed to support risk-based decision making, the biological properties to be measured should be based on the identification of assessment endpoints and of measures of effects that can be used to estimate those endpoints (Chapter 18). However, this does not preclude using multiple metrics in biological surveys. Any number of endpoints may be assessed and some common endpoints, such as species richness, combine metrics in a natural and comprehensible manner.

To be useful, biological surveys must include measurements or observations of habitat characteristics that influence biological properties. Guidance for obtaining habitat information is provided by the US EPA for bioassessment and rapid bioassessment, cited above. In addition, if the biological survey is conducted in response to previously identified environmental effects, it should include a survey of potential causal factors.

4.2 BIOLOGICAL ASSESSMENT

Biological assessments (or bioassessment) use the results of biological surveys to determine whether ecosystems are impaired and to define the nature and extent of the impairment. Defining an impairment requires that the assessors identify a reference state from which the impaired system departs (Section 18.2) and a degree of departure that is considered to constitute an impairment. Reference has been defined in several ways in bioassessments.

Historical reference: In many cases, the ideal state is considered to be the one that prevailed in the absence of humans or at least of Europeans. Historical reference states in the new world may be established from the records of the first European visitors and settlers or from natural records. Examples of the latter include pollen, remains of diatoms and other aquatic organisms in sediment cores from lakes or wetlands, and plant remains in pack rat middens (Cowgill 1988; Dixit et al. 1992).

Self-reference: If an impaired site was monitored before the occurrence of an impairment, the prior state of the site may be used as its own reference. Unless a long time series has been obtained, this approach provides poor estimates of variance. However, its relevance to the impairment is undeniable.

Local reference: Often reference conditions are found in undisturbed or uncontaminated locations near the impaired ecosystem. These may include upstream or upwind locations, tributaries in streams, nearby lakes, other embayments in a reservoir, and adjacent watersheds on a ridge.

Regional natural reference: A common approach to defining reference states in the United States is to define a region that is reasonably uniform ecologically, so that any forest stand or stream segment in the region can be compared to any other (Bailey 1976; Hughes et al. 1986; Omernik 1987; Klijn et al. 1995; Bailey et al. 1998). Within a region, natural undisturbed ecosystems can be used as references.

Regional acceptable reference: If undisturbed ecosystems of a particular type are rare or absent in a region, ecosystems that are least disturbed or otherwise considered to be the acceptable representatives of their type are used as references. This approach has been used for stream bioassessments in Ohio (Yoder and Rankin 1995a).

No reference: In some cases, bioassessments do not identify a reference. Rather they simply determine the distribution of a biological property across all sites and define some percentile (e.g., 10%) as dividing impaired and unimpaired. This approach is equivalent to grading on the curve.

The necessary degree of departure from the reference state to declare an ecosystem impaired may also be established in multiple ways.

Extraordinary events: Some events such as mass mortalities of fish or birds or local extinction of a native vertebrate species are manifestly significant departures from the usual state of an ecosystem. They may have natural causes, but the natural condition that caused such an event must itself be extraordinary and therefore potentially worthy of assessment.

Range of observations: If the reference condition is defined in terms of observations of natural undisturbed ecosystems, the range of observed parameter values over space or time represents the natural variability. Hence, any observed conditions outside that reference range may be considered an impairment.

Percentiles of observations: If the reference condition is defined in terms of observations of least disturbed sites, acceptable sites, or randomly chosen sites (i.e., no reference), some percentile of the distribution of observations may serve to divide impaired and unimpaired conditions. The 25th percentile of reference values has been used to define impairment in some states (Yoder and Rankin 1995a; Barbour et al. 1996).

Significant difference: Systems may be considered impaired if their condition has departed from an acceptable state in a way that is ecologically or societally significant. This significance may be established by law, regulation, or policy. For example, in the United States any killing or reduction in abundance of an endangered species is unacceptable. Alternatively, significant changes can be determined by consulting the decision makers. The US EPA has developed the Data Quality Objectives process for that purpose (Quality Assurance Management Staff 1994). Statistical significance is not a substitute for ecological or societal significance, although it is used that way in some programs (e.g., Borgmann et al. 2004).

The final step in biological assessment is to define the impairment in a useful manner. Simply declaring that an ecosystem is impaired is equivalent to defining a person as disabled. It is sufficient to prompt certain legal protections, but it does not indicate what is actually needed. Just as a paraplegic and a blind person need different assistance, an impaired ecosystem with deformed fish needs different management than one with no fish or one dominated by alien fish. If the ecosystem is declared to be impaired based on a multimetric index or a multivariate statistical technique, it is necessary to recover the constituent metrics. This disaggregation should be carried to the lowest available level. For example, the percentage of fish with deformities, erosion of fins, lesions, or tumors (DELT anomalies) is a metric of the IBI. However, because each of those types of anomalies has different causes, the DELT metric should be disaggregated. Similarly, the percentage of macroinvertebrates that are Ephemeroptera, Plecoptera, or Tricoptera (EPT taxa) is a metric in many invertebrate community indexes, but these organisms have different sensitivities to many agents, so it is important to examine the actual counts of taxa. It is also important to convert percentages to actual abundances. A decrease in the percentage of a taxon may be due to a decline in abundance of that taxon or an increase in abundance of other taxa. Next, the responses that characterize the impairment must be identified. Of the many responses that are measured in a biological survey, only a few will have changed in a way that results in a declaration of impairment. For example, Ohio defined the Little Scioto River as impaired based on the IBI and ICI, but the causal analysis characterized the specific impairment by increased relative weight and frequency of anomalies of fish and increased percentage of mayflies and tolerant invertebrate taxa, components of the indices (Norton et al. 2002).* Finally, the magnitude of the responses that constitute the impairment and their spatial and temporal distributions should be defined to the extent possible.

4.3 CAUSAL ANALYSIS

Paradoxically, we all know how to identify the causes of events and do so every day, but the concept of causation is so problematical that some philosophers, including Bertrand Russell (1957), have suggested abandoning the concept, and at least one ecologist suggested that, because causal mechanisms are complex, we should forget determining causes and focus on making empirical predictions (Peters 1991). We are, nonetheless, compelled to identify causes. No management action will be taken without some assurance that it will cause an improvement in the environment by eliminating or forestalling some cause of impairment.

The modern concept of causation derives from Galileo Galilei, who distinguished the teleological concept of causation from an empirical concept based on the relationship between events. He stated that a cause can be defined as a necessary and sufficient event: the effect must occur whenever and wherever the cause occurs and the effect must not occur in the absence of the cause. The strict empiricist, David Hume, argued that all we know of causation is association: we believe that A causes B because they are associated in space and time (contiguity), A precedes B (temporal succession), and A is always conjoined with B (consistent conjunction). This is similar to Galileo's empirical criterion, but Hume denied the concept of causal necessity and emphasized the weakness and subjectivity of causal inference. We might legitimately believe that the rooster's crow causes the sun to rise until the morning after the rooster's death. John Stewart Mill argued that, while observed associations provide a basis for hypothesizing a causal relationship, the necessary and sufficient nature of a cause cannot be demonstrated without experimentation. We must be able to manipulate the associations between putative causes and the effects (i.e., bind and gag the rooster) and observe a clear outcome. Because it is often impossible to control all potential causes or obtain a clear consistent outcome, we now commonly employ the probabilistic concept of causation based on statistical correlations developed by Karl Pearson. Because cancer is more probable in smokers, we say that smoking causes cancer. These probabilities may be derived from observation or experimentation. However, they provide a rather unsatisfying result: we know that there must be some reason that some smokers get lung cancer and others do not, but we must express our ignorance as a probability. In addition, because of confounding variables, measurement error, and misspecification, the probabilities are often unreliable.

The problem of determining causes from observations of uncontrolled systems can be illustrated by the problem of determining whether cigarette smoking causes lung cancer. The probability of contracting lung cancer is approximately 0.1 if you are a smoker and only 0.0005 if you are not (a relative risk of 200). More dramatically, the probability that a patient with lung cancer was a smoker is 0.87 (Dawes 2001). However, the great statistician Ronald Fisher pointed out that this dramatic association does not prove causation. As

---

*The selected effects violate the advice given above against results expressed as percentages or aggregate metrics (e.g., DELT anomalies), because the State of Ohio could not provide the raw data. This serves to emphasize the importance of retaining unprocessed data for reanalysis in later phases of an assessment.

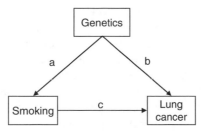

FIGURE 4.2 A conceptual model of the alternative causes of lung cancer in smokers. (a) makes people more likely to smoke; (b) a susceptibility gene for lung cancer also (c) Smoking causes lung cancer.

represented in Figure 4.2, a genetic factor that made people more likely to smoke might make them more susceptible to lung cancer, and that hypothesis would explain why so few smokers get cancer. A similar situation currently exists with respect to *Pfiesteria* spp., a marine dinoflagellate that has been associated with fish kills. Burkholder et al. (1995) have found *Pfiesteria* in waters associated with 52% of fish kills. However, the probability of a fish kill given the presence of *Pfiesteria*-like organisms is 12% (Stow 1999; Newman and Evans 2002). As with smoking, it is possible that other factors are involved. In this case, low dissolved oxygen or another pathogen may be causing the fish kills and the *Pfiesteria* may be opportunistic pathogens or even decomposers, or *Pfiesteria* may damage the skin predisposing the fish to other pathogens (Vogelbein et al. 2001) (Figure 4.3). Unknown or known factors such as Fisher's hypothetical gene, unidentified pathogens, and low dissolved oxygen are said to confound inferences concerning the possible cause being investigated.

Associations of causes and effects are generally assumed to be more reliable when observed in an experiment such as a toxicity test. It is worthwhile to consider why that is the case. Proper experimental designs include replicate systems, which assure that extraneous variance is minimized; controlled exposure to the hypothesized cause, which assures that the nature and level of the one unrandomized factor is known; and random assignment of those exposures to the individual systems, which assures that residual variance among the systems can be treated as noise. Hence, any nonrandom effects may be assumed to be caused by exposure to the treatment. Observational studies lack these three properties of good experiments. Samples of organisms above and below a pollution source are not randomly assigned to control or exposure groups, and, while they are replicate samples, they are not replicate

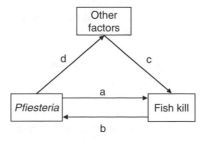

FIGURE 4.3 A conceptual model of the alternative causes of *Pfiesteria*-related fish kills. (a) *Pfiesteria* may directly cause the kills. (b) The kills may cause proliferation of *Pfiesteria*, which may feed on weakened or dead fish. (c) Other factors including low dissolved oxygen or other pathogens may cause the kills. (d) *Pfiesteria* may damage the skin of fish making them susceptible to other pathogens, particularly *Aphanomyces invadans*. Currently, causal model (a) seems most likely, but the issue is unresolved and controversial.

systems. Some studies have some but not all attributes of good experiments. For example, caged organisms placed above and below a source may be replicated and may be randomly assigned, but exposure is not fully controlled. That is, differences between the upstream and downstream location other than the pollutant emissions can confound the results. While it is clear that causation can be established in a well-designed experiment but not in a purely observational study, it is not clear how much confidence is gained by performing experiments that are not well designed. In addition, the design of experiments on complex ecological systems is not simple. For example, early experiments on the effects of plant diversity on production and other ecosystem properties included hidden treatments such as fertilization and a bias to high-production plants in high-diversity treatments (Huston 1997). Finally, the use of experimental results for causal analysis requires an extrapolation from the experimental system to the actual impaired system (see also Chapter 26). In sum, experiments can prove causal relationships within the experiment, so we know that the agent is capable of causing the effect, but not that the agent caused that effect in a real-world case.

Faced with a problem in causality, scientists often appeal to the concept of mechanism, i.e., we believe that A causes B if we know how A can cause B. What do we mean by mechanism? Simply put, it is associations at a lower level of organization than the effect to be explained. For example, a mechanistic explanation of how copper causes the local extinction of trout would involve changes in age-specific survival and fecundity of individual trout that are themselves associated with exposure to copper. A mechanistic explanation of reduced survival might involve damage to the gills and loss of the ability to maintain ionic balance, again associated with exposure to copper. In other words, mechanistic explanations are reductionistic. One of the sources of the power of science is coherence of evidence across levels of organization (this type of coherence is termed consilience; Wilson 1998a). The consistency of population responses with known organismal responses and of organismal responses with physiological responses provides some assurance that associations are likely to be causal, not coincidental. However, it must be borne in mind that these mechanisms are simply associations at a lower level, and they may themselves be based on flawed observations or may be irrelevant to the case at hand.

The potential roles of experimental and mechanistic associations can be illustrated by the cases of smoking and *Pfiesteria*. Experimental studies of smoking and lung cancer in animals could not be concluded because of the difficulty of inducing realistic exposures. However, mechanistic support has come from the identification of carcinogens in tar and other components of tobacco smoke (Figure 4.4). This analysis does not provide absolute proof that smoking causes lung cancer. In fact, the hypothesized mechanism may not operate in smokers or may not be sufficient to account for the observed cancer levels. However, as long as there is a plausible mechanism for smoking as a cause and no mechanistic evidence of the alternative, the hypothetical gene, the case for smoking is stronger. The case of *Pfiesteria* is more difficult. Because *Pfiesteria* kills fish in some investigations and not others, and, even when studies succeed in killing fish but other pathogens have not been conclusively eliminated, mechanisms are ambiguous. Mechanistic studies have also been inconclusive to date in that a toxin has not been isolated and characterized.

To summarize, Hume's contention that the evidence for causation is simply association is correct, but does not adequately acknowledge the differences among types of associations. Those from observational studies are less reliable than experimental studies. Some designs of observational studies are more reliable than others. Associations at lower levels of organization (mechanisms) are qualitatively distinct and therefore highly supportive of associations at the level of interest. This view implies that causation cannot be proven in epidemiological studies, but the strength of evidence for a cause can be analyzed and the relative strength of evidence for alternative causes may be compared.

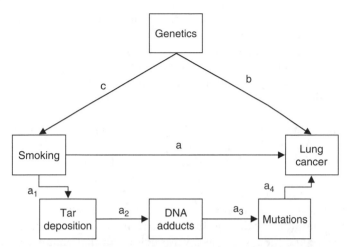

FIGURE 4.4 A mechanistically elaborated conceptual model of the alternative causes of lung cancer in smokers. One mechanism of lung cancer induction by cigarette smoking (a_1–a_4) involves the formation of DNA adducts of polycyclic and heterocyclic compounds, which results in mutations that cause cancer.

A methodology for assessing the causes of biological impairments is presented in the US EPA's stressor identification guidance (EPA 2000c; Suter et al. 2002a). This methodology was developed for application to specific sites, but it can be applied to site-independent causal analyses by ignoring the site-related aspects. It relies on two premises. The first is the comparison of alternative candidate causes. While it is not possible to prove a cause, it is possible to say which candidate cause is best supported by the evidence. The second is the application of the available data to every candidate cause in a consistent inferential process. Hence, the methodology is an instance of abductive inference, inference to the most probable explanation, which Josephson and Josephson (1996) summarize as follows:

1. D is a collection of data about a phenomenon.
2. H is a hypothesis that explains D.
3. No alternative hypothesis (H_A) explains D as effectively as H does.
4. Therefore, H is probably true.

4.3.1 Identifying Candidate Causes

Because the comparison of candidate causes is an essential feature of causal analysis, the identification of candidate causes is important not only to ensure that the true cause is included, but also to provide a basis for evaluating our relative confidence in the results of the analysis. This process requires that the assessors consider what might qualify as a cause and what potential causes should be considered candidates.

4.3.1.1 What is a Cause?

Strictly speaking, the state of an impaired ecosystem is caused by its prior state and by all external agents acting on it. While it is true that the presence of fish (the prior state) is part of the cause of a fish kill (i.e., it is a necessary condition), it is not the sort of cause that interests us. Agents that are potential causes in an ecoepidemiological study are causes that we hope to control either by preventative means to avoid future events or by interventions that restore an ecosystem to a preferred condition. Before we can influence environmental conditions, however, we need to identify the agent(s) directly harming the biological entity. After we have identified this proximate causal agent, it becomes necessary to identify how and why that causal agent came to be present in a way that was harmful. To this end, conceptual models are valuable for focusing the investigation first on the agent harming the biological entity and later on the routes that created that situation. Agents that are potential causes of interest are likely to have the following traits:

1. We are seeking anthropogenic causes. We would normally not perform a causal analysis unless it was suspected that an anthropogenic cause was involved. However, part of the investigation is to determine what is natural and what is due to human activity. To make this task even more challenging, many agents are naturally occurring. Therefore, the inappropriate introduction, frequency, duration, and level of the cause are important. For example, an impairment may be due to a combination of a metal and low pH. If the stream was naturally acidic and the metals came from an effluent, we would treat the metals as a cause and the low pH as a prior condition that enhanced the metal toxicity. However, if the metal levels were high due to watershed geology and the low pH resulted from an acidic effluent, we would consider the acidity to be the cause and the metals to be a prior condition (i.e., background). While causal analyses are performed because anthropogenic causes are suspected, natural events such as floods or droughts must be considered as alternative causes.
2. We are seeking causes that can be remediated. While it may be true to say that suburban development caused an impairment, it is not helpful. We will not remove the suburbs, but we may take actions to remediate specific agents. Therefore, assessors must determine whether the impairment is due to increased hydrological variance (flashiness), lawn fertilizers or pesticides, removal of riparian vegetation, road salt, or some other agent before we can determine the appropriate management action. Knowledge gained from ecoepidemiological studies about the causes associated with land use may inform true risk assessments. Although we may not remove a suburb, we may plan better suburbs.
3. We are seeking causes at an appropriate scale. For example, an entire watershed may have high suspended sediment loads that stress the aquatic community, but we would focus on that cause only if we were interested in explaining the state of the watershed as a whole. For an impaired reach in that watershed, we might focus first on a toxic effluent or a local habitat modification. Similarly, we would not consider normal climatic features a cause of an observed change in the ecosystem condition even if, like the lack of summer rainfall in California, they caused severe stress to organisms. We might, however, consider rare and extreme events such as droughts and hurricanes, because their temporal scale is appropriate.
4. We may be seeking more than one cause. Indeed, multiple causes are problematical. For example, algal production may be increased in a stream because of increased nutrient levels and increased light due to removal of riparian trees that shaded the stream. Both agents are necessary and neither one alone is sufficient. In such cases, the mutually acting agents should be considered together as a single cause. Such multiple-agent scenarios should be limited to the minimum set of agents to account for the effects (i.e., all agents are necessary and together they are sufficient).

Definition of causes also requires a decision about the degree of aggregation of agents. For example, we may treat all metals leached from mine wastes as the cause rather than the individual metals, or an entire effluent rather than its constituents. This aggregation is appropriate if we remediate the metals or effluent as a whole. However, it is often necessary to identify the individual constituents that must be removed. For effluents, this is readily done by using toxicity tests to confirm that the effluent is toxic and then performing a toxicity identification evaluation (TIE) to identify the toxic constituents (EPA 1991a, 1993a,b; Ankley and Schubauer-Berigan 1995).

4.3.1.2 Developing the List

The process of listing candidate causes is not fundamentally different from the processes of analyzing the evidence and characterizing the cause. In each case the assessors are looking for evidence of association. The differences are that the identification phase is less demanding of data quality or analysis, the potential causes are not compared, and potential causes may be included for political reasons even if evidence is lacking. A good strategy for completing the list of candidate causes is to prepare a list of potential causes based on a variety of sources and then develop from those a list of candidate causes to be analyzed. The following are some of the ways in which potential causes could be identified:

Consulting with stakeholders: Local governments, industries, environmental organizations, and other stakeholders are likely to have their own ideas about the cause of an ecological impairment.

Consulting with local environmental scientists: Local schools, consulting firms, and agencies may employ environmental scientists who are familiar with the impaired system or similar systems in the region and may be able to suggest potential causes.

Visiting the site: Members of the assessment team should always walk or float the site and note any signs of potential causes.

Consulting lists of pollution sources: State records of permitted effluents, Superfund sites, licensed waste disposal facilities, etc. are potential sources of potential causes.

Considering land uses: Land uses generate characteristic hazardous agents such as fertilizers, pesticides, herbicides, eroded soil, and manure from agricultural lands.

Considering past events: Spills and other discrete past events may continue to cause impairments long after their occurrence (Diamond et al. 2002).

Considering the impaired organisms, populations, or communities: To an experienced biologist, the nature of the observed changes in organisms, populations, or communities is suggestive of the potential cause of the impairment.

Using accessory data from the biological survey: Biological surveys commonly include measurements of habitat characteristics that may be indicative of causes of impairment.

Multivariate statistics may be used in a synoptic manner to find potential causes. However, it is important to remember that the most highly correlated parameters are not necessarily the most likely causes.

4.3.1.3 Developing Maps and Conceptual Models

Any site-specific assessment must develop a map of the important site features. For a stream, the map would show the stream-course, delimit the impaired reach, identify locations of effluents and sampling, indicate segments with channel modifications, and show transitions in riparian land use or cover. Analogous features would be shown in maps of other types of impaired ecosystems.

It is also important to develop conceptual models for all of the potential causal relationships (Chapter 17). These would show how each of the potential causes could directly or indirectly induce the effects that constitute the impairment. Some potential causes are consistent in that they may be portrayed in the same conceptual model. Such models may even suggest scenarios in which multiple agents are acting together. On the other hand, separate conceptual models may be developed for some potential causes because they suggest conflicting mechanisms or different organization of the system. For example, the lack of fish recruitment may be caused by toxicity-induced failure of the adults to spawn, absence of spawning substrate, or diseases of the early life stages, but not all three. A conceptual model can also show that a cause can have more than one mechanism of action. For example, nitrogen additions can result in ammonia toxicity or increased algal production, which may in turn cause changes in trophic structure, substrates embedded in algal mats, improved conditions for pathogens, or low dissolved oxygen levels. Hence, the development of a conceptual model can help to clarify the nature of the potential causes and their relationships to each other.

4.3.2   ANALYZING THE EVIDENCE

To identify causes in ecoepidemiology, evidence is assembled that supports or weakens the belief that a particular cause and effect relationship accounts for specific impairments. Evidence comes from existing or ad hoc data from the impaired site, from other sites, and from laboratory studies. These data can be analyzed to reveal five characteristics of causal relationships (Table 4.1). Analysis turns data into evidence by revealing relationships between the effect and the potential cause that have one or more of those causal characteristics. The characteristics can be used in each of the types of causal inference discussed in Section 4.3.3. In particular, the types of evidence that are evaluated in strength-of-evidence analysis (Section 4.3.3.4) are more specific aspects of these characteristics.

The smoking and *Pfieteria* examples and most other published examples of causal analyses are typically in the form of contingency tables and use associated concepts such as conditional probability. These analyses depend on dichotomous variables or, at most, a few categories, each of which may be individually associated. This works when causes are adequately characterized by their presence or absence and when effects are equally simple (alive or dead organism, present or absent species, etc.). Regulatory science tends to generate such dichotomies artificially if not naturally to classify systems as acceptable or unacceptable. However, most environmental data are based on count (e.g., number of fish) or continuous (e.g., concentrations) variables. For these cases, linear or nonlinear models may be derived. For example, the abundance of the mayfly *Hexagenia* may be regressed against the concentration of sediment ammonia. Such models have two advantages. First, by showing that effects change in a consistent way in response to exposure, they can clarify the likelihood that

---

TABLE 4.1
Characteristics of Causal Associations that are Used to Guide Analysis that Generates Evidence to Support or Weaken the Case for a Cause and Effect Relationship

Characteristic of Causal Relationships	Principle
Co-occurrence	An effect occurs where and when its cause occurs and does not occur in the absence of its cause
Sufficiency	The intensity or frequency of a cause should be adequate to produce the observed magnitude of effect
Temporality	A cause must precede its effect
Manipulation	Changing the cause must change its effect
Coherence	The relationship between a cause and effect must be consistent with scientific knowledge and theory.

an association occurs mechanistically rather than by chance. Second, they quantify a putative exposure–response relationship that can be compared to relationships from the laboratory or field experiments so that the sufficiency of the cause can be evaluated. For the same reason, it is important to derive exposure–response models for toxicity tests and other experiments used in causal analyses rather than using hypothesis testing to derive thresholds for statistical significance (Chapter 23). Additionally, using classifications to reduce count or continuous variables may artificially determine the outcome by increasing or decreasing correlations, depending on the criteria for classification. Finally, dichotomous classifications are often inappropriate, even for regulatory or remedial decisions. Often there is a range of possible decisions that require a range of assessment outcomes (e.g., area to be remediated, concentration of a chemical in an effluent, and permissible release rates).

Exposure–response correlations and models can often be improved by identifying and isolating confounding factors in either the receptors or the environment. For example, liver tumors in fish are more frequent in waters with high levels of polycyclic aromatic hydrocarbons, but also in waters with high frequencies of old fish (Baumann et al. 1996). Correction for age of fish would increase the consistency and potentially the biological gradient in the relationship of hepatic neoplasm frequency to polycyclic aromatic hydrocarbon levels. Similarly, a decline in fish species richness is a common measure of impairment, but the number of species generally increases with stream size (Vannote et al. 1980). Therefore, including stream size in the causal model could strengthen the evidence for association of a candidate cause with species loss.

4.3.2.1 Evidence of Co-occurrence

The basic evidence for causation is the spatial and temporal association of the candidate cause with the impairment at the site being assessed or at other sites (Table 4.1). The primary objectives of this analysis are to demonstrate that the cause occurs where and when the effect occurs (e.g., cadmium is detected at the site of the impairment), or, conversely, that they are not associated. The second objective is, as far as possible, to determine whether the nature or magnitude of the impairment changes with the magnitude of exposure to a candidate cause (e.g., as the concentration of cadmium decreases downstream of a source, the effects also decrease).

In general, exploration of the data from a site to find evidence of causation is best begun by plotting data related to sources, exposure, and effects on common spatial or temporal axes. Where spatial associations are complex, a geographic information system is useful. The demonstration of associations becomes difficult when time lags are involved or when either the affected organisms or the candidate cause move long distances, e.g., migrating salmon and birds that are exposed to different agents in different locations during different life stages. Various associations may be identified:

1. Observed co-occurrence of the causal agent and its direct effects
2. Observed absence of a necessary resource or habitat feature in the impaired ecosystem
3. Association of intermediate entities in the causal chain with the affected organisms
4. Presence of the contaminant, biomarker, or symptom in the affected organisms
5. Identified source and potential route of transport and exposure of affected organisms

As far as possible, the associations should be quantified. For categorical data, the frequencies or probabilities of associations are calculated. As discussed earlier for smoking and *Pfiesteria*, the seemingly simple task of quantifying the probability that a cause and effect are associated can be confusing and can result in misleading results. Assessors must be aware that simple joint probabilities differ from conditional probabilities and that sampling designs differ for each. For example, if we go to sites of fish kills and sample the water for *Pfiesteria*, we can readily calculate the conditional probability of *Pfiesteria* given a kill, because the sampling design is conditioned on kills. However, for a causal analysis we would like to know the conditional probability of a kill given the presence of *Pfiesteria*. A *Pfiesteria*-conditioned sampling program that identified sites as *Pfiesteria* present or absent and then monitored for fish kills could provide a good estimate of that probability, but would be relatively difficult to perform. One could, like Newman and Evans (2002), estimate that conditional probability by parameterizing Bayes's theorem from various studies, but with little confidence relative to a *Pfiesteria*-conditioned study (Section 5.3.2). Finally, a random sampling program could provide a good estimate of the joint probability of *Pfiesteria* and fish kills. Such joint probabilities are commonly calculated in environmental studies but are much less enlightening than the conditional probabilities (Dawes 1993, 2001). That is, knowing that only 0.2% of estuaries had a fish kill and *Pfiesteria* when randomly sampled does not tell us about causality. Hence, to sort out these correlations, it is necessary not only to define a contingency table for the candidate cause and effect to be associated, but also to understand that the conditioning of the data determines the correlations (Dawes 1993, 2001). In particular, observational studies that are conditioned on the antecedent (potential cause) are difficult to perform and yield lower probabilities than those conditioned on the effect, but they are the ones that are needed to understand causality (Dawes 1993).

4.3.2.2 Evidence of Sufficiency

The objective of analyzing data for sufficiency is to provide evidence that the cause is present in sufficient quantity or frequency at the site that the investigator would expect effects based on information from laboratory tests, field tests, or exposure–response relationships developed at other sites. This type of evidence is familiar to ecotoxicologists who combine measures of exposure from sites with exposure–response relationships from laboratory tests to estimate effects. For example, concentrations of chemicals measured in water are used in concentration–response models to estimate the frequency or magnitude of effects, which are then compared to the effects observed at the impaired stream. This analysis requires that the exposure metrics from the site be concordant with those in the exposure–response relationship (Chapter 30). More complex causal mechanisms, particularly those involving indirect causation, require more complex mechanistic models in order to establish plausible sufficiency. As models of causal processes become more complex, it becomes more difficult to judge whether an individual model provides an acceptable representation of the causes of ecological degradation at a site. In such cases, the best general strategy is to generate mathematical models of each proposed causal scenario and determine which model best describes the data (Hilborn and Mangel 1997).

4.3.2.3 Evidence of Temporality

A cause always precedes its effect. Evidence, showing with certainty that an effect preceded the causal agent, refutes the case for that agent. Likewise, evidence that the effect appeared shortly after the introduction of the cause strongly supports the case for that causal agent. However, evidence of a clear temporal sequence is uncommon, because it usually depends upon data collected over relatively long time scales, often long before the impairment is observed. Only measurements of the candidate causal agent (i.e., the proximate agent) should be used to evaluate temporal sequence: surrogates or measurements of other steps in the causal pathway are considered under other types of evidence.

Evidence of temporal sequence should be evaluated cautiously if multiple sufficient causes are present, as well as when the objective of the analysis is to identify all contributing causes rather than the most likely cause first. Under these circumstances, candidate causes occurring early in the time sequence may mask the effects of candidate causes occurring later, even though those candidates may contribute to the observed effects.

4.3.2.4   Evidence from Manipulation

When effects are diminished after a candidate cause is eliminated or reduced, it provides strong causal evidence (Table 4.1). Exposures may be reduced by manipulating the source. For example, cattle may be fenced from some locations where they have access to a stream, or an effluent may be eliminated for a time due to plant shutdown. These manipulations may be conducted at the site being assessed or may be conducted at other sites where the same type of source operates. Exposures may also be reduced by a regulatory or remedial action, which may be treated as a manipulation. Alternatively, experiments may be conducted that control the exposure of organisms or communities to potential causes, e.g., caging previously unexposed organisms at contaminated locations, placing containers of uncontaminated sediments in locations with contaminated water, or shading stream segments. These field experiments are often not replicated, so their results are potentially subject to confounding, but less so than associations derived from observations of the site. Finally, site media containing multiple chemicals or other agents can be brought into the laboratory and manipulated to eliminate or add different candidate causes. Then the manipulated media can be tested using laboratory organisms to determine which component causes the effects. These methods have been most extensively developed for the purpose of attributing causality among different chemicals in effluents (Chapter 24).

Valuable ancillary information is derived from the details of a manipulation, but the desired characteristic of the evidence is the fact that an effect was or was not altered by changing a putative cause. Evidence from manipulation of a cause can be so powerful that an unreplicated example at a site or even a similar site convinces most people even though the results could be coincidental. For example, demonstration that biological recovery followed a reduction of suspended solid loadings in the Androscoggin River in central Maine provided evidence that ended a contentious ecoepidemiological investigation in the Presumpscot River in southern Maine (EPA 2000c).

4.3.2.5   Evidence of Coherence

Knowledge of the manner in which a candidate cause could induce an effect can be used in three ways to generate evidence. First, intermediate steps in an indirect causal process may be identified and observed or measured. For example, if trout abundance is thought to be reduced by effects of forest pesticides on their aquatic insect prey, the reduced abundance of aquatic insects may be observed. Such intermediate steps in indirect causation should be found in the conceptual model. Second, whether effects are indirectly or directly induced, signs of lower-level mechanisms may be observed such as low growth rates or low fat stores in the trout. Third, knowledge of the mechanism of action of an agent may suggest effects other than induction of the impairment that may be observed or measured. For example, if pesticides are affecting trout through effects on insects rather than direct toxicity, fish that do not consume insects such as stone rollers should be abundant. These types of evidence are particularly useful when the ultimate effects of alternative candidate causes are similar, but they act through different causal mechanisms. It may be sufficient to document the occurrence of the intermediate step, but in many cases the level of the metric must be shown to be sufficient. For example, is the prey reduced sufficiently to account for the loss?

4.3.3   Characterizing Causes

We may follow Popper to the extent of claiming somewhat greater decisiveness for falsification than affirmation, but we cannot do without induction and affirmative tests.

Susser (1988)

The characterization phase of causal analysis uses the evidence developed in the analysis phase to determine the most likely cause. Four methods are discussed here: elimination of impossible causes, diagnosis, Koch's postulates, and strength-of-evidence analysis. Elimination is discussed first because it is most reliable and typically reduces the list of candidate causes for subsequent methods. Diagnosis is applied when the effect includes characteristic symptoms that can be used in a conventional medical or veterinary approach. For pathogens and chemicals, Koch's postulates can be applied. However, for the majority of causal analyses, application of elimination, diagnosis, or Koch's postulates will not result in a confident conclusion, and a strength-of-evidence analysis is performed to compare the evidence tending to support or refute each of the remaining candidate causes. These methods may be applied iteratively if they are not sufficiently conclusive on first application.

4.3.3.1   Elimination

Although causation cannot be proved, it can be disproved. As Popper and others have argued, no number of associations of putative cause and effect is as conclusive as an absence of association (Platt 1964; Popper 1968). That is, if cause *c* precedes effect *e* in 20 cases, we may hypothesize that *c* is a necessary cause of *e*. However, if in the 21st case *e* occurs without *c*, we have disproved that hypothesis. Note that Popperian disproof is not probabilistic, because probabilities cannot be refuted (Greenland 1988; Lanes 1988). If we eliminate a cause with 95% confidence, we have no more proof than if we confirmed a cause, so we are in the realm of considering the strength of evidence. However, we can use observations to logically eliminate candidate causes with the same certainty as deductive inferences (Greenland 1988; Maclure 1998). In particular, eliminative induction is commonly used by epidemiologists in outbreak studies to eliminate causes to which some cases were not exposed (e.g., if some of the people who got sick at the picnic did not eat the potato salad, it was not the cause). Candidate causes can be eliminated if:

- The effect began before the candidate cause
- There is no route of exposure for at least some of the affected species (e.g., the effect occurs upstream of the source)
- A link in the causal chain is missing (i.e., there are multiple mechanistic steps between the occurrence of the candidate cause and the effect, and at least one of those steps does not occur)
- The candidate cause occurs in reference ecosystems without the effect
- Elimination of the candidate cause does not eliminate the effect

Note that any of these inferences can be negated by incomplete or poor-quality information. For example, the effect may not occur in an exposed reference ecosystem because the reference is not well matched to the ecosystem of concern. If the assessors and reviewers are not fully confident that the candidate cause is eliminated by the evidence, the candidate cause should be retained for further analysis.

In theory, elimination could identify the cause of an impairment. If all possible causes have been identified and all but one have been eliminated, the remaining one is the true cause.

However, it is extremely difficult to identify all candidate causes with certainty. Therefore, the elimination step typically serves to reduce the number of candidate causes that are to be analyzed by other methods.

4.3.3.2 Diagnostic Protocols and Keys

In medical and veterinary practice, the causes of diseases are diagnosed by examining symptoms and determining which cause is indicated by the observed symptom set. In some cases, the diagnosis is aided by a dichotomous key or even an expert system. Similar approaches have been developed for diagnosing the causes of fish kills (Meyer and Barclay 1990), fish diseases (US Fish and Wildlife Service 2001), wildlife kills (Roffe et al. 1994), and wildlife diseases (US Geological Survey 1999). Texts and published reviews on veterinary, wildlife, fishery, and plant pathology also provide useful information on diagnosing the causes or disease, deformity, or death in nonhuman organisms. For example, a crinkle leaf of cotton is associated with manganese toxicity, and an accumulation of purple pigment in soybean leaves can signal cadmium toxicity (Foy et al. 1978). The diagnostic approach is useful when the impairment is defined in terms of effects on organisms that are available for examination or necropsy.

The concept of using symptoms to diagnose causes of impairments has been extended to higher levels of biological organization. Patterns of relative abundance of age classes and other population parameters may be used to diagnose the causes of population declines (Munkittrick and Dixon 1989; Gibbons and Munkittrick 1994). Similarly, it has been suggested that patterns of relative abundance of aquatic taxa may be used to diagnose the causes of impairments in aquatic communities (Yoder and Rankin 1995b; Norton et al. 2000; Simon 2002). None of these methods have proven to be sufficiently reliable to serve to identify the cause of ecological impairments. However, they may serve to help identify potential causes, which can then be analyzed by other methods (Section 4.3.1).

4.3.3.3 Koch's Postulates

In cases in which a chemical or pathogen is believed to cause the observed effects, a version of Koch's postulates may be used to organize the multiple lines of evidence. Koch's postulates provide a standard of proof for causation in medical and veterinary microbial pathology. They have been adapted to chemical toxicity in humans (Yerushalmy and Palmer 1959; Hackney and Linn 1979), air pollution effects on crops (Adams 1963) and forest trees (Woodman and Cowling 1987), and toxicity to ecological systems (Suter 1990, 1993a). The ecotoxicological version is as follows:

1. The injury, dysfunction, or other putative effects of the agent must be regularly associated with exposure to the agent and any contributory causal factors. This is Hume's now familiar requirement of consistency of association. However, the added requirement that other contributory factors such as appropriate levels of pH or temperature must also be associated is necessitated by the complexity of ecological causation.
2. Indicators of exposure to the agent must be found in the affected organisms. For pathogens, this means isolating the pathogen from the diseased organisms. For chemicals, it means finding elevated levels of the chemical, a metabolite, or a characteristic biomarker.
3. The effects must be seen when normal organisms are exposed to the agent under controlled conditions, and any contributory factors must contribute in the same way during the controlled exposures. This requirement is met by toxicity tests or by inoculating organisms with the isolated pathogen, thereby inducing the disease or other effect. Because of differences between laboratory and field conditions, exposure duration, life

stages, etc., application of this criterion requires some extrapolation (Chapter 26), but these should be minimized.

4. The same indicators of exposure and effects must be identified in the controlled exposures as in the field. For pathogenic diseases, this requires that the pathogen be isolated from the test organisms and shown to be the same as the pathogen that caused illness in the field. For chemicals, it can be met by demonstrating that body burdens, biomarkers, or some characteristic response occurs in the tests at levels similar to the field.

Koch's postulates are seldom satisfied in ecological studies, and they are applicable to a relatively narrow range of cases. However, they organize multiple lines of evidence in a way that can be considered a practical standard of proof. To serve as a proof of cause, the inferences must be based on sufficient high-quality data. For example, the associations for the first rule should be founded on well-designed and conducted field studies with an adequate number of cases spanning the area and range of ecosystem types within which the effect occurs. Hence, Koch's postulates are most useful when studies can be designed and conducted to satisfy them, or when an important ecological effect prompts multiple focused studies that together complete the postulates (Box 4.1). Otherwise, it is best to use all evidence in the strength-of-evidence analysis.

BOX 4.1
Application of Koch's Postulates to the Decline of Peregrine Falcons

In the 1950s and 1960s, populations of peregrine falcons precipitously declined in the United States (Cade and Fyfe 1970). Alternative causal hypotheses included shooting, collection by falconers, loss of prey, disease, and toxic chemicals (Hickey 1969). The evidence used to ultimately prove that DDT was the cause can be shown to satisfy Koch's postulates.

1. Association of DDT with the decline was difficult to establish. At first the effect was too vaguely defined to allow clear association. Changes in abundance can occur naturally or due to the alternative anthropogenic causes. However, the association of reproductive failure with eggshell thinning provided a specific effect that could be more clearly associated with potential causes (Hickey and Anderson 1968). In addition, the eyries that were no longer occupied were not spatially associated with DDT applications. However, at a larger scale, the reproductive failure of peregrines associated with eggshell thinning occurred in regions where DDT was used, after the introduction of DDT, and in all regions where there was significant DDT use but not in other regions. Hence, the association met all three of Hume's criteria for causal association: (a) spatial and temporal contiguity; (b) temporal succession; and (c) consistent conjunction.
2. DDT metabolites were found in thin-shelled peregrine falcon eggs and the adults that laid them (Hickey and Anderson 1968; Cade et al. 1971).
3. Standard avian test species were more resistant to DDT than falconiforms, so early test results were not supportive of the hypothesis. However, tests performed on falcons (kestrels) demonstrated effects at realistic exposure levels (Wiemeyer and Porter 1970; Lincer 1975).
4. The effects on tested kestrels, including eggshell thinning, were the same as those seen in peregrines in the field and occurred at similar body residues (Lincer 1975).

While this evidence is sufficient to demonstrate causation, other evidence was available that does not fit Koch's methodology. For example, analogy to the concurrent effects observed in brown pelicans and bald eagles supported the inference that DDT caused the decline of peregrine falcons. The use of strength-of-evidence analysis would allow incorporation of that evidence, and would include evaluation of the evidence for and against the alternative causes.

4.3.3.4 Strength-of-Evidence Analysis

While philosophers agree that we cannot prove causation, causes must be identified so that effects can be managed. Pragmatists have developed various approaches to meet this need. One is to use the professional judgment of an expert or team of experts as in the Surgeon General's Advisory Committee that resolved the issue of smoking and lung cancer in the United States (US Department of Health Education and Welfare 1964). An alternative is to evaluate all of the available types of evidence in a consistent manner. The best known of these is Hill's criteria (Hill 1965), which were expanded (Susser 1988) and adapted for ecoepidemiology (Fox 1991; Suter 1993a, 1998b; Beyers 1998). They were adapted for site-specific ecological impairments by the EPA (2000c). The criteria were further modified in response to user comments, and that version appears in the US EPA Causal Analysis/Diagnosis Decision Information System (CADDIS B; http://www.epa.gov/caddis) and here. This approach has been criticized, because none of the criteria prove causation and each can be misleading in some cases (Rothman 1986; Weed 1988). However, even if we acknowledge that the results of such analyses rely on expert judgment, it is at least judgment applied in a consistent and transparent manner that can determine which candidate cause is best supported by the available information.

4.3.3.4.1 Types of Evidence Used for Causal Inferences

The following types of evidence, which are equivalent to Hill's criteria, are divided into two major groups. Evidence using data from the case under investigation is evaluated first, because it is often the most compelling and highly relevant to the case. Then evidence from outside of the case, e.g., from other field or laboratory studies, is brought in to increase confidence that the observed associations are indeed causal. In practice, this step involves examining the evidence generated by associating measures of the potential causes and the effects (Section 4.3.2), and determining what specific type of evidence it is (Table 4.2 and Table 4.3) so that the strength of its contribution to the causal inference can be evaluated (see Table 4.4).

It is very unlikely that all the types of evidence described here will be available for a particular causal analysis. However, consideration of all types of evidence helps assessors to understand the inferential implications of the available information, and helps to direct any data collection toward generating critical evidence.

4.3.3.4.2 Types of Evidence that Use Data from the Case

Causal analysis should begin with an examination of evidence from the case at hand. For example, a field biologist might observe that effects occur when a particular candidate cause is present, but do not occur when it is absent, and that association may be quantified as a temporal correlation. The associations described in this section often provide the most compelling evidence for characterizing causes, and when confidence is sufficiently high, causes can even be confirmed or eliminated from further consideration.

Spatial or temporal co-occurrence. The biological effect must be observed where and when the cause is observed, and must not be observed where and when the cause is absent. For agents that flow downstream from a source, co-occurrence implies that effects occur downstream and not upstream of an identified source. For agents that do not flow, such as low habitat structure, co-occurrence implies that the effects occur in the same location as the agent, and not where it is absent. This consideration should be interpreted with caution when several sufficient causes may be present and when the objective of the analysis is to identify all potential and contributing causes. In this situation, causes occurring upstream may mask the effects of causes that only occur downstream.

TABLE 4.2
Types of Evidence That Use Data from the Case

Type of Evidence	Concept	Causal Characteristic
Spatial–temporal co-occurrence	The biological effect is observed where and when the causal agent is observed and is not observed in the absence of the agent	Co-occurrence
Evidence of exposure or biological mechanism	Measurements of the biota show that relevant exposure has occurred or that other biological processes linking the causal agent with the effect have occurred	Co-occurrence
Causal pathway	Precursors of a causal agent (components of the causal pathway) provide supplementary or surrogate evidence that the biological effect and causal agent are likely to have co-occurred	Co-occurrence
Exposure–response relationships in the field	The intensity or frequency of biological effects at the site increases with levels of exposure to the causal agent and decrease with decreasing levels	Co-occurrence
Manipulation of exposure	Field experiments or management actions that decrease or increase exposure to a causal agent decrease or increase the biological effect	Manipulation
Laboratory tests of site media	Laboratory tests of site media can provide evidence of toxicity, and toxicity identification evaluation (TIE) methods can provide evidence of specific toxic chemicals, chemical classes, or nonchemical agents	Manipulation
Verified prediction	Knowledge of the causal agent's mode of action permits prediction of unobserved effects that can be subsequently confirmed	Co-occurrence
Temporal sequence	The cause must precede the biological effect	Temporality
Symptoms	Biological measurements (often at lower levels of biological organization than the effect) can be characteristic of one or a few specific causal agents; a set of symptoms may be diagnostic of a particular cause if they are unique to that cause	Coherence

Evidence of exposure or biological mechanism. Measurements of the biota can show that relevant exposure to the cause has occurred, or that other biological mechanisms linking the cause to the effect have occurred. Evidence for exposure or mechanism may include measurements such as body burdens of chemicals, presence of parasites or pathogens, or biomarkers of exposure. For agents that do not leave internal evidence (e.g., siltation, some pesticides), it may be possible to observe behavior related to the mechanism (e.g., avoidance or convulsions) or to compare responses of organisms with different feeding or life history strategies that might discriminate among different mechanisms.

Causal pathway. Steps in the pathways linking sources to the cause can serve as supplementary or surrogate indicators that the cause and the biological effect are likely to have co-occurred. Data relevant to the hypothesized steps linking a candidate cause to potential sources can be used to assess the likelihood that the agent is present. These steps in the causal pathway serve as surrogates for the proximate cause when data on the causal agent itself are unavailable, and can bolster confidence that an agent is present. However, the absence of evidence of a causal pathway cannot be used to eliminate a cause, because there is always the possibility that an unknown source or pathway has generated the candidate cause.

TABLE 4.3
Types of Evidence that Use Data from Elsewhere

Type of Evidence	Definition	Causal Characteristic
Exposure–response from other field studies	The causal agent in the case is at levels that are associated with similar biological effects in other field studies	Sufficiency
Exposure–response from laboratory studies	The causal agent in the case is at levels that are associated with related effects in laboratory studies, which may test chemicals, materials, or contaminated media from sites contaminated by the same chemical, a mixture, or other agent as the case	Sufficiency
Exposure–response from ecological simulation models	The causal agent in the case is at levels that are associated with similar effects in mathematical models that simulate ecological processes	Sufficiency
Manipulation of exposure at other sites	At similarly affected sites, field experiments or management actions that alter exposure to a causal agent also alter the biological effects	Manipulation
Mechanistically plausible cause	The relationship between the causal agent and biological effect is consistent with known principles of biology, chemistry, and physics and with properties of the affected organisms and the receiving environment	Coherence
Analogous agents	Evidence that agents that are similar to the candidate causal agent in the case cause effects similar to the effect observed in the case is supportive of that candidate causal agent as the cause	Coherence

Exposure–response relationships from the field. The effect should increase with increasing magnitude or duration of exposure to the agent. This is the classic requirement of toxicology that effects must be shown to increase with dose, but it is also applicable to other types of causes. For example, if a high degree of cobble embeddedness from siltation is believed to cause reduced diversity of benthic invertebrates, diversity should decline along a gradient of increasing embeddedness. Other examples include demonstrating recovery of a community downstream of an outfall, or evidence that an effect decreases with concentration of an effluent or with increasing mean flow. Note that there may be nonlinearities in these gradients including apparent thresholds. This consideration requires that the effects be expressed as count or continuous data, rather than as pass or fail, so that the magnitude of effects can be evaluated. Regression analysis may quantify the gradient, and both high slopes and large correlation coefficients increase the strength of evidence.

Manipulation of exposure. This type of evidence refers to the manipulation of a cause by eliminating a source or altering exposure. The most compelling manipulations are controlled field experiments that involve eliminating or reducing exposure to a source (e.g., fencing cattle out of a stream), changing the level of an agent (e.g., adding large woody debris to a stream), or artificially inducing exposure to an agent (e.g., placing caged organisms at sites with varying levels of the suspected causal agent). Ideally, changes in both the causal agent and the observed biological effect should be measured before and after manipulation. The power of this type of evidence comes from the control of exposure achieved by deliberate manipulation of events, and even the potential for replication. However, uncontrolled experiments (e.g., elimination of effluent during facility shutdown) can also be useful.

TABLE 4.4 (Continued)
A System for Scoring the Types of Evidence Used for Causal Inferences

Type of Evidence	Finding	Score[a]
	Types of evidence that use data from the case	
Spatial/temporal co-occurrence	The effect occurs where or when the candidate cause occurs, or the effect does not occur where or when the candidate cause does not occur.	+
	It is uncertain whether the candidate cause and the effect co-occur.	0
	The effect does not occur where or when the candidate cause occurs, or the effect occurs where or when the candidate cause does not occur.	− − −
	The effect does not occur where and when the candidate cause occurs, or the effect occurs where or when the candidate cause does not occur, and the evidence is indisputable.	R
Temporal sequence	The candidate cause occurred prior to the effect.	+
	The temporal relationship between the candidate cause and the effect is uncertain.	0
	The candidate cause occurs after the effect.	− − −
	The candidate cause occurs after the effect, and the evidence is indisputable.	R
Exposure--response relationship in the field	A strong effect gradient is observed relative to exposure to the candidate cause, at spatially linked sites, and the gradient is in the expected direction.	+ +
	A weak effect gradient is observed relative to exposure to the candidate cause, at spatially linked sites, or a strong effect gradient is observed relative to exposure to the candidate cause, at non-spatially linked sites, and the gradient is in the expected direction.	+
	An uncertain effect gradient is observed relative to exposure to the candidate cause.	0
	An inconsistent effect gradient is observed relative to exposure to the candidate cause, at spatially linked sites, or a strong effect gradient is observed relative to exposure to the candidate cause, at non-spatially linked sites, but the gradient is not in the expected direction.	−
	A strong effect gradient is observed relative to exposure to the candidate cause, at spatially linked sites, but the relationship is not in the expected direction.	− −
Causal pathway	Data show that all steps in at least one causal pathway are present.	+ +
	Data show that some steps in at least one causal pathway are present.	+
	Data show that the presence of all steps in the causal pathway is uncertain.	0
	Data show that there is at least one missing step in each causal pathway.	−
	Data show, with a high degree of certainty, that there is at least one missing step in each causal pathway.	− − −
Evidence of exposure or biological mechanism	Data show that exposure or the biological mechanism is clear and consistently present.	+ +
	Data show that exposure or the biological mechanism is weak or inconsistently present.	+
	Data show that exposure or the biological mechanism is uncertain.	0
	Data show that exposure or the biological mechanism is absent.	− −
	Data show that exposure or the biological mechanism is absent, and the evidence is indisputable.	R

TABLE 4.4 (Continued)
A System for Scoring the Types of Evidence Used for Causal Inferences

Type of Evidence	Finding	Score[a]
Manipulation of exposure	The effect is eliminated or reduced when exposure to the candidate cause is eliminated or reduced, or the effect starts or increases when exposure to the candidate cause starts or increases.	+ + +
	Changes in the effect after manipulation of the candidate cause are ambiguous.	0
	The effect is not eliminated or reduced when exposure to the candidate cause is eliminated or reduced, or the effect does not start or increase when exposure to the candidate cause starts or increases.	− − −
	The effect is not eliminated or reduced when exposure to the candidate cause is eliminated or reduced, or the effect does not start or increase when exposure to the candidate cause starts or increases, and the evidence is indisputable.	R
Laboratory tests of site media	Laboratory tests with site media show clear biological effects that are closely related to the observed impairment.	+ + +
	Laboratory tests with site media show ambiguous effects, or clear effects that are not closely related to the observed impairment.	+
	Laboratory tests with site media show uncertain effects.	0
	Laboratory tests with site media show no toxic effects that can be related to the observed impairment.	−
Verified predictions	Specific or multiple predictions of other effects of the candidate cause are confirmed.	+ + +
	A general prediction of other effects of the candidate cause is confirmed.	+
	It is unclear whether predictions of other effects of the candidate cause are confirmed.	0
	A prediction of other effects of the candidate cause fails to be confirmed.	−
	Multiple predictions of other effects of the candidate cause fail to be confirmed.	− − −
	Specific predictions of other effects of the candidate cause fail to be confirmed, and the evidence is indisputable.	R
Symptoms	Symptoms or species occurrences observed at the site are diagnostic of the candidate cause.	D
	Symptoms or species occurrences observed at the site include some but not all of a diagnostic set, or symptoms or species occurrences observed at the site characterize the candidate cause and a few others.	+
	Symptoms or species occurrences observed at the site are ambiguous or occur with many causes.	0
	Symptoms or species occurrences observed at the site are contrary to the candidate cause.	− − −
	Symptoms or species occurrences observed at the site are indisputably contrary to the candidate cause.	R
	Types of evidence that use data from elsewhere	
Mechanistically plausible cause	A plausible mechanism exists.	+
	No mechanism is known.	0
	The candidate cause is mechanistically implausible.	− −

Continued

TABLE 4.4 (Continued)
A System for Scoring the Types of Evidence Used for Causal Inferences

Type of Evidence	Finding	Score[a]
Exposure–response from laboratory studies	The observed relationship between exposure and effects in the case agrees quantitatively with exposure–response relationships in controlled laboratory experiments.	+ +
	The observed relationship between exposure and effects in the case agrees qualitatively with exposure–response relationships in controlled laboratory experiments.	+
	The agreement between the observed relationship between exposure and effects in the case and exposure–response relationships in controlled laboratory experiments is ambiguous.	0
	The observed relationship between exposure and effects in the case does not agree with exposure–response relationships in controlled laboratory experiments.	−
	The observed relationship between exposure and effects in the case does not even qualitatively agree with exposure–response relationships in controlled laboratory experiments, or the quantitative differences are very large.	− −
Exposure–response from other field studies	The exposure–response relationship in the case agrees quantitatively with exposure–response relationships from other field studies.	+ +
	The exposure–response relationship in the case agrees qualitatively with exposure–response relationships from other field studies.	+
	The agreement between the exposure–response relationship in the case and exposure–response relationships from other field studies is ambiguous.	0
	The exposure–response relationship in the case does not agree with exposure–response relationships from other field studies.	−
	There are large quantitative differences or clear qualitative differences between the exposure–response relationship in the case and the exposure–response relationships from other field studies.	− −
Exposure–response relationships from ecological simulation models	The observed relationship between exposure and effects in the case agrees with the results of a simulation model.	+
	The results of simulation modeling are ambiguous.	0
	The observed relationship between exposure and effects in the case does not agree with the results of simulation modeling.	−
Manipulation of exposure at other sites	At other sites, the effect is consistently eliminated or reduced when exposure to the candidate cause is eliminated or reduced, or the effect consistently starts or increases when exposure to the candidate cause starts or increases.	+ + +
	At other sites, the effect is eliminated or reduced at most sites when exposure to the candidate cause is eliminated or reduced, or the effect starts or increases at most sites when exposure to the cause starts or increases.	+
	Changes in the effect after manipulation of the candidate cause are ambiguous.	0
	At other sites, the effect is not consistently eliminated or reduced when exposure to the cause is eliminated or reduced, or the effect does not consistently start or increase when exposure to the cause starts or increases.	− −

TABLE 4.4 (Continued)
A System for Scoring the Types of Evidence Used for Causal Inferences

Type of Evidence	Finding	Score[a]
Analogous agents	Many similar agents at other sites consistently cause effects similar to the impairment.	+ +
	One or a few similar agents at other sites cause effects similar to the impairment.	+
	One or a few similar agents at other sites do not cause effects similar to the impairment.	−
	Many similar agents at other sites do not cause effects similar to the impairment.	− −
	Evaluating multiple lines of evidence as a form of evidence	
Consistency	All available types of evidence support the case for the candidate cause.	+ + +
	All available types of evidence weaken the case for the candidate cause.	− − −
	All available types of evidence support the case for the candidate cause, but few types are available.	+
	All available types of evidence weaken the case for the candidate cause, but few types are available.	−
	The evidence is ambiguous or inadequate.	0
	Some available types of evidence support and some weaken the case for the candidate cause.	−
Reasonable explanation of the evidence	There is a credible explanation for any negative inconsistencies or ambiguities in an otherwise positive body of evidence that could make the body of evidence consistently supporting.	+ +
	There is no explanation for the inconsistencies or ambiguities in the evidence.	0
	There is a credible explanation for any positive inconsistencies or ambiguities in an otherwise negative body of evidence that could make the body of evidence consistently weakening.	−

[a]The number of + and − increases with the degree to which the evidence either supports or weakens the case for a candidate cause. When confidence is very high, evidence can be used to refute (R) or diagnose (D) a cause.

Laboratory tests of site media. Controlled exposure in laboratory tests to causes (usually toxic substances) present in site media should induce biological effects consistent with the effects observed in the field. This type of evidence is most commonly used to evaluate toxic substances in water, sediment, or effluents. For example, sediments from a stream flowing through an industrial area may be tested in the laboratory using organisms such as *Hyallela azteca* or *Chironomus riparius*. Such tests may be followed by TIE procedures designed to identify the specific toxic constituents of the media (US EPA 1991a,b, 1993a,b).

Verified predictions. Knowledge of a cause's mode of action can lead to prediction and subsequent confirmation of previously unobserved effects. The ability to make and confirm predictions is one of the hallmarks of a good scientific hypothesis. For example, if the proposed cause of a fish kill is drift of an organophosphate insecticide into a stream, we could make the specific prediction that cholinesterase levels would be reduced or the more general prediction that insects and crustaceans would also be killed, but organisms without cholinergic systems would be alive. If these predicted conditions are then observed at the site, it increases confidence in the causal relationship. Multiple predictions in both the positive and negative directions would strengthen this criterion (e.g., plants and protozoa would not be harmed, but arthropods would be).

Temporal sequence. A cause must always precede its effects. For example, a baseline monitoring study showing a productive trout population before a dam was built and a population decline following construction provides evidence that the dam caused the subsequent population decline. As with co-occurrence, this criterion should be applied with caution when several sufficient causes may be present and when the objective of the analysis is to identify all potential and contributing causes. In this situation, the causes occurring early in the time sequence may mask the effects of causes occurring later.

Symptoms. Biological measurements (typically at lower levels of biological organization than the effect) can be characteristic of one or a few specific agents. The presence of these symptoms can be used to diagnose that agent as the cause; conversely, the absence of known symptoms can weaken the case for a candidate cause. For example, necropsies of fish may reveal the presence of ovotestes, and vitellogenin in serum from males. These findings are unique characteristics associated with exposure to endocrine-disrupting chemicals. Confidence in this type of evidence is increased when a larger number of characteristic symptoms are observed, or when the observed symptoms are highly specific to few potential causes. Nonspecific effects are more difficult to diagnose, so this type of evidence is more helpful when impairments are defined as specifically as possible (e.g., decreases in specific insect taxa of concern, rather than decreases in total insect abundance). In the most confident cases, symptoms and other measurements are combined into diagnostic protocols (discussed earlier).

4.3.3.4.3   Types of Evidence that Use Data from Elsewhere

Although data from the case provide the core of causal analysis, most investigations benefit from bringing in knowledge gained from laboratory studies, past experiences, and observations in other similar systems. Virtually everything that is known about an impaired ecosystem and the candidate causes of the impairment may be useful for inferring causality. For example, one of the most useful types of evidence from elsewhere uses the exposure–response relationships developed from laboratory studies. Familiar examples are single-chemical, single-species toxicity tests. These associations are related to observations from the case.

Exposure–response relationships from other field studies. At the impaired sites, the cause must be at levels sufficient to cause similar biological effects in other field studies. The objective of analyzing exposure–response relationships is to provide evidence that organisms at impaired sites are exposed to the candidate cause at quantities, durations, or frequencies sufficient to induce observed biological effects. Although these relationships most frequently have been used to evaluate chemicals, a similar approach can be used for other agents, such as sediment, water, and temperature (Chapter 23). Consider increased levels of deposited fine sediments as a candidate cause of decreased mayfly (Ephemeroptera) taxonomic richness. In this example, 15% of the stream bottom is covered with a 1 mm thick layer of fine silt. If monitoring data from sites throughout the state shows that mayfly taxonomic richness declines steadily once the area of stream bottom covered by more than 1 mm of fine silt exceeds 10%, this would support the argument that fine silt is contributing to the declines in mayfly taxonomic richness. Exposure–response relationships from other field studies are most compelling when they are based on many studies providing exposure–response curves. Preferably these studies would be from similar ecosystems and the effects data would include the specific taxa showing impairment within the case.

Exposure–response relationships from laboratory studies. Given a known relationship between the candidate cause and the effect, would effects be expected at the level of the agent seen in the environment? Randomized, replicated, controlled laboratory tests are valuable experimental tools because they allow researchers to control the variability and potential confounding factors normally encountered under field conditions. These laboratory studies may test chemicals, materials, or media from other sites with contamination similar to that from the case. Relationships can be estimated using indicators of internal exposure (e.g., body burdens), but more commonly use measures of external exposure (e.g., aqueous concentrations). The comparison of environmental concentrations from the case under investigation to laboratory-derived concentration–response relationships is a common approach used in chemical assessments and provides strong evidence of causality if concentrations are higher than those causing a relevant response. Exceedence of criteria or standards does not necessarily imply causation, because those regulatory values are intended to be safe levels.

Exposure–response relationships from ecological simulation models. The cause should be at levels associated with effects in relevant mathematical models simulating ecological processes. As with exposure–response from laboratory tests and from other field data, model output is interpreted by comparison to exposure–response information from the case. Although this may be accomplished by comparing point data from the case (e.g., the aqueous concentration at which the number of benthic arthropod species is reduced by 35%) to model output, it is much more informative to compare trends from the case such as changes in the relative abundances of scrapers and collectors along a concentration gradient. Ecological simulation models can be especially helpful when a complex network of events influences the observed effect (Chapter 27 and Chapter 28). Manipulating one potential contributing factor at a time, different scenarios can be modeled to determine how expected effects change. As models of causal processes become more complex, it is increasingly difficult to judge whether an individual model adequately represents the causes of ecological degradation at a site. In such cases, the best strategy is to generate models of each proposed causal pathway, and determine which model best explains the site data.

Mechanistically plausible cause. Given what is known about the biology, physics, and chemistry of the candidate cause, the receiving environment, and the affected organisms, is it plausible that the effect resulted from the cause? It is important to distinguish a lack of information concerning a mechanism (e.g., the ability of chemical x to induce tumors is unknown) from evidence that a mechanism is implausible (e.g., chemical x is not tumorigenic). It is also important to carefully consider whether some indirect mechanism may be responsible.

Manipulation of exposure at other sites. This type of evidence is the same as manipulating exposure at the site under investigation (discussed earlier), but extrapolates the information from a similar situation. That is, at similar impacted locations outside of the case sites, field experiments or management actions that increase or decrease exposure to a cause must increase or decrease the biological effect. A key step in this type of evidence is explicit comparison of the manipulated sites and the impaired site in terms of physical conditions, biota, and other characteristics. Extrapolation as well as fate and transport models can be used to quantify differences between the sites, allowing for more accurate extrapolation of results to the impaired sites.

Analogous agents. Is the hypothesized relationship between cause and effect similar to any well-established cases? Hill (1965) used the criterion of analogy to refer specifically to similar causes. For example, a new pesticide with a structure similar to one that is in use may induce similar effects. The idea can be extended to other types of agents. For example, an introduced species that has similar natural history characteristics to one that had been previously introduced may have similar impact on an ecological system.

4.3.3.4.4   Evaluating Multiple Types of Evidence
Determining the most probable causal agent from several candidates requires the retention and weighing of much information. Table 4.4 presents a system for scoring and summarizing the types of evidence. When applying this table to specific cases, the available types of evidence are in the left-hand column. Each of the other columns presents results for a candidate cause. The rows show the appropriate number of +, −, or 0 symbols associated with the strength of evidence for each consideration evaluated for each candidate cause. Supporting narratives describe how the scores were obtained from the evidence.

The number of + and − increases with the degree to which the evidence either supports or weakens the case for a candidate cause. Evidence can score up to three pluses (+++) or three minuses (− − −). The maximum number recommended for a particular type of evidence depends on the likelihood that an association might be observed because of chance rather than because of the true cause. Alternatively, evidence may be scored for not applicable (NA) or for no evidence (NE). There are two other types of scores: refute (R) is used for indisputable evidence that disproves that the candidate cause is responsible for the specific effects; and diagnose (D) is used when a set of symptoms for a particular causal agent or class of agents is, by definition, sufficient evidence of causation, even without the support of other types of evidence.

The scores for a candidate cause should not be added. Adding would imply that each consideration is of equal importance and that the same types of data and evidence are available across all candidates. Rather, the scores should be used to identify the types of evidence that are most compelling. They are also used to evaluate the overall consistency of the evidence, and to identify whether any inconsistencies can be explained.

Consistency of evidence. Is the hypothesized relationship between cause and effect consistent across all of the available evidence? The strength of this consideration increases with the number of lines of evidence. When the candidate cause is consistently supported or weakened by many types of evidence, the confidence in the argument for or against the cause increases (Yerushalmy and Palmer 1959).

Explanation of the evidence. Does a mechanistic conceptual or mathematical model explain any apparent inconsistencies among the lines of evidence? For example, lead concentrations may be sufficient to prevent reproduction in fish, yet juvenile and adult fish are found at the site. This evidence may be coherent if reproduction is not occurring at the site, but juvenile or adult fish recolonize the site from unexposed locations. Another explanation may be that the measured total metal concentration is not 100% bioavailable. These explanations depend on the expertise and judgment of the assessors for their strength. They constitute a weak line of evidence, because of the possibility that post hoc explanations are wrong. However, the hypotheses may lead to experiments or predictions in future iterations of the causal assessment (e.g., testing the bioavailability of the metals), which could support stronger inferences.

The final determination is reached by comparing the evidence across candidate causes. All of the candidate causes must be compared to verify if there is more than one probable cause and to ascertain the level of confidence in the overall determination. Comparison across causes ensures that each candidate is treated fairly and that any bias in data collection and analysis is acknowledged. The comparison relies on professional judgment; there is no formula. In the best case, a probable cause or causes are identified, and the information is then communicated to managers and stakeholders. When evidence is sparse, the analysis may still identify the candidate cause with the strongest support relative to the others. In some situations, no cause is identified or the confidence in conclusions will be too low to support management action. However, even when this happens, it should be clear what information would be most useful to collect to increase confidence in the conclusions.

4.3.4. ITERATION OF CAUSAL ANALYSIS

If the cause is not identified or if confidence in the identification is inadequate, it may be because the impairment was not properly specified, the information was insufficient or misleading, or the cause may not have been listed as one of the candidates. These possibilities should be investigated and the causal analysis repeated if appropriate.

If the impairment was inadequately identified, there may actually be no impairment or the impairment may be different from that identified by the bioassessment or other source. Possible reasons may include exaggeration of observed effects, inappropriate survey techniques or statistical analyses, inappropriately defined reference conditions, or poor quality assurance in the biological survey, habitat survey, or chemical analysis. For example, a stream may be declared to be impaired because it was compared with the wrong set of reference streams or because the field crew could not distinguish species of sunfish. In addition, the impairment may be correct but not sufficiently specific. For example, the cause of peregrine falcon decline could not be determined until the impairment was defined as reproductive failure associated with thin eggshells (Box 4.1). In some cases, this problem may be corrected by reanalyzing the survey data, but in many cases it will require obtaining new data.

When the impairment has been adequately defined, additional observations or measurements may be performed to support or eliminate causal associations. The insight obtained during the prior causal analysis should be used to focus these studies on critical information. In particular, it should be possible to design experiments or observations that will potentially eliminate certain causes. The ability to design such critical experiments is a hallmark of successful science. However, such experiments are not always feasible. Alternatively, one may identify critical pieces of positive evidence such as body burdens or biomarkers that would strongly support one scenario and none of the others.

Finally, a prior causal analysis, reconsideration of the impairment, or collection of new data may reveal additional candidate causes. For example, an assessment of the causes of impairment of fish and mussel populations in the upper Clinch and Powell Rivers found that none of the land use studies explained much of the variance. Therefore, chemical contamination, which had not been characterized, was hypothesized to be a significant cause (Diamond and Serveiss 2001). New candidate causes should be included in an additional iteration of causal analysis.

4.4 IDENTIFYING SOURCES AND MANAGEMENT ALTERNATIVES

In most cases, characterizing the cause of an impairment also includes identifying the source. That is, the cause and its source are clearly associated, and in many cases knowledge of the source helps to lead to identification of the cause. For example, assessors may know that a sewage treatment plant exists in the impaired stream reach, which causes them to suspect that ammonia may have caused the impairment. However, that is not always the case. For example, we may identify polycyclic aromatic hydrocarbon (PAH)-contaminated sediments as a cause that could be remediated by dredging, but the sources of PAHs need not be identified and remediated if it is clear that they are not still operating. In an urban environment, the source may not be obvious, because potential sources are numerous. Some pollutants, such as suspended sediment and nutrients, typically have multiple sources. Methods for source identification are discussed in Chapter 19.

If the source is not identified along with the cause, identifying sources to be remediated is part of the process of developing alternative management actions (Figure 4.1). Management actions are typically developed by engineers or equivalent professionals (e.g., range managers, pest management practitioners, and stream restoration hydrologists). However, ecological assessors should be involved in ensuring that appropriate alternatives are considered and that the information needed to assess the risks from those alternatives is generated.

4.5   RISK ASSESSMENT IN ECOEPIDEMIOLOGY

In some cases, a causal analysis is sufficient to support a decision. If the effects are clearly unacceptable and there is one and only one acceptable alternative for remediating it, no further assessment is needed. However, in most cases, a risk assessment will be needed because there are remedial alternatives with uncertain consequences (i.e., risks). Those uncertainties may result from lack of knowledge of the efficacy of the alternative actions, of their potential undesirable consequences, or of the long-term response of the ecosystem (Chapter 33). Risk assessments may also be needed, because the bioassessment did not address all or the important endpoints. The bioassessment may address standard indicator metrics or indexes, excluding important organisms such as mussels and anadromous fish, or important ecosystem properties such as nutrient retention. Without knowledge of risks to important endpoint attributes, decisions are not well informed. These risk assessments may be conducted and communicated to the risk manager by the methods discussed in the following chapters. If biological surveys are used to monitor the success of a management action, the results may feed back into another round of bioassessment, causal analysis, risk assessment, and decision making (Figure 4.1).

4.6   SUMMARY

Ecoepidemiology has been described as retrospective risk assessment. However, it is helpful to distinguish cases in which the cause of existing ecological impairments is unknown (ecoepidemiology) from cases in which the ecological effects of a source or agent are unknown (ecological risk assessment) (Chapter 1). In particular, ecoepidemiology uses bioassessments to characterize the impairment followed by causal analyses to determine their causes. Ecoepidemiology feeds into conventional ecological risk assessment when remedial and restoration actions are designed and evaluated. The process comes full circle when postremedial monitoring identifies residual impairments or effects of the remediation.

5 Variability, Uncertainty, and Probability

We wish truth, and we find only uncertainty.

Blaise Pascal

Risk is conventionally understood to be related to the concepts of uncertainty and probability (Chapter 1), because risk is derived from a sense that the future is imperfectly or incompletely known. Uncertainty concerning the future may be due to inherent randomness (e.g., quantum indeterminacy), effective limits on knowledge of future conditions even in deterministic systems (chaotic systems or simply the limits of data collection and analysis), or simple lack of knowledge (ignorance). The concept of probability and the related concepts of uncertainty, variability, likelihood, error, credibility, etc. are sources of confusion and controversy, which lead to linguistic uncertainty. Most environmental scientists are aware of two schools of statistics with different concepts of probability: Frequentist and Bayesian. Many are not aware that there are other approaches such as information-based statistics (Burnham and Anderson 1998, 2001) or evidence-based statistics (Taper and Lele 2004), or that many statisticians consider their field to be in need of a conceptual revolution (Gigerenzer et al. 1989; Salsburg 2001; Royall 2004). This chapter will avoid most of these issues by sticking to approaches that are in reasonably common use. Those who are uncertain about probability should consult Hacking's (2001) marvelously clear text. Those who want specifics about how quantitative methods are used in risk assessment might consult a text on quantitative methods in risk assessment such as Vose (2000), Burgman (2005), and Warren-Hicks and Moore (1998).

5.1 SOURCES OF UNPREDICTABILITY

We speak of risks and analyze probabilities because we wish to predict the future but realize that we cannot know the future. This is due to variability and uncertainty and to unquantified or unacknowledged factors.

5.1.1 VARIABILITY

Variability is the property of sets of entities or events that differ in some significant way. Examples include daily rainfall, weights of red foxes, or concentrations of a chemical in an effluent. Variability may be observed and estimated, but it cannot be reduced, because it is an inherent property of the system. Observations of a variable trait result in a distribution of frequencies of that trait. The response to variability is to form probabilities or equivalent expectations concerning that trait. Variability is an objective property of entities and events, so objectivist concepts and frequentist methods dominate its analysis. However, subjective concepts and Bayesian methods are also used. In particular, not all variable traits are observed, so personal or expert judgment may be employed to estimate probabilities of variable traits.

5.1.2 UNCERTAINTY

Uncertainty is the lack of knowledge about a system. Unlike variability, it can be reduced by obtaining additional information. The responses to uncertainty are to form beliefs or suspend judgment until data can be generated. In probabilistic analyses, beliefs, like variability, should be expressed as distribution functions. However, they are commonly expressed qualitatively using terms such as likely, credible, and reasonably certain.

When mathematical models are considered to simulate a real-world phenomenon, and not simply describe a data set, issues of model uncertainty arise that go beyond simple lack of fit. Uncertainty concerning the parameters and forms of mathematical models is a difficult issue that is usually acknowledged without being quantified (Risk Assessment Forum 1996). Model uncertainty includes issues such as linearity, the inclusion of thresholds, the aggregation of species into trophic levels, and the inclusion of feedback processes. Like parameter uncertainty, model uncertainty can be reduced by research, but it is often more difficult than simply measuring a parameter. Research to resolve model uncertainties typically must address the mechanisms underlying the modeled processes. When there are multiple credible models, model uncertainty can be estimated from the variance among model results; otherwise it can be estimated from the variance due to changes in individual assumptions of a model (Gardner et al. 1980; Rose et al. 1991).

5.1.3 VARIABILITY UNCERTAINTY DICHOTOMY

The distinction between variability and uncertainty as aspects of unpredictability dates to the earliest writing on probability (Hacking 1975). The pair is more properly referred to as Aleatory (having to do with accounting) and Epistemic (having to do with how and what we know), but that terminology has not caught on (Hacking 2001). It is also tied to the two sources of probability: variability produces frequency-type probability and uncertainty produces belief-type probability, although the connection is not necessary or consistent. In particular, subjectivists treat all probabilities as beliefs, even those that arise from variability among entities or events.

The variability/uncertainty dichotomy has become conventional in environmental risk assessment (MacIntosh et al. 1994; McKone 1994; Price et al. 1996; Risk Assessment Forum 1997; Science Policy Council 2000; Linkov et al. 2001). It has practical as well as conceptual importance.

Variability is a property of the system while uncertainty is a property of the observer: This distinction is important to decision makers and stakeholders. In health risk assessment, this is reflected in the desire to know the range of responses of humans and the sources of that variability. Uncertainty concerning that variability is a secondary concern. Similarly, in ecological risk assessments we may be interested in the range of responses of populations or ecosystems due to variability in their exposure and sensitivity. We may then be concerned about our uncertainty concerning those estimates.

Variability is irreducible while uncertainty may be reduced by research: We cannot change the inherent variability in sensitivity to cadmium within a set of organisms or of species, but we can reduce our uncertainty concerning that variability by performing toxicity tests. Uncertainty about the magnitude of a constant such as the solubility of a chemical can be reduced as well. This distinction is clearly important when performing sensitivity analyses to decide how to allocate funds for measurement and research.

Estimates of variability can be directly verified but estimates of uncertainty cannot: Hence, if we estimate the distribution of a variable parameter, such as stream flow or invertebrate density, we can verify that estimate by taking measurements of flow or density over an appropriate temporal or spatial sampling frame. We can do the same for model-generated estimates. Consider, for example, a case in which we predict the proportion of fish species that will be affected if an effluent is released. We may estimate variability in exposure by toxicokinetic modeling and variability in sensitivity using a species sensitivity distribution. The resulting estimate of the proportion of species affected could be verified by monitoring if the effluent is permitted. In contrast, the uncertainty about a prediction cannot be verified, but a series of predictions may allow assessors to determine whether they have in fact over- or underestimated their uncertainties (Section 5.2.1).

These three aspects of the variability/uncertainty dichotomy correspond to different uses of assessments. These different uses could lead to treating different sets of parameters as variable or uncertain. For example, when performing a sensitivity analysis to prioritize a sampling or testing program, all parameters in the risk model that could be measured should be treated as uncertain, and uncertainty should be distinguished from irreducible variability. In contrast, to estimate a variable endpoint attribute, only those parameters that vary with respect to that attribute should be treated as variable. For example, flow varies over time, but, if the endpoint variable is the number of species responding, it would not be included as a source of variability, because it is the same for all species in a receiving stream.

5.1.4 Combined Variability and Uncertainty

The future is unpredictable because of the joint influence of variability and uncertainty. For example, we cannot predict the minimum flows in a stream because of variability in meteorology and hydrology and because of our uncertainty concerning the applicability of available data to the future (e.g., the climate is changing and the watershed is being developed). This combination has been termed total uncertainty.

The results of analyzing total uncertainty for an unreplicated event can be termed credibilities, following Russell (1948), to indicate that they are not probabilities in a conventional sense. An example is weather reports that present the credibility of rain tomorrow, the unreplicated event, based on uncertainty in measurements and models and the stochasticity of the atmosphere. If we are interested only in the credibility of an outcome, we need not distinguish variability and uncertainty. For example, if we wish to estimate the credibility that a drought will extirpate a reintroduced population of fish in a stream in the next 50 years, we might estimate and propagate all sources of variability and uncertainty together. For Bayesians, this concept of credibility is simply a case of degree of belief (Section 5.3.2).

5.1.5 Error

Error can be introduced at any stage in the generation and analysis of data including sampling, measurement, data transcription and analysis, model selection, and presentation of results. Generally, assessors deal with error by attempting to minimize it (Chapter 9) and hoping that it is inconsequential. Error, particularly measurement error, may, however, be large enough to significantly contribute to unpredictability (Sarda and Burton 1995). If measurement error is a particular concern or if it is important to estimate the true variability underlying a data set, statistical techniques are available to estimate the error and variability components of a data set (Zheng and Frey 2005). Alternatively, error may be estimated independently by quality assurance audits or equivalent studies. For example, from studies of interlaboratory variability in aquatic toxicity tests, error is generally in the range of

3× to 5× (i.e., the highest result in a set of replicate tests is three to five times as great as the lowest) but may be >10× depending on the test, toxicant, and set of laboratories (Johnson et al. 2005). To the extent that we are aware of, and estimate, error, it is a component of uncertainty.

The importance of error in science is illustrated by the experience of physicists, who often find that data that differ from the null model by three to six standard deviations are illusory, apparently because of errors that bias the results (Seife 2000). It seems likely that such erroneous results are common in environmental studies, given the routine use of a two sigma standard (95% confidence), the many opportunities for error, and the absence of independent replication of environmental studies.

5.1.6   IGNORANCE AND CONFUSION

Ignorance and confusion make the results of an assessment uncertain in ways that the assessors do not recognize. Ignorance is the source of unknown unknowns. That is, in some cases one is uncertain about some aspect of the problem without knowing it, because of ignorance that the issue even exists. For example, due to ignorance, some assessments of aquatic risks from low dissolved oxygen (DO) have been based on measurements of daytime DO, which is elevated by photosynthesis, rather than nighttime DO, which is depressed by respiration.

Confusion is the result of excess complexity of a problem. Often it results from attempts to mentally model a problem that requires a formal conceptual, mathematical, or logical model.

5.1.7   SUMMARY OF SOURCES

We can categorize input or output parameters in terms of the types of unpredictability that characterize them:

> None—we know the answer and it is constant.
> Variability—we know the answer but it varies.
> Uncertainty—we are unsure of the answer.
> Total uncertainty—we are unsure of the answer, but we know it varies.
> Error, confusion, or ignorance—we do not know the answer at all.

5.2   WHAT IS PROBABILITY?

Since the founding of the probability theory in around 1660, it has been applied to multiple concepts (Hacking 1975). It is best to think of probability as a unit that scales 0–1 and follows certain logical rules. For example, if A and B are independent random events, the probability of observing both A and B (the joint probability) is $p(A \& B) = p(A) \times p(B)$. That relationship holds no matter how the constituent probabilities are defined or derived.

Note that probability is not the only possible unit of risk. Alternatives include qualitative scales such as acceptable/unacceptable or highly likely/likely/as likely as not/unlikely/highly unlikely. They also include frequencies and relative frequencies, which are equivalent to some probabilities, but more easily understood (Gigerenzer 2002). However, probability is the standard unit of risk.

Like any unit, probability can be applied to diverse concepts and situations and can be estimated in various ways. While there are other concepts of probability—Good (1983) describes six—the most common are probabilities as expressions of frequencies and as expressions of beliefs. Since probability is used as the measure of uncertainty in predictions, the sources of probability correspond to the sources of unpredictability (Section 5.1). The correspondence is, however, rough and imperfect.

5.2.1 Types of Probability: Frequency vs. Belief

5.2.1.1 Frequency

Probability is intuitively understood as an expression of frequency. That is, we can summarize the results of 100 coin flips as the frequency of heads (54 out of 100) and infer that the probability of heads with this coin is 0.54. The frequency becomes a probability when it is normalized as 0–1 and when it is assumed that the frequency is an estimate of an underlying property, the probability of heads. This classical view of probability and statistics is easily accepted, because it reflects the human propensity to extrapolate from experience (a series of coin flips) to a general rule (the probability of heads), which may be applied to individual instances (the next coin flip). However, this approach runs into problems when we try to apply this experience to other coins or other coin flippers. Experimental science can be thought of as a means of making real-world variability analogous to flipping a coin. Experimental design involves creating replicate systems and randomly assigning treatments to them, so that the frequency of responses can be extrapolated to probabilities of response of such systems in the real world. Problems occur when attempting to assign experimentally derived probabilities to individual events in the complex and uncontrolled real world. However, even without experimental replication, some probabilistic parameters are routinely derived from frequencies. Examples include frequencies of failures of treatment plants and of parameters resulting from meteorology such as frequencies of wind direction or of low stream flows.

Probabilities are derived from frequencies, because of the associated variability in the system that results in uncertain outcomes. If there were only one possible outcome, we would not bother to express its probability as 1. To the extent that particular types of errors can be identified and expressed as discrete variables, they can be used to estimate probabilities. If, for example, quality assurance audits provide the frequency of an error, that frequency can be used to estimate the probability that a value in a data set is erroneous. In that case, we are treating variability in the data generation process as equivalent to variation in nature. Ignorance and confusion are effectively immune to the frequentist version of probability.

5.2.1.2 Belief

Clearly, probability is applied to situations in which a frequency is not its source. For example, I might say that drought probably reduced the abundance of San Joaquin kit foxes on the Elk Hills Petroleum Reserve in the 1980s, and I might even state that the probability (or more properly the credibility) is 0.8. That is not a repeated event, so I am expressing my degree of belief rather than appealing to a frequency. In addition, even when we can appeal to frequencies our statements of probability may be better thought of as degrees of belief rather than as instances from a set of repeated observations. For example, if we speak of the probability of an effect in a stream due to a proposed domestic wastewater effluent, what frequency is relevant? If we require similar streams in the same region and very similar effluents, we are unlikely to find enough cases to estimate a frequency. However, if we accept all streams and domestic wastewater effluents, the relevance to our particular case is doubtful because of the range of stream and effluent flow rates, stream ecologies, effluent contents, technological reliability, etc. None of the frequencies provides exactly the probability that we want. This is a serious conceptual challenge for frequentist statisticians, but not for subjectivists who equate all probabilities with degrees of belief. However, it is a serious practical problem for assessment of risks for a single event, no matter what statistical framework is used.

Beliefs should correspond to frequencies in the long term. If a risk assessor says that the risk that a particular effluent will cause local extinction of brook trout is 0.7, that prediction cannot be verified, because the extinction will or will not occur. There is no condition of 70% extinct. However, over a career, predicted effects should occur 70% of the time that a risk estimate of 0.7 was derived. That verification occurs for weather reports. Over the millions of weather predictions that have been made by the US National Weather Service, it rains on 70% of days at locations for which a 70% chance of rain is forecast. However, that sort of verification is not possible for ecological risk assessors, because there are too few predictions, methods and circumstances of prediction are too diverse, and predictions of significant risk result in management actions that negate the prediction.

Finally, subjectivists argue that even when observed frequencies are converted to probabilities, the conversion is based on belief rather than logic. We have no reason for certainty that the next ten coin tosses will be like the previous ones. However, that is the way to bet.

The concept that probabilities are expressions of belief is clearly more flexible than the frequentist concept. That is, beliefs concerning variability, uncertainty, error, ignorance, and confusion can all be expressed in units of probability.

5.2.2 TYPES OF PROBABILITY: CATEGORICAL VS. CONDITIONAL

Traditionally, statisticians have been concerned with estimating the probability of some event or hypothesis. An example is the probability of a fish kill below a treatment plant. This is a categorical probability $p(y)$. However, more commonly in recent years, the probability of some event given a prior event, or of some hypothesis given data, is of concern. An example is the probability of a fish kill given that a treatment plant has failed. This is a conditional probability $p(y|x)$. The basic formula of conditional probability is

$$p(y|x) = p(x \& y)/p(x) \tag{5.1}$$

where $p(x \& y)$ is the joint probability of x and y.

This distinction may not be obvious. The reader may say that the first example is conditioned on the existence of the treatment plant. However, if we treated that example as conditional, it would refer to the probability of a kill in a stream reach, given the probability that it has a treatment plant at its head. We would not be interested in that conditional probability, because we defined the assessment problem and delineated the reach based on the known existence of the plant. That is, $x = 1$, so $p(y|x)$ reduces to $p(y)$. The second example is more appropriately conditional, because we are interested in the probability of treatment failure, which is not constrained, and its influence on the probability of a fish kill.

Conditional probability is associated with Bayesian statistics, because Bayes' rule is a formula for estimating probabilities given prior beliefs and evidence. However, Bayesian statistics also involve concepts of subjectivism and updating that are not inherent in the concepts and calculus of conditional probabilities (Section 5.3.2).

5.3 WAYS TO ANALYZE PROBABILITIES

Some probabilities are simply calculated from the logical rules of probability. For example, if the probability that an out-migrating salmon will be killed by hydroelectric dams is 0.2 and by irrigation withdrawals is 0.08, the probability of being killed by these anthropogenic hazards is $(0.2 + 0.08) - (0.2 \times 0.08) = 0.26$, i.e., the probability of being killed by either minus the probability of being killed by both, because a fish cannot be killed twice. It is equivalent to the response addition model in mixtures toxicology (Section 8.1.2). More commonly, probabilities are estimated using one of the various forms of statistics.

5.3.1   FREQUENTIST STATISTICS

The dominant school of statistics termed frequentist is concerned with estimating the frequency of defined errors given an assumed model and an experimental or sampling design. This school of statistics has been primarily concerned with testing statistical hypotheses. In this context, "probability arises not to assign degrees of belief or confirmation to hypotheses but rather to characterize the experimental testing process itself: to express how *frequently* it is capable of discriminating between alternative hypotheses and how *reliably* it facilitates the detection of error" (Mayo 2004). Hence, it tells us about the probability of errors in our procedure and not, as many users believe, about the support for hypotheses provided by evidence. The latter is supplied by "evidential relation logics (whether Bayesian, likelihoodist, hypothetico-deductive or other)" (Mayo 2004). This aspect of frequentist statistics is the reason that your consulting statistician wants to be involved in designing your experiments. The hypothesis test is using the data to test the experimental design under the assumption that the null hypothesis is true.

Frequentist statistics actually belong to two schools: Fisher's testing of a null hypothesis using *p* values and Neyman–Pearson statistics, which estimates relative error rates, usually for a null hypothesis and an alternative. The standard approach to hypothesis testing in the natural sciences is a hybrid, which is interpreted in a way that is not strictly correct but which seems to work well enough for analysis of experiments (Gigerenzer et al. 1989; Hacking 2001; Salsburg 2001). In any case, statistical hypothesis testing is, in general, not appropriate to risk assessment (Box 5.1). Frequentist statistics are also applied to more useful analyses such as confidence intervals and statistical modeling techniques such as regression analysis (Section 5.5).

BOX 5.1
Hypothesis Testing Statistics

The statistical testing of null hypotheses is commonly perceived to be an objective and rigorous approach to scientific inference. Such tests are frequently used to inform environmental management, but these uses are nearly always inappropriate. Many publications have criticized the ascendancy of statistical hypothesis testing in applied science (Parkhurst 1985, 1990; Laskowski 1995; Stewart-Oaten 1995; Suter 1996a; Johnson 1999; Germano 1999; Anderson et al. 2000; Roosenburg 2000; Bailar 2005; Richter and Laster 2005). In fairness, it must be noted that some environmental statisticians still defend statistical hypothesis testing (Underwood 2000). The following are brief descriptions of a few of the problems with hypothesis testing.

We are not testing hypotheses: Hypothesis testing statistics were developed to test the reality of a hypothesized phenomenon by attempting to refute the null hypothesis that the phenomenon does not exist. "Most formal inference is not hypothesis testing but model construction, selection and checking (formal and informal), estimation of parameters and standard errors, or calculation of confidence regions or of Bayesian posterior distributions of parameters" (Stewart-Oaten 1995). This is particularly true in ecological risk assessment. We are not interested in testing the hypothesis that a chemical has no toxicity; we want to know what its effects will be at given levels of exposure. Similarly, we are not interested in testing the null hypothesis that two streams are identical; we know they are not. We want to know how they differ. Even when researching scientific hypotheses, it is better to compare genuine alternative hypotheses than to test a null hypothesis that nobody believes (Anderson et al. 2000; Taper and Lele 2004).

We are not interested in statistical significance: We are all taught, but often forget in practice, that statistical significance has no particular relationship to biological significance or societal significance. Real-world decisions should be based on real-world significance.

Continued

BOX 5.1 (Continued)
Hypothesis Testing Statistics

Statistical significance is subject to manipulation: If a responsible party wishes to avoid statistically significant effects in a test or field study, they can use few replicates, perform many comparisons, and use imprecise techniques so that the null will not be rejected. On the other hand, if environmentalists wish to find an effect, any difference can be statistically significant if the investigators are careful to keep method errors small and are willing to take enough samples. This is a result of the fact that statistical significance is about statistical designs, particularly the number of "replicates."

The relative degree of protection is biased by the statistics: If statistical significance is the criterion, more precisely quantified, more readily sampled, and more easily replicated life stages, species, and communities are more protected. For example, in biological surveys, a given proportional loss of species is more likely to be statistically significant in invertebrates than in fish, so biocriteria are less protective of fish (Suter 1996c). Some important ecological effects may never be detectible with statistical significance in practical or ethically acceptable studies (Roosenburg 2000).

Statistical hypothesis tests with field data are nearly always pseudoreplicated: The multiple samples of contaminated media and biota that are taken from an exposed environment and used as replicates in hypothesis tests are in fact pseudoreplicates (Hurlbert 1984). The message that environmental pseudoreplicates should not be treated as replicates has been widely heard but has not been appreciated. Few risk assessors would accept the results of a toxicity test in which one rat was treated with a chemical and another was not, even if replicate samples of blood from one rat had a "significantly" different hematocrit from the other. The reason that the test would not be accepted is that while the measurements were replicated the treatment was not. There are any number of reasons why the hematocrit of one animal might be lower than another's, which has nothing to do with the treatment. In studies of contaminated sites, the site is the "rat" and the waste, effluent, or spill is the unreplicated treatment. Ecological risk assessors should be at least as leery of testing hypotheses about differences between differently treated sites as they are about differently treated animals. There are more reasons why two ecosystems might differ than why two rats might differ. Note that pseudoreplication affects all estimates of variance, not just those in hypothesis tests. However, the issue is much more important for hypothesis tests because the variance is used to decide whether a treatment has an effect.

Statistical hypothesis tests with field data almost never randomly assign treatments: Even if we have true replication (e.g., by comparing sets of replicate streams with and without sewage treatment plants), field studies almost never randomly assign receptors to treatments. That is, we do not have a set of 20 streams and randomly pick 10 to have a sewage treatment plant each and 10 to have none. In fact, streams receiving treated sewage were picked by engineers and planners for nonrandom reasons and are likely to differ in ways that nullify any hypothesis test to determine the significance of differences.

Hypothesis tests invite misinterpretation: Even when testing a scientific hypothesis with an appropriate experimental design, the statistics are often misinterpreted. A common mistake is to accept a null hypothesis when one fails to reject it (Parkhurst 1985; Anderson et al. 2000). Null hypothesis tests assume that the null is correct and determine the probability of data at least as deviant from the null as the data obtained, given that assumption. Clearly a test cannot address its basic assumption. For example, in developing a model of the aquatic toxicity of polycyclic aromatic hydrocarbons (PAHs), it was assumed that the slopes of regressions of log LC_{50} vs. log K_{ow} are equal for all species (DiToro et al. 2000). The investigators tested the null hypothesis of equal slope for 33 species and, after correcting for multiple comparisons, accepted the null hypothesis and concluded that the slopes are all equal to a universal slope of -0.97. (The correction for $n = 33$ makes it very difficult to achieve a significant difference, but that is a less fundamental problem.) The conclusion that the slopes are effectively equal is defensible, but is not justified by the hypothesis test. To provide evidence for equality of slopes within a hypothesis testing approach, the investigators should have determined the deviation δ from the universal slope that would be biologically significant and then designed a study to determine whether that difference occurs. Those who recognize that they should not accept their null based on failure to reject are often tempted to perform a post hoc analysis of the power of the test. Such retrospective power analyses are, however, technically indefensible (Goodman and Berlin 1994; Gerard et al. 1998; Shaver 1999).

5.3.2 BAYESIAN STATISTICS

Although modern Bayesian statistical practices, like frequentist statistics, are concerned with making inductive inferences, Bayes himself was concerned with a problem in deductive inference. Classical deduction is represented by syllogisms such as *A* is greater than *B* and *B* is greater than *C*, therefore *A* is greater than *C*. This logic is beyond dispute until we apply it to the real world. Then we must admit that the magnitudes of *A*, *B*, and *C* are estimated with some degree of confidence. Hence, the inference is based on some prior degree of belief in the relative magnitudes, and even deduction must be admitted to be as subject to bias and error as induction. This position, that all inference is subject to uncertainty, is termed subjectivism.

Modern Bayesianism is characterized by three concepts, none of which is adopted by all Bayesians.

Subjectivism: Bayesian statisticians argue that we know nothing about underlying distributions, so we can say nothing about error rates in estimating true frequencies. All we know is the defensibility of beliefs or of changes in beliefs given evidence. Although there are personalistic Bayesians, subjective does not necessarily mean personal. Most Bayesians believe that their analyses provide a rational basis for interpersonal subjective beliefs.

Updating: Bayesian statistics provide an appropriate way to go about modifying our beliefs given new evidence. The presumption is that we have good reasons for believing what we believe (i.e., we are rational and our reasoning to date has been coherent). Bayes' theorem tells us how to go about that updating (we multiply by the likelihood of the new evidence given the hypothesis). Non-Bayesians respond that it often makes no sense in practice to update. For example, if we have an empirical model of plant uptake of soil contaminants and we have new data on plant uptake from the site that we are assessing, should we use the data directly or should we use it to update estimates from the model? Most assessors would favor using the data directly, because plant uptake is so dependent on site characteristics. As Dennis (2004) wrote concerning the conditioning of data analyses on prior beliefs: "Why dilute the Rothschild with Boone's Farm?" Bayesians can avoid this problem by using uninformative priors, and some Bayesians insist on uninformative priors. However, a nominally uninformative prior such as a uniform distribution is still an assumption that influences a Bayesian outcome. A more defensible but less common approach is to combine data from prior instances of the phenomenon being estimated as the prior (e.g., waste tanks of the same type or closely related species) (Myers et al. 2001; Goodman 2005).

Conditional probabilities: As discussed earlier, Bayesian statistics are associated with the calculation of conditional probabilities. Those who use Bayesian analysis of conditional probabilities, but do not subscribe to subjectivism, are called logical Bayesians. Bayes' rule appears in various versions of including versions for multiple conditions and for continuous variables. The basic version of Bayes' rule is the following formula:

$$p(\text{B}|\text{A}) = [p(\text{A}|\text{B})p(\text{B})]/p(\text{A}) \tag{5.2}$$

where $p(\text{B}|\text{A})$ is the probability of B given A. B can be a parameter value, a state of a system or anything else that has a probability, but often it is described as a hypothesis. Hence, the formula may be recast as

$$p(\text{hypothesis}|\text{evidence}) = [p(\text{evidence}|\text{hypothesis})p(\text{hypothesis})]/p(\text{evidence}) \tag{5.3}$$

where $p(\text{evidence}|\text{hypothesis})$ is the likelihood function and $p(\text{hypothesis})$ is the prior.

Bayesian analysis of conditional probabilities has become a favorite tool for addressing the controversial question, does *Pfiesteria piscidae* cause fish kills (Stow 1999; Brownie et al. 2002; Newman and Evans 2002; Stow and Borsuk 2003). Newman and Evans's (2002) version is

$$p(\text{fish kill}|\textit{Pfiesteria}) = [p(\textit{Pfiesteria}|\text{fish kill})p(\text{fish kill})]/p(\textit{Pfiesteria}) \tag{5.4}$$

As is so often the case with ecological assessments, problems come in parameterizing the formula. In this case,

  p(fish kill) is taken to be the daily rate of fish kills in the Neuse and Pamlico Rivers $= 0.081$.
  p(*Pfiesteria*) is from the frequency of *Pfiesteria* in samples from east coast sites $= 0.205$.
  p(*Pfiesteria*|fish kill) is from the frequency of *Pfiesteria*-like organisms in water samples at fish kills in the Neuse and Pamlico Rivers $= 0.52$.

Hence, by Bayes' rule,

$$p(\text{fish kill}|Pfiesteria) = 0.205 \tag{5.5}$$

Note that in this analysis, the prior, p(fish kill), is not really a prior belief. It is simply the number of fish kills divided by the number of observation days. A different answer would have been obtained if the authors had used their actual prior beliefs concerning the probability of a fish kill given an episode of *Pfiesteria* occurrence. Further, the use of a daily rate as the prior does not provide the desired probability per episode. These comments are not intended to discount the analysis, but rather to highlight the conceptual difficulties involved in just one simple term of the formula. Other controversies are associated with the other two terms.

Bayesian statistics provide an alternative to frequentist statistics. Bayesian analogs of mainstream frequentist techniques are available including Bayesian hypothesis testing, Bayesian confidence intervals, etc. Bayesians are common in academic statistics departments, and their methods are increasingly being applied in industry and government agencies for a variety of purposes (Box 5.2).

5.3.3 Resampling Statistics

The availability of computers has made it possible to estimate probabilities by repeatedly sampling from a data set or distribution, and analyzing the results of those samples. If one wishes to estimate the distribution of a variable from a sample (e.g., the distribution of fish weights from a sample of five fish), one might estimate a distribution by assuming a distribution function (e.g., log normal) and using conventional frequentist statistics. However, one might doubt that we know the appropriate function and may doubt that five fish are

---

BOX 5.2
Bayesian Statistics and Sampling Design

Conventional approaches to sampling design begin with a pilot study that allows one to estimate the distribution of the variable of interest. Then, if acceptable Type I and Type II error rates are provided by the decision maker, an adequate number and distribution of samples can be estimated. This approach is the basis for the quantitative portion of the US EPA Data Quality Objectives process (Quality Assurance Management Staff 1994). In the author's experience, this is impractical, because decision makers will not admit to having acceptable error rates. This problem is avoided with Bayesian analysis, because calculations are based only on the data obtained and not on possible values given the statistical model. More importantly, Bayesian updating lends itself to efficient sampling. One simply keeps collecting data until probabilities are adequately defined or time or money run out. This approach has revolutionized clinical trials Berry et al. 2002. One keeps adding patients to the trial and updating the probability estimates until it is sufficiently clear that the drug works, causes unacceptable side effects, or does neither. This approach gets results more quickly and maximizes benefits to participants.

enough to define it. Remarkably, we can generate an estimate of the underlying distribution by repeatedly sampling from the set of five weights, with replacement, and then determining the distribution of the mean, standard deviation, or other sample statistics from that set of samples. This is the nonparametric Bootstrap method (Efron and Tibshirani 1993).

Resampling can also help us estimate probabilities from models. If we have a model with multiple parameters that are defined by distributions, we may be able to solve the model analytically by variance propagation (Morgan and Henrion 1990). However, such analytical solutions are often impractical because of nonlinearities and correlations among parameters. The solution is to use Monte Carlo simulation, which involves repeatedly sampling from each parameter distribution, solving the model, saving the results, and then reporting the distribution of those results as the distribution of the modeled variable (Section 5.5.5).

5.3.4 Other Approaches

The number of ways of analyzing and expressing variability and uncertainty is large and growing. In addition to frequentist, Bayesian, and resampling statistics, they include interval arithmetic, fuzzy arithmetic, p bounds, fault-tree analysis, and possibility theory. The lack of discussion of these techniques should not be taken to imply rejection of them. It reflects a judgment that current conventional methods are appropriate and much more likely to be accepted by reviewers and decision makers than exotic methods.

5.4 WHY USE PROBABILISTIC ANALYSES?

The first step in a probabilistic analysis must be to determine motivation. The form and content of the analysis depend on the desired output of the analysis. However, most guides to uncertainty analysis assume a particular motivation and desired output and proceed from that assumption. Reasons include the following.

5.4.1 Desire to Ensure Safety

Because of variability, a realized effect may be considerably larger than the most frequent effect. Because of uncertainty, true effects may be larger than estimated effects or may occur more frequently. Therefore, if the goal of an assessment is to ensure that all credible hazards are eliminated or at least accounted for in the decision, variability and uncertainty must be incorporated into the analysis. This may be done in at least four ways:

1. One may determine that the uncertainties are so large and poorly specified that no quantitative uncertainty analysis is possible. In such a case, a risk management decision may be made that all members of an allegedly hazardous class of chemicals or other hazardous entities should simply be banned. This is known as the precautionary principle. Once the risk management decision is framed in that way, the output of the risk analysis is a conclusion that a chemical or technology belongs or does not belong to a banned category.

2. One may make conservative assumptions. For example, in human health risk assessments it is assumed that an individual drinks 2 L of water a day from a contaminated source for a lifetime, consumes fish caught in contaminated waters, consumes vegetables grown on contaminated soil irrigated with contaminated water, etc. Following this example, ecological risk assessors may assume that an entire population of a wildlife species occupies the most contaminated portion of a site. By hypothesizing levels of exposure higher than are credible for any real human or wildlife population, these

conservative assumptions ensure that exposure is not underestimated, even though the exposure is uncertain. The product of stringing conservative assumptions together is a "worst case" or "reasonable worst case" estimate of risk.

3. One may apply safety factors to the components or results of the assessment. These are factors (usually 10, 100, or 1000) that are applied to ensure an adequate margin of safety (Section 5.5.1). They are based on expert judgment and simple analyses of past cases. The output of analysis using safety factors is a conservative risk estimate. However, because of the way factors are derived and the way in which they combine multiple sources of uncertainty, the degree of conservatism that results from safety factors is unclear.

4. One may perform a formal quantitative uncertainty analysis and choose as an endpoint a probability of effects that is very low. For example, one might declare that the probability must be less than 0.01 that the likelihood of extinction is as high as 0.0001 over the next 50 years.

5.4.2 DESIRE TO AVOID EXCESSIVE CONSERVATISM

As already discussed, the desire to ensure safety has led to the use of numerous conservative assumptions and safety factors in risk assessments. Some risk assessors, regulated parties, and stakeholders have objected that the resulting margins of safety are excessive (Kangas 1996). One response has been to argue for reduction of number and magnitude of factors and conservative assumptions or their elimination (i.e., use best estimates). An alternative is to develop anticonservative factors to correct, at least in part, the compounding of conservatism (Cogliano 1997). Another approach is to replace uncertainty factors and conservative assumptions with estimated distributions of parameters and to replace the compounding of factors with Monte Carlo simulation (Office of Environmental Policy and Assistance 1996). If low percentiles of the distributions of risk estimates are used to ensure safety, this approach is not necessarily less conservative than traditional regulatory approaches.

5.4.3 DESIRE TO ACKNOWLEDGE AND PRESENT UNCERTAINTY

It is generally considered desirable to acknowledge and estimate the uncertainties associated with assessments. It is both safer to admit your uncertainties and more ethical than ignoring or hiding them. This is more the case with ecological risk assessments than with human health risk assessments, because estimated ecological effects are often detectable, and therefore a conservative deterministic estimate may be refuted by subsequent observations. A formal probabilistic analysis provides a clear and defensible method for estimating variability and uncertainty and justifying the estimates. However, many uncertainties are not estimated by conventional uncertainty analysis, such as the uncertainty associated with model selection or the uncertainty concerning assumptions about the future use of a site. Hence, presentations of uncertainty must include lists of issues, and qualitative judgments, as well as quantitative estimates.

5.4.4 NEED TO ESTIMATE A PROBABILISTIC ENDPOINT

Probably the least common reason for analysis of uncertainty in ecological risk assessment is the specification of a probabilistic endpoint by the risk manager. Probabilistic endpoints have been estimated by ecological risk assessors since the founding of the field (Barnthouse et al. 1982), but the impetus has come primarily from the assessors, not the risk managers. A conspicuous exception is population viability analysis, which estimates probabilities of extinction of species or populations given prescribed management practices (Marcot and

Holthausen 1987). Such analyses should be done in the event that a proposed action may pose a threat of extirpation. A more likely impetus for probabilistic endpoints are cases in which the ecological risks are driven by the probability of occurrence of an extreme event. Examples include the failure of a dam that holds a waste lagoon, an extremely wet period that brings contaminated groundwater to the surface, or the entry of a large flock of horned larks into a field recently treated with granular carbofuran. Finally, demands for probabilistic analysis may come from reviewers or from responsible parties. In any case, ecological assessment endpoints can be expressed as a probability, given variability or uncertainty in exposure or effects.

5.4.5 PLANNING SAMPLING AND TESTING

Ideally, the field and laboratory investigations that provide the data for risk assessments should be prioritized and planned on the basis of an analysis of uncertainty. The goal of the quantitative data quality objectives (DQO) process is to gather enough data to reduce uncertainty in the risk estimate to a prescribed, acceptable level (Chapter 9). This formalism is not directly applicable to ecological risk assessment, but one can still allocate resources on the basis of expected reduction in uncertainty. This use of uncertainty analysis requires a separate analysis of uncertainty that can be reduced by feasible sampling, analysis, or testing, rather than total uncertainty. For example, a model of mink and heron exposure to polychlorinated biphenyls (PCBs) and mercury was used in the Clinch River assessment to determine that the greatest source of reducible uncertainty in the exposure estimates was the distribution of PCB concentrations in water and sediment (MacIntosh et al. 1994). Analyses for this purpose are termed sensitivity analyses (Section 5.5.7).

5.4.6 COMPARING HYPOTHESES AND ASSOCIATED MODELS

Although conventional hypothesis testing has little place in ecological risk assessment (Box 5.1), statistics are needed to choose among alternative models, based on different hypotheses concerning the nature of the system, when mechanistic evidence is unclear. A simple common case is the selection of an exposure–response function to model a set of test data from among alternative functions that imply different characteristics of the response (linearity, thresholds, hormesis, etc.). Other cases involve selecting from among more complex alternative hypotheses such as compensatory, depensatory, and proportional responses of populations to mortality (Chapter 27) or bottom–up vs. top–down control of ecosystems (Chapter 28). Three approaches to model comparison are available:

1. One may choose the model that is best supported by the evidence. The most direct approach is to use likelihood ratios to compare the relative likelihoods of hypotheses given the available data (Royall 1997, 2004). This approach may be extended to include the goal of model parsimony (i.e., Occam's razor) by applying Akaike's information criterion. It is effectively the relative likelihood, normalized by the number of parameters in the model (Akaike 1973).
2. One may use knowledge of underlying mechanisms rather than the evidence provided by a data set to choose among hypotheses. This approach is particularly appealing when, as is often the case, data are not sufficient to confidently distinguish among models. For example, it is generally impossible to detect depensatory processes in population-monitoring data. However, one can estimate the influence of density-dependent predation and of difficulty in finding mates (the two mechanisms of depensatory responses) given the abundance and selectivity of predators and the abundance of mates in the ecosystem being assessed.

3. One may apply all plausible models. One can then present the range of results to the risk manager and stakeholders as an expression of model uncertainty (Section 5.1.2). Alternatively, one can combine them using Bayesian model averaging (Hoeting et al. 1999; Wasserman 2000). This is conceptually appealing, if information concerning underlying mechanisms is used to assign prior probabilities.

5.4.7 AIDING DECISION MAKING

Finally, the results of an uncertainty analysis may aid the risk manager in making a decision concerning the remedial or regulatory action. Decision analysis and some other decision-support tools require estimates of the probability of various outcomes, which must be derived by a probabilistic analysis of risks (Chapter 34). More generally, the additional information provided by an uncertainty analysis may lead to a better informed and more defensible decision, even without quantitative decision analysis.

5.4.8 SUMMARY OF REASONS

These reasons for evaluating variability and uncertainty are not mutually exclusive, so an assessor may have multiple motives. However, the chosen analytical method must be able to satisfy the most restrictive reason. For example, if one wishes to ensure safety, any analysis will do; but, if one wishes to ensure safety and present a full disclosure of uncertainties in the assessment, only a quantitative analysis will serve; and if one is using uncertainty analysis to help plan a program of sampling, testing, and analysis, only a quantitative analysis that distinguishes sources of uncertainty and variability will serve.

5.5   TECHNIQUES FOR ANALYSIS OF VARIABILITY AND UNCERTAINTY

This section presents six important classes of methods that are commonly used in ecological risk assessment.

5.5.1 UNCERTAINTY FACTORS

The most common technique for incorporation of uncertainty is uncertainty factors (also referred to as safety factors). These are numbers that are applied to either parameters of a risk model or the output of a model to ensure that risks are not underestimated. Most factors are based on expert judgment, informed by experience and simple analyses (Dourson and Stara 1983). For example, the NOAEL values used to calculate wildlife toxicity benchmarks are divided by a factor of 10 if they are based on subchronic studies, because of uncertainties concerning subchronic endpoints as estimators of chronic toxicity (EPA 1993e; Sample et al. 1996c). This factor is based on expert judgment that the threshold for chronic toxicity is unlikely to be more than a factor of 10 lower than a subchronic NOAEL. Most other uncertainty factors are also multiples of 10, reflecting their imprecision.

   In addition to the informality of their derivation, the chief complaint against uncertainty factors is the way they propagate uncertainty through a model. If a model contains four parameters, which are multiplied, and each has an associated uncertainty factor of 10, the total uncertainty is a factor of 10,000. This implies that in the case being analyzed, all things are simultaneously, individually as bad as they can credibly be. The uptake factor is much higher than has been observed, the organisms have extremely small foraging ranges, the endpoint species is much more sensitive than the test species, etc. To avoid obtaining absurdly extreme estimates when using this method of uncertainty analysis, one should estimate a maximum credible uncertainty (i.e., an overall uncertainty factor) in addition to factors for the individual parameters.

Uncertainty factors are operationally equivalent to the extrapolation factors discussed in Chapter 26. The distinction is simply that extrapolation factors account for identified systematic differences between measures of effect and assessment endpoints, while uncertainty factors account for uncertainties when systematic differences are not identifiable. For example, the subchronic tests discussed above are designed to be equivalent to chronic tests, so we do not expect them to be different, but we are uncertain of the truth of that assumption in any particular case. Therefore, an uncertainty factor is employed. If we knew that there was a predictable difference between subchronic and chronic test results, we might develop an extrapolation factor.

5.5.2 CONFIDENCE INTERVALS

Confidence intervals and their bounds are the most generally useful statistics for expressing variability or uncertainty. Although confidence intervals are often treated as equivalent to hypothesis tests, confidence and significance are distinct concepts within frequentist statistics (Hacking 2001). Unlike significance tests, confidence intervals allow us to estimate properties of the population from which a sample was drawn. Hence, their use reduces the sterility of performing hypothesis tests and reporting significance or lack thereof. Much more useful information is provided by presenting confidence intervals as interval estimates of a parameter or as a "best" (mean, median, etc.) estimate and confidence intervals on that estimate or on the data themselves. Confidence intervals on the data, given a model, are called prediction intervals (see, e.g., Figure 26.3). Information is increased by presenting multiple confidence intervals. That is, rather than presenting only the conventional 95% confidence interval, which has no particular relevance to environmental management, the 50%, 75%, 90%, and 95% intervals are presented (Figure 5.1).

Frequentist confidence intervals are, strictly speaking, statements about confidence in a method of sampling and its associated model (i.e., the method of estimating the interval is correct 95% of the time), not about the data (i.e., you cannot say that the true value falls

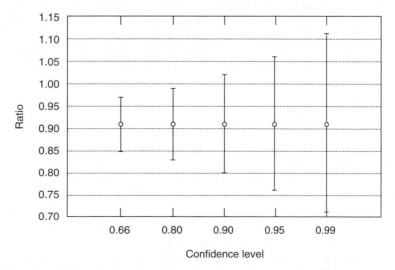

FIGURE 5.1 Sets of confidence intervals used to show how the uncertainty in a value (e.g., the ratio of exposure concentration to toxic concentration) increases as the required level of confidence increases.

in the interval with 95% confidence). To estimate confidence in sample-derived estimates themselves, you need Bayesian confidence intervals.

5.5.3   DATA DISTRIBUTIONS

Uncertainty often appears in the form of a distribution of realized values of a parameter. For example, concentrations of chemicals in repeated water samples or concentrations of chemicals at which individual organisms respond in a toxicity test have some distribution on the scalar, chemical concentration. These distributions can be used to estimate uncertainties concerning the mean water concentration during the period sampled, a future distribution of water concentrations, the concentration that will result in a fish kill, etc. As will be discussed, it is important to carefully consider the relationship between any data distribution and the distributions to be estimated in the assessment. In general, distributions can serve two functions. First, they can be used to represent the uncertainty or variability of a parameter in a mathematical model of exposure or effects. Second, when exposure or effect metrics are directly measured, the distribution of the measurements may directly represent the uncertainty or variability of exposure or effects.

An important decision to be made is whether to fit a function to a data distribution and, if so, which. Conventionally, one describes a distribution by fitting a mathematical function to it, such as the normal, lognormal, uniform, or logistic. The result is referred to as a parametric distribution function. However, if the distribution is not well fit by a parametric function or if the purpose is to display the actual distribution of the data, an empirical distribution function (EDF) may be used. In software for Monte Carlo analysis, EDFs are typically referred to as custom functions. One limitation of EDFs is that they do not describe the distribution beyond the data. This is a problem if the data set is not large and extreme values are of concern (e.g., estimating infrequent high exposure levels or responses of sensitive species). However, parametric functions may also poorly represent extreme values if they have infinite tails, if the extremes of the data are not symmetrical (e.g., the resistant organisms are more highly resistant than the sensitive organisms are sensitive), and if, as is typically the case, their fit is influenced primarily by the bulk of data near the centroid. Issues related to choosing and fitting functions are discussed in a book (Cullen and Frey 1999), a workshop report (Risk Assessment Forum 1999), and in two special issues of the journal *Risk Analysis* (vol. 14, no. 5; and vol. 19, no. 1).

Published distribution functions are available for some variables, either in prior risk assessments or open literature publications. For example, distributions of great blue heron exposure parameters are found in Henning et al. (1999), and such distributions may be expected in the future for other species. The following strategies may be employed to develop distribution functions:

• If the data set is large, one can employ statistical software packages to use statistical criteria to select the best function. In general, the best model provides the highest likelihood of the data, but other criteria such as least sum of squared deviations are also used and all should give the same ranking with good data. However, if there are few data points or the data are noisy, fitting algorithms may fail to give appropriate results. In addition, one must be aware of the fact that models with more parameters fit data better than those with fewer. Hence, one would not choose a three-parameter logistic model over a two-parameter logistic model unless the fit was significantly better, or the additional parameter had some mechanistic significance (e.g., to fit a variable maximum value). Akaike's information criterion, which is the log likelihood normalized by the number of parameters, provides an appropriate basis for comparing functions with different numbers of parameters fit to the same data set (Burnham and Anderson 1998) as it does for comparing mechanistic hypotheses (Section 5.4.6).

- One may choose a function to fit based on experience or on knowledge of the underlying distribution from which the data are drawn. For example, one may know from experience that stream flow rates are nearly always lognormally distributed when sufficient data are available to confidently define the distribution for a site. Therefore, one would use that function at a new site even though the data set is too small for its form to be clearly defined.
- One may choose a function based on the processes that generate the distribution. The addition of a large number of random variables results in a normal distribution due to the central limit theorum. The multiplication of a large number of random variables results in a lognormal distribution. Counts of independent random events result in Poisson distributions. Time to failure or to death of organisms results in Weibull distributions.
- One may use parsimonious strategies, i.e., include in the distribution nothing beyond what is known with confidence. If one feels that the shape of the distribution cannot be specified but the bounds can, a uniform distribution may be defined. If only the bounds and centroid can be estimated, they can be used to define a triangular distribution.
- Finally, if the form of the distribution is unclear or clearly does not conform to any simple function (e.g., is polymodal), an empirical distribution may be used. Even if the form is clear and conforms reasonably well to a function, empirical distributions may be preferable because they reveal the actual form and variability of the data. The only technical difficulty is the proper choice of bins to avoid excessive smoothing (too few bins) or irregularity (too many bins with too few data per bin).

For mechanistic reasons (multiplicative variance) and because many environmental data sets are approximately lognormal in shape, the lognormal distribution is the most commonly used distribution in human health and ecological risk analysis (Koch 1966; Burmaster and Hull 1997).

The selection of a distribution should proceed by a logical process of determining what functions are likely, based on knowledge of the type of data, the mechanism by which the variance in the data was generated, and the goodness of fit statistics. The function should not be selected by inappropriately applying hypothesis testing. It is common practice to assume a function and then test the null hypothesis that the data have the assumed functional form. However, this practice is inappropriate for two reasons. First, it is logically inappropriate to accept the null hypothesis when one has failed to reject it, although this is the most common conclusion drawn from such analyses. Second, it is inappropriate to apply a tool developed to prove the occurrence of treatment effects, where there may be reason to favor the null, to a problem in estimation, where there is not. Rather the assessor should choose the best distribution based on prior knowledge and relative goodness of fit of those functions that are logically plausible.

5.5.4 STATISTICAL MODELING

Statistical modeling (also called empirical modeling) is the use of statistical techniques to generate a predictive and potentially explanatory mathematical model of the relationship between one or more independent variables and a dependent variable of interest. The most obvious example in ecological risk assessment is the generation of exposure–response models from test data (Chapter 23). Other examples include quantitative structure–activity relationships to estimate fate and effects parameters (Chapter 22 and Chapter 26), models to extrapolate toxic responses between species, durations, etc. (Chapter 26), and stock-recruitment models (Chapter 27).

Statistical modeling, in its simplest form, resembles the fitting of distribution functions to data (Section 5.5.3). The function is the same, but in one case it is simply a description of the data and in the other it is a model of the process that generated the data. For example, one might interpret a log-normal function fit to the results of a fish toxicity test as simply the distribution of a parameter, such as the proportion dying, with respect to concentrations of the tested chemical. That is, the probability of death for fathead minnows has a log-normal cumulative distribution function with the median expressed as the LC_{50}. However, the fitted function is more usefully interpreted as a model of the responses of fish populations to the chemical that may be used to estimate responses in other circumstances. Clearly, we are more interested in toxicity test results as models. However, many other fitted functions are purely descriptive such as the distribution of concentrations of a chemical in soil at a contaminated site or the distribution of plant species richness in a forest.

These two interpretations have important implications for estimating uncertainty. The error statistics for a fitted function are, by definition, appropriate estimates of uncertainty concerning the function as a description of the data. However, those error statistics are not adequate estimates of the uncertainty associated with using the same function as a model of toxicity. In other words, departure of the data from the model is not an adequate estimate of the deviation of reality from the model. The uncertainty in the toxicity model should, depending on the use of the model, include interlaboratory variance, variance in water characteristics, variance among fish populations, or other variables.

At the other extreme of model complexity, statistical modeling blends into mathematical simulation modeling. That is, after variables have been independently estimated to the extent possible, the remaining variables in a simulation model may be estimated by fitting the model to data. This process is commonly termed model calibration. Increasingly, parameter estimation for complex models is performed using Bayesian techniques (Calder et al. 2003; Clark 2005).

5.5.5 MONTE CARLO ANALYSIS AND UNCERTAINTY PROPAGATION

When mathematical models are employed with multiple uncertain or variable parameters, appropriate error propagation techniques are required. Many risk models are simple enough to perform the propagation analytically (IAEA 1989; Morgan and Henrion 1990; Hammonds et al. 1994). However, the availability of powerful personal computers and user-friendly software packages for Monte Carlo analysis has resulted in the displacement of analytical solutions by that numerical technique. Monte Carlo analysis is a resampling technique that samples from the distributions assigned to each model parameter, solves the model, saves the solution, and repeats the process until a distribution of results is generated (Figure 5.2). Reviews and guidance for Monte Carlo analysis can be found in EPA documents (Risk Assessment Forum 1996, 1997), in relevant texts (Rubinstein 1981), and in a special issue of the journal *Human and Ecological Risk Assessment* celebrating the 50th anniversary of the technique (Callahan 1996).

5.5.6 NESTED MONTE CARLO ANALYSIS

As discussed above, situations involving risks can be thought of as involving both variability and uncertainty. Both contribute to the estimation of the probability that a specified effect will occur on a particular site, but in conceptually different ways. This distinction matters when one is estimating a variable endpoint (e.g., the probability of extinction of a species given variability in sensitivity among species) and wishes to estimate the associated uncertainties or

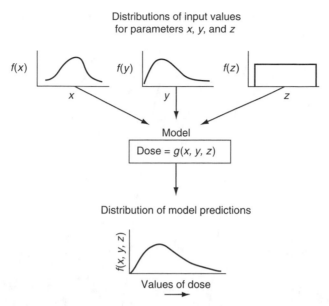

FIGURE 5.2 A diagram depicting the process of Monte Carlo simulation.

when one is using models to plan a sampling and analysis program and needs to distinguish reducible uncertainties. In such cases, the parameters of models should be divided into those that are well-specified constants, those that are uncertain constants, those that are variable but well-specified, and those that are variable and uncertain. The nested Monte Carlo analysis (also called two-stage or two-dimensional Monte Carlo analysis) is begun by assigning distributions to the inherent variability of the variable parameters (e.g., dilution flow in a stream), uncertainty distributions to the uncertain parameters including the uncertain variable parameters (e.g., the uncertainty concerning the true distribution of flow), and constants to the well-specified parameters (e.g., molecular weight of the contaminant). Monte Carlo analysis is performed by first sampling from the variability distributions and then sampling from the uncertainty distributions for the uncertain variables and constants, and solving the model. By iterating the sampling, one generates a distribution of the model output based on variability and a distribution of the percentiles of that distribution based on uncertainty. An example of output from such an analysis is presented in Figure 5.3. Examples of the use of nested Monte Carlo analysis are presented in (MacIntosh et al. 1994; McKone 1994; Price et al. 1996).

Although this nested analysis is computationally complex, the greater difficulty is the conceptual problem of deciding how to classify the parameters. As discussed above, the assessor must determine how variability and uncertainty relate to the goals of the assessment and use that knowledge to consistently apply the analytical techniques. A nested analysis increases the conceptual complexity, but it may increase the likelihood of performing the analysis appropriately by compelling a more thorough review of the problem.

Although the discrimination of variability and uncertainty is the most common use of nested Monte Carlo analysis, any category of probabilities may be nested. For example, one may be interested in separately analyzing variance among entities (organisms, populations, or communities) and variance among temporal intervals (days or years) or events (spills, pesticide applications, or species introductions). In particular, one might wish to distinguish variability among fields being treated with a pesticide from variability among applications due to factors such as time to first rainfall.

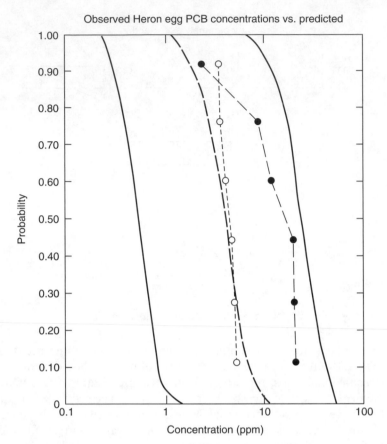

FIGURE 5.3 Complementary inverse cumulative distribution functions for concentrations of polychlorinated biphenyls (PCBs) in great blue heron eggs. The dashed central curve represents the expected variance among eggs from different nests (i.e., due to variance among females). The solid outer curves represent the 5th and 95th percentiles on that distribution based on uncertainty in the parameters. The dots connected by dashed lines are measured concentrations in heron eggs from two rookeries. (Redrawn from MacIntosh, D.L., Suter, G.W., II, and Hoffman, F.O., *Risk Analysis*, 14, 405, 1994. With permission.)

5.5.7 SENSITIVITY ANALYSIS

Sensitivity analysis estimates the relative contribution of parameters to the outcome of an assessment. It may be performed a priori or a posteriori. A priori sensitivity analyses determine the inherent sensitivity of the model structure to variation in the parameter values. That is, if one knows the model structure but not the parameter values or their variance structures, one can still determine the rate of change in the model output with respect to an input parameter at any nominal value of the parameter by calculating the partial derivative. More commonly, the individual parameter values are displaced by a prescribed small percentage from a nominal value, and the magnitude of change in the output is noted. The ratio of the change in the output to the change in the input variable is termed sensitivity ratio or elasticity. The model is more sensitive to parameters with higher sensitivity ratios (at least in the vicinity of the assigned value). This technique is applicable even when no quantitative uncertainty analysis is performed as a means of identifying influential parameters, and it has been recommended for determining which of the parameters should be treated probabilistically in the uncertainty analysis (Section 5.7). However, the sensitivity of the model to a parameter depends on the value assumed by the parameter (for models that are not fully linear) and its variance (for all models). Hence, the relative importance of parameters may differ from that predicted by an a priori sensitivity analysis (Gardner et al. 1981).

A posteriori sensitivity analyses determine the relative contribution of parameters to the model estimates. These analyses are typically done by performing a Monte Carlo analysis, recording the pairs of input and output values for each parameter, and regressing the input parameter values against the model output values. Greater slopes indicate greater sensitivity. Various specific techniques have been employed including single, multiple, and stepwise multiple regression (Bartell et al. 1986; Brenkert et al. 1988; IAEA 1989; Morgan and Henrion 1990). This sort of sensitivity analysis incorporates the expected values of the parameters and their assigned distributions due to variability or uncertainty. It can be used for communication of uncertainty in the risk estimates as well as to indicate which parameters of the model may be important to address in future iterations of the assessment. Good discussions of sensitivity analysis for risk models can be found in Iman and Helton (1988), Morgan and Henrion (1990), and Rose et al. (1991).

5.5.8 LISTING AND QUALITATIVE EVALUATION

Many uncertainties are not quantified because they are judged to be relatively trivial, either as a result of sensitivity analysis or based on judgment. In addition, some uncertainties are not quantified because they cannot be quantified in any reasonable manner or because the probabilities would not contribute to decision making. For example, uncertainty concerning future land uses cannot be quantified with any reasonable confidence and is more usefully treated as an assumption in the scenario than as a quantitative uncertainty in the risk models. Some uncertain parameters are not quantified, because they are set by policy. Finally, some uncertainties are not quantified because of limitations of time and resources. It is desirable to address unquantified uncertainties for the sake of openness (Section 34.2).

5.6 PROBABILITY IN THE RISK ASSESSMENT PROCESS

Unfortunately, uncertainty analysis techniques are often applied without sufficient thought to whether they make sense, given the goals of the assessment. In some cases, the results are misrepresented; more often they are presented ambiguously. As in other aspects of the risk assessment, it is appropriate to begin with a clearly defined assessment endpoint and determine the most appropriate way to estimate it (Chapter 16). For the analysis of uncertainty, one should begin by determining which of the cases in Table 5.1 is applicable. That is, must the assessment estimate a probability (e.g., probability of extinction) or a value (e.g., percent reduction in species richness), and are uncertainties of those probabilities or values to be quantitatively estimated (are they treated as determined or uncertain)? If probabilities derived from either variability or uncertainty are to be estimated, distributions must be derived.

The next step is to define those distributions, and that requires answering two questions:

- What is distributed?
- With respect to what is it distributed?

These questions need to be answered as clearly as possible. Risk distributions may result from variance or uncertainty in exposures, effects, or both.

TABLE 5.1
Types of Endpoints for Risk Assessment Categorized in Terms of Their Acknowledged Uncertainties

	Endpoint	
State of Knowledge	Single Value	Probability
Determined	Specified value	Probability from a specified distribution
Uncertain	Probability of an uncertain value	Probability from a probability of an uncertain distribution

Source: From Suter, G.W., II, *Guidance for Treatment of Variability and Uncertainty in Ecological Risk Assessment*, ES/ER/TM-228, Oak Ridge National Laboratory, Oak Ridge, TN, 1997.

5.6.1 DEFINING EXPOSURE DISTRIBUTIONS

In ecological risk assessments, exposure distributions are distributions of exposure metrics (e.g., concentration or dose) with respect to space, time, organisms, or belief. The specific type of space, time, or belief must be specified.

Space may be defined as arrays of points, linear units, or areas. Points are appropriate for immobile or near-immobile organisms such as plants or benthic invertebrates if the endpoint is defined in terms of individuals. For example, an assessor may be asked to determine whether any plants are exposed to toxic soils or to estimate the proportion of plants exposed to toxic soils. In those cases, one would estimate the distribution of point concentrations from the distribution of sample concentrations (assuming an appropriate sampling design). Streams are typically defined as linear units called reaches, and wildlife associated with streams such as kingfishers have territories defined in linear units. For example, an assessment of belted kingfishers would consider the distribution of exposures experienced in 0.4–2.2 km territories. Most wildlife are exposed within areas defined as territories or foraging areas. Other areas of exposure include distinct plant communities and distinct areas with a particular land use.

Time may be defined as a succession of instants, as intervals, or as incidents. Most samples are instantaneous, and distributions of such instants in time may be appropriate when temporal variance is purely random. However, few relevant exposures are instantaneous, so such distributions are most often of use when one is interested in estimating an average exposure over some period and its uncertainty. Most relevant exposures occur over some interval. For example, one may be interested in determining whether a chemical will cause an effect, which is known to require an exposure to relevant concentrations during a time x. One would then be interested in the distribution of concentrations over time intervals with duration x (i.e., moving averages). Another relevant interval is the seasonal exposure experienced by migratory species or sensitive life stages. That is, one would estimate the distribution of doses received during the seasonal interval when a life stage or species occupies the site. Finally, one may be interested in incidents that result in exposure or elevated exposure, such as storm events that flush contaminants into a stream or suspend contaminated sediments. These might be expressed as the distribution of concentrations over incidents of some specified duration or the joint distribution of concentration and duration of incidents.

Exposure may be distributed across individual organisms as in human health risk assessments, either because the endpoint is risks to individuals of an endangered species or other highly valued species, or because the endpoint is risks to populations expressed as the proportion of individuals experiencing some effect. Exposures of individuals may be distributed due to variance in the areas they occupy, the food they consume, or inherent properties such as weight or food preferences.

Distributions of the credibility of exposure arise when the distributions are defined in terms of uncertainties or some mixture of uncertainties and variances. For example, a polyphagous and opportunistic species like mink may feed primarily on muskrats at one site, fish at another, and a relatively even mixture of prey at a third, depending on prey availability. Hence, the uncertainty concerning the mink diet at a site may be much greater than the variance among individuals at the site, in which case the fractiles of the distribution of individual dietary exposures are credibilities rather than proportions of individuals.

5.6.2 DEFINING EFFECTS DISTRIBUTIONS

In ecological risk assessments, effects distributions are distributions of responses of organisms, populations, or communities (e.g., death, abundance, or species richness) with respect to exposure. It is necessary to specify the interpretation of the effects metric and the sort of exposure with respect to which it is distributed (see above). Four general cases will be considered: effects thresholds, exposure–response relationships from toxicity tests, distributions of measures of effects, and outputs of effects simulation models.

Effects thresholds are often defined by thresholds for statistical significance such as NOAELs or LOAELs. These values do not have associated variances or other uncertainty metrics and are conventionally treated as fixed values. However, while their inherent variance is unspecified, the uncertainty associated with extrapolating them between taxa, life stages, durations, etc. can be estimated (Chapter 26). The most common approach is uncertainty factors.

In conventional laboratory or field testing, organisms are exposed to a series of chemical concentrations, doses, or some other exposure variable, and the number of organisms responding or the magnitude of responses at each exposure level is recorded. Models are fit to those data that permit the calculation of either the exposure causing a certain level of the effect or the level of effect at a given exposure (Chapter 23). If the response is dichotomous (e.g., dead or alive) or dichotomized (e.g., the continuous variable weight can be converted to normal or underweight), a frequency of response can be treated as a probability of effect in the endpoint organisms (e.g., probability of dying). Alternatively, the frequency can be treated deterministically as the proportion of an endpoint population experiencing the effect (e.g., proportion dying). If one is concerned about the uncertainties associated with these results, one might begin with the variance in the model estimate (e.g., confidence intervals) or the variance of the observations around the model (e.g., prediction intervals). However, these variances are generally smaller than the variance among tests, which is a more relevant measure of uncertainty. The minimum variance among well-performed acute tests using the same protocol and species results in test endpoint values within $\pm$ a factor of 2 or 3 (McKim 1985; Gersich et al. 1986). However, ranges of results in less uniform test sets may be more than a factor of 10 (Section 5.1.5). In addition to this variance, which is inherent in the test, extrapolations between test species, life stages, response parameters, etc. should be represented by subjective uncertainty factors, empirically derived factors, or extrapolation models (Chapter 26).

Exposure–response relationships may also be derived from observational field data. That is, the observations of co-occurring levels of a biological variable and one or more environmental characteristics may be used to generate a model (Chapter 23). For example, the number of insect taxa in streams might be related to the total suspended solids concentration. Although these models are based on real-world data, they are not necessarily better representations of causal relationships. Field data have high inherent variance due to biological, chemical, and physical variability among sites and over time, have high error due to lower quality of measurements in the field, and, most importantly, are confounded due to correlations among the "independent" variables. In extreme cases, biological responses may be modeled as a function of a measured variable, when the true cause is an unmeasured variable. For example, episodic exposures to pesticides may be affecting streams in agricultural areas, but routine water quality monitoring seldom includes pesticides and is unlikely to detect those episodes. As a result, effects may be related to sediment rather than agricultural chemicals or to habitat rather than lawn chemicals. Hence, study design is a larger source of uncertainty in field-derived exposure–response models than the error and variance measures provided by statistics.

Finally, mathematical simulation models are used to estimate effects, particularly those on ecosystem properties or on populations mediated by population or ecosystem processes (Chapter 27 and Chapter 28). The uncertainties associated with these effects estimates are usually generated using Monte Carlo analysis.

5.6.3 ESTIMATING RISK DISTRIBUTIONS

Risk is a function of exposure and effects. If only one of these components of risk is treated as a distributed variate, the estimation of risk as a distributed variate is relatively straightforward. If the dose has been estimated to be distributed over organisms due to variance among individuals, and the effects are assumed to have a determinate threshold, the output of the risk characterization is the probability that an individual in the exposed population will receive an effective dose (Figure 5.4). However, if the exposure and effects are both estimated probabilistically, and the risk will be expressed as the joint probability, the concordance of the distributions must be ensured. If both distributions are derived as degrees of belief, concordance is not a problem; however, if they are based on variance, concordance should be assured. If one is estimating the probability that organisms of a particular species are affected, both the exposure and effects metrics must be distributed with respect to organisms. For example, if effects are distributed based on the variance among organisms of a species observed in a toxicity test, the variance in exposure should be limited to variation among organisms of a species due to weight, diet, water consumption, etc. Even though ecological risks have been

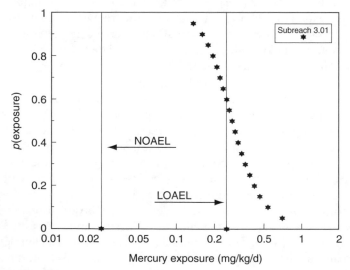

FIGURE 5.4 Inverse cumulative distribution function for exposure of rough-winged swallows to mercury, derived by Monte Carlo simulation of an oral exposure model. The vertical lines are the no observed adverse effect level (NOAEL) and the lowest observed adverse effect level (LOAEL). The probability that an individual would receive a dose greater than the LOAEL is 0.6.

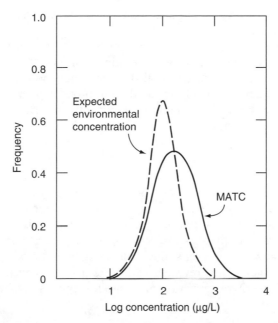

FIGURE 5.5 Probability density functions for a predicted *Salvelinus* maximum acceptable toxicant concentration (MATC) (solid line) and an expected environmental concentration (dashed line). (From Barnthouse, L.W. and Suter, G.W., II, *User's Manual for Ecological Risk Assessment*, ORNL-6251, Oak Ridge National Laboratory, Oak Ridge, TN, 1986. With permission.)

defined as the joint probability of an exposure distribution and an effects distribution since the earliest published methods (Figure 5.5), little attention has been devoted to determining and explaining what those probabilistic risks represent. Methods for addressing this issue are discussed in Chapter 30.

5.7 PARAMETERS TO TREAT AS UNCERTAIN

If the number of parameters in a risk model is large, or if research is required to define distributions, it is necessary to decide which parameters to treat probabilistically. The number of parameters is likely to be large in the early stages of a risk assessment when few data are available or in assessments that never perform much testing, measurement, sampling, or analysis. For example, once contaminant concentrations in plants have been measured, the plant uptake submodel of the wildlife exposure model can be eliminated. Hence, a multiparameter model is replaced with a single measured parameter. If a probabilistic analysis is performed that does not include all parameters, the following criteria should be used in the selection:

- If a probabilistic analysis is replacing an analysis that included uncertainty factors or conservative assumptions, the parameters to which those factors or assumptions were applied should be treated as uncertain.
- If the regulators, responsible parties, or other stakeholders have expressed concern that misspecification of a variable or uncertain parameter may affect the outcome of the assessment, that parameter should be treated as variable or uncertain.
- If the probabilistic analysis is performed in support of a decision, the parameters relevant to the decision must be treated as uncertain. For example, if the analysis is performed to

aid development of a sampling and analysis plan, the parameters that may be measured must be treated as uncertain.

- The EPA and others have specified that parameters determined to be influential by the sensitivity analysis should be treated as uncertain (Hansen 1997; Risk Assessment Forum 1997). This requirement should be applied when other more relevant criteria are not applicable. This requirement could cause the selection of parameters, such as the molecular weight of the chemical, that are not significantly uncertain or variable and are not directly relevant to the decision.

5.8   SUMMARY

This chapter provides an overview of issues related to probability that should allow the reader to understand the discussions and examples in the rest of the book. In particular, the reader should understand that there is no uniform practice of probabilistic ecological risk assessment. Rather, a variety of techniques may be applied depending on the goals of the risk assessment, the audiences for the results, the available data and models, and other considerations. These general issues will be revisited with respect to the development of the analysis plan (Chapter 18) and the characterization of variability and uncertainty (Chapter 34).

6 Dimensions, Scales, and Levels of Organization

Biologists think they are chemists,
Chemists think they are physicists,
Physicists think they are god,
God thinks he is a mathematician.

Anonymous

Scientific enterprises are organized in terms of multiple conceptual axes, three of which are important in ecological risk assessment: the first axis comprises levels of physical, chemical, and biological organization; the second, spatial and temporal scales; and the third, dimensions of the analysis. The location of the assessment on these axes determines how it will be performed and what results will be produced.

6.1 LEVELS OF ORGANIZATION

Scientists conceptually organize the universe in terms of a hierarchy of levels of organization. Those that are relevant to ERA are ecosystems, populations, organisms, and an amalgam of lower levels commonly referred to as suborganismal. While these levels seem to be simple, their use leads to much confusion in practice, particularly when defining assessment endpoints. Much of this confusion can be alleviated if we consider the attributes that characterize these entities.

The concept of organisms is relatively straightforward. Except for some highly colonial organisms such as corals, it is easy to identify an organism. Even when, as in turf grasses and aspens, organisms are connected, it is possible to identify tillers or trunks as organisms. The attributes of organisms that are used in assessment endpoints are also familiar (Chapter 16). They include mortality, growth, fecundity, and deformity. Where these attributes are of concern, we are dealing with organisms, even if the number of organisms dying or deformed is large. The confusion results in part from confusion between statistical populations and biological populations. That is, a set of organisms subject to a treatment level in a test make up a statistical population, although they respond as organisms; but because they are not a biological population, they do not display population responses.

Biological populations are a level of organization characterized by a set of interbreeding conspecific individuals with distinct attributes. Those attributes may be clarified by contrasting them with related attributes of organisms. Organisms die; populations have mortality rates and may become extinct. Organisms have fecundity; populations have immigration and recruitment. Organisms grow by adding mass; populations grow by adding organisms. Although attributes of populations are easily identified, populations as discrete entities are difficult to identify. Except in island-like cases such as fish in a pond, the boundaries of populations are ill defined because of movement of organisms and propagules. Modern population ecology is concerned with variance in density of organisms of a species and rates of movement of organisms among habitats more than with the attributes of isolated discrete populations. As a result, emphasis is placed on metapopulations of subpopulations in habitats of varying quality and exchanging members at various rates. Although traditional, discrete populations can seldom be identified, within any defined habitat patch or other area the population attributes of members of a species may be estimated.

Biotic communities, sets of populations that interact through trophic, competitive, and other relationships, are even more difficult to define in practice. Not only are the boundaries difficult to identify due to movement of organisms and propagules, but it is also often difficult to determine whether a particular set of organisms constitutes an instance of a particular community type. As one moves down a stream or through a forest, species composition changes but it is seldom possible to identify a transition from one community to another. As with populations, discrete communities can seldom be identified, except when some physical feature such as the shore of a lake isolates the organisms. Even then, the definition is muddled by birds that move among lakes or by differences between littoral and pelagic species compositions. However, attributes of communities are easily identified: e.g., the number of species (species richness); the relative number of plants, herbivores, and predators; and the food web linkage.

Ecosystems are the same level of biotic organization as communities; they are communities considered in their physical–chemical context. The difference is easily characterized by the attributes of ecosystems such as primary production, element cycling rates, and element export rates. It is often easier to define the bounds of ecosystems than those of communities. For example, while it may be difficult to define the boundaries of a forest community type, it may be easy to define a forest ecosystem in terms of watershed boundaries if the assessment is concerned with element dynamics. The physicochemical processes of ecosystems may be modified by anthropogenic agents directly rather than through effects on organisms. Examples include the modification of soil and water attributes by acid deposition, modification of nutrient cycling by nutrient deposition, and changes in temperatures due to emissions of greenhouse gases. However, it should be kept in mind that community and ecosystem ecology are simply different ways of looking at the same entities, and even the different models employed in the two fields are interconvertible (DeAngelis 1992).

Suborganismal levels of organization play an auxiliary role in ecological risk assessment. Biochemical, physiological, and histological attributes are very seldom used as ecological assessment endpoints. However, some argue that the health of organisms, as indicated by suborganismal responses, is an important endpoint, because healthy ecosystems require healthy animals (Depledge and Galloway 2005). Nevertheless, they do have roles to play in mechanistic studies of higher-level exposure and effects. For example, physiological rates and organ volumes are important parameters of toxicokinetic models (Section 22.9). Similarly, suborganismal responses can be important diagnostic traits in studies of observed effects in the field (Section 4.3).

It is relatively straightforward to define the attributes of the various levels of biological organization. It is also easy in most cases to identify an instance of an organism, but it is seldom easy to identify an instance of a population or community for an assessment. In a sense, the attributes are more real than the entities. However, since attributes of concern must be estimated within some bounds, populations and communities must be operationally defined. To do this we define assessment populations and communities. These consist of organisms within an area of interest such as the stream reach below an outfall and above the next confluence for a new source assessment, or the US corn belt for the assessment of a new corn pesticide. While assessment populations or communities are seldom defined in classic ecological terms, they should be defined using a mixture of pragmatism and ecological realism (Chapter 16). If the area of concern is a 2 ha contaminated site, one would not define an assessment population of golden eagles inhabiting that site.

Ecologists sometimes engage in arguments about the best level of organization for ecological risk testing and modeling. In general, assessors must focus on the levels of organization at which the endpoint entities are defined (Chapter 16). For example, if the endpoint is production of a game fish population, the assessment should focus on the population biology of that species. However, hierarchy theory teaches us that we must look up and down from our focal level (O'Neill et al. 1986). To understand the mechanisms that drive the population biology, we must examine organismal biology, including effects of contaminants, food requirements, and habitat use, on the constituent organisms. To understand the constraints on the population, we must examine community/ecosystem structure and dynamics. For example, models of population responses to toxicants may be based on the bioenergetics of individual organisms (Hallam and Clark 1983; Kooijman and Metz 1984), but the energy resources available for maintenance in the face of toxic effects are constrained by ecosystem processes.

Although the choice of organization level for assessing endpoints is a policy decision and therefore somewhat idiosyncratic, some generalizations can be drawn. Organisms constitute the lowest level of organization that is exposed to contaminants in the environment, is harvested, or otherwise interacts with agents in the environment in a way that is recognized as significant by humans (who are organisms themselves). The boundary between organisms and their environment divides the domains of environmental chemistry and toxicology, and organismal responses are the best-characterized component of toxicology. However, most nonhuman organisms are transitory on a human time scale, and, because organisms are consumed by humans, they are treated as expendable. Populations of macroorganisms typically persist on human time scales, and therefore are more amenable to management and conservation: we eat the fish but preserve the fish stock (at least in theory). Ecosystems provide the context without which neither organisms nor populations could persist. While humans tend to focus on organisms and populations, ecosystems are according to ecologists, in general, more important. The loss of the American chestnut is a tragedy, but it is largely unrecognized by the public because of the persistence of the eastern deciduous forest ecosystem of which it was a part. In fact, from an ecosystem perspective, the American chestnut is hardly missed because its place has been largely filled by other mast-producing trees, particularly the chestnut oak (Shugart and West 1977). This last example illustrates another generalization: while higher-level entities are generally seen as more important than lower-level ones, they are also more resistant to most agents. The most important reason for this generalization is functional redundancy. Despite the loss of a dominant species, forest ecosystem functions were maintained. Even agents that directly modify ecosystem attributes such as lake acidification have detectable effects on organisms before higher-level changes are seen (Schindler 1987). Analogously, the loss of sensitive or highly exposed organisms may not be reflected in population production or abundance because of compensatory survival or reproduction by the survivors. However, this generalization does not always hold. Compensation may not occur because no other species can fulfill a function performed by a sensitive species. The loss of some species-specific functions can significantly alter the structure and function of the ecosystem. Hence, providing protection at lower levels of organization is considered appropriately precautionary.

Some practical generalizations may also be made about levels of organization. It is easier to assess risks to organisms than to higher levels of organization. The biology of organisms is relatively well characterized, and data concerning responses of organisms to chemicals, radiation, noise, pathogens, and other agents are abundant relative to responses of

TABLE 6.1
Scales of Observation in Ecological Assessments

Spatial Scale	Temporal Scale	Entities	Functions
Organism-level tests (cc–m^3)	Hours–days	Organism	Survival Growth Reproduction
Microcosms (laboratory ecosystem tests) (cc–m^3)	Days–months	Organism Population Micro-community	Survival Growth Production
Field tests (m^2–km^2)	Days–years	Organism Population Community	Survival Growth Production
Environmental monitoring (m^2–km^2)	Years	Organism population Community	Growth Production Cycling

populations or ecosystems. The cost and effort required to perform tests or observations of responses increase at higher levels of organization, while the precision of the data and the ability to generalize them decrease. As a result, the Organization for Economic Cooperation and Development (OECD 1995) concluded that it is impractical to routinely assess chemicals at the ecosystem level. Clearly the size of the experimental system and the duration of the test increase greatly at higher levels of organization, unless macroorganisms are excluded in higher-level studies (Table 6.1). Similarly, ecosystem models are more complex than population models, and organisms seldom must be mathematically modeled. However, for some major classes of agents, generalizations concerning ecosystem responses have been empirically derived. For example, changes in primary production from addition of phosphate can be estimated, and we can predict that addition of insecticides to aquatic systems will decrease the abundance of arthropods and increase algal biomass and abundance of other invertebrates such as rotifers and snails. These issues are discussed in Chapter 27 and Chapter 28.

6.2 SPATIAL AND TEMPORAL SCALES

Spatial scale is roughly related to the level of biological organization. Organisms typically occupy less than the full range of the population to which they belong and ecosystems are usually defined at larger spatial scales than their constituent populations. However, there are numerous exceptions. Disperser organisms move among populations and migratory populations move among ecosystems and even continents. Further, the concept of spatial scale and level of organization is sometimes confused. For example, regions are sometimes treated as a level of organization although, because they have no functional attributes beyond those of ecosystems, they are functionally a spatial scale. Hence, one may treat a region as a single ecosystem for an assessment of nitrogen export from the Mississippi River watershed, or as an area occupied by loosely related ecosystems of various types as in an assessment of risks to California condors in the southwestern United States.

Similarly, temporal scales are roughly related to a level of organization and to a spatial scale. Populations outlast their constituent organisms, and ecosystems often outlast populations. However, an ecosystem may be destroyed and the populations that inhabited it may

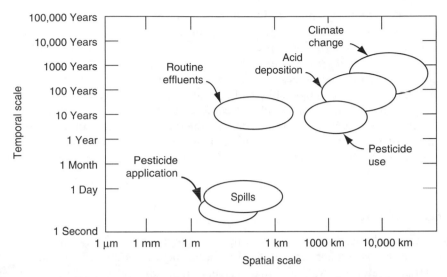

FIGURE 6.1 The distributions of various hazardous agents on spatial and temporal scales.

persist in other ecosystems. Similarly, there is a general relationship between the temporal scale of agents and their severity. That is, exposure levels that cause severe effects are allowed only rarely (e.g., during extreme low flows that provide little dilution) or as a result of accidents. At the other extreme, effects are allowed to persist for long periods only if their consequences are minimal. This relationship is the basis for the conventional dichotomy of acute (short-term and severe) vs. chronic (long-term and not severe) effects.

The appropriate spatial and temporal scales for an assessment depend on the spatial characteristics of the populations, ecosystems, or other endpoint entities and those of the agents being assessed and the interactions between them (Figure 6.1 and Figure 6.2). One way to organize this problem is to begin by defining a core area and time in which an agent is released or some other activity occurs and endpoint entities are potentially exposed, e.g., the

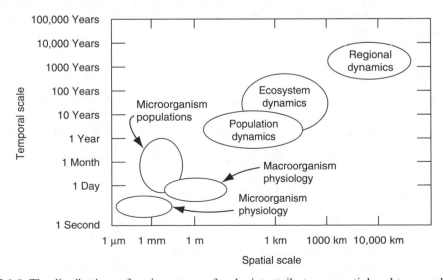

FIGURE 6.2 The distributions of various types of endpoint attributes on spatial and temporal scales.

area to which a pesticide is applied and the duration of the applications, a waste disposal site and the time until the waste is degraded, a clear-cut tract and the time until regrowth, or an impounded river reach and the life expectancy of the dam. One should then consider how the influence of events in that core area may be extended in space or time. Pesticides may drift during spraying, and both pesticides and waste chemicals may be distributed by surface water and groundwater. The hydrologic effects of a dam extend downstream at least to the next reservoir. Silt from a clear-cut may enter streams at high rates until vegetation and a litter layer are established. Finally, one should consider the range of the endpoint entities relative to the core areas of influence. If a specific site is involved, this analysis is best performed by creating a map of the distributions of the agent and the habitats of the endpoint populations or communities.

6.3   REGIONAL SCALE

Some ecological risk assessments have regional scales because the agent or the endpoint entity has regional scales (Hunsaker et al. 1990). Examples include assessments of acid deposition, climate change, water or pollutant management in a watershed, and management of a regional population (McLaughlin 1983; Salwasser 1986; Dale and Gardner 1987; Hunsaker and Graham 1991). These are like other ecological risk assessments discussed in this book, except that they analyze exposures and responses over a large area.

Regional assessments that address a region rather than a regional-scale agent or endpoint are special cases. Many are simply descriptive, quantifying the occurrence of ecological entities and attributes and of conditions that may affect them (Jones et al. 1997; Office of Research and Development 1998; Berish et al. 1999). Others attempt to analyze the multiple sources, agents, and endpoint entities in a region to rank the ecological risks if not estimate them (Harwell 1998; Serveiss 2002; Cormier et al. 2000; Landis 2005). They must often resort to unitless ranking because of the incommensurable nature of the risks to the various entities and attributes in a region as well as the difficulty of quantifying numerous exposure–response relationships. Such risk ranking does not provide the basis for particular regulatory or management decisions, because it does not address the magnitude of potential effects and therefore provides no basis for evaluating significance. However, it can provide background and context for particular regulatory or management decisions. For example, the result of the relative risk assessment for the fjord of Port Valdez, Alaska, "was a new and inclusive port-wide ecological risk perspective" (Wiegers and Landis 2005).

6.4   DIMENSIONS

Ecological risks are defined within a multidimensional state space. One of the most important tasks during problem formulation is to define the dimensionality of the risk assessment to be performed. In particular, the analysis of exposure and the analysis of effects must have concordant dimensions so that they can be combined to characterize risks (Chapter 30). The following dimensions should be considered.

6.4.1   ABUNDANCE OR INTENSITY OF THE AGENT

This dimension is most familiar to ecological risk assessors as concentration. It may be concentration of a pollutant chemical (e.g., mg/L benzene), of a prescribed mixture (e.g., mg/L total petroleum hydrocarbons), or of an effluent (e.g., % dilution of a refinery waste-water). For introduced species, it is the abundance of species in an ecosystem. For thermal energy, it is temperature; for incident solar radiation, $Watts/m^2$; and for ionizing radiation, millirems. Other agents have particular units to express their intensity (e.g., trawler hours). Some have no obvious units because they are dichotomous (e.g., either an area is paved or it is not) or they are effectively dichotomous (e.g., we need to know whether a host organism is infected, not the abundance of the pathogen).

6.4.2 TEMPORAL DURATION

Time is a continuous variable that may be defined for exposure or effects. The duration of exposure determines the likelihood that an effect will occur. Because effects are induced by processes of contact or uptake and succeeding processes of damage induction and accumulation, some duration of exposure is necessary for effects to begin; effects tend to increase with increasing exposure duration, but beyond some duration an equilibrium is achieved and the frequency and intensity of effects no longer increase. However, even if effects are constant, the duration of those effects is important. Various strategies are used to deal with the temporal dimension.

- Exposure duration is often treated as dichotomous: acute or chronic. This roughly corresponds to episodic and continuous exposures, but in practice the dichotomy is often problematical. Because of variance among chemicals and organisms in uptake, metabolism, and depuration rates, the duration of an "acute" episode is critically important. In addition, the duration that is effectively continuous depends on the life cycle of the species considered.
- For some agents such as explosions, pesticide applications, and harvesting, the exposure duration is negligible or irrelevant and the relevant temporal dimension is the recurrence frequency.
- For most episodic exposures to chemicals, both the duration of exposure and the recurrence frequency must be considered.
- If exposures are effectively continuous and constant (i.e., variance in concentration is small or rapid), duration of exposure and effects can be ignored in most cases.
- If concentrations are highly variable in time (i.e., exposures cannot be treated as either continuous or as a pattern of episodic events), time and concentration may be integrated by calculating a time-weighted average or temporal integral of concentration or by using a toxicokinetic model to estimate internal exposure (Chapter 22).

6.4.3 SPACE

The area exposed or affected is the least reported dimension of ecological risk. Ecological risks are often treated as occurring at a point such as the edge of the zone of initial dilution or the point of maximum observed concentration. However, it is important to know, for example, whether an accident killed more than half of the fish in a 10, 100, or 1000 m reach of a river. Space is particularly important with respect to physical disturbances. The area that is clear-cut, tilled, dredged, filled, inundated, paved, etc. is the primary dimension of exposure and of effects for those ecosystem-destroying activities. That is, the estimate of exposure is 5 ha filled and the estimate of direct effects is 5 ha of wetland destroyed.

6.4.4 PROPORTION AFFECTED

The proportion of organisms in a population, species in an ecosystem, or ecosystems in a region that is affected is the most commonly reported dimension of effects. It is a quantal variable derived from binary responses (e.g., living or dead) or a count of continuous responses that have been converted to binary form (e.g., above or below a birth weight of 50 g or a species richness of five).

6.4.5 SEVERITY OF THE EFFECTS

The severity of nondichotomous effects tends to increase with exposure. For example, the decrement in growth or fecundity of organisms, in productivity of a population, or in species richness or biomass of an ecosystem increases. Severity is a count or continuous variable.

6.4.6 TYPE OF EFFECT

As exposure to contaminants increases, the number of types of effects and their importance tend to increase. For example, effects on organisms might begin with abnormal behavior, and, as exposure duration increases, reduced growth, reduced fecundity, and finally death might occur. This categorical dimension does not lend itself to numerical scaling. The simplest approach is to develop standard categories of effects. For example, the Environmental Protection Agency (EPA 1982) has arrayed effects of air pollutants on plants as (a) no observed effects, (b) metabolic and growth effects, (c) foliar lesions, and (d) death. A more general scale developed for human health risk assessment is (a) no observed effects, (b) no observed adverse effects, (c) adverse effects, and (d) frank effects (Dourson 1986). Such categorizations may be analyzed using categorical regression or qualitative analyses (Section 23.2.5). Alternatively,

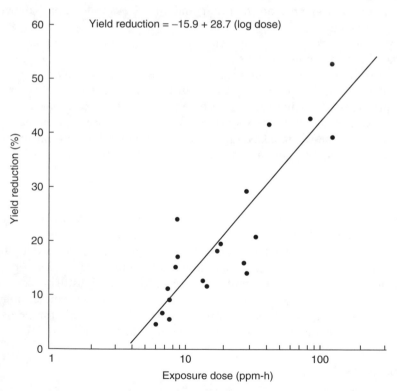

FIGURE 6.3 Percent reduction in yield of beans as a function of SO_2 dose (ppm-h). (From McLaughlin S.B., Jr. and Taylor, G.E., Jr., in *Sulfur Dioxide and Vegetation*, W.E. Winner, H.A. Mooney, and R.A. Goldstein, eds., Stanford University Press, Stanford, CA, 227, 1985. With permission.)

by assigning scores, such scales can be converted to a numerical scale that potentially facilitates modeling (Newcombe and MacDonald 1991; Newcombe and Jensen 1996).

6.4.7 WHAT TO DO WITH MULTIPLE DIMENSIONS?

All six of these dimensions are potentially relevant to assessing the risks to an ecological assessment endpoint, and one could estimate risks with respect to all of them. However, modeling risks in a six-dimensional state space is difficult, and the results are complex to communicate. The first strategy to deal with the dimensions is to eliminate those that are not important for estimating risks to the endpoint. For example, as already discussed, effects that result from either an instantaneous or continuous exposure can neglect exposure duration. Second, dimensions may be combined: e.g., the product of concentration and duration has been used as an expression of exposure (Section 23.2.3) (Figure 6.3). Third, one or more dimensions may be held constant: e.g., we may consider only one type of effect (e.g., mortality) and define one proportion responding as a threshold for concern (e.g., 10% mortality) so that the only effects variable is a severity dimension (e.g., proportion of years).

7 Modes and Mechanisms of Action

Although risk assessments may be entirely empirical, an understanding of the underlying mechanisms by which effects are induced can improve an assessment's efficiency and credibility. This requires employing the concepts of mode of action and mechanism of action, and extending them from toxicology to the analysis of effects of other agents. These concepts are important to the analysis of effects of mixtures (Chapter 8), development of quantitative structure–activity relationships (Section 26.1), use of toxicokinetic models to extrapolate among species (Section 26.2.7), and other aspects of ecological risk assessment.

7.1 CHEMICAL MODES AND MECHANISMS

Mechanisms of action (MoAs) are the specific means by which chemicals induce effects in organisms. Examples include narcosis, respiratory uncoupling, and inhibition of calcium uptake. Modes of action are more general and phenomenological; a mode of action implies a common toxicological outcome but not necessarily the same underlying mechanism. Examples include acute lethality, tumorigenesis, feminization, teratogenesis, and hatching failure. Note that the terms mechanism of action and mode of action are not used consistently in the literature, and there is no agreed dividing line between them.

MoAs are important in ecological risk assessment, because chemicals with a common MoA should behave similarly and may even be used interchangeably in some models. For example, they should be fit by the same quantitative structure–activity relationship (QSAR), they are expected to cause the same effect at the same molar concentration at a site of action, they should have a concentration-additive or dose-additive combined effect when they appear as a mixture, species are expected to have the same relative sensitivity to chemicals with the same MoA, if acclimation to a chemical occurs it is likely to occur with other chemicals with the same MoA, species sensitivity distributions and other exposure–response models are expected to have the same slope for chemicals with the same MoA, etc.

The potential advantages of mechanism-based assessment are not easily achieved. Most ecotoxicology is based on only whole-organism responses; without observations of suborganismal effects, the MoA is not apparent. (However, the organismal responses are the mechanisms by which effects on populations and communities are induced.) Further, exposures are expressed as external concentrations or applied doses, so the concentration at the site of action is unknown. Toxicokinetic modeling can estimate site of action concentrations, but they are not well developed for ecological species of concern and in most cases the site of action is unknown or uncertain. Many chemicals have multiple MoAs. Most information on MoA is derived from acute lethality tests, and the MoA may be different for nonlethal effects at lower doses and longer exposures (Slikker et al. 2004a,b). In addition, the MoA may depend on the route of exposure because of effects at the portal of entry or toxicokinetic differences such as first-pass hepatic metabolism. An example of a chemical with multiple MoAs is lead, which causes acute lethality and chronic stress in fish by acting on calcium ion channels on the gills (Niyogi and Wood 2004), but also causes neurotoxic effects in extended aqueous exposures, and constipation and gut impaction in dietary exposures (Woodward et al. 1994a,b). Another example is organophosphate pesticides, which are acutely lethal by cholinesterase inhibition, but some are also androgen receptor antagonists, possibly having chronic effects by that mechanism (Tamura et al. 2003). MoAs often do not correspond to chemical classes (Figure 7.1), so identification of MoAs requires specific studies (Russom et al. 1997).

Interest in developing and using knowledge of MoA in ecological risk assessments has been increasing (Escher and Hermens 2002, 2004). Mechanistic toxicology references have been available for mammals for some time (Boelsterli 2003), and are increasingly available for other taxa such as fish (Schlenk and Bensen 2001) and arthropods (Korsloot et al. 2004). Lists of modes and MoAs have been published (Russom et al. 1997; Wenzel et al. 1997; Escher and Hermens 2002). Although they are a good start, they are limited to common mechanisms of acute lethality.

Classifications are incomplete because of the extremely large number of potential MoAs. For example, for each of the many hormone receptors in the various animal and plant taxa, there are at least four MoAs: two involving the hormone receptors and two involving regulation. Agonists are chemicals that bind to, and activate, the receptor (i.e., they act like the hormone), and antagonists bind to, but do not activate, the receptor and block the hormone. They may be quite specific. For example, a chemical may be an agonist or

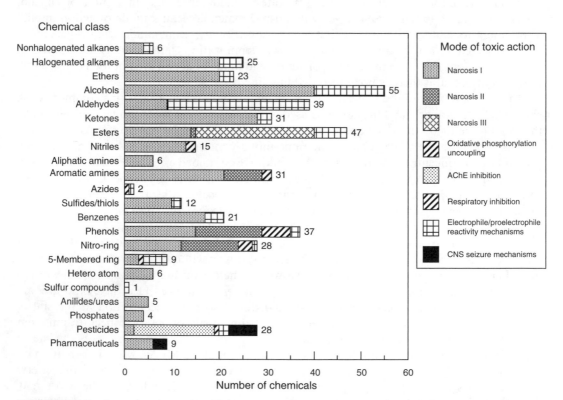

FIGURE 7.1 A chart showing that members of a chemical class may have different mechanisms of action (MoAs). (From Russom, C.L., Bradbury, S.P.S., Broderius, J., Hammermeister, D.E., and Drummond, R.A., *Environ. Toxicol. Chem.*, 16, 948, 1997. With permission.)

antagonist of estrogen receptors on different cells in an organism (Safe et al. 2002). Regulators are chemicals that affect the mechanisms that regulate the production of hormones. They may either upregulate or downregulate. Hence, the mechanisms of endocrine disruption alone are too numerous to address given current knowledge and resources.

One solution to this surfeit is to define a few categories of chemicals that have a simple common MoA [inert (baseline narcotic or narcosis I) and less inert (polar narcotic or narcosis II)] that can be used to predict toxicity of most organic chemicals and a few broad categories that can be used to estimate an interval within which toxicity will fall (reactive chemicals and specifically acting chemicals) (Verhaar et al. 1992, 2000). The simple four-bin categorization is popular in Europe for assessment of aquatic risks from organic chemicals (De Wolf et al. 2005).

A few MoAs and categories of mechanisms are discussed here because of their potential importance in the practice of ecological risk assessment:

Narcosis: Also termed baseline or inert toxicity, narcosis is the least toxic mechanism and is common to all organic chemicals. It results from the nonspecific disruption of cell membranes due to partitioning of the toxicant into the lipid bilayer. It causes death at a relatively constant membrane molar concentration and is fully reversible prior to death. It occurs in at least two distinct classes of chemicals. Baseline or nonpolar narcosis is caused by nonionic organic chemicals that partition to both storage and membrane lipids. Polar narcosis is caused by organic chemicals such as phenols and anilines with an ionic component, which causes them to partition more readily to the phospholipids of cell membranes than to storage lipids. Polar narcotics are more toxic on a body residue basis, because of this preferential partitioning to the site of action. Narcosis is the most common MoA; Russom et al. (1997) estimated that 71% of a sample of 225 industrial organic chemicals comprised narcotics in acute lethality tests of fathead minnows.

Disruption of ion regulation: Many metals induce toxic effects in aquatic organisms by disrupting ion channels on respiratory surfaces involved in the regulation of inorganic ion levels, particularly Na^+ and Ca^{++}. This is the basis for biotic ligand models (Chapter 26). This mechanism also includes disruption of channels involved in neural sodium and potassium pumps. In addition to heavy metals, disruptors of these channels include some neurotoxins produced as defensive chemicals. Well-known examples include cardioglycosides produced by foxgloves (*Digitalis* spp.) and tetrodotoxin from puffer fish.

Reactive oxygen generators: Some chemicals such as the herbicide paraquat and quinones such as tetrachlorohydroquinone can undergo redox cycling in which they reduce oxygen to reactive oxygen species such as superoxide anion, hydrogen peroxide, and hydroxyl radical. These reactive oxygen species cause nonspecific effects including degradation of membrane lipids and proteins.

Uncouplers: Some chemicals, including pentachlorophenol and dinitrophenol, uncouple the oxidative phosphorylation pathway that generates adenosine triphosphate (ATP) in mitochondria. Consequences of uncoupling include reduced useful energy (ATP), increased oxygen consumption, and excess heat generation.

Photosynthesis inhibitors: Some herbicides and other chemicals block photosynthetic electron transport.

Cholinesterase inhibition: Organophosphate and carbamate pesticides act by inhibiting the enzyme acetylcholinesterase, leading to the accumulation of acetylcholine in synapses, and overstimulation of the cholinergic nervous system.

Endocrine disruption: This is a category of mechanisms which includes chemicals that mimic endocrine hormones (agonists), that occupy hormone receptors without activating them (antagonists), and that act on the systems that regulate endocrine hormone production.

The best-known environmental endocrine disruptors are estrogen agonists and antagonists, but all hormone systems are potentially subject to disruption by chemicals in the
environment. Ecdysone-inhibiting pesticides such as diflubenzuron are examples of
ecologically significant endocrine disruptors.

Pheromone disruption: This category of mechanisms is analogous to endocrine disruption
but affects hormones involved in intraspecies communication. For example, endosulfan
inhibits reproduction in red-spotted newts by blocking the production of pheromones
(Park et al. 2001).

In many cases, the MoA for a chemical is identified in terms of inhibiting a particular organ
or organ system without knowledge of the specific mechanism. Hence, many toxicology texts
are organized in terms of organ systems (e.g., Schlenk and Bensen 2001). Such MoAs are
defined by where the effect occurs (e.g., hepatotoxicants) rather than how they occur (e.g.,
respiratory uncoupling). In particular, many chemicals have been identified as immunotoxic
in fish, but the mechanisms are often unclear (Rice 2001). Immunotoxicity serves to illustrate
the difficulty of interpreting MoAs based on organ system performance. While changes in
immune function clearly have implications for disease resistance and tumor suppression, it is
extremely difficult to translate a change in immune system condition into probabilities of
survival. Even when realistic experimental studies showing immunotoxicity can be related to
an actual disease outbreak, it is difficult to prove that the magnitude of outbreak was due to the
toxic effect. For example, more than a decade after the 1988 mass mortality of European
harbor and gray seals associated with phocine distemper infections, and after several observational and experimental studies, the role of immune disruption by polychlorinated biphenyls
(PCBs) has continued to be a contentious issue (O'Shea 2000a,b; Ross et al. 2000).

Some chemicals have both nutrient and toxic MoAs. All metals are toxic at sufficient
exposure levels. For macronutrient metals such as calcium, iron, potassium, and sodium,
toxic levels are so high that they are more commonly deficient than toxic. However, for many
micronutrients, including chromium, cobalt, copper, manganese, molybdenum, nickel, selenium, vanadium, and zinc, toxicity is a more common concern, particularly in industrialized
nations. A typical exposure–response relationship for these nutrient metals is illustrated in
Figure 7.2. Risk assessment is particularly difficult for elements such as selenium that have a

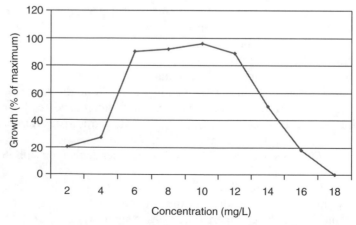

FIGURE 7.2 An exposure–response relationship of a micronutrient metal. Growth is inhibited by
deficiency at very low concentrations, and regulated over a wide range of concentrations resulting in
optimal growth; however, at high concentrations toxicity reduces growth.

narrow optimal range. If a precautionary approach is employed, benchmark levels for protection from toxicity may be set in the deficient range. Ammonia is another nutrient that commonly achieves toxic concentrations in the environment. In some cases, chemicals may have a nutrient MoA in organisms at all levels of exposure but have adverse effects on communities at high levels. Increased loading of ecosystems with nutrient elements, particularly nitrogen and phosphorous, cause changes in species composition and relative abundances which may be considered adverse. At high levels in aquatic ecosystems, the increased production can result in anoxia due to autotrophic and heterotrophic respiration that exceeds photosynthesis and aeration.

7.2   TESTING FOR MECHANISMS

Tests can be used to identify mechanisms by which chemicals might induce effects. In vitro tests for mechanisms cannot be used to estimate effects but can potentially allow the screening of chemicals so that testing can be focused on the most relevant tests of the most hazardous chemicals. Examples include the Ames *Salmonella* bioassay to detect mammalian mutagens and the fish liver slice test to detect estrogenic chemicals. Escher and Hermens (2002) provide in vitro tests for their classification of aquatic ecological MoAs. Such tests are currently not a component of regulatory testing schemes, although research to make them more acceptable is ongoing. Alternatively, conventional tests can be modified or supplemented to yield information on the MoA. For example, observations of the behavior of fathead minnows during 96 h lethality tests distinguished three syndromes that corresponded to narcosis I, narcosis II, and either cholinesterase inhibition or electrophile reactivity (Drummond et al. 1986; Drummond and Russom 1990; Russom et al. 1997).

7.3   NONCHEMICAL MODES AND MECHANISMS

The concepts of mode of action and MoA are applicable to other hazardous agents and may be clarified by those applications. A mode of action may be defined as a response at one level of organization that has a common implication at higher levels. For example, acute lethality is a common mode of action resulting from high levels of exposure to chemicals, harvesting, explosions, high temperatures (as in passage of organisms through a cooling system), crushing by vehicles, etc. Each of these examples has a different mechanism. This organism-level mode of action, however, has the same implication for population abundance or production and would be represented in the same way in a population model. Other examples of organism-level modes of action include reduced fecundity, increased developmental deformity, and reduced growth. Modes of action for populations include reduced abundance of older-age classes (from harvesting or slowly bioaccumulating chemicals), changes in sex ratio (from harvesting or endocrine disrupting chemicals), and local extinction (many mechanisms). Multiple mechanisms can induce these effects, but they have the same implications for the population's role in the community. Modes of action for ecosystems include reduced primary production, structural diversity, and nutrient retention. As with the other levels, all of these modes of action could result from multiple mechanisms. Careful consideration of MoAs is necessary when estimating risks from multiple activities and agents affecting a particular site or population (Chapter 8).

8 Mixed and Multiple Agents

Can any mortal mixture of earth's mould breathe such divine enchanting ravishment?

John Milton

Risk assessments often deal with single agents such as a new chemical, an exotic species, or a harvesting method. However, organisms and ecosystems are always exposed to mixtures of chemicals and to diverse and multiple hazardous agents. Often, things that are treated as single agents for regulatory purposes such as crude oil, polychlorinated biphenyls (PCBs), and pesticide formulations are actually mixtures. This chapter begins with a discussion of methods for addressing chemical mixtures and then presents an expansion of those methods to address combined risks from multiple activities and agents at a site.

This chapter is limited to assessment of mixed and multiple agents per se. An alternative approach to multiple agents is to measure or observe their effects in the field and then infer their causes, which may be mixed or multiple (Foran and Ferenc 1999). Such bioassessments, which are limited to existing and ongoing conditions, are presented in Chapter 4. Also, some discussions of multiple agents treat background environmental conditions such as temperature and light as part of the set of agents to be assessed. Here, the multiple agents to be assessed are limited to those that may be subject to regulation, remediation, or management, and environmental conditions are treated as cofactors.

8.1 CHEMICAL MIXTURES

Methods for estimating risks from chemical mixtures can be divided into those that use test results for whole mixtures and those that use test results for component chemicals. The choice depends primarily on three considerations:

Availability of effects data: Methods that employ whole mixtures require that a mixture be available for testing. Methods that are based on components require that the components have been tested in a consistent and appropriate manner or that the resources be available to perform the needed tests.

Complexity of the mixture: In general, the models that estimate effects of mixtures from toxicity data for components are feasible only for simple mixtures. If a mixture contains many chemicals, not only is it unlikely that suitable toxicity data will be available for each, but it is also difficult to defend assumptions about their combined effects.

Availability of exposure data: Whole-mixture methods require that the exposures be defined in terms of dilution in an ambient medium of a mixture with a sufficiently consistent composition. Processes such as differential partitioning or degradation of constituents may render test data for mixtures irrelevant. Component-based methods require a characterization of the chemical composition of the mixture to which organisms are exposed, derived by analysis of contaminated media or by modeling of the transport

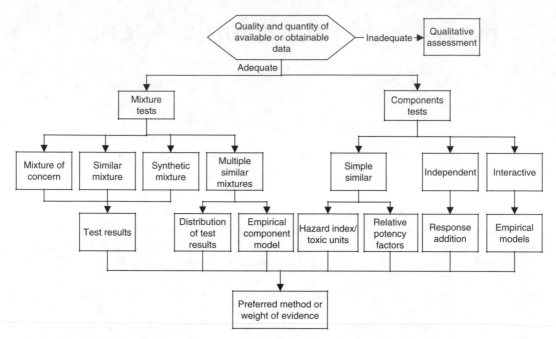

FIGURE 8.1 The different methods of assessment of chemical mixtures. (Highly modified from Risk Assessment Forum, Supplementary guidance for conducting health risk assessment of chemical mixtures, EPA/630/R-00/002, US Environmental Protection Agency, Washington, DC, 2000.)

and fate of a release that has been chemically characterized (Part III). Modeling is complicated by the effects of the mixture on the transport and fate of the constituents.

The options for assessment of mixtures are diagrammed in Figure 8.1. With the availability of exposure information and effects data concerning mixtures or their constituents, the quality of those data, and the possibility of performing new tests to provide necessary data, one may proceed to an assessment based on whole mixtures, constituents, or a qualitative assessment. If only a qualitative assessment is possible, it should explain why mixture risks cannot be quantified and also that the combined effects of a mixture are likely to be greater than those of the constituents.

This section focuses primarily on estimation of effects of mixtures, because that is the most difficult problem. However, estimating the transport and fate of mixtures presents its own problems. For example, the antibacterial properties of one chemical may slow the biodegradation of others in a mixture. Similarly, the fate of benzene dissolved in an aqueous effluent is quite different from its fate in petroleum, a nonaqueous liquid. It is important to assure that the effects analysis and exposure analysis produce consistent descriptions of a mixture at the point of exposure.

This section is consistent with the guidelines from the US Environmental Protection Agency (Risk Assessment Forum 2000). It is not inconsistent with guidance and practices in Europe and elsewhere. The basic concepts and models are the same, but terminology differs, and different approaches are emphasized, even among agencies in a nation.

8.1.1 METHODS BASED ON WHOLE MIXTURES

Some whole mixtures are materials such as pesticide formulations that have a consistent composition and may be tested, assessed, and regulated like an individual chemical. In such cases, assessors are primarily concerned with the fate of the mixture in the environment. If the constituents have highly diverse properties, the toxicity of the original mixture can quickly become irrelevant. Over time, such a mixture in the environment will be transformed from the original mixture, to a different but still similar mixture, and then to a mixture that is so dissimilar that it must be tested as a new mixture or assessed in terms of its constituents. An important example is the current situation with PCBs. Because all of the isomers in a commercial PCB mixture such as the Aroclors are relatively similar, their fates are similar; so it has been possible to relate mixtures in the environment back to the commercial mixture. However, in the years since PCBs were banned, the differences in partitioning and degradation rates among isomers have resulted in changed mixtures in the environment. There is now a lively debate between those who believe that toxicity should be estimated for the individual isomers and then combined to estimate mixture risks and those who believe that a concentration that is reported as Aroclor 1242 is still sufficiently similar to the tested commercial mixture. The debate continues because it is not clear whether our lack of knowledge about the toxicity and interactions of the constituent isomers or the changes in the mixtures are larger sources of uncertainty.

Other mixtures are undesigned and variable. They include combinations of wastes, pesticides, fertilizers, and other chemicals brought together at a location by various processes (e.g., the mixture of chemicals in the sediments of an urban estuary) or complex effluents from various industrial, commercial, and domestic processes. These mixtures differ from the materials discussed above in that their compositions are often unknown or poorly known and are highly variable over space or time. However, these mixtures may still be collected and tested, and, in fact, tests have been developed for that purpose (Box 8.1). These tests differ from conventional toxicity tests in that they are designed to be relatively inexpensive, so that they can be performed in large numbers, yet sensitive enough to be protective. For example, the standard toxicity tests for aqueous effluents and contaminated freshwaters in the United

---

BOX 8.1
Mixtures Risks to Chesapeake Bay Striped Bass

In situ toxicity tests with larval striped bass have shown that the waters of some spawning areas in Chesapeake Bay tributaries are acutely lethal and that these effects are associated with mixtures of metals (Hall et al. 1985). Logan and Wilson (1995) developed a predictive risk characterization method and used it to explain and extend the results of the tests. They defined risk under the assumption of simple similar effects as

$$P(\Sigma TU_i > 1)$$

or, since the data were log-normally distributed, as

$$P[\log(\Sigma TU_i) > 0]$$

Risks were estimated for six test sites. The means and variances of concentrations of five metals were taken from measurements at the sites. The toxic concentrations were LC_{50}s with an estimated variance of 0.018 for log-transformed values from the pooled variance in Barnthouse and Suter (1986). At the least and most metal-contaminated sites, the sum of toxic units was 0.17 and 3.27, respectively, and the risks were 0.00 and 0.99, respectively. The three sites with risks greater than 0.2 had greater than 85% mortality of striped bass larvae. However, for the sites with risks less than 0.2, mortalities ranged from 60% to 83%. The mortality at low risks suggests that either the caging or contaminants other than metals affected many larvae.

States are 7 d tests of fathead minnow (*Pimephales promelas*) larvae and the life cycle of the cladoceran *Ceriodaphnia dubia*. These tests may be performed periodically to determine the temporal variance in toxicity of effluents or ambient waters, whereas the conventional chronic tests would be prohibitively expensive and might not detect important short-term variance in toxicity. Similar soil or sediment tests can be used to determine the distribution of mixture toxicity across space (Figure 20.2).

In some cases, the tested mixture is the mixture of chemicals accumulated by organisms. The advantage of this approach is that the tested mixture is the mixture to which organisms in the field are internally exposed. It integrates spatial and temporal variance in exposure concentrations, bioavailability, and differential uptake and retention. This approach can be particularly useful in establishing causation (Chapter 4). For example, exposure to hydrocarbons extracted from blue mussels with reduced feeding rates collected from contaminated sites was shown to cause reduced feeding in clean mussels (Donkin et al. 2003). Similarly, Tillitt et al. (1989) tested chlorinated hydrocarbon mixtures extracted from eggs collected from colonies of cormorants and terns that had experienced reproductive impairments.

If adequate data are not available for the toxicity, biological oxygen demand, or other property of a mixture, test or measurement results for a similar mixture may be substituted. In such cases, the assessors must determine whether the mixtures are sufficiently similar, which

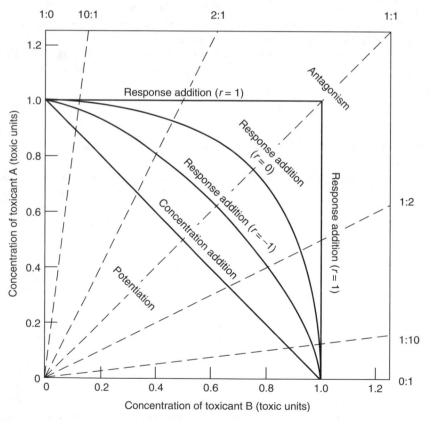

FIGURE 8.2 Isobologram (isobole diagram) for a quantal response to a mixture of two toxic substances. Dashed lines represent constant ratios of concentrations of the two chemicals. Solid lines are isoboles, lines of equal response, representing the set of concentrations of A and B at which the endpoint response (e.g., LC_{50}) occurs.

depends on the use of the data. For screening purposes, to determine whether a hazard requires more assessment, a roughly similar mixture may be adequate. For example, we may assume that all heavy crude oils or all polycyclic aromatic hydrocarbon (PAH) mixtures are similar, perhaps with a safety factor. However, in more definitive assessments in which the risks are near the margin of acceptability, the mixtures should be highly similar. The criteria for determining similarity include the proportion of components that are common to the mixtures, their proportions in the mixture, the presence of components in one mixture with a different mode of action (MoA), and the presence of components in one mixture from a different chemical class.

In some cases, data are available for multiple similar mixtures. For example, we may know the degradation rates of several heavy crude oils, the fathead minnow LC_{50}s for effluents from multiple chrome-plating operations, or the *C. dubia* chronic value for wastewater from a municipal treatment plant at multiple times. In such cases, the distribution of those properties could be used to estimate the same property for an untested heavy crude oil, chrome-plating effluent, or municipal effluent. If chemical composition data are available for the similar mixtures, a model may be developed using multiple regression to relate concentrations of the constituents to the toxicity or fate-related property. That empirical constituent model could then be used to estimate the properties of the mixture of concern, if it has been analyzed. However, that approach depends on the assumption that the concentrations of the constituents are independent.

Finally, tests may be performed with synthetic mixtures. That is, if the composition of a future effluent or other unavailable mixture can be estimated, the mixture can be approximated by mixing reagent chemicals in an appropriate medium. This mixture and, if appropriate, dilutions of the mixture may then be tested. Because stability of the mixture over time is not an issue, these tests may be performed with longer durations and more life stages than the short-term tests used with effluents and contaminated ambient media. This approach also provides the opportunity to systematically study the effects of variation in composition of the mixture or characteristics of the medium such as pH or hardness.

The results of mixture tests may be modeled as exposure–response relationships like individual chemicals. That is, if the mixture is tested at full strength and in dilutions, responses such as proportional mortality p may be modeled as

$$p = f(s) \tag{8.1}$$

where s is strength from 0 to 1 and f is the probit, logit, Weibull, or some other fitted function. In the field, dilutions may occur naturally due to dilution of an effluent plume, leachate, aerial deposition, or spill. In such cases, the contaminated media may be collected and tested on a transect or grid and tested. The results may be modeled as in Equation 8.1, with dilution defined by concentration of an index chemical or some other measure of contamination (e.g., total PAHs) relative to concentration in the source or maximally contaminated sample.

In some cases, only the full-strength mixture is tested. In such cases, one can only describe the differences in response from the control (e.g., clean water) or reference (e.g., upstream water). Statistical significance tests are commonly applied to these differences, but the usual warnings about the differences between statistical and biological or regulatory significance apply (Section 5.3).

8.1.2 METHODS BASED ON TESTS OF COMPONENTS

In many instances it is not possible to test the mixture of concern or, because the composition of a mixture is not stable, it is not feasible to assess risks based on tests of the mixture. In such cases, the toxicity and other properties of the mixture must be estimated from the properties

TABLE 8.1
Categories of Interaction of Pairs of Chemicals

	Similar Action	Dissimilar Action
Noninteractive	Simple similar (concentration or dose addition)	Independent (response addition)
Interactive	Complex similar	Dependent

Source: Adapted from Plackett, R.L. and Hewlett, P.S., *J. Royal Stat. Soc.*, B14, 141–163, 1952; Hewlett, P.S. and Plackett, R.L., *The Interpretation of Quantal Responses in Biology*, Edward Arnold, London, 1979. With permission.

of the individual chemicals. The occurrence of chemicals in mixtures can affect toxicity in two ways. First, the presence of one chemical may influence the kinetics of uptake, transport, metabolism, and excretion of other chemicals. Examples include the thickening of the mucous coating of fish gills by zinc and damage to kidney nephridia by cadmium–metalothionine complexes. A common form of kinetic interaction is the induction or inhibition of metabolic enzymes. Second, by jointly inducing effects, chemicals in a mixture can cause types or levels of effects that are different from those of any of the chemicals acting alone.

The combined effects of toxic chemicals can be categorized based on whether they interact and whether they have similar joint action (Table 8.1). Similar action can refer simply to having the same mode of action (e.g., reproductive failure) or more strictly to having the same mechanism of action in the same organ (e.g., inhibition of the shell gland) (Chapter 7). Chemicals may be assumed to have similar action if they are fitted well by the same structure–activity relationship (Section 26.1). Interactions include effects of one chemical on either the kinetics of the other or on dynamics, including modification of the site of action or combining to produce an effect that neither could produce alone. Both chemicals with similar action and dissimilar action may interact. In either case, the interactions may be potentiating (also termed synergistic or enhancing) or inhibiting (also termed antagonistic or attenuating). Models of simple similar action are commonly referred to as concentration- or dose-additive, and models of independent action are commonly referred to as response-additive.

Figure 8.2 is an isobologram, a diagram of the possible combinations of two chemicals, normalized as proportions of the concentrations producing a particular effect such as an LC_{50}. Isoboles are the lines of equal effect in that space.

Care should be taken when extrapolating knowledge of the nature of chemical interactions between taxa, life stages, or exposure patterns. In particular, most of the studies of chemical interactions use acute lethality as the test endpoint. Chemicals that have similar action for that brief exposure and simple response may act dissimilarly in chronic exposures where more mechanisms such as developmental or behavioral toxicity may come into play.

8.1.2.1 Simple Similar Action and Concentration Addition

If chemicals have the same mechanism of action and do not interfere with, or potentiate, each other, their joint action is simple similar and their combined effects may be estimated by concentration addition or dose addition. (In the remainder of this section, read "concentration" as "concentration or dose.") Simple similarity can be assumed if the mechanism of action of the chemicals and their uptake, transport, and elimination (i.e., toxicokinetics) are known to be the same based on mechanistic toxicology. For example, organophosphate insecticides, which inhibit cholinesterase, have simple similar combined toxicity. Simple similarity can be determined empirically by testing mixtures of the chemicals. That is, if different proportional mixtures of the chemicals are tested and the results fall on the concentration addition line of Figure 8.2, the interaction is simple similar addition. This mode of interaction is suggested by the observation that chemicals have parallel exposure–response curves. Finally, concentration addition models are commonly used as a default for mixtures containing a large number of chemicals with various mechanisms of action (Risk Assessment Forum 2000). This is based on the observation that the toxicities of many mixtures of chemicals are reasonably approximated by a concentration addition model (Alabaster and Lloyd 1982).

Concentration addition models are all based on the assumption that the chemicals in the mixtures are functionally the same, but differ in their potencies by a constant factor t. For two chemicals occurring at C_1 and C_2, and a response R such as proportional mortality, this model is

$$R_1 = f(C_1) \tag{8.2}$$

$$R_2 = g(C_2) = f(tC_2) \tag{8.3}$$

where f and g are concentration–response functions. Equation 8.3 shows how the factor converts the concentration of one chemical to an effective concentration of the other. As a result, the response to binary mixtures of simple similar chemicals (R_m) can be represented as

$$R_m = f(C_1 + tC_2) \tag{8.4}$$

For example, quantal responses (e.g., death or deformity) may be modeled by a log-probit function and for a set of n concentration-additive chemicals, the function would be

$$P_m = a + b \log(C_1 + t_2 C_2 + \dots t_n C_n) \tag{8.5}$$

where P_m is the probit of the response to the mixture, and a and b are fitted variables.

For a particular level of effect such as an LC_{50}, concentration addition models reduce to

$$\Sigma(C_i/LC_{50i}) = 1 \tag{8.6}$$

For example, if a mixture contains three similar chemicals, each at a third of its LC_{50} for a species, that mixture is expected to kill half of the exposed organisms. The concentration (or dose) divided by its effective concentration (e.g., C_i/LC_{50i}) is known to ecotoxicologists as a toxic unit (TU) following Sprague (1970). However, it is known more generally in risk assessment as the hazard quotient (HQ) (Chapter 23). Sums of HQs are called hazard indexes (HI):

$$\Sigma TU = \Sigma HQ = HI \tag{8.7}$$

If the HI is greater than 1, we would expect the prescribed effect to occur. This HI approach depends on having toxicity data for all chemicals in the mixture for an appropriate species and response (e.g., *Daphnia magna* LC_{50} values for each).

Minor constituents of a mixture should not be ignored; if chemicals act by the same mechanism, even those at very low concentrations contribute to effects (Deneer et al. 2005). A solution to the lack of data for some chemicals in a mixture is to use a quantitative structure–activity relationship (QSAR) model to estimate missing values (Section 26.1). If chemicals are additive, they have the same mechanism of action and therefore their toxicity should be predicted by the same QSAR. For example, QSARs using the octanol–water partitioning coefficient (Kow) as the independent variable have been used to estimate the acute lethality of PAHs to amphipods for a TU model of PAH mixtures in sediments

(Swartz et al. 1995). The approach has been extended to the estimation of sediment quality guidelines for PAH mixtures (DiToro and McGrath 2000).

If at least one chemical in a mixture of similar chemicals has well-characterized toxicity, the effects of the mixture can be estimated if their relative potencies (the t values in Equation 8.3 through Equation 8.5) can be estimated. The best-developed example is the dioxin-like chemicals. The toxicity of 2,3,7,8-TCDD is relatively well characterized for a number of vertebrates. Other halogenated dibenzodioxins and dibenzofurans, and some PCBs, have the same mechanism of action, which involves binding to the Ah receptor, but less potency. On the basis of data for toxicity, enzyme induction, and receptor binding, and with expert judgment, toxicity equivalency factors (TEFs) have been developed for the compounds that are known to be dioxin-like for mammals, birds, and fish (van den Berg et al. 1998). The TEF terminology is limited to the dioxin-like compounds, but equivalent factors termed relative potency factors (RPFs) may be developed for other chemical classes such as PAHs (Schwarz et al. 1995; Safe 1998; Risk Assessment Forum 2000), organophosphate pesticides (EPA 2003e), and chlorinated phenols (Kovacs et al. 1993). The general formula is

$$C_m = \Sigma RPF_i * C_i \tag{8.8}$$

where C_m is the mixture concentration expressed as concentration of the index chemical C_i. This normalized concentration can then be used in concentration–response models for the index chemical to estimate effects of the mixture.

The utility of concentration addition models is illustrated by a classic study of the acute lethality of 27 industrial chemicals to fathead minnows (Broderius and Kahl 1985). The hypothesis that these chemicals all acted by narcosis was affirmed by testing them individually and finding that the slopes of log-probit models fitted to test results of each chemical were approximately equal. That is, the concentration–response curves were parallel, and the relative potencies were indicated by the intercepts (Figure 8.3a). Hence, a single concentration–response function could be created by using RPFs to normalize all of the chemicals to the toxicity of 1-octanol (Figure 8.3b). The common narcotic mechanism of action was also suggested by common symptomology: after an initial excitatory phase, the fish darkened and became passive and if death occurred it was within hours. Finally, LC_{50} values from eight binary tests of these chemicals all fell along the concentration addition line of an isobologram (Figure 8.4).

A test of the use of literature values and assumed additivity to predict effluent toxicity is found in Bervoets et al. (1996). They found that the ΣTU for the four most toxic constituents based on 24 and 48 h LC_{50}s were predictive of 24 and 48 h LC_{50}s in tests of the effluent. However, the ΣTU based on chronic no-observed-effect concentrations (NOECs) overestimated the toxicity of the effluent as indicated by the NOEC. The latter result is not surprising given that the NOEC does not correspond to a level of effect and therefore does not provide a consistent TU for addition.

An approach similar to RPFs is used in life cycle assessments (Section 2.3). Ecotoxicity factors and other components of life cycle assessments are commonly normalized to some well-characterized chemical or agent (Guinee 2003). The primary purpose, however, is to create a common unit for comparison rather than to estimate the effects of a mixture.

8.1.2.2 Independent Action and Response Addition

Chemicals have independent action when they are toxicologically dissimilar and exposure to one chemical does not influence the effects of another. Hence, an organism will die of the

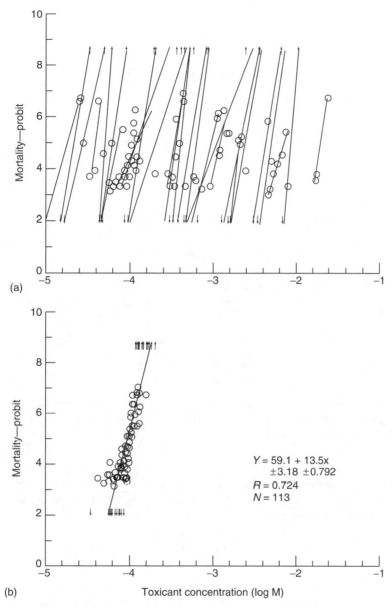

FIGURE 8.3 (a) Percent mortality in probits as a function of log chemical concentrations for 27 chemicals. Up and down arrows represent 100% and 0% mortality, respectively. (b) Plot of the data in part (a) normalized to the toxicity of 1-octanol. (From Broderius, S. and Kahl, M., *Aquat. Toxicol.*, 6, 307–322, 1985. With permission.)

effects of the chemical to which it is most sensitive relative to the concentration. Independent action is represented by response addition models, which sum the effects of each component. Response addition models are applied to mixtures of carcinogens in the United States and elsewhere, because it is assumed that at low concentrations each chemical in a mixture has a

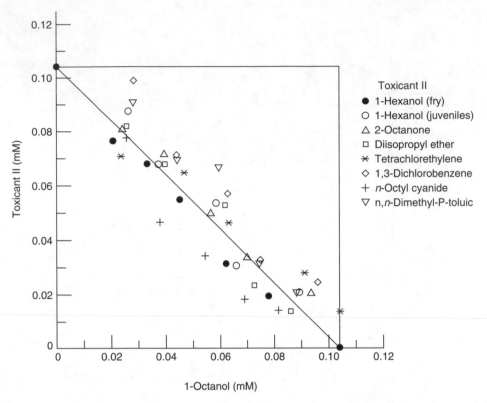

FIGURE 8.4 An isobologram of the tests results of mixtures of organic chemicals and octanol. The diagonal line indicates equal toxic potency on a molar basis. (From Broderius, S. and Kahl, M., *Aquat. Toxicol.*, 6, 307, 1985. With permission.)

small probability of causing a particular tumor type in a particular organ; other chemicals are unlikely to contribute but may cause another type of tumor in another location. A response addition model for two chemicals is

$$P(C_1 + C_2) = P(C_1) + P(C_2) - P(C_1) * P(C_2) \tag{8.9}$$

where $P(C_i)$ is the probability (or, equivalently, the rate or incidence) of a quantal response given concentration C_i. This is simply the probability of an effect that has two independent potential causes. The subtracted term accounts for the practical impossibility of two independent causes of the same effect (if one chemical kills you, the other cannot). The general formula for response addition is

$$P_m = 1 - (1 - P_1) * (1 - P_2) \cdots (1 - P_n) \tag{8.10}$$

To determine whether test results for a pair of chemicals are consistent with independent action, an appropriate exposure–response function is fitted to concentration–response data for each chemical, and the goodness of fit of mixture test data is determined for

$$P(C_1 + C_2) = F_1(C_1) + F_2(C_2) - F_1(C_1) * F_2(C_2) \tag{8.11}$$

where F_x is the fitted functions. Note that a different function may be fitted to each chemical (e.g., probit, logit, Weibull). If the response levels are small, the formula may be simplified to

$$P(C_1 + C_2) = F_1(C_1) + F_2(C_2) \tag{8.12}$$

If, as is usually the case in ecological risk assessment, we are interested in the frequency of an effect in a population rather than the probability of effects on an individual, the formula is the same (P becomes proportion rather than probability), but correlations of sensitivities to the chemicals must be considered. If the probabilities are perfectly negatively correlated (i.e., organisms that are most tolerant of one chemical are least tolerant of the other so that $r = -1$, Figure 8.2), the probability for the mixture is the sum of the independent probabilities, up to a maximum of 1. The other extreme of response addition is perfect correlation of sensitivity to the chemicals ($r = 1$), so organisms respond to the most toxic chemical and the other chemicals do not contribute to the frequency of response. If there is no correlation ($r = 0$), Equation 8.10 is unchanged.

Response addition models are seldom applied to populations because of the difficulty of determining the correlations of sensitivities and the inability to calculate higher-order associations from pairwise associations (Risk Assessment Forum 2000). One might be tempted to use a default assumption of $r = 0$, but there are no good biological reasons to expect that. Hence, if independent action is assumed or demonstrated, perfect positive and negative correlations could be used as bounding assumptions.

8.1.2.3 Interactive Action

Chemicals may interact in a variety of ways to increase or decrease toxic effects. Most of the described cases are antagonistic. For example, one chemical may inhibit the uptake of another, promote the metabolism of another by promoting enzyme expression, or compete for a binding site. However, synergism may occur. For example, injury caused by one chemical may decrease the rate of metabolism or excretion of another, thereby increasing its effective dose. No standard methods are available for modeling these interactive combined actions, and available test data indicating interactions are for pairwise tests that have no clear relevance to typical multichemical exposures (Risk Assessment Forum 2000; Hertzberg and MacDonald 2002). If the mixture cannot be tested, the best solution is to employ toxicokinetic and toxicodynamic models that mechanistically represent the interactions (Section 22.9) (Haddad and Krishnan 1998; Krishnan et al. 2002; Liao et al. 2002). However, suitable models and interaction data are seldom available for nonhuman species and chemicals of concern. Information from the literature may be used to adjust effects estimates ad hoc and generate a qualitative index of combined toxicity such as the mixture toxicity index (Konemann 1981a) or the interaction-based HI (Risk Assessment Forum 2000; Hertzberg and MacDonald 2002). However, if interactions are believed to be important, it is generally better to perform tests of the mixture (Chapter 24).

8.1.2.4 Multiple Chemicals and Multiple Species

These discussions of combined toxic effects have been framed in terms of conventional exposure–response relationships for individual organisms of a species. However, species sensitivity distributions (SSDs), the equivalent multispecies functions, may be treated in the same way (Traas et al. 2002; deZwart and Posthuma 2005) (Chapter 26). That is, if the SSDs for a set of chemicals are parallel or for some other reason are thought to have a common mechanism of action, concentration addition models may be employed. For example, a TU approach might be used as in Equation 8.6 and Equation 8.7, where the common effect might be the concentration causing some effect (e.g., LC_{50}) in half of the species, the median hazardous concentration for 50% of species (HC_{50}). Similarly, if the chemicals in a mixture are believed to act independently, the response addition models could be applied (Equation 8.10 and Equation 8.11). In place of the probability of effect on an organism [$P(C_i)$], one would use the potentially affected fraction (PAF_i).

8.1.3 INTEGRATION OF COMPLEX CHEMICAL MIXTURES

In some cases, more than one method is available for estimating the toxicity or other property of a mixture. The US EPA mixtures guidelines say: "all possible assessment paths should be performed" (Risk Assessment Forum 2000). If multiple estimates are obtained, the results of one method may be chosen or the evidence from all methods may be weighed. One method may be chosen when it is clearly superior to the others, and the uncertainty in its results is low. For example, testing the mixture of concern is generally preferable to using models based on assumed MoAs of the component chemicals. This generalization may be applied if the tests are of high quality and use species, life stages, and responses that are relevant to the assessment endpoint. However, the quality of tests may be marginal, tests of the components may be more relevant, mechanisms of action may be known, or for various other reasons, testing may not be superior or have low uncertainty. In such cases, judgment must be used to weigh the results in terms of their similarity and relative quality to derive a weight-of-evidence estimate of the response to the mixture. Even when only one method is used to generate the risk estimate, the results of other methods may be used to determine the assessor's knowledge of the nature of the toxic effects and confidence in the estimates.

One particular integration problem is the question of how to estimate the effects of combinations of heterogeneous mixtures of chemicals. Typically, when using a component-based approach, one begins by categorizing chemicals into classes with the same mechanism (or at least the same mode) of action and applying a concentration addition model to each class. The problem is that the combined effects of the classes are unknown. The conceptually conservative approach is to simply report the class results and the fact that their combined effects are unknown. If the effects of each group are sufficiently heterogeneous, the results may be acceptable. For example, if one class of chemicals causes tumors and another causes reduced fecundity at the estimated exposure levels, it is sufficient to report the estimated levels of each effect. However, in most cases, the decision maker will require an estimate of the combined effect. One alternative is to use concentration addition across all classes. Although it is unlikely that chemicals with different MoAs will be truly additive, concentration addition is the default assumption of some regulatory programs such as the risk assessment guidance for Superfund (Office of Emergency and Remedial Response 1991), and it provides good estimates of effects for many complex mixtures (Alabaster and Lloyd 1982). Another alternative is to use a response addition model to combine the effects of the chemical classes. This approach has been recommended in the Netherlands (Traas et al. 2002; deZwart and Posthuma 2005) and the EPA (2003b,e). In a study of toxicity to algae of 10 chemicals commonly co-occurring in sediments, use of a concentration addition model for the non-specifically acting components followed by response addition to include the others resulted in a better model than concentration addition for all components (Altenburger et al. 2004). This approach is based on the assumption that chemical classes with different MoAs will act independently. However, for nonspecific effects such as death or decrements in growth, toxic effects are unlikely to be fully independent. It seems advisable to apply both methods and report those results, along with the results for the individual classes. That is, to report (1) the concentration addition results for heavy metals, baseline narcotics, ionizing narcotics, cholinesterase inhibitors, etc.; (2) the results for the whole mixture from using concentration addition for all chemicals; and (3) results using concentration addition within classes and response addition to combine the classes.

These approaches require toxicity data for all components of the mixture. If this requirement is not met, one may simply identify a well-characterized representative chemical from each class and assume that the entire class is composed of that chemical (Suter et al. 1984; MADEP 2002).

8.2   MULTIPLE AND DIVERSE AGENTS

While chemical mixtures are a common risk assessment issue, they are relatively simple compared to the typical real-world situation of an ecosystem or region exposed to various anthropogenic agents associated with a set of activities. The activities may be coincidentally associated, as in the sewage treatment plant added to a watershed that already contains other sewage treatment plants, farms, industries, etc. In other cases, they are parts of a program that must be assessed (Table 8.2).

The logic for assessing the cumulative risks of multiple activities generating multiple agents is presented in Figure 8.5. It can be applied to a program or to a new activity that may significantly interact with existing activities. It is based on a series of logical steps that consider first the spatial and temporal relationship of the activities and agents and then the mechanistic relationships among the interacting agents (Suter 1999a). The process diagramed represents cumulative risk characterization. It presupposes that the activities have been characterized, endpoints selected, and the location characterized in the problem formulation, and that the analysis phase has defined the transport, fate, exposure, and exposure–response relationships for each agent. The characterization must be performed for each assessment endpoint (receptor), because the extent of overlap and the mechanisms of action will differ.

The logic of this process is based on attempting to combine the risks of each activity with the minimum of additional analysis. That is, first determine whether the multiple agents are effectively the same. If they are not, are the risks independent? If they are not independent, can they be added or otherwise combined? If effects are not additive, can exposures be added and the risks recalculated? Finally, if exposures are not additive, one must estimate the

---

TABLE 8.2
Examples of Programs that Involve Multiple Activities and Agents

Program	Example Activities	Example Agents
Pulp mill	Cutting native forest; planting plantation trees; cutting plantation trees; road construction; aqueous emissions; atmospheric emissions	Saws, construction equipment; herbicides, alien monoculture; silt; channel modification; polyphenolic chemicals; nitrogen oxides
Military training	Infantry maneuvers; tank maneuvers; artillery practice	Smokes and obscurants; noise; live ammunition firing; digging; tracked vehicle traffic
Cattle farming	Pasturage; hay production; feed production; feed lots	Trampling; nonnative vegetation; silt, pesticides, nutrients in runoff; manure
Coal-fired electrical generation	Coal mining; coal processing; coal combustion; condenser cooling; ash disposal	Acid drainage; fines in runoff; sulfur and nitrogen oxides, entrainment of aquatic organisms; chlorinated blowdown; metal leachate

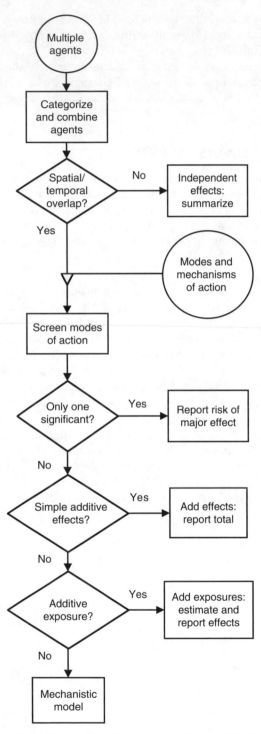

FIGURE 8.5 A diagram of a logical procedure for estimating the combined risks of multiple activities. (From Suter, G.W. II, *Hum. Ecol. Risk Assess.*, 5, 375, 1999. With permission.)

combined risks using a mechanistic model. This strategy depends on the spatial and temporal relationships among the activities as well as the MoAs of the agents.

This process was developed for cases of multiple activities. However, it can also be applied to a single activity that generates multiple agents by skipping the first two steps. There would be no need to combine agents or determine spatial and temporal overlap for a single activity.

8.2.1 CATEGORIZE AND COMBINE AGENTS

When characterizing the combined risks of multiple activities, it will often be the case that some of the activities have hazardous agents in common. For example, dairy and beef farms within a watershed will all have cattle that generate manure, trample streambeds, etc. Similarly, tank maneuvers, training in refueling or rearming tanks, and mechanized artillery training all involve driving tracked vehicles. These sets of agents may be combined with respect to crushing organisms and disturbing the soil.

8.2.2 DETERMINE SPATIAL AND TEMPORAL OVERLAP

If there is no spatial and temporal overlap of the activities or their consequences, the risks of the individual activities need not be integrated. The spatial and temporal extent of the program should have been defined in the programmatic problem formulation. However, this estimate of the temporal and spatial extent is completed for planning purposes and may have been shown to be inaccurate by subsequent assessment activities. The spatial and temporal extent of each activity should have been defined in the problem formulation in terms of a core area (the area in which the activity occurs) and influence area (the additional area to which the influences of the activity extend). At this step in the process, the core and influence areas for all activities must be overlain geographically and on a time line to determine whether the activities are spatially or temporally independent. Independence may be determined by asking the following questions:

- Are the activities performed in the same or overlapping areas?
- If not, do the agents generated directly or indirectly by an activity extend to a significant extent into areas where other activities occur?
- If not, is there significant interaction between the endpoint populations or ecosystems in the areas?

If the answer to these questions is no, the activities are spatially independent. If they are not spatially independent, temporal independence must be considered by asking the following questions:

- Are the activities performed at the same time?
- If not, do subsequent activities occur before the system has effectively recovered from prior activities?

If the answer to these questions is no, the activities are temporally independent. If the agents or activities are either spatially or temporally independent, the programmatic risk characterization consists of a summary of the risks from all of the component activities.

8.2.3 DEFINE EFFECTS AND MODE OF ACTION

If risks to a particular endpoint are a product of effects from multiple exposures, those effects must be combined in a common model that estimates the combined risks. This requires analysis and understanding of the MoAs and potentially the mechanisms of action underlying the effects and their relationship to the assessment endpoint (Chapter 7). Note that the distinction between endpoint effects and MoA is a matter of perspective. For example, if the effect is reduced abundance, the MoA might be mortality, emigration, or reduced reproduction. At a lower level of organization, the effect might be acutely lethal toxicity and the MoA might be narcosis, uncoupling of oxidative phosphorylation, asphyxiation, cholinesterase inhibition, etc.

The recommended strategy in these cases is to define the effects at as high and appropriate a level as possible given the definition of the endpoint and the available evidence (i.e., exposure–response models). For example, if an endpoint is survival of individuals of an endangered species, mortality is the highest-level relevant effect. One could simply add the number of individuals killed by each MoA, or estimate the number killed assuming independent action. In this way, risks from different MoAs can be estimated by a single model by treating them as one effect.

Even when multiple effects are involved, aggregation of MoAs to generate higher-level effects is an important first step. For example, if the endpoint is population abundance of an animal species, aggregation of the various mechanisms of action for lethality (the MoA) is appropriate. If other MoAs such as reduced fecundity or increased emigration were potentially significant, the resulting mortality would then be used along with those effects in a demographic model to estimate the population-level endpoint. See, for example, the combination of toxic effects on survival and fecundity of fish with the effects of harvesting on survival in Barnthouse et al. (1990).

Once the MoAs have been properly aggregated into common effects, the appropriate combined risk model may be identified and employed. For example, if tank maneuvers kill some tortoises, cause others to emigrate, and reduce fecundity of the remainder, a simple projection matrix or other demographic model could be used to estimate population reductions.

8.2.4 SCREEN EFFECTS

If only one effect is significant to the decision, it is not necessary to consider the combined effects. This simplification may occur for two reasons. First, if the magnitude of one effect is much greater than the others, the others may be ignored. In general, if the increment in risk obtained by including the minor effects is less than the uncertainty concerning the magnitude of the effect, the minor effects are negligible and can be screened out. For example, in the case of tank maneuvers and terrestrial herbivores, the effects of loss of food may be negligible relative to the direct effects of crushing for a rare species like the desert tortoise, which is not limited by food. However, for a species that is limited by food and is able to avoid being crushed, the direct effects may be negligible.

Second, if the magnitude of one or more individual effects is clearly so large as to be unacceptable, the risks from the combined effects need not be estimated. For example, if planned tank maneuvers are expected to kill half of the population of desert tortoises, there is no need to consider the additional risks from loss of food.

The decision to dismiss insignificant effects must be made on the basis of screening assessments. These are assessments that are performed using simple assumptions or models to determine whether it is appropriate to continue a line of assessment (Chapter 31). Simple screening methods and models are not available for most nonchemical hazards, but they will not be needed in most cases because the risk models are not highly complex and the number of effects to be screened is not large in most cases. Therefore, the screening assessments may be performed by characterizing the risks for each effect independently and then comparing the results to criteria for absolute or relative significance. These criteria must include the potential significance of indirect effects as well as direct effects.

8.2.5 SIMPLE ADDITIVE EFFECTS

The effects of activities on an endpoint are additive if they have the same mechanism of action and if they are independent. To return to the tortoise and tanks example, driving tracked vehicles, driving wheeled vehicles, and firing live artillery rounds all have acute lethality as an effect. Those effects may be simply added (i.e., the number killed is the sum of the numbers killed by each agent). If the probabilities are not small, it will be necessary to consider that you cannot cause the same effect twice. By analogy to Equation 8.9, for two activities a and b, the number of tortoises killed by both is

$$N_k = p_a N + p_b N - (p_a N * p_b N) \qquad (8.13)$$

where N_k is the number killed; p_a, the probability of being killed by activity a; N, the number of individuals in the exposed population; and p_b, the probability of being killed by activity b. The probabilities themselves would be functions of properties of the activities such as the number of tanks, the distance traveled, the season, and the time of day. The extension to more than two chemicals and to nonquantile responses should be obvious.

8.2.6 ADDITIVE EXPOSURES

Although effects are not additive, exposures may still be additive, as in the concentration addition models discussed above (Section 8.1.2). If the mechanism of action is the same so that the same or parallel exposure–response models are applicable, exposure levels or normalized exposure levels may be added.

Habitat losses can be considered to be exposure-additive. For example, if several activities eliminated habitat or made it unavailable, the losses may be added and the effects on a population or community may be estimated. A more complex case would be activities that each reduced the quality of habitat areas. By analogy with the RPFs for chemicals, one could devise factors that were proportional to the loss in habitat quality. The model for total reduction in habitat would be

$$UHE = \Sigma RHQ_i * a_i \qquad (8.13)$$

where UHE stands for undisturbed habitat equivalents; RHQ, for relative habitat quality with activity i; and a_i, for the area affected by activity i.

For example, if tank training is being expanded into a new 1000 ha area involving both construction of paved roads on 1 ha and designation of 600 ha for off-road tank operation, and if studies in the old training areas found that tank traffic reduced tortoise abundance by half, then the tortoise habitat equivalents will be

$$UHE = 0(1\,\text{ha}) + 0.5(600\,\text{ha}) + 1(399\,\text{ha})$$
$$UHE = 699\,\text{ha}$$

If undisturbed habitat in the new area supports 0.7 tortoises/ha, the 1000 ha area after conversion to tank training is estimated to support $0.7(699) = 489$ tortoises, a 30% reduction. The information for deriving RHQ values could be derived ad hoc, as in this example, or from the literature on habitat quality for the species of concern, including the Habitat Evaluation Procedure reports (FWS 1980).

Another common mechanism for which exposure addition might be assumed is agents causing reductions in energy resources. That is, if less food is available for heterotrophs or light for plants, if a contaminant causes energy costs for metabolism or repair, if animals must travel further between water, food, or other resources, or if plants must replace tissues lost from herbivory or harvesting, such energy costs might be summed. This example is hypothetical and would require some creativity because there is no common exposure unit like habitat area or chemical concentration. While common units are not necessary for exposure addition models, the functions are likely to be heterogeneous and would therefore need to be converted so that they are approximately parallel. Simulations using bioenergetic models might be more appropriate.

8.2.7 MECHANISTIC MODELS OF COMBINED EFFECTS

If the mechanisms by which multiple agents act are heterogeneous and particularly if they are interactive, a model must be selected that incorporates them all. This requires a consideration of the mechanisms underlying the effects, and selection of an appropriate mechanistic level for the model. In general, selection of higher-level mechanisms results in simpler models that are more easily implemented. However, interactions among the exposure and response processes often require a deeper level of mechanistic description. For example, combined risks to fish populations from contaminants and harvesting can be modeled using relatively simple demographic models by assuming that the contaminants reduce only survivorship, and fecundity and harvesting affect only survivorship (Chapter 27). However, if one wishes to estimate the influence of compensatory processes such as reduced mutual interference among survivors and increased food supply, one must incorporate mechanisms that affect individual fish (Rose et al. 1993 and Chapter 27). In an ecosystem-level example, the combined effects on phytoplankton and zooplankton production of exposure to a mixture of metals (As, Cd, Cu) and nutrient (N, P) enrichment were estimated (Moore and Bartell 2000). Concentration addition was used to estimate direct reductions in plankton production due to the metals; then the CASM model was used to estimate the combined effects of direct metal toxicity, direct nutrient promotion of primary production, and food-web propagation of effects on plankton taxa. In many cases, the extent to which these more deeply mechanistic models can be applied is limited by both scientific knowledge and by lack of needed site-specific information. Nevertheless, simple threshold models or empirical models that may have provided the best estimate of risks from an individual agent are likely to be inappropriate for the estimation of combined risks.

8.2.8 INTEGRATION OF COMPLEX SETS OF AGENTS AND ACTIVITIES

For cases involving multiple activities generating multiple agents and exposing multiple receptors, integration becomes more complicated because the decision structure is more complicated (Harwell and Gentile 2000). In order to determine risks to an endpoint like wood storks in the everglades, one must consider the effects of agriculture, urban sprawl, tourism, and climate change. However, these activities are not under the control of one agency and they are not legally or politically controllable to the same degree. Hence, risk characterizations must be tailored to particular decision makers and stakeholders by treating different activities and agents either as cofactors or as subject to control, and by adjusting the spatial and temporal scales to match constraints on the decision maker. Such tailoring not only makes the risk assessment more useful but can also reduce the uncertainty in the results by limiting the sources of variance and uncertainty to those that are relevant.

9 Quality Assurance

Sound environmental decision making requires forecasts that are ... perceived as credible, if credibility is deserved.

William Ascher (2006)

The quality of risk assessments is often a source of controversy. This is in part because of the role of risk assessments in contentious decisions and in part because standards for quality in risk assessments are not clearly defined. This chapter will discuss quality assurance in terms of the quality of input data, of the individual models, and of the risk assessment. It is important to emphasize that quality standards depend on the use to which the results will be put. In particular, data and models for a screening assessment that serves to identify the scope of subsequent assessments meet different standards from those for a definitive assessment that guides a remedial or regulatory action.

In the US Environmental Protection Agency (US EPA), information quality is defined by the following properties (Office of Environmental Information 2002):

Objectivity—information is accurate, reliable, and unbiased, and is presented clearly and completely.
Integrity—information has not been corrupted, falsified, or otherwise compromised.
Utility—information is useful for its intended purpose and user.

While most discussions of quality assurance deal with increasing accuracy, it is also important to avoid unnecessary accuracy, precision, or completeness. If risks can be shown to be acceptable or unacceptable with an analysis of direct exposures of one endpoint receptor using available data, the expenditure of additional resources on models or data generation would not increase the utility or even the overall quality of the assessment. Large data sets and complex models increase the opportunities for loss of integrity through errors or even deliberate corruption, and decrease the likelihood of detecting errors. Einstein is often quoted as having said that a theory should be as complex as necessary, but no more. The same advice applies to risk assessments.

Influential information is held to a higher standard of quality than other information. Influential information has a clear and substantial effect on the outcome of an important decision. To the extent that decision is influenced by the results of models, the influence of information can be determined by sensitivity analysis (Section 5.5.7). Otherwise, information is influential if it plays a prominent role in the assessment, particularly if it is likely to be controversial or contested. In general, parameters that have generally accepted default values or for which values are provided in official guidance are less controversial than those that are unique to the assessment.

9.1 DATA QUALITY

The data used in a risk assessment may come from three sources: primary data generated for
the assessment, secondary data drawn from the literature, and default values or assumptions.
Each source raises different data quality issues.

9.1.1 PRIMARY DATA

Quality of primary data is a relatively well-defined component of quality assurance for risk
assessment, i.e., sufficient and adequate data should be generated to provide risk estimates
that meet the needs of the decision maker. If a well-defined quantitative risk model is going to
be used and the decision maker is willing to quantitatively define his or her decision criteria
and acceptable error rates, the data needs can be defined statistically. The US EPA has
developed a procedure for this purpose called the data quality objectives (DQO) process
(Quality Assurance Management Staff 1994), which is outlined in Box 9.1. Its focus on linking
data quality assurance to the rest of the assessment and decision-making process makes the
DQO process a part of problem formulation (Chapter 18). Because of the complexity of
the process and the potentially controversial issues involved, it typically requires one or more
full-day meetings.

The DQO process can ensure that data collection provides the information needed to make
a defined, risk-based decision. However, the DQO process was designed for human health
risk assessment and has been difficult to apply to ecological assessments. Part of the problem
is simply the complexity of ecological risks relative to human health risks. It is difficult to
define a "bright line" risk level, like 10^{-4} human cancer risk, for the various ecological
endpoints. A probability of exceeding a "bright line" significance level is not even the best
expression of the results of an ecological risk assessment. In most cases, it is better to express
results as an estimate of the effects level and associated uncertainty (Suter 1996a; EPA 1998a).
In addition, ecological risks are assessed by weighing multiple lines of evidence, so the
uncertainty concerning a decision about the level of ecological risk is often not quantifiable.
The DQO process is directly applicable if only one line of evidence is used, and if, as in human
health risk assessments, one is willing to assume that the decision error is exclusively a result
of variance in sampling and analysis. Also, in my experience, risk managers have been
reluctant to identify a quantitative decision rule for ecological risks. This is in part because
there is little clear policy for decisions based on quantitative ecological risks. Finally, the
remedial decision is not actually dichotomous. There may be a number of remedial alternatives with different costs, different public acceptability, and different levels of physical
damage to the ecosystems. Therefore, the remedial decision typically does not depend simply
on whether a certain risk level is exceeded, but also on the magnitude of exceedence, how
many endpoints are in exceedence, the strength of evidence for exceedence, etc.

These issues, however, do not completely negate the utility of using an adaptation of the
DQO process for ecological risk assessment. Steps 1 through 4 of the process (Box 9.1)
correspond to conventional problem formulation. Therefore, even if only those steps
are completed, the risk managers and assessors should be able to develop assessment endpoints, a conceptual model, and measures of exposure and effects in a manner that leads to a
useful assessment because of the collaboration and the emphasis on the future remedial
decision. Further, even if the risk manager does not specify decision rules, he or she should
be willing to specify what effects are to be detected with what precision using what techniques.
Discussions of the use of the DQO process in ecological risk assessment can be found in
Barnthouse (1996) and Bilyard et al. (1997).

BOX 9.1
The Steps in the Data Quality Objectives Process

1. *State the problem*: Clearly specify the problem to be resolved through the remediation process. For example, the sediment of a stream has been contaminated with mercury and is believed to be causing toxic effects in consumers of fish. The ecological assessment endpoint entity is the local population of belted kingfishers.
2. *Identify the decision*: Identify the decision that must be made to solve the problem. For example, should the sediment be dredged from some portion of the stream?
3. *Identify inputs to the decision*: Identify the information that is needed in order to make the decision and the measurements and analyses that must be performed to provide that information. For example, the diet and foraging range of kingfishers, the relationship between concentrations of mercury in food and reproductive decrement in kingfishers, and the distributions of mercury concentrations in sediment.
4. *Define the study boundaries*: Specify the conditions to be assessed, including the spatial area, time period, and site. Use scenarios to which the decision will apply and for which the inputs must be generated. For example, the kingfisher population of concern is that occurring in the entire stream from its headwaters to its confluence with the river.
5. *Develop decision rules*: Define conditions under which an action will be taken to remove, degrade, or isolate the contaminants. This is usually in the form of an "if...then..." statement. For example, if the average production of the population is estimated to be reduced by 10% or more, the stream should be remediated to restore production.
6. *Specify acceptable limits of decision error*: Define the error rates that are acceptable to the decision maker, based on the relative desirability of outcomes. For example, the acceptable rate for falsely concluding that production is not reduced by as much as 10% is 10% and for falsely concluding that it is reduced by at least 10% is 25%.
7. *Optimize the design*: On the basis of the expected variance in the measurements and the exposure and effects models, design the most resource-efficient program that will provide an acceptable error rate for each decision rule. For example, on the basis of Monte Carlo analysis of the kingfisher exposure model, the species composition of the kingfisher's diet should be determined by 10 h of observation during each of four seasons for each bird inhabiting the stream or a maximum of six birds, and the mercury composition of the fish species comprising at least 80% of the diet should be determined twice a year in ten individuals with a limit of detection of 0.1 µg/kg.

Source: From Quality Assurance Management Staff, *Guidance for the Data Quality Objectives Process*, EPA QA/G-4, US Environmental Protection Agency, Washington, DC, 1994, with annotations previously published in Suter, G.W. II, Efroymson, R.A., Sample, B.E., and Jones, D.S., *Ecological Risk Assessment for Contaminated Sites*, Lewis Publishers, Boca Raton, FL, 2000.

In addition to determining that the necessary measurements are made using an appropriate statistical design, it is necessary to ensure that good methods are applied. For conventional measures such as chemical analyses and toxicity tests, this usually implies using standard methods. Sources of standard methods include regulatory agencies (e.g., US EPA, Environment Canada) and standards organizations (e.g., Organization for Economic Cooperation and Development, American Society for Testing and Materials, British Standards Institution, and International Organization for Standardization). Even when standard methods are not available or are inadequate (e.g., standard analytical methods are not sufficiently sensitive), they should be consulted for applicable components. For example, even if a novel analytical method is applied to environmental media, the requirements for documentation, holding times and conditions, chain of custody, trip and method blanks, replication, matrix spikes, etc., should be followed.

Finally, primary data must be adequately documented for quality control, both as basic input to the uncertainty analysis and to assure that the data cannot be defended if challenged. The process of reviewing the quality control documentation to determine the acceptability of the data is termed data validation. It involves determining whether the documentation is adequate, whether procedures were followed, whether the results are acceptable (e.g., whether results of duplicate analyses are sufficiently similar), and whether the results make sense. Examples of nonsense include reported concentrations far below the detection limit of the analytical method, unstable daughter products reported to be more abundant than relatively stable parent compounds or parent radionuclides, or survival rates greater than 100%. During validation, data that have deficiencies should be marked with qualifier codes, which the assessors must use to determine the usability of the data. A standard reference addressing these issues is the EPA guidelines on data usability (Office of Emergency and Remedial Response 1992). If the DQO process is used to design the data generation process, a data quality assessment is used to ensure that the decision can be made within the desired error constraints (Quality Assurance Management Staff 1998).

9.1.2   Secondary Data

Secondary data are those that were not generated for the assessment at hand. Hence, they were generated not to estimate a parameter of the assessment and were generated to attain quality criteria other than those of the assessors and manager. Much of the data in ecological risk assessments fall into this category. Secondary data may have problematical quality because of their inherent problems (i.e., they may be bad data) or because of the way they are used (i.e., they may be good for their original purpose but bad for the assessment). This section describes some of the ways of judging the quality of scientific studies and of judging the suitability of a datum for the use of an assessment.

The fundamental check for quality in science is reproducibility: if multiple scientists produce the same result, that result is reliable, but if a result cannot be reproduced, it is unreliable. However, both of these inferences may themselves be unreliable. Twenty scientists in Seattle, Washington, may independently obtain the same estimate of mean barometric pressure, but that does not make the result reliable for Billings, Montana. It also does not refute the differing result obtained by a single scientist in Denver, Colorado. In complex and poorly understood systems, it is likely that unrecognized and uncontrolled factors, like altitude in the example, will lead to conflicting results. The replicability criterion is even more problematical in cases of advocacy science. Differences in the way a toxicity test or other study is performed that are unreported or seemingly unimportant may alter results. The resulting failure to replicate a prior result can then be used to support charges of junk science. Hence, in the absence of well-understood systems and disinterested scientists, the reproducibility criterion must be applied with care.

The most common criterion for quality of published scientific results is peer review. Publication in a peer-reviewed journal is routinely taken as a guarantor of quality. However, my reading of the peer-reviewed literature has found clearly false results such as survivorship of over 200%, and 75% mortality among 10 treated organisms, in addition to the usual bad statistics, bad logic, and poor methods. This is not surprising. Given the large number of journals that require 10–40 papers per issue and the relatively small amount of funding for environmental science, even poor papers can find a home. Drummond Rennie, former editor of the *Journal of the American Medical Association*, was quoted as saying:

> There seems to be no study too fragmented, no hypothesis too trivial, no literature too biased or egotistical, no design too warped, no methodology too bungled, no presentation of results too inaccurate, too obscure and too contradictory, no analysis too self-serving, no argument too circular, no conclusions too trifling or too unjustified, and no grammar or syntax too offensive for a paper to end up in print. (Crossen 1994)

To a certain extent, the expectation of quality may be increased by using papers from highly selective journals. However, there is no scale of journal quality, and both of the quantitatively impossible results described above came from one of the most prestigious journals in environmental science. Journals usually use only two or three reviewers, and those few reviewers may not have the expertise to evaluate all aspects of the research. Further, they may not devote the time necessary to check the results of calculations, the appropriateness of assumptions, or other technical issues. Hence, it is incumbent upon assessors to be their own reviewers and check results. For influential data, it may be worthwhile to pay experts in the field to critically evaluate the data both in terms of its fundamental soundness and in terms of its appropriateness for the intended risk assessment use.

Another aid to appraising the quality of a study is disclosures of conflicts of interest. Although the source of a scientist's funding or his or her financial interest in a company should not affect the results of a test or measurement, they do, at least in some cases (Michaels and Wagner 2003). In a review of studies of calcium channel blockers, 96% of authors of supportive studies had financial relationships with the drug manufacturer vs. 60% of authors of neutral studies and 37% of authors of critical studies (Stelfox et al. 1998). In a survey, 15.5% of biomedical researchers admitted to altering a study in response to pressure from a sponsor (Martinson et al. 2005). Different types of funding create different pressures on research. Investigator-initiated grants generate relatively little pressure for bias in favor of the government agency or foundation that funded them. At the other extreme, if an industry or advocacy group develops the research project, shops it to investigators, and controls the publication of results, the pressure for bias is severe. Some journals require disclosure of conflicts of interest, and acknowledgments of funding sources or affiliations of the authors may suggest potential conflicts. This information alone should not be used to dismiss a study. Rather, it should serve to alert the reader to look for departures from standard methods and other potential sources of bias in the results.

Even when there are no financial interests or other overt sources of bias, investigators may be subject to "wish bias," the desire for important results even when the data are ambiguous. This has been recognized as a particular problem in epidemiology, where results are often ambiguous and causal criteria may be selectively and subjectively applied (Weed 1997). One proposed solution to wish bias is to take data from original sources but take inferences about causation or other implications of the data only from independent reviews (Weed 1997).

Secondary data must be evaluated for the quality of the methods and their implementation. Questions to be asked include:

- Was a standard method or other reliable method used?
- Were proper blanks and standards used with analyses, and were the results acceptable?
- Were proper controls or references used for comparisons?
- Were methods well documented, particularly any deviations from standard methods?
- Were the methods well performed (e.g., were control survival and growth high)?
- Are the statistical methods appropriate?
- Are the data presented in sufficient detail to check the statistics or apply a different analysis?

- Are the results logical (e.g., no survival rates >100%)?
- If the results are inconsistent with other studies of the phenomenon, are the differences explained satisfactorily?

This is not an exhaustive list of considerations. Each different use of a type of secondary data will inspire the user to consider different aspects of the methods, their implementation, and the reporting of results.

In addition to these general quality criteria, secondary data must be evaluated in terms of their suitability for the particular assessment use. That is, which of the data available to estimate a parameter are the best for that purpose, and, are they good enough? If we wish to estimate effects of chromium in brook trout, would we be better off with a low-quality study of rainbow trout or a high-quality study of fathead minnows, a more biologically dissimilar species? How much should differences in water chemistry or metal species be considered? Ideally, these choices should be based on a quantitative analysis of uncertainties that would consider both the inherent uncertainty of a literature value as an estimator of the parameter, and the quality of models that might be used for the extrapolations. However, this ideal is technically difficult, if not impossible, in many cases. Hence, it is often necessary to qualitatively or semiquantitatively evaluate the relative uncertainties associated with alternative data.

The goal of evaluating secondary data is not necessarily to identify the best number from the best study. If multiple acceptable values are available, it may be appropriate to use the mean of those values, a weighted mean based on an evaluation of their relative quality, or a distribution of the values for use in an uncertainty analysis (Section 5.5). If values differ due to some variable in the studies (e.g., water hardness or pH), the multiple values may be used to generate a model of the parameter as a function of those variables. An example is the model of metal toxicity to aquatic biota as a function of hardness used in the US ambient water quality criteria (Stephan et al. 1985).

To assure the quality of secondary data, assessors should document a priori criteria for data selection. Because risk assessments are often contentious, it is best to be able to say that you chose a value or set of values because they had certain quality attributes that the alternatives lacked. Guidance and criteria for data evaluation have been published by governmental and quasi-governmental agencies and synthesized by Klimisch et al. (1997). However, these documents are focused on the needs of routine assessments of individual new chemicals. Most ecological risk assessments will require ad hoc evaluation of secondary data quality.

9.1.3 DEFAULTS AND ASSUMPTIONS

Defaults are functional forms or numerical values that are assigned to certain models or parameters in risk assessments, based on guidance or standard practice, in the absence of good data. Examples include a default rooting depth of 10 cm for plant uptake, a default bioaccumulation factor of 1 for invertebrates in soils, or a default logistic exposure–response model. Assumptions are equivalent to defaults but are derived for a specific assessment rather than being taken from guidance. They may be complex, implying functional forms or sets of parameters. For example, an assumption in a pesticide risk assessment might be that a 100 m^3 pond receives runoff from a 1 ha treated field. Defaults are appealing, because they are easy to use and uncontroversial, at least to the organization that published them. They are of high quality in the sense that they are likely to be acceptable to the decision maker and may have withstood judicial challenge when used in prior assessments. However, almost any real data concerning the actual situation being assessed are likely to be more accurate than a default. Therefore, even relatively poor data are likely to have a high quality relative to defaults, in the sense of generating more accurate risk estimates. Ad hoc assumptions must be individually justified.

9.1.4   REPRESENTING DATA QUALITY

Funtowicz and Ravetz (1990) proposed a scheme for presenting data or results of numerical analyses for uncertain policy decisions. The components are numeral, unit, spread, assessment, and pedigree, so the scheme is termed NUSAP. Numeral is the number, set of numbers, or other elements expressing magnitude, e.g., 10, 1/8, 5 to 8, and third of ten. Unit is the base of the operations expressed by the numeral, such as kg/ha or $\$_{1998}$. Spread expresses the distribution of values that the numeral might assume, based on the data, and is the output of uncertainty analyses (Chapter 5). Expressions of spread include variance, range, or within a factor of x. Assessment is a more complex concept that relates to justified interpretations of the result including expectations of the values that the numeral might assume given all knowledge and beliefs of the assessor. It may be expressed in simple statistical terms such as confidence intervals if statistical uncertainty dominates, it may be qualitative such as high confidence that the true numeral value lies in a particular range, or it may be a description of a numerical result as conservative, highly conservative, optimistic, etc. Finally, pedigree describes the source of the numeral. This may be simply a citation of a published source, the identity and qualification of the expert who provided the numeral, or the agency that established a default value. The NUSAP system could be used to improve the communication of quality to decision makers and stakeholders (van der Sluijs et al. 2005). Even if it is not formally adopted, each of its components should be considered when choosing data or reporting their quality.

Numerical results in risk assessment are plagued by false precision. Although we are all taught the concept of significant figures, assessors often neglect it in practice. Consequently, results are often presented as more precise than the precision of the input parameters. In practice,

$$5000 + 13 = 5000$$

because the trailing zeros in 5000 are simply place holders to indicate that the 5 is applied to a particular order of magnitude (10^3), and 13 is lost in the imprecision of that number. This is the basic arithmetic of significant figures, and it should be applied to ensure that the results are not presented as more precise than the least precise input. The problem of false precision also occurs when the individual numerals in an assessment have high precision, but, because they are not directly applicable to the situation being assessed, the reporting of all available significant figures would be misleading. For example, we may know that the environmental half-life of a chemical in a lake was 332 h, but we may not know the lake-to-lake or year-to-year variance. When applying that number to another lake or even the same lake in a different year, we might, based on experience, judge that the value had only one significant figure. That is, we do not believe that the true value in a new application is likely to be between 331 and 333, but rather that it may be between 200 and 400. This use of appropriate significant figures is a form of truth telling about the precision of results. It is important to remember that the reduction of significant figures should occur after all calculations, to avoid rounding errors.

9.1.5   DATA MANAGEMENT

High-quality data from the best methods and sources may be corrupted during transcription, units conversion, copying, subsetting, aggregation, etc. A data management plan must be developed to ensure that the data do not become corrupted, that data descriptors (e.g., units and definitions of names for variables) are correct, that meta data (information describing the data such as sampling date and location) are correctly associated with the data, and that data are archived.

The creation, management, and use of compilations of data such as the US EPA ECOTOX database present particular problems. Data in the literature are presented in various formats and levels of detail, and methods differ in ways that make it hard to determine whether a particular result should be included in a database as an instance of a prescribed data type. Even without errors in data extraction and entry, two people may extract different results from a paper because of differences in interpretation. Hence, to assure the quality of a database, rules for data interpretation must be developed. Even then, two individuals should extract data from each source, the entered data should be compared, and differences should be arbitrated by an experienced expert.

When using databases, assessors should be cautious because not only may the entry be erroneous, but the primary source may not have adequate quality. The quality evaluations performed for the US EPA ECOTOX database and the Australasian Ecotoxicity Database (AED) focus on the adequacy of reporting methods and results rather than the actual quality of the methods and results (EPA 2002a; Hobbs et al. 2005). Hence, if a datum is influential, it should be checked against the original source. This check is to determine whether the datum is correct and to read the paper or report carefully to determine whether it is applicable to the assessment and is of sufficient quality.

9.2 MODEL QUALITY

A high-quality model is one that contributes to a high-quality assessment and a well-informed decision. This statement seems self-evidently true, but it hides a host of controversies about the use of models in risk assessment and environmental management. Models, whether complex mathematical simulations or statistically fitted functions, cannot be literally true or valid because they are inevitably abstractions of the natural system being modeled. However, it is unreasonable to abandon the concept of model quality and simply strive for consensus through a social process of confidence building as some have advocated (Oreskes 1998). Some models provide better predictions than others, regardless of their acceptability to stakeholders (Box 9.2). Hence the aphorism: all models are wrong, but some models are useful. Several considerations go into recognizing or developing a model that is likely to achieve the goal of a well-informed decision. The first four are pragmatic; the others are more conventional, technical, or procedural considerations. Together they constitute the criteria for model evaluation.

Useful output: A fundamental requirement is that the model must predict the parameters required by the assessment. An exposure model that predicts water concentrations is not adequate if effects are expressed in terms of body burdens, but it can be useful if a bioaccumulation module is added.

Appropriate application niche: The application niche is the range of conditions to which a model may be defensibly applied. A model should be designed or selected to have an application niche that includes the conditions being assessed. In quantitative structure–activity relationships (QSARs) and other empirical models, this is referred to as the model's domain.

Transparency: Proprietary models cannot be evaluated or modified to make them more appropriate for an application. Hence, the source code for computer-coded models should be made available so that the model structure and parameterization are accessible.

Endorsement: When performing an assessment for an agency, models that are endorsed by that agency are assumed to have sufficient quality for their prescribed uses. Models endorsed by other agencies or standard organizations may also be assumed to have sufficient quality if their authority is recognized by the decision makers and reviewers of the assessment.

BOX 9.2
A Comparative Test of Fate and Exposure Models

The International Atomic Energy Agency sponsored a comparative test of 11 environmental transport and fate models using data on the fate of ^{137}Cs in southern Finland from the Chernobyl accident (Thiessen et al. 1997). The modeling teams were supplied with information on ^{137}Cs concentrations in air and soil, meteorological conditions, and background land use and demographic information. They were asked to estimate, for specified time points over a 4.5 y period, ^{137}Cs concentrations in various agricultural products and natural biota, and daily intake and body burdens for men, women, and children. Model results were compared to measurements.

Reviews of model quality tend to focus on the structure of the model, the functional forms, and the assumptions concerning the structure and function of the system, but those were not the most important differences among models in this test. "The two most common reasons for different predictions, however, were differences in the use and interpretation of input information and differences in selection of parameter values" (Thiessen et al. 1997). Hence, modeling exercises are at least as dependent on the quality and clarity of the information provided by environmental biologists and chemists as on the quality of the modelers and their model.

Experience: A model that has been widely and successfully used is more likely to have acceptable quality. Such models have undergone a practical peer review.

Peer review: Traditional peer review is a weak basis for evaluating the quality of a model because of the difficulty of evaluating the performance of a model by reading a paper or report. However, if other practitioners do not have experience using the model, conventional peer review is better than no outside evaluation. Peer review requires transparency: users and reviewers should have access not only to the equations and parameters of the model but also to the assumptions and any publications from which they were derived.

Parameterization: If the parameters of a model cannot be measured or estimated with reasonable accuracy using available data or readily obtained data, it is a low-quality model.

Identifiability: Models should not disaggregate a parameter into numerous parameters when their effects on the system cannot be individually distinguished. For example, inclusion of multiple susceptibility states of organisms may make the model seem more realistic, but if those states are not identifiable in data for the population, they are likely to lower the quality of the model.

Mechanistic understanding: Mechanistic models are potentially more general and reliable than empirical models, if the mechanisms controlling the modeled systems are understood. Even empirical models benefit from mechanistic understanding, which informs the selection of model forms and independent variables. Hence, mechanistically supported models have, in general, a higher quality. However, mechanisms of ecological systems are often poorly understood, so mechanistic assumptions can be misleading if they are not themselves based on high-quality information.

Completeness: All models are incomplete, but a model that represents all of the major components and processes of the modeled system is more credible. For example, a wildlife exposure model that includes water and soil ingestion as well as food is more complete. However, complex models are difficult to implement and inclusion of poorly specified components is likely to decrease, rather than increase, accuracy.

Verification: Models are built by trying to represent some set of data about a system of the type being modeled. For example, data concerning the hydrology of a lake and the concentrations of a chemical in that lake may be used to develop a model of transport and fate of that chemical in any lake. At a minimum, the model should be able to represent that lake. If runs of the model indicate that nearly all of the chemical is in the sediment when in fact significant concentrations occur in the water and biota, verification of the model has failed. Note that verification is a weak criterion because the site data were used to derive the model. In particular, empirical models are, to a certain extent, self-verifying, because they are fitted to the data. Even mathematical simulations typically contain parameters that are derived from the data by a process called calibration. Verification may be quantified by goodness of fit or likelihood estimators.

Repeatability: If the modeled system's properties vary significantly from year to year, a model should repeatedly produce accurate output estimates by incorporating the variance. A classic example is the year-to-year variance in recruitment of fish populations, which depends on the previous year's stock and on physical conditions.

Corroboration: Models should produce acceptably accurate estimates for different instances of the type of system being modeled. For example, a lake model would be corroborated by using it to successfully estimate properties of several lakes that were not used in the model's development. This process is commonly referred to as validation, but it is just one particularly good way to evaluate a model and does not prove that the model is valid in any absolute sense (Oreskes 1998). All data sets have their own problems of quality and representativeness that prohibit their use as absolute standards of comparison. Corroboration of empirical models such as QSARs or bioaccumulation models is performed by using part of the data (the training set) to develop the model and an independent part of the data as a test set to determine that the model has some general applicability (Sample et al. 1999). As with verification, corroboration may be quantified by goodness of fit or likelihood estimators.

Predictive success: The highest form of model success is the demonstration of some previously unrecognized phenomenon that was predicted by the model. The classic example is the demonstration of the bending of light in a gravitational field, predicted by the general theory of relativity.

Robustness: A model is robust if it is insensitive to changes in underlying mechanisms. For example, density dependence may be caused by food limitations, space limitations, interference, or other mechanisms, each of which may be modeled by the same function.

Reasonable structure: In practice, the most important quality assurance checks for a model involve determining the reasonableness of its structure (Ferson 1996). Does it yield the correct dimensions, as expressed by the output units? For example, if a model is said to estimate concentration in mg/kg, do the model expressions reduce to those units? Monte Carlo analysis places particular constraints on structure. Is the correlation structure reasonable (e.g., if X is correlated with Y and Y is correlated with Z, is X correlated with Z)? Does the model include division by a distribution that includes zero? Does each instance of a variable have the same value in an iteration?

Reasonable behavior: The model should be checked to determine if the results are reasonable, particularly at extreme values of the input variables. For example, concentrations and abundances may not be negative, and predators may not consume more than their prey. In addition, aberrant model behavior can be revealed by examining results for simplified or trivial conditions. For example, Hertzberg and Teuschler (2002) showed that, for one recommended model of chemical interactions, when the interaction parameter is set to 1, the result is a constant, independent of mixture composition. Application of this criterion can require some scientific sophistication to recognize that a result is unreasonable.

While the last two criteria are absolute, the others are points to consider, not clear and absolute criteria. Some of them are apparently contradictory. In particular, completeness and mechanistic representation can make a model more difficult to parameterize and less robust.

These points are also primarily aimed at assessors who are choosing from existing models. In addition, there are quality assurance criteria for model development projects (EPA 2002g) that include tests to assure that the code is correct, that data transfers between component models function correctly, and that units balance These concerns are also relevant when assessors implement a published model in a computer or modify a model for an assessment. For example, if a published exposure model is implemented in a spreadsheet, results should be checked by hand calculations.

So far, this discussion of model evaluation is intended to determine whether a model is acceptable for a given use. An alternative approach is to choose from among a set of alternative models (distinct hypotheses about the structure and function of the system) the one that is the most probable, given the evidence. This can be done by using Bayesian decision theory to evaluate the probability of the models given prior information such as results of previous implementations of the models and a set of data concerning the system to be modeled. One can then calculate the posterior probability of each model using Bayes' theorem. This approach is seldom used in risk assessment of pollutants but has been employed in natural resource management (Walters 1986; Hilborn and Mangel 1997). Multiple models may have equivalent quality including approximate consistency with the available data. Hilborn and Mangel (1997) argued that, because all models are metaphors, we should expect to have multiple models of any complex system. If multiple models are plausible and presentation of multiple results is unacceptable, model averaging may be used, preferably weighted by the probability or predictive accuracy of the models (Burnham and Anderson 1998). Simple information-theoretic criteria (Akaike's or Bayes' information criteria) may be used for model selection or weighting (Burnham and Anderson 1998). The use of these approaches will be limited to cases in which differences in model assumptions or structure are critical to a management decision, and sufficient high-quality data are available to discriminate among the models. However, the probabilities generated by these methods are relative and do not provide estimates of the probability that a model is true or the best possible model. All of the models evaluated may be incomplete or biased in some important way.

The issue of model quality is complex and difficult. Good discussions of quality assurance for environmental models include Ferson (1996), Pascual et al. (2003), and Walters (1986).

9.3   QUALITY OF PROBABILISTIC ANALYSES

The quality of probabilistic analyses of uncertainty has been a major concern of risk assessors, stakeholders, and regulators. Sources of guidance on quality assurance include Burmaster and Anderson (1994), Ferson (1996), and Risk Assessment Forum (1996). The EPA has issued requirements for acceptance of Monte Carlo or equivalent analyses (Risk Assessment Forum 1997). Such guidance is more often cited than applied, in large part because the requirements are labor-intensive, in part because they require a real understanding of probability and the Monte Carlo technique, and partly because they inflate the volume of assessments (which are already criticized as excessively voluminous) with information that is incomprehensible to most readers. However, quality assurance must be performed and the results must be available, but not necessarily in the primary assessment report.

Both the EPA and Burmaster's requirements are clearly intended for use with human exposure analyses, but the EPA requirements are said to be applicable to ecological risk assessments as well. The following points are adaptations for ecological risk assessment of the EPA's eight requirements (Risk Assessment Forum 1997) plus nine bonus points to consider.

1. Assessment endpoints must be clearly and completely defined. It is particularly important to specify whether an endpoint is defined as a probability. If so, it is important to know what source of variance or uncertainty is of concern to the risk managers.

2. Models and methods of probabilistic analysis and associated assumptions and data must be clearly described. The disclosure called for in this condition is good practice whether or not the methods are probabilistic.
3. Results of sensitivity analysis should be presented as a basis for deciding which input parameters should be treated as distributed.
4. Moderate to strong correlations among input parameters should be identified and accounted for in the analysis. Correlations are common in risk models, and, if ignored, they inflate the output distribution. For example, body weights, feeding rates, and water consumption rates are all highly correlated. If correlations are believed to occur but cannot be estimated from available data, assessors should perform Monte Carlo simulations with correlations set to zero and to high but plausible values to determine their importance and present the results (Burmaster and Anderson 1994).
5. Each input and output distribution should be provided including tabular information and plots of probability density and cumulative density functions. The tabular presentation should include the following:

- Name of the parameter
- Units of the parameter
- If variable, with respect to what does it vary?
- Formula for the distribution of variability
- Basis for the distribution of variability
- If uncertain, what sources of uncertainty are considered?
- Formula for the uncertainty distribution
- Basis for the distribution of uncertainty

Distributions that are developed ad hoc may require considerable explanation. These may include the data from which they are derived or the elicitation techniques for expert judgments plus an explanation of how the data or judgments relate to the assumed sources of the variability or uncertainty. If the expert judgments of individuals are used to derive distributions, any information or logic that went into the judgment should be described as far as possible. Burmaster and Anderson (1994) indicate that a typical justification for a distribution would require five to ten pages.

6. The stability of both central tendency and the extremes of the output distribution should be noted and recorded. The requirement refers to the stability of the moments of the output distribution as the number of iterations of a Monte Carlo analysis increases. Most software packages provide criteria for termination of the analysis based on the stability of the output distribution.
7. Calculations of exposure and risk of effects using deterministic methods should be reported for comparison with the probabilistic results. The deterministic analyses may be performed using realistic or best estimate values for the parameters (e.g., the mean or medians of the parameter distributions) or assumptions and parameters favored by a regulatory agency. In some cases, discrepancies among conservative point estimates, best point estimates, regulatory estimates, and medians of probabilistic results will be quite large. The causes of these differences should be explained.
8. The expressions of exposure and effects must be concordant as well as individually making sense, given the site conditions and assessment endpoints (Chapter 30). Note that for distributions, this requirement goes beyond simple checking of units. The assessor must consider not only what is distributed but also with respect to what it is distributed.

9. As far as possible, use empirical information to derive distributions (Burmaster and Anderson 1994).
10. The correlation matrix must have a feasible structure, e.g., if parameters a and b are both strongly positively correlated with c, they cannot be strongly negatively correlated with each other (Ferson 1996).
11. Multiple instances of the same variable in a model must be assigned the same value in an iteration of a Monte Carlo analysis (Ferson 1996). This is a particular problem in stepwise or nested analyses in which different components of risk are estimated by separate simulations.
12. Care must be taken to avoid nonsensical values of input and output distributions. For example, negative values should not be generated for parameters such as concentrations or body weights, herbivore consumption rates should not exceed plant production rates, and contaminant concentrations should not exceed a million parts per million. This can be prevented by truncation, by the appropriate selection of the distribution, or by constraints on the relationships between variables.
13. In general, it is most important to treat the input parameters correctly (e.g., do not treat variables as constant), next most important to get the magnitude of variability or uncertainty right, and least important to get the form of the distribution right (e.g., triangular vs. normal).
14. For fitted distribution functions such as concentration–response distributions, species sensitivity distributions, and distributions of measures of exposure, goodness of fit statistics and prediction intervals should be reported as estimates of model uncertainty.
15. As far as possible, specify whether model assumptions introduce an identifiable bias. Examples include:

    • Assuming 100% bioavailability introduces a conservative bias.
    • Assuming independent toxic effects (i.e., response addition) introduces an anticonservative bias.
    • Assuming additive toxic effects (i.e., concentration addition) introduces a conservative bias.
    • Assuming that the chemical occurs entirely in its most toxic form introduces a conservative bias.
    • Assuming that the most sensitive species of a small number of test species is representative of highly sensitive species in the field introduces an anticonservative bias.

A bias does not mean that there is a consistent direction of error in every case. For example, strong antagonistic or synergistic effects could negate the bias associated with assuming additive toxicity. However, the bias is real because such effects are relatively uncommon. When possible, the influence of biases should be estimated. For example, the uncertainty from assuming that the chemical occurs entirely in its most toxic form can be bounded by presenting results for the least toxic form.

16. In general, model uncertainty cannot be well or reliably estimated because the range of models cannot be well defined. At least, model uncertainty should be acknowledged. The acknowledgment should list specific issues in model selection or design that are potentially important sources of error. The list should include any issues about which there was initial disagreement among the parties or issues about which there is no consensus in ecological risk assessment practice. When there are clear differences of opinion concerning assumptions, models should be run with each of the alternatives to determine their influence on the results.
17. Acknowledge that the quantified uncertainties are a small part of the total uncertainty concerning future events.

9.4   ASSESSMENT QUALITY

In addition to assuring the quality of the data and models that go into an ecological risk assessment, the quality of the assessment as a whole should be assured. The following are some important considerations when performing or evaluating an assessment.

Completeness: An ecological risk assessment should include all of the components specified in the US EPA or other applicable guidelines. In addition, it should include all aspects of the scenario and other aspects of the problem formulation. For example, if an assessment endpoint is abundance of cutthroat trout, the assessment must include an estimate of exposure of cutthroat trout to each of the contaminants or other agents of concern, corresponding exposure–response relationships, and a characterization of risks to cutthroat trout including uncertainties.

Expertise: The individuals who performed the assessment must have adequate qualifications and their qualifications should be reported. Assessors who have experience in ecological risk assessment and related sciences, appropriate degrees, or professional certification are more likely to produce a high-quality assessment.

Generally accepted methods: Generally accepted approaches and methods are more likely to be accepted. However, that should not inhibit needed innovations. When new methods are used, they should be compared to the generally accepted method so as to show how their use may affect the results. In the extreme, the use of standard methods assures a minimum quality. An example of standard method quality is the European Union System for the Evaluation of Substances (EUSES).

Transparency: Methods, data, and assumptions should be clearly described and their sources identified. However, this requirement can result in bulky documents that are difficult to read (Section 35.1).

Reasonable results: If results are unreasonable, the methods, data, and assumptions should be carefully scrutinized. However, apparently unreasonable results may not be false. For example, it has been suggested that the occurrence of wildlife on a site where estimated effects on wildlife are significant indicates that the assessment is unreasonable (Tannenbaum 2005a). However, a contaminated site may be causing reduced life span or fecundity, but populations will persist on the site because losses at sink habitats are replaced by individuals dispersing from source habitats. Similarly, estimates of significant risks at metal concentrations below regional background levels may be due to differences in form and speciation of the metal rather than unreasonable risk assessment.

In addition, the overall quality of an assessment can be assured by following a good process, by peer reviewing the assessment, and by comparing replicate assessments performed by different organizations.

9.4.1   PROCESS QUALITY

By assuring that an assessment is performed in a prescribed manner, one can assure that important components of the assessment or sources of information are not neglected and that appropriate reviews are performed. The frameworks for ecological risk assessment (Chapter 3) serve that function in a general way. For example, following the EPA framework and guidelines assures that the assessors will seek input concerning goals and will clearly define the assessment endpoints that they derive from those goals. Context-specific guidance documents assure process quality by detailing procedures. For example, the Interim Final Ecological Risk Assessment Guidance for Superfund specifies a procedure that includes six scientific or management decision points (Sprenger and Charters 1997). These are points in the process at which the risk manager meets with the risk assessors to review interim products and plan future assessment activities. Such frequent reviews and consultations have the potential to assure the quality and utility of an assessment. However, elaborate procedures can also be bureaucratic waste of time and effort if they become empty process checks. Process quality is discussed at length by Benjamin and Belluck (2002).

9.4.2 PEER REVIEW OF THE ASSESSMENT

The conventional means of assuring the quality of scientific products is peer review. Although risk assessments have not been routinely peer-reviewed, the practice is increasingly common. Peer reviews help to increase the quality of risk assessments, assure the risk manager of the quality of the assessment as input to the decision, and help to deflect criticism or legal challenges. Depending on the circumstances, peer reviews may be performed by consultants, academic scientists, or staff of the regulatory agency, responsible party, or stakeholder organizations. Reviews of risk assessments are increasingly performed by super peers, panels of distinguished scientists such as the US EPA Science Advisory Board or US National Research Council committees.

Because of the complexity of risk assessments and the need to assure the quality of the peer review, it is valuable to provide checklists and other guidance for the peer reviewers. These should include a list of components that should be found in the report and technical points to be reviewed such as the life history assumptions for endpoint species. A good example is provided by Duke and Briede (2001). In addition, it is valuable to communicate directly with the reviewers. Briefing the reviewers at the beginning provides a means of assuring that the reviewers understand the manager's charge to the assessors and the reasons for choices that were made during the assessment process as well as assuring that the reviewers understand their own charge. For example, reviewers may be charged with addressing the entire assessment process including the scope and goals, or, more commonly, they may be limited to the technical aspects of risk analysis and characterization. In the latter case, reviewers should be warned not to waste their time reviewing goals or other policy-related aspects of the assessment.

Peer review may also be improved by increasing the number of reviewers. Individual reviewers differ in their experience, expertise, level of attention, and dedication. For example, it is sometimes clear that a reviewer has skipped certain sections of a long report. A large number of reviewers (≥ 10) tends to assure that at least one reviewer will notice a particular error. Large sets of reviewers also make it possible to identify outlier opinions. If only two reviewers are used and they disagree, it is not possible to objectively determine which opinion is consistent with consensus in the field and which is eccentric. Even three reviewers may provide three different opinions.

9.4.3 REPLICATION OF ASSESSMENTS

As with scientific studies (Section 9.1.2), it is possible for risk assessments to be biased by the interests of the organizations performing or funding the assessment. Chemical manufacturers and other regulated parties allege that the US EPA and other regulatory agencies systematically bias their risk assessments so as to be excessively protective. Environmental advocacy organizations allege that both industry and government agencies use risk assessment to hide problems and avoid protecting the environment (Tal 1997). In part, these different perceptions are due to the degree of precaution applied in the selection of data and assumptions. They are also due to differences in how the problem is formulated. Should risks be assessed using attributes of organisms, populations, or ecosystems as endpoints? Should they be assessed for a site or a region? Should background concentrations be combined with concentrations attributable to the source when estimating exposures? These differences are partly due to the unconscious biases that creep in when a particular outcome is desired.

If the parties to a contentious environmental decision each perform their own assessments, it can clarify the differences among the parties, reveal errors in the assessments, determine which differences affect the decision, provide a basis for reaching a consensus on technical issues, and better inform the decision maker. Industries commonly perform their own risk assessments, particularly for new chemicals or products. However, they seldom make them public. Environmental advocacy groups have often opposed risk assessment, but they have been influential when they have performed and presented their own assessments (Tal 1997).

There are some potentially serious problems with the use of replicate risk assessments. Different groups have different levels of resources and expertise to apply to risk assessment. The performance and evaluation of multiple risk assessments could slow the assessment and decision-making process. Finally, rather than leading to consensus or increased understanding, replicate risk assessments could lead to incomprehensibly arcane arguments among dueling experts. Some of these problems could be avoided by the appointment of a neutral party or organization to mediate or judge the relative scientific merits of the alternatives. However, no institutional mechanisms for generating and comparing replicate risk assessments currently exist outside the legal system, and the courts choose winners and losers rather than trying to reach a scientific consensus.

When multiple assessments of a particular hazard are performed, for whatever reason, they provide an opportunity to improve quality. Each assessment group should examine the other assessments to identify differences in methods, data, and results, and determine whether they reveal weaknesses or errors in their own assessment that should be corrected. If the assessors from each group are brought together to discuss the reasons for the differences, the overall quality of the assessments can be greatly improved.

9.5 SUMMARY

Most available guidance on quality assurance for environmental scientists relates to the generation of high-quality primary data. However, for most ecological risk assessments, that is not the most important quality concern. Ecological risk assessors must evaluate the quality of secondary data, of models, of probabilistic analyses, and of the overall quality of their risk assessment.

Part II

Planning and Problem Formulation

Before everything else, getting ready is the secret to success.

Henry Ford

Before risks are analyzed, it is necessary to lay the groundwork by defining the problem and the means to solve it. This process is driven by the needs of the manager who will use the results of the assessment to make a decision, but it is constrained by the limitations of time, resources, and technical capabilities. The EPA's framework for ecological risk assessment (Section 3.1) distinguishes between planning, which involves the risk manager and potentially the stakeholders, and problem formulation, which is performed by the assessors alone. The intent was to comply with the National Research Council's (1983) mandate to keep policy and science distinct. However, the distinction seldom holds up in practice because the formulation of the problem also involves the interpretation of policy. The best example is the EPA's Data Quality Objectives (DQO) process that requires the risk manager to participate in the development of the analysis plan, the last step in problem formulation (Quality Assurance Management Staff 1994). Further, the distinction is not generally recognized outside the United States. Finally, this distinction is unlikely to convince skeptics that risk assessment is not biased by the policies of its sponsor. External peer review is a better guard against bias in the technical aspects of an assessment. Therefore, the planning of an assessment and formulation of the problem are presented here as an integrated process.

The distinction between routine assessments and novel assessments has more practical importance. Assessments of new chemicals and similar cases may follow standard policies and procedures. In such cases, goals, endpoints, and management options are set for all assessments, and little planning or problem formulation is required. The assessors need not consult with the risk manager for each assessment and, after assembling the data, may proceed to the analysis and characterization of risks. At the other extreme, site-specific assessments and assessments of novel agents such as acid deposition, a genetically modified fish, or global warming typically require an extended process of planning and problem formulation. Multiple meetings may be held with the stakeholders and the public, and the process may be iterated as assessors develop conceptual models, endpoints, and other planning products for review and revision. It is the responsibility of the risk manager to determine how routine or ad hoc the planning and problem formulation should be. However, it is the assessor's responsibility to inform the managers about cases in which routine policies and methods may not be applicable. For example, a new chemical may have physical or toxicological properties that suggest the need to consider novel endpoints or may have patterns of use that suggest the need for novel management options.

The success of risk assessments and the resulting management decisions depends on the quality of the planning and problem formulation process. If the problem formulation is done in a haphazard manner, the resulting assessment is unlikely to be useful to the risk manager. The process should be taken as seriously as the performance of toxicity tests or the creation of a hydrologic model and should be done with at least as much care. In some cases, the completion of a problem formulation may be sufficient to drive a management decision. That is, once the nature of the situation is delineated, the need for action or the appropriateness of inaction may be apparent without quantitatively assessing the risks.

10 Impetus and Mandate

Assessors should understand why they are performing an ecological risk assessment and the powers and limitations of the decision maker. These may be briefly and clearly defined, as in the regulation of new chemicals, or broadly and ill defined, as in the assessments of risks from climate change.

On the basis of the paradigm of the release of agents by sources resulting in exposures that cause effects, the impetus for assessment falls into one of three categories: sources and agents, exposures, or effects. Sources and agents are the most common and familiar. They include sources such as waste outfalls, new technologies or new resource management plans and hazardous entities such as chemicals, exotic organisms, and dredges. Such assessments support the issuance of permits or approvals to operate a facility, market a product, import an organism, release an effluent, harvest a resource, or build a structure. The usual management options are to approve, disapprove, or approve with restrictions.

Effect-initiated risk assessments are prompted by the observation of ecological impairments such as the fishless lakes that prompted the aquatic component of the National Acid Precipitation Assessment (Baker and Harvey 1984), or the deformed frogs that have prompted assessments of the risks to anurans (Burkhart et al. 2000). These ecoepidemiological assessments are characterized by the need to perform a causal analysis before the risks from alternative actions can be considered (Section 4.3).

Exposure-initiated assessments are the least common. For chemicals, a risk assessment may be initiated by the observation of elevated body burdens. For exotic organisms one may be initiated by observations of a new pathogen in an organism or new species in an ecosystem. Exposure-initiated assessments require determination of the sources as well as the risks of the effects. For example, the observation of high mercury levels in fish has led to risk assessments that determine the sources as well as the risks to humans and piscivorous wildlife (EPA 1995).

In addition, the EPA (1998a) identifies value-initiated assessments. These are cases in which a valued aspect of the environment is the subject of the assessment. No sources, agents, exposures, or effects are identified as requiring analysis. Instead, sources, agents, exposures, and effects must be hypothesized or identified through observation, measurement, or modeling. Examples of such impetus include rare or high-valued species or places such as parks or watersheds. A specific example is the Big Darby Creek watershed in Ohio, which was assessed because it has an exceptionally high-quality biotic community including rare and endangered fish and mussels (Cormier et al. 2000; Serveiss 2002). These value-initiated assessments must evolve into one or more assessments based on a conventional impetus, i.e., effects, exposures, or sources of agents must be identified before risks can be assessed.

11 Goals and Objectives

If you do not know where you are going, any road will take you there.

The White Rabbit to Alice (Lewis Carol)

The planning of an ecological risk assessment depends primarily on the goal of the management action to be supported. Most environmental laws in the United States provide rather vague goals such as "protect public health and the environment" or "protect and restore the physical, chemical and biological integrity of the Nation's waters." Agencies that implement laws should interpret those vague goals in more concrete terms that can be evaluated. For example, the International Joint Commission (1989) interpreted the biological impairment goal for Lake Superior thus: "The Lake should be maintained as a balanced and stable oligotrophic ecosystem with lake trout as the top aquatic predator of a cold-water community and with *Pontoporeia hoyi* as a key organism in the food chain." Such specific goals, called objectives, may apply to an entire regulatory or management program, or may be assessment-specific. A programmatic example is the European Commission's goals for their water quality objectives (WQOs). Accordingly, a WQO

- should be such as to permit all stages in the life of aquatic organisms to be successfully completed
- should not produce conditions that cause these organisms to avoid parts of the habitat where they would normally be present
- should not give rise to the accumulation of substances that can be harmful to the biota (including man) whether via the food chain or otherwise and
- should not produce conditions that alter the functioning of the ecosystem (CSTE/ EEC 1994)

Examples of appropriate management goals in the EPA's guidelines include "reduce or eliminate macroalgal growth" and "maintain diversity of native biotic communities" (EPA 1998a). Goals for site-specific or "place-based" assessments may be generated through workshops or other consensus-building processes. Goals for public lands or other natural resources are often contained in their management plans. In addition to these goals for a specific law or assessment, it may be possible to define national environmental goals. However, goal setting is probably the most inconsistent and ill-defined component of the ecological risk assessment process (McCarty and Power 2001). In any case, careful thought should be devoted to defining goals (Box 11.1). However derived, ecological goals provide the basis for identification of the assessment endpoints.

Some goals are defined in terms of desired attributes and require no comparison. Examples for fish species include: (1) the endangered species should persist for at least 50 years after the action, (2) the fishery should support a median yield of 100 MT, or (3) no kills should occur. However, it is often necessary to define goals relative to a reference condition. As discussed in

---

BOX 11.1
Goals for Urban Streams

A goal of the US Clean Water Act is to "protect and restore the physical, chemical and biological integrity of the Nation's waters." The biological integrity goal obviously requires some clarification. The most common approach is to define relatively undisturbed streams in the least disturbed watersheds within a region as having integrity and then to develop an index or other indicator to define a scale of loss of integrity relative to the undisturbed stream (Yoder and Rankin 1995b). However, since even developments that result in only 10% to 20% impervious surfaces in a watershed cause significant changes in stream communities, it is not possible to achieve that sort of integrity in urban streams. Rather, somewhat degraded standards such as "modified warm-water habitat" are created for such streams. An alternative would be to develop definitions of biological integrity for urban streams, based on what is desirable and possible in those conditions. This need not be an unambitious goal. It may require expensive projects and controversial regulations to eliminate combined sewer overflows, store, treat and slowly release storm water, eliminate toxicity and high nutrient levels in effluents, reduce residential pesticide and fertilizer use, and create habitat structure. However, the goal of a high quality urban stream rather than a somewhat less impaired stream could provide significant incentives and psychosocial benefits. This would require more than a change in the semantics of the goal. The metrics that define the degree of departure from an ideal undisturbed stream would not be the best metrics to define the departure from a high quality urban stream. For example, it may be impossible to reduce temperatures sufficiently to support trout and darters, but it may be possible to sustain an urban fishery of catfish and sunfish that differs from undisturbed streams in the region but has an integrity of its own. Hence, the goals would be to achieve design criteria for designated uses including recreation, flood control, aesthetics, and recreational fisheries, rather than minimal departure from a regional reference state. This would require the development of a practice of urban aquatic ecology that would be equivalent to urban forestry.

---

succeeding sections, the definition of reference is usually treated as a technical problem during the definition of assessment endpoints and the development of the analysis plan. However, as the urban stream example illustrates (Box 11.1), the choice of reference has policy implications. The results of management will be different if the goal is to achieve an attribute of undisturbed and uncontaminated ecosystems (e.g., wilderness steams) of some percentile of all ecosystems (e.g., the tenth percentile of all streams arrayed from highest to lowest quality), of a historical reference (e.g., the community composition reported in the first records), or of high quality urban streams. Hence, the bases for comparisons should be defined by decision makers during the goal setting process.

Ideally, the management goals would also specify the decision criteria. Thresholds for effects, three-part logics, cost-effectiveness, cost–benefit, net environmental benefit, or other decision criteria should lead to different risk assessments. For example, thresholds for acceptability may be based on any variety of metrics, but cost–benefit or net benefit analyses require that the expected changes in the environment be clearly specified and quantified.

12 Management Options

Environmental risk managers have limited range of options that are determined by legal and regulatory constraints and by practical constraints. For example, a US EPA remedial project manager for a Superfund site may mandate a variety of remedial actions, such as removal or capping of contaminated soil, but cannot order the purchase and protection or restoration of replacement ecosystems. Practical constraints include monetary limitations; public acceptance; and limitations of the available science and technology for remediation, restoration, or treatment. Within those constraints, the management options considered should be those that could potentially achieve the management goals. These options must include not only environmental goals but also human health goals and socioeconomic goals.

The management options determine the range of future conditions to be assessed. For a new chemical, that range may be simply the conditions with and without the use and release of that chemical. For pesticides, it could be the conditions resulting from the range of uses defined by the manufacturer and those associated with various restraints on use, such as no application within a prescribed distance of a water body. For forest fires it would include allowing the fire to burn or sending various sizes of fire crews and support teams, so the future conditions might include not only the extent of the fire, but also the effects of fire breaks, flame retardants, etc.

The result of consideration of management options should be descriptions of each alternative action that will be assessed and compared in the risk characterization. Hence, they must include sufficient detail to allow the analyses and characterization to be performed. For example, it is not sufficient to identify wastewater treatment as an option. Rather, the quality and quantity of the treated water must be specified as well as the likely frequency and consequences of treatment failures.

13 Agents and Sources

The subject of most risk assessments is an agent that has been or will be imposed on the environment such as a new pesticide, an effluent, a timber harvest, or an exotic species. In some cases, the impetus for the assessment is a new source of such agents. These include new technologies such as a new coal liquefaction technology, facilities such as a highway, activities such as off-road motorcycle races, plans such as a timber management plan, or even cultural pressures such as suburban sprawl that will impose multiple agents on the environment. The assessment must begin with a reasonably complete description of the agent and its source. That information is typically provided to regulatory agencies by the applicant for a permit, and the required information is specified in regulations. However, if the source is contaminated media, an established exotic species or equivalently existing ambient source, then it and its associated agents must be characterized by census, sampling, or analysis. If such data are not available, the characterization of sources and agents must be included in the analysis plan.

13.1 EMISSIONS

Conventionally the design of a facility, including any regulatory restrictions, determines what may be released from a facility, how it is released, and how frequently. This information is commonly known as the source term. Source terms typically estimate the characteristics of normal operations including variance in concentrations and release rates due to variance in raw materials, product mix, and conditions. However, they should go further and estimate the variance due to unintended releases. Chemicals and other agents may be released by spills and other accidents, by fugitive releases (e.g., leaks around gaskets or during pouring), or by operation during startup, shutdown, or upset conditions (e.g., when operating temperatures or pressures are not achieved or when switching between products). In addition, many chemicals get into the environment even though they are not intentionally released (PCBs) or even not intentionally produced (chlorinated dioxins).

In some cases, unintended releases are characterized by engineering risk assessments. These assessments evaluate the results of potential accidents such as tank ruptures or upset conditions such as power loss (Rasmussen 1981; Haimes 1998; Wang and Roush 2000). There are two basic types of analysis for accidents. Fault trees define the causal links from a defined failure (e.g., a pipe rupture resulting in a spill) back to all of the potential initiating events (e.g., corrosion, vandalism, impact of a vehicle). Event trees begin with an initiating event (e.g., a stuck valve) and trace all of its possible consequences (e.g., safe shutdown, pipe rupture). Probabilities are assigned to each node in a tree so that probabilities of a failure given an event or probabilities that an event caused a failure can be estimated. In some cases, standard models are available for assessing risks of unintended releases to the environment. For example, LANDSIM was developed to simulate release of contaminants from landfills in Britain with various design features (Environment Agency 1996). The importance of engineering risk assessment is highlighted by disastrous accidents such as the Exxon Valdez oil spill and the dike failure on the Tisza River in Romania that released cyanide and metals resulting in mass mortality of aquatic life.

13.2   ACTIVITIES AND PROGRAMS

In some cases, the subject of the assessment is an activity or program rather than a chemical or emission. Examples include forest management plans, military training exercises, mining activities, irrigation projects, or construction and operation of ports or other facilities. In such cases, the source characterization is an analysis of the activity to determine what hazardous agents will be employed or generated. It may appropriately begin by listing the phases of the activity ranging from, e.g., site exploration through construction and operation to decommissioning. Other activities my have little temporal phasing but may have distinct constituent activities that may be listed. For example, a military exercise may include driving tracked vehicles, live fire of weapons, aircraft overflights, and excavation of defenses. In either case, listing of the constituent activities or phases focuses the source characterization and assures that no major component of the activity is overlooked.

For each constituent activity or phase, the potentially hazardous aspects of the activity to be assessed should be listed. For example, driving a vehicle off-road involves crushing plants, striking animals, compacting soil, rutting the soil surface, and starting fires. For repeated activities such as forest management plans or military exercises, a checklist based on prior assessment experience may be used. For novel activities, lists must be generated by brainstorming exercises involving experts on the activity, experts on the site, and possibly stakeholders.

13.3   SOURCES OF CAUSES

If the impetus for the assessment is an observed biological effect, it is necessary to determine the cause of the effect and then the source of the causal agents (Chapter 4). In such assessments, the characterization of the source or sources is part of the assessment rather than an input to the assessment (Section 19.2).

13.4   PROPERTIES OF THE AGENT

In addition to characterizing the source, the problem formulation must identify the properties of the agent that are potentially relevant to the risk assessment. These include both exposure-related and effects-related properties. To develop the conceptual model it is necessary to know, at least qualitatively, what the agent does in the environment. Where does it go, what is its fate, and what are its effects on the various organisms and processes of ecosystems? Traditional human health and ecological risk assessments have focused on identifying the hazards and characterizing stressors. However, ecological risk assessments must recognize that they are assessing agents such as nutrients, fire, temperature, and habitat modification that are beneficial at typical levels and may be stressful only at exceptionally high or low levels. Hence, it is important to describe the range of effects of an agent and not simply those that are potentially harmful.

13.5   SOURCES OF INDIRECT EXPOSURE AND EFFECTS

The sources to be identified are usually the primary anthropogenic sources: pesticide releases, waste emissions, dams, explosives, etc. However, experienced ecological assessors know that the effects of those primary sources are themselves sources of secondary exposures and effects. For example, nutrients cause increased algal production, which is a source of organic material for aerobic decomposition and subsequent anoxia. Similarly, logging causes increased soil erosion that is a source of silt to streams, which causes reduced salmonid reproduction. These secondary and tertiary sources are best identified during the process of developing conceptual models (Chapter 17).

13.6   SCREENING SOURCES AND AGENTS

If a screening assessment has not preceded the assessment at hand (i.e., it is not a higher tier assessment), and the assessment is sufficiently complex, a screening assessment may be performed as part of the problem formulation to focus the assessment on sources or agents that present credible hazards (Section 3.3). The primary purpose of these assessments is to screen out inconsequential sources, agents, pathways, and receptors from further consideration. If all sources, agents, pathways, or receptors are screened out, the assessment may end without proceeding beyond the problem formulation stage. These screening assessments use existing data and simple conservative assumptions and may even be qualitative. For example, it may be determined that exposure to a source would require that contaminants flow uphill.

14 Environmental Description

Ecological risk assessments may address the environments of real places, generic places, or representative places. Real places include sites contaminated with wastes or spills, the sites of effluents or industrial facilities, forests or rangelands to be managed, or other actual locations with real physical and biological properties. The obvious advantage of real places is that their properties may be observed or measured. The disadvantage is that, in order to avoid obviously unrealistic results, one must expend one's time and resources to obtain observations and measurements from the site.

Generic places are abstractions of the sorts of places in which risks might occur. They are used for the assessment of new chemicals or technologies, which might be widely distributed or for which locations have not been selected. They are typically defined by choosing values for model parameters that are representative or conservatively representative of the array of real environments. For example, when modeling the dilution of a new detergent chemical in sewage effluents one may choose a reasonably conservative dilution factor like 10 (Beck et al. 1981), or might examine the distributions of receiving stream and effluent flows to define an average dilution factor or a dilution factor distribution. Hence, generic places can be defined by a set of default assumptions about the receiving environment for chemicals or other agents being assessed.

Representative places are real environments that are chosen to represent all places which will be exposed to an agent. For example, assessment of oil shale technologies and coal liquefaction technologies used representative environments in the Green River Formation of western Colorado and the Central Appalachian Coal Basin, respectively (Travis et al. 1983). The advantage of using representative places is that a degree of realism is assured. They also assure a greater degree of transparency than the use of generic environments. In the coal liquefaction example, the use of an eastern Kentucky environment makes it clear that the assessment does not address risks in the coalfields of Wyoming or Alaska. A representative place may be chosen as a realistic worst case. For example, an assessment of aquatic ecological risks from pyrethroid pesticides modeled exposures using the actual distribution of cotton fields and water bodies as well as other properties in Yazoo County, Mississippi (Hendley et al. 2001). That county was chosen through a systematic process based on the area of cotton fields, the area of water, and use of pesticides. Further, a representative environment may serve as a site for field tests or for collection of realistic media or organisms for laboratory studies. Finally, if a prototype of a proposed technology or instances of a similar technology are in operation, studies of that site can serve as the basis for defining a positive reference environment for the assessment of future facilities. For example, studies of effluents from an existing coke oven were used as the basis for assessing the risks of PAH emissions from a hypothetical future coal liquefaction plant (Herbes et al. 1978). Both technologies are likely to be close to coal mines and on rivers large enough for barges, so the physical and biological properties of the environment as well as the effluent chemistry were relevant.

The selection of one or more scales at which the assessment will be performed is a critical component of the environmental description (Chapter 6). When assessing a pesticide to use on corn, is the scale of assessment an individual corn field, agronomic regions such as the eastern corn belt, or the entire nation? For an aqueous effluent, is the mixing zone, the downstream reach, or the watershed the appropriate scale? For a timber sale, is it the area of the sale, the watershed within which it will occur, the National Forest, or the range of that forest type? Increasing the scale increases the likelihood of encompassing all significant direct and indirect effects. On the other hand, increasing the scale of the assessment beyond the area that is directly affected will tend to dilute effects. For example, the use of an organophosphate insecticide may result in the death of 50% of the birds in a field to which it is applied, but less than 1% of birds in the agronomic region, and a minuscule loss nationally.

The scale of an assessment may be matched to the scale of the agent or action being assessed, or to the scale of the ecological endpoints. For example, a treated field or an agronomic region would act as appropriate scales for assessing a pesticide based on its use. However, neither of them is likely to bear a natural relationship to the scales of biological processes associated with any potential ecological endpoints. Scales inevitably match when the action assessed is directed at the ecological endpoint, e.g., harvesting a fish stock. In some other cases, the scales match due to some coincidence of properties. For example, release of a chemical in the headwaters of a stream may contaminate the entire flood plain of that stream and, if the assessment endpoint is an attribute of the flood plain community, the scales of exposure and effect match. However, in most cases there is no apparent conjunction of the scale of an activity or agent and the endpoint entities. This situation is further complicated by the fact that the scale of ecological entities is difficult to define. Hence, the EPA (2003c) has developed the concept of assessment populations and assessment communities to provide a basis for pragmatically defining the scale of endpoints in a manner that is ecologically reasonable and consistent with the activity or agent being assessed (Box 16.6). This is equivalent to the concept of the reasonable maximally exposed individual in human health risk assessment; it is a way of defining the entity to be assessed that is practical and acceptably conservative without being unreasonable or unrealistic.

When a real place is described, it is necessary to actually define the bounds on the site. This involves applying the scale definition to specific geography. For example, if the reach scale is chosen for an assessment of an aqueous effluent, the reach may be defined as extending from the source to the next downstream tributary, a dam, or some other feature that significantly changes the ecology or the level of disturbance. This issue has been considered at length in risk assessment for contaminated sites, where the extent of the site must include not only the directly and indirectly contaminated area, but also the area that would be contaminated in the future if no action is taken (Suter et al. 2000). When assessing a large site or region, it is often necessary to divide it into units that are reasonably uniform with respect to ecology and the nature or levels of exposure to anthropogenic agents. Division of an area or water body into such units not only facilitates assessment by making it possible to uniformly apply a set of models and data, but also makes it possible to choose a management action that could be applied to the unit. For example, the Milltown Reservoir, Montana, Superfund site was divided into 12 units based on physiography and metal concentrations, and each was assessed as a potential remedial unit (Pascoe and DalSoglio 1994).

As soon as one has delimited the environment to be described one can consider what constitutes an appropriate description. It is important to avoid description for its own sake; the long species lists often found in environmental impact assessments are not needed, as they

BOX 14.1
Information Normally Included in the Site Description for Ecological Risk Assessments of Contaminated Sites (Suter et al. 2000)

Location and boundary: Latitude and longitude, political units, and boundary features (e.g., bounded on the south by the Clinch River) should be described.

Topography and drainage: Gradients of elevation and patterns of surface and subsurface drainage determine the hydrologic transport of the contaminants. Floodplains and other depositional areas are particularly important.

Important climatic and hydrological features: For example, if flows are highly variable due to infrequent storms or spring snow melt or if groundwater seasonally rises to contaminated strata, those features should be noted and characterized.

Current and past site land use: Land use suggests what sorts of contaminants may be present, what sorts of physical effects (e.g., soil compression) may have occurred, and what sorts of ecological receptors may be present.

Surrounding land use: Land use in the vicinity of the site determines to a large extent the potential ecological risks. A site in a city surrounded by industry will not have the range of ecological receptors of a site surrounded by forest.

Nearby areas of high environmental value: Parks, refuges, critical habitat for endangered species, and other areas with high natural value that may be affected by the site should be identified and described, and their physical relation to the site should be characterized.

Ecosystem (habitat) types: On terrestrial portions of the site, ecosystem types correspond to vegetation types. Aquatic ecosystems should be described in appropriate terms such as ephemeral stream, low-gradient stream with braided channel, or farm pond. In general, a map of ecosystem types should be presented along with a brief description of each type and the proportion of the site that it occupies. The map should include anthropogenic as well as natural ecosystems (e.g., waste lagoons).

Wetlands: Wetlands are given special attention because of the legal protections afforded wetlands in the United States. Wetlands on the site or receiving drainage from the site should be identified.

Species of special concern: These include threatened and endangered species, recreationally or commercially important species, and culturally important species.

Dominant species: Species of plants or animals that are abundant on the site and may be indicative of the site's condition or value should be noted.

Observed ecological effects: Areas with apparent loss of species or species assemblages (e.g., stream reaches with few or no fish) or apparent injuries (e.g., sparse and chlorotic plants or fish with deformities or lesions) should be identified.

Spatial distribution of features: A map should be created, indicating the spatial distribution of the features discussed above.

distract from the purpose of informing a decision. Rather, the description should serve two purposes. First, it should contain enough information for the risk manager and stakeholders to understand the situation. Second, it must be sufficient for the risk models to be developed and parameterized. Box 14.1 describes information that is typically included in descriptions of contaminated sites.

15 Exposure Scenarios

In addition to describing the place where exposure will occur, it is necessary to describe the circumstances under which it will occur (Oliver and Laskowski 1986). This is particularly important if an agent acts episodically, has significant temporal variability, or takes different forms in alternative futures. The scenarios define sets of hypothetical or actual conditions under which exposure may occur and for which risks will be characterized. Scenarios may be designed to represent typical cases, reasonable worst cases, the expected range of cases, or a set of plausible bad cases. If the life cycle of a chemical or product is assessed, scenarios must be developed for manufacturing, transportation, use, and disposal (Scheringer et al. 2001; Weidema et al. 2004). In a risk assessment of regional development, the development scenarios must be defined (Gentile et al. 2001). For contaminated sites, scenarios must address future movements of the contaminants, changes in land use, and changes in the biota. For example, ecological risk assessments of the Oak Ridge Reservation assumed that river otters and bald eagles would soon become reestablished on the site. The number and type of scenarios developed should be defined in consultation with the risk manager or should be based on established policy.

The need to define scenarios can be illustrated by considering proposed pesticide use. The applicant for a permit to market a pesticide must propose a crop that it would protect and the pest to be controlled, but other questions must be answered to define exposure scenarios. Examples include the following:

- What weather will be associated with the application? For example, will a rain occur shortly after application; if so, how much and how soon?
- What species of birds and how many birds use the field after application?
- Will the pesticide be applied more than once per year? At what interval?
- Will the pesticide be applied in sequential years?
- If the field is irrigated, how is it watered and how soon after pesticide application?

While it is tempting to devise a worst-case scenario, such cases may be so extreme that the assessment results are discounted by the risk manager. Even cases that have been observed to occur may be considered too extreme. For example, in the assessment of granular carbofuran, an observed bird kill involving hundreds of horned larks was mentioned but deleted from the quantitative analysis as an aberrant outlier (OPP 1989).

The alternative to defining scenarios would be to simulate all possible cases by assigning distributions to the model variables. That is, one could estimate the distributions of temperature, precipitation frequency, and intensity, soil texture, abundance of each species of bird that may occur in the fields, etc., for all areas and all times in which the pesticide may be used. Even without considering uncertainties in those variances, the results from Monte Carlo simulation of such a model would be distributions of exposure estimates that could span several orders of magnitude. Further, the correlation structure would be unmanageable, because the correlations among many parameters are not the same across the United States, Europe, or other continents or regions. The results of such an all-cases model would simply serve to suggest that the assessors could predict nothing. However, by appropriately defining scenarios, the assessors can focus on estimating relevant variance. For example, a manager might wish to know the year-to-year variance in aquatic exposures in a region given the variance in weather. Similarly, a manager might wish to know the variance in maximum exposure of stream communities in a region given uniform weather, but considering variance in distance from fields where the pesticide would be applied.

Various methods are used to derive future scenarios (Weidema et al. 2004):

- *Extrapolation methods* simply assume that conditions are constant (e.g., the current conditions and practices will continue) or current trends will continue. For example, one might temporally project trends in climate and in area devoted to a crop.
- *Dynamic simulation* uses mechanistic understanding of the forces changing the system and causal connections among components of the system that may be influenced by the proposed actions to estimate future conditions.
- *Cornerstone scenario methods* assume that the future is not predictable and attempt to define a set of scenarios that encompass the reasonably plausible conditions.
- *Participatory methods* use the judgments of experts and stakeholders to define scenarios.
- *Normative methods* define how the world should be and create scenarios that achieve those goals. For example, one might assume that a contaminated site is allowed to undergo natural succession, that endangered species which no longer occur in the region will be restored, or that all sources of pollution in a watershed will be eliminated.

Another use of scenarios is to hypothesize alternative futures that could result from different policies or different chaotic courses of history. Such scenarios are used as planning tools to help corporations, governments, or other entities predict the results of current actions in different alternative futures. An example is the four global development scenarios used by the Millennium Ecosystem Assessment (Millennium Ecosystem Assessment 2005; Ness 2005). An application to ecological risk assessment might be the development of endangered species recovery plans with scenarios that assume constant climate and climatic warming.

16 Assessment Endpoints

Assessment endpoints are an explicit expression of the environmental values to be protected, operationally defined as an ecological entity and its attributes (Suter 1989; EPA 1992a). The concept of assessment endpoints is similar to other concepts but the distinctions are important (Box 16.1). The process of selecting assessment endpoints can be thought of as a process for converting the risk manager's environmental goals into specific attributes of real entities, which can be estimated by measurement or modeling. This selection may be difficult or contentious because the number of ecological entities and attributes is effectively infinite. Endpoints may be defined for structural or functional attributes at the organism, population, community, or ecosystem levels of organization. Differences in the perceptions and values of the risk mangers, stakeholders, risk assessors, and public further complicate the process. For example, while the public is concerned about individual oiled birds following a tanker accident, assessors who are trained in ecology may insist that endpoints be defined in terms of avian population attributes. The criteria listed in Box 16.2 have been used to select ecological assessment endpoints. The first three are the EPA's criteria.

The selection of endpoints for an assessment requires balancing and reconciling of those criteria. In practice, policy goals and societal values are the dominant consideration. If an endpoint is not sufficiently important to influence the decision maker who, in a democracy, represents a society, it will be useless. Ecological relevance serves to augment and clarify the consideration of societal values. That is, if an endpoint attribute is not only societally important in itself but also influences other ecological attributes that are important, its significance is increased. However, an endpoint that has little ecological relevance must be included in the assessment if it has significant societal value. For example, neither the loss of peregrine falcons in the eastern United States nor their recent recovery has had any apparent influence on ecosystem attributes. Nevertheless, their societal value has justified the expenditure of millions of dollars for their restoration. The other criteria are practical screens. If a potential endpoint is not susceptible, it is irrelevant. If it has an inappropriate scale or is impractical, it must have a low priority. If it is valued but not operationally definable, it must be redefined.

Consistency with policy goals and societal values may be determined on the following basis:

- *Explicit goals*: As discussed in Chapter 11, the goals of an environmental management program should be identified at the beginning of the planning and problem formulation process. These goals are assumed to embody societal values. When clear goals are defined, the assessors must simply identify those entities and attributes for which risks must be defined to achieve the goals. However, clarification of goals is often necessary. For example, if the goal is to protect the production of Coho salmon, it is important to distinguish whether the goal is to protect the productivity of salmon for the fishery (which would include hatchery-produced fish) or of wild naturally spawning salmon populations.
- *Established policy and precedent*: If there is a published policy to protect certain environmental attributes or if an attribute has been the basis of regulatory or other management actions in the past, then it may be assumed to reflect societal values. Such precedents are the basis for the generic assessment endpoints discussed below and are

BOX 16.1
Related Concepts

While the concept of assessment endpoints is generally accepted in the practice of ecological risk assessment, other concepts serve an equivalent role in other contexts. It is often necessary for risk assessors to adapt these concepts. The following examples illustrate the necessary interpretation and adaptation of such concepts.

Uses: The term appears in the Great Lakes Water Quality Agreement, the US Clean Water Act, and elsewhere. While the term implies utilitarian values, it has been interpreted broadly. In particular, in implementation of the Clean Water Act, "use" has included use of the water by aquatic life as well as by humans.

Services (of Nature): Like "uses" this term is utilitarian. It commonly appears in the environmental literature to emphasize the many useful functions of ecosystems (Costanza et al. 1997; Daily et al. 1997). In Natural Resource Damage Assessments, injuries to natural resource services are identified and damages are paid to resource managers for loss of those services (Deis and French 1998; DOI 1986).

Indicators: Indicators are products of environmental monitoring programs that are intended to indicate the state of some property of the environment that is not measured but is of interest (National Research Council 1999). They may be indicative of an abstract property such as biotic integrity or ecosystem health or a real property that is difficult to measure such as mutation rate. Examples of indicators include the index of biotic integrity and the area of a nation that is forested. Indicators are used in local, regional, or national assessments of environmental status and trends (Office of Research and Development 1998; John Heinz III Center for Science Economics and the Environment 2002). If an indicator has value in itself, it may be used as an assessment endpoint.

BOX 16.2
Criteria for Selection of Assessment Endpoints for Ecological Risk Assessments (Suter 1989; EPA 1992a)

1. *Policy goals and societal values*: Because the risks to the assessment endpoint are the basis for decision making, the choice of endpoint should reflect the policy goals and societal values that the risk manager is expected to protect.
2. *Ecological relevance*: Entities and attributes that are significant determinants of the attributes of the system of which they are a part are more worthy of consideration than those that could be added or removed without significant system-level consequences. Examples include the abundance of a keystone predator species, which is relevant to community composition or the primary production of a plant assemblage, which is relevant to numerous ecosystem attributes.
3. *Susceptibility*: Entities that are potentially highly exposed and responsive to the exposure should be preferred, and those that are not exposed or do not respond to the contaminant should be avoided.
4. *Operationally definable*: An operational definition is one that clearly specifies what must be measured and modeled in the assessment. Without an unambiguous operational definition of the assessment endpoints, the results of the assessment would be too vague to be balanced against costs of regulatory action or against countervailing risks.
5. *Appropriate scale*: Ecological assessment endpoints should have a scale appropriate to the site or action being assessed. This criterion is related to susceptibility in that populations with large ranges relative to the site have low exposures. In addition, the contamination or responses of organisms that are wide-ranging relative to the scale of an assessment may be due to sources or other causes not relevant to the assessment.

Continued

BOX 16.2 (Continued)
Criteria for Selection of Assessment Endpoints for Ecological Risk Assessments (Suter 1989; EPA 1992a)

6. *Practicality*: Some potential assessment endpoints are impractical because good techniques are not available for use by the risk assessor. For example, there are few toxicity data available to assess effects of contaminants on lizards, no standard toxicity tests for any reptile are available, and lizards may be difficult to quantitatively survey. Therefore, lizards may have a lower priority than other better known taxa. Practicality should be considered only after evaluating other criteria. If, for example, lizards are included because of evidence of particular sensitivity or policy goals and societal values (e.g., presence of an endangered lizard species), then some means should be found to deal with the practical difficulties.

Source: From Suter, G.W. II, in *Ecological Assessment of Harardous Waste Sites: A Field and Laboratory Reference Document*, EPA 600/3–89/013, W. Warren-Hicks, B.R. Parkhurst, and Jr. S.S. Baker Jr., eds., Corvallis Environmental Research Laboratory, Corvallis, Oregon, 1989; US Environmental Protection Agency, Framework for ecological risk assessment, EPA/630/R-92/001, Washington, DC, Risk Assessment Forum, 1992.

---

often the basis for selecting assessment endpoints in practice. The primary problem with this approach is that it leads to continued neglect of potential endpoints, which have for various reasons been neglected in the past. For example, amphibians have received little attention even though many amphibians, particularly frogs, are attractive to the public and the recent decline in frogs has caused public concern.

- *Manifest societal values*: Some societal values are apparent to all. For example, society values the production of fish for food, trees for timber, and other marketable resources. Similarly, some populations and ecosystems are valued for their recreational uses. These utilitarian values are reflected in the use of phrases like services of nature (Daily et al. 1997) and use impairments (International Joint Commission 1989), to describe environmental goals. While market values fall in this category, other judgments about societal values may be controversial. Environmental assessors and managers often make ad hoc judgments about what the public values without evidence beyond their personal judgment. For example, the US Army BTAG (2002) wrote that "the public generally does not currently accept fungi, bacteria and species of invertebrates as appropriate assessment endpoints." They present no evidence to support this judgment or to refute the judgment of those assessors who have used earthworms, butterflies, mussels, crabs, and other invertebrates as endpoint entities. Ecological risk assessors should perform the research necessary to generate better evidence concerning public environmental values.

- *Departure from natural conditions*: A common assumption among environmental scientists is that mandates to protect the environment amount to a goal of preventing or minimizing any departure from a natural state. This implies that any attribute that may differ from the natural state is an appropriate endpoint. In some cases, the language of laws or regulations suggest that a natural state is desired and maintaining a natural state is clearly an appropriate goal for national parks, wilderness, or equivalent areas. However, most of the world is human modified. It may not be appropriate to determine what attributes best distinguish agricultural streams or suburbans from forest streams and make those our endpoints (Box 11.1). Further, natural systems are variable targets. If we deal with this by saying that anthropogenic effects can induce variance in ecological conditions as great as natural variability, then we may allow anthropogenic effects as great as those caused by a hurricane, drought, or glaciation (Hilborn 1996). Hence, departure from natural conditions is not an appropriate expression of societal values in situations where the society has chosen to convert an ecosystem to a modified condition or when a long temporal perspective is required. However, it is appropriate when, as in the US Clean Water Act, goals are defined in terms such as "biological integrity of the Nation's waters." Under that mandate, biocriteria are typically defined in terms of detectable changes from the least disturbed ecosystems of the same type.

16.1   ASSESSMENT ENDPOINTS AND LEVELS OF ORGANIZATION

The appropriate level of organization of ecological assessment endpoints has been an ongoing source of controversy. Much of that controversy occurs because people do not recognize that assessment entity and attribute may be defined at different levels of biological organization (Section 6.1). It is important to remember the conceptual formula:

Assessment endpoint = attribute + entity.

Examples include:

- Survival of a kit fox in a territory containing a former sump for arsenic-treated water— an organism attribute associated with a hypothetical individual organism in a specially protected population.
- Survival of juvenile kit foxes on Elk Hills—an organism attribute associated with the organisms in an assessment population.
- Population growth rate on Elk Hills—a population attribute associated with an individual assessment population.
- Mean population growth rate for all kit foxes—a population attribute associated with a set of populations.
- Number of indigenous mammal species on Elk Hills—a community attribute associated with an individual community.

These examples show that attributes at each level of biological organization can occur in an individual entity (an individual organism or an individual population) or in multiple entities (the organisms in a population or multiple populations within a region).

An attribute at one level may be applied to an entity at a different level of organizations. For example, death may be an attribute of an organism (e.g., death of an individual fish), to the set of organisms in a population (e.g., 50% mortality of rainbow darters in a stream), or a community (e.g., 50% mortality of fishes in a stream). However, the application of an organism-level attribute to a population or community of organisms does not make it a population-level or community-level attribute. Population and community responses are not simply sums of organismal responses. For example, the decline in population abundance is not simply the proportional mortality, because of compensatory and depensatory effects of density on survival, fecundity, or susceptibility to disease or predation (Chapter 27).

The relationship of entities and attributes can be clarified by comparing risk assessment endpoints for humans and ecological entities (Table 16.1). Human health risk assessments are intended to protect organism-level attributes of individual humans (e.g., a 5×10^{-4} cancer risk to the reasonable maximally exposed individual), but health risk assessments often also consider risks summed across the members of an exposed population so as to elucidate the magnitude of potential effects (e.g., an incremental risk of five cancers in an exposed population of 10,000). In contrast, ecological risk assessments seldom use entities at the organism level. Rather, organism-level attributes typically are associated with an assessment population or community (EPA 2003c). In the United States, true population-level attributes are not

TABLE 16.1
Examples of Assessment Endpoints for Human and Ecological Risk Assessments

Entities	Human Health Risk Assessment	Ecological Risk Assessment
Organism-level attributes		
Individual organism	Probability of death or injury (e.g., risk to the maximally exposed individual)	Probability of death or injury (e.g., risk to an individual of an endangered species). Seldom used.
Population of organisms	Frequency of death or injury, Numbers dying or injured	Frequency of mortality or gross anomalies, Average reduction in growth or fecundity
Population-level attributes		
Individual population	Not used	Extirpation, production, or abundance
Set of populations	Not used	Seldom used (e.g., extinction rate or regional loss of production)

Source: From Suter, G.W., Norton, S.B., and Fairbrother, A., *Integr. Environ. Assess. Manag.*, 1, 397–400, 2005. (With permission.)

considered in human health risk assessments, because individuals are to be protected, and an effect on a human population is sufficient to lower its abundance or production would not be countenanced. However, in ecological risk assessments, risks to abundance, production, extirpation and other attributes of nonhuman populations, or sets of populations are assessed.

16.2   GENERIC ASSESSMENT ENDPOINTS

The plethora of potential assessment endpoints calls for some method to define and narrow the field of candidate endpoints while ensuring that important candidates are not overlooked. One way of addressing this problem is to define generic endpoints that may be reviewed for their relevance to the assessment at hand and, if appropriate, adapted to the case at hand. Two approaches are described in this section: one is based on policy judgments and the other on ecological functional classification of organisms. While the approaches are conceptually distinct, they may be reconciled when defining assessment-specific endpoints.

16.2.1   GENERIC ENDPOINTS BASED ON POLICY JUDGMENTS

The most straightforward approach to developing generic endpoints for managers in an agency or organization is to develop a list that is judged to be suitable for their context and purposes. Examples include sets of generic endpoints for contaminated sites in Alaska (ADEC 2000) and on the Oak Ridge Reservation in Tennessee (Suter et al. 1994). Policy-based generic endpoints may even be developed by ecologists based on their interpretation of entities and attributes that must be protected to meet the goals of environmental laws or regulations. An example of generic endpoints for a specific policy goal, sustainable freshwater ecosystems, is presented in Box 16.3.

The US EPA reviewed its policies and precedents in using ecological endpoints to define a set of generic ecological assessment endpoints (GEAEs) (EPA 2003c; Suter et al. 2004). The GEAEs are not prescriptive, but serve to indicate what sorts of ecological assessment endpoints have been mandated or used successfully in decision making. The entities are broadly defined in terms of conventional levels of biological organization, so as to be broadly useful, but the attributes are more specifically defined (Table 16.2). They are meant to help the process of endpoint selection by indicating endpoints that may be presumed to have policy

BOX 16.3
Generic Endpoints for Sustainable Freshwater Ecosystems

Decrease in Biodiversity
This concerns negative effects on

1. Overall species richness and densities
2. Population densities of ecological key species that are

   • critical determinants in trophic cascades (e.g., piscivorous fish, large cladocerans)
   • "ecological engineers," that have a large influence on the physical properties of habitats (e.g., marcrophytes)

3. Population densities of indicator species

   • with a high "information" level for monitoring purposes
   • protected by law and regionally rare or endangered species

Impact on Ecosystem Functioning and Functionality
This concerns negative effects on

1. Biogeochemical cycles and energy flow
2. Water quality parameters (e.g., persistence of contaminants, increased toxic algae, oxygen depletion)
3. Harvestable resources (e.g., drinking water, fish)

Decrease in Aesthetic Value or Appearance of the Water Body
This can be caused by

1. Disappearance of species with a popular appeal
2. Visible mortality of individuals or fishes, frogs, water fowl, and other vertebrates
3. Taste and odor problems
4. Decrease in water transparency and symptoms of eutrophication (e.g., algal blooms)

Source: From Brock, T.C.M. and Ratte, H.T., in *Community-Level Aquatic System Studies: Interpretation Criteria*, J.M. Giddings, T.C.M. Brock, W. Heger, F. Heimbach, S.J. Maund, S.M. Norman, H.T. Ratte, C. Schafers, and M. Streloke, eds., SETAC Press, Pensacola, FL, 2002, pp. 33–41.

support in the US EPA. A similar set of generic endpoints is developed for the ecological assessment of pesticides in Canada (Delorme et al. 2005).

   Generic endpoints based on policy have the disadvantage of being based on the haphazard processes that generate public policy. While they tend to be incomplete and inconsistent, they have the advantage of being relevant to the decision processes from which they are derived.

16.2.2 FUNCTIONALLY DEFINED GENERIC ENDPOINTS

Ecologists often approach the problem of choosing generic endpoints in terms of the familiar organization of ecosystems in terms of trophodynamics, i.e., they explicitly or implicitly create a food web and define each node as an endpoint entity. Reagan (2002) has made this

TABLE 16.2
The US EPA's (2003c) Generic Ecological Assessment Endpoints[a]

Entity	Attribute	Identified EPA Precedents
Organism-level endpoints		
Organisms (in an assessment population or community)	Kills (mass mortality, conspicuous mortality)	Vertebrates
	Gross anomalies	Vertebrates
		Shellfish
		Plants
	Survival, fecundity, growth	Endangered species
		Migratory birds
		Marine mammals
		Bald and golden eagles
		Vertebrates
		Invertebrates
		Plants
Population-level endpoints		
Assessment population	Extirpation	Vertebrates
	Abundance	Vertebrates
		Shellfish
	Production	Vertebrates (game/resource species)
		Plants (harvested species)
Community and ecosystem-level endpoints		
Assessment communities, assemblages, and ecosystems	Taxa richness	Aquatic communities
		Coral reefs
	Abundance	Aquatic communities
	Production	Plant assemblages
	Area	Wetlands
		Coral reefs
		Endangered/rare ecosystems
	Function	Wetlands
	Physical structure	Aquatic ecosystems
Officially designated endpoints		
Critical habitat for threatened or endangered species	Area Quality	
Special places	Ecological properties that relate to the special or legally protected status	e.g., National parks, national wildlife refuges, Great Lakes

[a]Generic ecological assessment endpoints for which US EPA has identified existing policies and precedents, in particular the specific entities listed in the third column. Bold indicates protection by federal statute.

approach the basis for defining general assessment endpoints (GAEs). He specifies three trophic categories (producers, consumers, and decomposers), which are further divided into functional components that serve to define generic endpoint entities. He argues that even complex ecosystems can be reduced to 20 such components. While ecologists disagree on how to define functional components (e.g., Reagan aggregates detritivores and scavengers; I would not) and what to include (e.g., he leaves out parasites and parasitoids), the process of defining the components is useful for developing the conceptual model as well as developing assessment endpoints. The attributes for these generic endpoint entities are limited to a few functional properties such as food and habitat. As the functions are assumed to be important, the justifications of the value of functionally defined generic endpoints tend to become circular. For example, carnivores are said to be important because of their role in predation (Reagan 2002).

A related approach was used to develop exposure guilds for ecological risk assessment at Los Alamos National Laboratory (Myers 1999). Rather than defining general functional components, this approach used functional characteristics of species such as diet and foraging strategy that are believed to control exposure to contaminants. A statistical cluster analysis was then applied to the functional characteristics of species to define the exposure guilds.

16.2.3 APPLYING GENERIC ENDPOINTS

The process of developing assessment endpoints for an ecological risk assessment may be thought of as bringing together five types of information and answering questions related to each (EPA 2003c). Together, the questions address the criteria for ecological assessment endpoints. Generic endpoints constitute one type of information that answers one question:

- *Agent (stressor) characteristics*: What is susceptible to the stressor? For some agents, this question is straightforward. Benthic invertebrates are susceptible to dredging, birds are susceptible to granular pesticides, wetlands are susceptible to filling, and so on.
- *Ecosystem and receptor characteristics*: What is present and ecologically relevant? For site-specific assessments, this is the species, communities, or ecosystems at the site. For other assessments, the scenario should define the types of species, communities, and ecosystems that are likely to be exposed. For example, assessment of a new pesticide for corn would consider the species likely to be found in or adjacent to cornfields in the Midwest. In the absence of specific information about the particular importance of an entity, those that are present may be assumed to be ecologically relevant. The generic endpoints based on food webs or other functional relationships may be used to identify and organize these entities for endpoint selection.
- *Management goals*: What is relevant to the management goals? Statements of management goals should suggest the changes in attributes of ecological entities that would preclude achieving the goal.
- *Input by stakeholders*: What is of concern? If stakeholders are consulted or make their preferences known, their concerns about particular ecological effects should be considered. Although societal values at a national scale are reflected in government policies, values that are specific to a locale or resource are expressed by stakeholders.
- *Policies or precedents*: What is supported by policy or precedent? The GEAEs in Table 16.2 provide a set of entities and attributes that meet this criterion for the US EPA, which express national goals and values with respect to regulation of pollutants.

The answers to each of these questions would be a list of potential assessment-specific endpoints. None of the questions imply absolute requirements. For example, susceptibility to a novel agent may be unknown, and the concerns of stakeholders are often unknown and often do not include important potential endpoints. No generally applicable procedure is available for answering the questions. If consistency with US EPA policy and precedent is particularly important, one might go through the GEAE set and ask the other four questions with respect to each generic endpoint. Alternatively, all of the questions might be answered and the lists are integrated. In that case, the endpoints for a specific assessment may simply be those that are represented on most of the lists.

16.3 MAKING GENERIC ASSESSMENT ENDPOINTS SPECIFIC

In general, it is desirable to define the endpoints as completely as possible during the problem formulation so as to avoid ad hoc decisions during the analysis and the characterization phases. Some of the problems to be avoided in defining the endpoints are listed in Box 16.4. Also, if one of the generic assessment endpoints is chosen, or if the endpoint is broadly defined by the risk manager (e.g., community structure), it is necessary to define the entity and attribute more specifically.

Specific definition of the species and population: When, as is often the case, the assessment endpoint is defined in terms of attributes of organisms or populations of a species, it is necessary to select that species. In some cases, the value that defines the endpoint is associated with a particular species. Examples include endangered species, commercial, or recreational fisheries, and culturally significant species such as the bald eagle in the United States. In most cases, however, the endpoint values are not associated with a particular species. In such cases, it is necessary to choose one or more species that are representative of the endpoint (Box 16.5). An alternative, which is sometimes employed in land management, is to choose an umbrella species. They are species that have a high demand for large areas of relatively undisturbed habitat. It is assumed that if the umbrella species are protected, all other species will be protected as well. When endpoints are defined in terms of organisms or a population of a species it is necessary to define the spatial limits of the population (Box 16.6).

BOX 16.4
Common Problems with Assessment Endpoints

1. Endpoint entity is defined but not the attributes (e.g., bluebirds, rather than bluebird abundance).
2. Endpoint entity and attribute are mismatched (e.g., fish in Sierra Nevada streams, which naturally have few species, and species diversity).
3. Endpoint is a goal rather than an attribute (e.g., maintain and restore endemic populations).
4. Endpoint is vague (e.g., estuarine integrity rather than eelgrass abundance and distribution).
5. Endpoint is a measure of an effect that is not a valued attribute (e.g., midge emergence when the concern is production of fish which depends in part on midge production).
6. Endpoint is not directly or indirectly exposed to the contaminant (e.g., fish community when there is no surface water contamination).
7. Endpoint is irrelevant to the site (e.g., a species for which the site does not offer habitat).
8. Endpoint does not have an appropriate scale for the site (e.g., golden eagles on a 1000 m^2 site).
9. Value of an entity is not sufficiently considered (e.g., rejection of all benthic invertebrates at a site where crayfish are harvested).
10. Attribute does not include the value of the endpoint entity (e.g., number of species when the community is valued for game fish production).
11. Attribute is not sufficiently sensitive to protect the value of the endpoint entity (e.g., survival when the entity is valued for its production).

Source: Modified from US Environmental Protection Agency, Guidelines for ecological risk assessment, EPA/630/R-95/002F, Washington, DC, Risk Assessment Forum, 1998.

BOX 16.5
Representative Species

It is a common practice when selecting endpoints, particularly for wildlife, to designate a representative species (Hampton et al. 1998). That is, one may choose the meadow vole as a representative herbivore or red fox as a representative carnivore. This practice can lead to confusion unless it is clear what category of organisms is represented and in what sense the species are represented. For example, the meadow vole may represent all herbivores, all small mammals, all herbivorous small mammals, or all microtine rodents. A representative species may be representative in the sense that it is judged likely to be sensitive, because its activities are confined to the site (e.g., the vole rather than deer as representative herbivore), its behavior is likely to result in high levels of exposure (e.g., birds feeding on soil invertebrates rather than on herbivorous invertebrates), it is known to be inherently sensitive to the contaminant of concern (e.g., mink and PCBs), or it is predicted to be sensitive by application of extrapolation models (e.g., larger mammals based on allometric models). A species may also be representative in an ecological sense if it is the most abundant representative of the category of organisms on the site. Finally, a representative species may be chosen because it is particularly amenable to sampling and analysis or to demographic surveys.

The groups that the representative species represent are commonly defined in terms of higher taxa or broad trophic groups. However, if the characteristics that control exposure and toxicological or ecological sensitivity can be defined, endpoint groups may be defined by cluster analysis of those traits. This approach was applied to birds at Los Alamos, New Mexico, using only diet and foraging strategy to generate "exposure guilds" (Myers 1999). This approach is more objective than the typical subjective grouping of species, and the hierarchy of clusters provides a basis for increasing the level of detail in the analysis as the assessment progresses.

In general, it is not a good idea to select high-valued species as representative species, because it tends to confuse the roles of endpoint species and representative of a community or taxon. For example, if bald eagles occur on a site, they are likely to be an endpoint species protected at the organism level. If piscivorous wildlife as a trophic group is also an endpoint, then bald eagles might also be thought to serve to represent that group. However, because bald eagles cannot be sampled except under exceptional circumstances and they are not likely to be highly exposed due to their wide foraging area, it would be advisable to choose a species which is more exposed, more abundant on the site, or less protected as a representative (e.g., kingfishers or night herons). By using a different species to represent the trophic group, one could perform a better assessment of the trophic group and could clearly distinguish the two endpoints in the risk communication process.

When using a representative species, it is essential to determine how the risks to the represented category of organisms will be estimated. The method may range from assuming that all members of the category are equivalent to using mechanistic extrapolation models to estimate risks to all members of the category once risk to the representative species is demonstrated to be significant.

Source: From Suter, G.W. II, Efroymson, R.A., Sample, B.E., and Jones, D.S., *Ecological Risk Assessment for Contaminated Sites*, Lewis Publishers, Boca Raton, FL, 2000.

Specific definition of the community: When community endpoints are chosen, the community must be defined. For example, do we define the biota of a stream as a community or treat the fish, benthic macroinvertebrate, periphyton, zooplankton, and phytoplankton assemblages as communities? If we define the community as a food web, riparian species may be included. Communities are often defined in terms of the constituent assemblages, because the differences in sampling techniques inhibit quantification of the structure of entire communities. The spatial dimensions of a community are discussed in Box 16.6.

BOX 16.6
Defining Assessment Populations and Communities

As the conventional ecological meaning of "populations" and "communities" presents problems in practice, the US EPA (2003c) guidance on generic assessment endpoints introduces the terms "assessment population" and "assessment community." An assessment population is a group of conspecific organisms occupying an area that has been defined as relevant to an ecological risk assessment. An assessment community is a multispecies group of organisms occupying an area that has been defined as relevant to an ecological risk assessment.

Although ecological assessment endpoints often include population properties, such as abundance and production, and community properties, such as species richness, it is difficult to delineate populations and communities in the field. Recently, ecology has focused on temporal dynamics, spatial patterns and processes, and stochasticity that belie the notion of static, independent populations. An example of this is metapopulation analysis, which reveals that population dynamics are significantly determined by the exchange of individuals among habitat patches or differential movement across a landscape that continuously varies in suitability (Chapter 27) (Hanski 1999). Communities are subject to the same dynamics. For example, the species diversity of Pacific coral reefs is apparently determined by the availability of recruits from other reefs within 600 km (Bellwood and Hughes 2001). If the composition of coral reefs on islands, which would appear to be classic discrete communities, is in fact determined by regional dynamics, there is little chance of delimiting discrete communities in general.

Populations may be readily delimited if they are physically isolated within a broader species range (e.g., a sunfish population in a farm pond) or if the species consists of only one spatially discrete population (e.g., the endangered Florida panther, whose current range is restricted to southwest Florida). Otherwise, population boundaries are difficult to define because they are typically structured on multiple scales. Genetic analyses, which are needed to define discontinuities in interbreeding frequencies and thus to delimit populations, are not a practical option for most ecological risk assessments.

The practical problems are even greater for communities. Although the members of a population consist of a single species, it is not always clear whether a particular group of organisms constitutes an instance of a particular community type. This is because the species composition of communities varies over space and time.

To protect properties such as population production or community species richness, it is necessary to develop a pragmatic solution to these problems. An example of such a solution is the approach taken by the Nature Conservancy and NatureServe (formerly the Association for Biodiversity Information) to inventory and map biodiversity (Stein et al. 2000). Because it is not feasible to define discrete populations or communities, these organizations inventory and map occurrences of conservation elements, which may be defined at various scales, depending on the elements and circumstances. For example, a plant community occurrence may be "a stand or patch, or a cluster of stands or patches." An occurrence of a bird species would be defined differently, but analogously.

For individual assessments, the population or community entities to be protected must be defined during the problem formulation stage of risk assessment. These assessment populations and assessment communities should be defined in a way that is biologically reasonable, supportive of the decision, and pragmatic with respect to policy and legal considerations. For example, it would not be reasonable to define the belted kingfishers in a 20 m stream reach as an assessment population if that reach cannot fully support one belted kingfisher pair. On the other hand, even though the kingfisher's range is effectively continuous, it would not be reasonable to define the entire species as the assessment population, given that it ranges across nearly all of North America. Rather, it may be reasonable to define the kingfishers nesting on a watershed or a lake as an assessment population.

Continued

BOX 16.6 (Continued)
Defining Assessment Populations and Communities

Definitions of assessment populations may include nonbiological considerations as well. For example, for Superfund ecological risk assessments on the Oak Ridge Reservation of the US Department of Energy, populations of large terrestrial vertebrates were delimited by the borders of the reservation (Suter et al. 1994). This definition was reasonable not only because the Superfund site was defined as the entire reservation, but also because the reservation was large enough to sustain viable populations of deer, wild turkey, bobcat, and other endpoint species. Although the reservation is more forested than are the surrounding agricultural and residential lands, its borders are not impenetrable and are not ecologically distinct at all points. However, the pragmatic definition proved useful and acceptable to the parties. A similarly pragmatic approach would define an assessment community of benthic invertebrates as occupying the first fully mixed reach of a stream receiving an effluent.

The selection of a scale to define an assessment population or community involves a trade-off. If the area is large relative to the extent of the stressor, the effects of that stressor will be diluted. However, if the area is small, the assessment population or the community may be significantly affected but may seem too insignificant to prompt stakeholder concern or action by the decision maker. Hence, appropriate spatial scales must be determined during the problem formulation stage of individual risk assessments, taking into consideration both the ecological and the policy aspects of the problem; it must not be manipulated during the analysis to achieve a desired result.

Source: From US Environmental Protection Agency, Generic Ecological Assessment Endpoints (GEAEs) for Ecological Risk Assessment, EPA/630/P-02/004B, Washington, DC, Risk Assessment Forum, 2003.

Specific definition of structural attributes: Structural attributes such as abundance, deformity or number of species tend to be relatively clear. However, the definition often depends on the specific methods used to estimate the attribute.

Specific definition of functional attributes: Functional attributes are relatively seldom used, and their definitions tend to be unclear. Protection of wetland functions is the one functionally defined goal of the US EPA, but it is necessary to define which functions are appropriate in a particular case. They include nutrient retention, nutrient cycling, pollutant retention, water retention, groundwater recharge, provision of waterfowl habitat, etc. The choice may be made based on the circumstances. For example, water retention would be an important endpoint attribute for wetlands in watersheds that are subject to destructive floods. Similarly, nutrient retention would be important in watersheds such as those in the Mississippi and Neuse Rivers, which have contributed excess nutrients to coastal areas, causing anoxia.

Even a clear definition of an assessment endpoint as an entity and one or more attributes is often insufficient to support the assessment process. Particularly if the assessment will involve the measurements and observations, it is necessary to define the magnitude or frequency of effects that is considered potentially significant and therefore must be detectable or distinguishable from reference conditions with reasonable confidence.

16.4 ENDPOINTS BASED ON OBJECTIVES HIERARCHIES

When assessments are based on stakeholder processes, ecological risk assessments must address endpoints that are meaningful to those stakeholders. One means to develop such endpoints is to define and elaborate their goals through an objectives process (Reckhow 1999).

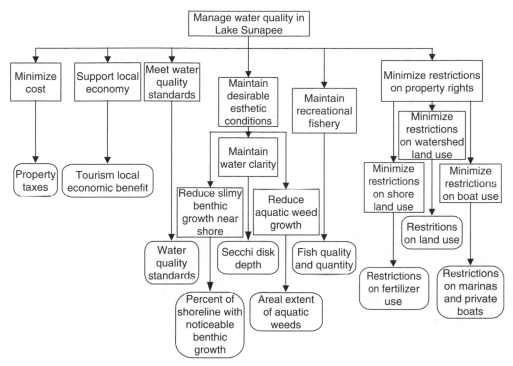

FIGURE 16.1 An example of an objectives hierarchy for an environmental decision. The overall objective appears at the *top*, the other *rectangles* are issue-specific objectives, and the *rounded rectangles* are specific operational objectives. (Redrawn and modified from Reckhow, K.H., *Hum. Ecol. Risk Assess.*, 5, 245, 1999.)

This technique, derived from decision analysis, begins with an overall objective and breaks down into constituent objectives until something is identified that can be measured or estimated (i.e., an endpoint). Figure 16.1 shows an objective hierarchy for water quality management in a lake that was experiencing eutrophication. Note that some are potentially appropriate endpoints for ecological risk assessment, and others relate to economic assessments, regulatory actions, and land use planning. This diversity of objectives is typical of multiattribute decision analysis (Chapter 36).

17 Conceptual Models

Effective use of information requires congruence between the scientific models presented and the decision maker's conception of the problem.

Anderson (1998)

Conceptual models are working hypotheses about how the hazardous agent or action may affect the endpoint entities (Barnthouse and Brown 1994; EPA 1998a). They include descriptions of the source, receiving environment, and processes by which the receptors come to be exposed directly to the contaminants and indirectly to the effects of the contaminants on other environmental components. They are conceptual in that they are purely descriptive, as opposed to implemented models that actually quantify the entities and relationships. They are hypothetical in that they represent an agreed description of the system to be assessed. Other conceptualizations of the system might differ in various ways such as having wider bounds (e.g., include the watershed and not just the lake) or more detail (e.g., include individual fish species rather than fish as a taxon), or even different concepts of exposure and response (Box 17.1).

17.1 USES OF CONCEPTUAL MODELS

Conceptual models have various uses in risk assessment:

Defining the risk hypothesis: The fundamental purpose for conceptual models in risk assessment is to define the conceptualization of how effects may be caused by the agent or action being assessed. This function is equivalent to the hazard identification phase of human health risk assessments.

Screening: The development of a conceptual model can serve to screen out impossible or implausible hazards. In some cases, a proposal that an agent poses a hazard to a receptor will seem plausible until it is conceptually modeled. That is, if there is no source, pathway, receptor, or mechanism for inducing harm, then that proposed hazard may be left out of the assessment.

Defining a mathematical model: The use of conceptual modeling in risk assessment is derived from the conceptual modeling of systems, as a first step in mathematically simulating the system (Jorgensen 1994). That is, the boxes represent both entities and state variables that describe them and the arrows represent both processes and functions that describe them (Figure 21.2 and Figure 27.1). Developing a simulation model then requires defining the variables and deriving the functions implied by the diagram. This use is relevant when an assessment is based, at least in part, on simulation modeling. It generally serves to indicate what must be modeled if it cannot be directly measured.

Causality: Causal analyses require a conceptual model to define the mechanisms underlying relationships between the observed effects and the hypothesized causes (Chapter 4). Unlike conventional risk assessments, these conceptual models may be developed by beginning with

BOX 17.1
Conceptual Models and the Assessment Framework

Two diagrams are critical to a risk assessment but are conceptually disjunct: the assessment framework (Figure 3.1) and the conceptual model. The framework portrays the process of performing a risk assessment, while the conceptual model portrays the system being assessed. The conceptual disjunction is best illustrated by the fact that the central concept of the conventional framework (Figure 3.1) is the distinction between exposure and effects, but there is no such dichotomy in a conceptual model of ecological risks. Instead, a conceptual model represents a chain of processes that cause changes in states, which in turn cause changes in processes. Every state is an effect of exposure to some causal process. When complex causal chains or networks are important to the risk assessment, it is desirable to use a framework that represents those processes (Section 3.2.4), so that the complexity of the system is not lost in the attempt to reduce it to a simple case of direct exposure and direct response. However, it is important to consider the nature of the cause–effect chain, even in simple toxicological risk assessments.

Consider Figure 17.1 as a simple generic conceptual model of direct toxicological risks to organisms. A chemical is released by a source, is transported and transformed in the environment resulting in a concentration in an ambient medium, uptake by organisms results in a dose, which is distributed, partitioned, and metabolized, resulting in an internal concentration, which results in binding to a membrane, receptor protein, or other biological structure (ligand), inducing proximate biological effects. Finally, the proximate effects of the bound chemical result in injury and healing processes which cause death, reduced fecundity, or other endpoint effect. What part of this diagram is exposed and what are its effects? One may cut the diagram at any point. For example, in a field test of a pesticide, the exposure is the source (e.g., an application of x g/m^2), and everything else in the diagram is aggregated into induction of the effects observed in the field test (e.g., y dead birds/m^2), so the cut is performed at a. Most commonly in ecological risk assessment, exposure is estimated by modeling or measuring concentrations in media (e.g., x mg/L), and everything downstream of cut b is aggregated into effects, as observed in a laboratory toxicity test. Similarly, if exposure is expressed as dose, the cut is made at c, and, if it is expressed as an internal concentration, it is made at d.

Where to make the cut between exposure and effects depends on practical and ideal considerations. The practical considerations are the availability of data and models. For example, we have good models of chemical transport and fate in water, good measurements of chemicals in water, good aquatic toxicity test methods, and abundant aquatic toxicity data, all expressed an aqueous concentration, so it is practical to make the cut at b. On the other hand, models to estimate concentrations in internal compartments of aquatic organisms and toxicity data for internal concentrations are uncommon and unfamiliar. The ideal consideration is that cutting exposure from effects at other points might reduce uncertainty and the costs of testing and assessment. For example, effective concentrations at the site of action are likely to be nearly constant for a particular mechanism of action and taxon (Section 24.2.5). Hence, if we could develop good models for the toxicokinetics of chemicals in a range of endpoint species and life stages, we might reduce the need for toxicity testing and extrapolation models and the conventional division between exposure and effects would be at cut d.

the observed effects and working backward. For example, if the effect to be explained is loss of brook trout from a stream, one may list the habitat requirements of the trout—such as spawning substrate, dissolved oxygen, and temperature; the factors that may negatively affect it—such as disease, harvesting, siltation, and toxicants; and create a web of potentially causal relationships. Those relationships may, in turn, be related to specific causal agents and their sources based on the knowledge of the stream and its watershed.

Heuristics: The creation of a conceptual model can act as a means for the assessment team to share knowledge and to come to a common understanding of the system to be assessed.

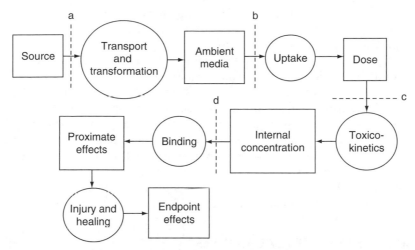

FIGURE 17.1 Generic conceptual model of toxicological risks to organisms, showing four points (*dashed lines* [a–d]) at which the causal chain may be divided into exposure and effects. The model is presented as alternating states (*rectangles*) and processes (*ovals*).

In particular, it can lead to the recognition of indirect effects, routes of transport, or other aspects of the problem that would not have come to mind without the discipline of diagraming the relationships.

Communication: The conceptual model can be a valuable tool for sharing the understanding of the system with risk managers and stakeholders. It cannot be expected that most people will be able to read and understand the mathematical formulation of a risk model or understand how the various models go together to generate the risk estimate. However, it should be possible to clearly convey the conceptual model that underlies the implemented models. Consideration of conceptual models allows the assessors, managers, and stakeholders to reach a common understanding of the structure of the system or, at least, to clarify the differences in their understanding. Disagreements about risk are often due to differences in conceptual models (Hatfield and Hipel 2002).

Life cycle definition: In product life cycle assessment and risk assessment of new chemicals or materials, it is important to develop a conceptual model of the origin, distribution, and fate so that routes of entry into the environment (i.e., sources) are defined (Weidema et al. 2004). For example, a chemical for fabric treatment may be released in effluent where it is manufactured, in laundry water, in sewage sludge, in dust, in landfills, and incinerator effluents.

All of these uses require that some care be devoted to developing the conceptual model. Too often conceptual models are vague or do not correspond to the assessment models. Conceptual models should contain all of the sources, endpoints, and pathways that are included in the assessment but no more. In some cases, it is necessary to simplify the system because of lack of data, models, or resources. In such cases, it may be desirable to present a conceptual model of the important features of the system and then a simplified model as implemented in the assessment. Vagueness can be avoided by considering whether the conceptual model could be handed to an assessment team for implementation with the assurance that everything important would be included in the assessment.

Conceptual models may be developed and used iteratively in the risk assessment process. First, following an initial site survey or other preliminary description of the problem, draft conceptual models can be developed as input to the problem formulation process. These models should be inclusive in that they should include all sources, receptor classes, and routes of exposure that are plausibly of concern. This preliminary conceptual model also serves as a conceptual model for an initial screening assessment performed to support the problem formulation process. During the problem formulation process, that preliminary conceptual model is modified to be more relevant to the decision. The model is simplified by eliminating (a) receptors that are not deemed to be suitable assessment endpoints, (b) routes of exposure that are not credible or important, (c) routes of exposure that do not lead to endpoint receptors, and (d) potential sources that are not deemed credible or important. In addition, the problem formulation process makes the conceptual model more specific by identifying particular endpoint species, by defining the spatial and temporal scale of the assessment, and by making other judgments. The results of the problem formulation process are presented in the conceptual models published in the analysis plan. Conceptual models for definitive assessments may be further modified based on the results of ongoing communications among the parties and on the results of the field and laboratory investigations or exposure modeling.

17.2   FORMS OF CONCEPTUAL MODELS

Conceptual models consist of a diagram and an accompanying explanatory text. Most conceptual model diagrams are presented as flow charts, but other forms are used as well. Pictorial models represent the components of the system as drawings of the entities involved (sources, media, and organisms), with arrows between them showing their relationships. They may be vague because it is not clear, e.g., whether a drawing of a mallard duck indicates that species, all ducks, all birds, all omnivores, or some other entity. However, pictorial models may be quickly and easily understood because images are readily interpreted than words. They can be made more useful by including appropriate labels and keys for the entities and processes (e.g., the conceptual model for excess nutrients in EPA [1999]). A useful compromise is the use of compartments in forms suggestive of the entities that they represent (Figure 21.2).

Maps are included in site-specific assessments to show boundaries and spatial relationships. They may also be used as conceptual models representing the spatial relationships among sources and receptors including flow paths. These map-based conceptual models are particularly useful when the exposure of receptors has not yet occurred but may occur in the near future due to the movement of contaminants from buried waste through groundwater to a stream.

Flow diagrams are the most generally useful form for conceptual models. The boxes may designate physical compartments in the environment that contain contaminants, endpoint entities, resources for an endpoint entity, etc. They may designate specific entities, such as a particular stream or waste tank, or classes of entities such as herbivorous small mammals. The clarity of the diagram can be increased by using different shapes to distinguish the categories of the entities, such as sources, media, receptors, and output to other units. They may even be in iconic forms representing the entities (Figure 21.2). The compartments are connected by arrows that designate processes such as flows of contaminants between compartments. In many cases, particularly if the processes are contaminant transfers, the nature of these processes is self-evident. However, if some of the relationships are not contaminant transfers (e.g., loss of habitat structure due to loss of plants) or if for some other reason the diagram would be hard to interpret, the arrows may be labeled or separate process boxes may be added between the compartment boxes. Numbered arrows or boxes allow the diagram to be clearly keyed to the narrative or to models.

Various conventions may be used to impart additional information in the conceptual model. For example, the amount of the contaminant transferred through particular routes may be conveyed by arrows of different thicknesses. Similarly, receptor boxes that are assessment endpoint entities may be distinguished from those that are not by shading, thickness of box outline, etc. Figure 17.2 shows a basic structure for a conceptual model

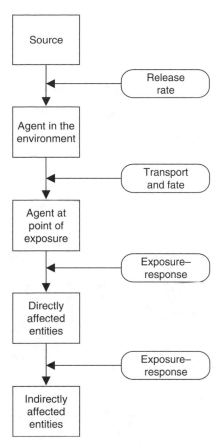

FIGURE 17.2 A highly generic conceptual model illustrating the most basic components of a conceptual model for release of an agent. *Rectangles* are states of entities and *rounded rectangles* are processes.

involving release of a chemical or other agents. It serves to illustrate both the basic components of conceptual models and the importance of identifying the functions that control the causal relationships among the components. Figure 17.3 shows a conceptual model for an activity, and illustrates the utility of structuring the model as alternating states and processes.

17.3 CREATING CONCEPTUAL MODELS

Because of the complexity of ecosystems, conceptual models for ecological risk assessment can be difficult to generate. In particular, important components or relationships are easily overlooked; it is also difficult to know when to stop in terms of bounds and detail. The following techniques may help:

Build from basic relationships: Walters (1986) recommends beginning with the basic relationship to be assessed and elaborating the variables and processes that link them, until apparent boundaries of the system are reached and further detail seems impractical or unnecessary. An example of an assessment for management of a fishery is presented in Figure 17.4. For a fishery manager, the fundamental relationship is between catch and employment. Elaboration shows that catch depends on stock size and harvesting effort, while employment depends on catch, harvesting effort, and fleet size. Further elaboration reveals that stock size depends on growth, recruitment, and natural mortality, each of which

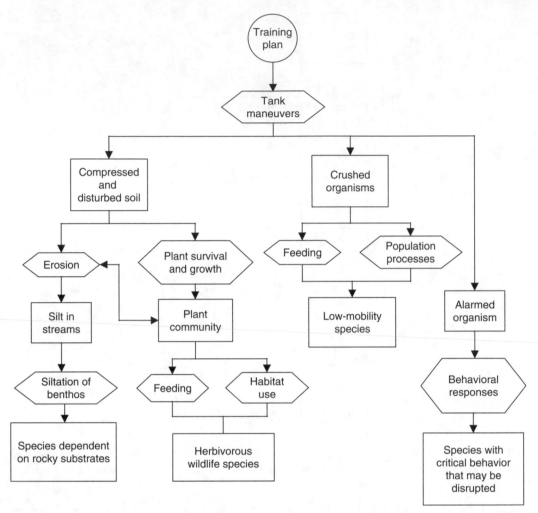

FIGURE 17.3 A generic conceptual model for an activity, tank training, that is a source of hazardous processes. The *rectangles* are states of entities and *hexagons* are processes linking those states. (Previously published in Suter G.W. II, *Hum. Ecol. Risk Assess.*, 5, 397, 1999. With permission.)

depends on ecological factors that cannot be quantified. Display of each step in the elaboration can make the logic of the model clear to the reader.

Break the model into modules: Often, when creating a conceptual model of a complex system, the modelers can get lost in the tangle of boxes and arrows. To avoid this, it is often helpful to define component modules of the system. In that way, relatively simple top-level and low-level models can be created for each of the modules (Suter 1999b). For example, a conceptual model of South Florida used a hierarchy of models of society, ecosystems (e.g., marl prairie and Biscayne Bay), and species (e.g., Florida panther) (Gentile et al. 2001). A more conventional example might be an assessment of a waste site that contributes leachate to a creek. The overall conceptual model of the system includes the biota of that creek as a module along with a waste module, groundwater module, terrestrial biota module, etc. The creek biota would be elaborated in a conceptual submodel with uptake of the contaminant by various species or trophic groups, movement of the contaminant through the food web, and primary and secondary effects (Figure 17.5). This approach lends itself to object-oriented

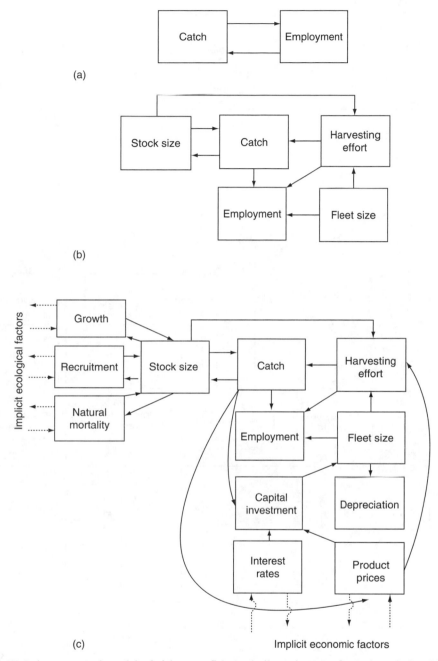

FIGURE 17.4 A conceptual model of risks to a fishery, built up in steps from the relationship that is fundamental to the decision. (Adapted from Walters, C.J., *Adaptive Management of Renewable Resources*, Macmillan, New York, 1986. With permission.)

programming that allows the independent creation and updating of component models. Modules may be defined in terms of distinct spatial units, distinct processes (e.g., eutrophication), or distinct types of entities (e.g., species or ecosystems). The conceptual models of the modules may be more detailed, or may represent the mechanistic processes at a lower level of

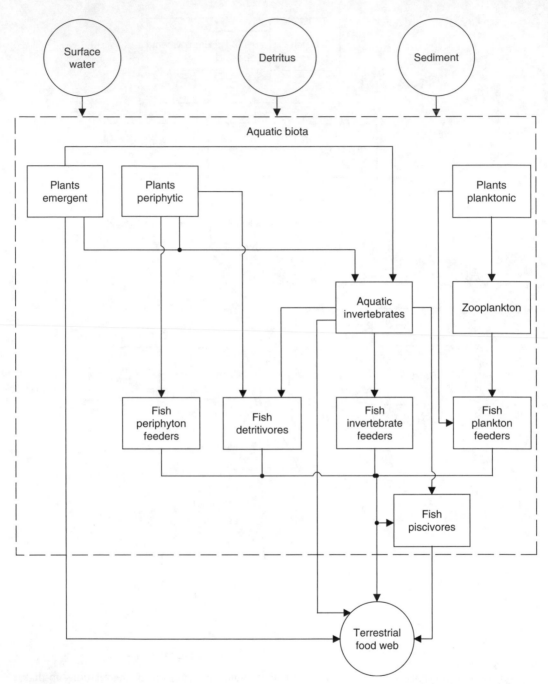

FIGURE 17.5 A stream biota module from a larger conceptual model of ecological risks from waste disposal. The module is connected to inputs from contaminated water, sediment, and detritus and is in turn connected to a terrestrial food web module through feeding by piscivorous and insectivorous wildlife.

organization. For example, a module for uptake of a chemical by an organism may simply show all of the routes of uptake, or may represent physiological toxico-kinetics (Section 22.9).

Linking of standard component modules: If certain sources, endpoint entities, or processes appear repeatedly in ecological risk assessments, then generic components can be created for

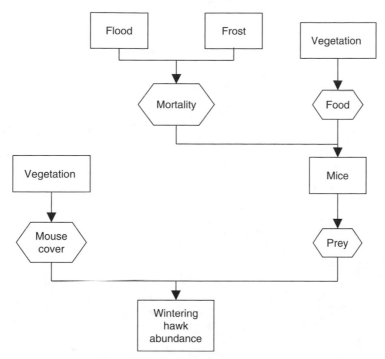

FIGURE 17.6 A receptor conceptual model for influences on the abundance of wintering hawks. (Modified from Andrewartha, H.G. and Birch, L.C., *The Distribution and Abundance of Animals*, University of Chicago Press, Chicago, 1984 and previously published in Suter, G.W. II, *Hum. Ecol. Risk Assess.*, 5, 397, 1999. With permission.)

repeated use (Suter 1999b). This is particularly likely when multiple risk assessments must be performed for the same site due to multiple activities or sources. These would fall into three categories. First, receptor conceptual models can be created for particular species or communities that are the objects of multiple assessments. The receptor models would represent all of the significant influences on the endpoint receptors (Figure 17.6). Conceptual models of this sort, termed influence models are important tools in ecology (Andrewartha and Birch 1984). Second, source conceptual models can be created for individual categories of sources or activities that generate the agents to be assessed (Figure 17.3). Examples might include buried wastes, coal fired power plants, logging operations, livestock grazing, or sewage treatment plants. The models would illustrate the release of chemicals, removal of resources, physical modification of ecosystems, and other activities that may affect the environment. If the source initiates a common causal chain, such as eutrophication from nutrient releases, it should be included in the generic source model. Third, site conceptual models would describe features of the site (actual or defined by a scenario) that can causally link the sources and endpoint receptors. These could include hydrologic models, atmospheric models, and food web models (Figure 17.7). The generation of a conceptual model for an assessment would involve selecting appropriate source, site, and receptor models and linking them (Figure 17.8). The linkage would involve identifying the logical relations among the three types of models. For example, release of any aqueous effluent or any runoff would link to the surface hydrology model, which would show how the materials moved through the streams, flood plains, etc., and any persistent chemicals would link to the aquatic food web models. Components of any model that do not provide a connection between source and receptors would be pruned. For example, the tank training source model (Figure 17.3) might be linked to a desert

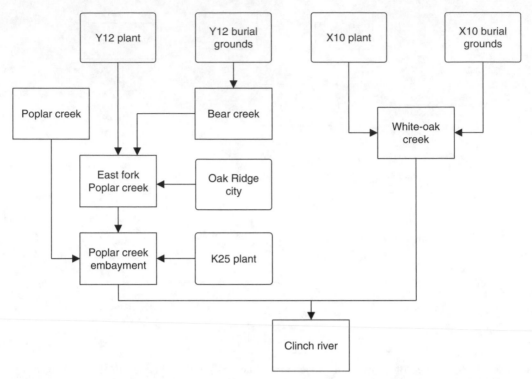

FIGURE 17.7 A site conceptual model, hydrologic transport of contaminants on the Oak Ridge Reservation, Tennessee. (Previously published in Suter, G.W. II, *Hum. Ecol. Risk Assess.*, 5, 397, 1999. With permission.)

pronghorn receptor model at all three terrestrial community links, but the population processes path would be pruned because pronghorns would not be crushed. Linkages that are peculiar to the assessment would be added.

Generic conceptual models: If the situation being assessed is repeated and sufficiently consistent, generic conceptual models may be used. Examples might include effluents from sewage treatment plants on small rivers in the upper midwestern United States, insecticide application to irrigated cotton, or release of surfactants from domestic laundry. It is important not to extend such generic conceptual models beyond their range of applicability, and to modify them as needed to reflect the peculiar aspects of individual cases.

Brain storming and pruning: One may begin by developing a model that includes all components and relationships that any member of the assessment team, manager, or stakeholder feels is relevant. One may then prune the model by deleting components that are not actually part of the system, relationships that are not possible (e.g., movement of contaminants into an isolated aquifer), components or relationships that would not influence the decision, and components or relationships that cannot be quantified and are not of sufficient importance to prompt research or testing (e.g., dermal exposure of reptiles). This approach is inefficient but is likely to serve the heuristic and communication functions discussed previously.

Representing implemented models: The fundamental purpose of conceptual models is to direct the selection and development of the implemented mathematical models of exposure and effects. Conversely, as those models are selected and developed, the conceptual model should be modified to represent the way that risks are quantitatively modeled. For example, if plant uptake from contaminated soil by both roots and leaves is simulated, it is important to

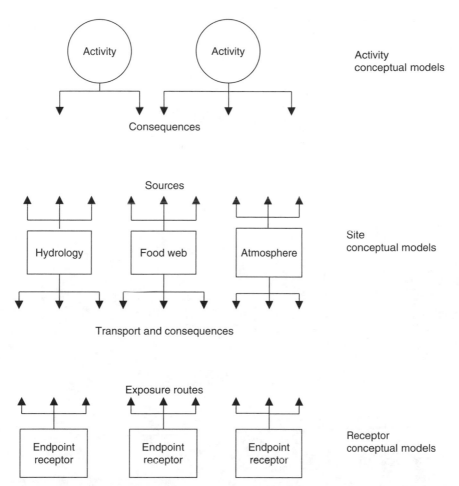

FIGURE 17.8 A representation of the process of linking modular conceptual models to create a complete conceptual model. (Previously published in Suter, G.W. II, *Hum. Ecol. Risk Assess.*, 5, 397, 1999. With permission.)

represent the two pathways. However, if an empirical soil–plant uptake factor is used, the conceptual model would show the soil–plant transfer without separating the routes.

17.4 LINKAGE TO OTHER CONCEPTUAL MODELS

The conceptual model for ecological risk assessment should be consistent with other conceptual models developed for assessments to inform the same decision. There will always be a human health risk model and sometimes an engineering risk model, or a socioeconomic model. At least, the conceptual models should be consistent. They should show the same sources, the same routes of transport, etc. In addition, the ecological risk model should link to the other risk models. Output from one should be input to another. One way to achieve this is to create an integrated conceptual model with engineering, human health, ecological, socio-economic, and common constituents. The common constituents would be modules such as a conceptual model of chemical transport and fate that provides input to multiple assessments. Such integration of conceptual models can lead to efficiencies in assessment (e.g., the human health and ecological assessors do not duplicate effort) and better-informed decisions because the results of the various assessments are consistent (Chapter 37).

18 Analysis Plans

A limited data base is no excuse for not conducting a sound risk assessment.
On the contrary, with less knowledge of a system, the need for risk assessment and management becomes imperative.

Haimes (1998)

The last step in problem formulation is the development of a plan for performing the assessment. It should define the data that will be collected, the models and statistical or logical analyses that will be applied, and the type of results that will be presented. It should also describe the methods that will be used to generate data and approaches that will be used to develop new models. It should define the milestones and completion date of the assessment, the level of effort, and the cost. It should also provide a quality assurance plan that specifies the measures taken to reduce uncertainty and the expected magnitudes of uncertainty in the results (Chapter 9). For routine assessments, such as those for new chemicals, the analysis plan may be defined by a standard protocol. However, for novel assessments and site-specific assessments, it will be necessary to develop a plan de novo. Typically, the analysis plan must be reviewed and approved by the risk manager, who will seek assurance that the planned work will provide the information needed for the decision, and by any others who may control the budget, staffing, and schedule.

The assessment team generates the analysis plan by synthesizing the other components of planning and problem formulation. They begin by considering what changes in what attributes of ecological entities (assessment endpoints) must be estimated, how those changes might be brought about (source descriptions and conceptual models) and under what circumstances (environmental descriptions and scenarios). Then they must determine, given the limitations of time and resources, the best way to estimate those changes. That is, given the combination of existing information and the possibilities for testing, measurement, modeling, and inference, what is the best combination of methods to apply for this assessment?

The analysis plan may include plans for tiering the assessment (Section 3.3). These plans include a screening assessment using existing data (if not already performed in the problem formulation), a screening assessment performed with data from preliminary sampling and analysis, and a definitive assessment with a full data set. Each tier should have its own analysis plan that describes the performance of that tier in some detail, and general plans for subsequent tiers. Analysis plans for screening tiers should assure that they provide the basis for the next tier's problem formulation, focusing the assessment on critical pathways, receptors, and locations, and providing a basis to design sampling or testing activities.

18.1 CHOOSING MEASURES OF EXPOSURE, EFFECTS, AND ENVIRONMENTAL CONDITIONS

Development of analysis plans is largely a matter of choosing the best measures of exposure, effects, and conditions. This is typically begun by identifying measures of exposure for each of the agents and exposure–response measures, which are appropriate to the measure of exposure and the endpoint entities and attributes. Types of measures of commonly assessed categories of agents in aquatic ecosystems are presented in Table 18.1 and are described in the following chapters. Ideally, the measures would be chosen so as to provide the most accurate and complete estimate of risks. The Data Quality Objectives (DQO) process, described in Chapter 9, is a method for identifying such ideal data sets. In fact, the DQO process may be considered an alternative to problem formulation, as a means of developing an analysis plan in cases that are as simple as health risk assessments. However, the selection process is often constrained by resource limitations, the desire to use standard or conventional measures, the need to accommodate existing data, and the data requirements of existing models. For example, the best measure of effect may be a time-to-death model,

TABLE 18.1

Associations Between Measures of Exposure and Measures of Effects from Controlled Studies for Different Types of Aquatic Stressors

Agents	Characterization of Exposure	Characterization of Exposure–Response
Chemical	External concentration in medium	Concentration–response or time–response relationships from laboratory or field
	Internal concentration in organism Biomarker	
Effluent	Dilution of effluent	Effluent dilution-response tests Field tests in plume
Contaminated ambient media	Location and time of collection	Lab or in situ tests using the medium
	Analysis of medium	Medium dilution-response Medium gradient-response
Habitat	Structural attributes	Empirical models (e.g., habitat suitability models)
Water level or flow	Hydrograph and associated summary statistics (e.g., 7Q10)	Instream flow models
Thermal energy	Temperature	Thermal tolerances Temperature–response relationships
Silt (suspended)	Suspended solids concentration	Concentration–response relationships from laboratory or field studies
Silt (bed load)	Degree of embeddedness, texture	Empirical siltation–response relationships from laboratory or field
Dissolved oxygen and oxygen-demanding contaminants	Dissolved oxygen	Oxygen concentration–response relationships from laboratory or field
Excess mineral nutrients	Concentration	Empirical concentration–response relationships from laboratory or field Eutrophication models
Pathogen	Presence or abundance of pathogens	Disease or symptoms
Nonindigenous species	Presence or abundance of the species	Ecological models (food web, energetic, predator–prey, etc.)

Source: Modified from US Environmental Protection Agency, Stressor Identification Guidance Document, EPA/822/B-00/025, Washington, DC, Office of Water, 2000; and Suter, G.W., II, Norton, S.B., and Cormier, S.M., A methodology for inferring the causes of observed impairments in aquatic ecosystems, *Environ. Toxicol. Chem.*, 21, 1101–1111, 2002. With permission.

but, if the only data for acute lethality is an LC_{50} and there are no resources for testing, then the LC_{50} is used, and the exposure data must be related to that test.

As suggested by the example, the analysis plan should assure that the dimensions of the problem are reflected in the measures and that the different measures are concordant (Chapter 23). The dimensions of exposure and effects for chemicals and similar agents are discussed in Section 6.4. Once exposure ends, time to recovery is another relevant component of the temporal dimension, but it does not relate to the risk of an effect. It becomes relevant when considering net benefits of an action or when comparing the total effects of heterogeneous alternative actions (Section 33.1.5). Hence, the choice of measures of exposure and effects should be based on an understanding of which of these dimensions must be quantified, how they are to be scaled and quantified, and how they are related (Chapter 6).

18.2   REFERENCE SITES AND REFERENCE INFORMATION

When performing a risk assessment for a site or region, information concerning the baseline state or a nominally uncontaminated or undisturbed state is needed for purposes of comparison and normalization. For example, one may need to know the concentrations of contaminant metals in uncontaminated media or the number of fish species in unchannelized streams. This is referred to as reference information.

Background is a special case of reference sites. Background sites are those that are believed to represent an uncontaminated or undisturbed state. In many cases, this ideal cannot be achieved. For example, the atmospheric deposition of anthropogenic mercury is ubiquitous. Therefore, one can identify negative reference sites for mercury with respect to a specific mercury source, but not true background sites. In practice, the distinction between reference and background is unclear. Nevertheless, background concentration is an important concept in risk assessment, because they are used to determine what concentrations of metals and other naturally occurring chemicals should not be remediated and are not even worth assessing (Chapter 31). Therefore, it is important, in the analysis plan, to identify sites that will be treated as effective background sites.

The sources of reference information must be clearly specified in the analysis plan. Several sources of such information can be recognized.

18.2.1   INFORMATION CONCERNING THE PRECONTAMINATION OR PREDISTURBANCE STATE

For the assessment of future disturbances or contamination, the proposed site may serve as its own reference. For example, if a dam or a wastewater treatment plant is proposed for a stream, then the current biota of the stream serves as a reference against which the predicted future conditions are compared. Surveys of the stream could provide the basis for that comparison, focusing in particular on the state of endpoint entities.

For the assessments of ongoing disturbance or contamination, studies may have been done at the site prior to the contamination or disturbance. Such historic information may have been developed for permitting or other regulatory purposes, for resource management, for research, for a monitoring program, for environmental impact assessments, or for evaluation by private conservation organizations. Hence, it is important to contact past landowners, local universities, regulatory and resource management agencies, and private conservation organizations to determine whether the information is available. However, such information must be used with caution. Information may not be of expected quality for risk assessment, it may be so old as to no longer be relevant, or it may have been obtained using inappropriate methods. For example, chemical analyses may not have achieved adequate limits of detection.

Some useful data may be truly historical. When agriculture, resource harvesting, fire suppression, or similar activities have long modified an ecosystem, reference information may come from the writings of early visitors, early photographs, or museum collections from early biologists. Such records have documented the conversion of grasslands to shrub lands and of open forests to closed canopy forests. However, it must be noted that these records reflect the management practices of indigenous or premodern peoples, not a natural state (Redman 1999).

18.2.2 MODEL-DERIVED INFORMATION

In some cases, no measurements or observations can supply adequate reference information. For example, there may be no undisturbed or uncontaminated streams in the area. In such cases, it may be possible to use a model to estimate characteristics of the uncontaminated state. In particular, habitat models may be used to estimate the biota of a site in the absence of contamination or physical disturbance (US Fish and Wildlife Service 1987; Wright et al. 1989). The results of such models are seldom sufficiently accurate to determine that a decline in abundance has occurred, but they can be used to indicate that species are absent or severely depleted that should be present in abundance.

18.2.3 INFORMATION CONCERNING OTHER SITES

The most common source of reference information is the study of reference sites, i.e., sites that resemble the site being assessed except for the presence of the disturbance or contaminants. Commonly, the reference is a single site. For streams and rivers, a single upstream site may be used and for terrestrial sites, a single location that is clearly beyond the contaminated or disturbed area is typically chosen. This approach is inexpensive, and, if exposure and effects are clear, it may be sufficient. However, a pair of sites may differ for any number of reasons other than disturbance or contamination. For example, upstream and downstream sites differ due to stream gradients. In addition, the use of a single reference site does not provide an adequate estimate of variance in the reference condition, because repeated samples from a single reference site are pseudoreplicates (Hurlbert 1984) (Box 5.1). Pseudoreplication is perhaps best explained here by a relevant example.

The San Joaquin kit fox population on the Elk Hills Naval Petroleum Reserve, California, crashed in the early 1980s, and it was suspected that the oil field wastes were responsible. A study was conducted, which analyzed the elemental content of fur from the foxes to determine whether they had elevated exposures to the wastes (Suter et al. 1992). The sponsor initially insisted that reference information be limited to analyses of fur from foxes trapped on the periphery of the oil field. However, a pilot study using that reference indicated that the concentrations of several elements were statistically significantly higher in the fur from the oil field than on reference fur. A subsequent study, which included areas away from oil fields and another oil field, found that the initial comparison had been misleading. The metal concentrations in fur from Elk Hills oil field foxes were not high relative to these other reference sites. Rather, the fur from foxes at the peripheral reference site was unusually low in metals. Without true replication of reference sites, an incorrect conclusion might have been drawn about the cause of the observed decline.

The best solution to this problem is the one employed in the Elk Hills study; obtain information from a number of reference sites sufficient to define the nature of and variance in the reference conditions. In addition to the obvious problem of increased cost, the chief problem in this approach is finding enough suitable sites.

Selection of reference sites involves two potentially conflicting goals. First, relevant properties of the reference sites should resemble those of the assessed site, except for the presence of the contaminants or disturbance. The relevance of properties of the sites depends on the comparisons to be made. For example, if a contaminated stream reach has a natural channel, but upstream reaches are channelized, that difference is probably not relevant to comparisons of metal concentrations, but it precludes using the upstream reach as a reference for population or community properties. Second, the reference sites should be independent of the assessed site and of each other. For example, if animals move between the sites, measurements of body burdens, abundances, or any other attribute involving those mobile organisms cannot be used for comparison, because the sites are not independent. In most cases, the sites that resemble the assessed site most closely are those that are adjoining or nearby, but those are also sites that are least likely to be independent of the assessed site.

The problem of deciding whether reference and assessed sites differ in any relevant way requires knowledge of physics, chemistry, or biology controlling the properties to be compared. For example, the occurrence of heavy metals in soils is a function of the cation exchange capacity, so that factor must be considered, but that relationship does not depend on the identity of the clay minerals providing the capacity. For ecological comparisons the relevant habitat parameters must be similar, but the parameters that must be similar depend on the taxa being assessed. For example, the fish species composition of a water body is affected by the diversity of habitat types, but is unlikely to be significantly affected by moderate differences in production. However, condition measures such as weight and length ratios are likely to be affected by production. As a result of considerations such as these, the availability of appropriate reference sites may constrain the measures of exposure and effect that are used. It is pointless to measure a property of a site if it is not possible to obtain relevant reference information against which to compare it. The EPA has suggested parameters that should be included when determining the similarity of sites, but that guidance cannot substitute for site-specific consideration of the relevance of differences in site characteristics to the types of comparisons among sites that will be part of the assessment (Office of Emergency and Remedial Response 1994).

In some cases, suitable criteria for similarity may already exist. The best example is the soil classification systems developed for the United States by the Natural Resource Conservation Service and elsewhere by other agencies. For the most part, a reference site with the same soil type as an assessed site is suitable for measures of soil properties such as metal concentrations. However, for other properties such as abundance of soil invertebrates, additional factors such as compaction, organic matter content, and vegetation must be considered. In any case, the soil type is a good starting point in selecting reference soils. Another example would be the use of the same stream characteristics that have been used to establish regional references (see below) to select reference stream reaches for local comparisons.

The magnitude of difference, i.e., necessary to declare a potential reference site to be acceptably similar to assessed sites, is determined by the magnitude of exposure or effects to be discriminated, and by judgment and experience. For example, experienced invertebrate ecologists know approximately how much difference in sediment texture is likely to cause a 25% difference in the species composition of the community. The magnitude of difference to be detected should be determined by interaction with the risk manager.

The independence problem requires that we consider not only factors that introduce common biases in the sites, but also factors that inappropriately reduce intersite variance. For example, it is inappropriate to simply establish reference sites at different distances upstream and to use them to estimate reference variance. The Oak Ridge Reservation's Biological Monitoring and Abatement Program addressed this problem by monitoring a suite of reference streams both on and off the reservation. This use of multiple reference streams establishes information equivalent to a regional reference, but more localized. There are technical and policy elements in the selection of a set of reference sites, because it is necessary to decide what variance is relevant and what properties must be independent. For example, all soils in Oak Ridge Reservation are slightly contaminated with mercury due to atmospheric deposition of mercury, used at the Y-12 plant, and coal combustion from nearby power plants. However, it was decided that low-level contamination was not relevant to determining which sites could be considered background, and therefore the lack of independence of mercury concentrations among reference sites on the reservation was judged to be irrelevant.

18.2.4 Information Concerning a Regional Reference

Regional reference information is a special case of reference information from other sites. It is derived from a population of sites within a region that encompasses the assessed site and that is deemed to be acceptably uniform and undisturbed with respect to the properties of interest. One example is the use of the US Geological Survey's summaries of elemental analysis of soils from the eastern and western United States as reference soil concentrations for sites in those regions (Shacklette and Boerngen 1984). Some states have compiled background concentrations for environmental media (Slayton and Montgomery 1991; Webb 1992; Toxics Cleanup Program 1994). Regional reference information is available for biotic communities in many states due to bioassessment programs (Davis and Simon 1995; Barbour et al. 1996). For example, the state of Ohio has been divided into ecoregions, and reference fish and benthic invertebrate community properties have been defined for each (Yoder and Rankin 1995). The use of regional reference information has advantages. Regional reference information is available from the responsible agency, eliminating the need for reference sampling; it is based on multiple independent sites so it accounts for variance; and, when the reference information is generated by a regulatory agency, it is likely to be acceptable to that agency.

Regional reference information has some disadvantages. One is that the data used to establish the regional reference may not be suitable for risk assessment. Detection limits for analyses of reference media may be too high or the quality assurance and quality control may not be up to the standards required by regulators or responsible parties. For example, the results of metal analyses of uncontaminated waters performed prior to the early 1990s were often too high because of inadvertent contamination during sample handling and analysis (Windom et al. 1991). And the other is that, the regional reference may not include the parameters that are needed to estimate the exposure or the assessment endpoint, as specified by the problem formulation. For example, the abundance or production of a particular fish species may be an assessment endpoint for a site, but regional reference values for fishes are commonly defined in terms of community properties. In addition, the ubiquity of contamination and disturbance may result in reference sites that are degraded. This is the case in Ohio, where the 25th percentile of the distribution of reference community indices is the threshold for acceptability (Yoder and Rankin 1998). That is, 25% of the best streams in Ohio are declared unacceptable. Finally, the risk managers often prefer site-specific reference information. This preference is particularly likely when the region is large, making the relevance of the reference information questionable.

18.2.5 Gradients as Reference

An alternative to the use of multiple reference sites to provide reliable reference information is the use of gradients. This is particularly useful where there is a gradient of contamination or where there is a biological or physical gradient. One use of gradient analysis is to establish a reference concentration of a contaminant in terms of the asymptotic concentration. This approach is particularly useful for sites where the extent of contamination is unknown.

An analogous use is to establish the natural gradient of stream community properties as a reference against which to compare the state of the community in assessed reaches. For example, fish species richness naturally increases along downstream gradients due to increased habitat quantity and diversity. Therefore, sampling fish on a gradient from well upstream of the site to well downstream can establish both the natural gradient and any deviations from it associated with the site. Gradient analyses eliminate the problem of balancing the need for similarity against the need for independence, by eliminating both. That is, if the relevant factors follow a gradient, then the lack of similarity and the lack of independence are incorporated into the gradient model. Gradients are relatively seldom used in ecological risk assessments, partly because their existence is not recognized, and partly because confounding disturbances or contaminant sources may disrupt the natural gradient.

18.2.6   POSITIVE REFERENCE INFORMATION

In addition to the conventional or negative reference information discussed above, positive reference information may be needed. Positive reference information concerns contaminated or disturbed sites that are not the sites being assessed. While the purpose of a negative reference is obvious, the uses of positive references are more diverse and difficult. The most common is the upstream reference on a stream that has sources of contamination upstream of the source being assessed. For example, Poplar Creek, Tennessee, is contaminated by various sources on the Oak Ridge Reservation that were assessed for possible remediation, but also by upstream sources including coal mines and a small city. Analyses and tests performed on water and sediment samples from Poplar Creek upstream of the reservation provided a measure of the level of contamination and toxicity to which Reservation's sources were an addition. Therefore, they provided an indication of the improvement that might be expected if remediation was implemented. Another use of positive reference is to determine whether a hypothesized cause of observed effects is plausible. For example, Poplar Creek receives mercury from its East Fork. If the effects observed in Poplar Creek were due to that mercury input, the same or greater effects with the same symptomology should have been observed in lower East Fork, where the mercury is less dilute. The fact that aquatic toxic effects were not observed in the East Fork indicates that mercury was at most a minor contributor to the observed effects in Poplar Creek.

Positive controls, tests of toxic levels of well-characterized chemicals, serve a similar function in toxicity testing. They should be used with site-specific toxicity tests to demonstrate that the tests, as performed, are sensitive to concentrations of chemicals that are known to cause relevant toxic effects. Cadmium chloride and other well-studied chemicals have been recommended as positive control chemicals, but for tests at contaminated sites, chemicals that are contaminants of potential concern at the site or at least have the same mode of action should be selected.

18.2.7   GOALS AS AN ALTERNATIVE TO REFERENCE

In some cases, no appropriate reference is available because contamination or disturbance is ubiquitous, because a desirable ecological condition no longer exists in the region, or because the desired state may have never existed (e.g., a high-quality urban stream—Box 11.1). In such cases, management goals serve in place of reference conditions as a standard for comparison. As discussed in Chapter 11, goals may be set in a variety of ways. However, if they are to serve in place of reference conditions, goals must be very specifically and clearly defined, like an assessment endpoint. For example, 30% eelgrass cover might be a goal for an estuarine bay and reproducing populations of three fish species might be a goal for an urban stream.

Part III

Analysis of Exposure

Ecological risks occur because organisms are potentially exposed to hazardous agents in a way that could result in effects. Analysis of exposure extends from the identification and characterization of sources (Chapter 19) to the estimation of levels of the agent in the environment by either sampling and analysis (Chapter 20) or transport and fate modeling (Chapter 21), and ends with the estimation of the actual exposure of organisms through contact and uptake (Chapter 22). Ultimately, the goal is to characterize exposure in a way that can be integrated with an exposure–response relationship to allow characterization of risk (Chapter 23).

The distinction between exposure and response may seem clear, but the location of the dividing line is largely a matter of convenience (Figure 17.1). At one extreme, one can treat a release or emission as the exposure (i.e., dose to the environment) and estimate an exposure–response relationship for that exposure. This is effectively the case for field tests or mesocosm tests in which a pesticide is applied at a set of rates, and responses are observed and recorded as a function of that rate (e.g., dead birds/ha given kg agent/ha). At the other extreme, exposure may be taken as occurring in a target tissue, or even a receptor molecule. In that case, exposure estimation requires the use of toxicokinetic models, and response, as a function of tissue concentration, is expected to be a constant for chemicals with a common mechanism of action. In most cases, the line is drawn at an intermediate point. In ecological assessments, exposure is most commonly expressed as a concentration in an ambient medium (air, water, sediment, or soil), and responses are expressed as functions of those concentrations (e.g., LC_{50}, NOEC). However, dose, density, temperature, area, and other units are used as appropriate.

19 Source Identification and Characterization

In most ecological risk assessments, the source is adequately characterized during the problem formulation (Chapter 13). For new facilities, for example, the process engineers will estimate the volume and composition of emissions. For new commercial materials such as pesticides and detergents, use scenarios estimate the release of the material during its use or in various waste streams. For waste repositories, estimates of risks of releases to the environment are provided by models such as LANDSIM (Environment Agency 1996). However, additional source characterization may be needed during the performance of an ecological risk assessment. In the simplest case, the source description provided by the manufacturer or process engineers may be inadequate. For example, the pH or temperature of an effluent may not have been provided, so the form of metals or the buoyancy of the effluent plume cannot be estimated. In such cases, the assessor's role is to go back to the source terms and request the needed information or to generate reasonable values. In some cases, assessors may be skeptical of the source information. If the source terms seem too good to be true or if they do not include unplanned, fugitive, or accidental releases, it may be desirable to compare them to information on releases from similar sources or of similar products. Otherwise, ecological risk assessors are most likely to be involved in characterizing the source if the environment modifies the source or if the source is unknown or uncertain.

19.1 SOURCES AND THE ENVIRONMENT

In some cases, characterization of sources involves defining the interaction between a human activity and a natural process. For example, assessing risks to streams from runoff of sediment from a clear-cut forest involves the natural processes of precipitation, runoff, and recovery of vegetation. Hence, the source characterization involves geohydrologic models with ecological components (Ketcheson et al. 1999).

Even for sources of chemical contaminants, natural processes may be involved in characterizing the source. For example, the release of buried wastes may involve slow processes of decomposition of containers and leaching of the waste constituents as well as the biological processes of burrowing animals and intruding tree roots (Suter et al. 1993). Some sources operate only during natural extremes, notably the spreading of wastes by floods. Materials located in flood plains or in holding ponds created by damming streams may be mobilized by floods causing aquatic and terrestrial exposures downstream. Flooding associated with hurricanes or other storms has repeatedly introduced animal wastes stored in "nondischarge" lagoons into the rivers of North Carolina (Wing et al. 2002). The occurrence of such sources can be estimated using a hydrological model in a geographic information system to estimate the extent of flooding by storms of prescribed intensity and then relate the extent of flooding to the distribution of waste lagoons. Such models could be extended by including the frequency distributions of storms with respect to intensity to create predictive risk models for new lagoons.

19.2   UNKNOWN SOURCES

When significant risks are caused by contaminated media, the sources of that contamination should be identified. In cases of contaminated water, air, and precipitation, remediation can occur only at the source. Even in cases of contaminated soil and sediment, which can be remediated at the site of exposure, sources should be identified to ensure that they are not recontaminated. Similarly, when risk assessments are based on a prior ecoepidemiological assessment of observed biological effects (Chapter 4), it is necessary to identify the source of the causal agents (EPA 2006c). Identification of sources of contamination, particularly when multiple possible sources are known, may be contentious. Even when the source seems obvious, if the responsible party is recalcitrant, it may be necessary to prove the source in a court of law. Hence, much of the guidance on identifying sources is contained in books and journals under the topic of environmental forensics (Murphy and Morrison 2002).

Source identification often begins by searching for possible sources. Depending on the pollutant and circumstances, relevant information may come from:

- Effluent permits
- Land use maps and aerial photos
- Interviews of local informants
- Historical records
- Surveys of the area

For example, if excess sediment is the causal agent, a survey of the watershed may reveal eroded and collapsed banks, tillage on steep slopes, construction, cattle in the stream or riparian zone, or down-cut channels. Information on where to look for sources may be found in generic conceptual models of pollutants, lists of potential sources (Office of Water 1999a,b), or databases of actual sources such as the Toxic Release Inventory (http://www.epa.gov/triexplorer/). Conceptual models are particularly useful when sources act indirectly. For example, suspended sediment may come from channel erosion, which is caused by changes in hydrology, so the ultimate source may be pavement that increases peak storm runoff. In some cases, it will be necessary to analyze effluents, wastes, or spilled material to establish that their composition makes them possible sources. If there is only one possible source and the issue is not contentious, it is sufficient to have identified the source.

When potential sources have been identified, assessors may be required to trace the connection between the point of exposure and the potential sources. Sometimes this step is as simple as tracing the flow of a stream. In other cases, the occurrence of the pollutant must be established by analyzing water or air along the path. Particularly when groundwater is involved, tracer dyes may be used to demonstrate a connection. In other cases, simple observations, such as a muddy tributary flowing from a farm, may be sufficient. When multiple sources are operating or when it is not clear that a source is sufficient to account for the level of contamination at the point of exposure, models may be used. They may be used conventionally to estimate transport and fate from a possible source (Chapter 21). Alternatively, transport and fate models may be used "in reverse" to identify a source or to apportion the exposure of a receptor at a location among multiple sources. Models for this purpose are termed receptor models (Gordon 1988; Scheff and Wadden 1993). An example is the use of atmospheric models with air parcel trajectories to establish that the sulfates that fell on the Adirondak Mountains of New York came from the power plants of the Ohio River Valley. The models for this purpose may be regression-based empirical models such as those that estimate soil and nutrient loss from particular land uses (e.g., the Revised Universal Soil Loss Equation). However, process models such as the watershed model BASINS (http://www.epa.gov/waterscience/basins/) and HSPF (http://water.usgs.gov/software/hspf.htm) are generally more useful because they can represent multiple sources and routes of transport and apportion responsibility among sources.

Sometimes, sources may be identified by analyzing the pollutants, using isotopic analyses, mixtures fingerprinting, and genetic analyses. Sources of some metals and other elements may be identified through isotopic analysis. When lead toxicity was found in Laysan albatross chicks, isotopic analysis determined that the source was lead paint chips from a decommissioned military base, not soil or food (Finkelstein et al. 2003). Lead isotope analyses were also used to distinguish lead shot from gasoline additives as sources of elevated lead levels in birds (Scheuhammer and Templeton 1998). Similarly, the contribution of sewage to the nutrient loading of a river was determined by analyzing nitrogen isotopes (DeBruyn et al. 2003).

For mixtures, sources may be identified by determining the similarity of the relative abundances of components of ambient mixtures to their relative abundances in potential source mixtures. This approach, termed fingerprinting, has been employed for polychlorinated biphenyls (PCBs) (Ikonomou 2002), petroleum hydrocarbons (Bence and Burns 1995), polycyclic aromatic hydrocarbons (PAHs) (Uhler et al. 2005), and atmospheric particles (Mazurek 2002). Fingerprinting may also be applied to less well-defined sources. For example, the profile of PAHs in Massachusetts estuaries was shown to indicate that urban storm water runoff was the primary source (Menzie et al. 2002). Chemical fingerprinting also showed that most of that PAH loading for four urban streams in the United States is due to parking lot sealants (Mahler et al. 2005).

Sources of pathogens or exotic organisms may be identified through traditional taxonomic methods or newer molecular genetic techniques (Simpson et al. 2002; EPA 2005b). Examples include the identification of sources of disease outbreaks such as sudden oak death, which came from imported rhododendrons, and distemper in sea otters, which came from domestic cats. Genetic techniques may also serve to identify sources of introduced nonpathogens such as fish that have been released or have escaped from cultivation.

When testing the toxicity of contaminated media, source identification must begin by determining the contaminants that cause the toxicity. Techniques for this purpose are called toxicity identification evaluation (TIE; Section 24.5).

19.3 SUMMARY

For most assessments, the role of ecological risk assessors is to assure that the source terms provided by others include the information needed to model transport, fate, and exposure. However, when environmental processes generate or mediate the source or when the source must be identified by tracing the pollutant back through the environment, ecological assessors and ecological models may be employed.

20 Sampling, Analysis, and Assays

Measure if you can; model if you must.

Colin Ferguson quoted by Thomp and Nathanail (2003)

This chapter discusses in general terms the activities that comprise the sampling of environmental media and biological materials for analysis of contaminants. Sampling and contaminant analysis are obviously relevant to situations in which contamination already exists. However, it is also relevant to assessments of future contamination. In such cases, sampling and analysis are needed to support transport and fate modeling by determining background concentrations and relevant characteristics of environmental media such as pH, hardness, and temperature (Chapter 21). These data are also relevant to analysis of effects. For example, aquatic toxicity data for metals should be chosen based partly on the similarity of test water chemistry to ambient water chemistry.

The specifics of sample collection techniques, sample preparation and handling, and analytical techniques can be found in texts on environmental chemistry and in guidance documents from the US Environmental Protection Agency (US EPA), other government agencies, and standard organizations such as the American Society for Testing and Materials and the American Public Health Association. Analytical methods for chemicals, microbes, and physical properties of water and sediment in the United States are available at the National Environmental Methods Index (http://www.nemi.gov). However, most of the technical guidance for environmental sampling and analysis is intended to support human health risk assessments. These techniques may not be appropriate for the estimation of ecological exposures. For example, analytical guidance for contaminated sites in the United States calls for total extraction and analysis of water and soil, but total extractable concentrations are typically much higher than bioavailable concentrations. Ecological risk assessors should, when possible, obtain and process samples that are relevant to the exposures of endpoint receptors. When that is not possible, the concentrations should be converted to more relevant estimates of exposure. These issues are discussed in Chapter 22.

20.1 SAMPLING AND CHEMICAL ANALYSIS OF MEDIA

Most of the funds and effort expended on studies of contaminated sites are devoted to the collection and chemical analysis of the abiotic media: soil, water, and sediment. Similarly, most of the guidance for site studies is devoted to media sampling and analysis. These activities should be performed as specified in the analysis plan, and the quality of the data should be verified before it is used in the risk assessment (Chapter 9). The issues to be addressed here are the general approaches to media sampling and analysis, particularly the summarization, analysis, and interpretation of the resulting data. These issues are particularly problematical when chemicals are detected in some, but not all, samples (Box 20.1). Specific issues with respect to using the measurements to estimate exposure are discussed in Chapter 22.

20.2 SAMPLING AND SAMPLE PREPARATION

Sampling should be performed in a way that produces a sample representative of the medium to which organisms are exposed and that does not modify the contaminant concentrations.

Soil samples should represent the range of depths to which organisms are exposed. As these vary greatly among taxa (Section 22.4), samples from multiple intervals should be obtained and analyzed so that the appropriate exposure can be estimated for each food web or endpoint species.

Sample preparation involves transforming raw samples into a form that can be chemically analyzed (Allen 2003). Initial sample preparation involves removing extraneous material (e.g., sieving or filtering), separation of phases (e.g., extracting pore water from sediment), stabilization (e.g., freezing to stop microbial processes or acidification of water to keep metals in solution), homogenization (e.g., mixing a soil sample or shaking a water sample), comminution

---

BOX 20.1
Handling Nondetects

Analytical data sets may include both reported concentrations (detects) and reported inability to detect the chemical (nondetects). Thus, the low end of the distribution of concentrations is censored. The problem is that nondetects do not signify that the chemical is not present, but merely that it is below the method detection limit (MDL) or quantitation limit. If a chemical is detected in some samples from a site, it is likely that it is also present at low concentrations in samples reported as nondetects. For screening assessments, this problem can be handled simply and conservatively by substituting the detection limit for the nondetect observations in order to estimate the distribution, or by using the maximum measured value as the estimate of exposure. However, for definitive assessments, such conservatism is undesirable. The most appropriate solution is to estimate the complete distribution by fitting parametric distribution functions (usually log-normal) using procedures such as SAS PROC LIFEREG or UNCENSOR (SAS Institute 1989; Newman and Dixon 1990; Newman et al. 1995). Alternatively, a nonparametric technique, the Product Limit Estimator, can be used to give more accurate results when data are not fitted well by the parametric functions (Kaplan and Meier 1958; Schmoyer et al. 1996). The US EPA provides guidance for analyzing data sets with nondetects that emphasizes simplicity of analysis (Quality Assurance Management Staff 2000).

The problem of censoring is exacerbated by the fact that method detection limits are not actually the lowest concentration that a method can detect, but rather the lowest concentration that can be reliably detected given a statistical criterion (Keith 1994). Therefore, an analytical laboratory may detect a chemical at 7, 9, and 11 μg/L in three samples but, if the MDL is 10 μg/L, the reported results are <MDL, <MDL, and 11 μg/L, respectively. Although the two lower concentrations in this example are more uncertain than the highest concentration, these measured values are clearly more accurate than the estimates generated by the methods discussed above. The best procedure from a risk assessment perspective would be to report all measured concentrations with associated uncertainties rather than allowing chemists to censor data that they deem to be too uncertain.

It should be noted that the methods for calculating detection limits and quantitation limits and even their conceptual definitions differ among users and can be highly contentious (Office of Science and Technology 2003).

(e.g., crushing or milling of solid materials), and subsampling. Initial preparation is followed by final preparation for the particular analysis, including extraction, digestion, and forming pellets. As with sampling, it is important to take steps to ensure that sample preparation does not significantly change the concentrations. This involves handling samples in a manner that is appropriate to the class of chemicals and method, and ensuring the cleanliness of equipment. Samples to be analyzed for volatile or labile chemicals should receive little preparation. Standard methods for chemical analyses often specify sample preparation methods, but they may not be appropriate for generating concentrations that are relevant to ecological exposures.

20.3   ENCOUNTERED DATA

At some contaminated sites, chemical concentrations in site media that were measured prior to the site investigation are available. The assessors must decide whether they should be used in the assessment. Although more data are generally better, encountered data may not be useful because of their age, quality, sampling techniques, or design. In general, the utility of encountered data must be determined by expert judgment. Considerations relevant to the age of the data include the rate of degradation of the contaminants, the rate of change in the rate of release of contaminants from the source, and the rate of movement of the contaminated media. Even if concentrations are declining, old data may be useful for screening assessments, because they provide conservative estimates of current concentrations. The quality of the data and the acceptability of the sampling methods must be judged in terms of the uncertainty that is introduced relative to the uncertainty from not having the data. For example, metal analyses that are performed without clean techniques may be acceptable if the contaminants of concern occur at such high concentrations that trace contamination of the sample is inconsequential. Another important consideration is the detection limits. Analyses with high detection limits may create misleading results in screening as well as definitive assessments.

20.4   SCREENING ANALYSES

Although the trend in practice is to employ specific analyses, screening analyses for classes of contaminants are still used which include analyses for total organic chlorine, total polycyclic aromatic hydrocarbons (PAHs), total hydrocarbons, gross alpha and beta radioactivity, and toluene-extractable organic matter (Thomp and Nathanail 2003). These can serve to identify hot spots and eliminate uncontaminated areas with respect to an entire class of contaminants. Hence, they can save effort and costs in analyses of specific chemicals. Some are conservative, which is acceptable in screening analyses. For example, toluene-extractable organic matter could include significant amounts of natural organic matter (Thomp and Nathanail 2003).

20.5   ANALYSIS OF COFACTORS

In addition to analyses of contaminant concentrations, analyses must be performed of the physical and chemical characteristics of the tested media that influence toxicity. These are particularly important when toxicity tests of the ambient media are performed, because the media may be unsuitable for the test organisms due to basic properties. For water, these include pH, hardness, temperature, dissolved oxygen, total dissolved solids, and total organic carbon. For sediments, they include particle size distribution, total organic carbon, dissolved oxygen, and pH. For soils, the same properties are measured, except that dissolved oxygen is omitted and water content (e.g., field capacity) and major nutrients (e.g., N, P, K, S) are added. For example, differences in plant growth between contaminated and reference soils may be due to fertility, pH, or texture rather than toxicity. Without information on these properties, the case for toxic effects cannot be defended.

Exposure analyses for ambient media toxicity tests require analyses of chemicals of potential ecological concern (COPECs) from samples that are representative of the tested material. Therefore, results of analyses that are performed independently of the test should be used with great caution. Aqueous concentrations are highly variable over space and time. Storm events or episodic effluent releases may cause aqueous concentrations to change significantly over the course of a 7 d static replacement test, potentially making the analysis of only one of the three tested water samples inadequate for exposure characterization. Soil samples are variable over space both vertically and horizontally. Therefore, exposures in a soil toxicity test may not be well characterized by analyses of samples that were collected from "nearby" or from a different range of depths. Sediments may be relatively stable in time, like soils, or may be mobile and therefore temporally variable.

At most sites, abundant analytical data are generated which must be summarized and presented. The data summarization must meet the needs of the risk characterization. Depending on the effects and characterization models, the data may be presented as means and variances, distribution functions, percentiles, or other forms. Care must be taken in statistical summarization to avoid bias. For example, because many sets of environmental data have skewed distributions that approximate the log-normal distribution, the geometric mean is commonly recommended. However, this results in an anticonservative bias when the value is used in calculations or interpretations that involve mass balance (Parkhurst 1998). For example, if fish are exposed to varying concentrations in water, the best exposure metric for calculating their body burdens is the arithmetic mean concentration. Use of the geometric mean would improperly minimize the influence of high concentrations on uptake.

In addition, the chemical data must be summarized for presentation to other members of the assessment team, risk managers, and stakeholders. The goal of these presentations should be to make important patterns in the data apparent. The best general approach to displaying relationships between parameters (e.g., stream flow and contaminant concentrations) is the conventional x–y scatterplot. Although maps are generally not as good as scatterplots for showing potentially causal relationships, they provide an important means of presenting spatially distributed data. The difficulty comes in converting data that are associated with points to areal representations. The simplest approach is to present the results on a map at the point where the sample was taken. The results may be in numeric form or as a glyph such as a circle with area proportional to the concentration. Alternatively, various geospatial approaches can be used to associate concentrations with areas. These may be discrete areas (e.g., Theissen polygon), isopleths (e.g., Kriging interpolation), or gradients (e.g., polynomial interpolation) (Figure 20.1). Discussions of data presentation for contaminated sites can be found in Stevens et al. (1989) and Environmental Response Team (1995). More technical guidance may be found in Goovarts (1997). This is an area in which a little creative thought can be useful. Good general guidance for data visualization is provided by Edward Tufte (1983, 1990, 1997).

Toxicity normalization provides a means of summarizing exposure data for numerous chemicals in an interpretable form. This is done by converting the concentrations (C) to toxic units (TUs), which are proportions of a standard test endpoint such as the *Daphnia magna* 48 h EC_{50}.

$$TU = C/EC_{50} \qquad (20.1)$$

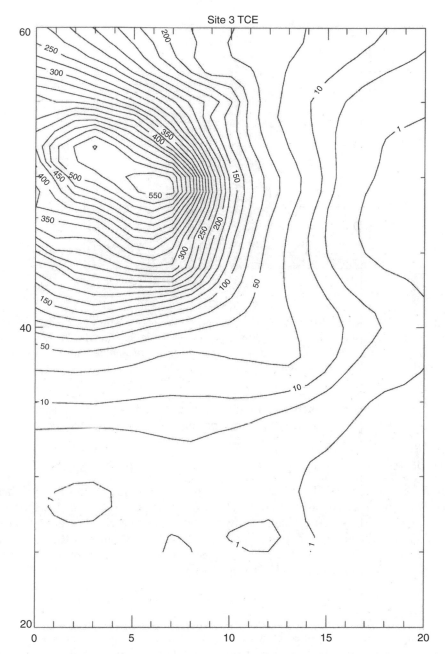

FIGURE 20.1 An example of a map generated by Kriging which, given a spatial array of chemical measurements, defines areas estimated to have chemical concentrations within prescribed ranges. (Provided by Yetta Jager, ORNL, and previously published in Suter, G.W., II, Efroymson, R.A., Sample, B.E., and Jones, D.S., *Ecological Risk Assessment for Contaminated Sites*, Lewis Publishers, Boca Raton, FL, 2000. With permission.)

TUs may be plotted as the values for each reach, subreach, transect, or other unit (Figure 20.2). The height of the plot is the sum of toxic units (ΣTU) for that location. The advantage of this approach is that it displays the contaminant concentrations in units that are indicative of potential toxicity rather than simply mass per unit volume. Therefore, one can

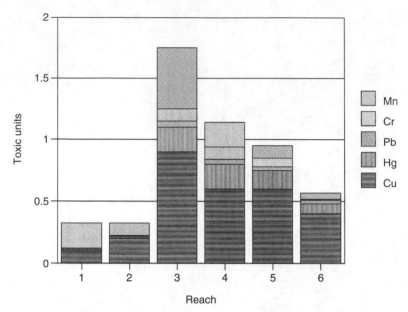

FIGURE 20.2 A display of the toxicity-normalized concentrations of metals in six stream reaches. The height of each bar is the sum of toxic units. (Previously published in Suter, G.W., II, Efroymson, R.A., Sample, B.E., and Jones, D.S., *Ecological Risk Assessment for Contaminated Sites*, Lewis Publishers, Boca Raton, FL, 2000. With permission.)

see which locations are most likely to pose significant risks, and which chemicals are likely to be major contributors to the toxicity. The purpose of this analysis is heuristic.

20.6 WATER

A common issue in analysis of ambient waters is the low concentrations of many chemicals, particularly when dissolved rather than total concentrations are desired. Detection of low concentrations requires not only high-quality analytical techniques, but also great care in sampling and sample handling to avoid trace contamination. Because of the general lack of ultraclean sampling and handling techniques, aqueous metal concentrations reported prior to the mid-1990s are generally unreliable (Benoit 1994).

20.7 SEDIMENT

Sediment samples may be obtained by a variety of dredges, coring devices, and, in wadeable streams or intertidal areas, by hand-held trowels or other devices. As with soil, the samples should be taken to a depth that is relevant to the organisms, usually a few centimeters. Another important consideration is the multiphasic nature of sediments. Apart from the basic distinction between the aqueous and solid phases, it may be important to distinguish sorbed material in the aqueous phase from the freely dissolved phase. The importance of these distinctions depends on the assumptions and models that are used to estimate exposures (Section 22.3).

The aqueous phase of sediments, pore water, is important as it is the most bioavailable phase for many chemicals and organisms. Extracting pore water from sediment samples can be labor-intensive, can require large amounts of sediment in order to obtain sufficient sample volume for multiple analyses, and can alter the form and speciation of the chemicals measured. The advantage is that measured pore water concentrations can be evaluated using the same techniques and effects data used for surface water. Measuring pore water concentrations is particularly useful for metals and ionic organic chemicals, because the particle–pore water partitioning mechanisms are complex and difficult to model.

20.8   SOIL

The concentration of contaminants in soil may be characterized in either bulk soil or soil solution. Sampling and analytical methods should be selected that are precise and easily related to measures of effects. The most direct and common approach is the collection and analysis of the bulk medium. Total concentrations of metals may be obtained by total digestion (e.g., $HF/HNO_3/HClO_4$), which removes elements bound to silicates, or by partial digestion (*aqua regia*), which removes all metals not firmly bound to silicates (Thomas 2003). Similarly, thorough extractions using organic solvents and heat allow the estimation of organic compounds (Hatzinger and Alexander 1995; Hendriks et al. 1995). These analyses have the reassuring feature of including the full extent of contamination as well as background concentrations of elements.

Aqueous extractions of soil may be designed to simulate the extraction processes of soil organisms. That is, the mass of the chemical extracted by an aqueous solution (somewhat less than the total) divided by the mass of the soil would approximate the bioavailable concentration. Appropriate procedures would depend on the organisms for which exposure is being estimated. Relatively mild extractions would be appropriate for root uptake, and stronger extractions would be expected to correlate with uptake by earthworms. Although many extraction procedures have been proposed, none has been demonstrated to be reliable for a variety of organisms, soils, and contaminants. For example, although concentrations of diethylenetriamine pentaacetic acid (DTPA)-extracted contaminants from soils sometimes correlate with those taken up by plants (Sadiq 1985) and earthworms (Dai et al. 2004), this estimate of bioavailability has been observed not to be valid for some metals (Sadiq 1985, 1986; Hooda and Alloway 1993; Dai et al. 2004) and for soils of varying pH (Miles and Parker 1979). As another example, three very different approaches for extracting bioavailable PAHs have been proposed in recent years, but none has demonstrated to be superior (Cuypers et al. 2000; Loibner et al. 2000; Liste and Alexander 2002).

Extractions intended to simulate mammalian gastrointestinal processes have been developed for human health risk assessment, but their range of applicability is unclear (Kelly et al. 2002). Because these extractions are believed to overestimate bioavailable concentrations in ingested soil, their results are referred to as bioaccessible concentrations.

20.9   BIOTA AND BIOMARKERS

Analysis of abiotic media provides a measure of external exposure to contaminants, but not internal exposure or exposure through trophic transfers, which require estimates of uptake from media and transfer between biotic compartments. In the absence of reliable models of uptake and transfer, internal exposures and trophic transfers can be estimated by collecting and analyzing biota from the contaminated site or from laboratory exposures to contaminated media. This approach has the advantage of avoiding the use of highly variable empirical models or unvalidated mechanistic models. However, analytical chemistry is expensive, and some chemicals are rapidly metabolized or may not accumulate to detectable levels. Similarly, body burden analyses are not feasible for some species such as those designated as threatened and endangered or those that do not currently occur on the site.

Care must be taken to ensure that the body burden analysis is relevant to the assessment. One issue is the treatment of unassimilated material. For example, if soil or sediment oligochaetes are not purged, the analysis may be dominated by chemicals in the gut contents that have not been incorporated. This may either overestimate or underestimate internal exposure of the worms and dietary exposure by vermivores, depending on whether the uptake factor (organism concentration/soil concentration) is less than or greater than one. However, for chemicals that are rapidly depurated following assimilation, long holding times for purging may result in underestimation of exposure. Although 24 h is the standard holding time to evacuate gut contents, as little as 6 h may be sufficient (Mount et al. 1999). The issue of unassimilated material also arises with contamination of the surfaces of leaves, and fur, feathers, and gut contents of wildlife. Decisions concerning this issue should be based on careful consideration of the actual mode of exposure of the endpoint organisms and of the exposure model used in the assessment. For example, if soil ingestion is included as a separate route in the exposure model, care should be taken to avoid incorporating soil into the chemical analysis of endpoint organisms or their food.

A second aspect of ensuring relevance of analyses to the risk assessment is selection of appropriate species, higher taxa (e.g., insects), or assemblages (e.g., benthic invertebrates) for sampling and analysis. This depends on the purpose of the sampling. In general, the purpose is either to estimate the dietary exposure of consumers (i.e., analyzing plants to estimate exposure of herbivores) or to estimate the internal exposure of endpoint organisms. In the first case, sampling should focus on the primary food organisms and on the parts that are consumed. In the latter case, the sampling should focus on the endpoint species or, if that is impractical, on a closely related species with similar habits. If the endpoint entity is a community or higher taxon, one may choose a representative species, representative set of species, or the entire group. To the extent that they can be identified and are relevant, the species that have the highest level of accumulation should also be selected. When other criteria are satisfied, organisms may be chosen on the basis of practical considerations such as ease of collection and body size.

A third aspect of ensuring relevant analyses is selection of appropriate components of the organisms for analysis. This requires first a decision as to whether to analyze the whole organism, some organ, or another component. Once again, the primary consideration is the relationship of the analysis to the mode of exposure. If one is interested in the dietary exposure of a grazing or browsing animal, the leaves of plants should be analyzed; for beavers, the bark and cambium of small branches; for granivores, the seeds. If the analysis is performed to estimate internal exposure of an endpoint receptor, one should perform the analysis that is appropriate for the exposure–response model.

If internal measures of exposure are to be employed in an assessment and site-specific field data are to be collected, it is important to know something of the toxicokinetics of the chemical of interest. Toxicokinetic data will provide an indication of whether the exposures are likely to result in detectable concentrations and which tissues should be sampled and analyzed. Tissue types most frequently sampled include liver and kidney, as they are the primary organs for metabolism and excretion, and are therefore likely to be adversely affected by contaminants. Brain is frequently analyzed for contaminants that are neurotoxic and accumulate in lipid. Chemical concentrations in eggs are widely used to evaluate the exposure of birds to contaminants that may be transferred through eggs (lipophilic chemicals occur in yolk) or are known to have adverse effects on development.

Some tissues are analyzed not because they are clearly associated with effects but because they are reservoirs for contaminants. Tissues such as bone and fat become reservoirs because of the chemical-specific affinities. Because most organochlorine contaminants are hydrophobic and lipophilic, they tend to accumulate in fatty tissues. Similarly, because lead and strontium are analogs for calcium, these inorganic contaminants tend to accumulate in bones. Contaminants in reservoir tissues may be mobilized episodically. Examples include mobilization of organochlorine chemicals from fat during starvation (e.g., hibernation or migration) or mobilization of lead associated with calcium mobilization during pregnancy or egg formation.

Other tissues, such as hair and feathers, are contaminant sinks. Chemicals in sink tissues cannot be reabsorbed, and are generally restricted to inorganic contaminants. Advantages of analyzing hair and feathers are that both tissues can be sampled nondestructively (without killing the animal) and that they may be sampled repeatedly from the same individual so that contaminant exposure may be tracked over time. Chemical concentrations in hair and feathers from various locations have been compiled and summarized (Huckabee et al. 1972; Jenkins 1979), but few data relate the concentrations to effects. Exceptions are provided by studies of lead and mercury in feathers (Burger 1995; Burger and Gochfeld 1997).

It is also necessary to consider the relevance of analyses of mobile organisms. Mobile organisms collected on a site may have spent little time on that site. To the extent that it is consistent with the endpoints of the assessment, organisms that are most associated with the site should be preferred, such as less mobile organisms and organisms with small home ranges. However, if the organisms of concern are not confined to the site, body burden analyses can still be relevant in that they realistically represent the proportional exposure of those organisms to the site and its contaminants. This rationale is applicable only if the organisms are not significantly exposed to sources of the contaminant outside the site.

In some cases, analysis of organisms from a site is not practical because the site is small or highly disturbed. In such cases, organisms can be exposed to the contaminated site media under controlled conditions. For example, at the Naval Weapons Station, Concord, California, plants and earthworms were exposed to site soils in the laboratory, and caged clams were exposed to site waters in the field (Jenkins et al. 1995). Similarly, earthworms were exposed in containers of soil at the Baird and McGuire Superfund Site in Holbrook, Massachusetts (Menzie et al. 1992). While providing consistent bioconcentration data, such studies can also provide information on toxicity.

An alternative to body burden analysis is analysis of biochemical biomarkers such as hepatic mixed function oxidase enzymes (Huggett et al. 1992). Biomarkers may be detected when the contaminant cannot be detected, and in some cases they may be measured without sacrificing the animal. For example, blood aminolevulinic acid dehydratase (ALAD) was used to estimate lead exposure in birds on the contaminated floodplain of the Coeur d'Alene River, Idaho, and liver lead concentrations were determined in a subsample of birds (Johnson et al. 1999). However, biomarkers tend to be nonspecific, to increase nonlinearly with increasing exposure levels (e.g., to decline at high exposures due to inhibited protein synthesis), and to vary with extraneous variables such as the animal's breeding cycle or nutritional state. In addition, few reliable exposure–response functions are available to relate biomarker levels to effects on organisms. For these reasons, biomarkers have been used much less than analysis of contaminant burdens in ecological risk assessments. However, one potentially important use is as bioassays (Section 20.10).

Body burdens and biomarkers of exposure must, in most cases, be related to concentrations in media to which the organisms are exposed. The derivation of such relationships requires sampling and analysis of the exposure media colocated with the sampled biota. A series of such analyses of colocated biological and media samples can be used to develop a site-specific uptake factor or other model. If the range of sites encompasses the range of contaminant levels, and if the uncertainty in the site-specific factor or model is sufficiently low, the factor or model can be used to predict body burdens or biomarker levels at locations where media samples, but not biological samples, have been analyzed. As media and biota concentrations may vary, samples should also be colocated in time. The acceptable interval between samples depends on the rate of variance of the biota and media, but the samples should not be taken in different seasons.

A wide variety of methods are available for the collection of biota samples for residue analyses, with sampling methods generally being medium- or taxon-specific. Common collection methods for taxa generally of interest in risk assessments are outlined in Appendix A of Suter et al. (2000) and in many other sources. General guidance on biota sampling is presented in Box 20.2.

20.10   BIOASSAYS

Bioassays are measures of biological responses that may be used to estimate the concentration or determine the presence of some chemical or material. Bioassays are seldom used since the development of sensitive analytical chemistry. One valuable use of bioassays is to determine the effective concentration of chemicals with a common mechanism of action. For example, the H4IIE bioassay provides a toxicity-normalized measure of the amount of chlorinated diaromatic hydrocarbons in the food of an organism (Tillitt et al. 1991; Giesy et al. 1994b). This use is analogous to the use of biomarkers to estimate internal exposure (Section 20.9), except that the goal is to estimate external response-normalized concentrations.

It has also been proposed that activity of contaminant-degrading microbes be used as a bioassay for bioavailable contaminant constituents (Alexander et al. 1995). A weak interpretation of this bioassay is that, if biodegradation has stopped, there is no more bioavailable chemical to cause toxicity. This conclusion requires the assumption that biodegradation has stopped because the residue is unavailable rather than because it is resistant to biodegradation. A stronger interpretation would be that bioavailable concentration is a function of biodegradation rate so that one could estimate exposure from measures of degradation. This idea requires the assumption that the availability of a chemical for degradation by microbes is proportional to availability for uptake by endpoint plants and animals. The use of microbial toxicity tests as measures of bioavailability or ecological effects is beyond the current state of practice.

Bioassays may be used more generally to screen contaminated media for toxicity (Loibner et al. 2003). That is, toxic responses may be used in place of chemical analyses in screening

---

BOX 20.2
Rules for Sampling Biota

- Take enough samples to adequately represent the variability at the site.
- Sample endpoint taxa for which internal measures of exposure are useful.
- Sample organisms or parts of organisms that represent the food of assessment endpoint species.
- Take samples of biota and contaminated media at the same locations and at effectively the same time.
- Take samples at reference and contaminated locations or on contamination gradients.
- Take samples from all sites at approximately the same time because chemical concentrations in organisms may vary seasonally.
- Be aware of the information that is lost when samples are composited.

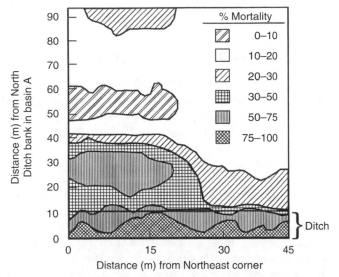

FIGURE 20.3 A map derived by Kriging of percentage of lettuce seed mortality in surface soil at the Rocky Mountain Arsenal. (From Thomas, J.M., Skalski, J.R., Cline, J.F., McShane, M.C., Simpson, J.C., Miller, W.E., Peterson, S.A., Callahan, C.A., and Greene, J.C., *Environ. Toxicol. Chem.*, 5, 487, 1986. With permission.)

assessments to discriminate areas that are significantly contaminated and require more study from areas that are not toxic and may be ignored. Bioassays used for this purpose must be sufficiently sensitive, but need not bear a defined relationship to the assessment endpoints. For example, tests of soil extracts using *Daphnia* or *Ceriodaphnia* spp. (sensitive organisms to many chemicals) may be good screening bioassays, but would not, in general, be accepted as predictors of effects on terrestrial organisms. However, tests that serve to estimate risks to assessment endpoints may further serve to define areas that require remediation without identification of the toxic chemicals (Thomas et al. 1986) (Figure 20.3). For example, it might be agreed in the problem formulation that any sediment that is acutely lethal to an amphipod would be dredged without further characterization or assessment, sediments with sublethal effects would be further characterized, but sediments with no effects would not be considered further. Tests used for this purpose must be sufficiently sensitive to the contaminants, reliable, robust to characteristics of the site media, and inexpensive relative to a full suite of chemical analyses.

20.11 BIOSURVEYS

Surveys of organisms (biosurveys) are used in ecological epidemiology to identify impaired communities (Chapter 4) and in ecological risk assessments primarily as a means of determining effects of contaminants (Chapter 25). However, they may also play a role in the analysis of exposure. Specifically, they can be used to determine whether a species or taxon is present in contaminated areas, what life stages are present, their abundance, and how long a migratory or otherwise transient species is present on the site. Without biological surveys, these presence and abundance parameters must be estimated using habitat models or assumptions. Biosurveys of contaminated sites provide estimates of these parameters for the current condition. Biosurveys of uncontaminated reference areas can provide estimates of these parameters for precontamination or postrestoration scenarios. Biosurveys may be conducted in conjunction with the collection of organisms for chemical analysis (Section 20.9), but care must be taken to ensure that sampling designs are adequate for both purposes.

20.12   SAMPLING, ANALYSIS, AND PROBABILITIES

In risk assessments, it is necessary to consider the distribution of the exposure metrics with respect to variability and uncertainty (Chapter 5). For example, if a data set consists of a set of cadmium concentrations in water, they may be time-averaged or instantaneous concentrations, at one or several points, at one or more times, which may extend over a season or a year, etc. Spatial or temporal variability may dominate the distribution of concentrations, or there may be significant uncertainty due to lack of information about the form of the measured cadmium. Distributions derived from these data need to be created and interpreted in such a way as to correspond to the distributions needed to estimate exposure of the endpoint receptors.

The most common probabilistic treatment of these data is to fit a distribution to the individual observations. However, these distributions often do not make sense as expressions of exposure. Assessors must determine the appropriate temporal and spatial units of exposure and how they are distributed given the definition of risk to the endpoint in question.

Concentrations in water may be treated as spatially constant (i.e., a fish community is assumed to occupy a pond or stream reach), and the critical variable is time. For effects of observed pulse exposures such as those that occur during storm events which flush pollutants into the system or failures of treatment or containment systems, the distribution of aqueous concentrations with respect to duration of the event should be derived. Choosing an appropriate duration for chronic exposures is more difficult. A default value is 7 days, the duration of the standard subchronic toxicity test developed by the EPA and employed at wastewater discharges and contaminated sites. The time period is based on the time required for most chemicals to induce effects on survival and growth of larval fish and survival and reproduction of planktonic crustaceans. Some chemicals induce effects on those organisms much more quickly, and some organisms, particularly larger ones, respond much more slowly. Ideally, the durations should be set at the time required to induce the endpoint response. The selection of that interval requires a careful study and analysis of the original toxicological literature. Once the duration has been chosen, the appropriate measure of variability is the variance in the x day moving average concentration, where x is the duration of the episodic exposure or the time required to induce chronic effects in routine exposures.

An additional concern is lack of correspondence between the measures of exposure and the appropriate exposure metric for the risk model. While it is generally recognized that measures of effects must be extrapolated, exposure extrapolations are less often acknowledged. Examples from ecological risk assessment include the following:

- Estimation of forage fish contamination from game fish analyses
- Estimation of whole fish concentrations from fillets
- Estimation of dissolved concentrations from whole water concentrations
- Estimation of undetected concentrations from limits of detection
- Estimation of pore water concentration from sediment concentration
- Estimation of annual average concentration from summer samples

In some cases, data and techniques are available to estimate the uncertainties associated with these extrapolations. Examples include maximum likelihood estimators for concentrations below detection limits and statistical models of the fillet to whole fish extrapolation (Bevelhimer et al. 1996). In other cases, expert judgment must be employed.

20.13   CONCLUSIONS

Sampling and analysis methods for contaminated environmental media are well developed and documented and routinely applied. However, their results are often less than optimal for use in ecological risk assessments. Ecological assessors need to be more involved in the design of sampling and analysis activities (Chapter 18) and must be insistent that their needs be met.

21 Mathematical Models of Chemical Transport and Fate

Donald Mackay and Neil Mackay

21.1 OBJECTIVES

The objective of this chapter is to describe the role that mathematical models can play in assessing the risks posed by chemicals in the environment. To fully appreciate this role, the ecotoxicologist should understand how these models are formulated, their strengths, weaknesses, and areas of applicability. This chapter first describes the process of model formulation and includes some basic concepts of environmental chemistry. This is followed by an example of a model of chemical fate in an aquatic system. Finally, available models and their application in North America and Europe in assessing risk in an ecological context are described.

The primary aim of the chemical fate and transport modeler is to gather available data on chemical properties, environmental conditions, and rates or quantities of chemical discharges, and then synthesize these data into a comprehensive statement of chemical fate, often as a mass balance. This statement enables the modeler to estimate the amounts and concentrations of a chemical that will be present in each part of the environment, the rates at which the chemical is degraded and transported from place to place, and therefore how long the chemical will reside in the environment in question. Ideally, these concentrations should be compared with available monitoring data to provide assurance that the model is "valid." The calculation, estimation, or prediction of the likely concentrations in various compartments requires a knowledge of the physical–chemical properties of the substance, notably partition coefficients between environmental media and especially parameters such as chemical reactivity, in the form of reaction rate constants or half-lives. It also requires information on prevailing environmental conditions such as hydrology, soil characteristics, and atmospheric processes.

It is important to appreciate that although concentrations in media such as air, water, and fish can be measured, it is usually not possible to measure *fluxes* such as rates of evaporation from water to air or uptake or metabolic degradation in organisms. As a result, monitoring provides only a partial, static "snapshot" of a dynamic system. The modeler's aim is to establish a more complete and quantitative picture of the dynamic behavior of the contaminant in the system. When the chemical's dominant environmental pathways and fate processes are understood quantitatively, a much clearer picture emerges of the behavior of the chemical and the nature of the contamination issue. This information can help to assess the risk of adverse effects, and evaluate actions that may be taken to reduce concentrations, exposures, and effects.

21.2 BASIC MODELING CONCEPTS

The assembly of this quantitative description is frequently referred to as the development of a "mass balance" model. The fundamental concept is illustrated in Figure 21.1 in which a volume in space of the environment is identified as a compartment, and a mass or material balance equation is written for chemical entering and leaving this volume. The volume may be water in a section of a river, a region of the atmosphere, or soil to a depth of 20 cm. It may even be an organism. It must have defined physical boundaries, and preferably be fairly homogenous in conditions within the phase envelope. The mass balance equation states that the inventory change of chemical quantity in the phase envelope (in kg/h) will equal the sum of the input rates to the phase envelope (in kg/h) less the sum of the output rates. The input terms may include flow in air and water, direct discharges, or formation from other chemical compounds. The outputs may include flows such as water from a lake, degrading reactions, and diffusion to other compartments. Antoine Lavoisier, the "Father of Chemistry," is credited with first enunciating this mass balance principle. It is no coincidence that he was also an accountant, as the same principle applies to money entering, leaving, and accumulating in a bank balance. It is now taken as axiomatic that the mass balance equation applies; thus the modeler's task is essentially to develop expressions, equations, or quantities for each term in the mass balance equation.

When assembling this mass balance, the modeler faces a number of decisions that require the exercise of judgment. The "art" of modeling can be viewed as the ability to make the "right" decisions when faced with alternative approaches or strategies. In particular, three aspects of the environmental system being simulated must be recognized and appropriately treated. These involve the nature of the emissions or "loadings" of the chemical and the scale of the system to be addressed.

21.2.1 EMISSIONS OR LOADINGS

Chemicals enter the environment as a result of a wide variety of human activities and in a large number of ways. Some chemicals such as pesticides and biocides are intentionally introduced into the environment. Others are introduced as a result of disposal following

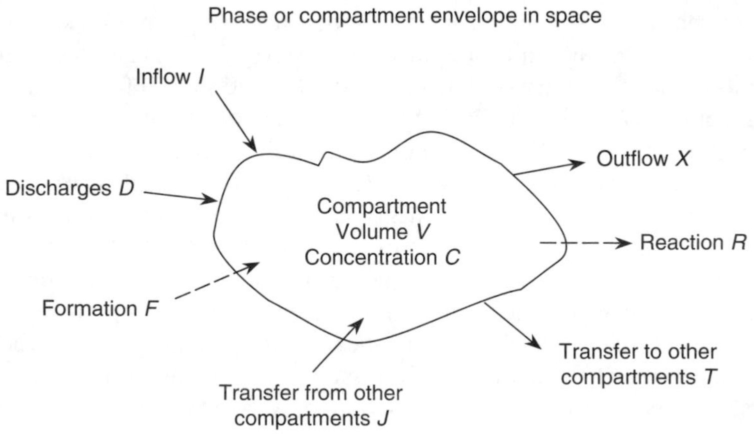

Phase or compartment envelope in space

Inflow I

Discharges D

Compartment Volume V Concentration C

Outflow X

Reaction R

Formation F

Transfer to other compartments T

Transfer from other compartments J

Inventory change = Inputs − Outputs

$$V\,dC/dt = I + D + F + J - X - R - T \text{ kg/h}$$

At steady state when $dC/dt = 0$

$$I + D + F + J = X + R + T \text{ kg/h}$$

FIGURE 21.1 The fundamental concept of a phase envelope or compartment and the corresponding dynamic and steady-state mass balance equations.

usage or production, e.g., pharmaceuticals, veterinary medicines, solvents, surfactants, fuels, and lubricants. Some are inadvertently formed during processes such as combustion, e.g., "dioxins" and polycylic aromatic hydrocarbons. This route or mode of entry into a given environment dictates the most appropriate methods for simulating its environmental fate and subsequent exposure to nontarget organisms. Two characteristics of the emissions are of importance:

1. Does the emission originate from a *point* or *nonpoint* (diffuse) source?
2. Is the emission *steady state* (continuous) or *non-steady-state* (pulse)?

21.2.2 POINT AND NONPOINT SOURCES

As the term suggests, point source loadings are spatially discrete and generally occur as a result of localized human activities such as disposal or accidental releases. In many cases the magnitude of point source loadings is determined by design specifications or can be measured (Chapter 19). In contrast, loadings from nonpoint sources are generally the result of more diffuse processes that occur throughout the environment such as volatilization from soil or water into air, or leaching to groundwater. The magnitude of these loadings into the environment is less easily defined because of dependence upon environmental processes that may themselves require modeling.

21.2.3 STEADY-STATE AND NON-STEADY-STATE SOURCES

Steady-state loadings are (or for pragmatic reasons are considered to be) continuous and constant in rate, whereas non-steady-state loadings are temporally discrete and occur sporadically over time as pulses. Steady-state loadings are often associated with human activities such as continuous production or discharge from industrial or municipal wastewater treatment plants. Examples of non-steady-state loadings include pesticide application or accidental chemical or fuel spills. Depending upon the complexity of the processes involved the scales of loadings can either be directly estimated from observations as in the case of chemical spills, or are themselves simulated, e.g., runoff following pesticide application.

21.2.4 IMPORTANCE OF SCALE

Determining the appropriate spatial and temporal scale is a fundamental component of any exposure and risk assessment (Chapter 6). As models are designed to support exposure and risk assessments at a variety of scales, the selection of an appropriate modeling tool for the task at hand is a vital first step. This is most often dictated by the two considerations discussed earlier that define how discrete loadings are in time and space. The scale at which we view the environment can make these decisions more subtle than one might imagine. Consider, for example, discharges of domestic surfactants from wastewater treatment plants. At the scale of an individual "reach" or "segment" of a river or water body the loadings would be considered point sources. Such loadings may also be investigated at a catchment or watershed scale along with loadings from other wastewater treatment plants. Depending upon the size of the catchment and the number of wastewater treatment plants, these loadings may be considered point sources or, more pragmatically, diffuse (nonpoint) sources. At a regional scale of a state or country involving several catchments because of the ubiquity of such discharges, these loadings are more likely to be considered diffuse sources and simulated accordingly. Examples of modeling approaches that illustrate the importance of scale are illustrated in Table 21.1.

TABLE 21.1

Examples of Modeling Approaches Applied at Different Scales when Simulating Discharges from Wastewater Treatment Plants

Scale of Assessment	Illustrative Modeling Approach	Simulation as Point or Nonpoint Source
Treatment plant	STP (Clark et al. 1995)	Point source
Waterbody "reach" or "segment" (100s of m)	EXAMS (Burns 2002)	Point source
Catchment or watershed (1 km^2 to 1000s of km^2)	GREAT-ER (GREAT-ER Task Force 1997)	Point source
Large lake (e.g., Ontario)	QWASI (Mackay 1989)	Point and nonpoint sources
Regional (1000s of km^2 to 100,000s of km^2)	EUSES (EUSES 1997)	Nonpoint sources

21.3 FORMULATING MASS BALANCE MODELS

There are five steps in formulating a model of the type illustrated in Figure 21.1: defining compartments, quantifying reaction rates, quantifying transport rates, acquiring emission data, and solving the equations describing the overall mass balance to obtain estimates of all chemical concentrations, masses, and fluxes.

21.3.1 DEFINING COMPARTMENTS

Step 1 definines the phase boundaries, envelopes, or compartments. Ideally the volume contained in each compartment or "box" should be well mixed and thus have a fairly homogeneous concentration. The mass of chemical in the compartment can then be expressed simply as the product of this concentration and the compartment volume. Examples are shallow pond water or surface sediments. If a more detailed and complex model is desired, it may be necessary to segment the water column into two: a surface layer and a deep layer. It may be desirable to treat particles in the water column as a separate phase, or they may be lumped into the water column. Obviously the greater the number of compartments, the greater will be the potential fidelity of the model to reality, but as the mathematics becomes more complex, more input data are needed, and the model tends to become more difficult to understand. If it is too complex, it is less likely to be widely used and it may suffer a lack of credibility. Figure 21.2 is a relatively simple, steady-state seven-compartment model of air, aerosols, water, fish, soil, bottom sediments, and suspended sediments, with a chemical (naphthalene) introduced into three compartments. Describing the processes that apply to the chemical as it enters and leaves each of these compartments requires estimations of the rates of 15 processes including rates of sedimentation, runoff from soil, evaporation, reaction, and advective flows.

21.3.2 REACTION RATES

Step 2 defines the various reaction rates experienced by organic chemicals through processes such as biodegradation, hydrolysis, and photolysis. For metals, degradation does not occur, although there may be speciation changes such as ferric to ferrous ion. This may be difficult because temperatures and other conditions such as the density and nature of microbial populations, sunlight intensity, and pH may vary diurnally and seasonally. It is not always clear what average conditions should be used. In some cases a distribution of values may be

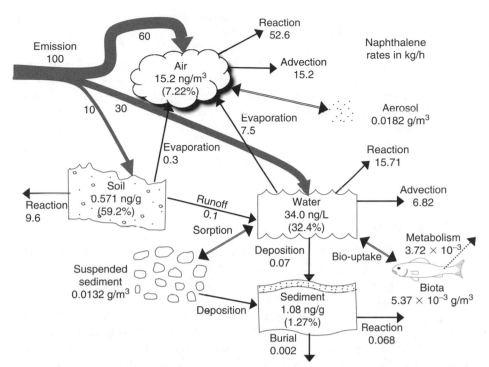

FIGURE 21.2 Illustrative mass balance diagram for the fate of naphthalene in a seven-compartment steady state Level III system. The concentrations and fluxes are obtained by solving the algebraic equations. Double arrows depict assumed equilibrium between compartments.

used. The most scientifically reliable and rigorous data are obtained from bench-scale experiments involving exposure of a chemical in a vessel to a known reactive environment under highly controlled conditions. The challenge is then to translate this physical–chemical information to environmental conditions. Experiments may also be conducted in larger microcosms or mesocosms in which environmental conditions are more closely simulated, but this results in some loss of control over the variables that determine the reaction rate. There is also a loss of reproducibility. It can be difficult to discriminate between the contributions of individual processes such as hydrolysis, evaporation, and biodegradation when all occur simultaneously.

Organic chemicals are usually subject to several competing degradation processes caused by reactive agents in the environment. These agents can be oxidants such as ozone or hydroxyl radicals, sunlight causing photolysis, both directly or indirectly, ionic species such as hydrogen or hydroxyl ions, and microorganisms including bacteria and fungi. Within an organism there may be metabolic conversion. The rate of reaction N is generally expressed as

$$\text{Rate} = N = VCxk_2 \text{ g/h}$$

where V is volume, C is concentration of the chemical, x is concentration of the reactive agent, and k_2 is a second-order rate constant. If environmental conditions are fairly constant, x will also be fairly constant and the terms x and k_2 can be combined as k, a first-order rate constant with units of reciprocal time (e.g., /h), and

$$N = VCk \text{ g/h}$$

The rate constant k is then the fraction of the chemical present that reacts during each unit of time. For example 0.01/h implies that 1% reacts each hour. It is more convenient to view these rate constants as half-lives, i.e., the time required for the amount of concentration to fall to half. This time $t_{\frac{1}{2}}$ can be shown to be $0.693/k$ where 0.693 is the natural logarithm of 2.0. Information on k or $t_{\frac{1}{2}}$ is obtained from laboratory experiments, from estimation methods often referred to as quantitative structure–activity relationships (QSARs) or from suitably interpreted field observations. Boethling and Mackay (2000) review many of these QSARs. Of particular importance are biodegradation rates in water and soil and reactions of the chemical with hydroxyl radicals in the atmosphere. These are often the principal mechanisms by which chemicals degrade. The persistence or longevity of the chemical is largely determined by these rates.

21.3.3 TRANSPORT RATES

Step 3 defines transport rates between the various media. A contaminant may be introduced into a medium such as soil, where it is relatively stable, but when conveyed by evaporation into another medium such as air, it can become subject to a fast degrading reaction. The chemical's lifetime in the environment may then be controlled, not by how fast it reacts in soil, but by how fast it evaporates from the soil to the reactive atmosphere. In the early days of environmental science the prevailing view was that most chemicals tended to remain in the medium into which they were discharged. For example, DDT applied to a soil would remain there indefinitely, subject only to soil degradation processes. It is now clear that most chemicals have the capability of migrating between all environmental media; thus application of DDT to soil will result in appreciable concentrations in the atmosphere that is exposed to that soil. Transport of this contaminated air will result in the contamination of distant areas, as in the case of DDT contamination of Arctic and Antarctic mammals, birds, and fish. Indeed, it was Rachel Carson in *Silent Spring* who first raised concerns about exposure of nontarget organisms to pesticide chemicals as a result of these intermedia transport processes (Carson 1962).

Transport processes can be classified into two general groups. The first is the advective or nondiffusive process (Figure 21.3), in which a chemical is conveyed from one medium to another by virtue of "piggybacking" on "carrier" material, i.e., moving between media for reasons unrelated to the presence of the chemical. Examples are deposition of chemical from the water column to sediment by attachment to depositing particles, and transport from the atmosphere to the soil or water in rainfall, in dust fall, or in wet deposition (dust scavenged from the atmosphere by rain). Ingestion of contaminated food by an organism is also a process of this type. These processes are readily quantified by multiplying the concentration of chemical in the migrating medium by the flow or transport rate of the medium. For example, if rain contains 10 ng/L of polychlorinated biphenyl (PCB) and is falling at a rate of 0.5 m/y, i.e., 0.5 m^3/m^2/y, on a region of 5 million m^2, the rain rate will be 2.5×10^6 m^3/y and the transport rate of PCBs will be a product of this rate and the concentration of 10 ng/L (expressed equivalently as 10 µg or 10×10^{-6} g/m^3). The PCB deposition rate will therefore be $2.5 \times 10^6 \times 10 \times 10^{-6}$ or 25 g/y. A calculation similar in principle to this can be done for other intermedia advective processes, including flow in water, air, and on solids, as well as ingestion in food by organisms and uptake by inhalation.

The second group is diffusive in nature and the chemical migrates between media because it is in a state of disequilibrium. For example, considering the two-compartment environment also depicted in Figure 21.3, benzene is present in water at a concentration of 1 mg/L and in air at 0.1 mg/L. The physical–chemical properties of benzene, specifically its air/water

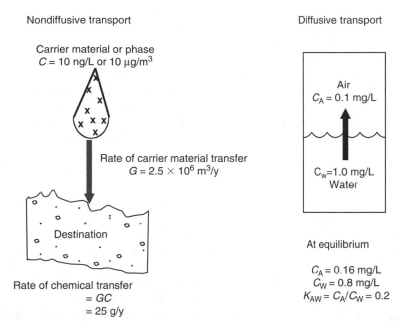

Nondiffusive transport

Carrier material or phase
$C = 10$ ng/L or 10 µg/m^3

Rate of carrier material transfer
$G = 2.5 \times 10^6$ m^3/y

Destination

Rate of chemical transfer
$= GC$
$= 25$ g/y

Diffusive transport

Air
$C_A = 0.1$ mg/L

$C_W = 1.0$ mg/L
Water

At equilibrium

$C_A = 0.16$ mg/L
$C_W = 0.8$ mg/L
$K_{AW} = C_A/C_W = 0.2$

FIGURE 21.3 Intermedia transport processes.

partition coefficient or Henry's law constant, dictate its concentrations when it is in equilibrium. This quantity is determined by benzene's solubility in water and vapor pressure. At equilibrium the ratio of air to water concentrations or the air/water partition coefficient is approximately 0.2 at the prevailing temperature. In this case the air is "undersaturated" with respect to the water. Benzene will thus diffuse from water to air until it establishes concentrations of perhaps 0.8 mg/L in water and 0.16 mg/L in air. From a physical–chemical viewpoint the chemical is striving to achieve a condition of thermodynamic equilibrium, or one in which the benzene has an equal chemical potential or fugacity in both media. It may or may not have time to achieve this equilibrium because of other input and output processes to or from the system, but the direction of diffusive transport is at least clear, and the rate of transport can be calculated.

It must be appreciated that this diffusion process is basically a manifestation of random mixing of benzene throughout the entire system. The benzene in the water does not "know" the concentration in the air, but the rate of loss of benzene from the water compartment to the air is proportional to the concentration of benzene in the water. Thus, that rate of loss will exceed the rate of gain from the air until such time as air and water concentrations adjust to give equal rates of upward and downward transport. It is common to write the equation for net diffusive transport in the following form:

$$N = k_M A(C_1 - C_2 K_{12})$$

where N is the transport rate (g/h), k_M is the transport rate coefficient (m/h) A is the area between the media (m^2), C_1 and C_2, are the concentrations in the two media (g/m^3), and K_{12}, the partition coefficient. At equilibrium, when there is no net transport, K_{12} equals C_1/C_2 or C_1 equals $C_2 K_{12}$ and the term $(C_1 - C_2 K_{12})$ thus becomes zero. This term is usually referred to as the "driving force" for diffusion or the "departure from equilibrium." Boethling and Mackay (2000) review these partition coefficients and recommend estimation methods.

These diffusive rate calculations can be applied to other combinations of media such as soil and air or water and sediment. The three key quantities are obviously: (i) the intermedia partition coefficients, e.g., K_{12}, which are thermodynamic quantities expressing the equilibrium condition and thus the extent of departure from equilibrium; (ii) a transport rate parameter analogous to k_M, which has the dimensions of velocity and can be viewed as a velocity with which chemical moves from one medium to the other; and (iii) the intermedia area A. A variety of methods are available for measuring, correlating, estimating, and predicting these intermedia transport coefficients. Full accounts are given by Thibodeaux (1996), Mackay (2001), and Schwarzenbach et al. (1993). The quantities involved are generally mass transfer coefficients, molecular diffusion coefficients, diffusion path lengths, or the thickness of boundary layers. Methods for estimating many of these qualities are conveniently reviewed by Lyman et al. (1982). The general experimental approach is to make measurements of intermedia transport rates, N, in controlled conditions in a laboratory in which A, C_1, C_2, and K_{12} are known, and then deduce k_M. It is then correlated with conditions such as turbulence in the atmosphere or water as reflected by wind speed or water current velocity. As it is difficult to make these measurements in the environment, there is often uncertainty that laboratory-to-environment extrapolation is valid.

In some cases the diffusing chemical may pass through two or more layers during its migration from one medium to the other. For example, benzene evaporating from water to the atmosphere must diffuse through near-stagnant boundary layers in the water and in the air. These diffusion processes occur in series and require application of the "two-resistance-in-series" concept, often referred to as the Whitman Two Film or Two Resistance Theory.

21.3.4 EMISSIONS

Step 4 defines the various discharges to the system, which may be from industrial or municipal sources, spills, deliberate applications of chemicals such as pesticides, leaching from groundwater, or use or disposal of consumer products. The total discharge rate, which is also referred to as the "loading rate," is often difficult to determine, at least for large regions in which there are multiple point and nonpoint sources, including deposition from the atmosphere and river flows. Some discharges such as pesticide applications must be treated as dynamic in time and space. The discharge or source term is critically important because it is the magnitude and distribution of the discharge that "drives" the model, establishes the magnitude of the concentrations, and therefore determines the exposure and ultimately the risk of adverse effects.

21.3.5 SOLUTIONS TO THE MASS BALANCE EQUATION

The modeler thus draws on information from a wide variety of sources when estimating rates of emission, transport, and transformation. For each loss process from a compartment, the rate can be expressed directly or as a function of chemical concentration, specifically the product of an estimated rate constant k and C, namely kC. Having defined each rate expression in the mass balance equation in Figure 21.1, the remaining task is to invoke the mass balance principle and solve the equations to obtain the desired concentrations. If there are n compartments there will be n unknown concentrations and n mass balance equations. Solution is then possible. The modeler now has several options.

A simple and valuable approach is to examine the constant or steady-state condition of the system. Often, environmental conditions are close to steady state as a result of long-term, continuous inputs of chemical. The inventory change terms on the left of the equation in Figure 21.1 are set to zero and the mass balance equations take the form of one or more

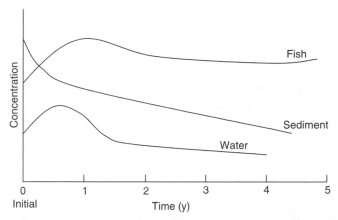

FIGURE 21.4 Illustrative time course of concentration changes in a lake subject to an increase then decrease in chemical input. The curves are obtained by solving the differential equations.

algebraic equations. These equations then express the concentrations that would prevail in the system if the input rate remained constant for a long period of time. Mathematical solution is accomplished by writing the set of algebraic equations and solving either by hand or matrix techniques. This solution, although not necessarily representative of actual conditions, is useful in that it demonstrates which processes and parameters are likely to be most important in determining the chemical's fate, concentration, and hence exposure. The modeler can then strive to establish more accurate values for the critical input parameters. For example, the reaction rate may prove to be so slow that it is relatively unimportant compared to a transport rate by evaporation. Further effort is then justified to obtain a more accurate evaporation transport rate. The quantities in Figure 21.2 were obtained in this way, and show that the important processes are reactions in air, water, and soil, advection from air and water, and evaporation from water.

The second, more rigorous but more demanding approach is to solve the set of differential equations as shown in Figure 21.1. For simple systems, analytical solutions are available, but often numerical integration techniques are used in which the initial or boundary conditions are first established, the change in inventory then deduced for a specified time increment or step, and the new set of concentrations for the system calculated. The equation is applied repeatedly to the new set of conditions. Care is necessary when selecting the time step and method of integration. The output from such a calculation as illustrated in Figure 21.4 is actually the time course of concentration changes in the system as a result of inputs which may be constant or may change with time. The disadvantage of this type of solution is that results are not easily generalizable to other conditions, as each solution is specific to the initial conditions selected, and of course, to the rates of chemical input.

21.3.6 COMPLEXITY, VALIDITY, AND CONFIDENCE LIMITS

Modelers strive to obtain models that describe real systems with high fidelity and provide accurate predictions of prevailing concentrations. This can result in complicated models of chemical migration in highly segmented or detailed multimedia environments involving hundreds of transport and transformation processes, and hence input parameters. So many parameters and assumptions may be built into the model that potential users become suspicious that there may be some hidden sensitivity, error, or even a mistake. Mathematically, the problem is that the "fit" involves a multitude of adjustable and poorly quantified parameters; thus the agreement with reality may be largely fortuitous. It may not be obvious which are the most sensitive, fate-controlling parameters. The key processes may be controlled by one or a few expressions or parameters that contain considerable error. Extrapolations to other chemicals or conditions may be unreliable. Frequently, the total rate of loss or "dissipation" of chemicals such as a pesticide is known fairly accurately from field data, and can be modeled accurately, but if it is the sum of several contributing processes such as volatilization and biodegradation, the individual components of this total may not be known accurately. It is important that field tests be designed to give the required process rates and that the model adequately expresses the rates of individual losses as well as the total rate of loss. A strong case can be made that the model should be only as complex as necessary, and should be sufficiently transparent that it is capable of being checked for intuitive reasonableness especially if it is to be widely used by non-modelers. (Chapter 9).

To an experienced environmental scientist the results should be in accord with what is expected from the chemical in question in the specified environment. Often the most useful aspect of the model is that it sorts the multitude of disparate processes into order of importance, and identifies the key processes. A very simple "back of the envelope" model can then be assembled containing only those key processes that can give a satisfactory, approximate simulation.

It is obviously desirable to compare the model results with reality in the form of monitored concentrations, often as a plot of predicted vs. observed values. Some adjustment of parameters may be needed to improve the fit. Some refer to this process of confirming the model to be correct as "validation." Others view "validation" as being too strong a word, as conditions can always be found in which the model is invalid. "Reconciliation", "corroboration"or "evaluation" may be better words. It is important that the modeler expresses all limitations of the model, e.g., it may be inapplicable to ionizing chemicals and it may be invalid at subfreezing temperatures. An expression of confidence limits is also desirable, for example, in the form that 95% of the estimates will be within a factor of 2.5. Such confidence limits or error estimates become very significant when calculated concentrations are to be compared with environmental standards or levels of toxicological concern when assessing risk.

Having set the stage, we now illustrate these concepts by developing and discussing a simple, illustrative model containing a variety of processes.

21.4 ILLUSTRATION OF A SIMPLE MASS BALANCE MODEL

21.4.1 THE SYSTEM BEING MODELED

To illustrate these concepts we treat a simple one-compartment aquatic model. The aim is to establish a mass balance for a specimen chemical in a lake subjected to a continuous direct discharge of chemical. The lake is 10 m deep and has an area of 10^6 m^2, the volume of water then being 10^7 m^3. Water flows into, and out of, this lake at a rate of 1000 m^3/h, the residence time of the water thus being 10,000 h or approximately 14 months. There is inflow of 75 kg/h or 50 L/h of suspended sediment (density 1.5 kg/L), of which 45 kg/h or 30 L/h is deposited on the bottom, and the remaining 30 kg/h or 20 L/h flows out of the system. The bottom (which we ignore here) consists of sediment. Chemical can react in the water column with a half-life of 289 d, i.e. a rate constant of 10^{-4}/h. It evaporates with an overall water–side mass transfer coefficient or diffusion velocity of 0.001 m/h. The chemical (molar mass 100 g/mol) has an air–water equilibrium partition coefficient K_{AW} of 0.01, a dimensionless particle–water partition coefficient K_{PW} of 5450, and a biota–water partition coefficient K_{BW} or bioconcentration factor of 5000. It is noteworthy that partition coefficients can be expressed as above in dimensionless form, e.g. (g/m^3 particle)/(g/m^3 water) or equivalently mg/L in both phases. Often the ratio is expressed as (mg/kg particle)/(mg/L water) and thus has units of L/kg. To convert from this dimensional form to the dimensionless form requires multiplication by the particle density. K_{PW} is thus 5450 (dimensionless) or 3633 L/kg. For biotic phases the density is usually 1.0 kg/L, i.e., equal to that of water; thus the values are numerically equal. The concentration of particles is 30 mg/L or 20 cm^3/m^3 or 20 ppm by volume and that of biota (including fish) is 5 ppm, again by volume. There is a constant chemical discharge of 40 g/h. The chemical is also present in the inflowing water at a concentration of 0.01 mg/L.

The aim of the model is to calculate the steady-state or constant concentration in the system of water, particles, and fish, and all the loss rates. We can undertake this calculation in two equivalent ways: first, a conventional concentration calculation, and second, a fugacity calculation.

21.4.2 CONCENTRATION CALCULATION

We set the total concentration of chemical in the water (including chemical present in particles and fish) as an unknown, C_W g/m^3. The various process rates are calculated in terms of C_W, summed, and equated to the total input rate, then solved for C_W, and finally the various process rates deduced. We use units of g/h for the mass balance calculation. The unknown could also be the dissolved concentration, rather than the total concentration.

21.4.2.1 Chemical Input Rate

The discharge rate is 40 g/h. The inflow rate is the product of 1000 m^3/h and 0.01 g/m^3 (i.e. 0.01 mg/L) or 10 g/h. The total input rate is thus 50 g/h.

21.4.2.2 Partitioning between Water, Particles, and Fish

The total amount in the water is $V_W C_W$, where V_W is the volume of water (10^7 m^3). But this contains 20 ppm by volume of particles, i.e. a volume of $20 \times 10^{-6} \times 10^7$ or 200 m^3, and similarly 5 ppm of biota or 50 m^3. The total quantity in the water is the sum of the quantities dissolved, in particles and in biota, namely:

$$V_W C_W = V_W C_O + V_P C_P + V_B C_B$$

The concentrations C_O, C_P, and C_B refer to dissolved, particulate, and biotic forms, respectively. The volumes are known, the concentration C_P is $K_{PW} C_O$ or 5450 C_O and C_B is $K_{BW} C_O$ or 5000 C_O. Substituting gives:

$$10^7 C_W = 10^7 C_O + 200 \times 5450 C_O + 50 \times 5000 C_O$$

$$10^7 C_W = C_O(10 + 1.09 + 0.25) \times 10^6 = 11.34 \times 10^6 C_O$$

It follows that C_O is $0.882 C_W$, C_P is $4800 C_W$, and C_B is $4400 C_W$, with 88.2% of the chemical being dissolved, 9.6% sorbed to particles, and 2.2% bioconcentrated in biota. Note that we used dimensionless partition coefficients K_{PW} and K_{BW}, i.e., ratios of (mg/L)/(mg/L). If the dimensional form is used, the volume of 200 m^3 is replaced by the mass of 300,000 kg and K_{PW} is 3633 L/kg; thus the amount is 300,000 $\times$ 3633 C_O/1000 or again 1.09×10^6 g. The 1000 converts L to m^3.

21.4.2.3   Outflow in Water

As the outflow rate is 1000 m^3/h, the outflow rate of dissolved chemical must be $1000C_O$ g/h or $882C_W$ g/h.

21.4.2.4   Outflow in Particles

In addition, there is outflow of 20 L/h (0.02 m^3/h) of particles containing $5450C_O$ g/m^3 of sorbed chemical, namely $109C_O$ g/h or $96C_W$ g/h, as C_O is $0.882C_W$. We assume that biota remain in the lake, but their outflow could be included if desired.

21.4.2.5   Reaction

The reaction rate is the product of water volume, concentration, and rate constant: $10^7C_W \times 10^{-4}$ or $1000C_W$ g/h. Note that we assume all chemicals in the system to be subject to the reaction. Different rate constants could be defined for dissolved, sorbed, and biotic forms.

21.4.2.6   Deposition to Sediment

The concentration on the particles is $5450C_O$ or $4800C_W$ g/m^3 particle. As the particle deposition rate is 30 L/h or 0.03 m^3/h, the chemical deposition rate will be $4800 \times 0.03C_W$ or $144C_W$ g/h.

21.4.2.7   Evaporation

The evaporation rate is the product of the mass transfer coefficient (0.001 m/h), the water area (10^6 m^2), and the concentration dissolved in water: $0.001 \times 10^6 \times 0.882C_W$ or $882C_W$. We assume that the air contains no chemical to create a "back pressure" causing diffusion from air to water. If this were the case, it would be included as another input term.

21.4.2.8   Combined Loss Processes

Combining the process rates, we apply the mass balance principle and assert that at steady-state conditions the sum of these loss rates will equal the total input rate of 50 g/h, hence:

$$50 = 882C_W + 96C_W + 1000C_W + 144C_W + 882C_W = 3004C_W$$

thus:

$$C_W = 50/3004 = 0.0166 \text{ g/m}^3 \text{or mg/L}$$

The dissolved concentration is thus 0.0146 g/m^3, that on the particles is 5450 times this or 80 g/m^3 particle, and as the particle density is 1.5 g/cm^3, approximately 53 mg/kg. This is also 0.0016 g/m^3 water. The concentration in biota will be 73 g/m^3 or mg/kg biota and is equivalent to 0.0004 g/m^3 water.

The process rates are thus:

Outflow in water	14.7 g/h	29%
Outflow in particles	1.6 g/h	3%
Reaction	16.6 g/h	34%
Deposition	2.4 g/h	5%
Evaporation	14.7 g/h	29%

These values total to the input of 50 g/h, thus satisfying the mass balance. The relative importance of the loss processes is immediately obvious. Figure 21.5 depicts this mass balance. There is now a clear and complete picture of the chemical's fate including concentrations, masses, and fluxes. The model thus provides the additional information about fluxes, whereas a monitoring program can only give concentrations.

21.4.3 FUGACITY CALCULATION

An alternative approach, which yields the same answer, is to use fugacity as the descriptor of chemical quantity rather than concentration. In this approach, partitioning of chemical between phases is expressed in terms of the equilibrium criterion of fugacity. Fugacity f can be considered the partial pressure of the chemical with units of pressure (Pa) and is a surrogate for concentration. It is related to concentration C mol/m^3 through a fugacity capacity or Z value, i.e., C is equated to Zf. These Z values are calculated from estimated or measured partition coefficients or indirectly from a chemical's physical properties of molecular weight, vapor pressure, solubility, and octanol/water partition coefficient, as well as environmental properties such as density and fraction organic content of the phases present. Full details are given by Mackay (2001). In this case, definition of Z values for the specimen chemical starts in the air phase where Z_A is $1/RT$ or 4.1×10^{-4} mol/m^3 Pa where R is the gas constant (8.314 Pa m^3/mol K) and T is absolute temperature (298 K). As a partition coefficient K_{12} is the ratio of Z values Z_1/Z_2, the Z for water can be calculated as Z_A/K_{AW} or 4.1×10^{-2}, for particles Z_P as $K_{PW}Z_W$ or 223, and for fish Z_B is $K_{BW}Z_W$ or 205.

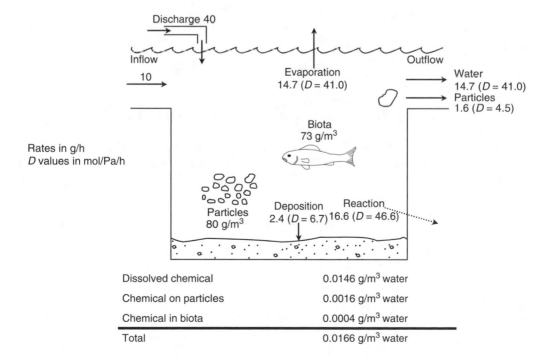

FIGURE 21.5 Steady-state mass balance of a chemical in a lake as described in the example in the text.

The total Z value for water, particles, and biota is the sum of these Z values, weighted in proportion of their volume fractions:

$$Z_{WT} = Z_W + 20 \times 10^{-6} Z_P + 5 \times 10^{-6} Z_B \text{ or } 4.65 \times 10^{-2} \text{mol/m}^3 \text{Pa}.$$

Diffusive transfer between air and water, air and soil, and water and sediment is again estimated from mass transfer coefficients, interfacial areas, diffusivities, and path length. Nondiffusive processes include air and water advection, wet and dry atmospheric deposition, leaching from groundwater, sediment deposition, burial and resuspension, and runoff from soil, and are estimated from flow rate, Z value, and fugacity. In each case the rate parameter is expressed as a "D value" such that the rate is the product of D and fugacity. Degradation in each compartment or subcompartment is included by means of a rate constant, again expressed as a D value.

The mass balance calculation can be repeated in fugacity format by first calculating Z values, then D values, then equating input and output rates as before. It is preferable to express rates in units of mol/h when doing fugacity calculations.

The D values (in units of mol/Pa·h) and fluxes (mol/h) are shown in Table 21.2.

The overall mass balance is then expressed in terms of the unknown fugacity of the chemical in the water f_W. The input rate of 50 g/h is also 0.5 mol/h. Equating inputs to outputs gives:

$$0.5 = f_W D_1 + f_W D_2 + f_W D_3 + f_W D_4 + f_W D_5 = f_W \, 139.8$$

thus $f_W = 3.58 \times 10^{-3}$ Pa

$$C_O = Z_W f_W = 1.46 \times 10^{-4} \text{ mol/m}^3 \text{ or } 0.0146 \text{ g/m}^3$$

$$C_P = Z_P f_W = 0.80 \text{ mol/m}^3 \text{ or } 80 \text{ g/m}^3$$

$$C_B = Z_B f_W = 0.73 \text{ mol/m}^3 \text{ or } 73 \text{ g/m}^3 \text{ or } 73 \text{ mg/kg}$$

The individual rates are Df and are identical to those calculated earlier. The D values give a useful and direct expression of the relative importance of each process as all are expressed in identical units.

TABLE 21.2
D Values and Fluxes

				Flux D_f (mol/h)
Outflow in water	$D_1 = G_w Z_w$	$= 1000 \times 4.1 \times 10^{-2}$	$= 41.0$	0.147
Outflow in particles	$D_2 = G_p Z_p$	$= 0.02 \times 223$	$= 4.5$	0.016
Reaction	$D_3 = V Z_{wt} K$	$= 10^7 \times 4.66 \times 10^{-2} \times 10^{-4}$	$= 46.6$	0.167
Deposition	$D_4 = G_D Z_p$	$= 0.03 \times 223$	$= 6.7$	0.024
Evaporation	$D_5 = K_M A Z_w$	$= 0.001 \times 10^6 \times 4.1 \times 10^{-2}$	$= 41.0$	0.147
		Total	$= 139.8$	0.500

21.4.4 DISCUSSION

The model provides invaluable information on the status and fate of the chemical in this lake system. Obviously it is desirable to measure concentrations in particles, water, and biota to determine if the model assertions are correct. If, as is likely, discrepancies are observed, the model assumptions can be reexamined to test if some process has not been included, or if the assumed parameters are correct in magnitude. When reconciliation is successful, the environmental scientist is in the satisfying position of being able to claim that the system is well understood. It is then possible to explore the effect of various changes in inputs to the system.

For example, if the discharge were eliminated, the water, particle, and biotic concentrations would eventually fall to one-fifth of the values calculated above because the input would now be 10 g/h instead of 50 g/h. The biota would now contain 14 mg/kg instead of 73 mg/kg. To achieve an ecologically based "target" of, for example, 5 mg/kg would require reduction of the input rate to about 3 g/h. The nature and magnitude of interventions necessary to achieve a desired environmental quality in the ecosystem can thus be estimated.

It may be judged that to protect the resident aquatic species from risk of adverse effects requires a dissolved concentration in water of less than 0.001 g/m^3 or mg/L. To achieve this target requires that the total inputs be reduced to 3.4 g/h.

Another issue is how long it will take for such measures to become effective. This requires solution of the differential mass balance equation, which in this case is

$$V_W dC_W/dt = \text{input rate} - \text{output rate} \quad \text{or} \quad V_W dC_W/dt = 50 - 3004 C_W$$

dividing by V_W of 10^7 m^3 gives

$$dC_W/dt = 50 \times 10^{-7} - 3004 \times 10^{-7} C_W = A - B C_W$$

If the initial water concentration is C_{WO} at time t of zero, this equation can be solved by separation of variables to give

$$C_W = C_{WF} - (C_{WF} - C_{WO}) \exp(-Bt)$$

where C_{WF} is the final value of C_W at t of infinity, and is A/B. C_W thus changes from C_{WO} to C_{WF} with a rate constant of B or a half-time of $0.693/B$ h. Here B is 3004×10^{-7}/h; thus the half-time is 2300 h or 96 d. The actual time course of the changing concentrations can be calculated.

In reality, the biota, especially fish, would respond more slowly because of the delay in biouptake or release, but the important finding is that within a year the system would be well on its way to a new steady-state condition. In fugacity terms, the equivalent of the rate constant B is the sum of the $D/V_W Z_{WT}$ terms and each individual process contributing to this overall rate constant is $D/V_W Z_{WT}$. It is thus apparent that water outflow, evaporation, and reaction have about equal equivalent influences on the rate or time of response, while deposition and particle outflow are relatively unimportant. Such information can be of considerable value when deciding on interventions to improve environmental quality and reduce risk.

It is possible to include a sediment compartment or another water compartment to provide more details. Figure 21.4 presented earlier illustrated the possible time-course of changing concentrations in such a system. If birds are exposed to chemical from fish consumption, it may be necessary to develop a separate bioconcentration or food chain model. The nature and detail of the modeling activity can be tailored to the exposure assessment needs. An example of a more complex aquatic model of PCBs in Lake Ontario has been given by Mackay (1989).

21.5   CHEMICALS OF CONCERN AND MODELS SIMULATING THEIR BEHAVIOR

More than 20 million chemicals have been identified by the Chemical Abstracts System of the American Chemical Society. Of these perhaps 100,000 are used in commercial quantities. Many are relatively innocuous inorganic substances but many are organic or contain "toxic" metals such as mercury or chromium. These chemicals find a wide variety of applications including pesticides, solvents, detergents, polymers, fuels, and medicinal products. They vary enormously in properties such as relative partitioning between environmental media and in degradability. Many are ionic and thus speciate between protonated and ionic forms depending on prevailing pH conditions. The web sites of most regulatory agencies provide lists of chemicals that are of concern to them, and handbooks such as those of Mackay et al. (2006) and Verschueren (1996) have compilations of chemicals and their properties. Models are available to simulate or predict the behavior of many of these chemicals, but extreme care is necessary to ensure that the model selected is appropriate and does not contain inherent assumptions that invalidate its application. For example, many models cannot treat ionizing substances.

In this section we describe some of the available environmental fate models that can be used for exposure estimation as part of risk assessment. No attempt is made to provide a comprehensive list, so this section should be viewed as merely a glimpse into the world of models. Many regulatory agencies such as the US Environmental Protection Agency (US EPA) publish lists of acceptable or recommended models. Examples of useful sources are listed in Table 21.3. References are given in this section to enable the reader to pursue the details of these models as desired.

21.5.1   GENERAL MULTIMEDIA MODELS

These models seek to simulate or predict the behavior of chemicals in a "multimedia" environment similar to that of Figure 21.2, consisting of air, water, soil, sediment, and biotic

---

TABLE 21.3
Useful Sources of Information on Models and Chemical Properties

Models and Modeling		Chemical Properties and Estimation Methods	
Jorgensen et al.	1996	Mackay et al.	2006
Mackay	2001	Baum	1997
Cowan et al.	1995a	Reinhard and Drefahl	1999
Turner	1994	Boethling and Mackay	2000
Thibodeaux	1996	Jorgensen et al.	1998
Clark	1996	Schwarzenbach et al.	1993
Nirmalakhandan	2002	Lyman	1982
DiToro	2001	Reid et al.	1987
Linders	2001	Fogg and Sangster	2003
Paquin et al.	2002	Howard and Meylan	1997

Websites
www.syrres.com/esc/on syracuse research corp.2003
www.epa.gov/crem
www.utsc.utoronto.ca/~wania
www.trentu.ca/cemc
www.rem.sfu.ca/toxicology

compartments such as vegetation, mammals, birds, fish, and other aquatic organisms. Each compartment is usually treated as a well-mixed "box" of material, homogeneous in composition. Subcompartments can be included such as aerosols in air and particles in water. The compartments are connected, with the chemical of interest having the opportunity of moving between them by processes of advection, diffusion, deposition, resuspension, and food intake. The models may contain segmented compartments such as different soil layers or soil types, e.g., agricultural and industrial. Steady-state and dynamic versions can be compiled.

Their primary value is their ability to describe how nonpoint sources of chemical will behave, the likely magnitudes of concentration in different media, the key transport and transformation processes, and the overall persistence or residence time of the substance. They are not suitable for accurate determination of concentrations on a local scale close to point discharges.

The applications of such models and a comparison of three such models are described in a monograph by Cowan et al. (1995b). The models can treat a range of complexity expressed as "levels."

21.5.1.1 Level I

Level I calculations describe a situation in which a fixed quantity of chemical is allowed to come to thermodynamic equilibrium within a closed and defined environment. There is no resistance to chemical migration between environmental phases, and each phase is considered to be homogeneous and well mixed. This calculation provides a general impression of the environmental media into which a chemical is likely to partition based principally on physical–chemical properties such as vapor pressure, octanol–water partition coefficient, and solubility in water.

21.5.1.2 Level II

Level II calculations describe a situation in which the chemical is continuously discharged at a constant rate and reaches both steady-state and equilibrium conditions at which the emissions are balanced by advective and degradation losses. Degradation half-lives must now be supplied. New information obtained includes the overall environmental persistence of a chemical and the relative importance of chemical degradation processes.

21.5.1.3 Level III

Level III calculations are the most useful for risk assessment. They introduce resistance to chemical migration between environmental compartments. The chemical is continuously discharged at a constant rate to one or more of the environmental compartments (e.g. air, water, soil, and sediment) and achieves a steady-state but nonequilibrium condition. Each compartment has a unique equilibrium status or fugacity. Additional input parameters that characterize the environment are required to calculate transfer rates between environmental media. The Level III calculation shows, as illustrated in Figure 21.2, the dependence of overall environmental fate on media of release, i.e., "mode of entry," and highlights the key intermedia transport pathways (e.g., air–water or air–soil exchange).

21.5.1.4 Level IV

Level IV unsteady-state (or dynamic) models can also be formulated to analyze short-term (e.g., seasonal) effects on environmental concentrations, or to determine the effect of changing chemical emissions over a period of years or decades. Figure 21.4 illustrates typical output of this level.

21.5.1.5 Fugacity Models

A series of fugacity-based models have been developed in the last two decades at the University of Toronto and Trent University by Mackay and coworkers. They are available from the Trent University Canadian Environmental Modelling Centre web site (www.trentu.ca/cemc).

The equilibrium criterion (EQC) evaluative model includes Level I through Level III and deduces the fate of a variety of specified chemicals in an evaluative environment of fixed properties (Mackay et al. 1996a–c). The aim is to provide a "benchmark" environmental fate profile for a chemical that can be compared with the fates of other chemicals in the same model world. The RAIDAR model of Arnot et al. (2006) extends the EQC model to include a natural and agricultural food web resulting in ecological and human exposure and risk.

21.5.1.6 CalTOX Model

The CalTOX model was originally developed by McKone (1993a) to describe the fate of chemicals in California, especially substances released from waste sites. It has a more detailed treatment of soil layers and is Level III in structure. Human exposure is also estimated.

21.5.1.7 Simplebox Model

The Simplebox model from the Netherlands is incorporated into the European Union System for the Evaluation of Substances (EUSES) assessment process (EUSES 1997). It is Level III in structure but has a nested configuration, with a small local region contained in a larger national region, which in turn is contained in a continental scale region. The EUSES program also calculates human exposure at these three scales.

21.5.1.8 Regional, Continental, and Global-Scale Models

Fugacity- and concentration-based models can be parameterized to simulate a region or a set of connected regions, states, or nations. Examples of such applications include the Chem-CAN model for Canada (Webster et al. 2004), the ChemFrance model (Devillers and Bintien 1995), the BETR North American model (MacLeod et al. 2001), the BETR World model (Toose et al. 2004), and the GloboPOP model (Wania 2003), which treats the entire planet and is used to assess long-range global fate and transport.

21.5.2 MODELS SPECIFIC TO ENVIRONMENTAL MEDIA

With increased specialization in environmental science and regulatory demands for more accurate and site-specific models, clusters of media-specific models have evolved which treat in more detail the atmosphere, water bodies (lakes, rivers, estuaries) and their underlying sediments, groundwater, soils, urban and indoor conditions, sewage treatment plants, and various organisms (including humans). The handbook by Jorgensen et al. (1996) lists many such models. Most national regulatory agencies have lists of "acceptable" models. Here we only briefly outline the general features of each cluster.

21.5.2.1   Plume Models in General

For the flowing media of air and water, it is frequently necessary to calculate the concentration of a chemical in a "plume" downwind or downstream of the discharge point. In its simplest form a dilution model can be applied in which the concentration in the flowing medium is E/G g/m^3, where E is the emission rate (g/h) and G is the flow rate (m^3/h). This is appropriate for narrow rivers in which mixing is rapid both vertically and horizontally from side to side. For wider rivers, estuaries, groundwater, and in the atmosphere the plume is not constrained and it expands vertically and horizontally. It is then necessary to calculate the changing plume dimensions. Essentially the flow term G increases and the concentration falls by dilution and diffusion. A common approach is to assume a Gaussian distribution of concentrations vertically and horizontally over the plume cross section. The distance parameters equivalent to the standard deviation of the distribution can be estimated from empirical correlations. The distribution of concentrations can be calculated from the maximum on the center line to the "edges" of the plume. Obviously, constraints may apply at surfaces such as the ground or river bottom, restricting diffusion. It is also possible to include an expression for degradation of the chemical as is done in river "die away" models, and for other loss processes such as deposition.

21.5.2.2   Atmospheric Models

Numerous air dispersion models with general or limited geographic applicability have been developed with the objective of deducing ground level concentrations and exposures from stack emissions. Most texts or handbooks on air pollution contain full descriptions of the basic principles underlying such models. The workbook by Turner (1994) is an excellent starting point.

The general approach is to first define the emission rate or strength of one or more sources in units such as kg/h. The emission may be continuous or as an intermittent "puff." As most discharges to the atmosphere are from stacks, it is usually necessary to deduce a plume rise height, yielding an estimate of the effective height at which contaminant is released. The chemical then blows downwind and disperses horizontally and vertically, diluting steadily at a rate controlled by the prevailing meteorology. A "map" of concentrations as a function of position can then be assembled. The equations employ standard deviation terms in the Gaussian distribution equation and are functions of wind speed and atmospheric stability. Ground-level concentrations downwind of the source are of most interest, as these control human and terrestrial ecosystem exposure.

Models range from simple application of Gaussian dispersion equation to complex, multisource models containing allowance for depositing particles and topographic features of the terrain. The use of such models is often written into legislations or regulations as a means of determining acceptable stack emission rates that will result in specified acceptable ground-level concentrations. Among these the Industrial Source Complex (ISC3) Dispersion Models of the US EPA and the commercial CALPUFF model are notable.

21.5.2.3   Aquatic Models

Water quality modeling is a highly developed subject with models being available for ponds, lakes, rivers, entire river basins, and estuaries. The approach is usually to segment the water into a number of connected "boxes" or to apply plume dispersion equations. Biotic components may be included to calculate concentrations in planktonic organisms, invertebrates, and fish. Models such as the WASP series, AQUATOX, and EXAMS are available from US EPA web sites (www.epa.gov/epahome/models). The QWASI model described early in the chapter is available from the Trent University web site cited earlier. Paquin et al. (2003) provide an

TABLE 21.4
Models of Aquatic Systems

Model	Scope	Reference
EXAMS	Exposure analysis modeling system, OSEPA	Ambrose 1997
WASTOX	Water quality analysis simulation of toxics	Connolly and Winford 1984
WASP4	Water quality analysis simulation program	Ambrose 1988
QWASI	Quantitative water–air sediment interaction fugacity model for lakes	Mackay et al. 1983; Mackay 2002
ROUT	GIS model applied to US rivers	Wang et al. 2000
GREAT-ER	European GIS river basin model	Feijtel et al. 1997, Boeije (1999)
DITORO	Sediment water exchange	DiToro 2000
SIMPLETREAT	Simpletreat (sewage treatment)	Stuijs 1996
STP	Sewage treatment plant	Clarke et al. 1995
TOXSWA	Edge of field systems	Adriaanse 1996, 1997

up-to-date listing of these models, especially for metals. Table 21.4 lists some of these models including wastewater treatment plant models.

Integral to these models are sediment compartments that often contain most of the chemical mass in the aquatic system. DiToro (2001) describes a number of such models. For metals there is usually a need to describe chemical speciation as a function of pH and the presence of other ions and dissolved organic matter. Speciation models are reviewed by Paquin et al. (2003).

Sophisticated river and river basin models have been developed in Europe and the United States to assess the fate of chemicals such as detergents or drugs that may be used domestically and are discharged to sewer and municipal treatment systems. They may incorporate GIS software to enhance the presentation of the results. An example is the GREAT-ER model (1997).

There is also a need to evaluate behavior in small, edge-of-field water bodies associated with agricultural uses of chemicals. One example of such a model is TOXSWA (Adriaanse 1996, 1997) developed for regulatory evaluations in the Netherlands and recently adopted as a standard regulatory tool for broader European evaluations under Directive 91/414/EEC.

21.5.2.4 Soil Models

The commonest application of soil models is to pesticides applied agriculturally. They are used for both scientific and regulatory purposes, primarily to provide an assessment of risk of leaching to groundwater, field drainage systems or runoff to surface waters. They may also be used to support more complex risk assessments for key terrestrial taxa such as soil invertebrates, birds and mammals. In principle, the aim is usually to quantify rates of degradation, leaching, runoff, and evaporation under the dynamic conditions following pesticide application. Models used in this context are described later in the section on pesticides. Also of concern is the fate of chemicals applied in sludge amendments to soil and of veterinary pharmaceuticals.

21.5.2.5 Fish Uptake and Food Chain Models

Because of the importance of the human and ecological exposure route resulting from fish consumption, considerable effort has been devoted to estimating chemical concentrations in fish. Contaminants may enter fish through the gills (bioconcentration) and, especially in the case of hydrophobic chemicals, by way of food (biomagnification). These models are

TABLE 21.5
Bioaccumulation Models

Name	Scope	Reference
GOBAS	Model of fish and food webs	Gobas 1993, 2003
AQUATOX	Aquatic fate toxicity model	Park 1998
FGETS	Food and Gill Exchange of Toxic Substance	Barber et al. 1991
BASS	Bioaccumulation and Aquatic System Simulator	Barber.craig@epamail.epa.gov
FISH	Fugacity model of fish	Mackay 2001
FOODWEB	Fugacity model of aquatic foodweb	Campfens and Mackay 1997
THOMANN	Model of fish and foodweb	Thomann and Connolly 1984
TOXSWA	Small-scale aquatic systems model	Adriaanse 1996, 1997
PEARL	Ricardian hydrology leaching model	Boesten and van der Linden 2001; Leistra et al. 2001
GeoPEARL	Spatially indexed hydrology leaching model	Tiktak et al. 2002, 2003, 2004

reviewed by Paquin et al. (2003), Gobas and Morrison (2000), and Mackay and Fraser (2000). Table 21.5 lists some of these models.

21.5.2.6 Miscellaneous Models

Models are also available for describing the fate of chemical and oil spills, contaminants in groundwater, in vegetation, in urban areas, in indoor environments, for remediation purposes and for processes within organisms (e.g. physiologically based pharmacokinetic (PBPK) models).

21.5.3 MODELS SPECIFIC TO CHEMICAL CLASSES

The models described above have been categorized according to the impacted environment such as air, water, or soil. In recognition of the need to regulate specific classes of chemical substances such as pesticides, the industrial and regulatory communities have developed a series of models that address the need for ecological and human risk assessment in the context defined by the nature of these substances and the way they are used. To illustrate this, we discuss models for only three classes of substances: agricultural pesticides, veterinary medicines, and biocides.

21.5.3.1 Agricultural Pesticides

This group of chemicals consists of fungicides, herbicides, and insecticides that are designed to eliminate or restrict the growth or infection of fungal, weed, or insect pests. For regulatory purposes, the scale of assessment is often the field or edge-of-field scale. Modeling may be carried out to assess the scale of exposure in soil, vegetation, surface water, sediment, groundwater, and air, although for ecological risk assessment the first four compartments are usually of primary interest. Table 21.6 lists some of these models recommended for use for various processes at a screening level, focusing on (primary) processes in the soil and (secondary) processes in the environment. Linders (2001) has compiled a valuable set of papers on these models.

Exposure experienced by soil organisms may be direct, following application "in crop," or indirect, typically as a result of spray drift. Two general exposure estimation methods are

TABLE 21.6
Pesticide Fate Models

Name	Scope	Reference
PELMO	Pesticide fate in soils	Klein et al. 2000
PRZM	Pesticide root zone model	Carsel et al. 2003; Mullins et al. 1993
SoilFug	Fugacity model of pesticide fate in soil	Di Guardo et al. 1994
MACRO	Model of pesticide fate in soils	Jarvis et al. 1994, 1995, 1996, 1998
GENEEC	Generic estimated environmental concentration based on PRZM	Parker et al. 1995
AGDRIFT	Model of pesticide drift	Spray Drift Task Force 1997

employed: models that simulate spray drift, and "look-up" tables that have been created from observations or models. In the United States horizontal transport and deposition field research has been used to support models such as AGDRIFT developed by the Spray Drift Task Force (Spray Drift Task Force 1997). In the European Union (EU) spray drift losses are most often assessed through use of "look-up" tables that condense the results of a large number of field-based horizontal deposition studies (Ganzelmeier et al. 1995; Rautmann et al. 2001). Within the EU regulatory arena it is necessary to assess "off-crop" impacts to nontarget arthropods. At present the spray drift tables discussed earlier are used within the ESCORT scheme (ESCORT 2001) to generate estimates of exposure on off-crop vegetation. It is, however, recognized that horizontal deposition estimates may give a poor representation of a combination of horizontal deposition and vertical interception by vegetation (Tones et al. 2001). "... interception by vegetation (Tomes et al. 2001)." ... Assessment of spray drift is just one example where a wide range of modelling approaches are available. Although such diversity is a healthy scientific state-of-affairs, this has led to regulatory uncertainty and confusion, e.g., "which model should be used as a basis for the regulatory assessment?" Within the European Union the need for standardisation of modelling frameworks to facilitate regulatory assessment of pesticides under Directive 91/414/EEC led to the establishment of a number of Working Groups under an initiative known as FOCUS. Regulatory and technical experts have reviewed modelling techniques and prepared shortlists of models recommended to support regulatory submissions, accompanying scenarios and supporting guidance on conduct and reporting of such assessments.

Loadings to surface waters for agricultural pesticides may occur as a result of a number of processes, the most important being drift, drainage, and runoff. Simulation models in use by industry and regulators in line with FOCUS guidance are listed below.

- Drift: Drift curves incorporated into the FOCUS SWASH based upon look-up tables and aspects of AGDRIFT (Spray Drift Task Force 1997)
- Drainage: MACRO, a Ricardian soil leaching model with macropore transport capability (Jarvis 1994, 1995, 1998; Jarvis et al. 1996)
- Runoff: PRZM, a capacity-based soil leaching and transport model with runoff simulation based upon the Universal Soil Loss Equation (Mullins et al. 1993; Carsell et al. 2003)

In addition to the need to assess loadings into surface water from these processes, it is also essential to consider the subsequent fate and behavior of chemicals within the receiving waters using aquatic models of the types outlined earlier. In the scheme established by FOCUS the loadings from each relevant process provide direct input for the TOXSWA model (Adriaanse 1996, 1997; Beltman and Adriaanse 1999a,b). TOXSWA is capable of simulating loadings from point source or distributed over a defined length of the water body. The simulated water body system is two dimensional and consists of two subsystems: a water layer containing suspended solids and a sediment layer whose properties (porosity, organic matter content, and bulk density) vary with depth. Within the version used to evaluate aquatic exposure for national registration in the Netherlands it is also possible to include partitioning interactions with macrophytes.

Within the United States assessments generally focus upon drift and runoff. In common with the regulatory assessment scheme in the EU, initial stages of exposure assessment are carried out with a set of conservative assumptions regarding loadings (generic estimated exposure concentration (GENEEC); Parker et al. 1995). Where appropriate, more sophisticated mechanistic modeling is then carried out employing models such as AGDRIFT and the PRZM runoff model already discussed. As outlined earlier, upon entry into surface waters it is necessary to consider the fate and behavior. Although various tools are available for this task, the most commonly used regulatory tool is the EXAMS modeling system developed (as in the case of PRZM and GENEEC) by the US EPA Office of Pesticide Programs.

21.5.3.2 Veterinary Medicines

This group of chemicals is of interest because they represent an "emerging" issue. Assessment and modeling procedures are still in the early stages of development. The two most significant classes from the perspective of scale of ecological exposure and risk are livestock and fish treatments.

Because livestock veterinary medicines are often repeatedly administered as whole herd treatments within intensive production systems, there are concerns surrounding the impact of potentially large quantities of active substance or active metabolites within excreta that are then applied to land, either directly in the case of pasture animals or indirectly in the case of spreading of manure or slurry produced by housed animals. Risk assessment concerns include impact on soil-dwelling organisms (such as microorganisms, earthworms, and any succeeding crops) and any subsequent exposure to aquatic organisms following runoff or drain discharge. There is increasing concern surrounding potential impacts on dung fauna and grassland invertebrate species following administration of veterinary medicines with insecticidal properties. In the United Kingdom a modeling system (VetPEC) developed for the Veterinary Medicines Directorate has provided a more complete mechanistic scheme for considering fate and behavior of veterinary medicines and their exposure in soil, groundwater, and surface waters following excretion.

Fish treatments (either freshwater or marine farmed fish) generally involve vaccines, "in-feed" treatments, and "bath" treatments. Depending upon the method of administration, these medicines enter the environment as pseudo-pulse doses (short-term administration) or pseudo-steady-state (long-term, chronic treatments). To assist with regulation of such formulations and agricultural practices, models are being developed as site-specific evaluation tools. Ultimately such models will require the capability of simulating partitioning of residues between organic matter (waste food, feces, suspended organic matter) and water.

There is still much work to be done to improve chemical fate representations in veterinary models. Sophisticated hydrological models including particle-tracking methods exist for simulating seabed deposition of treated fish feed and faeces (e.g., DEPOMOD, Cromey, 2002) but representation of chemical fate and behaviour remains generally very limited and is surprisingly inconsistent with the sophistication of other environmental process algorithms.

21.5.3.3 Biocides

This group of chemicals has been defined as chemicals or microorganisms, mixtures of either, or both, intended to control unwanted organisms such as animals, insects, bacteria, viruses, and fungi. Biocides present a unique set of risk assessment challenges as they are characterized by:

- Intentional introductions into the environment
- Both "point" and "nonpoint" introductions into the nontarget environment
- Diverse usage, disposal, and environmental release scenarios
- They are deliberately designed to be highly toxic

The range of product types covered by the Biocides Directive (98/8/EC) in the EU includes disinfectants and sanitizers used in drinking water, in public areas, and for veterinary purposes, preservatives used to protect wood, polymers, masonry and film, slimicides, molluscicides, rodenticides, antifoulants used on vessels, and even embalming fluids. In the United States, biocides are assessed under the auspices of the Federal Insecticide, Fungicide and Rodenticide Act (FIFRA) along with agricultural pesticides.

These Emission Scenario Documents have been developed for most biocides to provide a necessarily simplistic approach to exposure assessment based upon scenarios and conservative default assumptions of loadings and transport processes. Most calculations are simple enough to be carried out without computer models. Nonetheless, when simplistic exposure assessments suggest that exposure and risk may be unacceptable, more sophisticated modeling can be carried out to replace crude default assumptions, and thereby refine the risk assessment. Some of the modeling approaches discussed for other chemicals such as pesticides and veterinary medicines can be employed here.

21.5.3.4 Metals

Metals present a set of challenges that differ fundamentally from those posed by organic substances. Although the basic principles of mass balance modeling are identical, there has tended to be a disciplinary separation into those who study and model metals from those who address organics, and indeed from those concerned with radionuclides. Several key differences in focus are noteworthy.

Although organic molecules such as benzene are subject to degradation to carbon dioxide and water, metals are totally persistent. They may change from oxide to sulfide to carbonate, they can adopt different ionic forms such as ferrous and ferric ions, but the element is conserved. In this respect, metals are easier to model because the mass of the element does not change.

Metals ionize, and therefore their state and behavior in the environment is dictated, at least in part, by the prevailing acidity and redox conditions. This complicates the expressions describing transport in the environment including uptake by organisms. It is a much more severe problem for metals than for organics. As a result, considerable effort has been devoted to building chemical equilibrium models for aquatic systems, the most established being the MINEQL and MINTEQ series by Westall et al. (1976) which are available from the US EPA. These models predict speciation, adsorption, and precipitation of metal ions as a function of the presence of other cations and anions including natural organic matter.

To a first approximation, organic molecules partition into organic matter such as humic and fulvic acids and lipids in organisms by a solution or solvent mechanism. Because of this behavior, the octanol–water partition coefficient K_{OW} is effective for describing partitioning of a wide variety of organic substances to natural organic matter in soils, sediments, and aerosol particles as well as lipids in organisms. There is no analog for K_{OW} for metals. Partitioning tends to be highly specific to the metal in question and transport through membranes is often active in nature using ion-specific protein "pumps." Models have been developed describing metal partitioning to natural organic matter, the most notable being the Windermere Humic Acid Model (WHAM) of Tipping (1994).

The primary focus of metal models is the aquatic environment of water, sediments, and their resident biota. Only in rare cases such as mercury is transport to air important. The review by Paquin et al. (2003) provides a full account of fate, bioaccumulation, and toxicity models.

For both metals and organics the concept of bioavailability has proved crucial. This asserts that only a fraction of the total substance present is "available" to exert toxic effects. This fraction may range from 99% to 0.1%. For organics the fugacity approach automatically treats this issue. In the example given earlier, of the chemical in the water column only 88% is in free molecular form and thus "available." For metals the equivalent approach is the Free Ion Activity Model (FIAM) developed by Morel (1983). Campbell (1995) has discussed in detail the strengths and weaknesses of this model.

In summary, although metals differ from organics in many significant respects, these differences are more in process emphasis than of a fundamental nature. The same general modeling principles apply to both.

21.6   CONCLUDING THOUGHTS ON SELECTING AND APPLYING MODELS

Faced with the need to conduct an ecological risk assessment for a specific environment and specific chemicals, the assessor must evaluate how a model can assist the process and, indeed, if modeling is justified. If the risk arises from exposure to chemical substances that are of anthropogenic origin and there is a possibility that discharges can be reduced, it is likely that a model which relates discharge quantities to exposure concentrations will be useful. The extent of discharge reduction and the time response of the system can be estimated. For "new" situations in which a discharge is planned or expected it can be invaluable to forecast concentrations, even approximately. Many past "mistakes" such as the widespread contamination by PCBs could have been prevented if models had been applied prior to discharge.

In general, when an ecosystem is subject to a source of chemical and the properties and discharge rate of that chemical are known, even approximately, it is likely that a model will be useful for predicting concentrations or reconciling monitored concentrations. The model framework can then be used to determine the factors that control concentrations and exposures and to explore the implications of changing discharge rates. It should be noted that modeling can be relatively fast and inexpensive compared to ecological field investigations and chemical monitoring campaigns. As a tool, a model can then add considerable value to these conventional investigative tools.

Another problem facing the risk assessor is which model to use. Environmental situations vary considerably in nature and models are often specific to certain classes of chemical. The best strategy is to explain the nature of the situation to a modeler and seek advice on the preferred approach. An "off-the-shelf" model may be adequate, or a customized model may be required. The strengths and weaknesses of the model may not become apparent until it is actually applied. A valuable practice is to proceed from simplicity to complexity. The simplest available model is applied first, then its inadequacies are assessed with a view to introducing greater complexities only where they are most needed. This is essentially the principle expressed in Ockham's Razor that can be stated colloquially, in this context, as: "Don't make the model more complex than is dictated by the ecological risk assessment task being addressed."

22 Exposure to Chemicals and Other Agents

Exposure is the contact or co-occurrence of a contaminant chemical or other agent with a biological receptor (usually an organism, but possibly an organ, population, or community). The analysis of exposure estimates the magnitude of exposure of the receptors that constitute the endpoint entities, distributed in space and time. These distributions may be estimated by measuring the contaminants in media (Chapter 20) or by modeling the transport and fate of releases (Chapter 21). For new chemicals, technologies, organisms, or other agents, exposures due to permitting the release must be predicted based on expected releases. For contaminated or occupied sites, exposures due to current conditions and future conditions must be estimated. The boundaries of exposure modeling vary with context (see Box 17.1). Transport and fate models may include uptake and accumulation by some organisms, but often stop with concentrations in abiotic media. Commonly, the utility of chemical analyses at contaminated sites is limited by including only abiotic media or by limiting analysis of biota to those consumed by humans. Hence, estimation of exposure may require modeling of chemical accumulation and transport in food webs.

Exposure estimates should be appropriate for characterizing risks by parameterizing the exposure variables in the exposure–response models (Chapter 30). This requires that the exposure estimates address the same forms or components of the pollutants as the effects assessment and also have concordant dimensions. For example, the estimation of effects on plants may require that concentrations of chemicals in the aqueous phase of the soil be estimated, that concentrations be averaged over the rooting depth of the plants, and that the results be expressed as a median concentration and other percentiles of the empirical distribution of point concentrations. In contrast, the estimation of risks to wildlife due to soil ingestion may require total concentrations in surface soil, averaged over the foraging range of the species, expressed as the mean and standard deviation.

In all cases, the analysis of exposure must appropriately define the intensity, and the temporal and spatial dimensions of exposure. Intensity of chemical exposures is usually expressed as concentration in a medium that is in contact with a receptor, but dose and dose rate are also used. Equivalent intensity metrics must be developed for exposures to nonchemical agents (Box 22.1). Time is usually the duration of contact, but other relevant aspects of time are the frequency of episodic exposures and the timing of exposure (e.g., seasonality). The spatial dimension is usually expressed as area within which an exposure occurs or as linear distance in the case of streams. If contamination is disjunct (e.g., spotty), the spatial pattern may be important. Hence, the definition of exposure must be some measure of intensity with respect to space and time. A simple case is an average concentration over a site that does not vary within the time of concern.

The degree of detail and conservatism in the analysis of exposure depends on the tier of the assessment. Scoping assessments need only determine qualitatively that an exposure may

---

BOX 22.1
Intensity of Exposure to Nonchemical Agents

The following are some examples of potentially hazardous agents other than chemicals and corresponding expressions of intensity of exposure.

Exotic organisms: Exposure to exotic organisms is typically expressed as abundance (numbers per unit area or volume). However, for pathogens exposure is often expressed as dose or dose rate.

Noise: Noise pollution studies have typically used decibels as the measure of intensity, but it is based on human sound spectral sensitivity.

Aircraft: Aircraft overflights may startle and physiologically stress animals. Because both the noise and the sight of the aircraft are involved, the most generally useful expression of the intensity of exposure is the slant distance from the animal to the aircraft at closest approach (Efroymson and Suter 2001b).

Heat: Exposure to heated water or air is most commonly expressed as the temperature of the medium (°C).

Habitat modification: Exposure to habitat modifications is typically expressed as the area modified (e.g., hectares strip mined) or linear distance of stream modification (e.g., meters of stream channelized). However, some habitat modifications are characterized as changes in specific habitat variables (e.g., percent fines in silted stream riffles).

Light: Light pollution studies have used lux as the measure of intensity, but it is based on human spectral sensitivity (Longcore and Rich 2004). Some effects, such as bird collisions with illuminated structures, are related to position rather than intensity.

Radiation: The intensity of ionizing radiation is expressed as dose in Grays (formerly rad) or the absorbed dose equivalents in Sieverts (formerly rem). Both are most often used as dose rates (Gr/y or Sv/y).

---

occur by a prescribed pathway. Screening assessments must quantify exposure but should use conservative assumptions to minimize the likelihood that a hazardous exposure is inadvertently screened out. Definitive assessments should be realistic and therefore should treat the estimation of exposure and uncertainty separately. This may be done by estimating distributions of exposure or by estimating both the most likely exposure and upper-bound exposure.

The concept of bioavailability is critical to discussions of exposure to chemicals. Organisms are not equally exposed to every molecule in a medium. Some chemicals are sequestered by sorption to solid phases of the medium. Some forms of chemicals are more readily taken up than others. For example, methyl mercury is readily assimilated by aquatic and terrestrial organisms, but mercuric sulfide (cinnabar) is not taken up to any appreciable extent, and other mercury salts such as mercuric chloride and elemental mercury are taken up at moderate rates. Some chemicals readily change their form in response to environmental conditions. In particular, nonionic chemicals are generally taken up less readily than ionic forms and forms may change with the diurnal cycle of pH in productive waters. All of these phenomena are treated as aspects of bioavailability. One may speak of bioavailability in absolute terms: cinnabar is not bioavailable or dissolved cupric ion is bioavailable. However, it is usually more realistic to treat bioavailability as a continuous variable that is a function of the chemical species and environmental conditions.

This chapter begins with a discussion of the component activities that comprise an analysis of exposure. It then discusses issues specific to the individual environmental media: water, sediment, soil, and biota. Section 22.3 addresses exposure for particular taxa that are exposed to multiple media. The fourth topic is modeling of the uptake of contaminants by biota, primarily for estimation of food-chain exposures, but also for exposure–response models based on internal exposures. Wastes such as petroleum and its derivatives that are clearly a melange are discussed in Section 22.5, because they may appropriately be treated as a complex contaminant rather than as a collection of chemicals that are assessed separately. Section 22.6 finally discusses the presentation of results from an analysis of exposure.

22.1   EXPOSURE MODELS

After contaminant concentrations have been measured in ambient media or estimated using transport and fate models, it is necessary to estimate the actual exposure to contaminants using some sort of model of exposure. For most ecological risk assessments, the exposure is simply the concentration in water, sediment, or soil expressed as total concentration or as concentration in some phase (e.g., dissolved concentration) averaged over some time interval. Some sort of model is required because the measured concentrations must be related to the effective (bioavailable) concentrations by a conversion or simply by assuming that they are equivalent. Similarly, the instantaneous measured values must be related to some time interval. For example, one may assume constant concentrations for the period of concern (e.g., the life cycle of an endpoint species), constant concentrations for an episodic exposure, or some definable change in concentration (e.g., exponential decay). Alternatively, uptake within a particular exposure period (dose rate or external dose) may be required. Such doses are commonly calculated for wildlife as milligram of contaminant per kilogram per day, based on intake of contaminated food and media. Finally, exposures may be expressed as internal concentrations (internal dose) either in whole organisms or particular organs. If they are not determined by analyzing the organisms, those internal exposures may be modeled empirically (e.g., bioaccumulation factors) or mechanistically (i.e., linking a dose rate model with a toxicokinetic model). Although ecological exposure models are, in general, static, toxicokinetic models that can describe the time course of internal exposure have been developed for some chemicals and taxa.

22.2   EXPOSURE TO CHEMICALS IN SURFACE WATER

In most cases, ecological risk assessments of contaminated waters are based on measured or modeled aqueous concentrations of chemicals. In such cases, the major issues to be considered by the assessors are the appropriate averaging times and the forms of chemicals in water that must be measured or estimated.

Unlike soils and sediments, the concentrations of chemicals in water are often highly variable over relatively short time periods. The resolution of temporal issues in aqueous modeling or sampling and data reduction must be based on the variability of concentrations in the stream and the toxicokinetics and toxicodynamics of the chemicals and receptors. Human health risk assessments are usually based on the assumption that effects result from long-term (i.e., several decades) average exposures. Hence, chemical fate models and sampling and analysis plans based on human health concerns are designed to characterize those averages. In contrast, ecological effects may result from short-term (i.e., less than a week) exposures of relatively small organisms (e.g., algae, zooplankters, or larval fish) that rapidly reach equilibrium with highly mobile chemicals (e.g., metal ions). Hence, the model or sampling plan should include episodes of high concentration, and the analysis of exposure should include an analysis of the frequency and duration of such episodes.

The bioavailability of chemicals, particularly metals, is complex because of the many forms, including dissolved metal (either salts or free ions), particulate metal (e.g., associated with suspended clay), and metal complexed with dissolved material (i.e., organic colloids and colloidal hydrous metal oxides). Traditionally, regulatory organizations have required that total concentrations be used in ecological risk assessments for the sake of conservatism. Total concentrations can be useful for screening assessments, but metals associated with suspended particles clearly have little bioavailability and little relevance to typical toxicity tests, which are performed in clean water with highly soluble forms of the tested metal. Hence, the US Environmental Protection Agency's (US EPA) Office of Water recommended that assessments of effects of aqueous metals on aquatic biota be based on dissolved metal concentrations as determined by analysis of 0.45 μm filtered water (Prothro 1993). However, even dissolved concentrations are conservative in many cases, because they include metals complexed with dissolved materials as well as freely dissolved metals. Risk assessments should be based on the form that is best correlated with effects. For exposures of aquatic animals to metals in general, this appears to be primarily the free metal ion (Bergman and Dorward-King 1997).

It should be noted that use of total concentrations is not always conservative in practice. First, because the high levels of acid-extractable metals may cause analytical interferences, the limits of detection for metals may be greater for total concentration analyses, and therefore toxic concentrations of metals may not be detected. This apparently occurred in the assessment of Bear Creek in Oak Ridge, Tennessee, where copper was a chemical of concern in filtered samples but was not detected in total samples. Second, when comparing concentrations at contaminated sites to background concentrations, if the dissolved concentration is small relative to the total concentration, there may be a significant increase in dissolved concentrations relative to background concentration but no significant increase in the total. That is, the particle-associated background concentration may mask a relatively small but toxicologically significant increase in dissolved concentrations.

Speciation should be considered for both inorganic and organic chemicals that have multiple ionization states within the range of realistic ambient conditions. In general, non-ionic forms are more toxic, because they partition more readily from water to biota. Hence, unionized ammonia is more toxic than the ammonium ion, and unionized alcohols and phenols are more toxic than the ionized species. This rule does not apply to metals, particularly those that may have multiple ionic species within the range of ambient conditions. The expense of analyzing metal species is justified for metals of ecological concern that may occur at the site as multiple species that have significantly different toxicities or that are believed to occur predominantly as a single species that is different from the one assumed by regulators. Assessors should particularly consider speciation of arsenic, chromium, mercury, and selenium.

Forms and species of metals in water can be estimated from measured concentrations by applying metal speciation models (Bergman and Dorward-King 1997). The biotic ligand model (BLM), which is beginning to be used for water quality regulation by the US EPA, incorporates a version of the Chemical Equilibria in Soils and Solutions (CHESS) metal speciation model (Santore and Driscoll 1995; DiToro et al. 2001; Hydroqual 2003; EPA 2003a). The MINEQL+ speciation model may also be used (Schecher and McAvoy 1994). The BLM goes beyond CHESS to estimate concentrations of metals at the presumed site of action for acute lethality, the biotic ligands on the gills (Section 26.2.8). While speciation models are less reliable than analytical chemistry (e.g., ion-specific electrodes), they are more useful in this case, because the model accounts for competition between the toxic metal ion and other cations, particularly Ca^{++} and H^+ for both biotic ligands and sites on dissolved organic matter (Figure 22.1). The CHESS model is complex and requires 12 water quality parameters: pH, dissolved organic carbon (DOC), dissolved inorganic carbon, percent humic acid, temperature, major ions (Ca^+, Mg^+, Na^+, K^+, SO_4^-, Cl^-), and sulfide.

The BLM model includes only aqueous uptake, but at least in some cases, dietary exposure to metals is important (Meyer et al. 2005). In laboratory studies, of exposures equivalent to

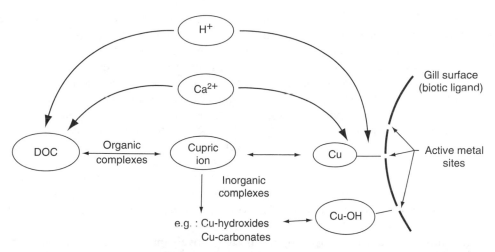

FIGURE 22.1 A diagram representing the partitioning of aqueous copper among abiotic and biotic ligands. (From EPA (U.S. Environmental Protection Agency), 2003 draft update of ambient water quality criteria for copper, EPA 822-R-03-026, Office of Watter, Washington, D.C., 2003. With permission.)

those in the Clark Fork River, Montana, and the Coeur d'Alene River, Idaho, diet was more important than direct aqueous exposure (Woodward et al. 1994a,b; Farag et al. 1999). Dietary exposures to metals may be important for aquatic invertebrates as well (Munger et al. 1999). There are no conventions for estimating aquatic dietary exposures or combining them with aqueous exposures, but toxicokinetic modeling (Section 22.9) has been suggested (Meyer et al. 2005).

The issue of bioavailability is relevant to organic chemicals as well as metals. Like metals, organic chemicals may bind to dissolved or suspended particles, making them less available for biological uptake. However, unlike metals, there is no guidance from the EPA to use concentrations of organic chemicals in filtered water to represent aqueous exposures. This is in part because the issue of bioavailability has not been considered important for aqueous organic chemicals.

In some cases, the physical properties of a chemical rather than its aqueous concentration and uptake determine its effective exposure. For example, ferric iron forms hydroxide and humic precipitates that settle on organisms and substrates limiting access to food, reducing habitat, and inhibiting respiratory exchange. Petroleum, aluminum flocs, organic particulate wastes, and other materials may similarly harm organisms without conventional uptake or toxic effects. Standard exposure metrics for these cases have not been developed, so they must be treated ad hoc.

22.3 EXPOSURE TO CHEMICALS IN SEDIMENT

Ecological risk assessments of contaminated sediments are typically based on chemical concentrations in bulk sediment or sediment interstitial (pore) water. The principal issues to be addressed are the heterogeneity of sediments and sediment contamination and the bioavailability of measured chemical concentrations (Ingersoll et al. 1997). It may be useful or necessary to also estimate contaminant uptake and trophic transfers.

Unlike surface water, sediment contaminant concentrations generally vary spatially (vertically and horizontally) more than temporally. The assessment of exposure must consider the spatial distribution of contamination in relation to the distribution of the receptors. Most sediment-associated organisms are exposed to surface sediment (e.g., the top 5 to 10 cm), rather than deep sediment. For example, the burrowing depth of sediment-dwelling insects and oligochaetes varied greatly among taxa and seasons, but seldom exceed 10 cm (Lazim et al. 1989; Charbonneau and Hare 1998). Typically, ecological assessors assume that the concentration reported for the topmost layer of a core or for a surface grab sample represents the exposure of benthic and epibenthic organisms at the sampled location. Careful consideration of the vertical dimension should be applied to species other than benthic invertebrates. For example, many centrarchid fish form nests in the sediment where the eggs and larvae develop. One might conservatively assume that they are exposed to epibenthic water that is equivalent to sediment pore water, but ventilation of the nest by the guarding male complicates the situation. If sediments are contaminated in an area that is heavily used by these species and risks are uncertain, a special effort to sample epibenthic water from the area of the nests may be justified.

Because most sediment-associated organisms are relatively immobile, it is not reasonable to assume that benthic organisms average their exposure to sediment contamination over large areas (e.g., entire stream reaches) or depth ranges (e.g., the top 50 cm of sediment). Rather, the median surface sediment concentration is an appropriate measure of the central tendency of sediment exposures for the benthic infauna in a given area (e.g., stream reach) and the maximum detected concentration is an appropriate conservative estimate of this exposure for use in contaminant screening.

Sediment contaminant concentrations at a given location generally vary little during the life cycle of most benthic invertebrates, since they can complete one or more life cycles in a year. Notable exceptions include estuaries in which physical–chemical characteristics of the overlying water that alter the partitioning of contaminants to and from the sediment compartment (e.g., salinity, dissolved oxygen, and hydrodynamics) vary on a biologically relevant time scale. The resulting variations in benthic exposures should be considered in the collection and evaluation of estuarine sediment data.

Even if the sediments are relatively stable, the settling of particles from the water column may require the collection of new data to replace the existing sediment data. The need for sediment data depends on the frequency and magnitude of scouring events, the rate of sedimentation, and the sources of sediment. For example, surface sediment in lakes and reservoirs may be scoured infrequently, but it may be buried by new sediment if the deposition rate is high. In any cases, the exposures may not have changed since samples were collected if the source sediment has not changed with respect to contaminant concentration or bioavailability. Finally, data for past samples or from buried sediments may provide a conservative estimate of current exposures if natural attenuation or up-gradient remedial actions are known to have reduced the flux of sediment-associated contaminants.

Two different expressions of sediment contamination are commonly used in ecological risk assessments: concentrations of chemicals in whole sediments and concentrations in sediment pore waters. The use of pore water is based on the assumption that chemicals associated with the solid phase are largely unavailable, and therefore sediment toxicity can be estimated by measuring or estimating the pore water concentration. This assumption is supported by much empirical data and is widely accepted by the scientific community (NOAA 1995). Sediment ingesters, however, may be more exposed to particle-associated chemicals, particularly hydrophobic chemicals, than other benthic species (Kraaij et al. 2002). Adams (1987) reviewed the feeding habits of benthic species and concluded that burrowing marine species frequently were sediment ingesters, but that, except for oligochaetes and some chironomids, few freshwater species were sediment ingesters. Still, those taxa can constitute most of the benthic assemblage in areas of fine sediment deposition.

For neural organic chemicals, bulk sediment concentrations may be adjusted for biological availability by normalizing to organic carbon content. The EPA has used this equilibrium partitioning (EqP) approach to develop sediment guidelines for nonionic organic chemicals (EPA 2000b). The dissolved chemical concentration in pore water can be estimated directly from the organic–carbon normalized sediment concentration. This EqP approach assumes that hydrophobic interactions with particulate organic carbon (POC) control the partitioning of nonionic organic chemicals between particles and pore water:

$$C_{pw} = C_s/(K_{oc}\,f_{oc}) \tag{22.1}$$

where C_{pw} = concentration in pore water; C_s = concentration in solid phase; K_{oc} = chemical-specific partition coefficient; f_{oc} = mass fraction of organic carbon (kilograms organic carbon per kilogram sediment).

This estimate is independent of the DOC concentration. The proportion of a chemical in pore water that is complexed to DOC (i.e., colloidal carbon) can be substantial. It is the free, uncomplexed component that is bioavailable and that is in equilibrium with the organic carbon normalized sediment concentration. Therefore, for highly hydrophobic chemicals and where there is significant DOC, the solid-phase chemical concentration gives a more direct estimate of the bioavailable pore water contaminant concentration than does the pore water concentration (EPA 2000b). Using the pore water chemical concentration to estimate the free pore water chemical concentration requires that the DOC concentration and partition coefficient be known. Dissolved chemical concentrations can be estimated from an extension of Equation 22.1, incorporating both the particulate-dissolved partitioning coefficient and a colloidal-dissolved partitioning coefficient (Burkhard 2000). For example, in the Elizabeth River, Virginia, the two-phase model did not fit the partitioning data for polycyclic aromatic hydrocarbons (PAHs) well, but a three-phase model that included dissolved organic matter did (Mitra and Dickhut 1999).

Another sediment phase that may be considered is black carbon, which comprises products of pyrolysis such as soot and related materials such as coal, shale, and tire rubber particles (see Burgess and Lohmann 2004). Its partitioning properties may be considerably different from other organic particles, but different black carbons may also have very different properties. Incorporation of black carbon into sediment partitioning models results in four-phase partitioning.

Where sufficient data are available, a site-specific partitioning model may be developed using locally derived coefficients. It can incorporate as many phases as are needed to explain variance in bioavailability across the site.

For some metals (cadmium, copper, lead, nickel, and zinc), bulk sediment concentrations may be adjusted for the lack of biological availability. Acid volatile sulfide (AVS) is a reactive pool of solid-phase sulfide that is available to bind metals and render that portion unavailable and nontoxic to biota (DiToro et al. 1992; Ankley et al. 1996). The AVS is extracted from sediment using hydrochloric acid. The metal concentration that is simultaneously extracted is termed the simultaneously extracted metal (SEM). When AVS is in excess of SEM (i.e., the SEM/AVS molar ratio is less than 1), toxicity is highly unlikely because the metal is in the form of bio-unavailable sulfides. The metals become potentially toxic (available) when the SEM/AVS molar ratio is greater than 1. In such a case, the molar SEM concentration in excess of AVS (SEM–AVS) is a better representation of exposure because it expresses the potentially available concentration (Hare et al. 1994). The EPA has used this approach to develop sediment guidelines for metal mixtures (EPA 2002e). One objection to this approach is that the excess SEM may be unavailable because of binding to other ligands. Another consideration is that AVS concentrations are highly variable in space and time (Luoma and

Fisher 1997). Sampling designs should account for variations in exposure with season and depth, to the extent practical. Other warnings and limitations associated with the AVS approach are presented by NOAA (1995).

Although unadjusted bulk sediment concentrations are relatively poor estimators of effective exposure, they are still commonly used in sediment exposure assessments (NOAA 1995). Collection and analysis methods for bulk sediment concentrations are standardized, and these measurements are often required for assessments of contaminated sites. Bulk sediment concentrations can be (1) compared to relevant effects concentrations, (2) used to focus sampling and assessment efforts on the contaminants and locations of greatest concern, and (3) used to estimate exposure of sediment ingesters to highly particle-associated chemicals.

Testing of overlying water concentrations is an alternative type of analysis that is potentially useful (Chapman et al. 1997). Near-surface pore water may be in equilibrium with overlying water if fine-grained sediment particles and organic carbon are effectively absent. Contaminated water is the primary exposure pathway in these sediments, because they are poor habitat for sediment ingesters. Sediment may also be a source of contaminants to the surface water. Epibenthic organisms may be primarily exposed to chemicals released into the overlying water in such situations.

Contaminant concentrations in benthic organisms are direct measures of exposure that can be compared with body burden-based effects concentrations for similar organisms (Section 24.2.5). Because the effective internal concentrations should be the same for sediment and water column species, it could help to alleviate the paucity of effects data for sediment invertebrates.

22.4 EXPOSURE TO CONTAMINANTS IN SOIL

Terrestrial plants, soil invertebrates, and soil microorganisms are continually exposed to chemicals in soil. Because they are immobile or nearly so at the scale that regulatory or remedial actions occur, their activity patterns (except for exposure depth) are not relevant for an ecological risk assessment. Exposure of wildlife to soil contaminants is discussed in terms of multimedia exposure (Section 22.8). Exposure to contaminated soils is much less well understood than exposure to water and sediments, and there is much less consistency in soil sampling, measurement, testing, and modeling. In part, this is because of less concern for contaminated soils and in part because soils are more complex. For example, they have a gas phase as well as liquid and solid phases and they undergo cycles of wetting and drying and of freezing and thawing. As a result, it is difficult to estimate exposures of organisms in contaminated soils in a way that allows the use of existing toxicity data.

The major soil exposure issues considered in the analysis plan are: (1) the appropriate soil or soil solution measurements, (2) the appropriate sampling depth for exposure of the endpoint organisms, (3) the applicability of existing data for uptake from soil, (4) soil characteristics that influence the biological uptake of contaminants, and (5) the expression of exposure in available exposure–response relationships for the contaminants of concern. This section describes issues that are generic to soil exposures and the following sections address issues specific to plants (Section 22.5), invertebrates (Section 22.6), and microbes (Section 22.7).

22.4.1 CHEMICAL ANALYSES TO ESTIMATE EXPOSURE

Soil chemical analyses, associated estimates of exposure, and measures of exposure used in exposure–effects relationships are presented in Table 22.1.

Because organisms do not extract chemicals so thoroughly as total extraction methods, total concentrations tend to overestimate exposure. Also, as with total concentrations in water, strong extractions of soil may obscure increases in bioavailable forms of chemicals and

TABLE 22.1
Alternative Methods of Soil Analysis and Associated Methods for Estimation of Exposure and Toxicity

Soil Analyses	Estimate of Exposure	Exposure in Effects Test
Analysis of a total extract	Total extractable concentration	Total extractable concentration in test soil Concentration added (spiked) to soil
	Solution-phase concentration modeled from total soil concentration	Modeled solution-phase concentration in test soil Aqueous concentration in a toxicity test in solution culture
	Total concentration normalized for soil factors that determine bioavailability	Soil concentration normalized for toxicity soil factors that determine bioavailability
Analysis of an aqueous extract	Concentration in aqueous extract (including mild acids for metals)	Aqueous concentration in toxicity tests in solution culture Modeled solution-phase concentration associated with test in soil (equilibrium partitioning or metal speciation)

Source: Adapted from Suter, G.W., II, Efroymson, R.A., Sample, B.E., and Jones, D.S., *Ecological Risk Assessment for Contaminated Sites*, Lewis Publishers, Boca Raton, FL, 2000. With permission.

may raise detection limits by increasing analytical interferences. An advantage of total extractions is that the results can be compared to similarly rigorous analyses of contaminated site soils used in toxicity tests, or, with less accuracy, to nominal concentrations of chemicals added to test soils.

Total chemical analyses are poorly predictive of toxicity; the fraction of the chemical that is bioavailable is highly variable, depending on soil and contaminant characteristics. Because of variation in soil properties that control the availability of chemicals to organisms, total concentrations in different soils or even the same soil at different times may result in very different levels of effects. Aged organic chemicals (i.e., chemicals that have persisted in soil for months to years) are less bioavailable to soil organisms than freshly added forms of the same chemicals (Kelsey and Alexander 1997; Ma et al. 1998). This is also the case for metal uptake and toxicity to soil invertebrates, microbes, and plants (Posthuma et al. 1998). The extractability of aged chemicals with some solvents has been observed to be associated with bioavailability to earthworms for uptake and to bacteria for degradation, but the rate and extent of sequestration varies markedly among soils (Chung and Alexander 1998). Therefore, no solvent extraction predicts bioavailability of diverse aged organic chemicals in a range of soils. Further, if a chemical has been taken up by plants, a large fraction may be immobilized in litter, dead roots, or humus (Banuelos et al. 1992). The following approaches to exposure estimation deal with the issue of bioavailability by either estimating the bioavailable component of contaminant concentrations in soil or by producing an estimate of exposure that is better correlated with toxicity than the total concentration.

22.4.1.1 Partial Chemical Extraction and Normalization

Soil pore water is commonly assumed to be the bioavailable component for plants and other soil organisms, but the measurement of chemicals in pore water is difficult and often imprecise

(Sheppard et al. 1992). It may be estimated by measuring total concentrations in bulk soil, and then estimating concentrations in soil pore water, as discussed above for sediments (Lokke 1994). Neutral organic compounds are assumed to be in equilibrium between the aqueous phase (pore water) and the organic component of the solid phase. If it is assumed that soil organisms are exposed solely to the aqueous phase, the estimated pore water concentrations can be used with data from aqueous toxicity tests (plants in hydroponic solutions, inverte-brates on blotter paper, or even aquatic invertebrates in water) to estimate risk. This EqP approach remains controversial when applied to sediments and is largely hypothetical for soils. Unlike in sediments, the variation in water content of soils can lead to saturation and other nonequilibrium dynamics, but these dynamics are probably important only for arid soils or during droughts. Further, plants take up chemicals directly from solid phases by contact exchange, although the rates are likely to be relatively small (McLaughlin 2001).

Multivariate methods of normalization may provide estimates of effective concentrations of metals and other inorganic chemicals. For example, Dutch reference values for various chemicals in soil were derived by normalizing values to a standard soil with 10% organic matter and 25% clay content using linear regression (VROM 1994). For example, a reference value for cadmium (R_{Cd}) in mg/kg is

$$R_{Cd} = 0.4 + 0.007(c + 3o) \tag{22.2}$$

where c is % clay and o is % organic matter (Van Straalen and Denneman 1989). How-ever, recent studies indicate that those equations must be expanded to at least include pH (Posthuma et al. 1998). Concentrations of some organic chemicals in soil could be normalized using organic matter alone. If, at a particular site, it could be shown that effective exposure concentrations for chemicals are a function of a set of soil properties, it would be possible to normalize soil concentrations across test soils and site soils.

Exposure estimates may be based on various aqueous extractions (Section 20.2.8), which can be compared with similar estimates of exposure from extraction of soil used in toxicity tests. Alternatively, extractions with dilute aqueous salts or acids could be assumed to approximate the concentrations in soil pore water, with the appropriate correction for dilution. With these extraction methods, comparison with aqueous toxicity test results would be possible, as in the sediment EqP approach. However, as discussed with respect to aged chemicals, no extraction method is generally applicable, so extractants must be evalu-ated and selected ad hoc (Sauve 2001).

An alternative approach for metals is to use mechanistic speciation models to estimate the free ionic concentration that is available for exposure and uptake, as in the aquatic biotic ligand model (Section 22.2). Results to date suggest that metals in soil compete most with Ca^{++} and H^+ for sites primarily on dissolved organic matter as well as biotic ligands (Weng et al. 2002). Water-saturated soils, as in wetlands, may behave more like sediments with sulfides dominating metal binding.

For a site contaminated with multiple inorganic and organic chemicals, the best alternative is often to measure concentrations of contaminants in soil, and relate them to the results of tests of that soil (Chapter 24). If toxicity data from the literature must be used, one may simply match the actual or adjusted measurement to the test conditions.

22.4.1.2 Input Form of the Chemical

The bioavailability of a chemical added to soil depends on the form in which it was added as well as the soil characteristics. If a metal is added to soil in inorganic salt form (typical for a laboratory toxicity test or accumulation assay), the added concentration of the element is likely to be more bioavailable than the same total concentration of an element in a contaminated field soil. For example, based on calculated soil–plant uptake factors, 10 of 15 metals added to soil as salts were more available than the background fractions (Cataldo and Wildung 1978). However, the variance in soil properties, biological properties, and forms of metals other than soluble salts prohibit simple generalizations (Efroymson et al. 2001). Finally, the speciation of chemicals may change after deposition in soils. For example, deposited hexavalent chromium from cooling tower drift in Oak Ridge was rapidly and almost completely transformed to the less bioavailable trivalent form.

24.4.1.3 Chemical Interactions

Interaction among chemicals is a potentially important factor in their uptake from soil that should be considered by assessors during risk characterization. For example, lead has been widely observed to increase cadmium uptake, but the results are not so clear for the effect of cadmium on the uptake of lead (Carlson and Bazzaz 1977; Miller et al. 1977; Carlson and Rolfe 1979). The amendment of soil with high concentrations of arsenic (50 mg/kg) increased the uptake of mercury by foliage of Bermuda grass (Weaver et al. 1984). The solubility of hydrophobic organic compounds in water is influenced by the presence of other such compounds (Eganhouse and Calder 1976). No existing models incorporate these interactions; they must be addressed by case-specific studies.

24.4.1.4 Nonaqueous Phase Liquids

When nonaqueous-phase liquids (NAPLs) such as petroleum or polychlorinated biphenyl (PCB) oils are present in soil, the bioavailable fraction of each chemical is not expected to be correlated with the total concentration in soil. Hydrocarbons or other constituents of NAPLs are divided among four phases: water, soil solid phase, soil gaseous phase, and NAPL. The most bioavailable fraction may be the aqueous portion, but EqP methods are not appropriate. When the concentration of a lipophilic chemical in the aqueous phase is close to saturation, and a NAPL is present, the aqueous concentration is independent of the total concentration in soil. Additionally, the NAPL fraction may be available to sorb to, and perhaps enter, plant roots, earthworm skins, and microorganisms. Measurement of the concentration of the dissolved fraction in soil water might approximate exposure. However, this type of measurement is not practical, since it is difficult to exclude NAPL from extractions of the aqueous phase. Further, petroleum and some other NAPLs may have a greater effect through their influence on the physical properties of soil than through the toxicity of their constituents. Thus, there is a high degree of uncertainty associated with the exposure of soil organisms to NAPLs, and the uncertainty should be acknowledged.

22.4.2 SOIL DEPTH PROFILE

The level of exposure of soil organisms to contaminants is defined partly by depth. The sampling depth is important because of the gradients of contaminant concentrations and of biological activity in soil. For example, if rooting depths are improperly estimated for plants, erroneous conclusions could be reached: (1) depth-averaged soil concentrations may be below toxicity benchmarks where effective root-zone concentrations are above, or (2) adverse effects observed during a vegetation survey might be related to different chemical concentrations than those actually in the effective rooting zone.

22.5   EXPOSURE OF TERRESTRIAL PLANTS

Chemicals are taken up by plants directly from soil and from air. Most contaminants are taken up passively from soil water in the transpiration stream, though nutrients such as copper and zinc are taken up actively from solution. The exposure of vascular plants to contaminants in soil is controlled by the distribution of roots in the soil profile, physicochemical characteristics of soil, and interactions among chemicals. In addition, physiological differences among plant species are responsible for differential accumulation of contaminants among taxa. At contaminated sites, the exposure of the plant community can be considered to be the distribution of exposures of individual plants in the area occupied by the endpoint community. However, if the risk manager desires to protect certain species, the spatial distribution of exposure is important.

22.5.1   ROOTING DEPTH

The optimal depth interval for sampling soil should be the interval where most feeder roots are found for plant populations or communities that are assessment endpoint entities or are the major food sources of endpoint herbivores. Ideally, the risk assessor should determine this depth by measurement. Roots may be weighed to determine the depth profile where rooting is most dense. The rooting depths of plants vary with species; nutrient and oxygen availability; soil water; soil temperature; presence of pathogens; soil pore size, distribution, and compaction (Foxx et al. 1984); and location of rock–soil interfaces (Parker and van Lear 1996). Thus, estimates of rooting depth at one site based on measurements at another are highly uncertain, but are better than no consideration of rooting depth.

  The sampling depth should include the interval containing most of the roots. Given that root densities often decrease exponentially with depth (Parker and van Lear 1996), and that chemical concentrations for surface-deposited contaminants decrease with depth, averaging soil concentrations over the interval from the surface to the maximum rooting depth would underestimate effective exposure. If samples are taken at multiple depth intervals, the surface intervals should be weighted more than subsurface intervals in the estimate of exposure. In the absence of multiple samples, the sampling depth should be less than the maximum rooting depth of the plant community if the chemical concentrations decrease with depth.

  Most of the root density for all biomes is in the top 30 cm of soil (Jackson et al. 1996). Thus, 30 cm is a good default estimate of the depth of plant root exposure to contaminants in soil. Of course, general information about biomes should be supplemented with site knowledge. A shallower depth may be more appropriate if the site is dominated by grasses. Among different biomes, grasses have 44% of their roots in the top 10 cm of soil (Jackson et al. 1996). Similarly, the variance in water availability with depth may alter the relative uptake of contaminants from different depth intervals. Also, if one soil depth must serve to estimate exposure to plants, soil invertebrates, and wildlife, a compromise depth for estimating all exposures may be necessary.

22.5.2   RHIZOSPHERE

Plants may influence the solubility of chemicals in the rhizosphere (the soil in the vicinity of roots), metals may be mobilized from solid phases by rhizosphere microbes, and the rate of degradation of organic chemicals in the rhizosphere may be higher than in bulk soil (Reilley et al. 1996; McLaughlin 2001). Thus, the solution concentration to which a plant is exposed may be different from that in bulk solution. It is not practical for assessors to measure chemical concentrations in soil water in the rhizosphere. However, it is important to be aware of this uncertainty in estimates of exposure of vegetation to contaminants.

22.5.3 WETLAND PLANT EXPOSURES

Wetlands straddle the line between water and soil exposures. The exposure of most wetland plants to chemical contaminants is assumed to be best represented by the concentration in the surface water, spring water, or groundwater that maintain the wetland condition. The rationale is that (1) concentrations of chemicals in soil are not necessarily in equilibrium with those in water moving through the wetland, and (2) roots are more directly exposed to concentrations in solution than in bulk soil. Research is needed to confirm this conjecture.

22.5.4 SOIL PROPERTIES AND EXPOSURE OF PLANTS

Soil physicochemical properties determine the forms of chemicals and their availability for uptake by plants. The similarity of site soils and test soils should be evaluated on the basis of these properties. Soil properties reported to affect the concentrations of inorganic chemicals in soil solution include pH, cation exchange capacity, organic matter, iron hydroxides, and particle-size fraction (Bysshe 1988; Sims and Kline 1991; He and Singh 1994; Jiang and Singh 1994; Weng et al. 2002). In addition, since passive uptake is a function of transpiration, uptake rates are increased by high soil water content and temperature and by low relative humidity (McLaughlin 2001). Hence, total concentrations of elements in soil may be normalized for these descriptors (Section 22.4.1).

Concentrations of nonionic organic chemicals in soil solution depend primarily on the soil organic matter content (Topp et al. 1986; Sheppard et al. 1991). That is, the exposure of plants to organic contaminants is likely to be higher in sand than in a muck soil, but commonly used soil partitioning coefficients (K_d) presume they are equal. However, the EqP models that are used for exposure to neutral organics in sediment have not been generally accepted for terrestrial plants.

22.5.5 PLANT INTERSPECIES DIFFERENCES

The physical or physiological differences that explain differences among plant species in accumulation of metals are largely unknown. In general, the uptake of inorganic chemicals varies much more among plant than animal species. Some plants are hyperaccumulators of metals, particularly selenium and nickel. Hence, concentrations in leaves or other plant parts are not likely to be useful measures of exposure to metals. Some evidence suggests that lipid content of plants may be useful in estimating their uptake and internal exposure to neutral organic chemicals (Bromilow and Chamberlain 1995).

22.5.6 PLANT EXPOSURE IN AIR

Plants may be exposed to chemicals in air from active atmospheric sources and from the volatilization of chemicals from contaminated soil and uptake by plant foliage. For pollutant gases and vapors, exposure is expressed as the concentration in air at canopy height. For chemicals taken up from soil via air, the soil concentration may be the best measure of exposure.

Soil contaminants primarily taken up by plants as vapors include high-molecular-weight, nonionic organic compounds such as DDT, dieldrin, endrin, and heptaclor, as well as PCBs. 2,3,7,8-TCDD is not transported in detectable amounts within the transpiration stream of herbaceous plants; aboveground plant parts are contaminated primarily from air (Trapp and Matthies 1997). Some forms of some inorganic elements are also transferred from soil to plants largely via air. In particular, most mercury in aboveground plant tissue is taken up as volatile, elemental mercury through the leaves, with little accumulation from the soil via the roots and transpiration stream (Bysshe 1988).

The atmospheric route of exposure is subsumed in empirical uptake models if concentrations of the chemical in air and soil are assumed to be in equilibrium and the soil is the only source of the chemical in the vicinity of the plant. Although mercury, for example, may be taken into the plant via air, a significant correlation between the concentration in soil and plant tissues may exist (Shaw and Panigrahi 1986). At or near background levels of organic chemicals, it is reasonable to assume that concentrations in air and soil are close to equilibrium (Trapp and Matthies 1997).

22.6   EXPOSURE OF SOIL INVERTEBRATES

This section primarily addresses exposure of earthworms because (1) more information is available on their exposure than on that of soil arthropods, (2) toxicity and internal concentrations in earthworms have been related to concentrations in soil, and (3) earthworms are more often identified as assessment endpoint organisms than other soil invertebrates. In addition, exposure of earthworms is generally assumed to occur primarily through the skin or through the gut in a manner equivalent to dermal exposure, while soil arthropods such as isopods, due to their exoskeletons, are probably exposed primarily through selectively ingested food, a more complex scenario (Van Brummelen et al. 1996). The exposure of earthworms to chemicals in soil is determined by several factors. These include: concentration of chemical in soil, depth of burrowing, material ingested, activity patterns, soil characteristics, and interactions with other contaminants. Although exposure to some chemicals may be related to the concentration of a chemical in soil water (van Gestel and Ma 1988; Janssen et al. 1997), more often the measurement that is taken is the total concentration of the chemical in soil.

22.6.1   Depth of Exposure and Ingested Material

The depth of burrows and the amount of time spent at each depth are factors that determine the exposure of earthworms to chemical contaminants. Depending on the species and conditions, earthworm burrows may be deep or shallow and horizontal or vertical (Lee 1985). Information on diets and burrowing depths of earthworm species in North America and Europe has been compiled in Lee (1985) and Suter et al. (2000). In general, earthworms are most abundant in near-surface soils (roughly 2 to 30 cm) with high organic content, and most tests simulate such conditions. Hence, the top 30 cm, which is a common default sampling depth for plant exposures, is also appropriate for earthworms. Species that burrow to extreme depths often feed on litter at the surface, so sampling at depths greater than 50 cm for earthworm exposure is unwarranted in most cases. If risks to earthworms are important and uncertain, studies should be conducted in the contaminated or potentially contaminated area to determine earthworm feeding and burrowing behavior.

Most arthropods inhabit the near-surface soil, but some are exposed to much deeper soils. Termites, for example, obtain clay-rich soil to build surface mounds and feeding galleries from deep soil horizons. Because of the relatively impermeable exoskeletons of arthropods, food may be a more important determinant of uptake than direct exposures to soil. For example, the uptake of PAHs by three isopod species was correlated to levels in humus and fragmented litter rather than fresh litter or mineral soil (Van Brummelen et al. 1996). Little is known about the exposure of microarthropods and free-living nematodes that are often highly abundant and diverse components of the soil fauna.

22.6.2 Soil Properties and Chemical Interactions

The exposure of earthworms to inorganic contaminants is dependent on soil chemical properties. Janssen et al. (1997) observed that the accumulation of several inorganic elements by earthworms was controlled by the same soil characteristics that affect the partitioning between the solid phase and pore water. Thus, they concluded that uptake is either from the soil pore water or a related route, but that bioavailability, as measured by the soil–earthworm uptake factor, is predictable using total metal concentrations supplemented with local soil characteristics. Soil factors that may determine exposure of earthworms to inorganic chemicals include pH, calcium content, cation exchange capacity, and organic matter (Corp and Morgan 1991; Saxe et al. 2001). The exposures of earthworms to at least some organic chemicals are dependent on the concentrations in soil water, so EqP may be used to estimate exposure and relate it to tests of worms exposed to chemicals in solution (van Gestel and Ma 1988).

22.7 EXPOSURE OF SOIL MICROBIAL COMMUNITIES

Microorganisms reside in distinct microenvironments of soil. Therefore, they may be exposed to a wide range of local concentrations, including concentrations in soil water, potentially higher concentrations at or near the surface of soil particles, or even droplets of organic liquids to which they may attach, such as petroleum or PCB oil. Because typical assessment endpoints for microorganisms are ecosystem-level microbial processes, the relevant exposure is a spatial and temporal average. The exposure that is typically used is the concentration in bulk soil, primarily because the available effects tests have used that measure. The concentration in soil water is an alternative measure of exposure, but most endpoints of interest to risk assessors (nitrogen transformation, enzyme activity) have not been tested in liquid culture in the presence of toxicants. Because the assessment endpoints of concern are the microbial processes that influence ecosystem dynamics, the microbes of primary concern are aerobic organisms in surface soil. Thus, the sampling depths that are selected for plants or soil invertebrates are reasonable for microbial processes.

22.8 EXPOSURE OF WILDLIFE

Wildlife exposures are likely to involve multiple media and multiple routes. Mammals, birds, reptiles, and amphibians may drink or swim in contaminated water, ingest contaminated food and soil, breathe contaminated air, or absorb contaminants through dermal contact. In addition, because most wildlife species are mobile, exposure is not restricted to a single location. They may assimilate contaminants from several spatially discrete sources. As a consequence, the accurate estimation of wildlife exposure requires the consideration of habitat requirements and the potential to move among habitats.

The contaminant exposures experienced by wildlife can be described using either internal or external measures of exposure. Internal measures of exposure are chemical concentrations in whole organisms or organs of the endpoint species. The latter may be target organs for toxic effects (e.g., liver, kidney, or brain) or nontarget organs that simply accumulate contaminants (e.g., fat or bone) and may be readily sampled (e.g., hair or feathers). Internal exposure measures have several advantages. They may integrate all exposure pathways through which the individual may have been exposed, average exposure over both time and space, indicate site-specific contaminant bioavailability (if data from the field are used), and eliminate exposure model error and parameter uncertainty. In addition, if contaminant concentrations are determined for target organs, exposure data may be directly related to toxicity. Internal measures of exposure may be obtained by field sampling and analysis (Chapter 20) or by modeling uptake (Section 22.9.5). External measures are concentrations in media or rates of contaminant intake measured or estimated at entry points to the animal, including the skin, lungs, and digestive system. These external measures may be used with simple models to estimate wildlife exposures.

22.8.1   EXPOSURE MODELS BASED ON EXTERNAL MEASURES

Wildlife may be externally exposed to contamination by oral, dermal, and inhalation pathways. Oral exposure includes the consumption of contaminated food, water, or soil. Dermal exposure occurs when contaminants are absorbed through the skin. Inhalation exposure occurs when volatile compounds or fine particles are inhaled. Total exposure is the sum of exposure by those pathways:

$$E_{total} = E_{oral} + E_{dermal} + E_{inhal} \qquad (22.3)$$

where E_{total} = total exposure from all pathways; E_{oral} = oral exposure; E_{dermal} = dermal exposure; and E_{inhal} = exposure through inhalation.

22.8.1.1   Dermal Exposure

Dermal exposure is generally assumed to be negligible for birds and mammals. While methods are available to assess dermal exposure to humans (EPA 1992b), necessary data to estimate dermal exposure are generally not available for wildlife. Additionally, feathers and fur of birds and mammals reduce the likelihood of significant dermal exposure by limiting the contact of skin with contaminated media. Therefore, dermal exposure is expected to be negligible relative to other exposure routes in most cases. However, if contaminants that have a high affinity for dermal uptake are present (e.g., organic solvents and some pesticides) and an exposure scenario for an endpoint species is likely to result in significant dermal exposure (e.g., burrowing mammals or swimming amphibians), dermal exposure must be considered. Pesticide sprays, particularly organophosphates that have been used as avicides by spraying perches, are an important example (Driver et al. 1991; Henderson et al. 1994).

Hope (1995) has recommended the use of two models to estimate daily dose from soil. The first assumes that an organism is exposed to all of the contaminant in the surface soil with which its skin is in contact. The second assumes that an organism is exposed to the soil that adheres to its skin and that not all of the contaminant is absorbed. The latter model, which is adapted from human health risk assessment, is

$$D = (A \times P \times S \times C \times F \times B)/W \qquad (22.4)$$

where D = daily dose (mg/kg/d); A = surface area of the organism (cm^2); P = proportion of the surface area contaminated; S = skin adherence factor (mg/cm^2); C = concentration of the contaminant in soil (mg/kg); F = conversion factor (10^{-3} kg/mg); B = bulk density of the soil (kg/cm^2); and W = body weight of the organism (kg).

Adherence factors and absorption factors are unlikely to be available for wildlife species; values for mammals may be borrowed from the human health literature. Hope (1995) recommends the using of the default human adherence factor of 0.52 ± 0.9 mg soil/cm^2 skin as a point of departure for other mammals.

Dermal exposure from pesticide applications is complex. The US EPA is developing models for birds that include direct spray exposure, dust and puddle bathing, foot contact with soil, and contact with sprayed foliage (OPP 2004).

The use of these dermal exposure estimates requires that dermal toxicity data be available (which is seldom the case) or that dermal exposure be converted to an equivalent oral exposure. The simplest assumption is that they are equivalent, but available data do not support that practice. The US EPA has developed an empirical model to perform that conversion for pesticide exposures of birds, but it is highly uncertain (OPP 2004).

22.8.1.2 Inhalation Exposure

Inhalation of contaminants by wildlife is frequently assumed to be negligible in wildlife risk assessments. This assumption is justified at contaminated sites by two considerations. First, because most contaminated sites are either capped or vegetated, exposure of contaminated surface soils to winds and resulting aerial suspension of contaminated dust particulates are minimized. Second, most volatile organic compounds (VOCs), which are the contaminants most likely to present a risk through inhalation exposure, rapidly volatilize from soil and surface water to air, where they are rapidly diluted and dispersed. Paterson et al. (1990) suggest that organic compounds with soil half-lives of <10 d are generally lost from soil before significant exposure can occur. Further, vapor diffusion is not a potentially significant exposure process if Henry's Law Constant is >24.3 Pa/m^3/mol (Wang and Jones 1994; Hope 1995). As a consequence, significant exposure to waste VOCs through inhalation is unlikely at most contaminated sites. When inhalation may be important at a contaminated site, the EPA (1993g) recommends consulting an inhalation toxicologist. However, models for inhalation exposure of a burrowing mammal in contaminated soil are provided by Hope (1995).

Some situations such as spraying and fumigating with biocides may result in significant inhalation exposure of wildlife (Driver et al. 1991). The US EPA has proposed models for inhaled doses of pesticide vapors and droplets by birds (OPP 2004). To convert the respiratory concentration to an equivalent oral dose, they recommend using the ratio of mammalian oral LD_{50} to inhalation LC_{50} as an equivalency factor.

22.8.1.3 Oral Exposure

Because exposures of wildlife through both the dermal and inhalation pathways are generally neglected, Equation 22.3 typically reduces to

$$E_{\text{total}} \approx E_{\text{oral}} \tag{22.5}$$

Oral exposure experienced by wildlife may come from contaminated food (either plant or animal), water, or soil. Soil ingestion may be incidental while foraging and grooming or purposeful to meet nutrient needs. In some cases, waste materials may also be directly ingested (e.g., oil from grooming or ethylene glycol in antifreeze). The total oral exposure is the sum of the exposures attributable to each source:

$$E_{oral} = E_{food} + E_{water} + E_{soil} + E_{direct} \qquad (22.6)$$

where E_{food} = exposure from food consumption; E_{water} = exposure from water consumption; E_{soil} = exposure from soil consumption; and E_{direct} = exposure from direct consumption.

Wildlife exposure estimates should be expressed in terms of a body weight–normalized daily dose, typically as milligrams of contaminant per kilogram body weight per day (mg/kg/d). Exposure estimates expressed in this manner may then be compared with similarly reported toxicological data presented in the literature. Models for the estimation of exposure from oral ingestion have been reported in the literature (EPA 1993f,g; Sample and Suter 1994; Hope 1995; Freshman and Menzie 1996; Pastorok et al. 1996; Sample et al. 1997). The basic form is

$$E_j = \sum_{i=1}^{m} (I_i \times C_{ij}) \qquad (22.7)$$

where E_j = oral exposure to contaminant (j) (mg/kg/d); m = number of ingested media (e.g., food, water, or soil); I_i = ingestion rate for medium (i) (kg/kg body weight/d or L/kg body weight/d); and C_{ij} = concentration contaminant (j) in medium (i) (mg/kg or mg/L). E_j represents the daily exposure averaged over the exposure duration, which at most waste sites is likely to be chronic, measured in terms of months or years.

Few wildlife species consume diets that consist exclusively of one food type. To meet nutrient needs for growth, maintenance, and reproduction and to cope with varying food availability, most wildlife species consume multiple food types. To account for differences in contaminant concentrations of different food types, exposure estimates should be weighted by the relative proportion of daily food consumption attributable to each food type and the contaminant concentration in each food type. Wildlife may also drink from different water sources and consume soils that differ in contaminant concentrations. Each food type, water source, and soil source may be included in a modification of Equation 22.7:

$$E_j = \sum_{i=1}^{m} \sum_{k=1}^{n} p_{ik} (I_i \times C_{ijk}) \qquad (22.8)$$

where n = number of types of medium (i) consumed; p_{ik} = proportion of type (k) of medium (i) consumed (unitless); and C_{ijk} = concentration of contaminant (j) in type (k) of medium (i) (mg/kg or mg/L).

It is generally assumed that chemical absorption by the wildlife species in the field is 100% (Sprenger and Charters 1997) or equivalent to that in the toxicity test species used to evaluate exposure (EPA 1993g). Because chemical forms used in toxicity tests are likely to be more bioavailable (e.g., test based on solutions of metal salts or organic chemicals in vegetable oils) than chemical forms at field sites, this assumption may result in overestimates of exposure. To address this issue, absorption factors may be included in oral exposure models (Hope 1995; Pastorok et al. 1996). These factors represent the proportion of the ingested dose that is actually absorbed through the gastrointestinal tract. Few absorption factors are available. Owen (1990) presents values for 39 chemicals. Empirical models for estimation of absorption of organic chemicals by birds and nonruminant mammals are presented by Garten and Trabalka (1983).

22.8.1.4 Spatial Issues in Wildlife Exposure

If the contaminated area is spatially heterogeneous with respect to either contamination or wildlife use or if the contaminated area provides some but not all habitat requirements (e.g., food but not water or shelter), the exposure model should be modified to include spatial factors. The most important spatial consideration is the movement of wildlife. Animals travel varying distances, on a daily to seasonal basis, to find food, water, shelter, and climatic conditions. The area encompassed by travels of nonmigratory species or of migratory species during the breeding season is defined as the home range (the term is used here to include territories). If the spatial units being assessed are larger than the home range of individuals of an endpoint species and provide the habitat needs of the species, the previously listed models are adequate. However, endpoint species often have home ranges that are larger than contaminated units, or the contaminated site may not supply all of a species' habitat requirements. In such cases, the wildlife exposure model must be modified.

If a contaminated spatial unit has similar habitat quality to the surrounding area but is smaller than the home range, use of the unit may be described as a simple function of site area. That is, one can assume that for wildlife that use the entire contaminated area, exposure is proportional to the ratio of the unit size to home range size:

$$E_j = \frac{A}{HR}\left[\sum_{i=1}^{m}\sum_{k=1}^{n}p_{ik}(I_i \times C_{ijk})\right] \qquad (22.9)$$

where A = area (m^2) contaminated; and HR = home range size (m^2) of the endpoint species. A/HR is commonly known as the area use factor and may be thought of as the probability that an individual of a wildlife species is in the contaminated area rather than another portion of its home range (Hope 2004).

Equation 22.9 implies that the entire habitat within a contaminated area is suitable and that use of all portions of the contaminated area is equally likely. Because many sites are industrial, agricultural, or otherwise highly modified in nature, it is unlikely that all contaminated areas provide suitable habitat. If it is assumed that use of an area is proportional to the amount of suitable habitat available in that area, Equation 22.9 becomes:

$$E_j = P_h\left(\frac{A}{HR}\left[\sum_{i=1}^{m}\sum_{k=1}^{n}p_{ik}(I_i \times C_{ijk})\right]\right) \qquad (22.10)$$

where P_h = proportion of suitable habitat in the contaminated area.

Another complication is the spatial heterogeneity of contaminants and habitat quality. These models are based on the assumption that either contaminants are evenly distributed, or wildlife forage randomly with respect to contamination on areas that constitutes habitat, so they are exposed to mean concentrations. Similarly, areas are assumed to be habitat or nonhabitat, but the quality of habitat may vary continuously across space or between a contaminated area and other portions of the home range. Similarly, contaminant levels vary in a manner that is not independent of habitat quality. For example, contaminant concentrations might be greatest near the center of a site, but the habitat quality might be highest near the edges. In such cases, it might be necessary to model the proportional contribution of each area with a distinct combination of contaminant level and habitat quality. The derivation of appropriate models for these cases is left to the reader.

As can be seen, if the distributions of contamination and habitat quality are complex, this approach to exposure estimation rapidly becomes ungainly. In such cases, it is advisable to implement the exposure in a Geographic Information System (GIS). A GIS is a tool to facilitate spatial modeling and mapping of results. Using a GIS, spatial units (cells) can be assigned contaminant concentrations and habitat characteristics such as food density, availability of water, cover, and bare soil. Occurrence of organisms of endpoint species may then be assigned probabilities based on habitat characteristics, and exposure may be calculated and summed across cells. If information on the distribution or movements of endpoint species (generated by radiotelemetry or censuses) is available, it may be combined with the habitat and contaminant data to provide a more accurate visualization of exposure. Examples of the application of GIS to exposure and risk assessments can be found in Banton et al. (1996), Clifford et al. (1995), Henriques and Dixon (1996), and Sample et al. (1996b). A spatially explicit exposure module (SEEM) has been developed for the US Army's risk assessment modeling system (ARAMS) (Wickwire et al. 2004). It allows simulation of the exposure of individuals in a wildlife population moving across a site.

All of these approaches require that assessors relate the ecological properties of lands that are contaminated or would be contaminated by a proposed action to habitat requirements of the endpoint species. If a habitat classification is available, land cover classes must be converted to equivalent habitat classes. An example from Britain is presented in Table 22.2. This example illustrates that different land cover classifications may be applied to the same area, and that land cover may not correspond to habitat classes. Alternatively, an expert in wildlife biology may judge which species occur in which land cover classes. Finally, the habitat qualities of contaminated lands may be evaluated directly, using the habitat evaluation procedure or some other source of habitat data (FWS 1980; Bovee and Zuboy 1988).

22.8.1.5 Temporal Issues in Wildlife Exposure

Temporal patterns of behavior may modify exposure by increasing or decreasing the likelihood of contact with contaminated media. Frequently wildlife behaviors are seasonally variable. Some foods may be available and consumed only at certain times of the year. Similarly, some habitats and certain parts of the home range may be used only in certain seasons. In addition, many species hibernate or migrate; by leaving the area or restricting their activity to certain times of year, their potential exposure may be dramatically reduced. All of these factors should be considered when evaluating contaminant exposure experienced

---

TABLE 22.2
The Relationship between Land Cover Classes and Avian Habitat Classes in the United Kingdom

Land Cover Classes[a]	Target Classes[a]	Bird Habitat Classes
Sea/estuary	Sea/estuary	Coastal sea
		Estuarine and coastal wetlands
Inland water	Inland water	Upland water
Beach/mudflat/cliffs	Beach and coastal bare ground	Coastal shore
		Coastal cliffs
Saltmarsh	Saltmarsh/intertidal vegetation	Estuarine and coastal wetland
Rough pasture/dune grass/grass moor	Lowland grass heath	Healthland
	Montane and hill grass	Mountain/moorland etc.

Source: From Dobson, S. and Shore, R.F., *Hum. Ecol. Risk Assess.*, 8, 45–54, 2002. With permission.
[a] Land cover classes apply to remote sensing data, while target classes apply to ground observations.

by wildlife, and exposure models should be adjusted accordingly. Comparison of exposure estimates generated for differing exposure scenarios aids in identifying the segments of population at greatest risk or times of year when risk is greatest.

Short-term patterns of exposure become critical in acute exposures, particularly when associated with pesticide applications. Many species have a diurnal pattern of foraging, which may result in a bimodal (morning and evening) pattern of entry into agricultural fields. The US EPA has proposed to incorporate this pattern into a probabilistic and dynamic model for acute lethal effects on birds (OPP 2004).

22.8.1.6   Exposure Modifying Factors

The models above imply that the endpoint species have uniform body size, metabolism, diet, home ranges, and habitat requirements. However, these properties may differ between juveniles and adults and between males and females. For example, because they are actively growing, metabolism (and therefore food consumption) is generally greater for juveniles than for adults. Diet composition may also differ dramatically between juveniles and adults of the same species. Similarly, the food requirements of reproductive females are generally greater than those for males. Because greater consumption implies greater exposure, contaminants may present a greater risk to these segments of the population. If a particular life stage or sex is likely to be highly exposed, it should be distinguished in the exposure assessment.

22.8.2   PARAMETERS FOR ESTIMATION OF EXPOSURE

To estimate contaminant exposure by terrestrial wildlife using the models described above, species-specific values for the parameters are needed. Because of large within-species variation in values for life history parameters, data specific to the site in question provide the most accurate exposure estimates and should be used whenever available. Because site-specific life history data are seldom available, published values from other areas must generally be used to estimate exposure. Summaries of wildlife life history information are available from multiple sources. Life history data for birds, mammals, reptiles, and amphibians may be found in the *Wildlife Exposure Factors Handbook* (US EPA 1993g), Sample and Suter (1994), and Sample et al. (1997). Wildlife exposure parameters are also available from state and national agencies such as the California Environmental Protection Agency (CAL/Ecotox; http//:www.oehha.ca.gov) and the Canadian Wildlife Service, and from professional societies such as the American Society of Mammologists and the American Ornithologists Union.

22.8.2.1   Body Weight

Metabolism and food and water consumption rates are functions of body weight. Larger animals consume more food and water than smaller ones. However, because larger animals have lower metabolic rates, smaller animals have higher food and water consumption rates per unit body weight. Hence, smaller animals experience greater oral exposure per unit body weight. In addition to sources cited above, wildlife body weights may be found in Dunning (1993) and Silva and Downing (1995).

22.8.2.2   Food and Water Consumption Rates

Field observations of food, water, and soil consumption rates can be realistic, but are unavailable for most wildlife species. As a result, media consumption rates from laboratory studies of wildlife species are used. Uncertainties associated with these data may be significant, because natural activity regimes and environmental conditions that influence metabolism and consumption rates are difficult to approximate in a laboratory setting.

In the absence of relevant data, food consumption values can be estimated from allometric regression models based on metabolic rates. Nagy (1987) derived equations to estimate food consumption for birds and mammals, including:

$$I_{df} = 0.235W^{0.822} \quad \text{Eutherian mammals} \tag{22.11}$$

$$I_{df} = 0.621W^{0.564} \quad \text{Rodents} \tag{22.12}$$

$$I_{df} = 0.577W^{0.727} \quad \text{Herbivores} \tag{22.13}$$

$$I_{df} = 0.492W^{0.673} \quad \text{Marsupials} \tag{22.14}$$

$$I_{df} = 0.648W^{0.651} \quad \text{All birds} \tag{22.15}$$

$$I_{df} = 0.495W^{0.704} \quad \text{Seabirds} \tag{22.16}$$

$$I_{df} = 0.398W^{0.850} \quad \text{Passerine birds} \tag{22.17}$$

$$I_{df} = 0.013W^{0.773} \quad \text{Insectivorous iguanid lizards} \tag{22.18}$$

$$I_{df} = 0.019W^{0.841} \quad \text{Herbivorous iguanid lizards} \tag{22.19}$$

where I_{df} = food ingestion rate (dry weight) in (g/d); and W = body weight (g live weight).

Variance parameters for these and the other allometric equations in this chapter can be found in the original sources or in EPA (1993g).

Food ingestion rates estimated using these allometric equations are expressed as grams of dry weight per day. Because wildlife species in the field seldom consume dried food, food consumption must be converted to fresh weight by adding the water content of the food. Percent water content of wildlife foods are available in Table 22.3 and the literature. Calculation of fresh food consumption may be estimated as:

$$I_{ff} = \sum_{i=1}^{m} \left(P_i x \frac{I_{fd}}{1 - WC_i} \right) \tag{22.20}$$

where I_{ff} = total food ingestion rate (kg food (fresh weight)/kg body weight/d); m = total number of food types in the diet; P_i = proportion of the ith food type in the diet; and WC_i = percent water content (by weight) of the ith food type.

Food ingestion rates may also be estimated based on the amount of metabolizable energy in foods and metabolic rates (EPA 1993g).

TABLE 22.3
Percent Water Content of Wildlife Foods[a]

Food Type		Percent Water Content		
		Mean	Standard Deviation	Range
Aquatic invertebrates	Bivalves (w/o shell)	82	4.5	
	Crabs (w/shell)	74	6.1	
	Shrimp	78	3.3	
	Isopods, amphipods			71–80
	Cladocerans			79–87
Aquatic vertebrates	Bony fishes	75	5.1	
	Pacific herring	68	3.9	
Aquatic plants	Algae	84	4.7	
	Aquatic macrophytes	87	3.1	
	Emergent vegetation			45–80
Terrestrial invertebrates	Earthworms (depurated)	84	1.7	
	Grasshoppers, crickets	69	5.6	
	Beetles (adult)	61	9.8	
Mammals	Mice, voles, rabbits	68	1.6	
Birds	Passerines (with typical fat reserves)			68[a]
	Mallard duck (flesh only)			67[a]
Reptiles and amphibians	Snakes, lizards			66[a]
	Frogs, toads	85	4.7	
Terrestrial plants	Monocots: young grass			70–88
	Monocots: mature dry grass			7–10
	Dicots: leaves	85	3.5	
	Dicots: seeds	9.3	3.1	
	Fruit: pulp, skin	77	3.6	

Source: From EPA (U.S. Environmental Protection Agency), *Wildlife Exposure Factors Handbook*, EPA/600/R-93/187, Office of Health and Environmental Assessment, Washington, DC, 1993. With permission.

[a]Single values indicate only one value available.

Water ingestion rates can be estimated for mammals and birds from allometric models (Calder and Braun 1983):

$$I_w = 99 W^{0.90} \quad \text{Mammals} \tag{22.21}$$

and

$$I_w = 59 W^{0.67} \quad \text{Birds} \tag{22.22}$$

where I_w = water ingestion rate (mL water/d); and W = body weight (g live weight).

22.8.2.3 Inhalation Rates

Allometric equations, based on body mass, have been developed to estimate inhalation rates of resting mammals (Stahl 1967) and nonpasserine birds (Laskowski and Calder 1971):

$$I_a = 0.5458 \ W^{0.80} \quad \text{Mammals} \tag{22.23}$$

and

$$I_a = 0.4089 \ W^{0.77} \quad \text{Nonpasserine birds} \tag{22.24}$$

where I_a = inhalation rate (m^3 air/d); and W = body weight (kg live weight). The similarity of the models for mammals and birds suggests that Equation 22.24 is likely to be suitable for passerine birds as well.

22.8.2.4 Soil and Sediment Consumption

Terrestrial vertebrates consume soil, either inadvertently while foraging (i.e., predators of soil invertebrates ingesting soil adhering to worms, grazing herbivores consuming soil deposited on foliage or adhering to roots) or while grooming. Soil may also be consumed purposefully to meet nutritional requirements. Diets of many herbivores are deficient in sodium and other trace nutrients (Robbins 1993). Ungulates, in particular, consume soils with elevated sodium levels, presumably to meet sodium needs. Because soils at waste sites may contain very high contaminant concentrations, direct ingestion of soil is a potentially significant exposure pathway. For example, field observations indicated that white-tailed deer consumed coal ash from an ash disposal basin on the Oak Ridge Reservation (Sample and Suter 2002). Because the sodium content of the coal ash was high (approximately 4 times higher than that of background soils), it was assumed that deer were consuming ash to meet their sodium needs. Exposure to metals in the ash was estimated by assuming that deer consumed a sufficient volume of ash to alleviate their sodium deficit. Ash consumption to meet sodium requirements was 7.3 times greater than estimated incidental ingestion. While it is typically assumed that chemicals in ingested soil are completely bioavailable, it is important to consider the digestion or extraction techniques used in the soil analysis (Section 20.8).

Models do not exist to estimate soil ingestion by wildlife, but measured rates are available for some species (Table 22.4).

These estimates of soil ingestion do not include passerine birds. However, a recent study of soil particles in the gizzards of birds that use agricultural lands can provide the basis for such estimates (Luttik and de Snoo 2004). The EPPO (2004) recommends a conversion factor from soil mass in the gizzard to a daily soil ingestion rate of 4.2.

Sediment ingestion is seldom considered, but it can be an important route of exposure of semi-aquatic wildlife, particularly for chemicals that are not bioaccumulative (Beyer et al. 1997). For example, swans have been killed by ingesting lead, white phosphorus, and other chemicals in sediments, while waterfowl feeding in the same areas that ingested little sediment were hardly affected (Henny 2003; Sparling 2003).

22.8.2.5 Home Range and Territory Size

Home ranges are the spatial areas occupied by individuals. These areas provide food, water, and shelter and may or may not be defended. Home range size is a critical component in estimating exposure (Section 22.8.1). Species with small home ranges may live exclusively within the bounds of a contaminated area and therefore may experience high exposure. Conversely, species with large home ranges may travel among, and receive exposure from, multiple contaminated areas.

TABLE 22.4
Measured Soil Ingestion Rates for Selected Wildlife Species

Reptiles	Species	Soil Ingestion (% of Diet)	Reference
Box turtle	*Terrapenne carolina*	4.5	Beyer et al. 1994
Eastern painted turtle	*Chrysemys picta*	5.9	Beyer et al. 1994
Birds			
Blue-winged teal	*Anas discors*	<2.0	Beyer et al. 1994
Ring-necked duck	*Aythya collaris*	<2.0	Beyer et al. 1994
Wood duck	*Aix sponsa*	11	Beyer et al. 1994
Mallard	*Anas platyrhynchos*	3.3	Beyer et al. 1994
Canada goose	*Branta canadensis*	8.2	Beyer et al. 1994
Stilt sandpiper	*Micropalama himantopus*	17	Beyer et al. 1994
Semipalmated sandpiper	*Caladris pusilla*	30	Beyer et al. 1994
Least sandpiper	*Calidris minutilla*	7.3	Beyer et al. 1994
Western sandpiper	*Calidris mauri*	18	Beyer et al. 1994
American woodcock	*Scolopax minor*	10.4	Beyer et al. 1994
Wild turkey	*Meleagris gallopavo*	9.3	Beyer et al. 1994
Mammals			
Opossum	*Didelphis virginiana*	9.4	Beyer et al. 1994
Short-tailed shrew	*Blarina brevicauda*	13	Talmage and Walton 1993
Nine-banded armadillo	*Dasypus novemcinctus*	17	Beyer et al. 1994
Black-tailed jackrabbit	*Lepus californicus*	6.3	Arthur and Gates 1988
Meadow vole	*Microtus pennsyvanicus*	2.4	Beyer et al. 1994
Cotton rat	*Sigmodon hispidus*	2.8	Garten 1980
White-footed mouse	*Peromyscus leucopus*	<2.0	Beyer et al. 1994
White-footed mouse	*Peromyscus leucopus*	1	Talmage and Walton 1993
Black-tailed prairie dog	*Cynomys ludovicianus*	7.7	Beyer et al. 1994
White-tailed prairie dog	*Cynomys leucurus*	2.7	Beyer et al. 1994
Woodchuck	*Marmota monax*	<2.0	Beyer et al. 1994
Feral hog	*Sus scrofa*	2.3	Beyer et al. 1994
White-tailed deer	*Odocoileus virginianus*	<2.0	Beyer et al. 1994
Mule deer	*Odocoileus hemionus*	<2.0	Beyer et al. 1994
Mule deer	*Odocoileus hemionus*	0.6–2.1	Arthur and Aldredge 1979
Elk	*Cervus elaphus*	<2.0	Beyer et al. 1994
Moose	*Alces alces*	<2.0	Beyer et al. 1994
Bison	*Bison bison*	6.8	Beyer et al. 1994
Pronghorn	*Antilocapra americana*	5.4	Arthur and Gates 1988
Red fox	*Vulpes vulpes*	2.8	Beyer et al. 1994
Raccoon	*Procyon lotor*	9.4	Beyer et al. 1994

Source: From Suter, G.W., II, Efroymson, R.A., Sample, B.E., and Jones, D.S., *Ecological Risk Assessment for Contaminated Sites*, Lewis Publishers, Boca Raton, FL, 2000. With permission.

Home range size is determined by habitat quality, prey abundance, population density, and other factors. In general, home range size decreases with increasing habitat quality. In contrast, home range generally decreases with increasing population density because antagonistic interactions with neighbors reduce movements.

In the absence of measured species-specific values, home range or territory size may be modeled based on body weight and trophic relationships. Jetz et al. (2004) produced the following relationships between body mass and home range in mammals:

$$HR_{herb} = 2.05\,(w)^{1.02} \tag{22.25}$$

$$HR_{omn} = 15.87\,(w)^{1.12} \tag{22.26}$$

and

$$HR_{carn} = 52.07\,(w)^{1.20} \tag{22.27}$$

where HR_{herb} = home range for herbivores (kg); HR_{omn} = home range for omnivores (kg); HR_{carn} = home range for carnivores (kg); and w = body mass (kg).

These home range estimates are corrected for group size. That is, when a home range is held by a group, its area is divided by the number of individuals in the group.

A strong positive relationship also exists between body mass and territory or home range size among birds (Schoener 1968). Predators tend to have larger territories than omnivores or herbivores of the same weight. Schoener (1968) provided a summary of home range or territory sizes for 77 species of land birds but not models that could be used to estimate territory size.

22.9   UPTAKE MODELS

Uptake models may be required for two purposes in ecological risk assessment. First, if exposure–response relationships based on internal exposure are used, internal concentrations must be estimated. This use is relatively uncommon but has many potential advantages (Escher and Hermens 2004). Second, modeling exposures of wildlife to contaminants require that the concentrations in food items be measured or estimated. In either case, direct measurement of chemical concentrations in biota is preferred, because it provides the best indication of actual bioavailability and bioaccumulation. However, measurement is not possible for future exposures and may not be feasible for contaminated sites because of time or budget limitations. Even when measurement is possible, the data must be converted to an empirical model (e.g., by regressing earthworm concentrations against soil concentrations) so as to interpolate or extrapolate to unmeasured concentrations, particularly those expected in the future. Hence, uptake models are used to estimate wildlife dietary exposures and also to estimate internal exposures of endpoint organisms.

Three general types of models exist for the estimation of contaminant concentrations in biota; however, none of these types of models is available for all biota types. They are: uptake factors, empirical regression models, and mechanistic bioaccumulation models.

Uptake factors are quotients of ratios of chemical concentrations in biota to concentrations in associated abiotic media. Uptake factors are also referred to as transfer coefficients or (particularly in aquatic studies) as bioconcentration factors (BCFs). In aquatic systems, uptake factors from studies that include exposure through food are called bioaccumulation factors (BAFs). Uptake factors from soil or sediment are often referred to as biota sediment/soil accumulation factors (BSAFs). Multiplication of an uptake factor by the chemical concentration in an abiotic medium produces an estimate of chemical concentrations in a tissue or organism. While uptake factors may be simple to use, variance and associated uncertainty in the estimates may be quite high. The factor is

$$UF = C_b/C_x \tag{22.28}$$

where UF = uptake factor; C_b = concentration in biota (mg/kg); and C_x = concentration in the contaminated medium (mg/L, mg/kg).

An implicit assumption in the use of uptake factors is that uptake is a simple linear function of media concentrations, with an intercept of zero. However, at least for inorganic elements in soil, uptake is commonly nonlinear with respect to soil concentration (Alsop et al. 1996; Sample et al. 1998, 1999a; Efroymson et al. 2001). Consequently, the use of uptake factors at highly contaminated sites may grossly overestimate actual concentrations in biota. In addition, it must be assumed that soil properties do not significantly affect uptake, at least for the range of soils used to derive and apply the coefficient. That is, properties such as pH and organic matter content are sufficiently similar that they need not be considered.

Empirical regression models are derived using concentrations in biota and abiotic media from contaminated sites. In general, regression models are preferable to simple uptake factors. First, physical and chemical parameters known to influence bioavailability and uptake of contaminants from media, such as pH, cation exchange capacity, and organic matter content, may be included in multiple regression models. The resulting models may explain more of the variability of the data and are likely to result in improved estimates of tissue concentrations by permitting more site-specific information to be included.

Second, regression models can address thresholds and nonlinearities in bioaccumulation. Because of saturation kinetics or equivalent processes, the rate of accumulation typically decreases at higher concentrations of contaminants. When uptake is known to be controlled by an enzymatically limited process, the parameters of the Michaelis–Menten equation may be estimated (McLaughlin et al. 1998). However, uptake is more usually modeled by fitting a power function:

$$C_b = a(C_x)^B \tag{22.29}$$

where a and B are fitted parameters.

While nonlinear regression methods may be used to fit these models to bioaccumulation data, it is easier to log-transform the data and perform simple linear regression analyses. Regression models based on log-transformed data, while linear in log-space, are nonlinear in untransformed space (Figure 22.2). The regression model for transformed data may be expressed as

$$\log(C_b) = a + B(\log C_x) \tag{22.30}$$

In place of these static empirical bioaccumulation models, one may, in principle, model accumulation using mechanistic mathematical simulations. The multimedia process models that are used to estimate transport and fate commonly contain biological uptake, but uptake is often represented simply (Chapter 21). Separate uptake models are commonly used when media concentrations are known from measurements at contaminated sites or when transport and fate models estimate only concentrations in abiotic media. When uptake models are dynamic, they are typically referred to as pharmacokinetic models because of their derivation for pharmacology but, in this context, they are properly termed toxicokinetic models.

When organisms take up a contaminant from one medium, such as water, the basic first-order toxicokinetic model is

$$dC_b/dt = k_u C_x - k_e C_b \tag{22.31}$$

where k_u and k_e are the uptake and elimination rate constants. At equilibrium, k_u equals k_e, the derivative is zero, the concentrations are constant, and static uptake factors or regression

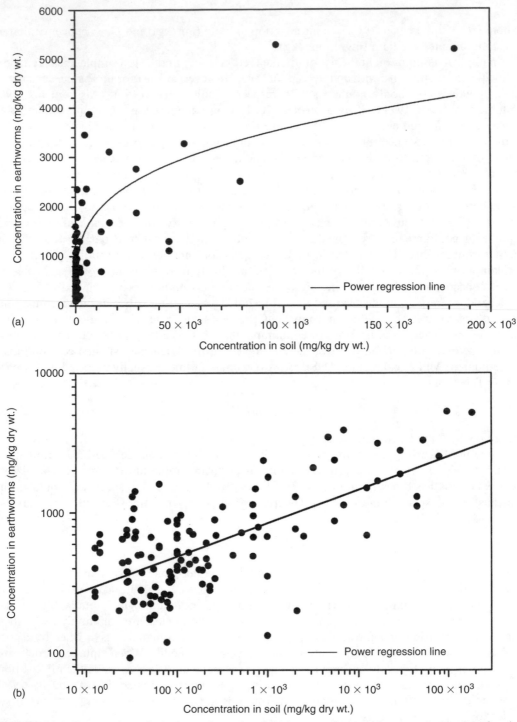

FIGURE 22.2 Scatter plots of (a) untransformed and (b) log-transformed bioaccumulation data for zinc from soil by earthworms. (From Suter, G.W., II, Efroymson, R.A., Sample, B.E., and Jones, D.S., *Ecological Risk Assessment for Contaminated Sites*, Lewis Publishers, Boca Raton, FL, 2000. With permission.)

models apply. This first-order model has been used to assess risks based on critical body residues and other toxicodynamics (Section 23.26) (Kooijman 1981; McCarty and Mackay 1993; Legierse et al. 1999; French-McCay 2002).

If organisms are assumed to be exposed to two media such as the solid and aqueous phases of sediments or soils, the first-order kinetic model is

$$dC_b/dt = (k_{u1}C_{x1} + k_{u2}C_{x2}) - k_eC_b \tag{22.32}$$

In an earthworm, for example, k_{u1} would be the uptake rate constant for passive absorption from the soil pore water (C_{x1}) and k_{u2} would be active digestive uptake from the solid phase (C_{x2}).

From these simple beginnings, toxicokinetic modeling can elaborate in a number of directions, depending on the needs of the assessment and their practicality given available information (Reddy et al. 2005). The rate constants might be treated as variable functions of environmental characteristics (e.g., temperature), organism characteristics (e.g., size and lipid content), or characteristics of the chemical of concern (e.g., solubility). The media concentrations may be dynamic. Higher-order kinetics may be employed, such as the Michaelis–Menten equation mentioned previously. Multiple compartments may be added to the environment or the organism.

Toxicokinetic modeling has a number of potential advantages. It can represent situations in which the environment or the organism is changing (e.g., variable emissions or an organism moving among areas of differing contamination). To the extent that differences in sensitivity of species and life stages are due to differences in kinetics, toxicokinetic models can substitute for toxicological extrapolation models (Chapter 26). In particular, if the effective concentration at the site of action of chemicals with a particular mechanism of action is constant, toxicokinetic modeling can potentially substitute for toxicity testing as well as modeling (Escher and Hermens 2004).

So far, toxicokinetic modeling has shown more promise than utility in ecological risk assessment. This is in part because static, equilibrium assumptions have been sufficient in many cases. Also, information concerning kinetics, both basic knowledge of processes and specific knowledge of rates and compartment characteristics, has been lacking. Finally, nearly all toxicity data are expressed as external exposure concentrations or administered doses rather than internal concentrations. Toxicokinetics can be better understood by applying a mechanistic understanding of chemical accumulation (Box 22.2).

22.9.1 AQUATIC ORGANISM UPTAKE

Commonly, chemical concentrations are measured in water but not biota. In such cases, it is necessary to model concentrations in aquatic biota if effects are to be estimated based on body burdens in aquatic biota or if risks to wildlife that feed on aquatic organisms must be estimated. The most common approach is to use BCFs derived from conventional laboratory studies of aqueous exposures or BAF derived from field studies that include all routes of exposure. The concentrations are most often total concentrations, but dissolved or freely dissolved aqueous concentrations may be used, and biological concentrations may be measured in fat or other specific tissues or may be lipid-normalized. The point of these adjustments is to create a BCF or BAF that is more nearly constant across different water chemistries or organism attributes.

BAFs are more realistic than BCFs, because they include dietary exposures and long exposure durations. However, the water concentrations for BAF derivation are often highly variable and poorly defined, dietary uptake is often unimportant, and, when diet is important,

BOX 22.2
Mechanisms of Chemical Accumulation

MacDonald et al. (2002) developed a unified mechanistic theory for the accumulation of chemicals in biota that includes four mechanisms.

1. *Solvent switching*: Media, including organisms or their constituents (e.g., fat) act as solvents for chemical contaminants, and the passive movement of the contaminant between media is termed solvent switching. Bioconcentration of hydrophobic chemicals occurs by solvent switching between water and the lipid phase of an aquatic organism; it is simulated by octanol/water partitioning.
2. *Solvent depletion*: When an organism ingests a contaminated food item, if the contaminant chemical is significantly less digestible than the food organism, the contaminant concentration increases. That is, as the fat that is the solvent for a hydrophobic organic chemical or the proteins that are the solvents for metals are digested, the solvent is depleted, raising the concentration in the gut and driving uptake. This is the mechanism by which biomagnification occurs in food chains.
3. *Filtering*: Filtering is the purely mechanical process of partitioning of particles. Examples include filter feeding and accumulation of aerosols on leaf surfaces.
4. *Pumping*: Organisms may actively accumulate, sequester, or excrete chemicals, particularly nutrient elements and analogous metals.

Bioaccumulation is the result of these mechanisms acting in combination.

BAFs are influenced by the particular structure and function of the ecosystems from which they were derived. BCFs are more precise and accurate, because they are derived in laboratory studies for which aqueous concentrations and conditions are well specified. Therefore, BCFs are often preferred unless (1) the uptake of the chemical of concern is known to have important dietary components, (2) the BCF is judged to be of low quality or relevance (e.g., the exposures were short relative to the uptake kinetics of the chemical as indicated by failure to reach an asymptotic concentration), (3) the BAF was derived from the site being assessed, or (4) the conditions under which the BAF was derived are similar to those at the site being assessed.

BCFs and BAFs from the literature may be unreliable substitutes for site-specific values. In the Trinity River, maximum BCFs from EPA criteria documents provided estimates of metal concentrations in fish that were within a factor of 10 of measured values in most cases (Parkhurst et al. 1996b). However, cadmium concentrations were consistently overestimated, and some predictions were in error by more than a factor of 1000. In the same study, geometric mean BCFs overestimated concentrations of pesticides in fish in nearly all cases, usually by more than a factor of 10.

When chemicals are so hydrophobic that they cannot be detected in water, or, for some other reason can be readily detected in sediment but not water, BSAFs can be derived for water column species. A generic method to predict BAFs for neutral organic chemicals in fish from sediment concentrations has been developed for regulatory use (EPA 2003d) and has been shown to give reasonably good estimates for PCBs in two ecosystems (Burkhard et al. 2003).

When empirical factors are not available, BCFs or BAFs must be modeled from chemical properties. A set of BCF models for fish has been developed for the US EPA Office of Pollution Prevention and Toxics (Box 22.3). These quantitative structure–activity relationships

(QSARs) are more broadly applicable than the models of neutral organics discussed below. They do not incorporate dietary considerations or lipid content, but the authors argue that those considerations are not needed. Like other QSARs for bioconcentration or bioaccumulation, these equations have large uncertainties (roughly $\pm 10\times$) and should be used for screening purposes or when specific data are unavailable.

22.9.1.1 Neutral Organics

The US EPA (2003d) recommends expressing accumulation factors for nonionic or sparingly ionic organic chemicals (henceforth, neutral organics) in terms of freely dissolved fraction in water and lipid fraction in organisms. A lipid-normalized BCF or BAF is one that has been divided by the fractional lipid content of the organism or analyzed tissue. While this method is widely accepted, it depends on the assumptions that concentration in an organism is proportional to its lipid content and the apportioning of a chemical to lipid does not depend on the amount of composition of the lipid. However, the partitioning of chemicals differs for structural and storage lipids, and the lipid fractions are not equally extracted by available methods (Randall et al. 1998). Further, storage lipid content varies greatly among species, individuals, sites, and seasons. These issues may become important in site-specific assessments.

The estimation of the freely dissolved concentration of neutral organic chemicals depends on the assumption of EqP with dissolved organic carbon (DOC) and particulate organic carbon (POC), so that the fraction freely dissolved (f_{fd}) is inversely proportional to the organic carbon concentrations and the associated partitioning coefficients:

$$f_{fd} = [1 + (POC \cdot K_{poc}) + (DOC \cdot K_{doc})]^{-1} \qquad (22.33)$$

If appropriate data are not available, the US EPA (2003d) estimates the partitioning coefficients as: $K_{poc} = K_{ow}$ with 95% confidence limits of $8\times$, and $K_{doc} = 0.08\ K_{ow}$ with 95% confidence limits of $20\times$.

QSARs are commonly used to estimate uptake of neutral organics by aquatic organisms. The conventional assumption is that body burdens of neutral organics are due to EqP between water and the lipid fraction of the organisms (Veith et al. 1979). The model most commonly endorsed by the EPA is

$$\log BCF = 0.79 \log K_{ow} - 0.40 \qquad (22.34)$$

Veith and Kosian (1983) indicate that this model's estimates have order-of-magnitude confidence limits for most chemicals and larger limits for chemicals with K_{ow} values greater than 6.5. This model works reasonably well for chemicals with short half-lives but underpredicts concentrations of persistent chemicals in field-collected fish (Oliver and Niimi 1985). Therefore, an analogous model was developed for bioaccumulation of PCBs and similar hydrophobic chlorinated organic compounds by (Oliver and Niimi 1988):

$$\log BAF = 1.07 \log K_{ow} - 0.21 \qquad (22.35)$$

For chemicals that accumulate to a significant extent via diet, bioaccumulation factors should be related to trophic level. To account for this, BCFs may be multiplied by food chain multipliers (FCMs = BAF/BCF) for the trophic level of the aquatic organism. National FCMs have been derived to estimate bioaccumulation of neutral organics in fish for the derivation of water quality criteria to protect fish-consuming humans from consumption of neutral organic chemicals (Table 22.5) (EPA 2003d). Chemicals with K_{ow} less than 4, and

BOX 22.3
A Method for Estimating Bioconcentration in Fish

Equations

Nonionic compounds

Log $K_{ow} < 1$	log BCF $= 0.50$
Log K_{ow} 1 to 7	log BCF $= 0.77$ log $K_{ow} - 0.70 + \Sigma F_i$
Log $K_{ow} > 7$	log BCF $= -1.37$ log $K_{ow} + 14.4 + \Sigma F_i$
Log $K_{ow} > 10.5$	log BCF $= 0.50$

Aromatic azo compounds
Log BCF $= 1.0$

Ionic compounds (carboxylic acids, sulfonic acids and salts, compounds with N of +5 valence)

Log $K_{ow} < 5$	log BCF $= 0.05$
Log K_{ow} 5 to 6	log BCF $= 0.75$
Log K_{ow} 6 to 7	log BCF $= 1.75$
Log K_{ow} 7 to 9	log BCF $= 1.00$
Log $K_{ow} > 9$	log BCF $= 0.50$

Compounds with $\geq C_{11}$ alkyl
Log BCF $= 1.85$

Tin and mercury compounds
Use the appropriate equation with the appropriate factor (below) or 2.0, whichever is greater.

Factors

The following correction factors are used as the F_i parameters in the equations above.

Compounds with an aromatic *s*-triazine ring (3 compounds)	-0.32
Compounds containing an aromatic alcohol (e.g., phenol) with two or more halogens attached to aromatic ring (17 compounds)	-0.40
Compounds containing an aromatic ring with a *tert*-butyl group in a position ortho to an –OH group (e.g., *tert*-butyl *ortho*-phenol) (6 compounds)	-0.45
Compounds containing an aromatic ring and aliphatic alcohol in the form of –CH–OH (e.g., benzyl alcohol) (4 compounds)	-0.65
Phosphate ester, O $=$ P(O–R)(O–R)(O–R), where R is carbon (one R can be H) (18 compounds)	-0.78
Ketone with one or more aromatic connections (18 compounds)	-0.84
Nonionic compounds with an alkyl chain containing eight or more –CH$_2$– groups (13 compounds)	-1.00
	(log K_{ow} of 4 to 6)
	-1.50
	(log K_{ow} of 6 to 10)
Compounds containing a cyclopropyl ester of the form cyclopropyl-C ($=$O)–O– (e.g., permethrins) (6 compounds)	-1.65
Compounds containing a phenanthrene ring (4 compounds)	$+0.48$
Multiply halogenated biphenyls and polyaromatics containing only aromatic carbons and halogens (e.g., PCBs) (19 compounds)	$+0.62$
Organometallic compounds containing tin or mercury (12 compounds)	$+1.40$

Source: Meylan, W.M., Howard, P.H., Boethling, R.S., Aronson, D., Pruntup, H., and Gouchie, S., *Environ. Toxicol. Chem.*, 18, 664–672, 1999.

organisms in trophic levels I (phytoplankton) and II (zooplankton) require no FCM. BAFs are estimated by multiplying BCFs from laboratory measurements or QSARs by FCMs, which are specific to trophic level and K_{ow} of the chemical.

22.9.1.2 Ionizing Organic Chemicals

Ionizing organics must be considered separately from neutral organics because their behavior is much more complex, involving multiple partitioning mechanisms, sensitivity to pH, and other water quality parameters. As a result, it is difficult to predict their accumulation, and factors or models empirically derived from the laboratory or field are recommended. However, models developed for nonionic chemicals may be applied to ionizing chemicals if water chemistry is such as to neutralize them. Organic acids exist almost entirely in neutral form when pH is two or more units below their pK_a, and organic bases exist almost entirely in neutral form when pH is two or more units above their pK_a (EPA 2003d). This rule of thumb is more useful for bases than acids, because most pK_as are smaller than ambient pHs (usually pH 6 to 9).

22.9.1.3 Inorganic and Organometalic Chemicals

Metals, metaloids, nonmetalic inorganic chemicals (e.g., cyanide), and their organic forms present significant difficulties when estimating bioaccumulation. They occur in multiple

---

TABLE 22.5
Food-Chain Multipliers (FCMs) for Trophic Levels 3 (Planktivorous Fish) and 4 (Piscivorous Fish) Based on the Gobas Food Web Model and Conditions in the Laurentian Great Lakes[a]

Log K_{ow}	Trophic Level 3	Trophic Level 4
4.0	1.2	1.1
4.5	1.7	1.3
5.0	3.0	2.5
5.5	5.8	6.6
6.0	9.8	15
6.5	13	23
7.0	13	24
7.5	11	18
8.0	7.6	7.2
8.5	3.6	1.5
9.0	1.4	0.21

Source: Condensed from EPA (US Environmental Protection Agency), *Methodology for Deriving Ambient Water Quality Criteria for the Protection of Human Health*, Technical Support Document Vol. 2 of *Development of National Bioaccumulation Factors*, EPA-822-R-03-030, Office of Water, Washington, DC, 2003. With permission.

[a]The original table contains FCMs with three or more significant figures for increments of a tenth K_{ow}. That level of precision is not justified by the accuracy of the data and model.

---

forms and valences that differ greatly in their accumulation kinetics. For example, selenium may occur in ambient water as inorganic selenite (+4) and selenate (+6) oxyanions, elemental selenium (0), and organocompounds of selenide (−2). In addition, metal accumulation is tied to ion regulation, and is highly variable. For example, metal accumulation varies by orders of magnitude among orders of aquatic insects because of differences in ion regulation strategies (Buchwalter and Luoma 2005). Hence, no simple rules exist for defining available forms of this class of chemicals or for the tendency of organisms to accumulate them. Rather, for each chemical, care must be taken to consider its various possible forms and their tendencies to bioaccumulate. If accumulation differs among trophic levels, trophic transfer factors (Reinfelder et al. 1998) or models from field studies are preferred.

Mercury accumulation is a complex and important case. A survey in the United States found that mercury levels in fish were strongly related to aqueous methyl mercury concentrations but not total aqueous mercury or sediment mercury (Brumbaugh et al. 2001). The model for length-normalized concentration in largemouth bass and equivalent predatory fish is

$$\ln y = 1.219 + 0.4923 \ln x \tag{22.36}$$

where y = length-normalized Hg concentration in age 3 fish fillets (mg/kg/m); and x = methyl Hg concentration in water (ng/L).

A slightly better model ($r^2 = 0.44$ vs. 0.38) also included percent wetland, pH, and acid volatile sulfide. Limiting the data to largemouth bass did more to improve the model ($r^2 = 0.51$), suggesting that species' identity is more important than conditions for methyl mercury uptake.

22.9.1.4 Aquatic Plants

Concentrations of chemicals in aquatic plants are rarely measured although they are a significant component of the diet of some fish, turtles, waterfowl, and mammals. Some bioconcentration factors for metals may be found in the literature (Garg et al. 1997; Gupta and Chandra 1998; Kahkonen and Manninen 1998; Kahkonen et al. 1998). In addition, environmental variables such as salinity may be important determinants of accumulation of heavy metals by macrophytes (e.g., cadmium by the sea grass *Potamogeton pectinatus*; Greger et al. 1995). Insufficient information is available to determine the relative importance of water and sediment in the contamination of rooted macrophytes or the possible nonlinearity of contaminant uptake.

22.9.1.5 Aquatic Toxicokinetics

If concentrations in aquatic biota must be estimated, and the assumptions of static models do not hold (e.g., aqueous concentrations are not sufficiently constant or multiple routes of exposure are important), it may be appropriate to model uptake dynamically. In particular, if equilibrium is not reached during the exposure, or if variance among species or life stages in anatomy, physiology, or biochemistry is important, toxicokinetic models should be used. Fish toxicokinetic models are similar to mammalian models except for gill uptake and excretion, which were modeled by Erickson and Stephan (1990) (Figure 22.3). The utility of fish toxicokinetic models is limited primarily by lack of knowledge concerning gut uptake and metabolic transformations (Nichols 1999). Toxicokinetic modeling permits the simulation of uptake by fish moving among sites with differing levels of contamination such as dredged material disposal sites (Linkov et al. 2002a).

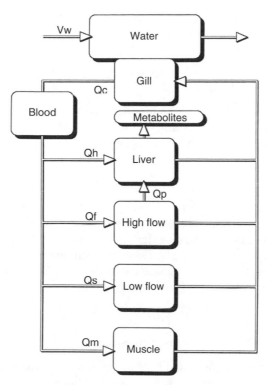

FIGURE 22.3 Diagram of fish toxicokinetics for organic chemicals taken up by the gills and metabolized by the liver. (Adapted from Barron, M.G., Mayes, M.A., Murphy, P.G., and Nolan, R.J., Aq. Toxicol., 16, 9–32, 1990. With permission.)

22.9.2 BENTHIC INVERTEBRATE UPTAKE

Body burdens of invertebrates exposed to contaminated sediments are used to estimate dietary exposures for predators of benthic organisms. For example, some flying insect-ivores (e.g., bats and swallows) forage over water and consume adult insects that have aquatic larval stages (e.g., mayflies and midges). These emergent insects can serve as an important vector for the movement of chemicals out of sediment deposits and into the terrestrial food chain (Larsson 1984; Currie et al. 1997; Froese et al. 1998). Similarly, diving ducks, shore birds, and some other wildlife species feed on benthic invertebrates in situ. Finally, dietary exposure to some chemicals may be important for some benthic feeding fish (Section 22.2).

Tissue concentrations must be estimated in the absence of measured values. BSAFs and models can be derived empirically from colocated sediment and biota samples. They have the same form, assumptions, and uncertainties as BAFs for water (Section 22.9.1). Unfortunately, they are highly variable, so BSAFs from the literature have little utility. As with contaminant uptake from water, uptake from sediments can be modeled based on partitioning with the aqueous phase, using the partitioning models in Section 22.3 with bioaccumulation models in Section 22.9.1. However, for at least some species and chemicals, uptake from ingested sediment may also be important. Toxicokinetic modeling of the results of a study of uptake of ^{109}Cd and ^{65}Zn by the oligochaete worm, *Tubifex tubifex*, showed that 9.8% of Cd and 52% of Zn uptake came from ingested sediment (Redeker et al. 2004). The behavior of invertebrates and anatomical features other than lipid content may also influence uptake in ways that deviate from EqP estimates (Belfroid et al. 1996). Nevertheless, toxicokinetic models are seldom used, although they have been available for over a decade (Landrum 1988).

22.9.3   TERRESTRIAL PLANT UPTAKE

Because plant uptake is highly variable among conditions and species, it is desirable, if possible, to sample and analyze chemicals. However, modeling contaminant accumulation in plants is part of many ecological risk assessments.

22.9.3.1   Soil Uptake

The simplest model for estimating the concentrations of chemicals in plants is the soil–plant uptake factor, the ratio of the concentration of a chemical in a plant or plant part to that in soil. Plant uptake factors may be derived for a site or found in the literature. As a last resort, for screening assessments, the US EPA recommends a default uptake factor of 1 for both organic and inorganic chemicals (Office of Solid Waste and Emergency Response 2003). Available evidence indicates that it is conservative (i.e., overestimates uptake) in most cases.

22.9.3.2   Empirical Models of Inorganic Chemicals

The relationship between the contaminant concentration in the plant and in soil is, in general, not linear. This is not surprising, since the uptake of ions in solution by plants has been observed to be more efficient at lower concentrations than at higher concentrations (Cataldo and Wildung 1978). Thus, above a certain soil contamination level, the concentration of metals in plants probably reaches a plateau or toxicity is observed. In particular, the uptake of plant nutrients, such as copper and zinc, would be expected to be regulated by the plant. Hence, uptake factors may overestimate accumulation at high soil concentrations and underestimate accumulation at low concentrations.

If uptake factors must be used, the associated uncertainty with soil–plant contaminant uptake factors can be represented by distributions of factors: for multiple species at a single site, for a single species at multiple sites, or for multiple species at multiple sites. The first type would require some site-specific measurement, and the second would require much existing data. The greatest uncertainty would be associated with the use of a distribution of uptake factors of multiple species at multiple sites to represent few species at a single site. Risk assessors may use these distributions as inputs to Monte Carlo simulations of wildlife exposure, though better models exist (regressions are described below). The range of the distribution for a particular chemical in a wide range of species at multiple sites can be 4 orders of magnitude or more. The distribution for selenium (Figure 22.4) spans about 3.5 orders of magnitude, with almost half of the uptake factors exceeding 1.

The concentration of several elements in aboveground vegetation (dry weight) is predictable using a log–log linear model fit to data from field studies of the uptake of chemicals from soil by plants (Efroymson et al. 2001). Studies in which salts were added to soil in pots were excluded as unrealistic. The models that were verified to be reliable are presented in Table 22.6. They are recommended by the US EPA for estimating both foliage and seed concentrations when deriving soil screening levels (Office of Solid Waste and Emergency Response 2003).

An example regression of selenium concentration in foliage and stems against the concentration in soil is shown in Figure 22.5. The relationship is fitted better by the log–log

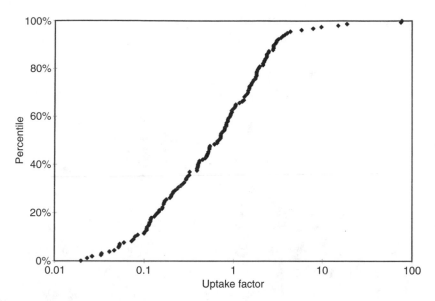

FIGURE 22.4 Cumulative distribution of soil–plant uptake factors for selenium from various published field studies. (From Suter, G.W., II, Efroymson, R.A., Sample, B.E., and Jones, D.S., *Ecological Risk Assessment for Contaminated Sites*, Lewis Publishers, Boca Raton, FL, 2000. With permission.)

relationship than the linear, zero–y–intercept relationship, which is equivalent to an uptake factor. The upper 95% prediction limit for the regression model is recommended for screening assessments (Efroymson et al. 2001).

The use of uptake factors or univariate regression models does not account for (1) soil properties, such as clay content, organic matter content, pH, cation-exchange capacity, or particle size, (2) plant properties, such as age, taxonomy, growth form, root depth, and evapotranspiration rate, (3) exposure time, and (4) other site characteristics, such as average

TABLE 22.6
Results of Regression of ln(Conc. in Plant) on ln(Conc. in Soil). Concentrations are mg/kg and Weights are Dry

Element	N	$a \pm SE$	$b \pm SE$	r^2	P Model Fit
Arsenic	122	-1.992 ± 0.431	0.564 ± 0.125	0.145	0.0001
Cadmium	207	-0.476 ± 0.088	0.546 ± 0.042	0.447	0.0001
Copper	180	0.669 ± 0.213	0.394 ± 0.044	0.314	0.0001
Lead	189	-1.328 ± 0.350	0.561 ± 0.072	0.243	0.0001
Mercury	145	-0.996 ± 0.122	0.544 ± 0.037	0.598	0.0001
Nickel	111	-2.224 ± 0.472	0.748 ± 0.093	0.371	0.0001
Selenium	158	-0.678 ± 0.141	1.104 ± 0.067	0.633	0.0001
Zinc	220	1.575 ± 0.279	0.555 ± 0.046	0.402	0.0001

Source: Bechtel-Jacobs, 1998. *Empirical Models for the Uptake of Inorganic Chemicals from Soil By Plants*, BJC/OR-133, Oak Ridge National Laboratory, Oak Ridge, Tennessee, 1998; Efroymson, R.A., Sample, B.E., and Suter, G.W., II, *Environ. Toxicol. Chem.*, 20, 2561–2571, 2001. With permission.
Model: ln(conc. in plant) = $a + b$[ln(conc. in soil)].

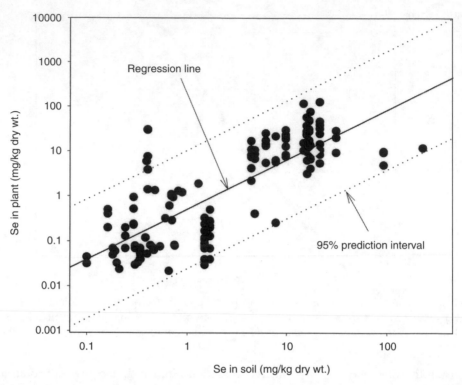

FIGURE 22.5 Scatter plot of plant concentration vs. soil concentration of selenium. The line represents a simple regression model of natural log-transformed data from published studies. (From Suter, G.W., II, Efroymson, R.A., Sample, B.E., and Jones, D.S., *Ecological Risk Assessment for Contaminated Sites*, Lewis Publishers, Boca Raton, FL, 2000. With permission.)

TABLE 22.7
Results of Regression of ln(Conc. in Plant) on ln(Conc. in Soil) and pH. Concentrations are mg/kg and Weights are Dry

Chemical	N	$a \pm$ SE	$b \pm$ SE	$c \pm$ SE	r^2	P Model Fit
Cadmium	170	1.152 ± 0.638^{NS}	0.564 ± 0.047^c	-0.270 ± 0.102^b	0.462	0.0001
Mercury	82	-4.186 ± 1.144^c	0.641 ± 0.062^c	0.423 ± 0.186^a	0.677	0.0001
Selenium	148	-8.831 ± 0.723^c	0.992 ± 0.050^c	1.167 ± 0.106^c	0.847	0.0001
Zinc	193	2.362 ± 0.440^c	0.640 ± 0.057^c	-0.214 ± 0.077^b	0.409	0.0001

Source: Efroymson, R.A., Sample, B.E., and Suter, G.W., II, *Environ. Toxicol. Chem.*, 20, 2561–2571, 2001. With permission.

Model: ln(conc. in plant) $= a + b$(ln(conc. in soil)) $+ c$(pH).
[a]$0.01 < p \leq 0.05$.
[b]$0.001 < p \leq 0.01$.
[c]$p \leq 0.001$.

temperature during the growing season. Many of the factors that control plant exposure to chemicals (Section 22.4) are pertinent here. Table 22.7 presents plant uptake models for metals that were significantly influenced by pH (Efroymson et al. 2001).

22.9.3.3 Empirical Models for Organic Chemicals

Studies of the accumulation of organic chemicals by plants from soil typically focus on the chemical characteristics that determine the soil–plant uptake factor, notably the octanol/water partitioning coefficient (K_{ow}). Less research is devoted to the role of soil characteristics or exposure time in determining uptake; the assumption is that the former are inconsequential and that the latter is not important, because equilibrium is achieved quickly. Similarly, existing published data are probably insufficient to evaluate whether the uptake factor is independent of chemical concentration in soil; thus it is unclear whether a nonlinear model of the form described for inorganic chemicals above would improve prediction of organic chemical concentration in plants. Accumulation models have been developed for estimating concentrations of organic compounds (many of them, pesticides) in plants from concentrations in soil. These are often referred to as partitioning models, because they are typically based on the chemical-specific octanol/water coefficients. Soil–plant uptake factor models for organic chemicals are presented in Table 22.8.

Most models in Table 22.8 relate the soil–plant uptake factor to the octanol/water coefficient. However, it is noteworthy that in some equations, the uptake factor is positively correlated with the octanol/water partitioning coefficient, and in other equations is negatively correlated. The correlation should be positive if soil water is used, but if the concentration of the compound in total soil is measured, the two-step partitioning process (soil to soil water and soil water to root) determines whether the correlation of uptake with the octanol/water partitioning coefficient is positive or negative (Scheunert et al. 1994). For this reason, K_{ow} may not be useful as a predictor of uptake by plants if the chemical, plant species, or soil is significantly different from those used to derive a model.

It should also be noted that the equations in Table 22.8 are not uniformly based on dry or wet weights of plants or soil. A common mistake in risk assessments is to combine an uptake model based on dry weight with a wildlife exposure that assumes that foods are expressed as wet weight.

Clearly the uptake models for organic chemicals require validation for a wide range of soils, plants, and chemicals. Alsop et al. (1996) compared concentrations of octachlorodibenzo-dioxin (OCDD) measured in oat plants with those predicted by the model by Travis and Arms (1988). The model, which was based on chemicals with log K_{ow} values between 1.75 and 6.15, underestimated the uptake of OCDD, with a log K_{ow} of 9.05, by a factor of almost 200 (Alsop et al. 1996). It is not known whether there was an air source of OCDD. For most contaminated sites, the models of Dowdy and McKone (1997) probably give the most accurate estimates of plant concentration for the broadest range of chemicals, soils, and plant species, though the results may be highly uncertain.

Even if the primary uptake pathway of some nonionic organic chemicals is not through the soil (see Section 22.4), correlations between soil and plant concentrations of organic compounds may exist where soil is the primary source of atmospheric chemical in the vicinity of the plant.

22.9.3.4 Surface Contamination

The assessor should be reminded that the equations in Table 22.6 through Table 22.8 do not include any contributions from atmospheric deposition; therefore, these models do not

TABLE 22.8
Equations Used to Estimate Soil–Plant or Soil Water–Plant Uptake Factors for Organic Chemicals

Equation	Chemical Parameter Required	Soil or Soil Water	Plant Part	Dry or Wet Weight	Chemicals Used in Derivation	Plants Used in Derivation	Reference
Log U_p = -0.578 log K_{ow} + 1.588	K_{ow}	Soil	Aboveground	Plant-dry, soil-dry	29 organic chemicals, including pesticides, Aroclor 254, TCDD, benzo(a)pyrene, and others	Multiple species from published studies	Travis and Arms 1988
Log U_p = 0.233 log K_{ow} + 0.971	K_{ow}	Soil water	Root	Plant-wet	5 chlorinated benzenes	Barley (no linear correlation for cress)	Scheunert et al. 1994
Log U_p = -0.24 log K_{ow} + 1.47	K_{ow}	Soil	Whole plant	Plant-wet, soil-dry	5 chlorinated benzenes	Barley	Scheunert et al. 1994
Log U_p = 0.391 log K_{ow} - 1.84	K_{ow}	Soil	Whole plant	Plant-wet, soil-dry	5 chlorinated benzenes	Cress	Scheunert et al. 1994
Log (U_p -0.82) - 0.77 log K_{ow} - 1.52	K_{ow}	Soil water	Root	Plant-wet	O-methylcarbamoyloximes and phenylureas	Barley	Briggs et al. 1982
Log U_p = -2385 log M + 5.943	M	Soil	Whole plant	Plant-wet, soil-dry	14 pesticides, chlorinated benzenes; benzene and pentachlorophenol did not fit	Barley	Topp et al. 1986a
Log U_p = -0.204 MCI + 0.589	MCI	Soil	Aboveground	Plant-dry, soil-dry	30 organic chemicals, including pesticides, TCDD, and other chlorinated organics	Multiple species from published studies	Dowdy and McKone 1997

U_p = soil–plant uptake factor (concentration of chemical in plant/concentration in soil); K_{ow} = octanol/water partitioning coefficient; M = molecular weight of the chemical; MCI = polar-corrected normal path first-order molecular connectivity index.

provide an estimate of the contamination on the plant surface. Atmospheric deposition may be important for atmospheric sources such as incinerators, and models are provided by the Office of Solid Waste (1999). In most risk assessments for contaminated sites, there is little interest in estimating current deposition of chemicals from air to plants. However, such a model is included in CALTOX (McKone 1993b). A simple model for deposition of wind-eroded contaminated soil is

$$C_p = F \times C_s \tag{22.37}$$

where C_p = concentration in plant aboveground parts (mg/kg); C_s = concentration in surface soil (mg/kg); and F = plant/soil coefficient. A proposed value of F is 3300 (McKone 1993).

Rainsplash may add contamination to the plant surface that is not accounted for in models of plant uptake. Rainsplash studies are designed to simulate tilled agriculture, so they are likely to overestimate rates for natural vegetation. An estimate of the contamination from rainsplash is not necessary if wildlife exposure models assume a reasonable rate of soil ingestion, since all routes of soil ingestion are included. The same model is proposed as for dry deposition (Equation 22.37) but with F of 0.0034 (fresh plant mass) (McKone 1993).

Another important route of plant external contamination is the deposition of sprayed pesticides. The processes of deposition, washoff, uptake, vaporization, etc. are combined in dynamic mass balance models of pesticides in agricultural systems (Ecological Committee on FIFRA Risk Assessment Methods 1999b).

22.9.3.5   Plant Tissue Type

The extent of accumulation of soil contaminants in plants often differs among tissue types. Metals are generally found at higher concentrations in foliage than in fruits and seeds (Sadana and Singh 1987; Jiang and Singh 1994). Thus, the uptake models described above probably provide conservative estimates of uptake by most reproductive tissues. In contrast, zinc has been observed to concentrate in wheat grain (Sadana and Singh 1987), and nickel may be remobilized from senescing tissues and accumulated in seeds (Cataldo and Wildung 1978). The US EPA assumes for screening purposes that all aboveground parts have equal concentrations (Office of Solid Waste and Emergency Response 2003).

Differences in water and lipid content in plant tissues may partly explain differences in contaminant uptake. For example, Trapp (1995) generalized the model of Briggs et al. (Table 22.8) to all plant species, using the water and lipid content of roots and stems. The relative accumulation of organic chemicals in various plant tissues depends on the route of uptake; volatile chemicals should concentrate more in the leaves, and soluble chemicals in plant roots (see Section 22.4).

22.9.3.6   Mechanistic Models

Mechanistic (toxicokinetic) models of plant uptake of chemical contaminants have been developed for few chemicals, soils, and plant species in few environments and the models have usually been developed in the laboratory without field validation. Models of uptake of metals from soil are derived from nutrient uptake models (Boersma et al. 1991; Barber 1995; McLaughlin 2001). Uptake is modeled as the sum of passive uptake of metals in the water replacing transpiration loss and selective uptake resulting in a concentration gradient and diffusion flow to the roots. Models of uptake of organic chemicals include passive uptake as well as partitioning processes. Mechanistic models for plant uptake of chemicals are not well established and have had difficulty accounting for the large observed differences in uptake among soils and plant species.

22.9.4 EARTHWORM UPTAKE

Earthworms are a significant pathway of contaminant transfer from soil, because they are more highly exposed through both ingestion and dermal contact than most other soil and litter invertebrates (Davis and French 1969; Ma 1994), and because they may represent a significant fraction of the diet of some vertebrates (e.g., shrews, moles, thrushes, woodcock). However, they are not always the most exposed terrestrial invertebrates. For example, snails may have higher metal concentrations because of their high assimilation efficiencies from food and their ability to take up metals through their feet (Hopkin 1989). At least in some cases, beetles with soil-dwelling larvae are much more contaminated than earthworms (Stansley et al. 2001).

Uptake factors for earthworms may be based on chemical concentrations in soil or in soil pore water. Uptake of inorganic contaminants is generally estimated using soil–earthworm uptake factors, while uptake of organic contaminants is more frequently estimated based on soil pore water. Soil–earthworm uptake factors are available in the literature for some chemicals (Suter et al. 2000) but may be highly variable because of soil properties and the form of the contaminant (Sample et al. 1999b; Peijnenburg 2001). While it can be assumed that chemical concentrations in soil and earthworms from field studies are at or near equilibrium, this assumption should be demonstrated for laboratory studies. Steady state is reached for a variety of organochlorine chemicals in 10 d (Belfroid et al. 1995) and for most metals within a few weeks (Janssen et al. 1997), but for Cd and some other chemicals, equilibrium is not achieved within the lifetime of a worm (Peijnenburg 2001).

Soil pore water–earthworm uptake models are commonly used with organic chemicals. They are consistent with soil–worm models if the solid, aqueous, and biotic phases of the soil are all in equilibrium. This EqP approach has been used for sediments by the EPA and others (Section 22.3). It is well supported for sediment invertebrates and is supported by some evidence for earthworms and possibly other soil invertebrates (van Gestel and Ma 1988; Lokke 1994; Ma et al. 1998). However, it has been suggested that the model may underestimate accumulation of a fraction of organic chemicals for which the dietary route is dominant (Belfroid et al. 1995). In addition to potentially making extrapolations between soils more accurate than soil/worm partitioning, the pore water model has the advantages of making available for use the large literature on water/biota partitioning factors (bioconcentration factors) and the QSARs for water/biota partitioning. However, it adds the burden of estimating soil pore water concentrations. The conventional formula for estimating soil pore water concentrations is

$$C_w = C_s/K_d \qquad (22.38)$$

where K_d = the soil (or sediment)/water partitioning coefficient (L/kg sediment); and C_w = water concentration (mg/L). Values of K_d are available from the literature for many metals and some organics but are highly variable among soils (Baes et al. 1984). If literature K_d values are used, this model is not expected to be more accurate than the use of soil–earthworm uptake factors from the literature. However, K_d values are available for some chemicals for which soil–earthworm uptake factors are not. For nonionic organic compounds:

$$K_d = f_{oc}K_{oc} \qquad (22.39)$$

or

$$K_d = f_{om} K_{om} \qquad (22.40)$$

or, from Karickhoff (1981):

$$K_d = 0.58\, f_{om} K_{oc} \qquad (22.41)$$

where f_{oc} = fraction organic carbon in the soil (unitless); K_{oc} = water/soil organic carbon partitioning coefficient (kg/kg or L/kg); f_{om} = fraction organic matter in the soil (unitless); and K_{om} = water/soil organic matter partitioning coefficient (kg/kg or L/kg).

This formula adjusts for the organic content of soil (expressed as either organic matter or organic carbon content), which is the major source of variance among soils in the uptake of neutral organic chemicals. This normalization makes the model more accurate than soil–earthworm uptake factors for neutral organic chemicals. For ionic organic chemicals, van Gestel et al. (1991) recommend correcting the coefficient (K_{oc} or K_{om}) by dividing by the fraction nondissociated (f_{nd}), which is estimated from:

$$f_{nd} = 1/(1 + 10^{pH - pK_a}) \qquad (22.42)$$

where pK_a = the negative log of the dissociation constant. When K_{oc} and K_{om} are both unavailable, they may be estimated from the QSAR developed by DiToro et al. (1991b):

$$\log_{10}(K_{oc}) = 0.983 \log_{10}(K_{ow}) + 0.00028 \qquad (22.43)$$

where K_{ow} = octanol/water partitioning coefficient (unitless).

Van Gestel et al. (1991) provide a formula for K_{om}:

$$\log_{10}(K_{om}) = 0.89 \log_{10}(K_{ow}) - 0.32 \qquad (22.44)$$

Values for K_{ow} are available in the literature for most organic chemicals, or they can be calculated from QSARs.

From these formulas, the earthworm concentration can be calculated as

$$C_v = K_{bw} C_w \qquad (22.45)$$

where C_v = concentration in earthworms (mg/kg fresh wt.); and K_{bw} = biota/water partitioning coefficient (L/kg organism). K_{bw} values for chemicals in earthworms may be assumed to be equivalent to bioconcentration factors for aquatic invertebrates from the literature. Alternatively, QSARs can be used to estimate this factor.

The model developed by Connell and Markwell (1990) for uptake by earthworms of 32 "lipophilic" organic chemicals (log K_{ow} 1.0 to 6.5) is

$$\log K_{bw} = \log K_{ow} - 0.6 \quad (n = 60, r = 0.91) \qquad (22.46)$$

These EqP models have been criticized for not incorporating gut uptake. Jager et al. (2003) responded by developing a three-compartment (soil, earthworm tissue, and gut contents) mass balance model. In studies of uptake of three organic chemicals by *Eisenia andrei* in artificial soil, their model provided better estimates of uptake but, at equilibrium, the results differed from EqP by less than a factor of 1.3. The authors concluded that the improvement was not significant given the uncertainties in risk assessments. However, if worms feed selectively on contaminated material such as leaf litter with pesticide residues, human sewage sludge, or manure from treated farm animals, EqP models are unlikely to be adequate.

Empirical models have been developed by regressing earthworm concentrations against soil concentrations. With the exception of models for TCDD and PCBs, all of these models have been developed for inorganic contaminants (Neuhauser et al. 1995; Sample et al. 1999b; Heikens et al. 2001; Peijnenburg 2001). Uptake of contaminants by earthworms may be influenced by soil parameters such as pH, organic matter, and cation exchange capacity. These parameters have been incorporated into multiple regression models, but the influential parameters differed among metals and species (Peijnenburg 2001). Model testing by Sample et al. (1999) indicated that, for models of multiple soils, inclusion of these additional soil parameters at best marginally improves estimates over those obtained from simple regression models.

Kinetic models of earthworm uptake could provide estimates of bioaccumulation that incorporate changes over time in soil concentration, soil conditions, and organism conditions. However, "despite the plethora of data detailing metal concentrations in earthworms collected from contaminated soils, relatively little is known about the accumulation and excretion kinetics of individual metals. Almost no kinetic information for other invertebrates has been reported" (Peijnenburg 2001). The same generalization applies to organic chemicals.

For any model, the risk assessor must verify the units of the estimate from the source literature. While earthworm concentration values estimated from the soil–earthworm uptake factors and the empirical regression models reported above are expressed as dry weight, those based on soil pore water uptake factors are expressed in terms of fresh weight. Because wildlife species do not consume dry food, all dry weight estimates must be converted to mg/kg of wet weight before they are employed in exposure estimation:

$$C_{wet} = C_{dry} \times P_{dry} \tag{22.47}$$

where C_{wet} = wet weight concentration; C_{dry} = dry weight concentration; and P_{wet} = proportion dry matter content of worm or other tissue. Water content of earthworms is reported to range from 82% to 84% (EPA 1993g).

Estimates generated by all methods reported above are for depurated worms. This should not result in an underestimate of exposure, because soil ingestion estimates, which include worm gut contents for vermivorous species, should be separately included as a source of exposure (Section 22.8). Concentrations are sometimes reported for undepurated worms, but there is no basis for correcting those values because of the variability in mass of ingested material.

22.9.5 TERRESTRIAL ARTHROPOD UPTAKE

Few models are available for the estimation of bioaccumulation by terrestrial arthropods. Examples of uptake factors from the literature are presented in Table 3.11 of Suter et al. (2000). While no empirical regression models are currently available for arthropods, some simple mechanistic models for arthropods have been developed (Kowal 1971; Van Hook and Yates 1975; Webster and Crossley 1978; Janssen et al. 1991). Because these models are parameterized for only a few species and contaminants, their use in risk assessments is problematic. Contamination of arthropods by pesticides is typically estimated by assuming that they are externally exposed to the spray, rather than accumulating the chemical indirectly from the environment.

22.9.6 TERRESTRIAL VERTEBRATE UPTAKE

To estimate the risks that contaminants present to wildlife that prey on terrestrial vertebrates, uptake models for birds, mammals, reptiles, and amphibians are needed. Uptake factors and models for reptiles and amphibians are not known, and those for birds and mammals are few. Examples of uptake factors for estimation of chemical concentrations in whole biota based on food or soil are summarized in Table 3.12 of Suter et al. (2000). Uptake factors have also been developed to estimate chemical concentrations in animal tissues based on concentrations in food. While these factors are generally applied as part of exposure assessments for human health, they may be of use in ecological risk assessments (Baes et al. 1984; IAEA 1994).

Uptake of lipophilic organic contaminants may be estimated based on chemical-specific K_{ow} values. Garten and Trabalka (1983) present diet–fat uptake factors for 93 organic chemicals for sheep, poultry, small birds, rodents, dogs, cows, swine, and primates. Because these uptake factors produce estimates in terms of mg/kg fat, the resulting estimates must be converted from mg/kg fat to mg/kg body weight using species-specific lipid content data.

Regression models based on field studies have been developed to estimate whole-body bioaccumulation from soil by small mammals in general and three trophic groups for 16 chemicals (Sample et al. 1998). These regression models have high variance, but they generally provide more accurate estimates of whole-body small mammal chemical concentrations than uptake factors in independent data sets. Soil screening levels are calculated using these models (Office of Solid Waste and Emergency Response 2003). These models have also been used to estimate body burdens in birds, based on the observation that trophically similar birds and small mammals from contaminated sites have similar whole-body concentrations (Beyer et al. 1985; Office of Solid Waste and Emergency Response 2003).

An alternative to these empirical and static models is the use of mechanistic and dynamic toxicokinetic models. For mammals and birds, toxicokinetic models are based on the physiologically based pharmacokinetic (PBPK) models derived for drug development and used in human health risk assessments. These mathematical simulation models estimate the internal dose of a chemical from external dose estimates by simulating the uptake, internal transport, partitioning among internal compartments, transformation, metabolism, and depuration of chemicals (Clewell et al. 2002). If effects are estimated from concentrations in specific tissues or organs, a toxicokinetic model with multiple internal compartments must be used (Figure 22.6).

Toxicokinetic models have been developed for the uptake of a few contaminants by terrestrial vertebrates. One example is a toxicokinetic model linked to a mass-based food web model for estimation of PCB bioaccumulation by piscivorous wildlife (Ram and Gillett 1993). Another example is the bioenergetically based one-compartment model of accumulation of PCBs by tree swallow nestlings (Nichols et al. 1995, 2004). This model was shown to accurately estimate body burdens of PCB congeners when dietary concentrations were set equal to those in food brought to the nestlings by adults (Echols et al. 2004; Nichols et al. 2004).

22.10 EXPOSURE TO PETROLEUM AND OTHER CHEMICAL MIXTURES

Contaminants seldom occur alone; therefore, nearly all exposure assessments must deal with mixtures (Chapter 8). However, most assessments address incidental mixtures of chemicals that are individually identified, analyzed, and modeled as described in the preceding sections. Other contaminant mixtures such as crude petroleum, petroleum-derived fuels, and PCB product mixtures (e.g., Aroclors®) are generally treated as a single material. Analysis of exposure to these complex materials can be divided into a set of distinct questions:

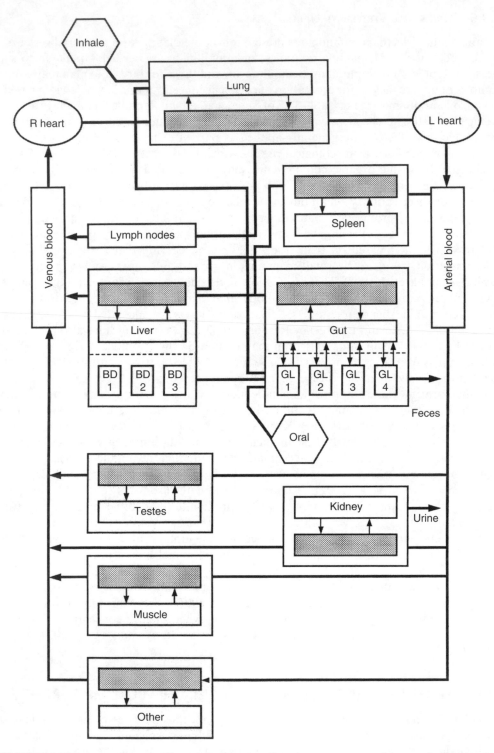

FIGURE 22.6 Diagram of mammalian toxicokinetics. Gray boxes represent the organ's blood volume. (Adapted from Menzel, D.B., *Environ. Sci. Technol.*, 21, 944–950, 1987. With permission.)

1. How can chemical mixtures be analyzed so as to generate useful fate and exposure properties?
2. How can media be analyzed to provide useful estimates of exposure?
3. Might analyses of biota be used to estimate dietary exposure or internal exposure?
4. Might bioassays substitute for chemical analyses?
5. How might current and future exposure and uptake be modeled from current concentrations?

The preliminary characterization of the released material should define as fully as possible the material released (e.g., North Slope Crude or PCB transformer oil). That characterization should be followed by analyses of the material in the contaminated media. Both the composition and concentration of the material should be determined. While the composition of the released material may be known, the weathering process may be quite rapid, so that by the time an assessment is conducted the composition may have changed considerably. The problem is deciding how to analyze the mixture in a way that both adequately characterizes the mixture and provides a basis for characterizing the fate, transport, and toxicological properties of the mixture. Three approaches are proposed: analysis of the whole material, analysis of chemical classes, and analysis of individual chemicals (Table 22.9).

The simplest approach is to analyze the whole material by determining total petroleum hydrocarbons (TPH), oil and grease, total PCB, or some equivalent metric of gross contamination. Such measures typically provide little basis for performing a risk assessment, because they provide little information about the composition and therefore the properties that determine potential fate and toxicity of the material. However, these gross metrics may be useful for determining the extent of contamination or the locations of the greatest contamination. Exceptions to this generalization are cases in which the composition of the whole material is well characterized, and the contamination is recent, e.g., synthetic lubricating oils and marketed PCB mixtures (e.g., Arochlor 1254®).

Rather than attempt to characterize the whole material, one may characterize constituent classes of compounds. Hydrocarbons are divided into aliphatics and aromatics; long and short chain aliphatics; one-, two-, three-, and more than three-ring aromatics, etc. In addition to hydrocarbons, petroleum contains metals, nitrogen, sulfur, and oxygen-containing organics, and other compounds. Analyses of these classes provide considerably more information than TPH or indicator chemicals, but in order to model transport and transformation of these chemical classes or to determine their toxicity, they must be associated with concentrations of particular chemicals. This can be done by identifying representative compounds for each class. A representative compound should be selected based on the following criteria:

TABLE 22.9

Methods for Analyzing Chemical Mixtures and Characterizing Their Physical, Chemical, and Toxicological Properties

Mixture Analysis	Property Characterization
Whole material	Whole material properties
Chemical classes	Properties of representative chemicals
	Distribution of properties of class
Individual chemicals	Properties of detected chemicals
	Properties of indicator chemicals

• It is an abundant member of the class of chemicals in the material being assessed.
• Data are available concerning its environmental fate and effects.
• It has greater than average (screening assessments) or average (definitive assessments) toxicity for its class, and greater than average (screening assessments) or average (definitive assessments) persistence and bioavailability.

Once representative chemicals are selected, the assessment can be performed by assuming that the entire mass of each class of chemicals is made up of that representative chemical.

A related approach to characterizing the properties of the mixture is to identify indicator chemicals (ASTM 1994). These are chemicals that are assumed to account for the major risks from the mixture. That is, if risks from those chemicals are acceptable, the risks from the whole mixture are acceptable. For risk assessments of petroleum products, ASTM recommends using benzene and benzo(a)pyrene [and, in cases where they are present, ethylene dibromide (EDB) and ethylene dichloride (EDC)] because of their carcinogenicity. Clearly, other criteria would need to be used to choose indicator chemicals for ecological risk assessments. Similarly, EPA Region 9 has a list of 28 PCB congeners that are of concern with respect to ecological risks (Valoppi et al. 1999).

Another sort of indicator chemical approach is the analysis of chemicals that are characteristic of the complex material. For example, the exposure of kangaroo rats to petroleum from a well blowout was confirmed by analyzing liver tissue for a set of PAHs that are characteristic of oil and coal (Kaplan et al. 1996). Vanadium, which is typically concentrated in petroleum, was found to be elevated in San Joaquin kit foxes from oil fields relative to foxes from other sites, suggesting that the foxes were exposed to petroleum in some form (Suter et al. 1992). This approach and the related concept of chemical fingerprinting (Chapter 19) are useful for confirming that exposure has occurred, that exposure is associated with a particular source, or that the material is being transported by a particular route. However, they are not adequate exposure metrics for risk assessment because they cannot be related to exposure–response relationships to predict toxicity.

If fate or effects data are available or can be estimated for several chemicals in a class, a more sophisticated approach could be used. The statistical distribution of the fate and effects properties of the members of a chemical class can be used in the assessment to represent the distribution of the properties in the entire class. For example, if water solubility values for several short-chain aliphatic hydrocarbons are found in the literature, a distribution fit to those data is an estimate of the distribution of water solubility for the class. Alternatively, if QSARs are available to estimate fate and effects properties, they could be used to estimate the parameter distributions for the class from the physical properties or structural characteristics of the individual chemicals in the class. For example, the water solubility of hydrocarbons can be estimated from their structure (Lyman et al. 1982; Mackay et al. 2000). Therefore, by specifying the structures of short-chain aliphatic hydrocarbons, one can estimate the solubilities of all members of the class, and the distribution of those individual solubilities is the solubility distribution for the class. If the relative abundances of the class constituents can be estimated, the distribution can be refined by weighting the observations (e.g., the individual solubility estimates).

Finally, a total analysis of the materials can be performed. For a petroleum-contaminated site, the analysis could include the hydrocarbons and other organic and inorganic compounds found in petroleum, as well as organic and inorganic additives (e.g., oxygenates), and the various chemicals that may occur in mixed wastes (e.g., drilling fluid components). This analytical approach offers the greatest flexibility in that all of the various exposure metrics discussed above can be reconstructed from a total analysis. In addition, if a site has been contaminated by chemicals other than the material of concern, a thorough chemical analysis may identify causes of toxicity that would not be revealed by a material-specific analysis. Finally, analytical quantitation limits are often lower for individual chemicals than for whole materials. For example, using EPA methods, quantitation limits for PCB congeners are 3 orders of magnitude lower than those for Arochlors and fall within the range of ecotoxicological benchmarks (Valoppi et al. 1999). However, it is likely that the majority of individual chemicals in any particular complex material will have unknown toxicity. Therefore, the risk characterization must be based on a subset of the detected chemicals.

22.11   EXPOSURE TO NATURAL EXTREME EVENTS

Ecosystems are exposed to, and affected by, extreme natural events such as floods, droughts, and fires. Methods for estimating the magnitudes and frequencies of extreme natural events are well developed for use in engineering design, insurance, and emergency planning (Haimes 1998; Reed 2001). The frequencies of extreme weather and hydrologic events are also an important input to many ecological risk assessments. For example, extreme low flows determine the minimum dilution of an aqueous effluent, extreme air stagnation events determine the maximum exposures to atmospheric emissions, and various extreme natural habitat disturbances such as floods or large fires contribute to the probability of extinction of rare species or small populations. In addition, management actions that are the subjects of ecological risk assessments may influence the frequency and intensity of nominally natural events. For example, channelization increases the frequency and intensity of downstream floods and timber management influences the frequency and intensity of forest fires.

The estimation of extremes is much more difficult than the estimation of means. Extremely long time series are required to obtain good estimates empirically. However, long time series are likely to be confounded by changes in land management, channel modifications, paving of watersheds, changes in measurement methods, and other trends or discontinuities. Alternatively, a system may be simulated or at least broken down into more frequent constituents. For example, the risk of an extreme flood may be the joint probability of a large snow pack, warm temperatures, and rain. Similarly, the risk of a severe fire may be the joint probability of high fuel load, dry conditions, and an ignition source. The large uncertainty associated with estimating extremes is increased by the need for data interpolation (e.g., between weather stations) or extrapolation (e.g., from one watershed to another). Hence, these assessments should be performed by specialists, and the ecological risk assessor should insist on a complete accounting of uncertainty.

22.12   EXPOSURE TO ORGANISMS

Exposure to introduced organisms must be estimated in order to estimate the probability of establishment and the probability of effects given establishment. For risk of establishment, exposure is expressed as the rate of arrival in potential habitat, sometimes termed the propagule pressure (Stohlgren and Schnase 2006). For risk of effects, exposure is generally defined by abundance, either in absolute terms or relative to indigenous species.

22.13   PROBABILITY AND EXPOSURE MODELS

Probabilistic environmental risk assessments for both human health and ecology have been primarily concerned with the estimation of multimedia exposure (Burmaster and Anderson 1994; Hammonds et al. 1994; Hansen 1997; Risk Assessment Forum 1997). Like humans, wildlife species drink potentially contaminated water, eat various proportions of foods, consume potentially contaminated soil, etc. These exposures are estimated using simple mathematical models that are subject to Monte Carlo analysis (Chapter 5). The trick in performing probabilistic analyses of these models is representing the parameter distributions in such a way that the results are those required by the risk characterization (Section 30.5). In general, the probabilities should be distributions of doses among individuals, credibilities of doses with respect to individuals, or both (i.e., from nested Monte Carlo analysis). Consider, for example, mink exposed to contaminants in a stream. The model is

$$E_j = \sum_{i=1}^{m} p_{ik} \left(\frac{IR_i \times C_{ijk}}{BW} \right) \tag{22.48}$$

where E_j = total exposure to contaminant (j) (mg/kg/d); m = total number of ingested media (e.g., food, soil, or water); IR_i = ingestion rate for medium (i) (kg/d or L/d); p_{ik} = proportion of type (k) of medium (i) consumed (unitless); C_{ijk} = concentration of contaminant (j) in type (k) of medium (i) (mg/kg or mg/L); and BW = body weight of the endpoint species (kg).

Body weight estimates may be obtained from the literature or summaries such as EPA (1993b), from which four distributions may be derived: (1) the distribution of weights of individual mink at an individual site (e.g., the mean and standard deviation of individual weights from a study), (2) the distribution of mean weights across sites (e.g., the mean and standard deviation of mean weights from multiple studies), (3) the distribution of individual mink weights across sites (e.g., the mean and standard deviation of individual weights from all studies), and (4) the distribution across sites of the moments of the distributions (e.g., means and standard deviations of means and standard deviations of individual weights across multiple studies). If one of the sites at which mink have been weighed is similar to the site being assessed, we may use the distribution of weights from that site to estimate the distribution of individual weights in the model (distribution 1). If none of the sites in the literature resemble the contaminated site, we may be tempted to use the distribution of means (distribution 2). However, assessments are concerned with the distribution of effects across the mink comprising a population at a site, not a hypothetical mean mink. Alternatively, one might use the distribution of individual weights across all studies to represent the distribution of individuals at the contaminated site, but that would inflate the variance by the amount of systematic variation in weight among populations (distribution 3). The most complete description of the situation would be to perform a nested Monte Carlo analysis with the distributions of the means and standard deviations (assuming normal or log-normality) from all studies that are believed to be potentially representative of the site (e.g., wild mink only) used to estimate the variance among individuals and the uncertainty concerning that variance (distribution 4).

Ingestion rates should be treated in the same manner as body weights. For example, if the mean and standard deviation of individual weights is taken from the study that is most similar to the site being assessed, the same should be done for ingestion rates.

Contaminant concentrations also require careful interpretation. The input data for this parameter are likely to be concentrations in individual fish collected at various points on the site. The variance in concentration among individual fish is not the appropriate measure of variance in this parameter unless some individual fish are so contaminated as to cause acute toxic effects. Rather, because effects are due to long-term exposures in nearly all cases, the relevant exposure metric is the mean concentration in fish. The next issue is over what set of fish should the mean be calculated? The appropriate mean is the mean across fish within a foraging range of a mink. This assumption leads to alternative expressions of the parameter, depending on the size of the spatial unit being assessed.

- If a unit is approximately as large as the foraging range of a mink, there is a single mean concentration. The variance is the variance on that mean, given sampling error, which can be estimated statistically if sampling is replicated. This can be taken as an estimate of the sampling variance among individual mink. The major uncertainty is the uncertainty concerning how representative a sample of fish taken by electrofishing or netting is of the sample taken by a foraging mink. That uncertainty may be estimated by expert judgment guided by the observed variation among fish of different species and sizes. Other uncertainties may be relevant depending on the sample preparation and analysis. For example, if the whole-fish concentrations are estimated from fillet concentrations, the uncertainty concerning that conversion may be important (Bevelhimer et al. 1996).
- If the unit is smaller than the foraging range of a mink, the procedure is as above, except that the fish from inside the unit and those from outside the unit must be treated as different dietary items in the exposure model.
- If the unit is much larger than the foraging range of a mink, the unit should be divided into subunits that correspond approximately to a foraging range. In most cases, a river or stream will already have been divided into subunits, termed reaches, based on differences in contamination and physical features. For the mink exposure model, one would simply aggregate or subdivide those reaches to approximate a foraging range. This approach requires the assumption that the bounds of a foraging range correspond to the bounds of the assessment's spatial units. This is not unreasonable for the purposes of the assessment, since the actual bounds are unknown and would vary from year to year, and the bounds of the units are relevant to the remedial decision. However, it is possible to eliminate the assumption about the bounds on foraging ranges. If there are sufficient sampling locations, one may use a moving average concentration where the average is of concentrations at sampling sites and the window within which they are averaged is the length of a foraging range. In either case, the most important variance is the variance among foraging ranges (i.e., among individual mink). The most important uncertainty is likely to be, as above, the uncertainty concerning the representativeness of the sample.

The treatment of variance in exposure depends on the exposed organism. For example, we assume that spatial variance is important to mink, but temporal variance is important to fish, even though they are both exposed to aqueous contaminants. The difference results from their modes of exposure. Fish are exposed directly to water by gill ventilation and are susceptible to effects of short-term variance due to spills, storms, and other such events that make fish kills a common occurrence on industrialized streams. Mink, in contrast, are exposed primarily through diet, which results in considerable temporal averaging of contaminant concentrations. Spatial variance is assumed to be relatively unimportant for fish, because they are assumed to reside within a reach, which is designed to have uniform contamination. In contrast, because mink have foraging ranges that may be large relative to reaches, and divide space among the members of a population, space is a critical variable in estimating risks to mink. Hence, for fish, temporal variance is much more important than spatial variance, but for mink, space is more important. These assumptions hold for most sites with which we are familiar, but they are not inevitable. For example, at some sites aqueous concentrations may be effectively constant over time (i.e., no significant episodic exposures), but may be highly variable in space relative to the range of a fish or fish population.

22.14 PRESENTING THE EXPOSURE CHARACTERIZATION

Since exposure characterization is an intermediate analytical stage in ecological risk assessment, its results should primarily be input to the next stage (risk characterization) rather than
voluminous text, tables, or figures. The results are presented as an exposure profile (EPA
1998a). It should summarize, for each endpoint receptor, the exposure pathways addressed so
as to ensure that all pathways included in the conceptual model have been analyzed. If not,
the conceptual model should be modified to remove pathways that have been determined to
be absent or insignificant. Risk assessors should not claim that their conceptualizations of the
exposure process are more complete than is in fact the case. The exposure profile should
describe how exposure was estimated for each pathway, summarize the results, and present
the associated uncertainties. If multiple lines of evidence are used in the risk characterization
of an assessment endpoint, proper exposure metrics should be derived for each. In general, a
tabular summary is an appropriate format for the exposure characterization. For each line
of evidence and endpoint, it would present exposure media, routes, point estimates of
exposure intensity (e.g., mean concentration, median dose, maximum concentration), distributions in space and time, and uncertainties (e.g., sampling variance, analytical precision,
model uncertainties).

When performing an exposure characterization for a preliminary phase of a multiphased
assessment (e.g., development of a definitive assessment plan) or when presenting preliminary
results to the public, it is often useful to graphically or cartographically summarize the
temporal and spatial distribution of contaminants. However, since presentations of distributions of exposure may be misleading or cause undue concern, results of the exposure
characterization should not be emphasized when risk estimates are available for presentation.

Part IV

Analysis of Effects

In the analysis of effects, assessors characterize the nature and magnitude of effects of chemicals or other agents as functions of exposure. Effects may be estimated by performing tests, by observing effects in the field, or by mathematically simulating effects. In the analysis of effects, effects data must be evaluated to determine which are relevant to each assessment endpoint, and then reanalyzed and summarized as appropriate to make them useful for risk characterization. Two issues must be considered.

First, what form of each available measure of effect best approximates the assessment endpoint? This issue should have been considered during the problem formulation (Chapter 18). However, the availability of unanticipated data and better understanding of the situation after data collection often require reconsideration of this issue.

Second, is the expression of the effects data consistent with the expressions of exposure? Integration of exposure and effects defines the nature and magnitude of effects, given the spatial and temporal pattern of exposure levels. Therefore, the relevant spatial and temporal dimensions of effects must be defined and used in the expression of effects. For example, if the exposure is to a material such as unleaded gasoline that persists at toxic levels only briefly in soil, effects that are induced in that time period must be extracted from the effects data for the chemicals of concern, and the analysis of field-derived data should focus on biological responses such as mass mortalities that could occur rapidly rather than long-term responses.

The degree of detail and conservatism in the analysis of effects depends on the tier of the assessment (Section 3.3). Screening assessments typically define the exposure–response relationship in terms of a benchmark value, a concentration or dose that is conservatively defined to be a threshold for toxic effects (Chapter 31). Definitive assessments should define the appropriate exposure–response relationship. Typically, this requires performing tests (i.e., controlled exposures to the agent of concern; Chapter 24) or field studies (Chapter 25) associating exposures and effects (Chapter 23). Because tests typically do not include all species and life stages of concern, extrapolation models are needed to estimate effects on attributes of organisms (Chapter 26), populations (Chapter 27), or ecosystems (Chapter 28). Because nearly all testing determines organism-level responses, extrapolations to organism-level endpoint attributes use simple assumptions or statistical models. In contrast, the extrapolation to population and ecosystem-level attributes requires an extrapolation across levels of organization that typically requires mathematical simulations.

23 Exposure–Response Relationships

> *What is there that is not poison?*
> *All things are poison, and nothing is without poison.*
> *Solely the dose determines that a thing is not a poison.*

> Paracelsus, translation by Deichmann et al. (1986)

Paracelsus's famous insight that the dose makes the poison implies that toxicologists must determine the relationship between the dose level and the toxic response. More generally, to assess the risk posed by any agent, it is necessary to determine the relationship between exposure and response. Exposure–response relationships are, in general, quantitative models of the form $r = f(e)$, where r and e are response and exposure metrics, respectively. However, they may be qualitative relationships such as: where introduced species e is present, native species r is extirpated. Hence, we may more generally state that we wish to estimate the expected response r given a specified exposure e, $E(r|e)$. Exposure–response relationships serve at least three purposes.

Estimation: If an exposure–response model is available, an appropriate estimate of exposure can be used with that model to estimate the response. Such estimates can be used in risk assessments to characterize risks from future contamination or in ecological epidemiology to determine whether observed levels of exposure are credible causes of observed impairments.

Benchmarks: If an exposure–response relationship is reduced to a point such as an EC_{20} or a Benchmark Dose Limit (these are test endpoints; Box 23.1), that value can be used to separate acceptable from unacceptable exposure levels. They are used as regulatory standards or as screening benchmarks, either directly or after adjustment with safety factors or other means (Chapter 29).

Communication: Stakeholders and decision makers are often unfamiliar with the ways in which effects change in response to changes in exposure levels. Rather, they tend to think in terms of dichotomies such as safe or unsafe. Hence, it is often important to present exposure–response relationships, particularly when complexities such as time to response, optimal exposure levels, or thresholds are involved.

Exposure–response relationships are expressions of the observation that effects are caused by associations of affected entities with causal agents. The associations may occur in a toxicity test or other experimental study (Chapter 24) or in observational studies (Chapter 25). In either case, the importance of the analysis comes from the assumption that by quantitatively modeling the association of exposure and response one can generalize to other cases in which that cause and the affected entity are associated. That is, if a chemical's 96 h LC_{50} for fathead minnows is determined from a laboratory test to be 2 mg/L, one would expect a fish kill to occur if that chemical occurred at 2 mg/L for at least 96 h in a stream with similar water

BOX 23.1
Terminology for Test Endpoints

Analyses of exposure–response relationships should aim to develop models of how responses change as exposure changes. However, results of tests or observations are commonly reduced to a point that is thought to provide a threshold or to summarize the results. The terminology for these values is inconsistent in practice and may be confusing. In this explanation, the intensity of agents is defined as concentration (C), since it is the most common unit in ecological risk assessment. However, one can substitute dose (D), time (T) or, more generally, level (L) for C in any of the terms.

If regression analysis has been used to develop a model that relates responses to exposure estimates, inverse estimation can derive an exposure level corresponding to a specified effects level (Figure 23.1). For quantal variables, those that are proportions of subjects displaying a dichotomous trait such as survival/death or presence/absence, these are termed ECp, the concentration causing the effect in proportion p. The median lethal concentration (LC_{50}) is a particular ECp. For continuous variables such as weight or eggs per female, these values are known as ICp, the concentrations inhibiting the response by proportion p. Other terms are used in particular circumstances. For example, the term infective dose (IDp) is used for tests of pathogens. For simplicity, ECp may be used for all of these values.

In human health risk assessment and some wildlife risk assessments, the term equivalent to ECp is the benchmark dose (BMD) (Crump 1984). A lower confidence limit on the BMD is termed a BMDL.

If hypothesis-testing statistics are used, two test endpoints are derived. The first is the lowest concentration causing an effect that is statistically significantly different from control or reference, the lowest observed effect concentration (LOEC). The second is the highest concentration that is below the LOEC, which is the no observed effect concentration (NOEC). If adverse effects are distinguished from those that are not considered adverse, LOAEC and NOAEC terminology is used.

chemistry. Hence, when developing exposure–response relationships, we must answer the question, what expression of the observed association between the causal agent and the effect of interest will allow us to make the most useful predictions of future effects?

The responding unit in nearly all toxicity tests and in many studies of biological responses to nutrients, heat, and other nontoxic agents is the individual organism. What proportion die, what is the average growth, etc.? However, responses of other entities such as experimental populations (e.g., algal tests), experimental communities (e.g., microcosms), and field populations and communities may be related to their levels of exposure. The most common responding unit in ecological risk assessment, after organisms, is species. Models relating the responses of individual species to exposure levels are termed species sensitivity distributions (Posthuma et al. 2001). However, since they are most commonly thought of as models to extrapolate from species to communities, they are discussed in Section 26.2.3.

Depending on the assessment problem, it may be useful to define an exposure–response relationship with respect to any number of the dimensions described in Section 6.3: space, time, intensity, severity, proportion responding, and type of response. Exposure–response relationships are often expressed as points, such as LC_{50}s or no observed effect concentrations (NOECs), but the most useful of the commonly available relationships is a line in two-dimensional space defined by one exposure metric (usually concentration or dose) and one response metric (usually severity or proportion responding) (Figure 23.1). These relationships

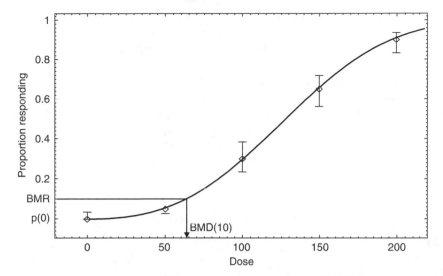

FIGURE 23.1 Exposure–response relationship with inverse regression. The benchmark dose (BMD) is derived from the benchmark response (BMR). Generated by the Benchmark Dose software.

are typically sigmoid. The slope or spread of the curve depends on the variance in sensitivity among the exposed units. Tests of very similar organisms, such as inbred laboratory rats, yield very steep curves, while dissimilar units, such as stream communities in a field study, yield much broader curves with respect to a particular range of exposure. Surfaces in three dimensions should be used much more often than they are, because we often want to know about responses to both the concentration and duration of exposure (Figure 23.2). Similarly, for a fish population model, we may need to know the relation to concentration of both the reduction in fecundity and the proportion of females exhibiting a given level of reduction, because the implications of a relatively uniform reduction in fecundity may be different from the same average effect due to sterility of part of the population and no effect on the rest. The next logical step is volume in four dimensions (e.g., concentration, duration, severity, and

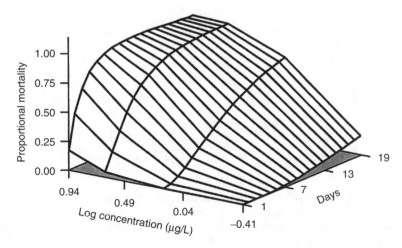

FIGURE 23.2 Toxic effects as a function of concentration, duration, and proportion responding.

proportion) or even five (add the distribution in space). More information is always desirable, but data become limiting. We can gather data on concentration, duration, severity, and proportion from a conventional toxicity test, but the conventional number of replicates would seldom be sufficient to statistically fit a four-dimensional model. However, such data may be displayed without fitting a function (Figure 23.3).

Often, risk assessors must settle for whatever standard or nonstandard expressions of the exposure–response relationship are available. This chapter discusses alternative approaches and associated issues so that assessors understand the types of relationships that they may be required to use and in the hope that they will have the opportunity to derive relationships from available data or even direct the generation of new data.

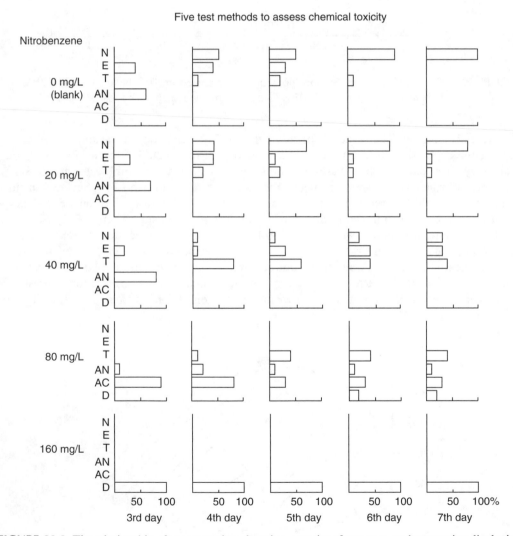

FIGURE 23.3 The relationship of concentration, duration, severity of response, and proportion displaying the response, shown as a set of severity vs. proportion responding relationships, arrayed on concentration and time axes. The responses are N = normal; E = eyespot; T = tetratopthalmic; AN = anopthalmic; AC = acephalic; and D = death. (From Yosioka, Y., Ose, Y., and Sato, T., *Ecotoxicol. Environ. Saf.*, 12, 15, 1986. With permission.)

23.1 APPROACHES TO EXPOSURE–RESPONSE

Exposure–response relationships can be derived in various ways depending on the amount and quality of data and background information that are available. Ideally, mechanisms are understood and can be used to model responses to exposures. At the other extreme, one may not be able to do more than report that a particular response occurred at a particular exposure level. Currently, the best available guidance for ecotoxicological exposure–response analysis is provided by Environment Canada (2005), but other organizations provide different guidance (ASTM 1996; OECD 1998, 2004; Crane and Godolphin 2000; Klemm et al. 1994; IPCS 2004).

23.1.1 MECHANISTIC MODELS

If the mechanisms by which an exposure causes a response are understood, a mathematical model that represents that relationship may be developed. These include toxicodynamic models of organismal responses (Section 23.3), population dynamic models (Chapter 27), and ecosystem models (Chapter 28). One advantage of these models is that their functional form is defensible on bases other than convention or goodness of fit. Another advantage of mechanistic models is their flexibility. If mechanisms are well understood, a mechanistic model may be used to simulate conditions outside the range of tests or observations. If mechanisms are fully specified, responses could be modeled from basic knowledge without any testing or observation, as in the application of physical laws. However, models in ecological risk are almost never purely mechanistic. They usually rely on empirical approaches to estimate parameter values, and, in the simplest case, they are equivalent to biologically plausible regression models. Hence, their range of applicability must be carefully considered (Chapter 9). Examples of mechanistic exposure–response models for organisms include the Dynamic Energy Budget model (DEBtox) (Kooijman and Bedaux 1996) and conventional toxicodynamic models (Section 23.3).

23.1.2 REGRESSION MODELS

If data are available for responses at multiple exposure levels, the best general approach to exposure–response modeling is statistical regression analysis. The generally preferred method for regression analysis is maximum likelihood estimation (Environment Canada 2005), but least-squares regression is usually effective. Either method provides confidence bounds, unless the error distribution is unclear, in which case bootstrap estimates should be used (Shaw-Allen and Suter 2005). One may either choose a single function and fit it to the data or fit multiple plausible functions and choose the one that provides the best fit. Functions may be chosen because they are the standard function for a particular use, because the form is appropriate for the data, or because it has an appropriate biological interpretation. The most commonly used function in ecotoxicology is the log probit (the linearized log-normal distribution), which is used to relate quantal data (e.g., proportional mortality) to the independent exposure variate (Figure 23.4). There are no standard models for continuous data; appropriate functions should be chosen, fitted to the data, and compared. When models are compared, their relative likelihoods are the appropriate metrics unless they differ in the number of fitted parameters, in which case Akaike's information criterion should be used (Section 5.4.6). In addition, plots of the data and fitted model should be inspected for their plausibility and for outliers. Finally, residuals should be plotted and inspected for patterns that suggest a systematic lack of fit or heterogeneity of variance.

Methods for fitting of exposure–response distributions to toxicity data are discussed by Kerr and Meador (1996), Moore and Caux (1997), Bailer and Oris (1997), and Environment

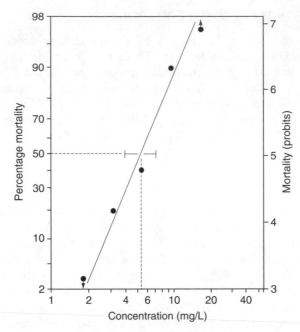

FIGURE 23.4 Results of an acute lethality test plotted as probits of response against the log concentration. The $LC_{50} = 5.6$ mg/L and the 95% confidence bound are plotted. (From Environment Canada, Guidance document on application and interpretation of single-species tests in environmental toxicology, EPS 1/RM/34, Ottawa, Ontario, 1999. With permission.)

Canada (2005). Software for regression analysis that can be used to generate exposure–response models can be found in any of the large statistical packages such as SAS, SPSS, and S+, and in R libraries. Commercial software packages specifically for analyzing toxicity test data include CETIS, TOXSTAT, and TOXCALC. Finally, government agencies have developed software that may be recommended for particular regulatory assessments. The US EPA has developed benchmark dose software that is particularly good for comparing alternative functions and calculating confidence bounds (http://www.epa.gov/ncea/bmdds.htm). Although it was developed specifically to calculate benchmark dose values for human health risk assessments, it is also useful for ecological risk assessments (Linder et al. 2004).

23.1.3 STATISTICAL SIGNIFICANCE

The traditional toxicity test endpoints for chronic tests, NOECs and LOECs derived by statistical hypothesis testing (Box 5.1), have low utility for ecotoxicology or ecological risk assessment (Hoekstra and van Ewijk 1993; Laskowski 1995; Suter 1996a; OECD 1998; Environment Canada 2005). Because they are based on statistical significance, these endpoints do not indicate whether the effect is, for example, a large increase in mortality or a small decrease in growth. The level of effect at an NOEC or LOEC is an artifact of the replication and dosing regime employed. As a result, NOECs and LOECs correspond to highly variable types and levels of effects (Suter et al. 1987; Crane and Newman 2000). They do not indicate how effects increase with increasing exposure, so the effects of slightly exceeding an NOEC or LOEC are not qualitatively or quantitatively distinguishable from those of greatly exceeding it. To estimate risks, it is necessary to estimate the nature and magnitude of effects that are occurring or could occur at the estimated exposure levels and associated uncertainties. Such estimates are supplied by the other approaches.

23.1.4 INTERPOLATION

When data are not adequate for statistical fitting of a model, linear interpolation may be employed (Klemm et al. 1994). Hoekstra and van Ewijk (1993) recommended using linear interpolation down from an observed effect of approximately 25%, because they felt that fitted functions are not reliable at low levels. This method is most accurate for approximately linear segments of exposure–response data and relatively small intervals between exposure levels. In most cases, log conversion of the exposure metric will increase the linearity. The US EPA standard method and program for linear interpolation are available in Norberg-King (1993).

23.1.5 EFFECT LEVEL AND CONFIDENCE

In some cases, the best that can be done with exposure–response information is to report the exposure level and associated effects level. If there is replication, geometric means and confidence limits should be calculated. This approach is appropriate when a test of a single exposure level and control is performed, as in tests of undiluted effluent or of a contaminated medium at a particular location. It may also be used when data do not permit regression, as when one treatment level produces partial mortality and all others cause 100% mortality, and one is reluctant to assume linearity for interpolation.

23.2 ISSUES IN EXPOSURE–RESPONSE

The modeling of exposure–response is a highly complex topic both because of the complexity and heterogeneity of causal relationships in ecology, and because the statistics is unsettled. The following issues are particularly important for ecological risk assessment.

23.2.1 THRESHOLDS AND BENCHMARKS

For regulatory standards or screening benchmarks, it is desirable to define points on the exposure distribution that are thresholds for significant effects; significant in this case means that, if the threshold is exceeded, some action should be taken. Thresholds for statistical significance are inappropriate for that purpose. Rather, one must choose a level of effects (p) that has legal, policy, or societal significance, but how? LC_{50}s have traditionally been reported, because values in the middle of the curves are estimated with greatest precision (Figure 23.4). Fifty percent mortality is clearly not a threshold effect. However, if the curve is sufficiently steep, so that there is little variance in the effective concentration relative to other sources of variance, the LC_{50} may be reasonably representative of partially lethal concentrations. However, a low effects level is generally desired for benchmarks. Because of concern for precision of the estimate, Environment Canada (2005) recommended that values less than EC_{10} not be used and that p not be within the range expected for control effects. OECD (1998) recommended that values from EC_5 by increments of 5 up to EC_{25} be determined routinely, and, if a mechanistic model is used, an EC_0 should be reported. This approach provides the decision maker with information to select a threshold effect based on policy and circumstances (e.g., the presence of important species). The approach would be enhanced by reporting confidence limits on each value.

If the effect in controls or reference areas is zero (or can be assumed to be zero plus error) and the exposure–response relationship has a lower threshold, the estimated intercept of the

x axis (EC$_0$) is an estimate of the biological threshold. Van Straalen (2002b) recommended using the HC$_0$ from species sensitivity distributions as community no-effect concentrations, using the uniform, triangular, exponential, or Weibull distribution. More conventional distributions with infinite tails (i.e., the normal and logistic) can be used if the number of organisms in the endpoint population or species in the endpoint community is specified (Kooijman 1987). If there are 100 species in a community, concentrations below the HC$_{01}$ (the first percentile of the SSD) are estimated to protect them all.

More commonly, the effects data display a nonzero threshold, which can be incorporated in the exposure–response model. Exposures up to some level produce responses equal to background (i.e., control treatments or reference sites), and higher levels produce increasing responses. Such cases may be fitted by a hockey-stick model, and the threshold is the exposure level at which the two segments meet (Figure 23.5). That is

$$\begin{cases} \text{Effect} = \text{Background} & \text{for } C < C_T \\ \text{Effect} = \text{Background} + \beta(C - C_T) & \text{for } C > C_T \end{cases} \tag{23.1}$$

where C_T is the threshold concentration and β is the slope. Examples of hockey-stick models in ecotoxicology include Beyers et al. (1994) and Horness et al. (1998). Beyers et al. (1994) found that hockey-stick thresholds were a factor of 2 to 4 lower than NOECs.

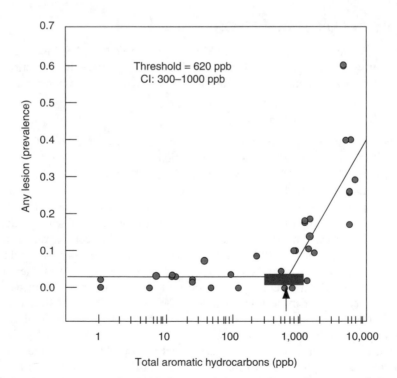

FIGURE 23.5 Hockey-stick regression of liver lesion prevalence in English sole as a function of total aromatic hydrocarbon concentration in sediment. The 95% confidence interval on the break point is plotted as a gray rectangle. (From Horness, B.H., Lomax, D.P., Johnson, L.L., Myers, M.S., Pierce, S.M., and Collier, T.K., *Environ. Toxicol. Chem.*, 17, 872, 1998. With permission.)

23.2.2 TIME AS EXPOSURE AND RESPONSE

Despite its importance, time is relatively neglected in exposure–response analyses. Variation in time may be more important than concentration; we may be concerned about fish swimming through a toxic plume or exposed to an episodic effluent (Brooks and Seegert 1977). In analyses of short-term toxicity tests, effects are conventionally reported for the point at which the test is terminated (e.g., a 96 h LC_{50}). For longer-term "chronic" tests, it is typically assumed that equilibrium exposure and maximum effects have been achieved, so duration is irrelevant. For many exposures, neither approach is adequate. One solution is to treat duration as the exposure metric. Figure 23.6 shows how revealing such relationships can be; BDE had no effect on egg production until day 10, when treated fish stopped spawning. The same functions may be used as for concentration or dose (e.g., fit the probit function to mortality vs. time data) (Environment Canada 2005). However, the associated confidence bounds are not accurate, because the same organisms are repeatedly observed over time, and therefore the observations are not independent replicates. More properly, time–event modeling approaches can be used. These techniques are not in common use, but appropriate procedures are available in software packages (e.g., LIFEREG in SAS, recommended by Newman and Aplin 1992) and guidance is available (Crane and Godolphin 2000; Crane et al. 2002).

These approaches treat time as duration, that is, an exposure continues for a certain discrete interval of time. However, if concentration or another measure of intensity is variable in time or if exposure episodes recur without sufficient time for recovery, duration is insufficient. Rather, exposure must be dynamically modeled. In ecotoxicology, toxicokinetic

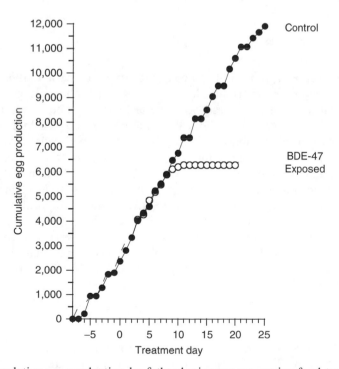

FIGURE 23.6 Cumulative egg production by fathead minnows consuming food treated with 2,2,3,3-tetrabromodiphenyl ether (open circles) and controls (solid circles). (From Muirhead, E., Skillman, A.D., Hook, S.E., and Schultz, I.R., *Environ. Sci. Technol.*, 40, 523, 2006. With permission.)

models may be used to estimate internal concentrations that are then used in internal exposure–response models (Section 23.3).

Time is a dimension of response as well as exposure. The duration of effects is seldom considered because the emphasis of regulation and risk assessment has been on determining whether an effect will occur. However, with increasing requirements for net benefit and cost–benefit analyses (Chapter 33 and Chapter 38), the time to recovery and other aspects of effects duration are increasingly important. In the simplest case, effects reach an asymptote during the exposure and cease shortly after exposure ends (Figure 23.7a). This is the implicit model behind the concept of chronic toxicity. However, even if the induction of effects ends when exposure ceases, recovery may be slow, so the duration of effects may be much longer (Figure 23.7b). In addition, effects may continue or even increase after exposure ceases because of time lags in the induction of overt effects; because body residues are remobilized at a later time due to metabolism of fat reserves or mobilization of bone during migration, hibernation, starvation, lactation, or the production of young; or because effects are expressed only during certain points in the life cycle (Figure 23.7c). Delayed effects are routinely reported only in single-dose wildlife toxicity tests, in which the need to wait for effects following exposure is obvious. Finally, effects may end before exposure due to acclimation or adaptation (Figure 23.7d). Use of the duration of exposure to estimate the duration of effects would be reasonable in case (a) but would underestimate effects in cases (b) and (c) and would overestimate effects in case (d).

23.2.3 COMBINED CONCENTRATION AND DURATION

Effects increase with exposure concentration or duration, so these two dimensions of exposure are somewhat interchangeable. The simplest expression of this relationship is Haber's rule:

$$Ct = k \qquad (23.2)$$

where C is concentration, t, time, and k, a constant exposure metric that is associated with a particular effect such as 50% mortality (log C and log t may also be used). This equation is quite handy in that it allows an assessor to apply test data for one duration to exposures of another duration. For example, if the fathead minnow 96 h LC_{50} is 10 mg/L, the concentration needed to kill the median minnow in 48 h is 20 mg/L. Haber's rule also allows an assessor to create an exposure metric, the product of concentration and time, which can be used to model response to data from exposures that vary in both concentration and time (Figure 6.3) (Newcombe and MacDonald 1991). Haber's rule does not apply to all chemicals or materials or to all effects, and should be restricted to relatively small temporal extrapolations. For reviews of these issues see Gaylor (2000), Bunce and Remillard (2003), and SAB (1998).

When sufficient data are available for responses at different times and concentrations, it is advisable to determine whether a nonlinear concentration–time relationship fits the data better than Haber's rule. For a fixed effect (e.g., the LC_{50}), Miller et al. (2000) recommend a simple power law:

$$C^{\alpha} t^{\beta} = Ct^{\gamma} = k \qquad (23.3)$$

where γ equals β/α. Even better, a surface can be fit to data for variable concentration, time, and response (either proportion responding or severity) (Figure 23.2) (Sun et al. 1995; Newcombe and Jensen 1996). Unfortunately, few reports of toxicity tests contain data for responses at any time other than the end of the test.

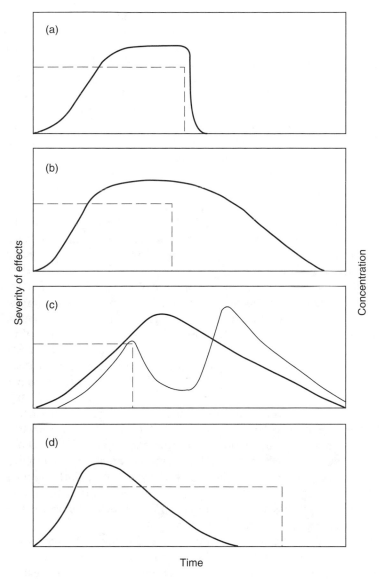

FIGURE 23.7 Toxic effects as a function of duration of effects (solid lines) contrasted with duration of exposure (dashed line). (a) Effects rapidly decline following cessation of exposure. (b) Induction of effects ceases after cessation of exposure, but recovery requires significant time. (c) Effects are induced after cessation of exposure due to lagging responses (thick line) or delayed responses (thin line). (d) The system adapts to the exposure before its cessation.

23.2.4 NONMONOTONIC RELATIONSHIPS

Monotonic exposure–response models are the norm in toxicology: as exposure increases, response increases. However, nonmonotonic models may arise for a variety of reasons.

Nutrients: Low levels of nutrients cause deficiency effects but, like other chemicals, nutrients cause toxic effects at sufficient exposure levels. Similarly, some elements that are often toxic contaminants are also micronutrients, which may be deficient at sufficiently low concentrations (Cu, Cr, I, Co, Mo, Se, and Zn). In addition, some nonchemical agents such as precipitation are effectively ecosystem nutrients. The approach to assessment of effects of these elements on humans and other organisms is relatively straightforward (IPCS 2002). The goal is to maintain intake in a region between deficiency and toxicity in which organisms of the endpoint species may maintain a desired level of nutrition, thereby maximizing some measure of performance such as growth (Figure 7.2). The issue becomes more difficult when community or ecosystem endpoints are considered. An excess of nutrient elements in aquatic ecosystems results in eutrophication, which is unesthetic and causes loss of fish and other animals due to anoxia. Oligotrophy is also generally undesirable due to low productivity of fish or other aquatic resources. Hence, one might assume that mesotrophy is the goal, analogous to intermediate levels of nutrient elements in an organism. However, there are exceptions; many alpine lakes and other ecosystems that are naturally oligotrophic are appreciated for the clarity of their waters, and their communities are adapted to low nutrient levels. Hence, the division of an exposure range into beneficial and adverse segments depends on the prior adaptations of the system and the goals of the environmental managers.

Intermediate disturbance: Many agents that physically disturb ecosystems are often beneficial at intermediate levels, but deleterious at high or low levels. Examples include fire, flooding, wind, and freezing temperatures. These agents are effectively equivalent to nutrients, but their direct effects, even at low levels, are deleterious. Their beneficial aspect results from the stress adaptation of the ecosystems involved. For example, when fire is suppressed, prairies may be replaced by woodland or shrubland. However, sufficiently high frequencies of fire diminish the diversity and productivity of a prairie and reduce its rate of recovery.

Hormesis: Radiation and some chemicals appear to have a stimulatory or protective effect on organisms at low exposure levels even though they are not nutrients (Calabrese and Baldwin 2000). Hormesis is thought to result from the organism's overcompensation for toxic effects. This results in J-shaped functions, because mortality or other adverse effects diminish with increasing dose before increasing (Figure 23.8). Although test results that suggest hormesis are common (Calabrese and Baldwin 2001), the reality and mechanism of the phenomenon are controversial. For example, apparent hormesis in fish toxicity tests may be due to reduced aggression resulting from toxicity.

Hormone-like chemicals: Because hormones are signaling agents involved in feedback mechanisms that maintain homeostasis, it is possible for a low level of an endocrine-disrupting chemical (EDC) to have greater effects than a higher level (Welshons et al. 2003).

23.2.5 CATEGORICAL VARIABLES

Because they are inherently qualitative, categorical data present problems for quantitative exposure–response modeling. Some dimensions may be expressed categorically because quantitative data are unavailable. For example, a quick survey of stream macroinvertebrates may simply classify the streams as having high, moderate, or low species richness. Other categorical dimensions such as acute and chronic durations are simply traditional. Some dimensions are categorical to combine disparate data, particularly when different studies are combined, or to compare chemicals with different reported effects (Teuschler et al. 1999). For example, a severity scale of (a) no observed effects, (b) no observed adverse effects, (c) adverse effects, and (d) frank effects may be used to place effects on various organs, species, or

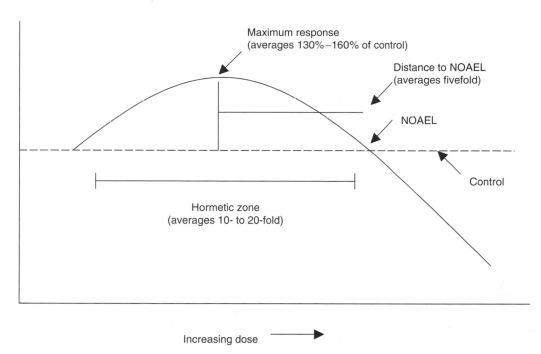

FIGURE 23.8 An illustration of the characteristics of a hormetic exposure–response relationship. (From Calabrese, E.J., *BELLE Newslett.*, 7, 1, 1998. With permission.)

ecosystems on a common scale. Finally, the type of effect (e.g., survival, growth, fecundity, and behavior) is unavoidably categorical.

The most common assessment problem involving categorical data is determining how categorical effects (i.e., types of response, categorical proportion, or categorical severity) change with exposure. This may be done simply by plotting the various types of responses using different symbols and then drawing boundaries between the types by eye (Figure 23.9). However, a technique called categorical regression quantitatively relates categorical responses to exposure levels (Dourson et al. 1997; Haber et al. 2001). By assigning scores to the categories, the probability of each category can be modeled as a function of an exposure, or a prescribed probability of each response (Figure 23.10). Software for categorical regression is available from the EPA (2005a).

Finally, one may treat a categorical scale of responses as a numerical variable, and regress it against exposure. The most prominent ecological example is the scale of 14 response types used in reviews of suspended sediment effects on fish (Newcombe and MacDonald 1991; Newcombe and Jensen 1996). The utility of this approach depends on defining the categories in such a way that they form a linear scale that is well correlated with exposure. Newcombe and Jensen (1996) generated response planes (severity score vs. log sediment concentration and log duration) with r^2 in the range of 0.6 to 0.7.

23.2.6 Exposure–Response from Field Data

Measurements of biological effects in the field can be used to generate exposure–response models. The advantage of these models is that they are based on exposure–response relationships from the real world. The disadvantages stem from the fact that exposure is not

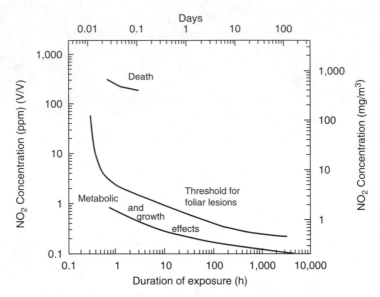

FIGURE 23.9 Categorization of effects of NO$_2$ on plants arrayed with respect to concentration and duration of exposure. (From EPA (US Environmental Protection Agency), *Air Quality Criteria for Oxides of Nitrogen*, EPA-600/8-84-026f, Research Triangle Park, North Carolina, 1982. With permission.)

controlled in the real world, so it is often poorly defined, it may not include the desired range of exposure levels or conditions, and the ecosystem is exposed to various anthropogenic and natural agents simultaneously. If the ecosystems being studied are sufficiently isolated, one can be reasonably certain that only one agent is significantly affecting the system and the

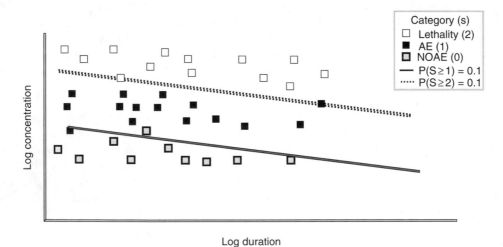

FIGURE 23.10 Categorical regression used to relate exposure concentration to exposure duration for three categories of effects: lethality, adverse effects, and no observed adverse effects. The two regression lines are for probability of 0.1 of severity greater than, or equal to, category one (adverse) and for category two (lethality). Generated using CatReg software.

same exposure–response functions that are used with toxicity tests may be employed. Examples might include acid mine drainage to a stream in a forested watershed or an illegal waste dump in a natural area. If a few agents are affecting the system and they are all measured, multiple regression may be appropriate.

The association of bird kills with pesticide applications is a type of case in which other causes may be neglected. Mineau (2002) used logistic regression to model the probability of occurrence of bird kills in fields treated with cholinesterase-inhibiting pesticides. He used data compiled from 181 studies of 35 pesticides. To make a single model for all of the pesticides, he created exposure parameters that reflect the exposure normalized to the potential routes of exposure. The primary predictive variable was the oral toxic potential, which is the fifth percentile of the SSDs for avian oral LD_{50}s per square meter in the application. This variable is equivalent to toxic units (Chapter 8), but expresses toxicity per unit area rather than per unit concentration or dose. Other contributing variables were the dermal toxicity index and, for inhalation, Henry's law constant. This approach provided remarkably good models for field crops, forests, and pastures, given the range of chemicals, species, application methods, and conditions. From the multivariate logistic models, Mineau calculated application rates that would result in a 10% probability of a bird kill for each of the 35 pesticides.

Griffith et al. (2004) developed models that estimate benthic macroinvertebrate community characteristics from metal concentrations in water or sediment (Figure 23.11). They reduced the multiple metal concentrations to a single dimension by using the sum of toxic units (Chapter 8), with toxicity expressed as the ambient water quality criteria or sediment threshold effects level. They dealt with threshold effects by using segmented regression, which is equivalent to hockey-stick regression but with the slope of the lower segment not constrained to zero. The break point was set to zero on the log scale, which is equivalent to a hazard quotient of 1, the expected threshold.

Because of the complexity of factors influencing communities in the field and their inherent variability among sites, it may be desirable to isolate the toxic response by testing contaminated media from the field. Smith et al. (2003) and Field et al. (2002, 2005) have used logistic regression to model the probability of amphipod toxicity of field sediments, given concentrations of multiple chemicals. They have explored various ways to deal with the multiple chemicals including stepwise multiple regression and (because of multiple collinearity) combined variables derived by principle components analysis and hazard quotients. Currently, they recommend simply modeling the probability of toxicity for all chemicals individually using appropriate data from all of North America and, to predict toxicity at a site, using the model that estimates the highest probability (Field et al. 2005). This approach seems to imply that one chemical at each site dominates sediment toxicity, but it may be that one chemical is usually a better representative of toxicity than a linear combination of chemicals or the average of probabilities across chemicals.

If numerous agents are contributing to the impairment of organisms, populations, or communities, conventional regression analysis will model the average effects of independent variables plus all other agents, but we often wish to estimate the effects of an agent acting alone (Figure 23.12). If we plot a biological response variable against levels of an agent of interest measured at numerous field sites, we will typically see a cloud of points, with a roughly linear upper edge for toxicants or a humped edge for nutrients or other agents with an optimum exposure level (Figure 23.13). The upper boundary represents the maximum value achieved by the biological response variable given the level of the agent. Points below the boundary are assumed to be reduced by co-occurring stressors. The upper boundary, which may be thought of as the response when the independent variable is the limiting factor, is

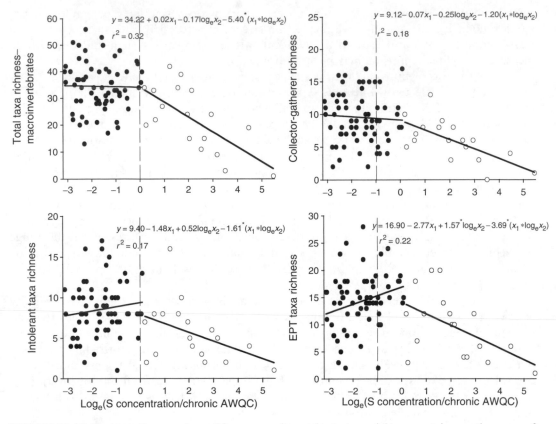

FIGURE 23.11 Segmented regression of four macroinvertebrate taxa richness metrics on the sums of ratios of Cd, Cu, Pb, and Zn concentrations to their chronic national ambient water quality criteria (AWQC). (From Griffith, M.B., Lazorchak, J.M., and Herlihy, A.T., *Environ. Toxicol. Chem.*, 23, 1786–1795, 2004. With permission.)

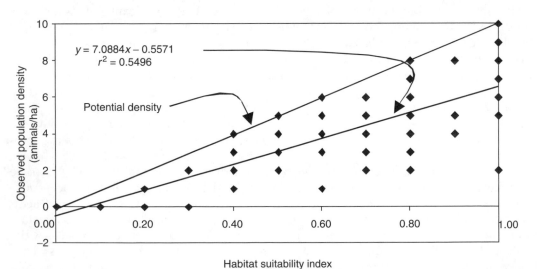

FIGURE 23.12 Relationships between population density and a habitat suitability index. The line fitted by linear regression estimates the typical density at a particular habitat suitability. The upper line, fitted by eye, estimates the maximum density given that the population is limited only by habitat suitability. (From Kapustka, L.A. *Hum. Ecol. Risk Assess.*, 9, 1425, 2003. With permission.)

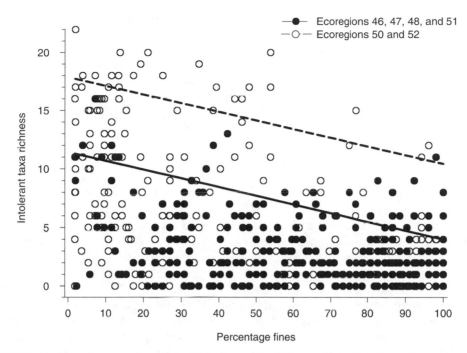

FIGURE 23.13 Quantile regression of the 90th percentile of the number of intolerant invertebrate taxa against percent fine sediment for Minnesota streams in two sets of ecoregions. The figure illustrates the variance among regions in the effects of siltation. (Courtesy of Michael Griffith. With permission.)

termed the limiting function. It is estimated by quantile regression, fitting a regression line for a high quantile of the variance of the response with respect to the level of exposure (e.g., 90%). This can be done by asymmetrically weighting the positive and negative deviations in least-squares regression. The technique was developed in economics (Koenker 2005) but has recently become popular in ecology (Cade and Noon 2003).

These examples serve to illustrate the complexity of modeling exposure–response in the field where multiple contaminants and other habitat variables all influence responses as well as some of the diversity of approaches that have been employed. None of these approaches have been sufficiently tested in real assessments to know which provides the best predictions. In addition to the obvious problem of multiple natural and anthropogenic causes and the lack of control of exposure, inherent problems of sampling artifacts must be considered (von Stackelberg and Menzie 2002). In large part, the choice of modeling approach depends on the assessor's conceptualization of the system. For example, the segmented regression approach of Griffith et al. (2004) is based on the assumptions that the toxicity of metals is concentration-additive and has a threshold, but that in those metal-contaminated streams, other contaminants or habitat variables have negligible effects. However, examination of the data (Figure 23.11) could also support the assumption that other agents are acting and that there is no threshold, so quantile regression could be used. Statistics alone cannot resolve this choice. Rather, the choice of modeling approach should be based on general scientific understanding and knowledge of the particular system being addressed.

23.2.7 RESIDUE–RESPONSE RELATIONSHIPS

Single chemical toxicity tests may be used to develop exposure–response relationships based on internal exposure measures (residues, also called body burdens) rather than external exposures (media concentrations or administered doses). In theory, this approach offers considerable advantages. Chemicals cause toxic effects in the organism, so measures of internal exposure should be more predictive of effects than measures of external exposures (McCarty and Mackay 1993; Escher and Hermens 2004). Estimation of effects from residues potentially bypasses most of the variance among sites, species, and individuals associated with the physical, chemical, physiological, and behavioral processes that control intake, uptake, and retention of chemicals. This approach may be particularly relevant to chemicals that may be significantly accumulated by aquatic biota through food intake as well as direct exposure to the chemical in water.

Internal exposure–response functions can be derived for body burdens like those for external exposures. For example, the test endpoint equivalent to the LC_{50} is termed the median lethal residue (LR_{50}). However, it is commonly assumed that the variance among organisms and even species is relatively small and equilibrium conditions are achieved, so a single threshold value suffices. These are usually termed the critical body residue (CBR).

Since residue–response relationships are not available for most chemicals, the approach has been extended by assuming equivalent potency for chemicals with the same mechanism of action. That is, all chemicals acting by the same mechanism of action should be effective at approximately the same molar concentration at the site of action (Escher and Hermens 2004). If all internal compartments (e.g., muscle, fat, and blood plasma) are in equilibrium and have roughly the same relative size across individuals and species, the absolute or adjusted whole-body effective concentration would be the same for all chemicals with the same mechanism of action. Finally, if all individual molecules of chemicals with the same mechanism of action have the same potency, effective molar concentrations should be constant. These assumptions underlie the compilation of estimated CBRs for eight groups of chemicals in fish presented in Table 23.1; these are supported by some research. For example, polycyclic aromatic hydro-carbon (PAH) whole-body residues were found to be effectively equal at LC_{50}s for multiple species (DiToro and McGrath 2000). Hence, these thresholds may be used as a first approximation to estimate whether measured residues of organic chemicals with known mechanisms of action are likely to be associated with acute or chronic effects.

Like all toxicity benchmarks, these should be used with caution, and the original sources consulted before using these values to estimate risks. The CBR values may be applied to field data for most chemicals and species in long-term exposures, because the measured body residues may be assumed to reflect equilibrium with the environment. However, that is not the case for brief or episodic exposures or for chemicals with very slow kinetics. For example, CBRs for 2,3,7,8-TCDD varied 122-fold when measured at the time of death in fathead minnows (Adams 1986). This variation was apparently due to nonequilibrium toxicokinetics in laboratory tests of different durations. Similarly, DiToro and McGrath (2000) found that their generalization about equal CBRs for LC_{50}s did not apply to NOELs, which may be due to problems inherent in that type of test endpoint or to toxicodynamic issues (see Section 23.3).

It is unlikely that all chemicals with a common mechanism of action have exactly the same potency, but relative potencies are seldom known. Relative toxicities of doses or external exposure concentrations do not estimate potency because of all the kinetic factors discussed above. When relative potency factors are available, as they are for the dioxin-like chemicals (Section 8.1.2), they can be used to estimate the effective internal concentration of chemicals.

If the mechanism of action is unknown or not included in Table 23.1, one may assume that an organic chemical's toxicity is at least as great as for chemicals acting by baseline narcosis (Chapter 7). Since all organic chemicals have at least that level of toxicity, body residues of any organic chemical of 0.8 mmol/kg (the upper limit for chronic narcosis; Table 23.1) or greater are clearly indicative of chronic toxicity in fish. However, since chemicals may have

TABLE 23.1
Summary of Modes of Toxic Action and Associated Estimates of Critical Body Residue in Fish[a]

Chemical and Effect	Estimated Residue (mmol/kg)
Narcosis	
Acute (summary)	2 to 8
Chronic (summary)	0.2 to 0.8
Acute (octanol, MS222)	1.68 or 6.32[b]
Polar narcosis	
Acute (summary)	0.6 to 1.9
Acute (2,3,4,5-tetrachloroaniline)	0.7 to 1.8
Chronic (summary)	0.2 to 0.7 (chronic/acute = 0.1 to 0.3)
Chronic (2,4,5-trichlorophenol)	0.2
Acute (aniline, phenol, 2-chloroaniline, 2,4-dimethylphenol)	0.68 or 1.76
Respiratory uncoupler	
Acute (pentachlorophenol)	0.3
Acute (2,4-dinitrophenol)	0.0015 or 0.2
Chronic (pentachlorophenol, 2,4-dinitrophenol)	0.09 to 0.00015 (chronic/acute = 0.1 to 0.3)
Chronic (pentachlorophenol)	0.094
Chronic (pentachlorophenol)	0.08
Acute (pentachlorophenol, 2,4-dinitrophenol)	0.11 or 0.20
AChE inhibitor	
Acute (malathion and carbaryl, chlorpyrifos)	0.5 and 2.7
Acute (chlorpyrifos)	2.2
Acute (aminocarb)	0.05 and 2
Acute (parathion in blood)	0.13 to 0.2
Chronic (chlorpyrifos)	0.003
Acute (malathion, carbaryl)	0.16 or 0.38
Membrane irritant	
Acute (benzaldehyde)	0.16
Acute (benzaldehyde)	2.1 or 13.2
Acute (acrolein)	0.0014 or 0.94
CNS convulsant[c]	
Acute (fenvalerate, permethrin, cypermethrin)	0.002 to 0.017
Acute (fenvalerate, permethrin, cypermethrin)	0.000048 to 0.0013
Acute (endrin in blood)	0.0007
Acute (endrin)	0.0018 to 0.0026
Acute (endrin)	0.005
Chronic (fenvalerate, permethrin)	0.0005 and 0.015
Respiratory blockers	
Acute (rotenone)	0.0006 to 0.003
Acute (rotenone)	0.008
Acute (rotenone)	0.0009 or 0.0028
Dioxin (TCDD)-like	
Lethal (TCDD)	0.000003 to 0.00004
Growth/survival (TCDD)	0.0000003 to 0.0000008
Early life stages, lethal (TCDD)	0.00000015 to 0.0000014
Early life stages, NOAEL (TCDD)	0.0000001 to 0.0000002

Source: Reprinted from McCarty, L.S. and Mackay, D., *Environ. Sci. Technol.*, 27, 1719, 1993. With permission.

[a]The rainbow trout used in this study weighed 600–1000 g; the other data presented are largely for small fish, sometimes in early life stages, that typically weighed less than 1 g. Most estimates were converted from mass-based data.
[b]The two values represent residues estimated by two different methods.
[c]Includes three subgroups characterized by strychnine; fenvalerate and cypermethrin; endosulfan and endrin.

more powerful specific modes of action, concentrations less than 0.2 mmol/kg (the lower limit for chronic narcosis; Table 23.1) cannot be assumed to be nontoxic.

Interpretation of residues of metals is more problematic. Because of the nutrient role of many metals and the numerous processes that control metal uptake, depuration, distribution, and sequestration, effective concentrations are highly variable (McCarty and Mackay 1993; Bergman and Dorward-King 1997). In particular, organisms have evolved a variety of mechanisms for regulating exposure by sequestering metals in granules or insoluble precipitates, in inactive tissues such as hair and exoskeletons, and bound to regulatory proteins (e.g., metallothioneins and phytochelatins). Hence, internal as well as external concentrations of metals include fractions that are not bioavailable. The biotic ligand model (BLM) (discussed in Section 23.3) addresses these issues in a limited context by assuming that death occurs in a species at a particular concentration of metal–ligand complexes on the surface of gills, termed the median level of accumulation (LA_{50}) (Meyer et al. 1999; EPA 2003a). However, effects of metals occur at different internal concentrations for dietary vs. aqueous exposures and effects of dietary metals on the gut may occur without any bioaccumulation (Meyer et al. 2005).

In addition to the summary values in Table 23.1, residues associated with effects in individual aquatic toxicity tests may be found in the literature, but the endpoints are not standardized. A review of such data is presented in Jarvinen and Ankley (1999). Effective residues for a variety of chemicals in sediments are presented in the Environmental Residue–Effect Database (http://www.wes.army.mil/el/ered/index.html).

The use of chemical concentrations in plant tissues to estimate effects may be advantageous. Measurement of tissue concentrations permits the assessor to bypass the very large differences in bioavailability of chemicals in different soils as well as interspecies differences in uptake. For example, phytotoxicity of metals in soils of low organic matter is not a good predictor of the toxicity of metals in sludge-amended soils. Chang et al. (1992) developed empirical models relating concentrations of copper, nickel, and zinc in crop foliage to growth retardation.

Although residue–effects data are usually obtained from the literature, it is also possible to generate them from field data collected for assessment of contaminated sites. As part of biological surveys, animals or plants may be collected, examined for signs of toxic effects, and subjected to chemical analysis. A function relating residues to the severity or frequency of observed effects may be developed, or a maximum residue associated with no observable effects may be established. This approach is potentially more reliable than the use of residue–effect relationships from the literature, but must be used with care. For mobile species, the time that the collected individuals have spent on the contaminated site must be considered. In addition, it must be realized that the most sensitive individuals and species may have been eliminated from the site by toxic effects, leaving only resistant organisms. These two phenomena may interact. That is, the loss of individuals to toxicity may result in immigration of relatively uncontaminated individuals and eventually to the evolution of resistant local populations.

An assessment of the Seal Beach Naval Weapons Station used residues in a somewhat unconventional manner that could be applied elsewhere. Because of the concern that persistent organic chemicals were reducing tern reproduction, the assessors collected tern eggs that failed to hatch and analyzed them for the chemicals of concern (Ohlendorf 1998). If those chemicals were responsible for reproductive failure, concentrations would be elevated relative to reference populations, and they would be similar to those found in controlled studies that demonstrated reproductive effects. In this case, the analysis of biological materials was used to investigate the cause of apparent effects (Chapter 4) rather than to estimate the exposure of the population.

23.3 TOXICODYNAMICS—MECHANISTIC INTERNAL EXPOSURE–RESPONSE

If the response to a particular internal concentration is not constant across exposure durations, species, or life stages, the induction of effects must be modeled. If the rates of induction of damage or repair are modeled, these are toxicodynamic models (Ramsey and Gehring 1980; Lee et al. 2002). The basic toxicodynamic models represent reversible and irreversible binding of a chemical to a receptor.

Reversible binding is typical of baseline narcotics and other chemicals that cause an effect when they reach a critical concentration, but effects short of death are reversible. Uptake and release from the receptor are represented by the same first-order model used to represent uptake (Equation 22.31). The receptor may be a specific organ or tissue or may simply be represented by the entire organism, in which case the concentration associated with an effect is the CBR. The CBR for aquatic lethality is the product of the incipient LC_{50} (the LC_{50} at equilibrium or effectively infinite time) and the bioconcentration factor (k_u/k_e) (Section 22.9). For shorter exposures, a kinetic formula is required:

$$LC_{50}(t) = CBR/[(k_u/k_e)(1 - e^{-ket})] \tag{23.4}$$

where t is the exposure duration (Lee et al. 2002). These CBR models require that the exposure concentration and bioavailability be constant, that organism weight be constant, and that there be negligible dietary uptake or biotransformation.

Irreversible binding is typical of organophosphate pesticides and other chemicals that cause an effect when they bind a critical proportion of the receptor sites. Such dynamics are represented by the critical target occupation (CTO) model (Legierse et al. 1999). The organophosphates are metabolized to oxon analogs that covalently bind to the neurotransmitter acetylcholine, which is thereby inhibited. Death occurs when a particular proportion of acetylcholine is bound and inhibited, the CTO. Since the binding is irreversible, the CTO occurs at the critical area under the reaction curve for the toxicant and receptor, not a critical concentration. Hence, this is also known as the critical area under the curve (CAUC) model (Verhaar et al. 1999).

These two models (CBR and CTO) can be derived as extreme cases of a more general damage–repair model (Lee et al. 2002). This model combines a first-order toxicokinetic model and a first-order damage–repair model:

$$dA/dt = k_a R - k_r A \tag{23.5}$$

where A = accrued damage (dimensionless); k_a = rate of accrual of damage (kg/mmol h); R = tissue residue (mmol/kg); and k_r = rate of repair (1/h).

Lee et al. (2002) found that this model fits lethality data for amphipods exposed to PAHs better than the CBR (equivalent to $k_r = \infty$) or CTO (equivalent to $k_r = 0$) models. It had been commonly assumed that a constant CBR would apply to PAHs. In fact, the CBR continued to decline after amphipods achieved steady state, apparently because damage continued to accumulate.

These toxicodynamic models are semimechanistic extensions of toxicokinetic models (Section 22.9) (Figure 23.14). They are based on assumed mechanisms but are derived in practice by empirical curve fitting to accumulation and toxicity data. Toxicodynamics could, however, be much more complicated and genuinely mechanistic. Models that simulate multistep processes of effects induction at the molecular level are being developed for human health risk assessments. Such models will be particularly useful for chemicals like dioxins or endocrine

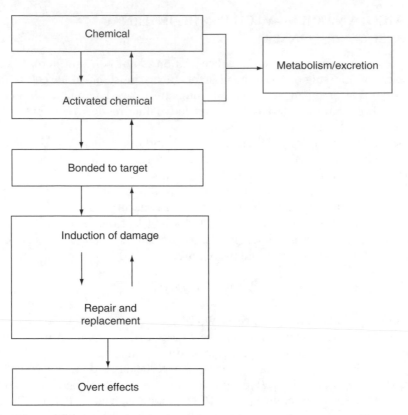

FIGURE 23.14 A generic conceptual toxicodynamic model.

disruptors that involve signaling systems. Genomics, proteomics, and metabolomics, in combination with computational biology, promise to provide the basis for simulating cells and even whole organisms at the molecular level.

23.3.1 TOXICODYNAMICS OF METALS ON GILLS

The biotic ligand model is a toxicodynamic model for effects of metals on the gills of aquatic organisms. It is atypical in that it is not linked to an external exposure model rather than a toxicokinetic model. The BLM assumes that the site of action is certain enzymes or ion channel proteins (biotic ligands) on the surface of gills, so the site of external exposure and effects induction is the same. These biotic ligands compete for metal ions with abiotic ligands including dissolved organic matter, hydroxides, chlorides, sulfides, and carbonates (Figure 22.1). Hence, the BLM consists of a metal speciation model to estimate free ion concentrations of the toxic metal as well as other ions that compete for ligand sites (primarily calcium, magnesium, sodium, and hydrogen) and a gill surface interaction model. Although the model does not depend on a specific mechanism of action (MoA), it appears that lethality is associated with loss of sodium or calcium due to loss of ion channel function. The BLM is described in Paquin et al. (2002), DiToro et al. (2001), and Niyogi and Wood (2004). BLM software is available (Hydroqual 2003) and has been used to develop proposed water quality criteria for copper (EPA 2003a).

The toxic effects portion of the BLM is an equilibrium model of toxic metal loading of the biotic ligands. Loading is determined by the binding affinities of the competing ions for the biotic ligands (the binding constant—log K) and the binding site density (B_{max}). These values were determined by studies of fish gills in short-term exposures such as Playle et al. (1993). It is assumed that the loading associated with a particular effect is constant. In particular, an LC_{50} has a corresponding concentration of the metal–ligand complex, the median lethal accumulation (LA_{50}). When using the BLM software, one provides an LC_{50} for a species–metal combination and the water chemistry for that test. The software then calculates an LA_{50} for the metal and species. This value can then be used to calculate a dissolved LC_{50} for any ambient water chemistry.

Although the BLM is a major advance in ecotoxicological modeling, it has significant limitations. In particular, it is currently limited to acute lethal effects of a few metals (Ag, Cd, Co, Cu, Ni, and possibly Pb) in freshwater. The BLM has not been successfully applied to nonlethal effects from sustained exposures, because the more complex toxicokinetics and dynamics make links between residues and responses elusive (McGeer et al. 2002). Extension to chronic exposures will also require determining which cases involve other toxic mechanisms of action. For example, lead appears to act on calcium and sodium transport in acute lethality studies, but is neurotoxic in long-term aqueous exposures and causes paralysis of intestinal peristalsis and other intestinal effects in dietary exposures (Chapter 7). Differences among species in LA_{50} values may be large and require more species-specific studies (Taylor et al. 2003). Extension of the BLM to saltwater requires dealing with very different ion chemistry. Other complexities to be considered include nonequilibrium uptake kinetics, effects of acclimation, differences in dissolved organic carbon properties, and temporal variance in water chemistry. However, these issues are active areas of research.

23.4 INDIRECT EFFECTS

Ecological risk assessments have been consistent with human health risk assessments in emphasizing direct toxic effects of contaminants. However, because nonhuman organisms are much more subject to indirect effects such as habitat modification and reductions in the abundance of food species than humans are, indirect effects should be included in more exposure–response models. The term indirect effects refers to effects that result when a contaminant directly affects an entity (population, community, or ecosystem), and that direct effect becomes a hazardous agent with respect to an assessment endpoint entity. Hence, the indirect effect is a response to exposure to a direct effect. Indirect effects of chemical contaminants result from effects on trophic and competitive relationships, such as reduced abundance due to toxic effects on food species. In addition, indirect effects due to habitat alteration should be considered. For example, the toxicity of chemicals to earthworms in a pasture may result in soil compaction, which could inhibit seed germination and result in other adverse effects to plants. Decomposition of organic contaminants in soil, surface water, or sediment could cause depletion of oxygen and reduced availability of nitrogen, adversely affecting endpoint species and processes. In contrast, after decomposition of petroleum is mostly completed, plant production may actually be greater due to improved soil structure, nitrogen availability, or other factors (McKay and Singleton 1974; Bossert and Bartha 1984). Like direct toxic effects, habitat-mediated effects of contaminants may depend on the magnitude of exposure. For example, at high exposures, petroleum and other nonaqueous-phase liquids may fill soil pores that would otherwise be habitat for microorganisms and mesofauna and that provide gas exchange for plant roots and soil macrofauna.

These indirect effects should have been identified in the conceptual model and their relationship to exposure should be quantified as far as possible. However, because of the complexity and heterogeneity of ecosystems, it is difficult to list all potentially important indirect effects, much less estimate them. When effects must be estimated from laboratory tests, assessors often can do little more than present qualitative relationships, for example, the reduction in abundance of aquatic insects will reduce growth and fecundity of fish. When the results of microcosm, mesocosm, or field tests are available, they can be used to empirically estimate the indirect effects or, for less selective agents, the combined direct and indirect effects. Biological surveys of contaminated or disturbed areas can potentially reveal indirect effects, but because the exposures are uncontrolled and unreplicated, indirect effects are difficult to distinguish from direct effects or biological variability in such studies. Alternatively, simple assumptions can be made such as an x% loss of wetlands will result in an x% reduction in the abundance of species that depend on that community for any of their life stages. Finally, ecosystem models may be used to estimate the consequences for all endpoint taxa of toxic effects on all modeled components of the exposed ecosystem (Chapter 28). As more and more indirect effects are considered, the uncertainties associated with exposure and effects models and measurements are compounded.

24 Testing

The toxicological literature is vast and largely trivial.

Moriarty (1988)

Toxicity tests are studies in which organisms, populations, or ecosystems are exposed to a chemical or mixture to determine the nature of effects and the relationship between the degree of exposure and effects. Similar tests may be performed for other agents. Examples include pathogenicity tests and tests of responses to physical/chemical conditions such as pH, dissolved oxygen, and suspended sediment. Tests resemble experiments in that the degree of exposure is controlled, exposures are replicated, assignment of replicate test subjects is randomized, and extraneous variance among replicates is minimized. They differ from classic scientific experiments in that they are intended to establish a functional exposure–response relationship. A test should determine not only that zinc can kill fish, or even that it kills them by disrupting ion exchange in the gills, but also how the proportion of fish killed increases with concentration or duration of exposure. This chapter describes tests of individual chemicals or materials, tests of contaminated media, and field tests.

The treatment of testing in this chapter is somewhat cursory, since the goal is to familiarize assessors with the types of test data being generated, not to teach them to perform tests. Detailed procedures are published by governments and standards organizations, which are cited in the appropriate sections. More detailed reviews can be found in ecotoxicology texts (Calow 1993; Rand 1995; Hoffman et al. 2003) or appropriate government documents such as Anderson et al. (2003).

24.1 TESTING ISSUES

Conventional toxicity tests determine effects on organisms and are divided into two classes, acute and chronic. Acute tests are those that last a small proportion of the life span of the organism (<10%) and involve a severe effect (usually death) on a substantial proportion of exposed organisms (conventionally 50%). Acute tests also usually involve well-developed organisms rather than eggs, larvae, or other early life stages. Chronic tests include much or all of the life cycle of the test species and include effects other than death (most often, growth and fecundity). Chronic test endpoints are typically based on statistical significance, so the proportion affected and the magnitude of affect may be large or small. In addition, there are many tests that fall between these two types that are termed subchronic, short-term chronic, etc. They typically have short durations but include sublethal responses. A prominent example is the 7 d fathead minnow test, which includes growth as well as death and includes only a small part of the life cycle, but that part is a portion of the larval stage (Norberg and Mount 1985).

In general, tests with longer durations, more life stages, and more responses are more useful for risk assessment, because they provide more information and because exposures in the real world are often sustained. However, if exposures are brief, acute or subchronic tests may be preferred. Examples include exposures of transients such as migratory waterfowl or highly mobile species that may use a site in transit or episodic exposures such as overflow of waste ponds, applications of pesticides, effluents generated during treatment failures, blow-down of cooling water, or flushing of contaminants into surface waters by storms.

The following are general recommendations for selecting tests of chemicals or materials. Other issues specific to tests of particular media or of other agents are addressed subsequently:

Standardization: In general, choose standard tests. Standard test protocols have been developed or recommended by governments (Keddy et al. 1995; EPA 1996b), and standard organizations (APHA 1999; OECD 2000; ASTM 2002). Most extrapolation models for relating test endpoints to assessment endpoints require standard data (Chapter 26). In addition, results of standard tests are likely to be reliable because methods are well developed, QA/QC procedures are defined, and test laboratories are likely to conduct standard tests routinely. However, nonstandard tests should be used when particular assessment-specific issues cannot be resolved by standard test results. Some effects such as effects of estrogenic chemicals on sexual development or the behavioral effects of lead are not observed in standard tests. Also, assessments may require tests of important local species or at least species that are relevant to a location. In particular, the biota of certain ecosystem types such as arid ecosystems are not well represented by standard tests (Markwiese et al. 2001).

Duration: Choose tests with appropriate durations. Two factors are relevant. The first is the duration of the exposures in the field. If exposures are episodic, as is often the case for aqueous contamination, tests should be chosen with durations as great as the longest episodes. The second factor is the kinetics of the chemical. Some chemicals such as chlorine in water or low-molecular-weight narcotics are taken up and cause death or immobilization in a matter of minutes or hours. Others have very slow kinetics and require months or years to cause some effects such as reproductive decrements. For example, tests of effects of polychlorinated biphenyls (PCBs) on mink have shown much greater effects on kit production and survival in the second year than in the first (Restum et al. 1998; Hornshaw et al. 1983).

Time course: Time to response is a neglected aspect of ecotoxicology and other components of applied ecology. In some cases, particularly episodic or accidental releases, variation in the duration of the exposure is more important to the risk estimate than concentration, which may be relatively constant or uncontrollable. In such cases, it is highly desirable to use data that report responses at multiple time points. For example, a report of a 96 h acute lethality test would be more useful if it reported the proportion surviving at each exposure level at 3, 6, 12, 24, 48, 72, and 96 h.

Response: Choose tests with appropriate responses. In particular, if an apparent effect of the contaminants has been observed in field studies, tests that include that effect as a measured response should be used. More generally, chosen tests should include responses that are required to estimate the assessment endpoint. The most common response parameters in toxicity tests are survival, fecundity, and growth, and most ecological effects models use one or more of these responses as parameters. Physiological and histological responses are generally not useful for estimating risks, because they cannot be related to effects at higher levels. However, if they are characteristic of particular contaminants, they can be useful for diagnosis (Chapter 4).

Media: Prefer tests conducted in media with physical and chemical properties similar to site media or media characteristic of the exposure scenario. For example, if assessing a pesticide for use on cotton, use tests in soils similar to those used for cotton production.

Organisms: Prefer taxa that are closely related taxonomically to the endpoint species and life stages that are likely to be exposed. If an assessment endpoint is defined in terms of a community, one may either choose tests of species that are closely related to members of the community or use all high-quality tests in the hope of representing the distribution of sensitivity in the endpoint community (Chapter 26). Species, life stages, and responses should also be chosen so that the rate of response is appropriate to the duration of exposure and kinetics of the chemical. In general, responses of small organisms such as zooplankters and larval fish are more rapid, because they achieve a toxic body burden more rapidly than larger organisms. Therefore, if exposures are brief and if those small organisms are relevant to the assessment endpoint, tests of small organisms should be preferred over larger organisms that are no more relevant. However, such tests may not be appropriate if, for example, the endpoint is fish kills. Finally, choose taxa and life stages that are known to be sensitive to the agent being assessed.

Multiple exposure levels: Studies that employ only a single concentration or dose level plus a control are seldom useful. If the exposure causes no effect, it may be considered a no observed effects level (NOEL), but no information is obtained about levels at which effects occur. If the exposure causes a significant effect, it may indicate that a reduction in exposure is required, but the necessary magnitude of reduction cannot be determined. Studies in which multiple exposure levels were applied allow an exposure–response relationship to be evaluated and thresholds for effects to be determined. Consequently, studies that applied multiple exposure levels are strongly preferred.

Exposure quantification: To correctly interpret the results of toxicity tests and to apply these results in risk assessments, the exposure concentrations or doses should be clearly quantified. Ideally, the test chemical should be measured at each exposure level; measured concentrations are preferable to nominal concentrations.

Chemical form: Correct estimation of exposure requires that the form of toxicant used in the test be clearly described. For example, in tests of lead, the description of the dosing protocol should specify whether the dose is expressed in terms of the element (e.g., lead) or the applied compound (e.g., lead acetate). Tests of chemicals in the forms occurring on the site are preferred. This is particularly important for chemicals that may occur under ambient conditions in multiple ionization states or other variant forms that have differing toxicities.

Statistical expressions of results: The traditional toxicity test endpoints for chronic tests, NOELs and lowest observed effect levels (LOELs), have been used to establish benchmarks or criteria (Chapter 29), but they have low utility for risk assessment, because they are based on statistical significance rather than biological significance (Chapter 23). To fully estimate risks, it is necessary to estimate the nature and magnitude of effects that are occurring or could occur at the estimated exposure levels. To do this, exposure–response relationships should be developed for chemicals evaluated in ecological risk assessments.

These criteria may conflict in some cases, because the best test data for one criterion may not be the best for another. Therefore, assessors must judge their relative importance to the particular assessment, and apply them accordingly.

24.2 CHEMICAL OR MATERIAL TESTS

In ecological risk assessments, effects data for single chemicals, organisms (e.g., an exotic parasitoid), or materials (e.g., gasoline, silt) may be obtained from tests performed ad hoc (primary data), but are usually obtained from the literature or from databases (secondary and tertiary data). One useful tertiary source is the EPA ECOTOX database (http://www.epa.gov/medatwrk/databases.html), which contains toxicity data for aquatic biota, wildlife, and terrestrial plants. Reviews such as those produced by R. Eisler for the US National

Biological Service are also useful tertiary sources (www.pwrc.usgs.gov/new/chrback.htm). Test data generated for an assessment (primary data) are relevant by design, but when data are taken from the literature or from reviews (secondary or tertiary data), assessors must select those data that are most relevant to the assessment endpoints and that can be used with the exposure estimates, as discussed in the previous chapter. However, because the variance among chemicals is greater than the variance among species and life stages, any toxicity information concerning the chemicals of interest is potentially useful. If no toxicity data are available that can be applied to the assessment endpoints (e.g., no data for fish or no reproductive effects data), or if the test results are not applicable to the site because of differences in media characteristics (e.g., pH or water hardness), tests may be conducted ad hoc. If combined toxic effects of multiple contaminants are thought to be significant, and if appropriate mixtures are not available in currently contaminated media, synthetic mixtures may be created and tested, or combined effects models may be applied (Chapter 8).

Test data from the literature have biases that should be understood by ecological risk assessors. Assessors must be aware of these biases when test data are used to derive toxicity benchmarks or exposure–response models for chemicals. Potential sources of bias in test data include:

- *Form*: The forms of chemicals used in toxicity tests are likely to be more toxic than the dominant forms in the field. For metals the tested forms are usually soluble salts, and organic chemicals may be kept in aqueous solution by solvents. In oral dosing tests, organic chemicals are often dissolved in readily digested oils.
- *Species*: The test species may not be representative of the sensitivity of species native to the site.
- *Media*: The standard media used in toxicity tests may not be representative of those at a particular contaminated site. For example, aqueous tests typically use water with moderate pH and hardness with little suspended or dissolved matter, and soil tests typically use agricultural loam soils or similar artificial soils.
- *Conditions*: Laboratory test conditions are less variable and may not be representative of field conditions (e.g., optimum temperature, sieved soil, or constant moisture).

24.2.1 AQUATIC TESTS

More test data are available for aquatic biota than any other type of ecological receptors (Table 24.1). In general, flow-through tests, which constantly renew the test water, are preferred over static-renewal tests, which renew the water periodically, and those in turn are preferred over static tests, which do not change the water. Flow-through tests maintain constant concentrations, whereas concentrations may decline significantly in static tests or even static-renewal tests due to evaporation, degradation, sorption, etc. However, static tests may be appropriate for extremely short-duration tests. The most abundant type of test endpoint is the 48 or 96 h median lethal concentration (LC_{50}). Life cycle tests that include survival, development, and reproduction provide the most generally useful data, but they are largely restricted to short-generation invertebrates because of the expense of long tests. For fish, early life stage tests of survival and growth are most commonly used, based on the presumption that early life stages are the most sensitive (McKim 1985). However, reproduction is often the most sensitive life stage (Suter et al. 1987), and, even if embryos or larvae are the most sensitive stage, maternal transfer may be an important route of exposure. These concerns may be addressed by short-term fish reproduction tests (Ankley et al. 2001), but for some chemicals, only a full life cycle test will reveal the effects on reproduction of long-term exposures of the adult fish.

TABLE 24.1
Examples of Standard Aquatic Toxicity Tests Published by the US EPA or the American Society for Testing and Materials (ASTM)

Taxon	Type	Reference[a]
Fish	96 h LC_{50} (juvenile or adult)	EPA/660/3-75-009
		EPA/600/4-90/027F
		EPA/712-C-96-118
		ASTM E729-96, -88
Fish	Early life stage survival and growth (egg through juvenile)	ASTM E 1241-97
		EPA/712-C-96-121
Fish	7 d larval survival and growth	EPA/600/4-91/002
		EPA/600/4-95/136
		EPA/600/4-91/003
Macroinvertebrates	48–96 h LC_{50}	EPA/660/3-75-009
		ASTM E729-96, -88
Mysid (saltwater crustacean)	Life cycle test	EPA/712-C-96-166
		ASTM E 1191-97
Daphnia	Life cycle test	EPA/712-C-96-120
		ASTM E 1193-97
Ceriodaphnia	7 d survival and reproduction	EPA/600/4-91/002
Algae	96 h growth	EPA/712-C-96-164
		ASTM E 1218-97a

[a]EPA method reports may be obtained by searching www.epa.gov for the listed report number.

ASTM methods may be purchased by standard number from www.astm.org.

Dietary exposures may be important contributors to toxicity for bioaccumulative organic chemicals and metals, but are seldom tested. This is in part because of the difficulty of culturing or collecting contaminated food organisms or of realistically contaminating artificial diets. It also reflects a lack of general acceptance of the importance of aquatic dietary exposures. Hence, most aquatic dietary toxicity studies have been concerned with demonstrating the reality and nature of the problem rather than generating relevant exposure–response relationships (Meyer et al. 2005).

Currently, the most popular freshwater test organisms in the United States are fathead minnows (*Pimephales promelas*) and daphnids (*Daphnia* spp. and *Ceriodaphnia dubia*). The most common saltwater organisms are sheepshead minnows (*Cyprinodon variegatus*) and the mysid shrimp (*Americamysis bahia*). Test results for algae (often *Selinastrum capricornutum*) and aquatic plants (often duckweed, *Lemna gibba*) are less abundant than for aquatic animals. These tests have short durations (72 to 96 h), but they include multiple generations of vegetative reproduction. Further, plant tests usually report growth (e.g., cell or frond number) or production (e.g., carbon fixation rates), which are often applicable to assessment of risks to ecosystems.

24.2.2 SEDIMENT TESTS

Selecting representative sediment tests and test results is complicated by the interactions among the multiple phases (i.e., particles, pore water, and overlying water) of the sediment system. Sediment tests may be conducted using whole sediment or aquatic tests may be used to represent one of the aqueous phases. Test selection depends on the expected mode of exposure, and more than one test type may be appropriate. Spiked sediment tests consist of the addition of known quantities of the test chemical or material to a natural or synthetic sediment to which the test organism is exposed. Spiked sediment tests provide an estimate of effects based on all direct modes of exposure, including ingestion, respiration, and absorption. Hence, toxicity to sediment ingesting organisms may be best approximated by spiked sediment tests. The primary disadvantage is the uncertain applicability of the exposure–response results to any particular field sediment or even to the distribution of field sediments. Aqueous tests are most appropriate if interstitial or overlying water is believed to be the primary exposure pathway for the toxicants and receptors at a site.

Aqueous tests are much more common than spiked sediment tests, but few aqueous tests use benthic species. Conventional aqueous tests and data are used to evaluate aqueous-phase exposures of benthic species, based on data suggesting that benthic species are not systematically more or less sensitive than water column species (EPA 1993d). For nonionic organic chemicals, aqueous concentrations and sediment concentrations can be interconverted by assuming equilibrium partitioning between the aqueous phases and the organic matter in the solid phase (Section 22.3).

An adjustment is also available for some sediment metals, based on the acid volatile sulfide (AVS) component of sediment (Section 22.3). However, it does not serve to estimate aqueous concentrations or effects.

When spiked sediment tests are used, physical and chemical properties of the test media are particularly important for evaluating chemical toxicity. Characteristics of the sediment (e.g., organic carbon content and grain size distribution) and of water (e.g., dissolved organic carbon, hardness, and pH) can significantly alter the speciation and bioavailability of the tested material. In site assessments, tests in sediments similar to the site media should be preferred. Regression models could be derived to account for confounding matrix factors (e.g., grain size or organic carbon content) (Lamberson et al. 1992). However, such models are species- and matrix factor-specific and would need to be developed on a case-by-case basis. The test method also can affect exposure. For example, chemical concentrations and bioavailability can be altered by the overlying water turnover rate, the water/sediment ratio, and the oxygenation of the overlying water (Ginn and Pastorok 1992). For all of these reasons, spiked sediment tests are relatively uncommon.

24.2.3 SOIL TESTS

There are relatively few standard soil tests, and the body of published toxicity data is small relative to water and sediment tests. In particular, few organic chemicals other than pesticides are represented. Standard methods using spiked soil or solutions are available for vascular plants (mainly crops) and earthworms from the US Environmental Protection Agency (US EPA) (OPPTS 850 test guidelines), American Society for Testing and Materials (ASTM; Committee E47 standards), European Union, and others. A variety of additional tests have been developed, mostly in Europe (Donkin and Dusenbery 1993; Donker et al. 1994; van Gestel and Van Straalen 1994; Kammenga et al. 1996; Heiger-Bernays et al. 1997; Lokke and van Gestel 1998). These tests generally treat soil as a medium in which particular species are exposed, rather than as an ecosystem. Tests of effects on field soils with associated communities are described in Section 24.5.3.

Tests in both spiked soil and aqueous solutions may be useful for assessing risks from soil contaminants. The relevance of published tests in soil to the assessment of risks to soil organisms seems self-evident, but because soil properties are highly variable and greatly influence toxicity, the toxicity in any other soil may be quite different. For example, Zelles et al. (1986) found effects of chemicals on microbial processes to be highly dependent on soil type. Unrealistic extremes of that variance should be eliminated by excluding data from tests in quartz sand, peat, or vermiculite, unless toxicity of chemicals mixed with these materials is demonstrated to be similar to that in natural soils. Tests conducted in solution have potentially more consistent results than those conducted in soil. Toxicity observed in inorganic salt solution may be related to concentrations in soil extracts, estimated pore water concentrations, or contaminated springs in which wetland plant communities are located. It has even been proposed that aquatic toxicity test results could be used to estimate the effects of exposure of plants and animals to contaminants in soil solution (van de Meent and Toet 1992; Lokke 1994), but that practice is not generally accepted.

Test endpoints for soil tests are less standardized than those for aquatic or wildlife tests. Plant tests most commonly include germination or growth. Invertebrate tests most commonly include survival but sometimes include reproduction. Tests of litter-feeding earthworms may not be representative of those that ingest soil, and vice versa. Similarly, although pollution-induced community tolerance (PICT) (Rutgers et al. 1998) has been used as a toxicological endpoint, it is not always clear that microbial communities that have become altered in their tolerance of contaminants are indicative of a decrease in the rate of a valued microbial process (Efroymson and Suter 1999).

24.2.4 ORAL AND OTHER WILDLIFE EXPOSURES

Terrestrial and semiaquatic wildlife are exposed through oral, dermal, and inhalation routes, and by intergenerational transfers. Tests exist for each of these routes, but oral tests to estimate the effects of toxicants in food, water, or other orally consumed materials are most common. These tests are employed primarily with birds and mammals.

For dietary tests, test animals are allowed to consume food or water ad libitum that has been spiked with the test material. The amount of food consumed should be recorded daily so that the daily dose can be estimated. A potential problem with dietary tests is that animals may not experience consistent exposure throughout the course of the study. For example, as animals become sick (e.g., due to toxicity), they are likely to consume less food and water. They may also eat less or refuse to eat if the toxicant imparts an unpleasant taste to the food or water or if the toxic effects induce aversion. These problems are sufficiently serious for cholinesterase-inhibiting pesticides and some other chemicals that the use of dietary tests is questionable (Mineau et al. 1994).

In oral tests, animals receive periodic (usually daily) toxicant doses by gavage (i.e., esophageal or stomach tube) or by capsules. The chemical is usually mixed with a carrier such as water, mineral oil, or acetone solution to facilitate dosing. Oral tests may include a single dose to simulate an isolated and brief exposure, or daily doses to simulate a continuous or long-term exposure. They provide better-defined exposures and can be representative of oral exposures other than food or water consumption such as incidental soil ingestion or oral ingestion of oil or other materials during grooming.

The choice of carrier used for oral or dietary tests has been shown to influence uptake by binding with the toxicant or otherwise influencing its absorption. For example, Stavric and Klassen (1994) reported that the uptake of benzo(a)pyrene by rats is reduced by food or water but facilitated by vegetable oil. Similarly, uptake of inorganic chemicals varies dramatically between tests with food and water. Chemicals are generally taken up more readily from water than from food.

Results of most dietary toxicity tests are presented as toxicant concentrations (mg/kg) in food or water. These data can then be converted into doses (mg agent/kg body weight/d) by multiplying the concentrations in food or water by food ingestion rates and dividing by body weights either reported in the literature or presented in the study (Section 22.8).

Standard methods for avian acute, subacute, and reproductive oral toxicity tests have been developed (Table 24.2). In general, risks to mammalian wildlife are assessed using the same tests of laboratory rats and mice that are used in human health risk assessments, but wild mammal tests may be required when the particular issues suggest that those tests may not be adequate.

Dermal and inhalation test results may be found for rodents and sometimes other species in the mammalian toxicity literature. The methods developed for laboratory test species may be adapted for mammalian wildlife (see US EPA guidelines OPPTS 870.3465 and .3250). Methods for testing dermal or inhalation exposures for birds, reptiles, or amphibians must be developed largely as needed.

Effects of developmental toxicants on birds and other oviparous species are readily tested by egg injection. For example, effects of dioxin-like compounds, for which embryo development is critically sensitive, have been tested by injecting the eggs of chickens and other species (Hoffman et al. 1998).

24.3   MICROCOSMS AND MESOCOSMS

Microcosms and mesocosms, together termed model ecosystems, are physical representations of ecosystems that contain multiple species and usually multiple media and that may be replicated. Microcosms are small enough to be maintained in the laboratory. They include everything from a mixed microbial culture in a beaker to aquaria and small artificial streams. Mesocosms are larger, more complex, and located out of doors. They include artificial streams and ponds and enclosed areas of terrestrial, wetland, and shoreline ecosystems. Microcosms and mesocosms have similar purposes and overlap in terms of size and complexity. In general, the purposes for conducting tests in these systems are to (1) provide realistic fate and exposure kinetics by including degradation, sorption, and uptake; (2) include all routes of exposure; (3) expose a large number and variety of types of organisms; (4) allow secondary effects due to species interactions; and (5) allow the operation of ecosystem processes. The biotic components of microcosms may be specified so as to improve replication and understanding of responses (Taub 1969, 1997). More commonly, the microbial and invertebrate components of these systems are provided by the process of enclosure or by inoculation from a natural ecosystem. For example, a pond microcosm or mesocosm might be inoculated with sediment and water from a pond or lake that contains microbes and invertebrates. Prescribed numbers of fish, amphibians, or other large organisms may then be added to the inoculated mesocosm or large microcosm.

The use of model systems in the assessment and regulation of chemicals has been a source of ongoing controversy. Advocates argue that if the goal is to protect ecosystems, one must test ecosystems (Cairns 1983), but they have disagreed about the designs. Opponents argue that these simplified systems do not capture critical properties of real ecosystems (Carpenter 1996; Schindler 1998). Even advocates recognize that different ecosystems respond in qualitatively different ways, which are only sometimes reflected in model systems (Harrass and Taub 1985). Hence, after decades of development and advocacy, these test systems are still rarely used and have had relatively little influence on environmental decision making. Many issues have contributed to the controversy, some of which are discussed here.

Ecosystem definition: Microcosms or mesocosms have been said to represent ecosystems in the sense that a rat represents mammals and a fathead minnow represents fish. That is, ecosystem responses like changes in net production or species number are thought to be reasonably consistent across flasks, ponds, and rivers. On the other hand, ecosystems are

TABLE 24.2
Selected Standard Oral Toxicity Methods for Birds

Test Type	Test Species	Duration	Exposure Route	Test Endpoint(s)	Reference[a]
Acute	Northern bobwhite, Mallard	14 d	Gavage	Mortality, intoxication	OPPTS 850.2100
Subacute dietary	Northern bobwhite, Japanese quail, Mallard, Ring-necked pheasant	5 d exposure, 3 d post	In diet	Mortality, intoxication, other effects	OPPTS 850.2200 ASTM E857-87
Reproduction	Northern bobwhite, Mallard	10 week	In diet	Adult mortality Eggs laid Egg fertility Egg hatchability Eggshell thickness Weight and survival of young	OPPTS 850.2300

[a]EPA methods reports may be obtained by searching www.epa.gov for the listed report number. ASTM methods may be purchased by standard number from www.astm.org.

much less consistent in terms of their components and function than classes of organisms, and might not be expected to respond as consistently to a particular exposure.

Size: Model systems vary greatly in size. Smaller systems provide more replication at less cost, but larger systems can support more types of organisms and perhaps better represent ecosystems of concern.

Composition: No model system can support the full range of taxa and trophic levels of a real lake, river, or forest. However, it is not clear how much simplification is allowable before a microcosm or mesocosm fails to be a model of ecosystems of concern.

Type of response: Are model systems best used to realistically expose organisms, elucidate species interactions, measure ecosystem properties, document recovery, or something else?

Model systems vs. system models: Mathematical models of ecosystems (Chapter 28) and physical models (microcosms and mesocosms) potentially serve the same purpose in ecological risk assessment. While physical models are clearly more realistic, mathematical models are more flexible in terms of being able to represent a variety of systems, states of the systems, and conditions of exposure, and they are much less expensive to implement. Physical models require a major assumption; they represent the real ecosystems of concern. Mathematical models require an equivalent major assumption (e.g., that the response of the system can be represented by a trophodynamic model) plus many assumptions associated with their equation forms and parameter values.

Most of these issues are a result of the difficulty in defining ecosystems compared to organisms. When testing fishes or birds, one does not need to decide how many livers or eyes to include, how big the tested piece should be (use the whole organism), or even what the basic responses are (there is a consensus for survival, growth, and reproduction). This is because organisms are clearly defined entities, not just representatives of a level of organization. Hence, assessors must define assessment communities but not assessment organisms (Box 16.6). While some of the problems have been addressed by developing standard designs, these fundamental issues have led to problems of interpretation that have severely limited the utility of these systems for regulation or management. In particular, the US EPA developed a standard aquatic mesocosm for tier IV testing of pesticides (Tuart 1988), but dropped the requirement for these tests in 1992. This decision was based on the judgment that field tests had not significantly improved the bases for most registration decisions over laboratory tests, in part because of the ambiguities in interpreting their complex and highly variable results. Also, it was judged that post-registration field monitoring could serve the need for field results in most cases (Tuart and Maciorowski 1997). Problems in interpreting results of microcosms and mesocosms were addressed by a recent workshop (Giddings et al. 2002). The recommendations include the following points that are particularly relevant to assessors:

Design: Tests should be designed to provide exposure–response relationships, not just to test a particular predicted exposure level.

Types of endpoints: Structural and functional endpoints are generally equally important, but species structure is primary, and functional endpoints alone do not protect biodiversity.

Recovery: Initial effects should not be considered unacceptable if population recovery occurs in an acceptable time, and they do not cause adverse indirect effects.

Models: Ecological models for extrapolation to real ecosystems or to other exposures should be developed.

Data for extrapolation: Biological and physical conditions of the actual ecosystems to be simulated by the model systems should be determined to aid extrapolation.

Scenarios: The agricultural landscape or other landscape context should be used to design reasonable exposure scenarios.

Endpoints: Regulatory authorities must develop goals and assessment endpoints that test systems can be designed to support.

Training: Tests in model systems are difficult to conduct and their responses are complex and difficult to interpret, so guidance, training and tools are required.

An important additional recommendation is that users of model systems should state their assumptions (Clements and Newman 2002). Like mathematical models, physical models are simplifications of real systems, and those simplifications imply assumptions about what is and is not important to understanding the response of the real systems being simulated. For example, an aquatic microcosm may require the assumptions that planktivorous fish are not important to understanding effects on plankton and that macrophytes are dominant components of the ecosystems being assessed.

Aquatic model systems range from flasks in the laboratory to artificial ponds and streams (Graney et al. 1994, 1995; Kennedy et al. 2003). The following are major types of model ecosystems:

Standard aquatic microcosms: This system is assembled in a flask from sterile sand and water, ten species of algae, five species of zooplankters, and a bacterium plus inadvertently introduced microbes (Taub and Read 1982; OPPTS 1996a; Taub 1997).

Pond microcosms: Pond water and sediment, with associated microbes and invertebrates, are placed in flasks, aquaria, or tanks, sometimes with macrophytes but seldom with fish (Giddings 1986).

Pond mesocosms: Ponds ranging from 0.04 to 0.1 ha are dug and filled with sediment and water from a real pond or lake. Macrophytes and fish may then be added. These systems have often been used to study the fate and effects of pesticides in the United States and Europe (Tuart and Maciorowski 1997; Campbell et al. 2003).

Artificial streams: Indoor or outdoor channels may be treated with once-through or recycling water. Unlike pond mesocosms these systems have not been standardized and they range from small channels that support algae and invertebrates to in-ground channels with pools and riffles that are large enough to support fish (Graney et al. 1989).

Littoral enclosures: Replicate systems are created by enclosing 50 m^2 portions of a pond or lake and adjoining shoreline (Brazner et al. 1989; Lozano et al. 2003).

Limnocorrals: Portions of a littoral ecosystem are enclosed by a large plastic bag or cylinder suspended from a floating platform and anchored to the bottom. They vary in volume from 1000 to 100,000 L (Graney et al. 1995).

Microcosm tests of the soil community and processes such as decomposition incorporate indirect effects of chemical addition as well as direct toxic effects. In the United States, standard test of soil function determines effects on respiration, ammonification, and nitrification in sieved soil (Suter and Sharples 1984; OPPTS 1996f), or nutrient retention, respiration, and plant production in soil cores (Van Voris et al. 1985; OPPTS 1996b). Other soil microcosms are used to test effects on soil community structure (Parmelee et al. 1997).

Soil microcosm tests usually focus on changes in the rates of soil microbial processes, which, however, may increase or decrease in response to a chemical exposure. Some metals are nutrients at low concentrations and most organic chemicals are microbial substrates. For example, the antibiotic streptomycin reduces bacterial abundance but increases overall soil activity by serving as a carbon and nitrogen source (Suter and Sharples 1984). Further, reductions in some soil processes such as litter decomposition may be desirable or acceptable in particular ecosystems (Efroymson and Suter 1999). A litter layer is esthetically desirable, reduces erosion, and is important to successful germination of some trees, but introduced earthworm species are destroying litter layers in forests of the northeastern United States. Therefore, if soil processes are assessment endpoints, it is desirable to determine the relevant exposure–response relationship and to understand the ecosystem context of the processes.

Mesocosm studies of wildlife are much less common. Even more than aquatic mesocosms, they are primarily used to study effects of realistic exposures, secondarily to reveal secondary effects, and very little to show population-level effects. For example, Dieter et al. (1995) placed mallard ducklings in littoral mesocosms to evaluate effects of aerial application of the organophosphate insecticide phorate on waterfowl in prairie wetlands. In another study, Barrett (1968) evaluated the effects of the carbamate insecticide carbaryl on plants, arthropods, and small mammals in 1 acre old-field enclosures.

24.4 EFFLUENT TESTS

Standard toxicity tests have been developed for determining the acceptability of aqueous effluents and are widely used in effluent permitting in the United States. Although conventional acute lethality tests may be used, short-term chronic tests using short-lived species or subchronic tests using sensitive life stages are used (Table 24.3). These tests are unique in the extent to which they have been validated against biosurvey data (Mount et al. 1984; Birge et al. 1986; Norberg-King and Mount 1986; Dickson et al. 1992, 1996). In those studies, the 7 d fathead minnow and *C. dubia* tests have been found to be predictive of reductions in the species richness of aquatic communities. As a result of this intensive development and validation, these tests are widely used beyond the regulation of aqueous effluents, and many laboratories are available to conduct them. Other species and taxa are used outside the United States (Herkovits et al. 1996). Effluent tests may pass or fail, i.e., only the undiluted effluent or only one critical dilution may be tested to determine acceptability. However, it is preferable to test a series of dilutions to establish the exposure–response relationship (EPA 2002b). Most effluent tests are static-renewal, but static tests may be used if the effluent has little oxygen demand and the toxic constituents are not lost through volatilization or other processes. If tests are performed on site, flow-through effluent tests are possible.

Effluent tests may also be used to identify which components of the contaminant mixture are responsible for effects, a process called Toxicity Identification Evaluation (TIE) (EPA

TABLE 24.3
Standard Procedures Used to Test the Toxicity of Effluents and Ambient Waters[a]

Species	Life Stage	Response	Duration (d)	Medium[b]
Sea urchin	Eggs and sperm	Fertilization	0.3	SW
Daphnia sp.	Juvenile	Immobilization	2	FW
Bivalve mollusk	Larvae	Mortality, shell development	2	SW
Fish[c]	Juvenile	Mortality	4	FW/SW
Algae (*Selenastrum capricornutum*)	Cell culture	Growth	4	FW
Ceriodaphnia dubia or *Mysidopsis bahia*	Juveniles–adults	Immobilization, fecundity	7	FW/SW
Fish[c]	Larvae	Mortality, growth	7	FW/SW
Fish[c]	Embryo–larvae	Mortality, deformities	7	FW/SW
Algae (*Champia parvula*)	Culture	Sexual fecundity	7	SW

[a]Test protocols are found in (EPA 2002b,h,i) and ASTM standards.
[b]FW = freshwater; SW = saltwater; FW/SW = protocols are available for species from both media.
[c]The standard freshwater fish in the United States is the fathead minnow (*Pimephales promelas*) and the saltwater fish is the sheepshead minnow (*Cyprinidon variegatus*) or inland silverside (*Menidia beryllina*).

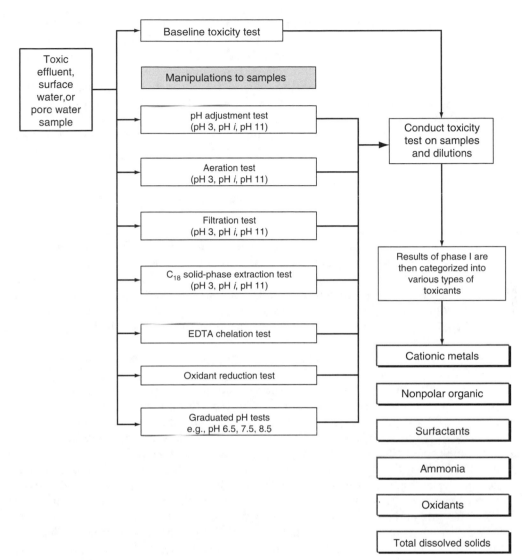

FIGURE 24.1 A logic diagram for toxicity identification evaluation (TIE) for acutely toxic aqueous samples. (From EPA (US Environmental Protection Agency), *Methods for Aquatic Toxicity Identification Evaluations: Phase I Toxicity Characterization Procedures*, 2nd ed., EPA-600/6-91-003, US Environmental Protection Agency, Duluth, MN, 1991a. With permission.)

1993a,b; Norberg-King et al. 2005). In TIE, the toxic constituents of a mixture are identified by removing components of a mixture and testing the residue, fractionating the mixture and testing the fractions, adding components of the mixture to background medium and testing those components, or other techniques (Figure 24.1).

24.5 MEDIA TESTS

The toxicity or other adverse properties of ambient media can be tested in at least three ways. In the least-used technique, contaminated biota are brought into the laboratory and tested. This technique is appropriate if the contaminant is persistent and bioaccumulated, or if it is known to cause persistent injury. For example, herring eggs from areas exposed to spilled oil and from unexposed areas were brought into the laboratory, and their hatching rates and frequencies of abnormalities recorded (Pearson et al. 1995). By far the most common approach is to bring contaminated and reference media into the laboratory for toxicity testing. This is a very active area of ecotoxicology, and test methods have been developed for ambient waters, sediments, soils, and biota. Methods for testing aqueous effluents (Section 24.4) have been adapted to testing ambient waters and media (Norberg-King et al. 2005). Methods specifically recommended for use at contaminated sites in the United States and Canada may be found in Office of Emergency and Remedial Response (1994b) and Keddy et al. (1995).

In assessments of contaminated sites, testing the contaminated media from the site has several advantages relative to testing individual chemicals in laboratory media:

1. The bioavailability of the contaminants is realistically represented. Because of sorption, formation of complexes, and other processes that reduce the availability of a chemical for uptake by organisms, the toxic effects of a particular concentration of a chemical may be highly variable. In particular, standard single chemical toxicity tests are conducted under conditions that tend to maximize bioavailability, so toxicity values from the literature may be conservative. Media toxicity tests can reduce or eliminate this source of uncertainty by conserving the bioavailability of the contaminants to which organisms are exposed on the site.

2. The forms of the contaminants are realistic. The toxicity of chemicals depends on their form including the ionization states and co-ions for metals and other ionizable chemicals. Typically, the forms of contaminants at a site are unknown. Even when known, the predominant forms found at the site may not be those for which toxicity data are available. Media toxicity tests can reduce or eliminate this source of uncertainty by conserving the forms of the contaminants to which organisms are exposed on the site.

3. Combined toxic effects are elicited. Few sites are contaminated by only one chemical, and the toxic interactions of chemicals are seldom well known. In addition, the interactions depend on the form of the chemicals which is itself problematical. Media toxicity tests can reduce or eliminate this source of uncertainty by retaining the combination of contaminants in the forms and proportions that occur at the site.

4. The effects of contaminants for which few or no relevant test data are available are included. Ecotoxicological testing has focused on pesticides and metals, not the industrial chemicals found at many contaminated sites. Even for metals and pesticides, the taxa and responses of interest may not have been tested. Media toxicity tests greatly reduce or eliminate this source of uncertainty by including all contaminants to which organisms are exposed on the site in a test that has been chosen to represent the endpoint response.

5. The type of effects may be determined. The specific effects of the mixture may not be predictable from available knowledge of the effects of the components. The test can be designed to determine the occurrence of effects that are relevant to the assessment endpoint.

6. The spatial distribution of toxicity can be determined (Figure 20.3). The extent of the area to be assessed or remediated and the priority to be assigned to different sources or receptor ecosystems can be more appropriately determined on the basis of the distribution of toxicity than from the distribution of individual chemical concentrations.

7. Remedial goals may be determined. Toxicity can provide a better basis for defining media and areas to be remediated than chemical concentrations can.

8. The potential for achieving the level of anthropogenic effects specified in the assessment endpoint can be determined. In some cases, because of upstream or background contamination, it may be uncertain whether site remediation will significantly improve the ecological condition of the receiving system. Demonstrated toxicity of upstream water and sediments can provide a better basis for this determination than chemical concentrations can.

9. The efficacy of remedial actions can be determined. In many cases, toxicity provides a better basis for defining whether additional remediation is needed than chemical concentrations can.

For reasons such as these, media toxicity testing has been recommended by the EPA for use at contaminated sites (Office of Emergency and Remedial Response 1994c). However, the qualifiers in the statements above point to the following limitations in media toxicity testing:

1. The medium may be modified by collection and preparation for toxicity testing. This has been a particular concern for testing of sediments, which may lose their physical structure and oxidation state during collection, sieving, and storage. Soils and water may be modified as well.

2. The forms and concentrations of chemicals may be modified by sample collection and processing. These changes may result from the changes in the medium just discussed or from direct effects such as changes in speciation due to adjustment of pH or addition of salts to make the medium acceptable to the test species, or from loss of volatile chemicals to air or loss of chemicals from solution due to sorption to the walls of the sampling and testing containers.

3. The samples may be unrepresentative. This problem also occurs in sampling for chemical analysis, but may be more severe for media testing, because typically fewer samples are tested than are analyzed.

4. Most media toxicity tests have short durations and few response parameters are recorded relative to conventional chronic toxicity tests.

5. The cause of the toxicity is unknown. Toxicity may be due to one or more contaminants in the tested medium, so it may not be clear what remedial actions are needed. In some cases, apparent toxicity may be due to extraneous factors such as chemical or physical properties of the medium or disease. For example, it was necessary to UV sterilize water from the Oak Ridge Reservation for fathead minnow larval tests because of an unidentified pathogen.

6. Apparent toxicity may be due to the choice of inappropriate reference locations. For example, relatively rapid growth may be due to high nutrient levels in reference media rather than toxicity in site media.

These limitations do not negate the considerable advantages of media toxicity testing. The first three can be minimized by care in the collection and handling of samples and in the conduct of the tests. The fourth point requires analysis and interpretation of the results during the risk characterization, as with other test data.

The fifth problem requires applying TIE to contaminated ambient media as described above for effluents (Section 24.4). An example is the use of TIE to demonstrate that low concentrations of nickel were responsible for the toxicity to *Ceriodaphnia* of water from Bear Creek, Tennessee (Kszos et al. 1992). Methods are also being developed for sediments and pore water (Ankley and Schubauer-Berigan 1995; Burgess et al. 2000; Norberg-King et al. 2005). Standard TIE methods are not available for soils, but could be developed ad hoc. This may be most readily done by spiking background soil. For example, toxicity of soils from the

Lehigh Gap, Pennsylvania, to isopods were correlated with concentrations of several metals, but tests of soils spiked with individual metals showed that zinc was responsible (Beyer and Storm 1995). Extension of the TIE process to include other properties of tested media could solve the sixth problem.

Both control and reference media should be tested along with the contaminated media. Control media are laboratory media that are known to be appropriate for the test species; they support nearly maximal rates of survival, growth, and reproduction of the test species. Their characteristics are usually prescribed in standard test protocols. Reference media come from the vicinity of the contaminated site, and are physically and chemically similar to test media except that they do not contain the site contaminants. Reference media include waters and sediments collected upstream of the site or soils from the same soil series as the contaminated soils, but not contaminated by the site. If upstream reference media are contaminated by an upstream source or if local soils are contaminated by a source other than the site (e.g., historic use of an arsenical pesticide or atmospheric deposition from a smelter), it may be desirable to obtain reference samples outside the range of those sources (a clean reference as opposed to the local reference). The control tests determine whether the test was conducted properly using healthy organisms. The local reference tests provide the basis for determining how much toxicity the site adds to proximate media. If a separate clean reference is used, it provides the basis for determining whether the differences from controls are due to contaminants or to properties of the media such as pH or texture. For example, water from Poplar Creek on the Oak Ridge Reservation was toxic to Japanese medaka embryos, water immediately upstream was a little less toxic, and water several kilometers upstream, above a municipal waste water treatment plant, was not toxic (equivalent to controls).

As in any form of toxicology, the best evidence for toxic effects is provided by demonstration of an exposure–response relationship. This can be done by testing samples collected on a contamination gradient or by testing a dilution series. An obvious example of the former is sampling and testing waters in a gradient downstream of an effluent or contaminated site. Often, particularly for soils, contamination gradients do not occur on a site. In such cases, an exposure series can be created by diluting the contaminated medium with a clean reference medium. It is obviously important to ensure that factors such as nutrient levels or texture do not confound the toxic effects by carefully matching the dilution medium to the test medium. Finally, an exposure–response series can be established by spiking site or reference media with the chemicals of concern. In such studies, it is important to match the forms of the test chemicals to those of the site contaminants. For soils and sediments, it is also appropriate to age the media to establish more realistic bioavailability (Heiger-Bernays et al. 1997). In addition to establishing that toxicity is responsible for observed effects, an exposure–response relationship can be used to define remedial goals by establishing what level of the site's contaminant mixture has acceptably low levels of effects.

Conventionally, media test data are analyzed using hypothesis testing statistics. Responses of each tested medium are determined to be either statistically significantly different from those in reference or control media, or not. If an exposure–response relationship can be established by gradients or dilution series, a function may be fitted to the data and used to estimate exposure levels that cause prescribed levels of effects (LC_{50}, EC_{10}, etc.). If, as is nearly always the case, there is a mixture of contaminants, exposure must be expressed as concentration of a representative chemical or some metric of aggregate concentration such as total petroleum hydrocarbons. If there are no exposure gradients, assessors should at least report the level of effects relative to reference or controls in the tests of the individual samples and the associated variance among replicates (e.g., proportional mortality $= 0.22 \pm 0.12$) (Box 24.1).

BOX 24.1
Replication of Tested Media

Care must be taken when analyzing media toxicity data to understand the nature of the replicates. Often, a sample is taken from a water body or site, subsampled in the laboratory, and the subsamples used as replicates for testing. Such laboratory replicates incorporate variance among containers within a test but not variance in the material. If the intent is to characterize the toxicity of soil in an area of the site or in a stream, such laboratory replicates are pseudo-replicates. A separate sample should be taken for each replicate from the area or water body to be characterized. Note that the same consideration applies to replication in effluent testing.

In some cases, the site medium may be unsuitable for the test organisms to survive, grow, or reproduce. In those cases, one has the option of adjusting the medium or changing the test species. Adjusting the pH, hardness, or other physicochemical property is problematical because of the potential effects on the form, bioavailability, or toxicity of the contaminants. Such adjustments should not be performed when the medium properties are unsuitable due to properties of the waste. For example, leachate from the S-3 ponds on the Oak Ridge Reservation was highly acidic and had high metal concentrations, so it would have been inappropriate to adjust the pH of test waters from the receiving stream. However, in cases where the medium is naturally unsuitable for the test species, and the adjustment will not affect the chemical state of the contaminants, adjustments may be appropriate. Alternately, use of a test species that is appropriate to the medium is conceptually more appealing. Particularly if the chosen species is resident on the site, use of a species characteristic of the type of ecosystem that is being assessed increases the apparent relevance of the test results. However, a standard test species may not be available that is adapted to the site medium and it may be difficult to develop testing procedures for a nonstandard species.

Tests for specific media are discussed below and test protocols are summarized in accompanying tables. In general, the tests performed at contaminated sites follow standard protocols from the US EPA, ASTM, or other organizations. Standard media tests have the advantage that they are reasonably reliable, can be performed at reasonable cost by many laboratories, are acceptable to most regulators, and have known sensitivities. The most common deviation from standard protocols is the substitution of local species. However, some thought should be applied to determining whether nonstandard tests may be more appropriate. For example, neither the acute lethality tests nor the "subchronic" 7 d larval toxicity tests can detect effects on fecundity or early development of fish. Any chemical that has a primary mode of action that involves disruption of endocrine control of the formation of gametes or development of embryos would not be adequately tested by those methods. A reproductive test might be more appropriate (Ankley et al. 2001). Another example would be chemicals with very slow uptake kinetics that would not be adequately tested unless that test was long enough for equilibrium between the test organisms and the medium to be attained. In such cases, longer than standard test durations may be needed. Finally, the relationship of the test species and responses to the assessment endpoints must be considered. For example, if some property of a reptile or amphibian population is the assessment endpoint, none of the standard tests are suitable.

24.5.1 CONTAMINATED WATER TESTS

The tests that were developed for testing of aqueous effluents are also commonly used for tests of ambient waters (Section 24.5 and Table 24.3). Specifically in the United States, they are recommended for assessments of Superfund sites (Office of Emergency and Remedial Response 1994a) and surface waters (EPA 1991b). Assessors using tests of contaminated water must consider the variance in aqueous contamination. The three water samples used in a 7 d static-replacement test may vary considerably in their composition and toxicity, so effects may be due to one particularly toxic sample out of three or effects may not occur because of one particularly clean sample. Of course, further variance occurs at longer time scales. Hence, it is important to conduct analyses of water in conjunction with the tests. Further, it is important to consider the processes that contaminate the water and their influence on toxicity. If contaminant levels are likely to be elevated during low flows due to poor dilution, storm events due to runoff, spray drift following pesticide application, or episodic releases of effluents, water for testing should be collected at those times rather than on a schedule.

24.5.2 CONTAMINATED SEDIMENT TESTS

Tests of contaminated sediments are in a less advanced stage of development than aqueous media tests. Relatively few protocols have been standardized (Table 24.4) and they have not been thoroughly validated against biosurvey data. However, short-term tests with marine and estuarine amphipods and polychaetes have been extensively used to evaluate the relative toxicity of coastal sediments (Long et al. 1995; MacDonald et al. 1996). Standard protocols for freshwater sediments in the United States are available for the amphipod *Hyalella azteca* and the midge *Chironomus tentans*. The selection of test organisms depends, in part, on their sensitivity to the site contaminants and tolerance to ecological conditions such as salinity and grain size. For example, *H. azteca* can be used in tests of estuarine ($\leq 15\%$ salinity) but not marine sediments, whereas *Rhepoxynius abronius* can be used only in marine sediments.

TABLE 24.4
Standard Procedures Used to Test the Toxicity of Ambient Sediments. The Tests Listed Include those that have been Recommended by the EPA for Contaminated Sites

Species	Life Stage	Response	Duration (d)	Medium[a]
Chironomus tentans	Larvae	Mortality and growth	10	FW
Hyalella azteca	N/A	Mortality and growth	10	FW/ME[b]
Amphipod, marine sp.[c]	N/A	Mortality, emergence, and reburial	10	ME
Polychaetes[d]	Recently emerged juveniles and young adults	Mortality	10	ME
Polychaetes[d]	Recently emerged juveniles	Mortality and growth	20–28	ME

Source: Office of Emergency and Remedial Response, *Catalog of Standard Toxicity Tests for Ecological Risk Assessment*, EPA 540-F-94-013, US Environmental Protection Agency, Washington, DC, 1994a. With permission.

[a]FW = freshwater sediment; ME = marine or estuarine sediment; FW/ME = protocols are available for species from both media.

[b]*H. azteca* can be tested in estuarine sediments up to 15% salinity.

[c]The standard marine or estuarine amphipods in the United States are *Rhepoxynius abronius*, *Eohaustorius estuarius*, *Ampelisca abdita*, *Grandidierella japonica*, and *Leptocheirus plumulosus*.

[d]The standard marine or estuarine polychaetous annelids in the United States are *Neanthes arenaceodentata* and *N. virens*.

While sediments are more stable than waters, seasonal changes in sediment chemistry (e.g., redox potential) may modify the bioavailability and toxicity of sediment-associated contaminants. For example, toxicity and bioconcentration tests of clams (*Mya arenaria*) were performed with a mixture of metals in water designed to simulate the interstitial water of Narraganset Bay (Eisler 1995). The mixture was lethal at simulated summer temperatures but not winter temperatures. Hence, it is advisable to conduct seasonal tests and to consider environmental characteristics that may modify toxicity.

Sediment tests are generally conducted using whole sediment samples and benthic infauna or epibenthic fauna. Sediment interstitial water can be tested using standard aquatic toxicity test methods. The pore water is extracted from the bulk sediment sample and, typically, used in tests with invertebrates. This approach can help identify the mechanisms of exposure for benthic infauna (i.e., respiration of contaminated pore water, ingestion of contaminated sediments, or both). This knowledge can be used to plan further sampling and interpret the results of other analyses (e.g., benthic invertebrate surveys). The disadvantage is that the extraction and testing processes can alter the form and bioavailability of the contaminants. Sediment elutriates may be formed by mixing sediments with water and then filtering or centrifuging. Elutriates may be tested to estimate effects of dredging, dredge spoil disposal, or other activities that suspend sediments in the water column. Guidance for collecting, handling, storing, and testing sediments can be found in Marine Protection Branch (1991) and Office of Water (2001).

24.5.3 CONTAMINATED SOIL TESTS

Most tests of contaminated soils use seedlings of vascular plants or earthworms. Some standard soil tests are presented in Table 24.5, and the tests for single chemicals can be adapted to contaminated soil testing (Section 24.2.3). A review can be found in Wentsel et al. (2003).

When tests are performed in soil from a contaminated site, there is less need for normalization of the chemical concentrations than when using literature values. However, care must be taken to match reference soils to contaminated soils in terms of chemistry, texture, and nutrient status. Particularly for growth and reproduction endpoints, tests may be highly sensitive to soil properties. Therefore, it is desirable to test soils from multiple reference locations to estimate the natural variation. Where variation is significant, it can be reduced by normalization of concentrations. For example, soil metal concentrations were pH-normalized to reduce variation among locations at a metal mining and milling site in Anaconda, Montana (Kapustka et al. 1995). If site and reference soils have low organic matter or inorganic nutrient levels, it may be necessary to amend or fertilize them in order to support the test organisms, to achieve reasonable growth in the reference soils, or to bring site and reference soils to the same levels.

As with single chemical toxicity tests (Section 24.2), the choice of test organisms and toxicity endpoints should balance practicality of standard tests with the need to be relevant to the assessment endpoints. If possible, invertebrate species should be representative of assessment endpoints in function, taxonomy, trophic level, life history strategy, and route of exposure to toxicants (Spurgeon and Hopkin 1996). Tests of earthworms are relevant to the soil invertebrate community in most nonarid continental ecosystems. *Eisenia fetida*, the most common test organism in soil toxicology, has about average sensitivity to toxicants among earthworms (Laskowski et al. 1998a).

Most phytotoxicity tests of contaminated soil use crop species. For Superfund assessments, lettuce is listed as the standard species of the seed germination and root elongation assay, though the use of other species is permitted (Greene et al. 1988; Kapustka 1997). LeJeune et al. (1996) tested the tree crop, hybrid poplar, as a surrogate for spp. and *Populus* spp. at the

TABLE 24.5
Standard Procedures Used to Test the Toxicity of Ambient Soils

Type of Organism	Life Stage	Response	Duration (d)	Medium	Reference
Earthworm[1]	Adult	Mortality	14	Soil	Greene et al. 1988
Earthworm	Adult	Reproduction	35	Soil	ISO 1997
Plant[1]	Seed	Germination	5	Soil	Greene et al. 1988; Linder et al. 1992
Plant[1]	Seed	Root elongation	5	Elutriate	Greene et al. 1988; Linder et al. 1992
Plant	Seedling	Mortality and vegetative vigor	20–90	Soil	Linder et al. 1992
Plant	Seedling	Weight	45	Soil	Linder et al. 1992
Plant	Life cycle	Reproduction and growth	28–44	Elutriate	Linder et al. 1992

[1]Recommended by the EPA for use at contaminated sites, in Office of Emergency and Remedial Response, *Catalog of Standard Toxicity Tests for Ecological Risk Assessment*, EPA 540-F-94-013, US Environmental Protection Agency, Washington, DC, 1994a.

Clark Fork River, Montana. Many crops grow well in the laboratory, and they have replicable toxic effects, because seed of standard strains can be used.

Because ecological risk assessors are interested in the toxicity to native endpoint species, it may be advantageous to test them. For example, standard plant species have different root morphology, development patterns, and carbon allocation patterns, that may make results of a root elongation test more or less relevant to assessment endpoint species. Kapustka (1997) offers advice to the assessor who plans to use nonstandard test species. A few of his rules are: choose nominal performance standards (e.g., percent germination), characterize variability using reference soils, and use one or more standard species in addition to the nonstandard species. In general, organisms used in toxicity tests should be laboratory cultured or at least started from seeds or eggs to obtain uniformity of history, age, and physical condition. If the local species cannot be raised in the laboratory, they should be maintained in the laboratory for at least 2 weeks prior to testing, and control organisms should be maintained and observed beyond the end of the test to assure their good condition (Laskowski et al. 1998b).

Soil for toxicity tests should be sampled to an appropriate depth for the endpoint plant, invertebrate, or community. Appropriate depths would be the depths within which most exposure occurs. These are also depths within which most plant roots and invertebrate activity occurs. In one of the Clark Fork River studies described above, a deeper depth interval was used for tests with hybrid poplar than with alfalfa, lettuce, and wheat (LeJeune et al. 1996).

24.5.4 AMBIENT MEDIA TESTS WITH WILDLIFE

As with chemicals, contaminated media may be tested for oral, dermal, or inhalation exposures of wildlife. In general oral tests are most applicable for risk assessment purposes. Contaminated food, soil, or water from the site of interest may be collected and fed or orally administered to test animals. If dermal or inhalation exposure pathways are considered critical at a particular site, toxicity tests for these pathways may be performed. Standard methods for ambient media toxicity tests for wildlife do not exist. Methods for these tests may be developed ad hoc or by modification of standard chemical test methods (Section 24.2.4).

While ambient media toxicity tests with wildlife are not widely performed, examples do exist (Table 24.6).

Media toxicity tests for wildlife are rare, largely because of the expense of housing, feeding, and caring for adequate numbers of test animals. Obtaining sufficient contaminated food materials to maintain test animals for the duration of the study may also be difficult. For example, the mink toxicity study performed as part of the Clinch River Ecological Risk Assessment (Halbrook et al. 1999a) required more than 2000 kg of contaminated fish to maintain 50 mink for 6.5 months. In addition, many wildlife species reproduce only once a year and have multiyear generation times, so reproductive tests are time-consuming.

24.6 FIELD TESTS

The most direct approach to ecological toxicity testing is to treat real ecosystems. These tests may consist of treating and monitoring replicate areas of an ecosystem type (i.e., plot studies) or distinct ecosystems like the experimental lakes area in Canada (Schindler 1974, 1987; Schindler et al. 1985). If a site is already contaminated or disturbed, tests may involve caging, penning, or planting organisms along a gradient of contamination or disturbance, or at matched sets of contaminated or disturbed locations and reference locations. These approaches, termed field experiments, field-testing, or in situ testing, are relatively easy for immobile organisms such as plants and relatively distinct ecosystems such as ponds. They are more difficult for organisms that are mobile and forage for food and for ecosystems that are large or involve flowing water. Field-testing may be highly realistic, in that the organisms can be subjected to realistic conditions and variation in exposure. However, such studies are subject to the effects of variation among sites in conditions other than contamination and to loss of the study due to vandalism, predation, or extreme conditions. In addition, cage effects may modify the sensitivity of the organisms. Finally, field tests may involve deliberately injuring an ecosystem. Advocates of field-testing argue that microcosms or mesocosms provide replication at the expense of realism (Schindler 1998). While the US EPA no longer requires field tests for registration of pesticides, it does encourage field tests as part of postregistration monitoring (Tuart and Maciorowski 1997).

Some methods are intermediate between testing and monitoring. Dosing of replicate plots or ponds is clearly testing, and the sampling of naturally occurring organisms from contaminated or disturbed sites is clearly monitoring. However, the study of molluscs confined in contaminated and uncontaminated bays or of birds using nest boxes in areas with different levels of contamination includes some but not all aspects of a test. Organisms may be randomly assigned to locations, but the contamination or disturbance is not randomly assigned to those locations and the locations may not be replicated, so the study may be confounded. Such studies are included here for the pragmatic reason that they require manipulation of the system by an investigator.

24.6.1 AQUATIC FIELD TESTS

Aquatic field tests are most often performed by confining organisms in the water column or sediment of contaminated systems and reference systems (Chappie and Burton 2000). A good example of this technique is the use of caged mussels or clams to measure the uptake of contaminants and associated effects on survival and growth (Jenkins et al. 1995; Salazar and Salazar 1998; Donkin et al. 2003). These tests are sufficiently common and well developed to have a standard guide (ASTM E: 2122-01). Bivalves may be suspended in the water column, placed in trays on the sediment, or even placed in the intertidal zone. In situ tests of other

TABLE 24.6
Examples of Ambient Media Toxicity Tests to Evaluate Effects of Environmental Contaminants on Wildlife

Test Species	Reason for Test	Contaminants of Concern	Test Media	Toxicity Test Endpoint	Reference
Mink	Determine toxicity of Great Lakes fish to wild mink	Organochlorine pesticides, PCBs, dioxins	Diets containing carp from Saginaw Bay, MI	Mortality, reproduction, hematology, liver pathology, bioaccumulation	Heaton et al. 1995a,b
Mink	Determine toxicity of fish downstream of a US DOE facility to wild mink	PCBs, mercury	Diets containing fish from Poplar Creek, TN	Reproduction	Halbrook et al. 1999
Least shrew	Determine toxicity of metals in sewage sludge to secondary consumers	Cadmium, copper, lead, zinc	Diets containing earthworms from a sewage sludge-treated site	Growth, bioaccumulation	Brueske and Barrett 1991
Mallard, Ferret	Determine toxicity of weathered Exxon Valdez crude oil to seabirds and sea otters	Weathered crude oil	Weathered Exxon Valdez crude oil by capsules, gavage, or incorporated into diets	Mortality Food avoidance Organ pathology Reproduction	Stubblefield et al. 1995a–c
Mute swan, Canada goose, mallard	Determine toxicity of contaminated sediments to tundra swans	Lead	Coeur d'Alene River sediments in diet	Multiple physiological	Beyer et al. 2000

Source: Modified from Suter, G.W., II, Efroymson, R.A., Sample, B.E., and Jones, D.S., *Ecological Risk Assessment for Contaminated Sites*, Lewis Publishers, Boca Raton, FL, 2000. With permission.

aquatic organisms is limited by size, mobility, and food requirements. Small benthic invertebrates may be used to test sediment toxicity using trays with mesh bottoms and perforated covers (Chappie and Burton 1997; Tucker and Burton 1999). The suitability of fish for long-term exposures is limited by their mobility and food requirements. However, individuals of small species and young fish may be used in short-term exposures such as tests of storm runoff (Newbry and Lee 1984; Hall et al. 1988). Fish eggs in mesh bags may be a good alternative for longer exposures (Hiraoka and Okuda 1984). Similarly, amphibian eggs and larvae may be experimentally confined in the field (Linder et al. 2003). Such tests may reveal interactions among agents and conditions and address agents other than chemicals such as suspended sediment or temperature.

Field tests may also involve treating replicate aquatic ecosystems such as streams, small lakes, or ponds, or segments of an ecosystem. Such tests were commonly conducted in the past to determine effects of pesticides (Giles 1970; Jeffrey et al. 1986). Exposures of caged organisms may be conducted in conjunction with these ecosystem exposures (Clark et al. 1986). They can provide realistic exposure levels and conditions, observations of indirect effects and effects on properties of populations and ecosystems, and observations of recovery. However, these tests typically have few replicates and high variance among replicates.

24.6.2 Field Tests of Plants and Soil Organisms

Field-testing with soil organisms or communities is rare and methods are not well standardized. An example of field-testing is the placement of worms for 7 d in contaminated soil in plastic buckets buried at the locations where the soil was collected (Menzie et al. 1992). This study determined that highly toxic soils occurred as veins through site drainage areas. Carabid beetles have been tested in field pens on pesticide-contaminated soils (Heimbach et al. 1994). Field tests may be performed for processes as well as organisms. Effects on soil function may be tested in the field by using introduced substrates and monitoring their loss, by measuring respiration or nitrogen transformation, or by measuring enzyme activities. For example, the loss of tensile strength of buried cotton strips was used to test the effects of wood preservatives (Yeates et al. 1994). The most common method involves burying bags of leaf litter or crop residues in contaminated soils or, in natural systems, placing them on the litter layer. The soils may be at contaminated sites (Strojan 1978) or may be experimentally contaminated to test a pesticide or other chemical (Rombke et al. 2003). Field-testing methods for soils have been evaluated by Linder et al. (1992) and Wentsel et al. (2003).

Field tests of pesticides may involve spraying fields at expected application rates. Properties of plants, soil organisms, or soil functions may be measured and related to those rates. For example, the US EPA test guidelines for phytotoxicity (OPPTS 850.4025, 850.4300) include injuries and effects on production of natural vegetations, crops, or lawns. Soil or plant responses may also be determined incidentally in larger-scale field experiments for effects on wildlife.

24.6.3 Wildlife Field Tests

The primary advantage of wildlife field tests is that they allow ambient conditions to influence effects, and therefore may provide a more realistic measure of actual toxicity at a site or at sites with similar conditions. Due to the great mobility of most wildlife species, field tests are problematic for chronic exposures of most species and are generally suitable only for species with small home ranges. Some wildlife field tests involve creating penned areas of habitat that are treated to achieve defined contaminant levels or that have prior contamination. Since most such studies involve treating the pens, they are discussed above as mesocosms

(Section 24.3). However, the same approaches could be applied to a contaminated or experimentally treated site.

Tests of acute effects of pesticides on wildlife lend themselves to field tests; the treatment of a field results in realistic conditions and scale, and the movement of organisms on and off the field is realistic. Such tests may involve application of the substance to actual fields or forests followed by monitoring of birds, bees, or other nontarget organisms (OPPTS 1996a–d). Avian field tests of pesticides have served to reveal effects that were previously unknown, confirm effects that were suggested by observations, disprove effects that were suggested by laboratory studies, and demonstrate secondary effects such as reduced survival of young due to loss of insects (Blus and Henny 1997). Because these organisms are mobile and, unlike aquatic organisms, soil organisms, or plants, are not immersed in a contaminated medium, it is important to confirm exposure by analysis of gut contents, body burdens, or biomarkers (Balcomb et al. 1984). Dead organisms should be necropsied to determine the cause. Radio tagging allows investigators to determine the extent of use of the contaminated or disturbed area and to assure recovery of the test subjects.

Field tests of avian reproduction can be facilitated by using nest boxes to attract cavity-nesting birds. They may be attached to posts or trees within or around an area that is or will be contaminated or disturbed. Because cavities are frequently a limiting resource for cavity-nesting birds, nest boxes are likely to be occupied. Once the birds become established, effects on behavior and survival of adults and on the number, diet, growth, and development of young can be studied. Nest boxes are used by a variety of species including starlings, bluebirds, tree swallows, wood ducks, barn owls, and kestrels. Guidance for the use of nest boxes for the studies of starlings at contaminated sites is presented in EPA (1989). Nest boxes have been employed to evaluate risks to birds from application of insecticides to agricultural fields or forests (Robinson et al. 1988; Pascual 1994; Craft and Craft 1996), PCBs and heavy metals at a Superfund site (Arenal and Halbrook 1997), and lead along a highway (Grue et al. 1986).

24.7   TESTING ORGANISMS

Testing of biocontrol agents, generically engineered crops, and other exotic organisms is analogous to testing chemicals. Potentially susceptible organisms or ecosystems are exposed in the laboratory or field to the test organism at defined levels, and responses are recorded. In the United States, biocontrol agents must be tested for their ability to attack or infect nontarget organisms (OPPTS 1996g). Environment Canada has developed a set of ecological tests for new microbial agents (McLeay et al. 2004). Such tests are analogous to toxicity tests, including tests of plants, invertebrates, and vertebrates exposed orally, by injection, by inhalation, and by exposure to microbes in water. Reported responses include infection, pathogenicity, symptoms of toxins, and conventional responses (survival, growth, fecundity). In some cases, tests must be developed ad hoc. For example, testing for effects of *Bt* corn on monarch butterflies involved feeding studies of larvae on milkweed leaves with defined levels of pollen (grains/cm^2) (Stanley-Horn et al. 2001). Genetically engineered organisms such as *Bt* corn and enhanced *Rhizobium* are usually field-tested before approval (McClung and Sayre 1994). Such tests are essentially field trials of the organism in its agricultural use, but with monitoring of fate and potential effects on nontarget organisms. Because effects of organisms are more diverse than those of chemicals and because of their potential to multiply and spread, it is important to base tests of organisms on a careful problem formulation, which considers the possible activities of the organism with respect to species and ecosystems other than the targets.

24.8   TESTING OTHER NONCHEMICAL AGENTS

Because ecological risk assessment may be applied to any hazardous agent or process, various testing methods are needed to provide exposure–response relationships. Examples of hazards to be tested include harvesting methods, water storage and diversion, construction of utilities such as roads and pipelines, farming practices, and ecosystem management practices such as burning and mowing. Since tests are simply experimental applications of an agent, this implies the adaptation of ecological experiments to assure that the results define the relationship between the level of exposure and effects on endpoint entities and attributes. Examples include rain exclusion experiments to test effects of climate change (Yarie and Van Cleve 1996), watershed studies of forestry practices (Coweeta, Hubbard Brook, Walker Branch, etc.), forest exposures to CO_2 to determine effects of elevated atmospheric levels (Zak et al. 2003), and manipulation of dam-regulated hydrology (National Research Council 1999).

24.9   SUMMARY OF TESTING

Exposure–response tests are the core of ecological risk assessment. This is true because only controlled, replicated, and randomized exposures provide assurance that the observed associations between exposure and effect are causal. However, that assurance is limited to the test itself. The use of test results to estimate risks requires an extrapolation from test conditions to real, uncontrolled field situations, and this is discussed in the next three chapters.

25 Biological Surveys

About thirty years ago there was much talk that geologists ought only to observe and not theorize, and I well remember someone saying that at this rate a man might as well go into a gravel-pit and count the pebbles and describe the colours. How odd it is that anyone should see that all observations must be for or against some view if it is to be of any service.

Charles Darwin

Biological surveys of effects include a variety of techniques for enumerating and characterizing biological populations and communities so as to relate them to exposure to some agent. In the simplest case, the measure of effect for the biological survey is an estimate of the assessment endpoint. In such cases, the effects analysis consists of summarizing the data in such a way as to define the relationship of effects to exposure. Examples include plotting the species richness of the soil microinvertebrate assemblage against an exposure axis such as kilometers from a source, soil compaction, or concentrations of a particular chemical. The US Environmental Protection Agency (US EPA) has recommended the use of biological surveys for ecological risk assessment of contaminated sites when feasible and appropriate (Office of Emergency and Remedial Response 1994b; Sprenger and Charters 1997) and in assessments of water quality (EPA 1991b).

A frequent problem in the use of biological surveys is that the entities and properties measured bear an undefined relationship to the assessment endpoints. They are often referred to as indicators or surrogates without defining what they indicate or for what they are surrogates. If the measures of effect do not directly estimate the assessment endpoint, the relationship between them must be clearly characterized by risk assessors. For example, if data are available for stream macroinvertebrates and the assessment endpoint is some property of the fish community, the relationship between them must be characterized in terms of the trophic dependence of fish on invertebrates, the relative sensitivity of fish and invertebrates, the similarity of their exposure, and other relevant properties. Clearly, this difficulty should be avoided in the problem formulation by selecting measures of effects that correspond as nearly as possible to the assessment endpoint (Chapter 18).

The following points should be considered when deciding whether biological surveys are appropriate for analysis of effects in an assessment.

Scale: Highly mobile organisms and the populations and communities that include them are seldom appropriate for biological surveys of a site. For example, a survey of breeding birds was conducted on the East Fork Poplar Creek flood plain in Oak Ridge, Tennessee, but it contributed nothing to the results of the ecological risk assessment. Territorial birds are highly mobile and are nearly always space limited, so all sites that contain physically suitable habitat are quickly occupied whatever the longevity or reproductive success of the resident birds may be. However, if the goal of an assessment is to estimate risks to a regional population or risks from an agent that acts at regional scales, mobility is not a constraint.

Interpretation: In order to interpret the variation observed in results of biological surveys, the properties measured must be stable and consistent across similar sites, in the absence of contamination or disturbance, relative to the magnitude of effects that is considered significant. For example, population densities of microtine rodents are notoriously variable across time and space, varying by orders of magnitude in the absence of any anthropogenic effects. In contrast, properties of stream fish communities are relatively stable and are commonly used to detect anthropogenic effects by comparing exposed communities to reference.

Difficulty: Clearly, biological surveys are inappropriate if they are costly and time consuming, are likely to fail due to the difficulty of proper execution, or if the necessary conditions for success are unlikely to occur. For example, determining the reproductive success of kingfisher populations has proved to be quite difficult, but the reproductive success of birds that nest colonially and in the open is relatively simple (Henshel et al. 1995; Halbrook et al. 1999a). Similarly, the fish communities of wadeable streams can be easily quantified with great accuracy, but the abundances of fish populations and communities of large bodies of water cannot be quantified with sufficient accuracy or precision for many assessments.

Appropriateness: Techniques employed must be suitable for the species or community, season, and habitat of interest and should produce results that meet the objectives of the risk assessment.

Technical expertise: In some cases, the expertise or experience needed to perform a particular survey is not available. In such cases, the need for technical expertise can be reduced by a simple change in the survey techniques or endpoint. For example, technicians who can identify benthic invertebrates to species are in short supply, but identification of families may be sufficient and individuals with very little training can sort invertebrates into higher taxa without knowing their names.

Consequences of the survey: Biological survey may cause unacceptable injury to the sampled population or ecosystem. The destructive sampling of rare species is an obvious example.

Data relevance: Data not generated by the assessment program should be used if pertinent and of adequate quality. However, care must be taken to appropriately analyze and interpret them. For example, fish survey data have been collected by the Tennessee Valley Authority for the purpose of comparing the quality of their reservoirs. These data were used to determine that the Oak Ridge Reservation had not altered the fish community of Watts Bar Reservoir relative to other reservoirs in the system, but they could not be used to infer risks at the scale of embayments, which was the scale of remedial actions (Suter et al. 1999).

25.1   AQUATIC BIOLOGICAL SURVEYS

Aquatic biota surveyed for waste site and water quality assessments may include periphyton, plankton, fish, and benthic macroinvertebrates (Office of Emergency and Remedial Response 1994b; Gibson et al. 1996; EPA 1996a, 1997a, 1998b; Barbour et al. 1999). The choice of assemblage and sampling method depends on the endpoints and habitat characteristics. Care should be taken to ensure that the survey locations capture the variation in exposure while recognizing the scale of the system relative to the habitat requirements and mobility of the surveyed organisms.

Habitat quality information is critical to the ability to discriminate between contaminant effects and natural variability. They must be accounted for in the survey design and should be quantified to the extent possible for all sites. The relevant habitat factors depend on the types of organisms being surveyed. For example, photosynthetically active radiation is important for algal and periphyton surveys, cover type and stream structure are important for fish surveys, and water chemistry (e.g., pH, hardness, and conductivity) is important for all assemblages.

The common use of biological surveys in aquatic systems, particularly streams, presents both advantages and disadvantages for risk assessors. The primary advantages are that methods are well established, and the expertise to perform surveys is commonly available. In addition, in states such as Ohio where bioassessment programs are underway, community types are already classified, reference conditions are established, and criteria for injury are defined (Yoder and Rankin 1995b, 1998). The primary disadvantage is that the metrics and indices chosen for monitoring or regulatory programs may not be appropriate for a contaminated site risk assessment and remediation. In particular, the biotic indices that are commonly used are designed to discriminate sites with common sorts or disturbances, particularly organic enrichment (Karr and Chu 1997). They are relatively insensitive to toxic effects (Dickson et al. 1992; Hartwell et al. 1995). Biological surveys for risk assessments should focus on measures of effect that are sensitive to the agents being assessed and are sufficiently valued to support a remedial or regulatory action.

25.1.1 PERIPHYTON

Algae and other aquatic plants are much less often included in biological surveys than fish or invertebrates. Except for estuarine species, aquatic macrophytes are more likely to be considered noxious weeds than valued endpoint entities. However, periphyton have long been used as indicators of stream quality because they are ubiquitous, constitute the base of most lotic food chains, are in direct contact with water, are sessile, are sensitive to a wide range of stressors, respond quickly to changes in water quality, are more stable than phytoplankton, and are easily sampled (Rosen 1995). Periphyton have the practical advantages of being easily associated with a site and being easily collected by scraping from natural or artificial substrates. However, it is often difficult to demonstrate to decision makers that a change in algal community properties is adverse, particularly when pollution and disturbances commonly increase algal production due to increased light or nutrient levels.

Periphyton samples can be collected from natural substrates, or from artificial substrates (e.g., frosted glass slides), which are placed in the water for a set period of time (e.g., 2 to 4 weeks) and then removed for analysis. The principal advantages of artificial substrates are ease of use, repeatability of measurements, reduced variability of taxonomic composition, and relative abundance. However, they are selective for particular species and the results may not be representative of the entire periphyton community. Both approaches have their supporters and detractors in the regulatory community (Rosen 1995), and the choice of natural or artificial substrates should be made in cooperation with the relevant agencies.

Periphyton communities vary widely in response to microhabitats even within the scale of individual sampling units, especially on rocks or other natural substrates. One can reduce variability by compositing multiple samples collected from a single type of habitat within a stream reach. For the assessment of water quality, collecting only from riffles and runs in streams is generally sufficient and periphyton communities in these habitats (particularly in those with current velocities of 10 to 20 cm/s) are less variable than in pools and edge habitat (Rosen 1995). Also, sampling the periphyton of soft substrates is more difficult and time consuming than sampling hard substrates.

The measures of effect may be structural or functional. Structural measures include measures of taxonomic composition and measures of standing crop (biomass). Common measures of taxonomic composition are species richness and relative abundance. Enough periphyton cells should be counted to ensure that uncommon species are included (Rosen 1995). Taxonomic identification should be at least to genus for soft algae and to species for diatoms (Rosen 1995). Because diatoms are common, abundant, and relatively easily identified by their silicaceous frustules, they are often counted to the exclusion of soft algae. Standing crop is measured as *chlorophyll a*, ash-free dry weight, cell counts, and cell volume. Each method has limitations, but all are generally acceptable for comparisons between exposed and reference sites. Indices of structural characteristics include diversity indices and indices of similarity between sites. Although useful in conjunction with other structural measurements, such indices should not be used as the only measure of periphyton structure.

The functional measure used for periphyton is primary productivity. The most common and widely accepted methods for estimation of primary productivity are based on the production of oxygen (O_2 method) or the uptake of radioactive carbon (^{14}C method) (Rosen 1995). Choose the method that best fits the budgetary, logistical, and quality requirements of the assessment. The O_2 method is inexpensive, relatively simple to perform, and readily used in the field. The advent of microelectrode technology has simplified and improved the measurement of oxygen production. The ^{14}C method is more expensive, more complicated to perform, and much less amenable to field use than is the O_2 method. However, it is a direct measure of primary productivity and is more sensitive than the O_2 method (Rosen 1995).

Physicochemical parameters to be measured and controlled in selecting reference sites include substrate composition, current velocity, temperature, photosynthetically active radiation, dissolved oxygen, conductivity, alkalinity, hardness, and nutrients.

25.1.2 PLANKTON

Plankton are the algae (phytoplankton) and small invertebrates (zooplankton) suspended in the water column with little or no ability to resist currents. Plankton are traditionally used as indicators of water quality in lakes and saltwater ecosystems. They are ubiquitous, are in direct contact with the water, are sensitive to a variety of stressors, respond quickly to changes in water quality, and have a direct impact on water quality. However, phytoplankton species composition and abundance are highly variable over periods of a few days, so they are seldom useful as measures of effects except for long-term changes in nutrient loading.

Plankton may be collected from discrete depths or be integrated over a range of depths or horizontal distances, depending on the expected distribution of the stressor(s). Methods include nets, pumps, and bottles. Their selection depends on the target organisms, target depths, and desired sample quality. Measurements include species richness, relative abundance, and community indices (e.g., diversity and similarity). Phytoplankton are often used as the sole representative of the plankton community. They are sufficiently diverse to permit the evaluation of a variety of stressors. It is especially important to collect physical data and water samples for analyses (temperature, photosynthetically active radiation, dissolved oxygen, conductivity, alkalinity, hardness, contaminants, and nutrients) in conjunction with plankton samples; otherwise, it is difficult to associate exposure with effects in large open bodies of water.

25.1.3 FISH

Biological surveys commonly include fish, because the value of fish is generally acknowledged and fish respond to a variety of aqueous contaminants. In addition, fish have practical advantages; their environmental requirements are well known, they integrate effects at lower trophic levels, and identification is relatively simple. Collection methods include electrofishing, nets, and traps. Method selection depends on the habitat characteristics and study design. Relevant habitat characteristics include cover type, stream structure, flow rate, pH, hardness, alkalinity, conductivity, and temperature.

Streams are typically sampled using electrofishing or, less commonly, seining methods. High-quality estimates of species presence and abundance can be obtained by blocking the upstream and downstream approaches with nets and then repeatedly sampling the reach. The resulting attributes are expressed per unit area, rather than per unit effort, which is less precise for this method. Electrofishing, seines, or hoop nets are used for large streams and rivers, whereas electrofishing, gill nets, fyke nets, and subsurface trawls are best used for lakes and marine environments. Boat-mounted electrofishing units are used in portions of large rivers and lakes that are sufficiently shallow such as shorelines and embayments. The inability to restrict fish movement in open bodies of water results in relative measures of fish community metrics (i.e., numbers per unit effort). Stationary nets are highly selective; the results should not be compared to results obtained with other sampling techniques.

Commonly in the United States, fish community properties or indices that combine several properties are used as measures of effect. The properties may include the number of species, the number of trophic groups, the abundance of species or trophic groups, the biomass of species or the community, and size distributions. The indices may include conventional diversity indices or arithmetic combinations of heterogeneous variables, most notably the Index of Biotic Integrity (IBI) and its derivatives (Karr et al. 1986). These indices are preferred by many state agencies because they are used in water quality management programs (Simon and Lyons 1995). They have many disadvantages as effects measures in risk assessment that can be largely mitigated by disaggregating the index to its component metrics (Suter 1993b, 2001). Properties of individual fish populations are less commonly used as endpoints in surveys, but they would be appropriate where game, commercial, rare, or otherwise particularly valued species are present. Appropriate population properties include abundance, size distribution, and production. The only commonly used properties of individual fish are frequencies of gross pathologies and anomalies. These are easily noted while counting and measuring fish from a community survey and are often of concern to the public and risk managers. The EPA recommends species richness and relative abundance as fish survey metrics for contaminated sites (Office of Emergency and Remedial Response 1994b).

Because fish are mobile, attention must be paid to the range of movement relative to the scale of contamination or disturbance. For this reason, fish surveys are used more in streams, where movement is relatively limited, than in lakes or estuaries. Where movement is a problem, it may be desirable to focus on species such as sunfish that are relatively sessile rather than on community properties that may be influenced by highly mobile or schooling species such as shad.

25.1.4 BENTHIC INVERTEBRATES

Benthic macroinvertebrate communities are commonly surveyed for ecological assessments because they are ubiquitous, important components of aquatic food chains, in direct contact with water or sediment, relatively immobile, and sensitive to a wide range of agents.

Benthic invertebrates in streams are frequently collected from cobble substrates in riffles and runs. The techniques are well established and the results can be compared with many other similarly sampled sites (DeShon 1995). These riffle communities are exposed to water-borne contaminants and conditions such as temperature but have relatively little exposure to sediment-associated contaminants. Benthic invertebrates in riffles are exposed primarily by respiration of contaminated water, whereas benthic invertebrates in sediment depositional areas are often immersed in the contaminated sediment and may ingest sediment. Respiration of overlying water may still be an important pathway for sediment-dwelling organisms, especially for those that ventilate their burrows (e.g., *Hexagenia* mayflies), but not to the exclusion of sediment-associated pathways, which include respiration of sediment pore water.

Surveying riffles but not pools can produce misleading results, as revealed by a survey of paired riffle and pool surveys in multiple streams in Tennessee (Kerans et al. 1992). In many instances the results for both riffles and pools correctly classified the streams regarding human impacts (based on a fish community index). However, when the classifications differed between riffles and pools, the results for pools were nearly always "correct" in the sense of being consistent with the classification based on the fish community index. Hence, benthic invertebrate communities in sediment depositional areas may be surveyed in addition to the riffle communities if any of the contaminants of concern are likely to be particle associated. The exception is streams in which sediment depositional areas constitute a relatively small fraction of the habitat. For example, the benthic invertebrate communities in sediment depositional areas of Upper East Fork Poplar Creek in Oak Ridge, Tennessee, were not surveyed, because such areas constituted less than 5% of the total available habitat (DOE 1995). In this case a preliminary stream survey was conducted to measure the size, distribution, and total surface area of deposited fine sediments. This proved to be a very useful tool for selecting assessment endpoints, habitats, and exposure pathways for a detailed analysis.

Survey methods vary in rigor from qualitative (e.g., sampling all habitats with a D-frame net) to semiquantitative (e.g., sampling for a specified time or distance with a kicknet) to quantitative (e.g., sampling $0.1 \, m^2$ with a Surber sampler). Kerans et al. (1992) compared the results of quantitative (Surber and Hess samplers) and qualitative (sampling all habitats with D-Frame net and hand picking) surveys for multiple streams in Tennessee. The quantitative surveys consisted of three to eight replicate samples per site, whereas the qualitative survey consisted of a single composite sample with collection time limited to 2 h. The qualitative surveys failed to detect human impacts that were detected by the quantitative surveys, probably due to the lack of replication. Thus, the assessor should select methods and survey designs that are quantitative and replicated within sites. A preliminary site evaluation may be limited to qualitative and semiquantitative surveys to establish the presence or absence of certain groups of invertebrates and provide qualitative taxa richness and semiquantitative abundance estimates. However, a definitive risk assessment should include quantitative, replicated estimates of community metrics. Definitive assessments should also consider including qualitative surveys of all habitats when the quantitative samples are collected using artificial substrates (DeShon 1995), because artificial substrates are selective and may not be representative of rare taxa or the actual taxa richness at a site.

Data from benthic invertebrate surveys typically consist of counts of individuals of species or higher taxa and, in some cases, biomass. One may simply use the numbers or biomass of individual taxa as the results. Alternatively, from these data, species richness (or taxonomic richness if some taxa are not identified to the species level) or other diversity metrics such as evenness, total numbers, and biomass can be derived. They may be aggregated into multimetric indices such as Ohio's Invertebrate Community Index (DeShon 1995). The total abundance of Ephemeroptera, Plecoptera, and Trichoptera (EPT taxa) is also a common metric. However, it is based on the sensitivity of these taxa to organic loading and siltation and may not be relevant to site contaminants. For example, the nominally sensitive ephemeropteran *Hexagenia limbata* was so abundant and so contaminated with mercury and polychlorinated biphenyl (PCB) in Poplar Creek that it posed a risk to its predators (Baron et al. 1999). In addition, some waters are unsuitable for EPT taxa, even in the absence of contamination. The EPA recommends biomass, species richness, density, diversity, and relative abundance as benthic invertebrate survey metrics (Office of Emergency and Remedial Response 1994a). Functional measures are seldom used, but may be assumed to be related to these structural measures (Clements 1997).

Kerans and Karr (1994) evaluated 18 attributes of benthic invertebrate communities as indicators of biological condition in streams. The authors conclude that all of the attributes should be used because they appear to be responsive to different human impacts (Kerans and Karr 1994). This is a reasonable approach for contaminated sites, provided the possible explanations for the status of each attribute are considered (see Kerans and Karr 1994 for examples). However, emphasis should be placed on metrics that are related to the assessment endpoint. Carlisle and Clements (1999) found taxa richness measures to be the most sensitive and statistically powerful metrics for evaluating metal pollution in Rocky Mountain streams. Abundance attributes were generally found to be insensitive to metal pollution or highly variable. The species richness of mayflies, which are generally sensitive to metals, is particularly noteworthy (Clements 1997).

Population or organism properties are seldom considered in benthic invertebrate surveys. In some cases, however, abundance of a particular sensitive and valued species may be an endpoint. One such endpoint was the abundance of the widgeon clam (*Pitar morrhuana*) at Quonset Point, Rhode Island (Eisler 1995).

It is important to determine the texture, organic matter content, depth of overlying water, and any other habitat properties that might influence the benthic invertebrate community at sampled locations. Elevated ammonium concentrations are particularly common and likely to result in toxicity that is unrelated to site contaminants. Even at a highly contaminated site, habitat variables are likely to explain more of the variance in invertebrate community properties than contaminant concentrations (Jones et al. 1999).

Spatial variability, rather than temporal variability, is the primary concern for sediment contaminants and sediment characteristics. This is especially true in slow-flowing systems with relatively stable sediments. Samples for sediment analysis should be collected as close to the biological survey sampling points as practically possible. Ideally, subsamples of the sediment included in each benthic survey sample, including replicates, should be analyzed for contaminants and sediment characteristics. This is rarely practical for contaminant analyses, but sediment characterization is relatively simple and inexpensive. Recommended sediment quality characteristics include grain size (percent sand, silt, clay), organic carbon content, ammonia, and pH. Quantitative measurements such as grain size fractions should be preferred over subjective and qualitative designations such as sandy or mucky. This allows the assessor to better compare results within and among studies. It also expands the risk characterization techniques available to the assessor. For example, the benthic invertebrate assessment for the Clinch River included multiple regression analyses of the benthic survey data with both contaminant and habitat characteristics as explanatory variables (Jones et al. 1999).

In addition, water quality may influence benthic communities and can vary significantly through time. Water samples should be taken such that representative exposures can be estimated.

25.2   TERRESTRIAL BIOLOGICAL SURVEYS

Terrestrial biological surveys are much less common than aquatic surveys as input for ecological risk assessments and there are no survey-based soil or air quality criteria like the aquatic biological criteria in the United States. The methods are less well developed and typically must be developed ad hoc or adapted from resource management or research methods.

25.2.1   SOIL BIOLOGICAL SURVEYS

Soil communities are surveyed less often than aquatic communities, even though there are fewer inherent difficulties in obtaining soil samples. Ecological risk assessments rarely use surveys of soil invertebrates, microorganisms, or soil processes. However, examples can be found in Menzie et al. (1992) and Jenkins et al. (1995). Approaches to surveying soil biota include: (1) collecting samples of soil and extracting taxa in the laboratory; (2) extracting organisms in the field, e.g., with mustard solution; and (3) trapping organisms using pitfall traps. The second method is the least quantitative, as it is likely to extract organisms to variable depths. Although most surveys have focused on invertebrates (Paine et al. 1993; Pizl and Josens 1995), microbial community properties, element transformations, and litter accumulation have also been surveyed (Jackson and Watson 1977; Strojan 1978; Tyler 1984; Beyer and Storm 1995). The abundance and composition of soil biota are highly dependent on soil characteristics (Nuutinen et al. 1998), so risk assessors must carefully determine that reference locations are appropriate.

25.2.2 WILDLIFE SURVEYS

Many methods are available for the collection of field data for wildlife populations, involving direct observation, trapping, vocalizations, track counts, netting, and attractants (Bookhout 1994; Heyer 1994; Wilson 1996; Suter et al. 2000). These methods may produce data that are useful in ecological risk assessments and may help elucidate the presence, nature, and magnitude of effects. Wildlife surveys may generate presence/absence, abundance, and age structure data as well as food habits information for exposure modeling. By the comparison of these data between the contaminated site and one or more reference sites, effects attributable to contaminant exposure may be differentiated from population fluctuations or habitat alterations that result from other causes. As noted previously, colonial nesting birds lend themselves to surveys for effects of contaminants or other agents (Giesy et al. 1994b; Henshel et al. 1995; Ludwig et al. 1996; Halbrook et al. 1999b; Custer et al. 2003).

Wildlife surveys differ from plant and invertebrate surveys in the extent to which they incorporate necropsy of organisms that are found dead, debilitated, or moribund (US Geological Survey 1999). This is because the techniques are available and because wildlife ecology is more focused on organisms. A good example is the necropsy of waterfowl performed as part of the assessment of lead mining in the Coeur d'Alene basin (Henny 2003). Although necropsy of opportunistically collected organisms is suggestive, a data set that is useful for assessment will usually require collection of other organisms from the site of concern and reference sites to determine the distribution and frequency of pathologies and body burdens.

25.2.3 TERRESTRIAL PLANT SURVEYS

Because vegetation provides the habitat for all inhabitants of terrestrial communities, it is important to survey and map vegetation on contaminated or disturbed sites, even before the problem formulation. In addition, if plant populations or communities are assessment endpoints, biological surveys may be an appropriate line of evidence for estimating risks. Because plants are immobile, they are clearly associated with a localized environment, and are easily sampled. However, few ecological risk assessments have been based on plant survey data. Guidance has been provided by the EPA (Environmental Response Team 1994b, 1996). The Agency recommends density, coverage, and frequency metrics as measures of effects for plant populations and communities. Examples are provided by Galbraith et al. (1995) and LeJeune et al. (1996), who took transect measurements of percent cover of tree, shrub, forb, and grass species to aid in the estimation of risks to the plant community in the Clark Fork River floodplain and Anaconda site in Montana. Similarly, surveys of vascular plants, mosses, and lichens showed severe effects in zinc-contaminated areas of the Lehigh Gap, Pennsylvania (Beyer and Storm 1995).

Because plants are valued and ecologically important for their primary production, measures of plant growth or production may be particularly useful for sites with contaminated soils or shallow, contaminated ground water. Tree coring is recommended by the EPA as a means to measure effects of contaminants on tree growth (Environmental Response Team 1994c). The width of annual growth rings may indicate the effects of contaminants, but because of the confounding effects of drought, frost, and other environmental factors, the interpretation should be performed by an experienced dendrochronologist. When vegetation is herbaceous, the EPA recommends that growth be determined by repeated clipping and weighing of the aboveground plant parts (Environmental Response Team 1994a).

It should be emphasized that few risk assessment schedules permit the repeated sampling of vegetation over long periods of time. The usefulness of a vegetation survey depends on whether observed effects can be related to measures at reference sites or reference (precontamination) points in time. Although one tree ring sample provides a time series (and each tree is its own control), the discernment of effects from herbaceous plant clippings generally requires multiple temporal samples. Thus, detrimental effects on production of the forest understory, old fields, or grasslands are not usually evident from a single vegetation survey.

If unhealthy plants or unvegetated areas are observed, the following question should be asked to determine the usefulness of the survey: can factors other than contaminants explain the brown foliage or other adverse response? These factors could include seasonal patterns, nutrient deficiency, insect herbivory, salt from winter applications to roads, acid rain, ozone, drought, grazing pressure, fire, or changes in hydrological patterns associated with the development of adjacent land. For example, when adverse impacts on forest trees were observed within the Bear Creek Watershed on the Oak Ridge Reservation, it was unclear whether dead trees were the result of contamination or altered hydrology associated with logging a neighboring area. Occasionally, specific toxic symptoms may be associated with particular contaminants. For example, "crinkle leaf" of cotton is associated with manganese toxicity, and an accumulation of purple pigment in soybean leaves can signal cadmium toxicity (Foy et al. 1978). However, these symptoms do not necessarily apply to other species, and most symptoms of toxicity such as stunted growth and chlorosis are common to many toxicants and nutrient deficiencies (Skelly et al. 1990).

Basic soil data should be obtained during the vegetation survey. These characteristics include major plant nutrients, pH, organic matter content, particle-size distribution, bulk density, and salinity, where relevant. One or more of these factors might explain differences in plant parameters at different locations.

25.3 PHYSIOLOGICAL, HISTOLOGICAL, AND MORPHOLOGICAL EFFECTS

Monitored effects on biochemical, physiological, or cellular properties of an organism that are indicative of toxic effects are commonly referred to as biomarkers. Their use has been inhibited by the fact that few of them are clearly related to the overt effects that constitute assessment endpoints in most ecological risk assessments. Although it has been proposed that remedial goals be based on elimination of any detectable biomarker response (Depledge and Fossi 1994), regulators do not normally take action on the basis of enzyme induction, even for humans.

Biomarkers of effects may play a supporting role in ecological risk assessments. In particular, biomarkers that are characteristic of a particular chemical, class of chemicals, or mode of action can support the inference that apparent effects are caused by particular contaminants (Chapter 4). For example, aminolevulinic acid dehydrogenase (ALAD) activity in the blood of waterfowl was used to diagnose lead toxicosis (Henny 2003). Even damage that is not particularly diagnostic can be useful if it can be even qualitatively related to population-level responses. For example, histological damage to the gonads of largemouth bass in Poplar Creek embayment in Oak Ridge, Tennessee, supported the inference that the low abundance and species richness of fish was due to toxic effects rather than habitat properties. When biomarkers are used to support inferences concerning causation, it is important to associate their levels or frequencies with contaminant concentrations.

Gross pathologies such as tumors, lesions, and skeletal deformities have played a more important role in ecological risk assessments than biochemical biomarkers. They are a common source of public concern, particularly where they occur in sport or commercial fish. Frequencies of gross pathologies are easily determined when fish are collected for chemical analysis or for biological surveys. Pathologies that are characteristic of chemicals or chemical classes can also contribute to attributing causation to both the pathologies themselves and any population or community effects.

25.4   UNCERTAINTIES IN BIOLOGICAL SURVEYS

Biological surveys potentially provide direct estimates of effects at sites receiving various levels of exposure. The primary uncertainties to be considered in such cases are sampling variance and biases in the survey results as estimators of the assessment endpoint. Sampling variance is estimated by conventional statistics. Biases must, in general, be estimated by expert judgment.

More difficult uncertainties arise when biological survey results are used to estimate effects at sites other than those surveyed. Such estimates may require interpolation or extrapolation. An example of interpolation would be the use of fish surveys at certain locations in a stream to estimate effects at locations lying between sampled locations. This might be done by algebraic interpolation, by spatial statistics, or by process modeling. An example of extrapolation would be the use of survey data for one contaminated stream to estimate effects on another stream with the same contaminant. At minimum, the uncertainty in such extrapolations would be equal to the variance among fish communities at uncontaminated sites (i.e., upstream reference communities in the case of interpolation or regional reference communities in the case of extrapolation). Additional uncertainty results from variance in the effective contaminant exposure due to differences in chemical form, patterns of temporal variance, etc.

25.5   SUMMARY

Biological surveys are used to determine whether a site is biologically impaired or to estimate exposure–response relationships for a site assessment or for a watershed or region. They are inherently realistic, but because exposures and conditions are uncontrolled, apparently causal associations are often misleading (Chapter 4). In addition, because of inherent variability of populations and communities, imprecision of most methods, typically small numbers of samples, and general lack of time series, effects must typically be large before they can be confidently detected. Hence, it is important to avoid accepting negative results without determining whether the methods used could detect levels of effects that are important to decision makers or stakeholders. Therefore, it is important to use biological surveys along with laboratory test results and modeling so that its realism can be weighed against the greater clarity and sensitivity of other lines of evidence (Chapter 32).

26 Organism-Level Extrapolation Models

Life was certainly simpler in the old days... when we could evaluate risk with a safety factor.

Doull (1984)

In most cases, ecological risk assessments must be based on exposure–response data for species, life stages, levels of organization, and responses other than those specified by the assessment endpoints. In some cases, no exposure response data are available for the agent of concern for any relevant organisms. For example, the endpoint is brook trout production and we have only a fathead minnow LC_{50} or only the structure of the chemical. Hence, it is necessary to use extrapolation models based on assumptions or statistical analyses to extrapolate to the endpoint species or community or to estimate parameters for population or ecosystem models (Chapter 27 and Chapter 28).

26.1 STRUCTURE–ACTIVITY RELATIONSHIPS

It is axiomatic that the biological effects of a chemical are a function of its structure. Hence, empirical models have been used to predict the pharmacological effects of drugs, the intended toxic effects of pesticides, and unintended toxic effects of these and other chemicals from their structural properties. These models, termed structure–activity relationships (SARs), are used to estimate effects when test data are unavailable. SARs may be qualitative, but the most useful models are quantitative (QSARs). (QSARs are also used to estimate fate-related properties; Chapter 22.)

SARs may be used to classify chemicals or to predict their toxicity. Chemicals may be classified in terms of whether they possess some property such as carcinogenicity, estrogenicity, or teratogenicity. This information may be used to reject a chemical during development or registration or to design tests by identifying potential test endpoints. Chemicals may also be classified in terms of their mode or mechanism of action (MoA; Chapter 7). If a chemical or class of chemicals is fitted by a QSAR that is associated with a particular MoA, it is likely to share that MoA. This information may be used as a basis for deciding that a concentration addition or dose addition model may be used to estimate the effects of mixtures of those chemicals (Chapter 8). MoAs include narcosis, respiratory uncoupling, and acetylcholinesterase inhibition. Predictions of toxicity from QSARs may be used directly to quantitatively estimate risks from untested or unreliably tested chemicals. Manufacturers use them to determine early in the process of chemical development whether a new chemical is likely to have significant toxic properties. They are used in a variety of regulatory contexts internationally, but are typically limited to screening applications (Cronin et al. 2003). For example, in the US EPA, QSARs are used to screen industrial chemicals to determine whether testing is required (Nabholz et al. 1997; Zeeman 1995).

26.1.1 CHEMICAL DOMAINS FOR SARs

One of the more conceptually difficult aspects of SAR development and use is the identification of the domain, the range of chemicals from which a SAR should be derived and to which it may be applied. The most common approach is to identify a chemical class that is considered to be homologous such as aliphatic hydrocarbons or phenols. Congeneric chemicals are homologous sets that have a common functional group, termed a toxiphore. Examples of toxiphores include amine, hydroxy, sulfhydryl, and carboxyl groups. Examples of domains defined by the US EPA for screening industrial chemicals include aliphatic amines, dinitro benzenes, and phthalate esters (Nabholz et al. 1997). Artificial intelligence–based algorithms have been used, in place of expert judgment, to generate chemical classes based on toxiphores (Klopman et al. 2000).

Alternatively, domains may be defined in terms of the MoA of the constituent chemicals (Drummond et al. 1986) (Chapter 7). This is more reliable than use of chemical classes, because the members of a class may have different mechanisms of action (Russom et al. 1997). Hence, a chemical may belong to a well-defined class, but its toxicity may not be predicted by the model for that class. For example, phenols are often described as having a narcosis MoA (i.e., low toxicity), but most phenols have other modes of action that are more toxic to fathead minnows (Figure 7.1). The difficulty in this approach is that a study must be performed to identify the MoA. Some studies are capable of identifying multiple modes of action by reporting multiple physical and physiological responses in whole organisms (Bradbury et al. 1989). Others identify specific modes of action such as Ah receptor or estrogen agonists, often in vitro (Wenzel et al. 1997; Schmieder et al. 2000). In vitro approaches must be used with caution because of the potential for metabolism to alter the MoA. In addition, a chemical may have more than one MoA, and the identified MoA may be secondary to an unidentified mode.

26.1.2 APPROACHES FOR SARs

The basic approach is expert judgment. Judgment is always required, but in some cases it has been used to develop systems of rules for assessing the activity of chemicals (Walker 1993; Karabunarliev et al. 2002). Judgment may be used quantitatively, but more often is used qualitatively to predict that a chemical will have a particular property or to assign it to a category (e.g., likely to bioaccumulate). A quantitative example is the use of judgment to decide what tested chemical is most similar to the untested chemical; that chemical's activity data are then used as surrogates for missing data.

The most common approach to QSAR development is regression modeling. Toxic end-point values such as fathead minnow 96 h LC_{50}s for a set of chemicals are regressed against some property of the chemical. The most common property in ecological QSARs is the octanol/water partitioning coefficient, K_{ow}. It is broadly useful because, for a set of organic chemicals with the same MoA, toxicity may be largely determined by the rate of uptake, which is in turn determined by hydrophobicity. A classic example is Konemann's (1981b) model of 14 d LC_{50} for neutral organic chemicals and fish:

$$\log(1/LC_{50}) = 0.87 \log K_{ow} - 4.87 \qquad (26.1)$$

Subsequently, Veith et al. (1983) showed that for 96 h LC_{50}s, the model must be nonlinear at high values of K_{ow}, because the slow uptake of the high–molecular weight chemicals inhibits the attainment of a lethal internal level in an acute exposure:

$$\log \text{LC}_{50} = -0.94 \log K_{ow} + 0.94 \log (0.000068 \ K_{ow} + 1) - 1.25 \tag{26.2}$$

These models describe toxicity by the baseline narcosis MoA in fish (Section 7.1).

A generic regression model for toxicity of organic chemicals is:

$$\log (1/C) - a(\text{hydrophobic}) + b(\text{electronic}) + c(\text{steric}) \tag{26.3}$$

where C is concentration and a, b, and c are fitted parameters (Hansch and Fujita 1964; Walker and Schultz 2003). Hydrophobicity is usually represented by K_{ow}; electronic properties may include charge, pKa, quantum chemical descriptors, or others; and steric properties include size and shape descriptors. Concentrations are usually expressed in terms of moles rather than mass, because, for a particular MoA, effects are more related to the number of molecules potentially reaching receptors than their mass.

26.1.3 STATE OF SARs

The current state of practice in ecotoxicological SARs is to use simple statistical approaches to relate effects to external exposure metrics (usually concentration) as in toxicity tests (Walker and Schultz 2003). This practice has been summarized in the US EPA ECOSAR software, which contains more than 100 SARs for more than 40 chemical classes (http://www.epa.gov/oppt/newchems/21ecosar.html). Further development is needed to include more chronic effects and effects on more taxa, particularly terrestrial organisms. In addition, greater acceptance of QSARs for regulatory purposes will require more extensive and consistent quality assurance including greater transparency, better defined endpoints, molecular descriptors, and domains, and more mechanistic bases (Eriksson et al. 2003). One effort in that direction takes a multivariate approach employing a large number of molecular descriptors to classify chemicals as having similar or dissimilar modes of action and then develops QSARs for prediction of effective levels (Vighi et al. 2002).

Future developments are likely to involve the development of a computational ecotoxicology analogous to current practices in pharmacology. SARs will be used to derive the parameters of toxicokinetic and toxicodynamic models, which simulate the uptake, metabolism, distribution, excretion, and effects of chemicals (Yang et al. 1998). In advanced versions, molecular modeling of potentially toxic chemicals and the various receptors in organisms should allow the estimation of binding energies, which could be used to predict specific effects, as in drug design (Raffa 2001). This is more difficult for toxicologists than pharmacologists, because the receptors are seldom specified a priori. The Ah receptor, which is a target of dioxin-like chemicals, is one of the few exceptions (Mekenyan et al. 1996).

26.2 EFFECTS EXTRAPOLATION APPROACHES

Most analyses of effects begin with a small set of data related to a few responses of a few species and life stages and perhaps a few ecosystem responses. Somehow, assessors must extrapolate from those few data to the entities and responses that constitute the assessment endpoints. Assumptions, factors, and statistical models are most commonly used for this purpose. Increasingly, mechanistic models are used to extrapolate to population and ecosystem level endpoints (Chapter 27 and Chapter 28). In such cases, the effects parameters of the models must be estimated from the available data, and the same assumptions, factors, and statistical models are used for that purpose. Numerous and diverse extrapolation models have been developed, but their application has been somewhat haphazard, and there is no consensus about which are appropriate.

This section presents the major approaches for developing extrapolation models and then discusses the models that are used for particular media and taxa. Although each of the approaches is applicable to any assessment, different extrapolation approaches are used in different contexts because of the constraints of available data and differences in the traditions of the different groups of toxicologists.

26.2.1 CLASSIFICATION AND SELECTION

It may be assumed that the endpoint species, life stages, and responses are equal to those in the most sensitive reported test or in the test that is most similar in terms of taxonomy or other factors. This process of classification and selection of test endpoints is the simplest and most commonly used extrapolation method. Sufficient similarity must be judged on the basis of some classification system. For example, plants are often classified by growth form, and the EPA has classified freshwater fish as warm-water and cold-water species (Stephan et al. 1985). However, species are most commonly classified taxonomically. Studies based on correlations of LC_{50}s of species at different taxonomic distances indicate that for both freshwater and marine fishes and arthropods, species within genera and genera within families tended to be relatively similar, which suggests that they can be treated as equivalent, given testing variance (Suter et al. 1983; LeBlanc 1984; Sloof et al. 1986; Suter and Rosen 1988). The same conclusion was reached for terrestrial vascular plants (Fletcher et al. 1990). Taxonomic patterns of sensitivity have been important in practice. For example, the observed levels of DDT/E in peregrine falcons and bald eagles did not appear to be sufficient to account for reproductive effects, until testing was done on a member of the same order (Lincer 1975). Other considerations in selection include the quality of the test, similarity of the test conditions to assessed field conditions, and relevance of measured responses.

In effect, this approach implies that the most sensitive or most relevant test organisms and conditions are surrogates for the endpoints in the field. Surrogacy is a concept with applicability to ecological risk assessments beyond toxicology. For example, when assessing a proposed introduction of a biocontrol agent, one would consider whether the tested potential nontarget hosts were adequate surrogates for the species, life stages, and exposure conditions that could occur in the field.

The advantage of this approach is its simplicity. One must choose the most appropriate test results in any case, and, by not applying any extrapolation model to the data, one avoids both effort and controversy. The disadvantage is that the available test data are seldom credible surrogates for the assessment endpoint response. Data selection is most defensible when based on a strong body of evidence. For example, one can use the threshold dose for developmental failure of chicken embryos to estimate risks to birds from dioxin-like chemicals, because a relatively large body of evidence indicates that it represents the critical response in birds, and chickens are a sensitive species (Giesy and Kannan 1998).

26.2.2 FACTORS

The next most common extrapolation method is to multiply or divide a test endpoint by a numerical factor. These factors are referred to as assessment factors, extrapolation factors, safety factors, and other terms. This extrapolation method may be treated as a formal extrapolation model

$$E_e = aE_t \tag{26.4}$$

where E_e and E_t are the effective exposure levels for the endpoint and test, respectively, and a can be statistically estimated (Section 26.2.5). However, in practice, the factors used in the

TABLE 26.1
Assessment Factors Used to Estimate Concern Levels in the Assessment of Industrial Chemicals by the US EPA Office of Pollution Prevention and Toxics. The Lowest Toxic Value is Divided by the Appropriate Factor to Set a Level of Concern for Exposures in the Environment[a]

Available Data	Assessment Factor
Limited (e.g., only one acute LC_{50} from a QSAR)	1000
Base set acute toxicity (fish and daphnid LC_{50} and algal EC_{50})	100
Chronic toxicity	10
Field test	1

Source: From Zeeman M.G., in G. Rand, ed., *Fundamentals of Aquatic Toxicology: Effects, Environmental Fate, and Risk Assessment*, Taylor & Francis, Washington, DC, 1995.

[a] A more complex set of factors, derived from this set, is used in Europe (CEC 1996).

regulation of chemicals are based on experience and judgment (Table 26.1) (OECD 1992, Zeeman 1995). The mathematical form is chosen for its simplicity, and multiples of ten are used to express order-of-magnitude precision. These factors have been often criticized, but they have been useful to regulatory assessors and have withstood legal scrutiny.

Sometimes, multiple factors are used. One might, for example, use a factor for interspecies differences, an acute/chronic factor, a laboratory/field factor, etc. These are nearly always treated multiplicatively

$$E_e = a_1 a_2 \cdots a_n E_t \qquad (26.5)$$

for the *n* extrapolations incorporated. Multiplicative chains of safety factors imply that everything will go wrong together: the test species is maximally resistant relative to the endpoint species, there is a particularly large acute/chronic ratio, field conditions are particularly conducive to toxicity, etc. Because of this conservatism, such chains of factors are less often used than formerly. However, they have the advantage over single integrative factors of clarifying what extrapolations are incorporated, and their potential conservatism may be appropriate for screening assessments or when precaution is particularly desirable.

The primary advantage of factors is their ease. We can all divide by 10, 100, or 1000 in our heads. However, unlike simple data selection, factors allow adjustment for data inadequacies. The use of factors is widely accepted, and some have argued that there is no evidence that more sophisticated extrapolation models perform any better (Forbes and Forbes 1993; Forbes et al. 2001). Their primary disadvantages are that they are largely subjectively derived, and their use results in a value that is considered a safe level but is not clearly associated with a particular effect (Fairbrother and Kapustka 1996; Chapman et al. 1998). Hence, they are best used in screening assessments (Chapter 31).

26.2.3 SPECIES SENSITIVITY DISTRIBUTIONS

Species sensitivity distributions (SSDs), which were developed to estimate water quality standards that would protect some proportion of species, are increasingly used as extrapolation models in ecological risk assessments (Posthuma et al. 2001). SSDs are exposure–response models that are fitted to responses of species rather than organisms as in conventional toxicology (Chapter 23) (Figure 26.1). A percentile of the distribution of test endpoint values for

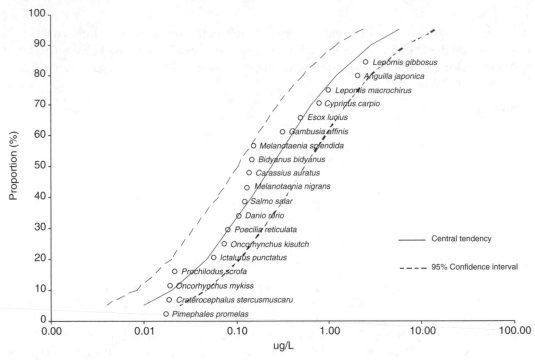

FIGURE 26.1 An example of a species sensitivity distribution (SSD), a logistic function relating the proportion of aquatic vertebrate species responding to copper in soft and warm water (<60 mg/L and >15°C). (Provided by Patricia Shaw-Allen. With permission.)

various species can be used to represent a concentration or dose that would affect that percentage of the exposed community. For example, if the 96 h LC_{50} values for fish exposed to a chemical are normally distributed (m_t, s_t), half of fish species in the field would be expected to experience mass mortality after exposure to concentration m_t within 96 h. This approach was developed for deriving water quality criteria independently in the United States (Stephan et al. 1985) and the Netherlands (Kooijman 1987). It has been repeatedly recommended as an ecological risk assessment technique (OECD 1992; Suter 1993a; Baker et al. 1994; Parkhurst et al. 1996a; EPA 1998a; Ecological Committee on FIFRA Risk Assessment Methods 1999a,b).

When SSDs are used to estimate levels of effects given an exposure level or to estimate the level of exposure corresponding to a level of effect, logistic or other functions are usually fitted to them. The choice of function makes relatively little difference if the data are well ordered (OECD 1992). Distributions may be fitted and percentiles calculated by any statistical software used for exposure–response modeling (Section 23.1.2). If used to support risk estimates based primarily on site-specific data or to support a causal analysis, an empirical distribution is simpler and adequate in most cases. Empirical distributions may even provide better numerical estimates, because many data sets are not fitted well by parametric functions (Newman et al. 2000).

The aggregation or partitioning of species in an SSD has been a topic of some debate. It is common practice in Europe to include all species, but the US EPA uses only multicellular animals to derive SSDs for water quality criteria, and others have advocated disaggregating as much as possible (e.g., fish, arthropods, other invertebrates, and algae). Aggregation of taxa in a common distribution provides more data with which to define the model. However, different taxa have different sensitivities, particularly for chemicals such as pesticides with defined modes of action. In extreme cases, this leads to manifestly polymodal distributions (Figure 26.2). Hence, knowledge of MoA and taxonomic relationships should be used to determine whether and how to disaggregate a data set.

Because of their growing popularity, and because they are more technically sophisticated than data selection or factors, SSDs have been subjected to detailed critiques of their conceptual bases and practical implementation (Forbes and Forbes 1993; Smith and Cairns 1993). These concerns range from the practical (e.g., the minimum number of species) to the conceptual (e.g., reasonableness of using a set of single species tests to represent a biotic community), and are discussed at length in Suter et al. (2002).

An advantage of SSDs is that they make use of all relevant and reasonably standard test results. Further, an SSD can be readily interpreted as representing the distribution of responses of species in a community or a taxon. The chief limitations of this method are

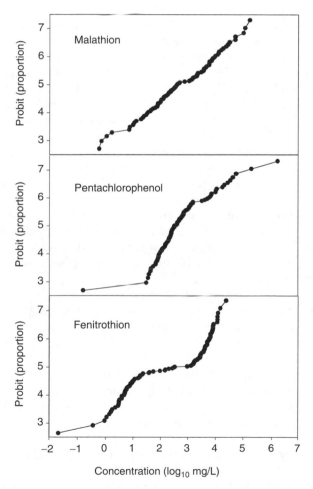

FIGURE 26.2 Species sensitivity distributions (SSDs) illustrating issues related to fitting standard functions. The Malathion SSD is linear on a log probit plot so it is fitted by the standard log-normal distribution. Pentachlorophenol has more highly sensitive and insensitive species than expected. Fenitrothion is bimodal, suggesting different modes of action or qualitatively different kinetics in different taxa. (From Newman, M.C., Ownby, D.R., Mezin, L.C.A., et al. in *Species Sensitivity Distributions in Ecotoxicology*, L. Posthuma, G.W. Suter II, and T.P. Traas, eds., Lewis Publishers, Boca Raton, FL, 2002. With permission.)

the requirement that enough species have been tested to define the SSD and that they be representative of the receiving community. When deriving water quality criteria, the EPA requires at least eight species from eight different families and they must be distributed across taxa in a prescribed manner, but regulatory assessments in the Netherlands may use as few as four species in an SSD. Relatively few chemicals have enough chronic toxicity data to establish the chronic SSD. Ingenious methods have been developed for developing SSDs when few data are available (Aldenberg and Luttik 2002; deZwart 2002). However, these approaches add significant uncertainty, so alternative extrapolation methods should be considered.

Another potential problem is that, if the media or the test conditions are variable and influential, the distributions will include extraneous variance. That is, the distribution is broader than the species sensitivities, because it includes variance due to conditions and test protocols. In fact, one extreme interpretation of SSDs is that they represent testing error, and the inherent differences in sensitivity among species are negligible (Van Straalen 2002a). Hence, attention must be paid to sources of variance and interpretation of SSDs. For aqueous toxicity extraneous variance can be low. The test methods and endpoints for aquatic toxic effects are reasonably consistent, so methodological variance should be relatively low. In addition, variance in test water chemistry is relatively low, particularly when hardness and pH normalization or speciation modeling are used for metals and ionizable compounds, so physical variance should be relatively low. For some chemicals, data are sufficiently abundant that SSDs can be derived for defined ranges of conditions (Shaw-Allen and Suter 2005). However, for both sediments and soils test data are sparse and the testing and survey methods and the endpoints are highly variable, the media have highly variable textures and chemistries, and reliable normalization methods are not available. Therefore, the physical and methodological variances may be significant contributors to the effects distributions in sediments and soils. The methodological variance is extraneous; the physical variance is an actual property of soils and sediments, and could be thought of as extraneous as well. However, if one takes an ecosystem perspective, the distributions resulting from the combination of biological and physical variance can be thought of as distributions of benthic ecosystem sensitivity, soil–plant system sensitivity, etc. It would be highly desirable to disaggregate those sources of variance by standardizing methods and by normalizing soil and sediment concentrations (Chapter 22).

Although SSDs were developed to address chemicals, they can be adapted to assess other agents. For example, the percentage of wildlife species behaviorally responding to aircraft overflights was related to slant distance to the aircraft (Efroymson and Suter 2001b).

SSDs may be interpreted in two ways (Suter 1993a, 1998a; Van Straalen 2002a). First, they may be interpreted as distributions of the probability that a species will be affected at a particular concentration. Hence, at a concentration of 100 μg/L, the probability of effects on any exposed aquatic species is 0.28 (Figure 26.3). Second, they may be treated as an estimate of the distribution of sensitivities of species in the exposed community. Hence, at a concentration of 100 μg/L the proportion of the community affected by the exposure is 0.28 (Figure 26.3). The results are a probability in the first case when the endpoint is effects on a population, but in the second case, when the endpoint is a community property, the result is deterministic.

The distinction may be clarified by analogy to exposure–response curves from conventional single species toxicity tests. The percentiles of those curves can be interpreted as probabilities of effects on individuals or as proportions of exposed populations. The former interpretation, which is used in human health risk assessments, is probabilistic, like the population-level interpretation of SSDs. The latter interpretation, which is more characteristic of ecological risk assessments but is also used in human health assessments, is deterministic, like the community-level interpretation of SSDs.

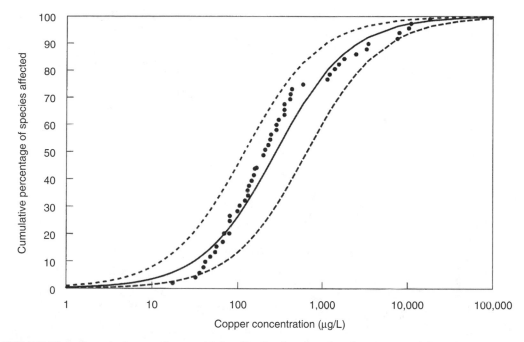

FIGURE 26.3 Cumulative species sensitivity distribution function for acute toxicity of copper. The curves are a logistic model fitted to the data points and upper and lower 95% prediction limits, generated using the Water Environment Research Foundation (WERF) software. (From Parkhurst, B.R., Warren-Hicks, W., Cardwell, R.D., Volosin, J., Etchison, T., Butcher, J.B., and Covington, S.M., *Aquatic Ecological Risk Assessment: A Multi-Tiered Approach*, Project 91-AER-1, Water Environment Research Foundation, 1996b With permission.)

In ecological risk assessments for the Oak Ridge Reservation, the aquatic assessment endpoints were defined at the community level and the endpoint properties included reductions in species richness or abundance (Suter et al. 1994). Hence, the appropriate interpretation is that the percentiles of the distribution are estimates of the proportion of species affected. In assessments of risks to aquatic communities, actual measurements of fish species richness and abundance as well as toxicity tests of ambient waters may be available. In such cases, the SSD is used to determine which chemicals are likely causes of any observed toxicity or community degradation rather than to estimate risks. For attributing cause, empirical distributions have been used and uncertainty analysis has not been judged important (Figure 26.4). However, if the risks are characterized on the basis of analyses of the SSD, uncertainty should be quantified.

The interpretation of SSDs is further complicated if uncertainty concerning the distributions is considered. Uncertainties concerning the percentiles of SSDs have been used to calculate conservative environmental criteria (Kooijman 1987; Aldenberg 1993; Aldenberg and Slob 1993) and to estimate risks as probabilities of effects (Parkhurst et al. 1996a,b). For a population endpoint, the percentiles of the distribution about the species sensitivity distribution are credibilities of a probability of effects on the endpoint population, i.e., we partition variability among species from uncertainty concerning that variable parameter. Consider Figure 26.3; the probability that a species will be affected at 100 µg/L is 0.27 with a credibility of 0.5 (the central line), but the upper 95% confidence bound indicates that there is a credibility of 0.025 (i.e., half of 5%) that the probability of effects is 0.46 or greater. This is conceptually equivalent to the output of a nested Monte Carlo simulation of an exposure

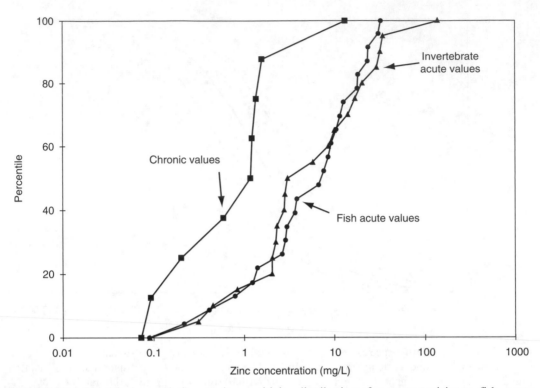

FIGURE 26.4 Empirical cumulative species sensitivity distributions for acute toxicity to fish, acute toxicity to aquatic invertebrates, and chronic toxicity to fish and invertebrates combined for zinc. (From Suter, G.W., II, Efroymson, R.A., Sample, B.E., and Jones, D.S., *Ecological Risk Assessment for Contaminated Sites*, Lewis Publishers, Boca Raton, FL, 2000. With permission.)

model in which variability in exposure is partitioned from uncertainty. However, the uncertainty is simply the residual variance from the fitting of the distribution function to the toxicity data. This is a very incomplete estimate of the actual uncertainty (Suter 1993a). For a community endpoint, the percentiles of the distribution could be probabilities based on the variability among communities or credibilities based on the uncertainty concerning the distribution as a representation of community response. However, as in the population-level interpretation, the variance from the fit is an incomplete estimate of the actual uncertainty. The total uncertainty is a result of biases in the selection of test species, differences between laboratory and field sensitivities, and differences between the laboratory responses and the endpoint properties. Subjective uncertainty factors can be employed to estimate total uncertainty, but these should be based on a careful consideration of the data in the distribution and their relationship to the site-specific endpoint. A factor of 10 can be considered minimal. Research is needed to develop more objective estimates of the uncertainties concerning risk estimates derived from these distributions.

26.2.4   REGRESSION MODELS

Regressions of one taxon on another, one life stage on another, one test duration on another, one level of organization on another, etc. can be used to extrapolate among taxa, life stages, durations, or levels of organization. This approach is extremely flexible and quantitatively rigorous but is seldom used. Regression models for aquatic extrapolations are presented in Table 26.2. More extensive discussions and examples of these methods can be found in the

TABLE 26.2
Linear Equations for Extrapolating from Standard Fish Test Species
to All Bony Fish (Units are log μg/L)

Test Species	Slope	Intercept	N	Mean	F_1	F_2	PI[a]
Pimephales promelas	1.01	−0.30	354	2.77	0.45	0.0006	1.31
Lepomis macrochirus	0.96	0.17	500	2.52	0.49	0.0005	1.37
Oncorhynchus mykiss	0.99	0.29	480	2.42	0.38	0.0004	1.20
Cyprinodon variegatus	0.97	0.03	51	1.25	0.58	0.0085	1.49

Source: From Suter, G.W., II, *Ecological Risk Assessment*, Lewis Publishers, Boca Raton, FL, 1993a.

[a]PI, the 95% prediction interval at the mean, is log mean $Y \pm$ the number in this column.

literature (Suter et al. 1983, 1987; Barnthouse and Suter 1986; Sloof et al. 1986; Holcombe et al. 1988; Suter and Rosen 1988; Calabrese and Baldwin 1993; Mayer et al. 2004).

Regression models provide an alternative to SSDs when data are few. If a test endpoint for a standard test species is available, the distribution of the endpoint for all fish species can be estimated from the equations like those in Table 26.2 that regress all fish species against a standard test species for multiple chemicals (Barnthouse and Suter 1986; Suter et al. 1987; Holcombe et al. 1988; Suter and Rosen 1988). The equations estimate the mean of log LC_{50} for saltwater fish from *Cyprinodon variegatus* LC_{50} or for freshwater fish from the standard freshwater species. The 95% prediction interval (PI) at the mean is log mean $Y \pm$ PI. The PI is estimated from the variance in LC_{50}s for other species (Y) at a given LC_{50} for a standard test species (X_0):

$$\text{var}(Y|X_0) = F_1 + F_2(X_0 - X)^2 \qquad (26.6)$$

Since the second term of the variance is relatively small, the PI at the mean is a reasonable estimate of the PI for all Y. That is, 95% of fish responses would be expected to fall within approximately ± 1.3 log units or approximately a factor of 20 of the log-normal mean fish response estimated from the equations.

A set of regression models for acute lethality to aquatic animals and terrestrial birds and mammals with supporting software, Interspecies Correlation Estimations (ICE), have been recently published (Asfaw et al. 2003). They are more convenient and are based on larger data sets than prior intertaxa regression models.

These intertaxa regression models were derived using data for a variety of chemicals. While this makes them generically useful, it also increases the variance. It is likely that the models could be made more precise by limiting them to a single mode of action, since relative sensitivities vary among chemical classes (Vaal et al. 1997).

26.2.5 TEMPORAL EXTRAPOLATION OF EXPOSURE–RESPONSE MODELS

As exposure duration increases, the concentration or dose necessary to kill an organism declines to some minimum, the incipient lethal level. Hence, acute lethality tests can be used to estimate thresholds for effects in sustained exposures by extrapolating fitted exposure–response models to effectively infinite time (Mayer et al. 1994). A curve fitted to effects concentrations or doses at multiple times (e.g., 24, 48, 72, and 96 h LC_{50} values) will approach an asymptote or reach an effectively infinite duration such as the maximum life span of the organism. If one also extrapolates to a low response level (e.g., LC_{01}), the corresponding concentration or dose may be interpreted as a safe level with respect to lethality for sustained exposures. Software—Acute to Chronic Estimation (ACE)—is available to perform such analyses using linear regression analysis, multifactor probit analysis, or accelerated life testing theory (Ellersieck et al. 2003). This approach has the advantage of estimating a threshold for chronic lethality without any chronic testing, but it requires that data for multiple acute durations be available. It also requires the assumption that the shape of the curve does not change at longer durations.

If data from multiple durations are not available, one may assume a relationship. The simplest is Haber's rule, which assumes that the product of dose or concentration (C) and time to effect (t) is a constant (k) for a particular effect. Therefore, for any exposure duration, the effective concentration or dose is:

$$C = k/t \qquad\qquad (26.7)$$

This formula is commonly used for temporal extrapolation as well as to reduce the dimensionality of exposure–response models (Section 23.2.3), although it often poorly fits actual temporal data and should not be used for extrapolations beyond a narrow range. The need for this warning is obvious. Some concentrations are too low to cause an effect no matter how long the exposure. A proposed alternative is Ostwald's formula:

$$C = k/t^a \qquad\qquad (26.8)$$

where a is a fitted constant. However, that formula requires either data fitting as in the ACE approach or a presumption that the exponent derived for a similar chemical with the same mode of action is applicable.

26.2.6 FACTORS DERIVED FROM STATISTICAL MODELS

Most factors are derived by expert judgment based on experience or simple reviews of relationships among general types of data (Section 26.2.2), but factors may also be derived by data analysis and associated with a particular extrapolation. Sloof et al. (1986) used the PIs around regression models to derive uncertainty factors. Calabrese and Baldwin (1993) applied this approach to previously developed extrapolation models (Suter et al. 1983, 1987; Barnthouse and Suter 1986; Suter and Rosen 1988). Results for acute–chronic extrapolations for defined chronic responses and intertaxa extrapolations are shown in Table 26.3 and Table 26.4, respectively. The reader should note that this method retains only the highly conservative 90%, 95%, or 99% upper bound estimate of effects levels and not the best estimate.

The intertaxa extrapolations require some explanation. Suter et al. (1983) developed an approach for extrapolating between any test species and reference species that involved aggregation of species within taxonomic hierarchies. Using a large data set of aquatic acute toxicity data, congeneric species were regressed against each other and then aggregated; next, genera confamilial were regressed against each other and then aggregated; after that families within the same order were regressed against each other. This process continued up to a regression of the phylum vertebrata against the arthropoda. The increasing PIs on these regressions as the taxonomic distance increased was used to demonstrate that toxicological similarity is related to taxonomic similarity. Calabrese and Baldwin (1993) used a later version of the regressions for fish taxa to reduce the regressions and PIs to 95% and 99% uncertainty factors for each taxonomic relationship by calculating confidence intervals on the set of PIs for pairs of orders of fish (Table 26.5). Calabrese and Baldwin (1994) later suggested that these generic factors were applicable to taxa other than fish, including humans. For

TABLE 26.3
Uncertainty Factors for Extrapolations from Acute Lethality to Specific Chronic Effects in Fish

X Variable	Y Variable	n	90%	95%	99%
			Uncertainty Factors — Confidence Interval		
LC_{50}	Hatch EC_{25}	31	26	50	198
LC_{50}	Parent mort EC_{25}	28	18	32	106
LC_{50}	Larval mort EC_{25}	89	18	31	93
LC_{50}	Eggs EC_{25}	42	32	64	228
LC_{50}[a]	Fecundity EC_{25}	26	26	50	206
LC_{50}[a]	Weight[b] EC_{25}	37	28	53	188
LC_{50}[a]	Weight/egg EC_{25}	14	91	246	2247
Mean			34	75	467
Weighted mean			27	55	265

Source: From Calabrese, E.J. and Baldwin, L.A., *Performing Ecological Risk Assessments*, Lewis Press, Boca Raton, FL, 1993. With permission.

[a] Regression analysis from Suter et al. (1987).
[b] Decrease in weight of fish at end of larval stage.

TABLE 26.4
Taxonomic Extrapolation: Means and Weighted Means for the 95% and 99% Prediction Intervals (PIs) for Uncertainty Factors Calculated by Calabrese and Baldwin (1994)[a]

X Variable	Y Variable	n	95% PI	99% PI
			Uncertainty Factor	
Taxonomic extrapolation: species within genera				
Salmo clarkii	*S. gairdneri*	18	9	13
S. clarkii	*S. salar*	6	6	10
S. clarkii	*S. trutta*	8	6	8
S. gairdneri	*S. salar*	10	7	11
S. gairdneri	*S. trutta*	15	4	5
S. salar	*S. trutta*	7	5	8
Ictalurus melas	*I. Punctatus*	12	5	7
Lepomis cyanellus	*L. macrochirus*	14	6	9
Fundulus heteroclitus	*F. majalis*	12	6	8
Mean			6	10
Weighted mean			6	7
Taxonomic extrapolation: genera within families				
Oncorynchus	*Salmo*	56	5	6
Oncorynchus	*Salvelinus*	13	4	5
Salmo	*Salvelinus*	56	5	7
Carassius	*Cyprinus*	8	4	6
Carassius	*Pimephales*	19	7	9

Continued

TABLE 26.4 (Continued)
Taxonomic Extrapolation: Means and Weighted Means for the 95% and 99% Prediction Intervals (PIs) for Uncertainty Factors Calculated by Calabrese and Baldwin (1994)[a]

			Uncertainty Factor	
X Variable	Y Variable	n	95% PI	99% PI
Cyprinus	Pimephales	10	7	10
Lepomis	Micropterus	30	8	11
Lepomis	Pomoxis	8	9	13
Cyprinodon	Fundulus	12	6	8
Mean			6	8
Weighted Mean			6	8

Taxonomic extrapolation: families within orders

Centrarchidae	Percidae	47	10	14
Centrarchidae	Cichlidae	6	4	6
Percidae	Cichlidae	5	13	24
Percidae	Esocidae	11	9	13
Atherinidae	Cyprinodontidae	32	7	9
Mugilidae	Labridae	12	55	78
Cyprinodontidae	Poecillidae	12	3	5
Mean			14	21
Weighted mean			13	18

Taxonomic extrapolation: orders within classes

Salmoniformes	Cypriniformes	225	20	27
Salmoniformes	Siluriformes	203	39	51
Salmoniformes	Perciformes	443	12	16
Cypriniformes	Siluriformes	111	11	15
Cypriniformes	Perciformes	219	32	43
Siluriformes	Perciformes	190	63	83
Anguiliformes	Tetraodontiformes	12	13	18
Anguiliformes	Perciformes	34	25	34
Anguiliformes	Gasterosteiformes	8	16	24
Anguiliformes	Atheriniformes	46	9	12
Atheriniformes	Cypriniformes	7	501[b]	786[b]
Atheriniformes	Tetraodontiformes	46	13	17
Atheriniformes	Perciformes	148	25	33
Atheriniformes	Gasterosteiformes	36	20	27
Gasterosteiformes	Tetraodontiformes	8	20	30
Gasterosteiformes	Perciformes	33	32	43
Perciformes	Tetraodontiformes	34	25	34
Mean			24	32
Weighted mean			26	35

[a]Values in this table are similar to, but differ from, those in Barnthouse et al. (1990) due to differences in the algorithm used, particularly the use of ordinary least squares regression by Calabrese and Baldwin (1994).
[b]Not included in calculations.

TABLE 26.5
Upper 95% Uncertainty Factors Calculated for the 95%
and 99% Prediction Intervals in Table 26.4

Level of Taxonomic Extrapolation	Prediction Interval	
	95%	99%
Species within genera	10.0	16.3
Genera within families	11.7	16.9
Families within orders	99.5	145.0
Orders within classes	64.8	87.5

Source: From Calabrese, E.J. and Baldwin, L.A., *Environ. Health Perspect.*, 102, 14, 1994. With permission.

example, when extrapolating between a mouse test and equivalent effects on a mammalian carnivore (order Carnivora), one would divide the mouse test endpoint by 64.8 to be 95% certain of including the carnivore species 95% of the time (Table 26.5).

An alternative approach was developed for the calculation of tier II water quality values (Chapter 29). These values were derived by applying resampling statistics to the data sets used to derive water quality criteria to obtain distributions of the ratios of the lowest concentration in a small sample of toxicity data to the criteria values. Factors were derived from these ratios that should protect 95% of aquatic invertebrate and fish species with 80% confidence. This method is best used to develop conservative screening benchmarks.

26.2.7 ALLOMETRIC SCALING

After factors, the type of quantitative extrapolation model used most commonly by human and wildlife toxicologists is allometric scaling. These models are based on the assumption that all members of a taxon have the same response to a chemical, but they differ in size and in processes that are related to size. The most commonly used allometric model is a power function of weight:

$$E_x = aW^b \qquad (26.9)$$

where E_x is the effective dose or concentration at some organism weight W. This model has been adopted by toxicologists because the metabolism and excretion of drugs and other chemicals are approximated by that function. The EPA and others have used the three-fourth power for piscivorous wildlife (EPA 1993f; Sample et al. 1996c). Allometric scaling may be applied to aquatic species (Patin 1982), but it is used almost entirely for wildlife extrapolations. Reviews of the theory and application of allometric scaling are provided by Fairbrother and Kapustka (1996), Davidson et al. (1986), and Peters (1983).

Allometric scaling is simple to apply, and it has a stronger scientific basis than uncertainty factors. If a toxic dose (D) and the body weights of both the test and endpoint species are known and an appropriate scaling factor is selected, the toxicity value for the wildlife species may be calculated (Sample et al. 1996c):

$$D_w = D_t \left(\frac{bw_t}{bw_w}\right)^{1-b} \qquad (26.10)$$

Confidence in allometric scaling is limited because current models are based on a few chemical classes (i.e., mammalian values are based primarily on drugs, and avian values are based primarily on cholinesterase-inhibiting insecticides). In addition, avian models are based on acute lethality. Because allometric scaling factors can vary widely among chemicals (Mineau et al. 1996), and because the toxic mode of action differs for acute and chronic exposures to the same chemical, the current practice of applying the same scaling factors for all chemicals and types of exposures may produce inaccurate estimates (Fairbrother and Kapustka 1996).

26.2.8 Toxicokinetic Modeling for Extrapolation

Toxicokinetic modeling, which is used to estimate body burdens and internal exposure levels (Section 22.9), also provides a means of extrapolating among species or life stages based on differences in physiology and the volumes of organs and other compartments. Toxicokinetic models are used to extrapolate rodent test data to humans (Clewell et al. 2002), but have seldom been used in ecological risk assessments. The conceptual approach is shown in Figure 26.5. It begins with the results of a conventional toxicity test, such as a laboratory rat reproductive test, expressed as the administered dose in mg/kg/d of the chemical of concern that causes a prescribed effect, such as 20% reduction in the number of viable pups. A toxicokinetic model is then used to estimate the corresponding internal exposure for the female rat. This may be the concentration in a particular compartment such as blood but is more likely to be a whole body concentration (mg/kg body weight). The internal concentration may be a peak concentration for a single dose or the equilibrium dose for a continuous exposure. This must be converted into an equivalent effective internal dose for females of the endpoint species (e.g., the internal concentration causing 20% reduction in the number of viable mink pups). Toxicodynamic models could be used to represent the induction of effects by the internal exposure (Section 23.3), but in practice, even in human health risk assessments, the effective internal concentrations are typically assumed to be equal. A toxicokinetic model for the endpoint species is then used to convert the internal concentration to an administered dose. This dose can then be related back to ambient concentrations in food or abiotic media using exposure models (Section 22.9).

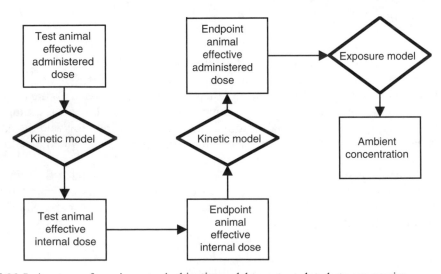

FIGURE 26.5 A process for using a toxicokinetic model to extrapolate between species.

The primary problem with this approach is that it is data-intensive, and most of the physiological parameters needed for most species are unavailable or poorly developed. However, despite data limitations, Fairbrother and Kapustka (1996) suggest that the use of even simple toxicokinetic models may significantly increase accuracy of interspecies extrapolations. One solution to data limitations is to use allometric relationships to estimate the needed parameters (Clewell et al. 2002). Another strategy is to focus on a few parameters that are likely to vary among species or organisms in a way that influences toxicity. For example, variance in the toxicity of chemicals to fish may be largely controlled by weight, the ratio of gill area to weight, and, for hydrophobic chemicals, by fraction of weight composed of fat (Lassiter and Hallam 1990). That is, when exposed to hydrophobic chemicals, big, fat, sluggish fish are most likely to survive.

Toxicodynamic models that relate internal exposure to response (Section 23.3) also have the potential to contribute to extrapolation among species and life stages. However, they are much less developed for that purpose than toxicokinetic models.

26.2.9 MULTIPLE AND COMBINED APPROACHES

These various potential methods for extrapolating toxic effects among species and life stages have traditionally been applied independently and not in a systematic manner. Fairbrother and Kapustka (1996) suggested that less reliance be placed on a single approach (e.g., allometric models) for all species and chemicals and that multiple approaches be applied to the problem of wildlife extrapolation.

To address all of the important extrapolations in an assessment, it is often necessary to combine approaches. For example, SSDs are used to extrapolate from individual species to communities, but the test endpoints used in the SSDs often do not represent the assessment endpoint. Hence, the individual species values for the model may be adjusted. For example, in calculating the acute ambient water quality criteria, the US EPA applies a factor of 2 to the 5th percentile of the SSD of acute values to extrapolate from 50% mortality to a small percentage (Stephan et al. 1985).

26.3 EXTRAPOLATIONS FOR PARTICULAR BIOTAS

26.3.1 AQUATIC BIOTA

If, as is often the case, the endpoint property for the aquatic biota is species richness or diversity, SSD is an obvious choice of extrapolation model. Modeling results show that continuous exposure to concentrations equal to the chronic value (CV) for a species can cause extinction of that species (Barnthouse et al. 1990). Therefore, the proportion of species for which the CV is exceeded by long-term exposures can be assumed to approximate the proportion of species lost from the community. In addition, because toxicity data are relatively abundant for aquatic organisms, it is often feasible to derive such distributions for individual chemicals. As discussed above, the SSD approach is widely accepted, because it is used for the derivation of water quality criteria. If responses are known to be a function of water chemistry, the individual test endpoints should be normalized to appropriate water chemistry before defining the distribution.

If the endpoint is a property of a particular population rather than a community, the extrapolations using SSDs are performed or interpreted differently. SSDs are still useful because they can be interpreted as probability distributions for effects on an individual species (Section 26.2.3).

Alternatively, one can extrapolate between species by using the appropriate intertaxa regression models or the uncertainty factors derived from them (Table 26.4 and Table 26.5).

That is, if one wanted to predict the toxicity of a chemical to brook trout (a salmonid) from test data for fathead minnow (a cyprinid), one could divide by 20 to be 95% certain of not underestimating the sensitivity of brook trout (or any other salmonid). If the desired taxonomic regression is not available, the appropriate generic factor (which would be 26 in this case of an interorder extrapolation) would be applied. These two approaches for estimating effects on particular species or taxa (SSDs and taxonomic regressions) have different weaknesses, and it is not clear which works better in practice. However, the taxonomic regressions and the factors derived from them require test data for only one species, so they are more generally useful. The factors are conservative and may estimate effects levels that are below background. For estimation of probabilities of effects, one should use the original regression models to estimate means and variances (see Table 7.4 in Suter 1993a).

Allometric models may also be used to extrapolate to specific aquatic species (Patin 1982; Newman and Heagler 1991; Newman et al. 1994). This approach has not been accepted in practice, because taxonomic differences have been perceived to be more important than size. At least, it seems likely to be useful for extrapolation among taxonomically similar species.

Acute–chronic extrapolations may be made with regression models or factors. Acute–chronic regression models for aquatic fish and invertebrates are presented in Suter (1993a), and factors derived from them are presented in Table 26.3. These factors are based on including the CV or EC_{25} with 95% or 99% confidence. Alternatively, CVs can be estimated with 80% confidence of not overestimating their value using a factor of 18 (Host et al. 1991). Calabrese and Baldwin (1993) recommend generic 95% and 99% uncertainty factors of 50 and 200 for acute–chronic extrapolations, based on the weighted means in Table 26.3. Analysis of a European data set yielded various acute/chronic factors including a 90% confidence LC_{50} to no observed effect concentration (NOEC) factor for organic chemicals and aquatic animals of 24.5 (Lange et al. 1998). Any of these factors are adequate if one is trying to conservatively estimate a chronically toxic concentration of a chemical in a screening assessment or to support an assessment based primarily on other lines of evidence. If acute lethality data are available for at least three exposure durations, extrapolation to low response rates and long durations can be used to estimate lethal thresholds (Section 26.2.5). If compelled by circumstances to estimate risks to aquatic organisms using only an LC_{50}, one of the regression equations in Suter (1993a) or Sloof et al. (1986) with its associated uncertainty should provide a better estimate of the chronic effects threshold than a conservative generic factor.

In some cases, multiple extrapolations are required including those between taxa and life stages. Such multiple extrapolations may be dealt with by chains of factors or by chains of regression models (Barnthouse et al. 1990; Calabrese and Baldwin 1993; Suter 1993a). Thresholds for lethality obtained by temporal extrapolation (Section 26.2.5) may be converted to estimates of thresholds for reproductive and other sublethal effects by multiplying by a factor of 0.1 (Ellersieck et al. 2003). However, methods for multiple extrapolations are not explained in detail here.

Mechanistic models, particularly the biotic ligand model (Section 22.2), are beginning to be used in aquatic ecological risk assessments. They can be used to extrapolate among water chemistries and, potentially, among species and life stages.

26.3.2 BENTHIC INVERTEBRATES

Species or community sensitivity distributions can be derived for toxicity of individual chemicals to benthic invertebrates based on published test results. In the case of exposure of benthic invertebrates to sediment pore water, the effects distributions are the same as the SSDs for aquatic biota. The use of aqueous data to evaluate effects on benthic species is based on data suggesting that benthic species are not systematically more or less sensitive than water column species (EPA 1993a).

In the case of exposure of benthic invertebrates to chemicals in whole sediment, the effects distributions are for species/sediment combinations and community/sediment combinations. This is necessary because it is not possible to adequately control for the effect of sediment characteristics, including co-contaminants in field-collected sediments, on toxicity. The most prominent examples of effects distributions for benthic invertebrates are those used to derive screening benchmarks for sediment-associated biota (Long et al. 1995; MacDonald et al. 1996). The effects in those distributions include taxa richness, diversity, density, mortality, growth, respiration, behavior, and suborganismal effects. As a result, those distributions only indicate an unspecified level of an unspecified effect. This is adequate for screening purposes, but not for definitive risk characterization. For definitive assessments, such nonspecific distributions can be parsed into distributions of thresholds for specific effects. For example, Jones et al. (1999) developed distributions of community-level effects and lethality from the sediment toxicity data presented in MacDonald et al. (1996) and Long and Morgan (1991).

26.3.3 WILDLIFE

Wildlife risk assessors have followed health risk assessors in using uncertainty factors (e.g., Banton et al. 1996, Sample et al. 1996c, Hoff and Henningsen 1998). As in health risk assessments, uncertainty factors are used to account for specific extrapolations such as interspecies, acute–chronic, laboratory–field, lowest observed adverse effect level/no observed adverse effect level (LOAEL–NOAEL). A general extrapolation model for wildlife has been proposed by Hoff and Henningsen (1998):

$$D_w = D_t/(UF_a * UF_b * UF_c * UF_d) \tag{26.11}$$

where D_w represents the estimated critical chronic dose for an endpoint wildlife species, and D_t is the literature-derived toxicity value for the test species. UF_a accounts for intertaxon variability and can range from 1 if the test and wildlife species are the same to 5 if the test and wildlife species are in the same class but in different orders. Uncertainty in study duration is represented by UF_b, which ranges from 1 to 15 for the range from chronic to acute. UF_c accounts for the type of toxicity data available and ranges from 0.75 for NOELs to 15 for severe or lethal effects ($\gg ED_{50}$). Finally, UF_d addresses other modifying factors (e.g., species sensitivity, laboratory–field extrapolation, intraspecific variability) and may range from 0.5 to 2. Hoff and Henningsen (1998) recommend reporting quantitative risk results only if total UF < 100. For total UF > 100, only qualitative (e.g., presence/absence, low, medium, high) estimates of risk should be reported. As with other uses of multiplicative factors, this model suffers from inappropriate error propagation and poorly defended factors. However, it is equivalent to current practice in human health risk assessment.

Allometric scaling has been commonly applied to wildlife as well as humans. Factors of 0.66 to 0.75 have been used for extrapolating from laboratory test species to humans and wildlife (EPA 1992c, 1993e; Sample et al. 1996c). Use of either the 0.66 or 0.75 scaling factor is conservative for humans and most endpoint mammalian wildlife in that large species such as deer are estimated to be more sensitive than the rodents, which are typically used in mammalian toxicity testing. However, small wild mammals are estimated to be less sensitive than laboratory rats or dogs. Common avian test species such as chickens and mallard ducks are much larger than most birds.

Use of allometric models for birds with the same exponents as mammals was supported by allometric models of avian physiology and pharmacology (Peters 1983; Pokras et al. 1993). In contrast, allometric regression analyses of 37 pesticides on 6 to 33 species of birds found that for 78% of chemicals the exponent was greater than 1 with a range of 0.63 to 1.55 and a mean of 1.1 (Mineau et al. 1996). Because of that mean and because scaling factors for the majority of the chemicals evaluated were not significantly different from 1, a scaling factor of 1 appears to be appropriate for interspecies extrapolation among birds (Sample and Arenal 1999). However, because of the apparently chemical-specific character of scaling factors, Sample and Arenal (1999) provided such factors for 138 chemicals for acute lethality to birds and 94 to mammals. In the absence of a scaling factor for the chemical of interest or for a similar chemical, they recommend generic factors of 1.2 for birds and 0.94 for mammals. Scaling factors should be developed for chronic exposures and for effects other than lethality.

SSDs have been recommended for assessment of risks to birds (Baril et al. 1994; Ecological Committee on FIFRA Risk Assessment Methods 1999b). However, sufficiently large data sets are limited primarily to acutely lethal doses of pesticides. When data are insufficient, safety factors derived from SSDs may be used (Baril et al. 1994). These factors are the geometric means, across chemicals, of the ratios of $LD_{50}s$ for common test species to the TLD_5 (the 5th percentile of the avian SSD) (Table 26.6).

Regression models are also potentially useful for wildlife. Intertaxa regressions are available for avian and mammalian wildlife species and families (Asfaw et al. 2003; Mayer et al. 2004). Because there are no avian test results for most chemicals, regressions of available avian data against rat data allow the use of the much more abundant data for laboratory rats. For example, regressions of ring-necked pheasant $LD_{50}s$ and mallard duck $LD_{50}s$ against rat $LD_{50}s$ for 24 organophosphate cholinesterase-inhibiting pesticides were used to assess risks to birds from disposal of organophosphate cholinesterase-inhibiting chemical warfare agents (Sigal and Suter 1989). The result for mallards was:

$$\log \text{ mallard } LD_{50} = 1.33(\log \text{ rat } LD_{50}) - 0.58 \qquad (26.12)$$

($r^2 = 0.47$) indicating that mallards are more sensitive than rats to these chemicals. Pheasants, however, were about equal in sensitivity to rats.

26.3.4 SOIL INVERTEBRATES AND PLANTS

Regression models were developed for pairs of plant taxa by Fletcher et al. (1990). For each of 16 chemicals, $EC_{50}s$ of between 7 and 36 plant species were compared. The variation in sensitivity ranged from 3.5-fold for linuron to 316-fold for picloram. Out of almost 300 chemical–plant species combinations, 59% of $EC_{50}s$ varied by less than a factor of 5 from other $EC_{50}s$ for the same chemical. Plants that were closely related, taxonomically, had similar sensitivities to the same chemical. No trends in the relative sensitivity of various species, genera, or families to different chemicals were observed. In this study interspecies variability in toxicity was much higher than the variability associated with the extrapolation from greenhouse to field. Hence, when assessing risks to a particular plant species, it is apparently more important to use data for a similar species than for similar conditions. However, these results are based on foliar applications of herbicides.

Since assessment endpoints for soil invertebrates and plants usually include community properties, SSDs would seem appropriate. The OAK Ridge National Laboratory (ORNL) benchmarks for exposure to contaminants in soil used SSDs for earthworms and plants (Efroymson et al. 1997a,b) (Figure 26.6). Because the variance of toxicity among soils can be significant and cannot be factored out, soil type is a source of variability in these distributions. Hence, the points in Figure 26.6 are species–soil type combinations, and the distribution is the distribution of effective concentrations across species and soils. As a

TABLE 26.6
Safety Factors for Commonly Tested Bird Species. An LD_{50} for One of These Species Divided by the Geometric Mean Safety Factor Should Be Protective of 95% of Avian Species with the Indicated Reliability

Safety Factor	Pheasant	Mallard	Bobwhite	Japanese Quail	Red-Winged Blackbird	Starling	House Sparrow	Common Grackle	Rock Dove	Red Partridge	Grey Partridge
Geometric mean	16.8	10.9	15.2	17.1	5.87	19.8	10.7	9.26	13.1	21.6	10.3
Maximum	298	113	141	174	18.7	1250	43.9	48.42	55.2	87.8	79.8
Minimum	2.00	2.12	2.40	3.10	2.28	3.76	2.13	2.13	3.51	10.5	3.58
n	22	22	16	22	22	22	22	22	22	7	7
%Reliability[a]	41	36	38	50	41	41	45	50	50	43	29

Source: Modified from Baril, A., Jobin, B., Mineau, P., and Collins, B.T., *A Consideration of Inter-Species Variability in the Use of the Median Lethal Dose (LD_{50}) in Avian Risk Assessment,* No. 216. Hull, PQ, Canadian Wildlife Service, 1994. With permission.

[a]Percentage of chemicals for which the mean safety factor is sufficient to obtain 95% protection.

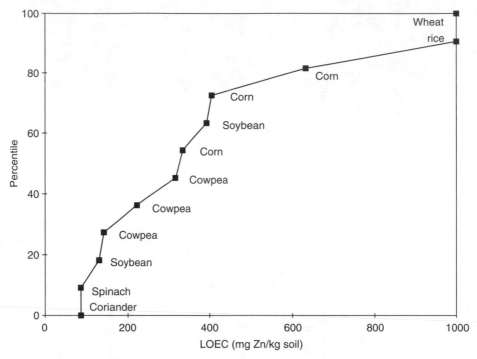

FIGURE 26.6 Cumulative distribution of LOECs for plants exposed to zinc in soil. Effects included are changes in the mass of whole plants of plant parts. (From Suter, G.W., II, Efroymson, R.A., Sample, B.E., and Jones, D.S., *Ecological Risk Assessment for Contaminated Sites*, Lewis Publishers, Boca Raton, FL, 2000. With permission.)

result, single plant species that are tested under different conditions have different LOECs in the distribution. An untested species in a particular soil may be assumed to be a random draw from the distribution, or the distribution may represent the proportion of species in a plant community that is likely to be affected by a particular concentration of a chemical given uncertainty concerning the influence of soil type. That uncertainty is reduced in the Netherlands by normalization to reference soil (Sijm et al. 2002).

26.3.5 SOIL PROCESSES

In the Netherlands, distributions of toxicity values for microbial processes and enzymatic activity are used, along with SSDs, to derive regulatory values for soil (Crommentuijn et al. 2000; Sijm et al. 2002). Studies of the same process in different soils are treated as separate observations, i.e., each soil ecosystem is considered equivalent to a species in an SSD. The NOEC and ECx concentration are normalized to standard soil to minimize the effects of soil chemistry on bioavailability and toxicity.

26.3.6 WATER CHEMISTRY

The properties of ambient and test waters such as salinity, pH, and hardness can influence the forms of chemicals to which organisms are exposed and the sensitivity of aquatic organisms to the exposure. The differences between test waters and waters to be assessed may be addressed in various ways. The simplest is to use data from tests conducted in waters that are similar to site water, but this raises the question of sufficient similarity. The question is particularly stark for the case of extrapolation from freshwater to saltwater. Few toxicity data are available for saltwater relative to freshwater, but there are few test data from freshwater sediments relative to estuarine sediments. Analyses have shown that the differences are small relative to the differences among species (Klapow and Lewis 1979; Hutchinson et al. 1998; deZwart 2002), but these differences may be important (Stephan et al. 1985; Hutchinson et al. 1998).

The second approach is to perform a standard test in site water and in standard test water and use the ratio of the results to adjust a water quality criterion or other test data for the effects of site water chemistry. Guidance for deriving and applying this water effect ratio is provided by the US EPA (Office of Science and Technology 1994, 2001).

A third approach is to model the influence of water chemistry. One example is the use of empirical models to adjust metal toxicity data for water hardness using regressions of LC_{50}s against test water hardness (Stephan et al. 1985; Pascoe et al. 1986). Metal speciation models have been used with the biotic ligand model to adjust the toxicity of certain metals for differences in the chemistry of freshwaters (Section 22.2).

26.3.7 SOIL PROPERTIES

The risk assessor should be aware that bioavailability in soil from the contaminated site may be substantially different from the bioavailability in published soil tests. As stated in Section 22.4.1, aged organic chemicals are typically less available and less toxic to biota than organic chemicals freshly added to soil in published toxicity tests (Alexander et al. 1995); thus the toxicity at the contaminated site may be overestimated if a published toxicity test of a chemical freshly added to soil is emphasized too heavily in the assessment. The risk assessor can make adjustments to observed toxic concentrations to account for differences in soils or chemical speciation. The variance in toxicity among natural soils may be reduced by normalizing the test soil concentrations to match normalized site soil concentrations (Section 22.3) (Sijm et al. 2002). Or free metal activities in soil solution may be estimated, potentially improving the precision of toxic thresholds for plants, soil invertebrates, or microbial processes (Sauvé 2001). The assessor may be more liberal in including tests in screening assessments (e.g., in the derivation of screening benchmarks) than in definitive assessments. In definitive assessments, soil type and chemical speciation should be factors in decisions about the acceptability of data.

26.3.8 LABORATORY TO FIELD

Many studies have been conducted attempting to relate conventional laboratory toxicity test results to responses in the field. Unfortunately, most of them have been intended to test the validity of laboratory toxicity data rather than to generate extrapolation models. The simplest formulation of the validation problem is, do classifications of field sites as impaired or unimpaired correspond to the classifications predicted from standard laboratory tests? The results of these attempts at validation are ambiguous at best, largely because field studies do not provide "true" results as a standard for comparison. Field experiments and biological surveys are highly inconsistent in their design and endpoints; they are pseudoreplicated, poorly replicated, or unreplicated; they typically include only one season; although they often include multiple responses, they miss many taxa, attributes, and processes; and they tend to be insensitive, so that when effects are found they are often severe (Neuhold 1986; Chapman 1995; LaPoint 1995; Luoma 1995; deVlaming and Norberg-King 1999). Because mechanistic understanding is usually lacking, it is unclear whether deviations from expectations are due to failure of the laboratory tests as predictors or to factors that are not relevant to evaluating the validity of a test. In addition, biological surveys are subject to confounding by differences among sampling locations other than the laboratory-tested toxicants.

Despite the difficulties, reviews of comparisons of laboratory and field studies typically conclude that when ambient dilution water or sediment is used, laboratory toxicity is usually indicative of field effects, and even tests with standard laboratory water are generally related to field effects (deVlaming and Norberg-King 1999; Long 2000). When laboratory and field results disagree, it is not clear that the laboratory test is erroneous. Even when they agree, it is not clear that the agreement is due to actual mechanistic correspondence. For example, field effects may be due to low dissolved oxygen rather than toxicity. Better comparisons are possible when consistent endpoints are used (e.g., percentiles of SSDs for invertebrates in laboratory and ditch mesocosm tests), but even then generalizations about the validity of laboratory tests are elusive (van den Brink et al. 2002).

An alternative approach to the relationship between laboratory and field test results is to simply regress the former against the latter. Sloof et al. (1986) regressed NOECs from aquatic mesocosm tests (NOEC$_e$) against the lowest reported single species acute LC$_{50}$ or EC$_{50}$ and against the lowest reported single species chronic no effect concentration (NOEC$_s$) for the same chemical and obtained:

$$\log \mathrm{NOEC}_e = -0.55 + 0.81 \log \mathrm{LC}_{50} \qquad (26.13)$$

$$\log \mathrm{NOEC}_e = 0.63 + 0.85 \log \mathrm{NOEC}_s \qquad (26.14)$$

The units are µg/L, the correlation coefficients are 0.77 and 0.85, respectively, and the arithmetic-scale PIs are ± 0.86 and 0.35, respectively. The authors concluded that the most sensitive responses in the mesocosms are more sensitive than the most sensitive acute lethality test but less sensitive than the most sensitive single species chronic test. Emans et al. (1993) derived a model similar to Equation 26.14, using different data selection criteria. They concluded that organisms in field conditions responded at concentrations similar to those that affect species in the laboratory. Although these models would not accurately predict effects in particular cases, they could be used to suggest the approximate range within which effects would be expected in field systems. They also suggest that although more species are exposed in the field and indirect effects and complex exposures occur in the field that do not occur in simple laboratory aqueous tests, the chemicals that are most toxic in the laboratory are also most toxic in the field and thresholds for effects are not greatly different, so toxicity in the ecosystem context is not completely unpredictable.

When developing empirical models of field effects, it is important to select field data that are measures of the assessment endpoint, because different community and ecosystem responses have very different sensitivities. Even different community metrics derived from the same data set can give very different relative sensitivities. For example, the sediments of the Louisianian province were most affected of three EMAP coastal provinces based on a benthic index score, least affected based on species richness, and intermediate based on infaunal abundance (Long 2000).

One can certainly imagine more sophisticated and potentially more predictive approaches to empirically modeling the relationships between laboratory and field responses. In particular, the use of chemical speciation models or toxicokinetic exposure models to normalize field and laboratory contaminant concentrations, the use of tests with more life stages and responses, and the use of body burdens as measures of exposure are likely to improve predictions. However, more progress may be obtained using laboratory data with mechanistic models of exposure (Chapter 22), population dynamics (Chapter 27), and ecosystem processes (Chapter 28).

26.4   SUMMARY

Since the first edition of this book was published in 1992, the situation has changed little with respect to extrapolation models for estimating risks to attributes of nonhuman organisms. The use of SSDs has become common, but otherwise most assessments still rely on selection of most relevant data and the occasional factor. There is no consensus that this actually represents good practice or that a particular alternative is preferable. Looked at positively, it gives assessors considerable freedom to select the approach that seems best to them, or to develop new approaches.

27 Population Modeling

Lawrence W. Barnthouse

If all significant environmental decisions could be based on predicted or observed effects of chemicals or other agents on organisms, there would be no need for this chapter. However, there are many important decisions for which knowledge of organism-level effects simply is not enough. Some species may, because of their life history or because of their greatly reduced abundance, be more at risk than others, given the same sensitivity of individuals. Mortality to certain individuals is unavoidable, in which case risk managers may be interested in the amount of mortality (or growth reduction) that can be tolerated by the exposed species. It may be necessary to know whether the combined effects of agents affecting several different life stages may reduce the abundance of populations or lead to increased risk of extinction. It may be important to forecast the rate of recovery of populations following an accident or a remedial action.

The toxicity tests and extrapolation models discussed in Chapter 24 and Chapter 26 are insufficient for addressing these problems. From a population viewpoint, the death or impairment of an organism is meaningless, because most organisms die after brief lives (on a human time scale) and few organisms achieve their full reproductive potential or maximum growth. Ecologists have long known that natural populations of many organisms frequently are subjected to extreme environmental variations that cause mass mortalities, and many species are composed of isolated pockets of organisms that occasionally become extinct, and are later reestablished. Modern ecological theory views disturbance and instability as normal, and the constant conditions observed in laboratory experiments to be highly unrealistic.

Many questions of interest in risk assessment relate to effects on the abundance, production, or persistence of populations. Responses of populations cannot be predicted from toxicity tests alone. The response of a fish population, for example, to a contaminant exposure will depend on the spatial pattern of exposure as related to the distribution of individuals in time and space, on the magnitude of other impacts that are imposed (including especially harvesting by fishermen), and the inherent capacity of the population to "compensate" or to evolve in response to exposure. The response of a soil invertebrate population periodically exposed to pesticides depends not only on spatial patterns of exposure and dose–response relationships, but also on the reproductive capacity of the population and on availability of nearby sources of immigrants that can replace organisms killed by the exposure.

Interest among both ecotoxicologists and risk assessors in techniques for predicting responses of populations to chemical exposures has grown rapidly in recent years. Between 1980 and 1990, for example, the journal *Environmental Toxicology and Chemistry* published only three papers dealing with effects of chemicals on populations. This same journal published more than 50 such papers between 1996 and 2005. The increase reflects both renewed awareness on the part of scientists and management agencies that population-level effects are important and can be quantified, and the emergence of new techniques for quantifying population dynamics, many drawn from the emerging field of conservation biology.

For small, short-lived species (e.g., microbes, cladocerans, and some other small arthropods), effects of chemicals on critical population parameters can be measured directly using laboratory experiments. However, these organisms are unrepresentative of the great majority of populations of interest in ecological risk assessment. In a few rare cases, effects of chemicals on the same parameters have been estimated from long-term field studies (e.g., Barnthouse et al. 2003). However, most population-level ecological risk assessments have and will continue to involve using mathematical models of populations to link organism-level experimental data to population-level responses. The objective of this chapter is to show how such models can be developed and applied.

The problem of quantifying population-level responses to death or impairment of organisms is not new. Fish and wildlife managers have struggled for decades with the problem of defining the effects of harvesting, habitat modification, and disease on the abundance and stability of exploited populations. Recently, conservation biologists have developed novel methods for quantifying the effects of environmental variability, habitat fragmentation, and reduced population sizes on the persistence of rare and endangered species. The methods used by resource managers and conservation biologists provide a useful frame of reference for assessments of other sources of stress, including toxic chemicals.

This chapter provides basic definitions, briefly describes some of the models that are especially relevant to assessing ecological effects of toxic chemicals, and discusses some specific applications. Readers interested in a more in-depth treatment of the principles of population biology are advised to consult the excellent textbooks and reviews that are available. Most modern undergraduate-level ecology textbooks (e.g., Begon et al. 1999; Krebs 2002) explain the fundamentals of population analysis. However, many of the most readable and thorough accounts are older. Among these, the most widely cited (and most frequently consulted by the author of this chapter) is the classic text by Andrewartha and Birch (1954). The texts of choice for fish population studies are Hilborn and Walters (1992) and Quinn and Deriso (1999); Bolen and Robinson (2002) provide a good general discussion of wildlife population biology. An overview of the new subdiscipline of "metapopulation biology," which is of substantial relevance to ecological risk assessment, is provided by Hanski and Gilpin (1996). For the mathematically inclined, Caswell (2001) provides a thorough exposition of the theory of matrix population models. Readers interested in developing computer simulation models of populations should consult Swartzman and Kaluzny (1987; unfortunately now out of print), or Jorgensen and Bendoriccio (2001).

Other authors have also written about population-level ecological risk assessment methods. Particularly noteworthy are the books by Newman (2001) and Pastorok et al. (2002). These authors provide useful alternative perspectives concerning issues addressed in this chapter.

One significant topic that is not discussed at all in this chapter is population genetics. Historically, the two principal branches of population biology, demography and population genetics, were intimately related, and many of the pioneers in theoretical population biology made fundamental contributions to both. Some, but not all, of the modeling approaches discussed in this chapter are used in population genetics as well as demography. Impacts of chemicals on the genetic composition and fitness of populations are topics that have been discussed in the ecotoxicological literature (Forbes 1999; Newman 2001); however, genetic impacts on populations have not yet been an issue in any major environmental controversy and no guidance has been developed concerning how risk assessments should or could address these impacts. Until ecological risk assessments begin to more routinely consider population genetics, it seems premature to include the topic in this chapter. Thus, the scope of this chapter is limited to demography.

27.1 BASIC CONCEPTS AND DEFINITIONS

The fundamental objective of population biology is to infer characteristics of groups of organisms (i.e., the populations) from organismal characteristics. The population characteristics of interest include total numbers or biomass, rate of population growth or decline, age, size, sex, or genotypic composition of the population, and the probability that the population will persist into the future. Only a subset of these characteristics is likely to be important in any single assessment. Managers of exploited populations may be interested in the number of organisms available to be harvested; conservationists may be interested in the probability of extinction of a population of a given size. These population characteristics are simply collective expressions of the state and fate of constituent organisms, such as reproduction rates, growth and development rates, and probabilities of death. The individual characteristics are, in turn, governed by (1) innate processes such as development and senescence; (2) the effects of the physical environment; (3) interactions with other organisms; and (4) deliberate or unintentional actions by man.

27.1.1 POPULATION-LEVEL ASSESSMENT ENDPOINTS

Population studies address a variety of endpoints of interest in risk assessment. The most basic of these are the endpoints traditionally used in management of exploited populations: total population density or biomass, age or size distribution, and sustainable rate of harvest. In resource management, models relate these endpoints to management actions such as harvest quotas, size limits, or harvest season lengths. Due to concerns about the conservation of endangered species, population biologists have recently formulated new models that contain environmental or demographic stochasticity and spatial heterogeneity, to address endpoints related to persistence of populations in variable environments. These models are used to estimate frequencies or probabilities of extinction within a given time period or expected time to extinction, as functions of population size, size of habitat required, or degree of habitat fragmentation.

27.1.2 IMPLICATIONS OF LIFE HISTORY FOR POPULATION-LEVEL ECOLOGICAL RISK ASSESSMENT

Are some species inherently more vulnerable to environmental stress because of their life history? Are some life stages more important than others to the survival of a population? To date, few ecotoxicologists have attempted to address the influence of life history on the vulnerability of populations to toxic chemicals. Exceptions include Barnthouse et al. (1990), who investigated effects of chemical exposures on two fish species with contrasting life histories and harvesting patterns; Spromberg and Birge (2005), who performed a similar investigation of five general fish life history types; and Calow et al. (1997), who developed a quantitative framework for relating life history traits to population-level effects of chemical exposures. However, the influence of various life history traits on population growth rates was a major topic of theoretical research during the 1960s and 1970s (see review by Stearns 1977). Both theoretical analysis and management experience have shown that long-lived vertebrates such as large mammals, predatory birds, and whales are more sensitive to mortality imposed on adults than are short-lived, highly fecund organisms such as quail and anchovies. Conversely, short-lived species are often vulnerable to short-term catastrophes that affect critical life stages. Populations in which survival or reproduction are strongly related to the density or abundance of the population should, on theoretical grounds, be less vulnerable than populations with a low degree of density-dependence. Qualitatively, it seems clear that the response of a population to a toxic chemical is influenced by the preexisting patterns of natural environmental variability, the age-specific survival and reproduction of the organisms, and the intensity and duration of exposure.

27.1.3 Representation and Propagation of Uncertainty

Chapter 5 discusses a variety of sources of uncertainty that are of interest in ecological risk assessment. Sources potentially important for population analysis include (1) environmental variability in time and space; (2) variations in sensitivity among individuals and life stages; and (3) stochastic birth and death processes. Interindividual and inter-life-stage variability are discussed in Chapter 26. Stochastic birth and death result from the fact that each organism has an indeterminate life span, even if the average life span for the population can be very precisely estimated. In practice, random birth and death processes, usually termed "demographic stochasticity," are important only in small populations (e.g., 50 individuals or fewer). Even for small populations, Goodman (1987) showed that this source of uncertainty is usually quite small compared to environmental variability.

Temporal environmental variability is readily incorporated in population models. Both periodic and stochastic variations have been studied. The principal mathematical tools available include time series analysis and stochastic modeling. These approaches may be used to quantify environmental variability (e.g., to estimate a periodic function of some important driving variable such as temperature or rainfall), or to estimate probability distributions for temporally varying population parameters (e.g., mortality rates). Many of these techniques are mathematically complex and well beyond the scope of this book. Time series analysis is widely used in economics and engineering as well as in ecology; Brockwell and Davis (2003) discuss many of the widely used procedures and software packages. Stochastic population models are discussed by Caswell (2001), and book-length treatments have been written by Nisbet and Gurney (1982) and Tuljapurkar (1990). The problem of fitting stochastic population models to time series data is still an active area of research in population biology, so none of the available textbooks provides a fully up-to-date treatment. For the purpose of typical risk assessments, however, theoretical elegance is often unnecessary. Techniques for using Monte Carlo modeling population-level consequences of environmental variability and parameter uncertainty have been available for more than 20 y (e.g., O'Neill et al. 1982; Barnthouse et al. 1990; Bartell et al. 1992). The popular RAMAS population and ecosystem modeling software (available from Applied Biomathematics, Setaucket, New York, web address http://www.RAMAS.com/) was specifically designed for this purpose.

Spatial variability can now readily be addressed using metapopulation models, i.e., models in which populations are represented as sets of semi-isolated subpopulations, each with potentially different rates of reproduction and mortality that are linked through the processes of immigration and emigration. Although the term "metapopulation" was first used in the 1960s (Levins 1969), most of the methods used to model metapopulations have been developed since 1990. As noted above, Hanski and Gilpin (1996) provide a good introduction to the theory of metapopulations. The most widely used software for metapopulation modeling is the VORTEX model described by Lacy (1993); however, a metapopulation version of RAMAS is also available. A related class of models, termed "spatially explicit" models represents populations either as groups of organisms or as individual organisms interacting on a spatial grid (e.g., Liu 1993; Turner et al. 1994). These techniques provide entirely new ways of quantifying ecological risks of spatially heterogenous chemical exposures; an example is discussed later in this chapter.

27.1.4 Density Dependence

Population regulation has been a fundamental problem in population biology since its inception. What prevents populations with high reproductive rates from increasing without bounds? How do fish and wildlife populations persist in the face of intensive exploitation by humans? The simplest answer to these questions is that in many, and perhaps in most or all, populations, either mortality, reproduction, or both change with the size of the population. When numbers are high, mortality increases and reproduction decreases; when numbers are low, mortality decreases and reproduction increases. Many empirical studies have documented effects of density on growth or reproduction of organisms; the widespread existence of density-dependence at the level of organisms is beyond doubt. However, the importance of these mechanisms for stabilizing populations and ensuring their persistence in variable environments is still widely debated. Rose et al. (2001) discussed the difficulty of quantifying effects of density-dependence in fish, the taxonomic group that has been the most intensively studied. In spite of the fact that harvested populations of fish could not sustain themselves if survival or reproduction was not strongly density-dependent, it continues to be very difficult to detect and quantify density-dependence in specific fish populations. In addition, populations of many types of organisms appear to be stabilized through dispersal of organisms between habitat patches, so that the population as a whole may persist indefinitely even though the subpopulations inhabiting individual patches frequently become extinct (den Boer 1968; Wu and Loucks 1995). Regardless of the mechanisms involved, it seems clear that some form of density-dependence, acting either within populations or between interacting subpopulations, is necessary to ensure the persistence of species (Murdoch 1994; Lande et al. 2002).

It is not always necessary to build density-dependence into population models used for management or risk assessment. Fish and wildlife managers have been reasonably successful with density-independent models, provided that the prediction horizon is short and that the population changes modeled are relatively small. For long-term predictions, however, explicit incorporation of density-dependence is usually necessary to provide realistic simulations. Projections from purely density-independent models inevitably grow either to infinite size or decline to zero, even without the imposition of anthropogenic agents such as harvesting or toxic chemicals.

27.2   APPROACHES TO POPULATION ANALYSIS

A variety of approaches to population analysis have been developed over the past several decades. This section provides an overview of the principal methods, with an emphasis on their conceptual relationships and past applications. Representative case studies involving toxic chemicals are presented in Section 27.3.

27.2.1   POTENTIAL POPULATION GROWTH RATE

The simplest approach to population analysis is quantification of the population growth rate. The theory in its present form was developed by Lotka (1924), Fisher (1930), and Cole (1954). Before the 1960s, analysis of the relationships between life history traits and population growth was the only approach to population dynamics used outside fisheries management. The approach requires only a compilation of (1) the fraction of organisms surviving from one age to the next, and (2) the average number of offspring of an organism of a given age. Define l_x as the fraction of organisms surviving from birth to age x, and m_x as the average number of offspring produced by an organism of age x. Suppose the maximum age or organisms in the population is n years. If l_x and m_x are constant, these parameters uniquely determine the relationship between reproduction, mortality, longevity, and population growth. This relationship is mathematically described by Equation 27.1:

$$\sum_{x=1}^{n} e^{-rx} l_x m_x = 1 \qquad (27.1)$$

The parameter r, the coefficient required to make the left side of Equation 27.1 sum to 1, has been termed the "Malthusian parameter," the "intrinsic rate of natural increase," or the "geometric rate of increase." In this chapter it will be termed the instantaneous rate of population change. If r is greater than 0, the population will increase indefinitely. If r is less than 0, the population will decline toward extinction, and if r is exactly 0 it will remain unchanged. It can be shown that if undisturbed, the age composition of this population will converge to a "stable age distribution" in which the fraction of organisms in each age class is the same from each generation to the next. Once this state is achieved, both the population as a whole and the number of organisms in each age class will grow (or shrink) exponentially with time:

$$N_t = N_0 \, e^{rt} \qquad (27.2)$$

where N_t = population time at time t, and N_0 = population size at time 0.

Equation 27.1 and Equation 27.2 are often expressed in alternative forms:

$$\sum_{x=1}^{n} \lambda^{-x} l_x \, m_x = 1 \qquad (27.3)$$

$$N_t = N_0 \, \lambda^t \qquad (27.4)$$

where $\lambda = e^r$ = the finite rate of population change.

The value of r and changes in r that might relate to the decline or extinction of a population are important for population management. Many authors have investigated the sensitivity of r to changes in fecundity or mortality. Mertz (1971) showed that because of its very low reproductive rate, the California condor population was extremely vulnerable to increased mortality of adults due to hunting. Mertz also concluded that management actions designed to increase reproductive success in this population were unlikely to improve the prospects for recovery. Unfortunately, Mertz's prediction proved correct and the California condor became extinct in the wild about 10 y later. During the 1980s, numerous authors used similar methods to assess the viability of northern spotted owl (Dawson et al. 1987; Lande 1988).

In ecotoxicology, the potential population growth rate approach is now frequently used to interpret results of chronic toxicity tests performed using cladocerans and other small, short-lived species (e.g., Daniels and Allan 1981; Gentile et al. 1983; Meyer et al. 1986; Walthall and Stark 1997; Kuhn et al. 2001; Salice and Miller 2003). Measurements of daily survival and reproduction obtained from these tests are sufficient to obtain estimates of r. Changes in r resulting from exposure to toxicants can be used as a relative index of chronic effects on populations. Although the calculated values of r cannot be directly extrapolated to the field, this approach to test data interpretation has the advantage of combining information on survival and reproduction into a single index. Forbes and Calow (1999) compiled a list of 41 such studies, including a total of 28 species and 44 chemicals. In addition to being used to assess risks to populations of the tested species, test-derived estimates of r for multiple species have been proposed as a potential method for deriving water quality criteria that protect aquatic communities (Forbes et al. 2001; discussed later in this chapter).

27.2.2 Projection Matrices

Age-structured or stage-structured projection matrices are an important extension of the potential population growth rate approach. The simplest matrix model is the linear "Leslie matrix" (Leslie 1945; Caswell 2001). The Leslie matrix contains exactly the same information as the potential population growth rate model, but the information is expressed in matrix form. The change in abundance of each population in time can be represented by the matrix equation:

$$N(t) = LN(t - 1) \qquad (27.5)$$

where $N(t)$ and $N(t-1)$ are vectors containing the numbers of organisms in each age class $(N_0, \ldots, N_k)$, and L is the matrix defined by

$$L = \begin{pmatrix} s_0 f_1 & s_1 f_2 & s_2 f_3 & \cdots & s_{k-1} f_k & 0 \\ s_0 & 0 & 0 & \cdots & 0 & 0 \\ 0 & s_1 & 0 & \cdots & 0 & 0 \\ 0 & 0 & s_2 & \cdots & 0 & 0 \\ \cdots & \cdots & \cdots & \cdots & \cdots & \cdots \\ 0 & 0 & 0 & \cdots & s_k & 0 \end{pmatrix} \qquad (27.6)$$

where s_k = age-specific probability of surviving from one time interval to the next, and f_k = average fecundity of an organisms at age k.

The Leslie matrix can also be expressed in a graphical form (Figure 27.1a) in which the different age classes are depicted as nodes and the survival and reproduction parameters are depicted as arrows connecting the nodes.

The matrix analog to Equation 27.4 is

$$N(t) = L^t N(0) \qquad (27.7)$$

where $N(0)$ = age distribution vector at time 0, and L^t = the matrix L raised to the power t.

Leslie (1945) showed that any population growing according to Equation 27.7 will converge to a stable age distribution, after which it will grow according to

$$N(t) = \lambda^t N(0) \qquad (27.8)$$

The parameter λ, which in matrix algebra is termed the "dominant eigenvalue" of the matrix L, is the same finite rate of population change that appears in Equation 27.3 and Equation 27.4.

Alternatively, a matrix of life stages and stage transitions (as opposed to ages and age transitions) can be constructed. Caswell (2001) presents a detailed discussion of the mathematics of stage-classified models. In addition to representing survival and reproduction rates, the coefficients of a stage-classified model can represent probabilities of transition from one size class or life stage to the next.

For example, in a stage-based alternative to the Leslie matrix, an organism alive during any time step might either remain in the same stage or size class during the next time step or transition to the next class. The transition matrix for a model of such as population could be

$$\mathbf{A} = \begin{pmatrix} P_0 & F_1 & F_2 & \cdots & F_k \\ G_0 & P_1 & 0 & \cdots & 0 \\ 0 & G_1 & P_2 & \cdots & 0 \\ 0 & 0 & G_2 & \cdots & 0 \\ \cdots & \cdots & \cdots & \cdots & \cdots \\ 0 & 0 & 0 & G_{k-1} & P_k \end{pmatrix} \tag{27.9}$$

In this matrix, the elements along the diagonal (P_i) are probabilities that an organism will survive and remain in the same class; elements along the subdiagonal (G_i) are probabilities that an organism will survive and transition to the next class. A graphical version of a stage-based model is provided in Figure 27.1b. The projection equation for this matrix is

$$\mathbf{N}(t) = \mathbf{A}^t \, \mathbf{N}(0) \tag{27.10}$$

Ecological examples of stage-based models include Sinko and Streifer (1967, 1969), Taylor (1979), Law (1983), Law and Edley (1990), De Roos et al. (1992), and many other papers cited by Caswell (2001). These models appear especially useful for plants and invertebrates, which have complex life cycles in which population dynamics are more strongly influenced by size and developmental stage than by age. Emlen and Pikitch (1989) used stage-classified models of generalized vertebrate populations to analyze the sensitivity of different types of vertebrate life cycle to mortality imposed on different stages. Munns et al. (1997) developed a

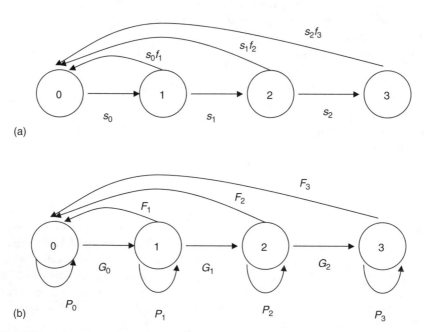

FIGURE 27.1 Life cycle graphs corresponding to age-based (Leslie) and stage-based projection models. (a) In the age-based model, organisms surviving at each time step transition to the next age group. No organisms survive beyond age group 3. (b) In the stage-based model, organisms surviving at each time step may either transition to the next stage or stay in the same stage. In both models, reproduction occurs immediately following the age–stage transition.

stage-based model of a mummichog population exposed to dioxin and polychlorinated biphenyls (PCBs).

The relative influence of different life history characteristics on the population growth can be calculated using a property termed "elasticity" (deKroon et al. 1986; Caswell 2001). Elasticity is a measure of the proportional sensitivity of λ to each element of the population transition matrix (a_{ij}):

$$e_{ij} = (a_{ij}/\lambda)(\partial\lambda/\partial a_{ij}) \tag{27.11}$$

The elasticities of all the elements (a_{ij}) of a population projection matrix sum to 1.0. Spromberg and Birge (2005) and Forbes et al. (2001) used the elasticity index to compare the relative influence of different life history characteristics on the population growth rates of species with different life history strategies.

For a population that is *not* at a stable age distribution, Equation 27.7 or Equation 27.10 can be used to predict the abundance and age distribution of the population over the next few time intervals, given any initial age distribution. Otherwise, as long as the coefficients of the matrices **L** or **A** are viewed as constant parameters, the matrix projection approach is essentially equivalent to the population growth rate approach. The real power of the matrix representation of population dynamics is its flexibility. The coefficients of **L** or **A** can be viewed not as constants, but as random variables or functions of environmental parameters.

These modifications permit entirely new types of analyses, many of substantial value for risk assessment. In the strictly linear and deterministic models discussed so far, the only endpoint that can be addressed is the future trend of abundance: will the population increase or decline in the future, and how sensitive is the rate of increase or decline to changes in life history parameters? Modification of the matrix approach in the ways described below permits assessment of any endpoint for which an operational definition can be formulated.

The remainder of this section deals with matrix models in which the elements are variables rather than constants. One straightforward modification is to make the coefficients random variables. Stochastic matrix models are generalizations of age- or stage-based models, in which one or more of the matrix coefficients are assumed to be a random variable. In very small populations, substantial fluctuations in abundance, and even extinction, can occur simply because of the random nature of birth and death processes. The most important application of these models for ecological risk assessment is in quantifying the probability of extinction, either due to strictly demographic and environmental stochasticity or due to stress (e.g., excessive harvesting, toxic chemical exposure, or periodic catastrophic mortality) imposed on stochastically varying populations.

The theoretical literature on stochastic matrix models emphasizes the use of analytical mathematics to obtain general results. In applications to specific populations, however, numerical simulations using Monte Carlo methods are usually adequate and are much easier to perform. Barnthouse et al. (1990) used Monte Carlo simulation to evaluate the combined effects of environmental variability and uncertainty of chemical toxicity on the abundance and risk of extinction of two fish populations. Snell and Serra (2000) describe a similar approach to quantify impacts of chemical exposures on rotifer populations influenced by short-term environmental variability and episodic catastrophes. Software designed for easy development of stochastic matrix models is now readily available; however, in many cases special software is not needed. Equation 27.3 and Equation 27.5 can be readily implemented on a spreadsheet, and Monte Carlo analysis of spreadsheets can be easily performed using add-on programs such as Crystal Ball®.

Another common modification of the basic matrix model is to make the coefficients density-dependent, i.e., to make one or more of the vital rates a function of the number of individuals either in the entire population or in some of the age classes. The purpose of incorporating density-dependence is to account for the fact that, in spite of fluctuations in abundance, populations in nature are always bounded within limits (Section 27.1.4).

As a practical matter, explicit incorporation of density-dependence in population models is often necessary to obtain realistic population projections. Inspection of Equation 27.2, Equation 27.4, Equation 27.8, and Equation 27.10 shows why. For deterministic models, unless r is *exactly* 0 (λ *exactly* 1), the model population must either grow without bounds or decline to zero. Inclusion of stochasticity does not alter this behavior. Given enough time, any density-independent model population will either grow to infinite size or become extinct. For this reason, most models intended to simulate population behavior over more than one generation incorporate density-dependence. Two common functions for density-dependent survival (s) are the Beverton–Holt function:

$$s = c_1/(1 + c_2 n) \tag{27.12}$$

where c_1, c_2 = constant parameters, and n = population size; and the Ricker function:

$$s = \alpha e^{-\beta n} \tag{27.13}$$

where α, β = constant parameters.

These functions are introduced here simply as convenient examples, because both have been extensively studied and are easy to apply to age- and stage-structured models. Both functions were originally developed for application to fish populations; explanations of their derivations and uses can be found in Hilborn and Walters (1992) or in Quinn and Deriso (1999). The key parameters of these two functions are the constants c_1 in the Beverton–Holt model and α in the Ricker model. They represent the maximum possible values of the survival rate s. When the population size (n) is very low, s is approximately equal to c_1 (Beverton–Holt model) or α (Ricker model). As n increases, s decreases in both models (although at a different rate in each model). In the fisheries literature, the parameters c_1 and α are sometimes termed estimates of "compensatory reserve," because they are measures of the capacity of populations to grow at low population sizes. Populations with high compensatory reserves (i.e., populations that grow rapidly at low population sizes) are more resilient to environmental disturbances, and can often sustain higher rates of harvesting, than can populations with low compensatory reserves (Christensen and Goodyear 1988; Rose et al. 2001).

The functional forms of Equation 27.12 and Equation 27.13 do not reflect specific biological processes, and there is no way to directly measure the critical parameters (c_1, c_2, α, and β). Moreover, there is no way to determine a priori which functional form is appropriate for a given population. In principle, it should be possible to determine which model is appropriate by fitting both models to a time series of data on population abundance. In practice, the available data are rarely sufficient to unambiguously distinguish between the two models. For the purpose of risk assessments, uncertainty due to differences between alternative models of the same process can usually best be dealt with by performing alternative simulations using all of the alternatives. If the results are not affected by the choice of models, the most convenient model may be selected; if the results are highly dependent on the choice of models, it may be better to perform parallel simulations using both.

Predictions about future population behavior or response to stress can be extremely sensitive to variations in functional forms and parameter values. For this reason, management applications of density-dependence have had mixed success and are often highly controversial (Rose et al. 2001). As in most other applications of mathematical models in risk assessment, these difficulties are minimized if the models are used for comparative purposes rather than for the purpose of predicting future states of nature. For example, Barnthouse et al. (1990) used stochastic, density-dependent population models to examine the relative influence of uncertainties concerning life history, harvesting mortality, environmental variability, and chemical toxicity on the predicted responses of fish populations to chemical exposures. They found that uncertainties related to the characteristics of the exposed populations were negligibly small compared with uncertainties related to laboratory-to-field extrapolation of toxicity test data.

27.2.3 Aggregated Models

Many readers of this book will be familiar with other kinds of population models, such as the logistic model and stock-recruitment model. These approaches to population modeling differ from the models discussed above in that all organisms are aggregated into one or two components such as "population size" or "parents and offspring." The logistic model has a long history in population biology, and is discussed in all college-level ecology textbooks. It is perhaps the simplest model that simulates the growth and stabilization of populations. A form of the logistic model, known as the Shaeffer surplus production model, is used in fisheries management. Good discussions of surplus production models and of the various forms of stock-recruitment models can be found in Hilborn and Walters (1992). Because of its extreme simplicity, many applied population biologists view the logistic model and its variants as being of little practical value. However, as long as precise numerical predictions are not required, the logistic model may be used as an approximation to more complex models. The model is usually expressed in differential equation form:

$$\frac{dN}{dt} = rN\frac{(K - N)}{K} \tag{27.14}$$

where K = population carrying capacity.

In integral form, the model can be expressed as

$$N_t = \frac{K}{1 + e^{rt}} \tag{27.15}$$

When the size of the population is very small relative to its carrying capacity, the rate of growth of the population is close to the maximum possible rate r, which is the same intrinsic rate of population growth defined in Equation 27.1. The rate of population growth declines as the population grows, and approaches zero as the population approaches the carrying capacity, K. The logistic model can also be expressed as a recursive discrete-time equation, which may be useful for simulating recovery of a population from repeated disturbance events:

$$N_{t+1} = N_t + rN_t\left(\frac{K - N_t}{K}\right) \tag{27.16}$$

A disturbance is simulated by eliminating a fraction of the population present on the day of the disturbance. The population immediately begins growing again toward carrying capacity, until the next disturbance occurs. Barnthouse (2004) used both the continuous and the discrete forms of the logistic model to estimate approximate population recovery times for various aquatic organisms following simulated mortality due to agricultural chemical applications. Nakamaru et al. (2002) used a stochastic version of the continuous logistic model to quantify the influence of DDT exposure on the probability of extinction of herring gull populations. Snell and Serra (2000) used a variant of the discrete logistic model to quantify the influence of generalized chemical exposures on the probability of extinction of rotifer populations. These studies are discussed later in this chapter.

27.2.4   METAPOPULATION MODELS

Many species, and in particular most terrestrial species, do not exist as continuous interbreeding populations. Instead, they consist of subpopulations inhabiting patches of suitable habitat interspersed among patches or regions of unsuitable habitat. All of these patches are subject to environmental variability. Small populations frequently become extinct, but new populations can be established in empty habitat patches by colonists arriving from other patches. This view of species as "metapopulations" was first formalized by Andrewartha and Birch (1954), although they did not use the term. Levins (1969) is credited with developing the first formal metapopulation model. He formulated a simple relationship between the fraction of habitat patches occupied by a species at any given time ($p(t)$), the rate of extinction of occupied patches (e), and the rate of production of propagules from each occupied patch (m). At any time t, the number of propagules produced is equal to the rate of production per occupied patch multiplied by the fraction of patches occupied. If each propagule has an equal probability of dispersing to occupied and unoccupied patches, a fraction equal to $(1 - p)$ of the propagules will colonize unoccupied patches. At the same time, a total number of patches equal to ep would become extinct. The rate of change in p at any time would be determined by the equation:

$$\frac{dp}{dt} = mp(1 - p) - ep \qquad (27.17)$$

The equilibrium frequency of occupied patches (p^*) is determined by the ration of the extinction and colonization rates:

$$p^* = 1 - e/m \qquad (27.18)$$

If extinction is more likely than dispersal (i.e., e is larger than m), extinction of the metapopulation is inevitable. This result is intuitively obvious, even without the model. What is not obvious, however, is that if e and m are nearly equal, the fraction of occupied patches can be expected to be very small, even if the rate of dispersal of propagules from occupied patches is very high. Under this circumstance, random variations in extinction and colonization rates can cause metapopulation extinction, even if under constant conditions the metapopulation would persist.

The above model is too simplistic to be of much value in the management of real populations. However, the fundamental processes and variables considered in the model, i.e., dispersal, extinction, and percent occupancy of available habitat, are central issues in conservation biology. In the 1980s, conservation biologists turned to metapopulation theory as a means of designing preservation strategies for vertebrate species that, although once widespread, were becoming restricted to isolated subpopulations because of increasing habitat fragmentation. Levins' original model has been extended to include influences of local population size, local population structure, spatial dispersal patterns, interspecies interactions, and population genetics. Hanski (1999) has provided an excellent overview of metapopulation ecology, including both theoretical and empirical aspects.

Metapopulation biology provides ecological risk assessors with both a conceptual framework and modeling techniques for addressing the effects of spatially variable chemical exposures on populations inhabiting spatially heterogeneous environments. Maurer and Holt (1996) used a metapopulation model to demonstrate that pesticide applications can endanger the regional persistence of species by reducing the pool of sites available for colonization. Spromberg et al. (1998) developed a generalized metapopulation model based on an extension of Equation 27.17 and used it to examine the influence of the spatial arrangement and connectivity of patches on the response of the metapopulation to a toxic chemical that affects one of the patches. Chaumot et al. (2002, 2003) used a multipopulation extension of the Leslie matrix approach to model the impacts of cadmium discharges on a brown trout metapopulation inhabiting a hypothetical river network. This study is discussed in more detail later in the chapter.

27.2.5 INDIVIDUAL-BASED MODELS

Ultimately, the health of a population is no more than a collective expression of the health of the individual organisms. The models discussed are at best abstractions that capture the general (we hope the essential) features of the biology of the organisms. Some, like the potential population growth rate model or the density-dependent Leslie matrix, are basically bookkeeping devices with which the deaths and births occurring during a given time are tabulated, while the biological mechanisms responsible for reproduction and mortality are ignored. All organisms within a given class (however defined) are assumed to be indistinguishable. Clearly, all organisms are *not* indistinguishable, and variations between individuals can have substantial influences on the responses of populations to anthropogenic stresses or management actions. Recognition of these problems has led to interest in "individual-based" models, i.e., models in which population dynamics are represented in terms of the physiological, behavioral, or other properties of the individual organisms. The general procedure is to develop a model of the individual organism to whatever level of detail is required, and then to infer the properties of the population as a whole either by analytical solution of equations or by numerical simulation of the activities of hundreds or thousands of individual organisms (Figure 27.2).

Individual-based models have made important contributions to understanding successional patterns in forests (Huston and Smith 1987), comparing the structure and development of different forest types (Shugart 1984), and predicting the effects of environmental stress on forest composition (Dale and Gardner 1987). A substantial number of applications to fish populations have also been published (Sperber et al. 1977; Adams and DeAngelis, 1987; DeAngelis et al. 1990; Madenjian and Carpenter 1991; Rose and Cowan 1993; Rose et al. 2003).

There are two broad approaches to developing individual-based models, of which one emphasizes Monte Carlo simulation and the other, analytical solutions to equations. The subtleties of the approaches and criteria for choosing one over the other have been discussed by Caswell and John (1992) and DeAngelis and Rose (1992). Elegant examples of the analytical approach have been published by McCauley et al. (1990) and Hallam et al. (1990). The principal advantage of the analytical approach is that results obtained are general, easy to verify, and easy to understand. The level of detail, however, must be compromised to achieve analytical tractability. In practice, the published biological applications of analytical individual-based models all deal with relatively simple organisms such as *Daphnia*.

The most widely known analytical individual-based models have emphasized physiological characteristics such as metabolism, growth, and chemical toxicodynamics. McCauley et al. (1990) developed a model of *Daphnia* growth and reproduction based on energetics and used the model to predict time-dependent changes in the age and size structure of *Daphnia*

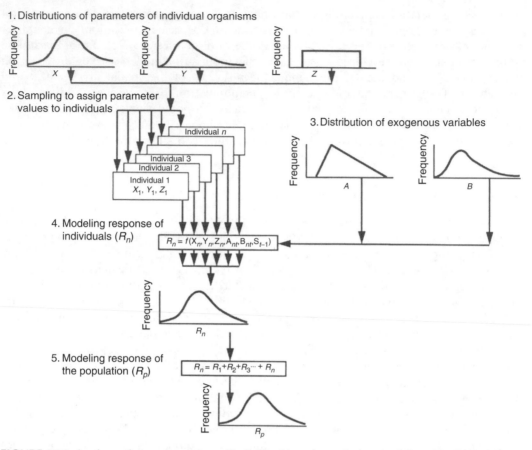

FIGURE 27.2 A schematic representation of individual-based population modeling. X_n, Y_n, and Z_n are characteristics of an individual organism n such as size or leaf area. A_{nt} and B_{nt} are characteristics of the environment experienced by individual n at time t such as temperature, pollutant concentrations, and prey availability. S_{t-1} is the state of the organism at the previous time step. R_n is the response of individuals such as death or maturation. R_p is the response of the population such as abundance or harvestable biomass.

populations in response to changes in food availability. Work on metabolism and toxicody-namics was pioneered by Kooijman and Metz (1984), who examined the influence of toxic chemicals on metabolism and population growth using *Daphnia* as a model organism. Hallam and Lassiter extended this approach to include (1) a thermodynamically based model of the uptake of contaminants from aqueous media, and (2) a definition of death in terms of the internal dissolved contaminant concentration within an organism (Hallam et al. 1990; Lassiter and Hallam 1990). Kooijman (2000) used the principles developed in these early studies as the basis of a formal framework for physiologically structured population modeling termed "Dynamic Energy Budget" (DEB) modeling. The DEB approach links physiological charac-teristics of the organisms to the growth rates and age distributions of populations, and also the exposure concentrations and modes of action of toxic chemicals to the physiological characteristics of the individuals. A software package (DEBtox; Kooijman and Bedaux 1996) has been developed for the purpose of estimating the chemical effects parameters used in DEB models from toxicity test data collected using standard Organization for Economic Cooper-ation and Development (OECD) protocols.

The DEB approach is intended for application to aquatic organisms with relatively simple life cycles, inhabiting homogeneous environments. The models are very general in form, and analysis of the models emphasizes analytical investigation of the equations. An entirely different approach has been used to develop individual-based models of organisms with more complex life cycles, inhabiting more complex environments. DeAngelis et al. (1991) and Rose and Cowan (1993) developed models of fish populations that include metabolism, growth, foraging behavior, and prey selection as functions of the life stage and age of the fish. The approach followed in developing both of these models was to use the existing extensive literature on bioenergetics, reproduction, and foraging of individual fish, coupled with exhaustive evaluation of the life history of specific fish species, to develop detailed models of each life stage from egg through to reproductive adult. Population-level consequences of changes in the physiology, behavior, or reproduction of individual fish were inferred by brute-force simulation of the birth, growth, and death of hundreds or thousands of individual fish. The models were calibrated to extensive data sets collected for specific fish populations. This approach was later used by Jaworska et al. (1997a) to model the effects of simulated PCB exposures on young-of-the-year largemouth bass in southeastern US reservoirs. Jaworska et al. (1997b) used individual-based models of walleye and yellow perch populations to test whether causes of adverse changes in populations could be inferred from observed patterns of abundance, growth, and age structure. Rose et al. (2003) used an individual-based model of an Atlantic croaker population to link experimental data on the effects of PCB exposures on fecundity, egg survival, and larval predator avoidance ability in this species to population-level effects. This study is discussed in greater detail later in the chapter.

The metapopulation model VORTEX, originally described by Lacy (1993) and subsequently applied to a wide variety of endangered vertebrate populations, is fundamentally an individual-based model. The core of VORTEX is a stochastic model of the birth, growth, movement, reproduction, and death of each animal present in a population. The growth, decline, or extinction of a population is calculated by simulating the fate of each animal and its offspring for multiple generations. Estimates of the probability of persistence and expected time to extinction of the simulated populations are obtained by performing multiple runs in which random values of key parameters are drawn from prespecified statistical distributions.

Within the last decade, as the availability of Geographic Information System (GIS) technology has expanded, ecologists have used this technology as a basis for a new class of individual-based models termed spatially explicit models. In spatially explicit models, organisms are distributed over a realistic landscape composed of habitat patches of different types and suitability for utilization by the species of interest. The spatially explicit approach permits ecologists to integrate theory and observations on foraging behavior and reproduction in individual animals, relate these to specific measurable habitat characteristics, and infer influences of habitat change on populations.

Thorough and well-tested models of this type have specially been developed for populations of ungulates foraging in Yellowstone National Park (Turner 1993; Turner et al. 1994) and for the population of Bachmann's Sparrow nesting on the US Department of Energy's Savannah River site. Recently, an individual-based model of skylarks utilizing an agricultural landscape in Denmark has been used to compare the relative influences of pesticide applications and land use change on the abundance and persistence of this species (Topping and Odderskær 2004). This study is discussed in greater detail later in the chapter.

27.3 APPLICATIONS TO TOXIC CHEMICALS

Most of the modeling approaches (DEB modeling is an important exception) discussed were developed to address theoretical problems, to manage exploited populations, or to aid in the conservation of endangered species. However, interest in applying these approaches to ecological risk assessment of toxic chemicals has grown over the past few years. In addition to numerous papers published in the scientific literature, three recent international workshops have addressed potential ecotoxicological applications of population models (Kammenga and Laskowski 2000; Baird and Burton 2001; Barnthouse et al. 2006). The following case studies provide examples of these applications.

27.3.1 QUANTIFYING UNCERTAINTIES IN INDIVIDUAL-TO-POPULATION EXTRAPOLATIONS

Barnthouse et al. (1987, 1988, 1990) developed a series of models that directly link toxicity test data to fish population models, and then used the combined models to evaluate the ecological implications of toxicity test data. Although this work was published nearly 20 y ago, it is still relevant today, for at least two reasons. First, it provides examples of applications of some of the extrapolation approaches discussed in Chapter 26 of this book. Second, it explores the relative uncertainties inherent in extrapolation of laboratory test data to effects on populations in the field in a way that has not yet been duplicated or superseded by more advanced methods.

Two different approaches to population modeling were used in these studies. In the first two papers of this series, estimates of the survival and reproduction parameters used in the Leslie matrix were used to calculate an "index of reproductive potential." The index was defined (Barnthouse et al. 1987) as the expected contribution of a female recruit (a 1-y-old female fish, in fisheries science terminology) to future generations of recruits, taking into account (1) her annual probability of survival (s_i), probability of being sexually mature (m_i), and age-specific fecundity (f_i); and (2) the probability that a spawned egg will hatch and survive to age 1 (s_0). The reproductive potential of a 1-y-old female recruit is given by

$$P = s_0 \sum_{i=1}^{n} s_i f_i m_i \qquad (27.19)$$

Although Equation 27.19 contains the same parameters found in the Leslie matrix (Equation 27.6), the reproductive potential index is not used to calculate the future abundance or age composition of a population. Instead, it is used as a relative measure of the effect of changes in mortality or fecundity on a population, expressed as a fractional reduction in reproductive potential (R_s):

$$R_s = (P - P_s)/P \qquad (27.20)$$

where P_s is the reproductive potential index in the presence of a stress that reduces survival, fecundity, or both. The value of P_s is calculated from

$$P_s = s_0(1 - C_m) \sum_{i=1}^{n} s_i (1 - C_r)^{i-1} f_i C_i m_i \qquad (27.21)$$

where C_m = probability of stress-induced mortality during the first year of life; C_r = probability of stress-induced mortality for 1-y-old and older fish (assumed to be equal for all age classes); and C_f = proportional reduction in fecundity due to stress (assumed equal for all reproducing age classes).

The reproductive potential index was originally used to assess impacts of power plant cooling systems on fish populations (Barnthouse et al. 1986). Since the mid-1980s, a variant of the index termed the "spawning stock biomass per recruit" index has been widely used in marine fisheries management (Goodyear 1993).

The reproductive potential approach, like the density-independent Leslie matrix, cannot account for natural environmental variability or density-dependence. To explore the influence of these processes on responses of fish populations to toxic chemicals, Barnthouse et al. (1990) developed density-dependent, stochastic matrix projection models for two especially well-studied populations: the Gulf of Mexico menhaden population and the Chesapeake Bay striped bass population. The models employ conventional projection matrices, but with the survival coefficient for young-of-the-year fish (s_0) containing both density-dependent and randomly varying components. Estimates of the coefficients were obtained from published abundance, age structure, and mortality statistics for these two populations.

Survival of young-of-the-year fish was calculated using

$$s_0 = e^{-\alpha + R_i \sigma - 0.5\sigma^2 - \beta N_0} \tag{27.22}$$

where α = expected annual instantaneous rate of density-independent mortality; σ = standard deviation of α; R_i = a unit random normal deviate; and β = coefficient of density-dependence. Effects of chemicals on young-of-the-year are incorporated by replacing α in Equation 27.19 with

$$\alpha' = \alpha - \ln(1 - C_m) \tag{27.23}$$

where C_m = fraction of young-of-the-year expected to die from effects of chemical exposure. Effects of chemical exposure on fecundity were incorporated by multiplying each age-specific fecundity rate (f_i) in the population matrix by a fecundity reduction factor.

The contaminant effects factors were estimated from standard life-stage-specific toxicity data using concentration–response models and extrapolation models (inter-life-stage and interspecies) described earlier in this book. These procedures were used to develop exposure–response relationships that explicitly incorporate three types of uncertainty in lab-to-field extrapolations: test variability, species-to-species uncertainty, and acute-to-chronic uncertainty. The concentration–response function used in these analyses was the logistic model:

$$P = e^{a+BX}/(1 + e^{a+BX}) \tag{27.24}$$

where P = fractional response of the exposed population; X = exposure concentration; and a and B = fitted parameters with no direct biological interpretation. When fitted to concentration–response data, the logistic function has a sigmoid shape similar to the probit model. Concentration–response data sets were fitted to Equation 27.24 using nonlinear least-squares regression. Uncertainty concerning the shape and position of the concentration–response function, as reflected in the variances and covariances of a and B, were represented graphically as confidence bands surrounding the fitted functions.

Concentration–response functions specific to each life stage were combined to produce integrated functions that express the effects of chemical exposures, including uncertainty, on population-level response variables. An example is provided in Figure 27.3, which shows a concentration–response function for brook trout exposed to methylmercuric chloride (data from McKim et al. 1976), with female reproductive potential as a response variable. The maximum acceptable toxicant concentration (MATC) for the data set (calculated as the geometric mean of the no observed effect concentration (NOEC) and the lowest observed effect concentration (LOEC) and also known as the chronic value) corresponds to a 55% to 70% reduction in brook trout reproductive potential. This result, and other similar ones presented in Barnthouse et al. (1987), demonstrated that MATCs calculated from life cycle

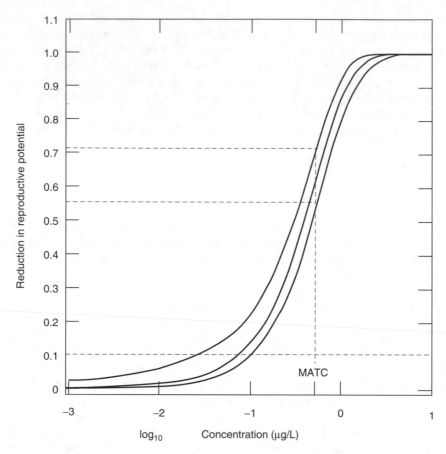

FIGURE 27.3 Concentration–response function and uncertainty band for the reduction in female reproductive potential of brook trout exposed to methylmercuric chloride. The lower dashed line denotes the 10% effect level (EC_{10}). The two upper dashed lines denote the 90% confidence band for the effects level associated with the maximum acceptable toxicant concentration (MATC). (From Barnthouse, L.W., Suter, G.W., II, Rosen, A.E., and Beauchamp, J.J., *Environ. Toxicol. Chem.*, 6, 811, 1987.)

toxicity tests often correspond to surprisingly high population-level effects and cannot in general be construed as ecological effects thresholds.

Barnthouse et al. (1987, 1988, 1990) used the above models for a variety of purposes. Extrapolated population responses were compared to MATCs derived from the same data sets, demonstrating that MATCs are not equivalent to population-level no-effects thresholds (Barnthouse et al. 1987, 1988). Comparisons of uncertainties associated with different test endpoints showed that fecundity responses are substantially more variable and introduce more uncertainty into risk assessments than do mortality responses (Barnthouse et al. 1988). Comparisons of uncertainty introduced in different extrapolation steps and toxicity test types were used to quantify the relative value for risk assessment of different toxicity testing strategies (Barnthouse et al. 1990). Comparisons between responses of menhaden and striped bass populations showed that, for typical screening-level assessments, uncertainties related to life history, environmental variability, and harvesting intensity would be negligible compared to toxicological uncertainties resulting from the use of quantitative structure–activity relationships and short-term toxicity tests to predict long-term population responses.

27.3.2   Life History–Based Ecological Risk Assessment

Calow et al. (1997) and Forbes et al. (2001) proposed an approach for assessing population and community-level risks based on a combination of toxicity test data and simplified life history models. The model used in this approach was first described by Calow and Sibly (1990). It assumes a highly simplified life history involving only two stages: juveniles and adults. It has only five parameters: the fractions of juveniles surviving to first breeding (S_j); the fraction of adults surviving between breeding events (S_i); the time from birth to first breeding (t_j); the time between breeding attempts (t_a); and the number of offspring produced per individual at each breeding (n). Each organism is assumed to have a potentially infinite life span. Only female organisms are included in the model. Under these assumptions, Calow and Sibly (1990) derived a simple formula for calculating the finite population growth rate (8) from the discrete form of the fundamental population growth rate equation (Equation 27.3). The procedure involves reformulating the fundamental equation as an infinite sum, recalling that the survivorship term (l_x) in the equation can be expressed as the product of survival fractions from birth to age x:

$$1 = \sum_{x=1}^{\infty} \lambda^{-x} l_x m_x = \lambda^{-t_j} S_j n + \lambda^{-(t_j+t_a)} S_j S_a n + \lambda^{-(t_j+2t_a)} S_j S_a^2 n + \cdots$$

$$= n S_j \lambda^{-t_j} \left(1 + \lambda^{-t_a} S_a + \lambda^{-2t_a} S_a^2 + \cdots \right) \tag{27.25}$$

Substituting $y = \lambda^{-t_a} S_a$:

$$1 = n S_j \lambda^{-t_j} \left(1 + y + y^2 + y^3 + \cdots \right) \tag{27.26}$$

Because both λ^{-t_a} and S_a are limited to values between 0 and 1, y must also be a number between 0 and 1. A theorem from the mathematics of infinite series states that for any number y between 0 and 1:

$$\sum_{x=1}^{\infty} (1 + y^x) = \frac{1}{1 - y} \tag{27.27}$$

Therefore, Equation (27.22) can be restated as

$$1 = \frac{n S_j \lambda^{-t_j}}{1 - y} = \frac{n S_j \lambda^{-t_j}}{1 - \lambda^{-t_a} S_a} \tag{27.28}$$

which can be rearranged as

$$1 = n S_j \lambda^{-t_j} + S_a \lambda^{-t_a} \tag{27.29}$$

Calow et al. (1997) used this model to illustrate the influence of chemical toxicity on species with different general life history types: semelparous species (species that reproduce only once before dying, so that $S_a = 0$); moderately iteroparous species ($S_a = 0.5$); and strongly iteroparous species ($S_a = 0.9$). They examined the effects of chemicals that affected juvenile survival, reproduction, or adult survival on species with these life history types. Not surprisingly, reductions in juvenile survival or reproduction per breeding event had the greatest effect on λ for semelparous species and the least effect on strongly iteroparous species. Reductions in adult survival, on the other hand, had the greatest effect on strongly iteroparous species and the least effect on semelparous species. The relative durations of the time to first reproduction (t_j) and the interval between reproductive events (t_a) also affected the responses of the model populations to reductions in survival or reproduction. The shorter the value of t_j relative to t_a, the greater is the sensitivity of λ to reductions in survival or reproduction.

The model described above is obviously quite simplistic, but it is also highly adaptable. Basic information on time-to-maturity, reproductive rate, and longevity are available for most types of organisms. Provided that dose–response models describing the relationships between chemical concentrations and effects on survival and reproduction can be developed, models relating chemical concentrations to changes in λ for different general life history type can be defined. Given information on the frequency distribution of different life history types present in a given ecosystem, "life history sensitivity distributions," analogous to the species sensitivity distributions (SSDs) described in Section 26.2.3, can be constructed.

Forbes et al. (2001) expanded on this idea in a subsequent paper, in which they compared hypothetical aquatic life protection criteria derived from population growth rate analysis to criteria derived using conventional toxicology-based approaches. In this paper, the authors defined four life cycle types representative of major organisms found in aquatic ecosystems: benthic invertebrates, fish, zooplankton (daphnids), and algae. They applied the life history model described above to each life history type, assuming that the long-term population growth rates (λ) for each organism type were approximately 1.0. They then investigated the influence of small changes in juvenile survival (S_j) on the population growth rate, using the elasticity index defined earlier in this chapter. Elasticities measure the proportional contribution of each life cycle parameter to the overall population growth rate. The authors found that the population growth rate of benthic invertebrates was the most elastic to small changes in juvenile survival, and that the population growth rate of daphnids was the least elastic. This means that, for any given decrease in survival caused by exposure to a contaminant, the corresponding reduction in λ would be greatest for the benthic invertebrate life cycle and least for the daphnid life cycle.

The authors then simulated a community composed of a mix of species with the four life history types, and compared protection strategies based on several conventional toxicity-based approaches to a protection strategy based on protection of population growth rates. The analysis assumed that the contaminant in question affected only juvenile survival (S_j). First, they developed a hypothetical dose–response function so that changes in juvenile survival could be translated into corresponding contaminant concentrations. For each life cycle type, they calculated the contaminant concentration corresponding to 10% reduction in λ. This value was assumed to be a safe concentration for each life cycle type, i.e., a population-level NOEC.

They then developed a series of scenarios in which both the relative toxicities of a hypothetical chemical to organisms with different life cycle types, and the relative contributions of different life cycle types to the total community, were varied. Variability in sensitivity among species within the same life cycle type was simulated by assuming a log-normal distribution of NOECs within each group. Monte Carlo simulation was used to calculate (1) the average NOEC for the community, i.e., the concentration at which the average value of δ for all species was reduced by 10%; and (2) a 95% protection level for the community, i.e., the concentration at which only 5% of the species-specific NOECs is exceeded. These values were compared with protective concentrations developed using conventional toxicity-based standard-setting toxicological methods: an application factor method and an SSD-based method. Forbes et al. (2001) found that the protective concentration based on a 10% reduction in λ was, in general, substantially higher than the concentrations calculated using the application factor or the SSD. Hence, conventional risk assessment approaches appeared to produce protective estimates of environmentally safe contaminant concentrations, and in some cases these approaches appeared to produce substantially overprotective concentrations. There were, however, some circumstances (e.g., a preponderance of species with a sensitive life cycle type and also high chemical sensitivity) in which the conventional approaches might produce underprotective concentrations. Forbes et al. (2001) concluded that more research on the relative contributions of different life history types to aquatic communities is needed to refine environmental protection protocols so that they provide truly protective, but not overprotective, estimates of safe exposure limits.

27.3.3 QUANTIFYING IMPACTS OF CHEMICAL EXPOSURES ON RISK OF EXTINCTION

Snell and Serra (2000) used an empirically derived, stochastic model of temperate rotifer populations to simulate the impacts of a hypothetical chemical exposure on the long-term risk of population extinction. Their model is based on a density-dependent variant of the exponential growth model:

$$\mathbf{N}_t = \mathbf{N}_{t-1}e^{r_t} \qquad (27.30)$$

Rather than being a constant as in Equation 27.2, the growth rate (r_t) is a function of previous population density and environmental variability. The authors obtained values for the parameters of the growth rate function by analyzing a time series of population density data for a natural rotifer population. Under most circumstances, rotifers are parthenogenetic (i.e., all individuals are female, and produce eggs that hatch into other females that produce only female eggs). At high densities, however, a fraction of the offspring produced comprises "mictic" females capable of sexual reproduction. Unfertilized eggs produced by mictic females hatch as male rotifers capable of fertilizing mictic females. The eggs produced from sexual reproduction are "resting" eggs, which fall to the sediment. Resting eggs are subject to mortality during both summer and winter, and they are the only life stage that survives through the winter. In the model simulation, the growing season for rotifers is assumed to last for 240 d, after which all individuals in the water column die. The following spring, the population in the water column is reestablished from resting eggs that hatch from the sediment. Only 10% of resting eggs hatch during any given spring, and those that do not hatch remain in the sediment as an "egg bank" that can potentially repopulate the water column for several years, and provide insurance against catastrophic events that may kill all rotifers in the water column.

Persistence and extinction in the model of Snell and Serra (2000) is determined by the density of resting eggs. If, during any year, the density of resting eggs falls below a critical value, the population is assumed to become extinct. The authors calculated extinction risks by simulating the fate of 1000 populations, all starting from the same initial population size, over a period of 100 y. They found that, even without chemical exposures, there is about a 5% extinction risk for a rotifer population over a 100 y period. The authors simulated three types of perturbations that could increase extinction risks: a continuous reduction in r caused by a chronic chemical exposure, an intermittent reduction in r caused by an intermittent chemical exposure, and a series of catastrophes occurring with different frequencies.

Effects of chemical exposures were simulated by multiplying each value of r_t by a constant fraction. The authors found that even a continuous reduction in r as small as 10% raised the extinction risk for a rotifer population to approximately 20%, and that a reduction in r of 30% or greater raised the risk to nearly 100%. The influence of intermittent reductions in r was simulated by assuming that r_t is reduced by 25% on a percentage of days ranging from 20% to 100%. Extinction risks were lower in the case of intermittent reduction than in the case of continuous reduction; however, reducing r_t on only 40% of days still increased the 100 y extinction risk to 30%. Simulated chemical exposures also reduced the abilities of the model rotifer population to survive periodic catastrophes. In the absence of chemical exposures, a catastrophe frequency of once every other year increased the risk of extinction from 5% to about 25%. If, in addition, a 10% continuous reduction in r occurs due to chemical exposure, the extinction risk is raised to 60%.

According to the authors, the underlying reason for the sensitivity of their model to chemical exposures is the dynamics of the resting egg bank. A minimum number of resting eggs must hatch in the spring so that the density of parthenogenetic rotifers can increase rapidly to the threshold density required for sexual reproduction. Reducing r reduces the rate of growth toward the threshold, and also the number of resting eggs that can be produced before the end of the growing season. If insufficient resting eggs are produced, the size of the egg bank begins to fall, reducing the number available for hatching the following spring. This further extends the time required for growth to the sexual reproduction threshold, again reducing the number of resting eggs produced. Eventually, the egg bank declines below the extinction threshold.

Snell and Serra (2000) concluded, based on their analysis, that even small reductions in r, including levels that are often presumed to be safe in toxicity tests, can pose significant extinction threats to natural rotifer populations. They argued for increased use of population-based methods in risk assessments.

Nakamaru et al. (2002) used a different modeling approach to quantify extinction risks in a herring gull population exposed to DDT. These authors modeled population growth and extinction using a stochastic version of the logistic model (Equation 27.14):

$$\frac{dN}{dt} = rN\left(\frac{K - N}{K}\right) + \sigma_e \xi_e(t) \circ N + \sigma_d \xi_d(t) \cdot \sqrt{N} \qquad (27.31)$$

In this equation the terms N, r, and K are defined in the same way as before, as population size, intrinsic rate of population growth, and carrying capacity. The additional terms in the equation represent environmental ($\sigma_e \xi_e$) and demographic ($\sigma_d \xi_d$) stochasticity. Environmental stochasticity is simply random variation in environmental factors affecting population growth. Demographic stochasticity is random variation in population due to the fact that, in stochastic models, births and deaths of individual organisms are random events. The symbols ($\circ$ and $\cdot$) used in the equation denote mathematical operations performed, using the "Stratonovich calculus" and the "Ito calculus," respectively. Readers interested in what these operations are and how they are performed should consult a good textbook on stochastic differential equations and be prepared for some heavy reading. The benefit of using a model like this rather than performing hundreds or thousands of Monte Carlo simulations is that mathematicians (in this case Hakoyama and Iwasa 2000) can derive an integrated equation that describes the influence of key model parameters on the risk of population extinction. The relationship between extinction time, carrying capacity, and environmental stochasticity is defined by

$$T = \frac{2}{\sigma_e^2} \int_0^K \int_0^\infty e^{-R(y-x)} \left(\frac{y + D}{x + D}\right)^{R(K+D)+1} \frac{1}{(y + D)y} dy dx \qquad (27.32)$$

The term R represents the more complex term $2r/(\sigma_e^2 K)$ and the term D represents $1/\sigma_e^2$. The mean time to extinction (T) is defined in units of generation time. Hence, Equation 27.31 can be used to calculate the mean number of generations a population will persist before becoming extinct, given estimates of r, K, and σ_e^2.

How does DDT affect the risk of extinction of a herring gull population? Nakamura et al. (2002) addressed this question by modifying the parameters r and K in Equation 27.32 to include a reduction caused by DDT exposure:

$$r' = r - \alpha \tag{27.33}$$

$$K' = (r - \alpha)\, K/r \tag{27.34}$$

Although the mathematics may be abstruse, the conceptual basis of the model can be very simply summarized: mean time to extinction is a function of population growth rate, carrying capacity, environmental variability, and DDT exposure. DDT decreases both population growth rate and carrying capacity, thereby shortening the expected time to extinction.

The authors used data from herring gull populations in Long Island, New York, to illustrate how the model can be used. They used observed doubling times of newly established herring gull populations and measures of interannual variability in long-established populations to derive estimates of r and σ_e^2. They estimated a range of carrying capacities (K), corresponding to large and small herring gull populations. They estimated effects of DDT on r and K using a multistep procedure that considered both age-specific fertility in female herring gulls and influence of DDT exposures on female fertility. Using this information, together with published estimates of DDT biomagnification factors and historical concentrations of DDT in Long Island Sound, Nakamura et al. (2002) estimated that DDT exposures during the 1960s reduced r in herring gulls by approximately 20%.

Because herring gull populations are fairly large, and interannual variability in herring gull abundance is small compared to many other types of animals, Nakamura et al. (2002) found that the expected time to extinction for both exposed and unexposed herring populations was quite long. For example, for unexposed populations ranging in size from 100 to 100,000 adult females, mean extinction times ranged from 10^5 to 10^{30} generations. Since the generation time of herring gulls is approximately 8 y according to the authors, this means that a population consisting of only 100 birds would be likely to persist for approximately 1 million years.

Perhaps more interesting than the extinction risks themselves are the calculations performed by Nakamura et al. (2002) to compare risks due to DDT exposures with risks caused by habitat disturbance. The authors argued that, because chemical exposures and habitat disturbance both reduce the carrying capacity of a population, their model could be used to compare both types of disturbance in common units of reduced time to extinction. The method for making the comparison is quite simple in concept (although not so simple mathematically). For a specific value of α, corresponding to a specific concentration of DDT in the environment, Equation 27.32 is used to calculate the change in time to extinction caused by DDT exposure. Assuming that the carrying capacity of a population is directly proportional to the size of its habitat, one then uses a version of the model in which only K is reduced (rather than both r and K as in the case of DDT exposure) to calculate the reduction in K required to produce the same change in time to extinction.

The authors found that the equivalent habitat loss for any given DDT concentration is strongly dependent on the size of the exposed population. For example, according to Table 2 in Nakamura et al. (2002), a water-column DDT concentration of 0.1 ng/L would cause the same change in time to extinction for a population of 100 herring gulls as would a 50.5% reduction in habitat area. For a population of 100,000 birds, a reduction in habitat area of 96.5% would be required to produce the same change in time to extinction.

Nakamura et al. (2002) qualified their results on the grounds that many of their parameter values were only approximations and that herring gulls are a very abundant species with a very low risk of extinction at present. They argued, however, that their method could be used to calculate mitigation requirements (e.g., habitat protection or enhancement) needed to protect endangered species exposed to DDT or other hazardous substances.

27.3.4 QUANTIFYING IMPACTS OF CHEMICALS ON METAPOPULATIONS

Maurer and Holt (1996) investigated the impacts of pesticide exposures on spatially distributed populations in which some subpopulations reside in habitats that are exposed to pesticides and other subpopulations reside in unexposed or "safe" habitats. If migration of organisms occurs between the safe habitats and the exposed habitats, it is possible in theory for the exposed habitats to become "sinks" that reduce the size of the population in the safe habitat, perhaps ultimately resulting in the extinction of the entire population. Typical agricultural landscapes consist of mosaics of vegetation types, some of which are treated with pesticides and others of which are not. If applying these chemicals to only a fraction of the vegetation patches in a region could still cause extinction of nontarget populations throughout the entire region, protocols for field-testing of pesticides would need to consider this possibility. The authors used two alternative metapopulation models to determine conditions under which this phenomenon could occur.

The first model used by Maurer and Holt (1996) is based on two coupled discrete-time equations that describe births, deaths, and migrations within and between safe and exposed habitat patches:

$$N_s(t) = N_s(t-1) + r_s N_s(t-1) - mN_s(t-1) + mN_e(t-1) \qquad (27.35)$$

$$N_e(t) = N_e(t-1) + r_e N_e(t-1) + mN_s(t-1) - mN_e(t-1) \qquad (27.36)$$

The terms r_s and r_e are net rates of population growth within safe and exposed habitats, respectively, accounting for both births and deaths. Since organisms in the exposed habitat should be declining in number, the growth rate (r_e) in that habitat is assumed to be negative. The term m refers to the rate of migration of organisms between habitats, assumed to be the same for both habitat types. The change in number of organisms within each habitat from time ($t-1$) to time (t) is thus equal to the number present at time ($t-1$) plus the number born in the habitat or migrating in, minus the number dying or migrating out.

The growth rate of the entire population in both habitats can be calculated by writing Equation 27.35 and Equation 27.36 as a matrix equation and then using the methods of matrix algebra (Caswell 2001) to find the dominant eigenvalue. The resulting population growth rate is given by

$$\lambda = \frac{2 + r_s + r_e - 2m + \sqrt{(r_s - r_e)^2 + 4m^2}}{2} \qquad (27.37)$$

As long as the rate of population decline in the exposed habitat is smaller than, or equal to, the rate of population growth in the safe habitat (i.e., $|r_e| \leq r_s$), λ will be greater than 1 and the population will increase. However, if the rate of decline in the exposed habitat is greater than the rate of growth in the safe habitat, the population will decline to extinction if the migration rate (m) is too high. If $|r_e| > r_s$, the threshold value for m, above which the population will decline to zero, is given by

$$m = \frac{r_s r_e}{r_s + r_e} \qquad (27.38)$$

To account for the possibility that density-dependent survival or reproduction might stabilize the population and permit persistence under conditions in which the above model predicts extinction, Maurer and Holt (1996) investigated an alternative model, based on the logistic equation:

$$\frac{dN_s}{dt} = r_s N_s \left(1 - \frac{N_s}{K_s}\right) - mN_s + mN_e$$

$$\frac{dN_e}{dt} = r_e N_e + mN_s - mN_e \qquad (27.39)$$

Note that density-dependence is assumed to occur only in safe habitat. The population in the exposed habitat would, as in the density-independent model, decline toward zero in the absence of immigration from the safe habitat.

In this model, because of density-dependence, the total population will stabilize at an equilibrium value determined by the values of the growth rate, migration rate, and carrying capacity parameters. The population persists if and only if this equilibrium population size is greater than zero. Maurer and Holt (1996) found that the condition for an equilibrium population size greater than zero is exactly the same as the condition for a positive value of λ in the density-independent model.

These results, according to the authors, have important and counterintuitive implications for the design of pesticide application programs. The authors noted that in both models the likelihood that the total population will be able to persist is a decreasing function of the migration rate between habitats. Hence, the ability of organisms to replenish exposed habitats through migration actually increases the risk that a regional population inhabiting a mosaic of treated and untreated habitats will become extinct. Persistence is more likely if little or no migration occurs between treated and untreated habitats. Moreover, the likelihood of persistence declines as the population growth rate in the safe habitat declines. The implication of this result is that species with very high maximal growth rates, including many pest species, can persist in the face of localized pesticide applications under conditions in which species with lower growth rates, including nontarget vertebrate species, might become extinct. On the basis of their analyses, Maurer and Holt (1996) concluded that typical approaches to pesticide risk assessment that emphasize measurement of effects in laboratory studies and test applications to individual fields are inadequate because they fail to consider the spatial structure of exposed populations.

Chaumot et al. (2002, 2003) used a multiregion matrix population model to investigate the responses of a hypothetical spatially distributed population of brown trout to cadmium discharges affecting a river network. In both studies, the trout population was distributed among 15 hierarchically organized compartments representing a network of first- through fourth-order stream segments. Three life stages were represented: alevins (age 0 trout), juveniles, and adults. Trout were assumed to migrate seasonally between compartments, with spawning occurring in the first-order segments, and the different age groups of trout distributed during the nonspawning season according to age distributions observed in field data. In the first paper (Chaumot et al. 2002), all trout in any compartment had the same probability of migrating during the spring season. In the second paper (Chaumot et al. 2003), the complete mixing assumption was relaxed to account for the observation that some trout did not migrate, and both a spring and a fall migration season were modeled. In the first paper, survival and reproduction were represented in an extended version of the Leslie matrix (Equation 27.6):

$$L = \begin{bmatrix} 0 & 0 & \mathbf{FP}_H \\ \mathbf{S}_1 & 0 & 0 \\ 0 & \mathbf{S}_2 & \mathbf{S}_3 \end{bmatrix} \tag{27.40}$$

In Equation 27.40, the elements $\mathbf{S}_1$, $\mathbf{S}_2$, and $\mathbf{S}_3$ are diagonal matrices containing compartment-specific survival rates for each life stage. The element $\mathbf{FP}_H$ is a matrix containing compartment-specific fecundities and migration probabilities. The corresponding matrix in the second paper is similar, but includes additional terms resulting from the more complex representation of the trout life cycle.

In both papers, effects of cadmium on brown trout were modeled using dose–response data from laboratory toxicity tests. Life-stage-specific concentration–response curves were derived using a logistic regression approach similar to the method used by Barnthouse et al. (1987, 1988, 1990). The matrix elements were then modified by multiplying the stage-specific fecundity and mortality rates by reduction coefficients derived from the concentration–response functions. The first paper considered only chronic exposures; the second considered both chronic and acute exposures.

In both studies, the authors used the elasticity analysis to determine the effects of cadmium exposure on the population growth rate, as a function of the location and intensity of a hypothetical discharge. In addition, they calculated the effects of hypothetical discharges on the age structure and spatial distribution of trout. In the second paper, the authors showed how discharges of cadmium into different levels of the stream hierarchy could have greatly different impacts on the spatial distribution of trout while having exactly the same effect on the population growth rate.

The authors made no claims concerning the management implications of their results, other than arguing that their approach could be adapted to a wide variety of river network types, be made highly site-specific, and be used to model other stresses in addition to toxic chemicals.

27.3.5 INDIVIDUAL-BASED MODELS

Topping and Odderskær (2004) described an individual-based model of a skylark population inhabiting an agricultural landscape in Denmark. The purpose of the study was to evaluate the influence of pesticides in relation to weather and agricultural practices as influences on skylark populations.

The landscape component of the model is a spatially explicit GIS-based system that includes three types of farms, each of which has its own characteristic crop rotation pattern, and detailed rules for simulating the sequence of activities (e.g., pesticide application, watering, plowing, sowing) that occur on each farm. Roadside vegetation, hedgerows, and other noncultivated areas are also included in the model landscape.

The skylark component of the model simulates the behavior of individual birds, including territory establishment, foraging, nest-building, incubation, and rearing. Each bird is modeled as an "agent" that engages in various behaviors contributing to survival and reproductive success according to sets of decision rules that take into account the bird's size, age, location, nest status, and other characteristics. The influence of weather on skylark reproduction is simulated by treating bad weather conditions (cold, wind, and heavy rain) as categorical variables (i.e., either present or absent during a given half-day time step), with the probability distributions of bad weather events determined from local meteorological records. Skylark foraging success is influenced by vegetation structure, which is determined by vegetation type and growth rate (explicitly simulated in the landscape model), and by the available insect biomass within a given vegetation type.

Topping and Odderskær (2004) developed parameter sets for the model based on landscape data for a specific region in central Jutland, Denmark, and evaluated the impacts of pesticide application, field size, crop heterogeneity, and weather on the skylark population present in the modeled region. Calculated population-level endpoints included total population size, number of breeding pairs, and total number of birds successfully fledged.

The pesticides and herbicides typically used in the modeled region have relatively low direct toxicity to birds. These chemicals were assumed to affect skylarks indirectly, through reductions in the availability of arthropod prey. Field size and crop heterogeneity affect skylarks by influencing the diversity of vegetation types and growth stages present in the territory of a typical bird, which in turn influence arthropod abundance and foraging success. Weather acts directly on the birds, influencing both incubation time, fledgling mortality, and foraging success.

Skylark populations were simulated over five 11 y weather cycles under four combinations of pesticide application patterns (used and not used) and field sizes (large and small). Impacts of crop heterogeneity were evaluated by assuming that only a single crop (spring barley) is cultivated throughout the region.

Results of the simulation showed that, although pesticide applications affected skylark abundance, the effects of pesticides were small compared with effects of landscape structure and weather. Doubling the average field size reduced mean skylark abundance over the 55 y simulation period by 37%. Applying pesticides to the fields using label-specified application rates and frequencies reduced skylark abundance by only about 4%. Variations in weather conditions between years resulted in annual variations in fledgling production between +19% and −13% compared with long-term average fledgling production. Assuming that only barley is grown in the region resulted in dramatically lower skylark abundance, including extinction of some or all model skylark populations.

The authors concluded that, although pesticide use has potentially adverse impacts on skylark populations in central Denmark, agricultural practices have a much greater influence. Intensification of farming practices, including increased farm size and decreased diversity of crops appear to be a more significant threat.

Rose et al. (2003) used an individual-based model to link laboratory observations of effects of PCBs on Atlantic croaker to field data on coastwide abundance trends in this species. The experiments providing the laboratory observations measured effects of PCBs on female fecundity, egg survival, larval swimming speed, and larval predator avoidance ability. The dynamics of the coastwide population were simulated using a matrix projection model similar to those discussed elsewhere in this chapter. Adults were assumed to spawn in the mid-Atlantic bight; larvae were assumed to migrate to nursery areas in North Carolina and Virginia, and to return to the ocean as juveniles.

The effects of PCBs on the behavior of Atlantic croaker larvae, which occur on time scales of seconds to hours, were linked to effects on the coastwide population, which occur on time scales of months to years, using a model of the feeding, growth, and mortality of individual Atlantic croaker larvae. The laboratory experiments measured effects of PCBs on larval avoidance behavior and swimming speed. Larvae exposed to PCBs through maternal transfer from PCB-exposed adult females were found to swim more slowly and respond less actively to simulated predator attacks than were unexposed larvae. Rose et al. (2003) used a statistical method referred to as a "regression tree" to translate the responses observed in the experiment to a probability that a larva encountering a predator would escape predation. Reduced swimming speed, in addition to reducing the probability that a larva would escape predation, would be expected to reduce the rate at which the larva would encounter zooplankton prey organisms. Hence, PCB exposure would be expected to result in reduced prey consumption, slower growth, and potential mortality due to starvation. The individual-based model used by

Rose et al. (2003) simulates the daily activities of a larval Atlantic croaker from hatching to transformation to the juvenile stage. It includes a bioenergetics submodel, a foraging submodel, and a predation submodel. Parameters for these submodels are derived from a variety of laboratory and field studies of larval fish. This same approach has been used in a number of other individual-based fish population models (e.g., DeAngelis et al. 1991; Rose and Cowan 1993; Rose et al. 1999). PCB effects were incorporated in the model through effects of predator avoidance and swim speed on the daily rates of mortality and growth. The daily mortality and growth rates, in turn, were used to modify the mortality and stage duration parameters used in the matrix projection model.

The matrix projection model was calibrated by adjusting the juvenile-stage mortality rates so that, in the absence of PCB exposures, the population would be stable, with interannual variability in juvenile abundance being similar to the variability observed in long-term monitoring data collected in Virginia and North Carolina. PCB exposures were simulated assuming that juveniles reared in North Carolina nursery areas are exposed during juvenile development, with effects expressed when the females spawn 1 to 2 y later. Fecundity and egg survival in the model were reduced by values observed in the laboratory experiments. The growth, development, and survival of PCB-exposed larvae were followed during the estuarine nursery phase of the life cycle, with the predator avoidance and swim speed parameters reduced according to the results of the regression tree analysis.

Two exposure scenarios were evaluated. In the first, females were assumed to be impaired only during their first spawning, and to completely depurate their PCBs during that spawning event. In the second, females were assumed to be impaired throughout their entire lifetime.

Rose et al. (2003) found that, under the first-time-only scenario, the predicted effects of PCBs on the long-term abundance of Atlantic croaker were negligible. Under the lifetime impairment scenario, long-term average abundance was about 10% lower than the baseline. In addition to demonstrating that their model could be used to link behavioral toxicity data to long-term effects on a population, the authors suggested that in the future, a better-verified version of the model could be used to quantify cumulative impacts of multiple stresses, such as PCB exposures combined with increased harvesting.

27.4   FUTURE OF POPULATION MODELING IN ECOLOGICAL RISK ASSESSMENT

The examples discussed in this chapter cover a wide range of applications of population models in ecological risk assessment. The first edition of this textbook suggested that projection matrices, stochastic extinction models, and individual-based models would in the future be used in ecological risk assessments performed to support Superfund assessments, pesticide risk assessments, natural resource damage assessments, and other types of regulatory activities. Research applications are clearly widespread, but applications of population models in chemical risk assessment and management are still relatively uncommon.

The US Environmental Protection Agency's (US EPA) recent guidance on endpoints for ecological risk assessments (EPA 2003) identifies several population-level endpoints (extirpation, abundance, and production) as being relevant to the agency's assessments; however, the agency's primary focus is still on organism-level attributes such as morphological anomalies, survival, reproduction, and growth (Chapter 16). The agency justifies this emphasis on the grounds of legal requirements, regulatory precedents, and practicality. Laws and regulations are beyond the scope of assessment science; however, practicality is an issue that can be directly addressed through research, demonstration, and guidance development.

A recent workshop on population-level ecological risk assessment (Barnthouse et al. 2006) identified a number of steps that could be taken to increase the use of population-level methods in ecological risk assessments. The most obvious of these steps is development of guidance documents explaining the availability, use, and interpretation of common modeling tools. Such guidance would cover the selection of models suitable for different types of assessment problems, methods for parameter estimation, and rules for model use and interpretation. Guidance on field data collection relevant to population-level assessment would likely also be needed. Beyond technical guidance, broader guidance intended to inform risk managers, and stakeholders concerning how, and under what circumstances, population-level assessment tools can lead to better environmental decisions. The workshop report includes a framework for population-level ecological risk assessment analogous to the well-known framework in the EPA's guidelines (EPA 1998a), and also contains recommendations for incorporating population-level considerations in risk management decisions.

Training and education are also important. Risk assessment practitioners would benefit from enhanced opportunities for training in population ecology theory, empirical field and laboratory methods, and GIS technology. Beyond training programs, perhaps the best single educational activity would be actual application of population models in one or several high-profile assessments. There can be no doubt that the widespread development and use of metapopulation models in conservation biology were enhanced by the use of these models to assess impacts of habitat fragmentation on the northern spotted owl. Similar case-specific applications involving environmental chemicals would educate both risk managers and the assessment practitioners themselves concerning the benefits, limitations, and proper use of population models in ecological risk assessments.

Even though population models are not yet being routinely used in ecological risk assessments, the integration of ecotoxicology and population biology that was envisioned by the authors of this book 20 y ago has occurred. Approaches originally developed for use in resource management and conservation biology and even for purely theoretical purposes are now being applied in ecotoxicology. Equally important, young scientists and assessment practitioners are entering the field with the training and expertise needed to understand, use, and advance these approaches.

Future editions of this book, or maybe successors to this book, will very likely discuss concrete regulatory applications of population models, and will provide specific recommendations concerning the uses of these models based on successes and failures in on-the-ground applications.

28 Ecosystem Effects Modeling

Steven M. Bartell

We must seek to model somehow the passing of a butterfly along with the growing of a tree.

Allen and Starr (1982)

28.1 AN ECOSYSTEM PARADIGM

The ecosystem remains a fundamental conceptual unit not only in basic ecological research, but also in environmental assessment and management. The limitations of organism-level ecological assessments raised decades ago remain (NRC 1981; O'Neill and Waide 1981; Kimball and Levin 1985). To address these limitations, ecologists, environmental toxicologists, and risk assessors have continued to develop, apply, and evaluate methods and models for characterizing ecosystem-level risks (Pastorok et al. 2002). Ecosystem modeling continues to contribute importantly to assessing ecological risk.

This chapter defines and describes ecosystem risk assessment and emphasizes the use of ecosystem models for estimating risk. In this presentation, *ecosystem models* include physical models (e.g., micro-, mesocosms), network analytical models, and compartmental simulation models. Physical models are discussed briefly. The intent is to underscore the similarities in issues (e.g., model structure, scale, initial conditions) that must be addressed in effectively using physical and mathematical models to characterize risk. Network analytic techniques (e.g., flow analysis, loop analysis) are mentioned because they offer a largely unrealized potential application in ecosystem risk assessment. Not surprisingly, the majority of the presentation addresses ecosystem simulation models as tools for assessing risk.

The chapter identifies and describes several ecosystem simulation models (AQUATOX, CASM, and IFEM) available for assessing ecological risks posed by chemical contaminants and other agents. These models have been developed with ecological risk assessment as a principal modeling objective. This chapter does not present an exhaustive list of ecosystem models that might be used to assess risk. Pastorok et al. (2002) comprehensively reviewed existing ecosystem models that might be adapted for estimating ecological risks, and the interested reader should consult this reference. Nevertheless, criteria for selecting among existing models are presented within this chapter. Following a discussion of the relative strengths and limitations of ecosystem models in assessing risk, the focus shifts to adapting available ecosystem models and developing new models for ecosystem-level risk assessment.

The development of practical capabilities in assessing ecosystem-level risks cannot proceed independently from the continuing evolution of ecosystem concepts and theory (e.g., Golley 1993; O'Neill 2001). Ecosystems are not simply places and the term "ecosystem" should not be used in risk assessment simply to denote habitat. Perhaps the most significant contribution of ecosystem theory to modern ecology includes the recognition of important biotic–abiotic feedback mechanisms that determine the dynamics of system structure and function. That is, physical–chemical factors dictate the nature and kinds of organisms that can inhabit a specific area or volume. In turn, the effects of resident organisms on those factors can result in subsequent habitat conditions that preclude the continued occupation by those organisms and open the area for new inhabitants. The essence of *ecosystem* lies not in habitat, but in the biotic–abiotic feedback mechanisms that strongly influence system dynamics and response to disturbance. Ecosystem-level risk assessments ought to rightly focus on risks to these feedback mechanisms.

Another conceptual contribution of ecosystem theory lies in the recognition of the significance of asymmetry in functional relationships among organisms and between biotic and abiotic components of ecosystems. Not all structures, processes, and interactions are of equal importance at all times or locations. Ecosystem dynamics integrate spatial–temporal shifts in interaction strengths (e.g., competition, grazing, predation) among participating organisms within a changing physical–chemical context. Seasonal changes in the relative importance of "bottom–up" and "top–down" control of production in aquatic systems provide one example of such asymmetry (e.g., Bartell et al. 1989).

Asymmetry in ecosystems is important to risk assessment. Characterization of ecosystem asymmetry can provide insights to the selection of endpoints for risk assessment (Chapter 16) and suggest relevant scales in time and space (Chapter 6) to guide the development of a conceptual model (Chapter 17). Knowledge of relevant scales of exposure can be used to identify corresponding species and ecosystem processes that appear important in relation to the hazardous agents.

Ecosystem asymmetry facilitates the simplifying assumptions that provide a basis for less comprehensive descriptions (e.g., population models) used to estimate risks. If a population model provides accurate estimates of measured population fluctuations, it is because the simplifying assumptions underlying the population model are congruent with the overarching ecosystem asymmetries relevant to the population model.

Properly structured ecosystem models afford the opportunity to incorporate biogeochemical asymmetries and examine their implications for estimating risk. For example, positive effects of nutrient enrichment on ecological production can mask the detrimental effects of simultaneous exposure to toxic chemicals (e.g., Breitburg et al. 1999; Riedel et al. 2003). Differential affinity for nutrients and susceptibility to toxic chemicals can together determine how any added productivity will be apportioned among the food web components. Such opportunity to explore asymmetry in ecosystem structure and function is absent from other ecological modeling constructs (e.g., organism models, population models, landscape models).

28.2 ECOSYSTEM RISK ASSESSMENT

Following from the proposition that asymmetric functional relationships and biotic–abiotic feedback control mechanisms are key concepts that distinguish ecosystem ecology from other levels of ecological inquiry, ecosystem risk assessment should correspondingly focus on changes in functional relationships and feedback mechanisms caused by single or multiple agents. In the proper parlance of ecosystem theory and risk assessment, ecosystem risk assessment ought to emphasize assessment and measurement endpoints specifically related to effects on the flow of energy, cycling of materials, strengths of competitive, grazing, and predator/prey interactions, and corresponding implications for system structure (e.g., species composition, community structure), function, and stability (e.g., resistance, resilience). These are the kinds of endpoints that cannot be addressed by assessments that focus on organisms or populations.

In practice, ecosystem risk assessment tends to emphasize population-level effects characterized within a dynamic physical–chemical context. Explorations into the relative contribution of direct and indirect effects on population risks do, however, provide examples of ecosystem models used to address some of the biogeochemical (minus the geochemical) asymmetries that connote ecosystem risk.

28.2.1 Ecosystem Assessment Endpoints

The specification of ecosystem-level endpoints follows logically from the preceding consideration of ecosystems. Suter and Bartell (1993) identified four kinds of ecological effects that can be observed in ecosystems, but are not possible for an organism or a population: (1) the effect of an agent on the nature of ecological interactions (e.g., predation, competition) among resident populations; (2) indirect effects that propagate through organisms sensitive to the agent and subsequently impact organisms not directly affected (e.g., reduced abundance of a predator resulting from toxic effects on prey); (3) alterations in the trophic structure or number of species; and (4) alterations in ecosystem function, including production, decomposition, and nutrient cycling. Suter and Bartell (1993) distinguished between assessment of population-level effects (i.e., effects 1 and 2) in an ecosystem context and true ecosystem-level effects (i.e., effects 3 and 4).

One of the key ecosystem concepts concerns feedback control mechanisms between biotic and abiotic system components. Abiotic factors (e.g., soil chemistry) can importantly determine the growth and establishment of species adapted for the existing conditions—abiotic factors determine ecological structure. Subsequently, the biological activities of resident organisms can modify the abiotic conditions to the point that the resident species can no longer tolerate the modified conditions and different species adapted to these conditions can replace the current inhabitants—here, biological activity determines the abiotic environment and eventual ecological structure. Thus, alterations in evolved patterns of abiotic–biotic feedback control mechanisms could pose serious threats to ecosystem integrity. Alterations in such patterns would in theory constitute important ecosystem-level endpoints. Yet, these sophisticated ecosystem endpoints have seldom been included in ecological risk assessments. Ecosystem models can be used to address these kinds of endpoints.

28.3 ECOSYSTEM SIMULATION MODELING

For this discussion, ecosystem simulation models refer to those ecological models which include both biotic state variables that describe one or more primary producers and consumers and one or more abiotic state variables or processes that are functionally linked to the biotic state variables. Their models should demonstrate functional interrelationships, e.g., grazing, predation, and competition, expressed between the producer and consumer state variables. The abiotic factors, e.g., light or nutrient limitation of primary production, should influence the expression of the biological and ecological interaction represented by the model. Importantly, the biological and ecological processes included in the model should permit the modification of the abiotic state variables (e.g., nutrient uptake or remineralization influencing dissolved or soil nutrient concentrations).

An ecosystem risk assessment model should be spatially defined. The modeled temporal dynamics should be specified over some spatial scale, e.g., a square meter or hectare for models of terrestrial ecosystems or similar volumetric scales for aquatic ecosystem models. Recent advances in ecosystem modeling have produced spatially articulated models wherein a single description of biotic and abiotic structures and interactions are defined repeatedly for multiple locations that are functionally interconnected by the flow of water, energy, or materials (e.g., Costanza et al. 1990; Bartell and Brenkert 1991; Voinov et al. 1998).

Several features of ecosystem models strongly recommend them for assessing ecological risks. The structural and functional complexity of ecosystem models provides risk assessors with tools to estimate both direct and indirect effects. The implications of differential susceptibility to chemical and other agents developed from single species tests can be explored in the context of system-level effects on structure and function. For example, ecosystem models such as CASM (Bartell et al. 1999, 2000) and AQUATOX (Park and Clough 2004) that define multiple populations within individual trophic guilds can be used to examine the indirect effects of chemicals on competitive and predator/prey interactions. Normally inferior competitors may gain the upper hand if their counterparts prove more sensitive to a chemical. Populations of prey species might increase substantially if their predators succumb more quickly to exposure to a chemical or other agent. Apart from costly and time-consuming field manipulations, ecosystem models provide the only means to address these kinds of direct and indirect effects that can propagate throughout complex ecological systems.

Ecosystem models can address ecological risks posed by simultaneous exposure to multiple agents of differing kinds. For example, CASM can be used to estimate risks posed by a combination of several toxic chemicals (organic and inorganic), nutrient enrichment (N, P, Si), sediment loading, depletion of dissolved oxygen, and fishing pressure, if necessary. Spatially explicit ecosystem models can also examine the implications of spatial patterns in habitat degradation and loss, as well as the effects of regional pollution and climatic change.

More detailed and explicit representation of ecological structure and function suggest that ecosystem models are more realistic descriptions of complex ecological systems (Pastorok et al. 2002; Bartell et al. 2003). Ecosystem models provide the capability to address subtle, but important, ecosystem-level endpoints, such as energy flow, nutrient cycling, alterations of abiotic–biotic feedback control mechanisms, and system stability (i.e., resistance and resilience). Ecosystem models emphasize the description of ecological systems as complex networks that propagate cause and effect, where the network complexity partly reflects current description and observations relevant to the system of interest. The network complexity also results from the biases introduced by the model makers and the specific nature of the assessment. The currency of flows through these complex networks can be energy (i.e., joules) or its material equivalents (e.g., carbon, dry mass, nutrients). Ecosystem models, as representative of complex and ecologically realistic networks, can be used to examine the probable ecological implications of even subtle alterations in these kinds of flows, whereas models of organisms or populations, given their structural limitations, cannot.

Ecosystem models can, in addition to addressing multiple and complex assessment endpoints, potentially provide insights for risk management and decision making that cannot be obtained using models of organisms or populations. The strictly empirical parameters of statistical models (e.g., regression coefficients) usually defy interpretation in relation to management practices; the models may prove accurate in estimating risk, yet provide little utility to managers who desire to use the models to reduce or mitigate risks. Similarly, the highly aggregated parameters of some population models (e.g., the carrying capacity, K) are difficult to use in managing risk. The process-level equations and parameters generally characteristic of ecosystem models are directly interpretable in terms of physical, chemical, biological, and ecological phenomena that underlie the model. This detailed level of description provides information that can be used to develop and evaluate the likely success of alternative management actions.

28.3.1 PHYSICAL ECOSYSTEM MODELS

Physical model ecosystems (e.g., microcosms, mesocosms, whole-system manipulations) provide an alternative approach to characterizing ecological risk (Section 24.3). The appeal of these "tangible" ecosystem models is not surprising. Risk can be characterized using the results of controlled and replicated experiments: organisms can be counted; chemistry can be analyzed; and variability in responses can be quantified. These attributes engender a perception of reality associated with physical ecosystem models.

At the same time, it should be remembered that the derivation and use of physical ecosystem models are subject to many of the same assumptions, limitations, and sources of uncertainty as their mathematical counterparts. In constructing or excising physical ecosystem models, decisions must be made concerning scale (i.e., physical dimensions) and how much ecological structure should be included and measured (Gardner et al. 2001). Initial values of the physical model "state variables" must be determined. The environmental context (e.g., light, temperature, precipitation) for physical models has to be defined or simply used as an uncontrolled regime defined by local conditions. All of the sources of bias and imprecision involved in sampling frequency, sample collection, sample processing, and data management are inherent to the use of physical ecosystem models. In addition, the resources required to use physical models routinely limit the number of replicates, and associated variability among replicate models can be substantial. Finally, the results of the physical models must be interpreted within the context of the ecosystem that they are intended to represent.

28.3.2 ECOSYSTEM NETWORK ANALYSIS

Ecosystems can be conveniently described using networks (e.g., Figure 28.1). The "box and arrow" schematic illustrations of ecosystem structure and function have been basic to ecological instruction for decades (e.g., Odum 1971) and practicing ecologists are familiar

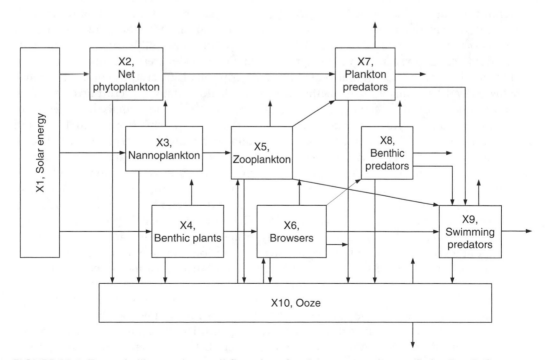

FIGURE 28.1 Example "box and arrow" flow chart for the ten-compartment Cedar Bog Lake ecosystem model. (Redrawn from Williams, R.B., in *Systems Analysis and Simulation in Ecology*, Vol. I, Patten, B.C., ed., Academic Press, New York, 1971. With permission.)

with such conceptual network models. Beyond mere illustration, more formal network models and network analyses of compartmental descriptions of ecosystems provide powerful tools for qualitative and quantitative understanding of ecosystem structure and function. These ecosystem network tools can help risk practitioners to characterize ecological risks. The following sections briefly outline several of these network ecosystem models.

Qualitative network analyses (e.g., "loop" analysis) can be used to describe the stability of ecosystems in the absence of detailed quantification of interactions among system components (Levins 1974). Simple knowledge of positive, negative, and neutral interactions among system components can be used to develop an interaction matrix, sometimes referred to as a "competition matrix" (Levins 1974; Lane and Collins 1985). Relationships among system components are designated by signs: $(+,-)$ denotes a predator–prey relationship; $(-,-)$ indicates competition between two components. The system is described as a directed graph, wherein each component is represented as a circle and interactions are described by connecting arrows. Mathematical analysis of the numbers and locations of these various signed interactions throughout the matrix can provide information concerning the general stability properties of the system. Relevant to risk assessment, the methods can be used to examine the stability implications of anthropogenic changes in the nature of the interactions or removal of a system component (e.g., Lane and Collins 1985). The sensitivity of these analyses to initial system specification and the overall qualitative nature of the approach have minimized the application of loop analysis since its introduction into ecology. In addition, the quantitative magnitudes of component interactions are likely more important than the overall qualitative network structure in determining the stability of ecological networks. However, Ortiz and Wolff (2002) have recently revitalized the use of this method. They apply loop analysis to evaluate the qualitative stability properties of north-central Chilean coastal marine benthic systems in relation to harvesting pressures on scallops. This recent application serves as a reminder that qualitative methods of ecosystem analysis are possible and that these kinds of analyses remain potentially useful as tools in contemporary assessment of system-level ecological risk. Given the incomplete and sparse data routinely available for ecosystem characterization, this ecosystem modeling approach might provide useful results for decision making based on more easily obtained qualitative description of structurally complex systems.

In the early 1970s, quantitative methods of network flow analysis were borrowed from economics and applied to ecological systems (Hannon 1973). Flow analysis was further developed for quantitative description of nonsteady-state ecosystems (e.g., Finn 1976) and subsequently elaborated for detailed characterization of hierarchical structure within ecological networks (Patten et al. 1976). Central to flow analysis is the construction of an input–output or production matrix (Figure 28.2), wherein the quantitative flows among all interconnected system components are estimated along with inputs to, and losses from, these components. The inputs, outputs, and intercompartment flows can be measured or produced by corresponding dynamic models (Bartell 1978). The production matrix provides a detailed description of system state and patterns of energy or material flux, either at steady-state or as a snapshot at a selected point in time for nonsteady-state conditions. Normalization of flows among components to inputs (outputs) provides quantitative information concerning the change in each component associated with a unit increase or decrease in input (output). Additional calculations can be made to estimate the total flow of energy or material through the system over a specified time scale. It is further possible to estimate a ratio of flow that is recycled to total system throughflow. This ratio quantifies the cycling efficiency of the system represented by the production matrix (Patten et al. 1976). Total system throughflow and cycling efficiency could serve as higher-order, ecosystem-level endpoints in ecological risk assessment.

To/from	x_1	x_2	..	x_n	Inflow	Outflow	Row Σ
x_1	φ_{11}	φ_{12}	..	φ_{1n}	z_{10}	0	T_1
x_2	φ_{21}	φ_{22}	..	φ_{2n}	z_{20}	0	T_2
.	.	.		.	.	.	.
.	.	.		.	.	.	.
x_n	φ_{n1}	φ_{n2}	..	φ_{nn}	z_{no}	0	T_n
Inflow	0	0	..	0	0	0	0
Outflow	y_{01}	y_{02}		y_{0n}	0	0	Σy
Column Σ	T_1	T_2	..	T_n	Σz	0	TST

FIGURE 28.2 Generalized production or flow matrix.

The main challenge in using flow analysis to describe ecosystem dynamics lies in accurately quantifying inputs, outputs, and flows among system components. This challenge increases nonlinearly as the number of components in the production matrix increases. The necessary values can be provided by detailed field studies (i.e., physical ecosystem models). Alternatively, dynamic models can be used to provide the value needed to develop a production matrix for the system of interest (e.g., Bartell 1978). In this case, the methods of flow analysis are used as another means to summarize modeled system dynamics. This integration of dynamic models with flow analysis can be used to assess potential effects on higher-order ecosystem endpoints (e.g., total system throughflow, cycling efficiency).

Flow analysis can be used to characterize ecosystem risk if exposure–response relationships are developed to quantify changes in flows, inputs, or outputs in relation to the agent of interest.

Figure 28.3 shows a production or flow matrix developed for the Cedar Bog Lake ecosystem. The values reflect the system inputs, outputs, and flows corresponding to the network compartmental diagram of this ecosystem. The balance between inflows and outflows characterizes a system that is in a state of dynamic equilibrium. The steady-state dynamics can also be inferred from the linear differential equations with constant coefficients used to describe this system (see Section 28.3.3).

To/from	x_2	x_3	x_4	x_5	x_6	x_7	x_8	x_9	x_{10}	Inflow	Outflow	Row Σ
x_2										13.1		
x_3										21.4		
x_4										61.4		
x_5		0.68										
x_6			0.21									
x_7	1.01			0.90								
x_8				0.17	0.48							
x_9					0.28	0.19	0.07					
x_{10}	8.79	15.32	45.99	5.13	0.20	0.30	0.07	0.15				
Inflow										0	0	0
Outflow	3.30	5.40	15.20	3.90	0.52	1.42	0.34	0.56	65.35	0	0	95.99
Column Σ										95.90	0	~96

FIGURE 28.3 Flow matrix based on the Cedar Bog Lake model.

Recent methods for network analysis have been derived in part to address the inability to completely define a network for any single ecosystem or comprehensively quantify the production matrix in flow analysis. These newer methods (ECOPATH/ECOSIM, NTWRK) focus on equilibrium conditions and attempt to identify plausible network structures that are compatible with available information that quantifies flows among specified system components (Christensen and Pauly 1992; Monaco and Ulanowicz 1997). These methods have been used to assess effects of selected agents on aquatic systems (e.g., Pauly et al. 2000); however, these studies have not been performed within a formal framework for ecological risk.

The preceding network methods for describing ecosystems are offered as complementary analyses to the more common ecosystem compartment simulation models described in Section 28.3.3. Importantly, the network analyses can be used to characterize changes in the patterns and magnitudes of energy flow or material cycling in systems subjected to pollution. Even the more qualitative method of loop analysis can provide some insights concerning alterations in the stability properties of ecosystem networks.

28.3.3 COMPARTMENT MODELS

The term "ecosystem model" has become more routinely associated in risk assessment with complex compartment simulation models (Pastorok et al. 2002). The ecosystem compartment model includes some calculus that describes time-dependent changes in the values of the model state variables as functions of changes in the inputs of abiotic factors and the values of the biotic state variables. Traditionally, ecosystem models have been formulated as coupled differential (or difference) equations, where one governing equation is defined for each modeled state variable, biotic and abiotic (e.g., Smith 1969; Patten 1971; Park et al. 1974). The values of the state variables change through time as a function of the model equations that describe ecological and environmental processes internal to the model, e.g., nutrient-dependent primary production, temperature-dependent grazing, predator/prey relations, and decomposition. Modeled temporal dynamics can also result from the time-varying input values of external environmental factors (e.g., temperature, nutrient loading, toxic chemical concentrations).

The temporal dynamics of each state variable defined in the ten-compartment model of the Cedar Bog Lake can be described by the following system of coupled differential equations (Williams 1971):

$$X_1' = 118,625 \, \text{cals cm}^{-2} \, \text{y}^{-1}$$
$$X_2' = f_{12} - X_2 \left(\rho_2 + \mu_2 + \varphi_{27} \right)$$
$$X_3' = f_{13} - X_3 \left(\rho_3 + \mu_3 + \varphi_{35} \right)$$
$$X_4' = f_{14} - X_4 \left(\rho_4 + \mu_4 + \varphi_{46} \right)$$
$$X_5' = \varphi_{35} X_3 + \varphi_{10,5} X_{10} - X_5 \left(\rho_5 + \mu_5 + \varphi_{57} + \varphi_{59} \right)$$
$$X_6' = \varphi_{46} X_4 + \varphi_{10,6} X_{10} - X_6 \left(\rho_6 + \mu_6 + \varphi_{68} + \varphi_{69} + \lambda_6 \right)$$
$$X_7' = \varphi_{27} X_2 + \varphi_{57} X_5 - X_7 \left(\rho_7 + \mu_7 + \varphi_{79} + \lambda_7 \right)$$
$$X_8' = \varphi_{68} X_6 - X_8 \left(\rho_8 + \mu_8 + \varphi_{89} + \lambda_8 \right)$$
$$X_9' = \varphi_{59} X_5 + \varphi_{69} X_6 + \varphi_{79} X_7 + \varphi_{89} X_8 - X_9 \left(\rho_9 + \mu_9 + \lambda_9 \right)$$
$$X_{10}' = \mu_2 X_2 + \mu_3 X_3 + \mu_4 X_4 + \mu_5 X_5 + \mu_6 X_6 + \mu_7 X_7 + \mu_8 X_8 + \mu_9 X_9 - X_{10} \left(\rho_{10} + \varphi_{10,5} + \varphi_{10,6} + \lambda_{10} \right)$$

Although simpler in form than other aquatic ecosystem models (e.g., CASM, AQUATOX), the Cedar Bog Lake model demonstrates features that are characteristic of these kinds of models. The model indicates losses to respiration (ρ_i), mortality losses to the "ooze" or sediments (μ_i), and physical losses from the system (λ_i) for each compartment i. Trophic transfers from compartment i to j are designated by the φ_{ij} terms. The same terms common to two or more equations define the functional interconnections between the model state variables as illustrated by the "arrows that connect the boxes"' in Figure 28.1. Indirect effects manifest themselves through these functional interconnections.

The critical requirement in selecting or developing an ecosystem model for assessing ecological risks is the derivation of a functional relationship between exposure to the agent and an associated response of one or more of the model process formulations or state variables. In addition, the exposed process formulations and model state variables should map onto the endpoints of interest for the assessment.

28.3.4 EXISTING ECOSYSTEM RISK MODELS

Pastorok et al. (2002) reviewed the ecological modeling literature and identified only three ecosystem models that were developed to estimate ecological risk. These models essentially extrapolate the results of laboratory toxicity assays to anticipated effects in complex aquatic systems. Given the emphasis on aquatic toxicity testing, it is not surprising that terrestrial ecosystem models are absent from the list of available models. Each of the available models is briefly described. The interested reader is referred to the provided references for detail.

28.3.4.1 AQUATOX

AQUATOX (Park and Clough 2004) models the fate and effects of toxic chemicals, nutrients, and sediments in a variety of aquatic systems, including lakes, ponds, streams, and reservoirs. This model estimates risks posed by these agents on modeled populations of aquatic producers (e.g., phytoplankton, periphyton, submersed aquatic plants) and consumers (e.g., zooplankton, benthic invertebrates, and several functionally defined guilds of fish). AQUA-TOX addresses lethal and sublethal effects on important ecological processes, including photosynthesis, consumption, reproduction, and mortality. Such toxic effects are integrated to estimate the impacts of chemicals on the daily sizes of modeled aquatic populations. Toxic effects are modeled in relation to biologically available chemical, which is determined by modeled chemical transport and fate processes (e.g., sorption, hydrolysis, volatilization, photolysis). The model has been developed within a user-friendly interface to facilitate site-specific applications. Necessary input data include nutrient, sediment, and toxic chemical loadings to the system of interest; general limnological characteristics of the site; growth characteristics of each modeled population; and sensitivity of each population to the agents of concern.

28.3.4.2 CASM

The Comprehensive Aquatic Systems Model (CASM) is a bioenergetics-based compartment model that describes the daily production of biomass (carbon) by populations of aquatic plants and animals for an annual cycle. CASM permits site-specific specification of food web structure and delineation of daily values of surface light intensity, water temperature, and nutrients (N, P, Si) that determine rates of photosynthesis of modeled plant populations. The model provides for as many as 30 populations of phytoplankton, periphyton, and macro-phytes. Up to 40 populations of zooplankton, benthic invertebrates, decomposers, and fish can be specified. Modeled populations can be defined taxonomically or functionally. The model was designed originally to examine theoretical relationships between food web structure, nutrient cycling, and ecosystem stability (DeAngelis et al. 1989). Since its adaptation for risk estimation, CASM has been applied to generic assessments for rivers, lakes, and reservoirs in Canada (e.g., Bartell et al. 1999), as well as smaller lakes in central Florida (Bartell et al. 2000). It has also been implemented for site-specific assessments of ecological risk posed by chemicals in Lakes Biwa and Suwa, Japan (Naito et al. 2002, 2003). CASM has been designed for probabilistic risk estimation using Monte Carlo simulation and characterizes risk as the probability of specified decreases in the annual production of each modeled population. It has also been adapted to assess more site-specific risks posed by pesticides in littoral ecosystems (i.e., the Littoral Ecosystem Risk Assessment Model (LERAM); Hanratty and Stay 1994).

28.3.4.3 IFEM

The Integrated Fates and Effects Model (IFEM) is an integration of the toxic effects model Standard Water Column Model (SWACOM) (Bartell et al. 1992) and a polycyclic aromatic hydrocarbon (PAH) fate model Forecasting Ocean Assimilation Model (FOAM) (Bartell et al. 1981). It combines environmental fate processes, bioaccumulation, bioenergetics descriptions of growth, and toxicity data to estimate the probable effects of PAHs on the production dynamics of lotic ecosystems (Bartell et al. 1988). Sublethal toxic effects of accumulated PAHs are modeled for 11 representative populations of aquatic plants and animals; toxic effects and risk are estimated as a function of a dynamic body burden. Body burden reflects the differential uptake, metabolism, and depuration of PAHs. Available PAH is determined by loading rate and environmental fate processes (dissolution, photolysis, sorption, volatilization). Fate process rates can be estimated using quantitative structure–activity relationships developed for PAHs. The data demands of the IFEM have thus far permitted only assessment of risks posed by naphthalene (Bartell et al. 1988).

28.4 MODEL SELECTION, ADAPTATION, AND DEVELOPMENT

Because few ecosystem risk assessment models are available (Pastorok et al. 2002), risk assessors interested in applying ecosystem models will likely be challenged to (1) adapt an existing model or (2) develop a new model. The following discussion addresses these challenges.

28.4.1 MODEL SELECTION

The first step in selecting an existing model is to identify candidate ecosystem models. Pastorok and Akçakaya (2002) recommended nine criteria for evaluating the potential selection and application of ecological models, including ecosystem models, for assessing risks posed by toxic chemicals. Their recommendations included six technical and three regulatory criteria:

Technical criteria
   1. Model realism and complexity
   2. Relevance of ecological effects addressed by the model
   3. Flexibility
   4. Characterization of uncertainty
   5. Degree of development, consistency, and validation
   6. Ease of parameter estimation

Regulatory criteria
   1. Acceptance among regulators
   2. Credibility
   3. Resource efficiency

TABLE 28.1
Brief Description of Nine Criteria for Selecting an Ecological Model
for Adaptation to Risk Assessment

Criterion	Description
Technical	
Model realism	The model includes ecological structure and processes known to be important in determining the dynamics of the ecosystem of interest in the assessment. Model assumptions are realistic in relation to ecological understanding of the system.
Relevance of ecological effects	The kinds of model calculations (e.g., change in biomass, trophic structure, energy flow, material cycling) can be easily mapped onto one or more of the ecosystem-level assessment endpoints.
Flexibility	The model can be implemented for systems similar to its original derivation without major restructuring or reformulation of governing equations, major alteration of external forcing functions, or redefining model parameters and outputs.
Characterization of uncertainty	The model has been developed to explicitly describe and include potential sources of bias and imprecision in its calculus. Model outputs reflect the uncertainties propagated through model calculations (e.g., distributions, intervals, fuzzy numbers).
Development, consistency, and validation	The physical manifestation of the model (e.g., spreadsheet, commercial software, custom program) has become essentially error-free (i.e., "debugged," verified). The user is warned of potentially erroneous input values in model applications. The model has been compared with observations from systems similar to the ecosystem of interest; model biases have been characterized.
Parameter estimation	Model input values can be estimated from commonly available data. Model parameters have clear ecological or toxicological interpretation.
Regulatory	
Acceptance	The model is frequently used by the regulatory community or the models results are routinely accepted as useful by regulators and decision makers.
Credibility	Previous model applications have been peer-reviewed and accepted by the technical community; the model has been widely published and it is generally familiar to ecological modelers.
Resource efficiency	The time and effort required to adapt the model to a particular assessment does not discourage selection of the model.

Source: Summarized from Pastorok, R.A. and Akçakaya, H.R., in *Ecological Modeling in Risk Assessment—Chemical Effects on Populations, Ecosystems, and Landscapes*, R.A. Pastorok, S.M. Bartell, S. Ferson, and L.R. Ginzburg, eds., Lewis Publishers, Boca Raton, FL, 2002. With permission.

Pastorok and Akçakaya (2002) discuss each of these reasons for selecting an existing model in useful detail. These criteria are briefly annotated in Table 28.1.

28.4.2 Model Adaptation and Development

If an ecosystem-level assessment is necessary or desired, the risk assessor might be able to adapt an existing model or might be forced to develop a new model. The following sections describe some of the key issues to be addressed in either situation. More in-depth treatments concerning the development of ecological models were provided in some of the earlier references on ecological modeling (e.g., Patten 1971, 1972; Levin 1974; Hall and Day 1977; Halfon 1979; Shugart and O'Neill 1979). The detailed instructions for model building and application provided by these pioneers remain largely relevant today. Recent treatments include Odum and Odum (2000) and Swartzman and Kaluzny (1987). Additionally, the works edited by Hall and Day (1977) and Halfon (1979) include case study applications of ecosystem models to environmental problems, although these studies predate "ecological risk assessment" per se. Decades prior to the formalization of ecological risk assessment (EPA 1992a), ecosystem scientists were examining the usefulness of ecosystem models in assessing human influences on natural systems (e.g., Loucks 1972).

In adapting an existing model or developing a new model, the risk practitioner must address model structure, model process, scaling, exposure–response relationships, necessary input data, model results, and model performance.

28.4.2.1 Model Structure

Model structure refers to the ecological entities that are represented in the model. In the modeling lexicon, structure defines the state variables in the model and there will be one governing equation for each state variable. The "boxes" in an ecosystem "box and arrow" flow diagram identify the model state variables and thus model structure. Examples of biotic state variables in an ecosystem model include the numbers, biomass, or energy equivalent of two or more kinds of organisms. Concentrations of particulate organic matter and dissolved inorganic nutrients (e.g., N, P) are examples of abiotic state variables in ecosystem models. Importantly, the model must contain state variables that correspond to the assessment endpoints identified in the problem formulation.

In adapting or developing an ecosystem model for risk assessment, the risk assessor must examine the feasibility of incorporating ecological structure germane to the assessment if it is not already present. In adapting a model, this might mean adding structure. Adding structure to an existing model requires an evaluation of compatibility with other state variables already in the model. In developing a new model, the necessary structure can be designed at the outset. The challenge will then take the form of deciding how many other state variables (i.e., additional structures) are needed to describe the dynamics of the variables corresponding to endpoints with sufficient accuracy and precision to usefully characterize risk.

28.4.2.2 Governing Equations

In addition to the ecological structure of ecosystem models, structure might also refer to the mathematical structure, i.e., the calculus of the model. For example, many traditional ecosystem models have been designed as sets of coupled differential or difference equations. Methods of calculation can range from simple algebra to analytical calculus to sophisticated schemes for numerical integration. Recent developments in ecosystem modeling have used numerical algorithms and cellular automata or combinations of automata and various forms of equations to determine how the values of the state variables change in space or time.

The equations or mathematical formulations that govern the calculations of the model should be compatible with the nature of the endpoints. For example, if an endpoint is the biomass (e.g., dry weight, carbon) of a particular population, the calculus of the governing equations should be in the same units. Otherwise, a conversion will be required (e.g., kcals to grams carbon) if the model calculations are performed in other units. These kinds of conversions can serve as a source of inaccuracy or imprecision in model performance.

28.4.2.3 Scaling

Here scale refers to the spatial and temporal dimensions explicit to the model. In space, the model describes some spatial subset (i.e., extent) of the biosphere defined by the selected ecosystem boundaries. The spatial resolution (i.e., grain) within this extent defines the smallest spatial unit represented by the model (e.g., 1 m^3). Parallel concepts apply in the temporal domain. Temporal scales include the duration (e.g., 1 y) of the model calculations and the temporal resolution (e.g., daily values of the state variables).

Four scales are important in adapting or developing an ecosystem model for ecological risk assessment. The *ecological scales* relevant to the ecological structure(s) of concern are fundamental to useful model application in support of risk characterization. The ecological scale will have been already determined for models being adapted for risk assessment. The risk practitioner will have to evaluate the scale of the model in relation to the scales appropriate to the assessment at hand. It might prove feasible to rescale an existing model, depending on the specification of the state variables and formulation of the governing equations. It is generally easier to "scale up" or aggregate structure and process than to add finer resolution to an existing model. In developing a model, the model builders can apply basic knowledge of organism life history and previous observations of their ecology to identify spatial–temporal scales that are appropriate for the assessment and compatible with the scales of the exposures. Similar understanding of important physical–chemical processes, augmented by local or regional data, can be used to characterize spatial–temporal variability in environmental forcing functions that must be integrated with biological and ecological scales in the selection of an overall scale for a new ecosystem risk assessment model.

The characteristic spatial–temporal *scales of the agents* must be factored into selecting among existing models, adapting a model, or developing a new model. For example, if the agent is a toxic chemical, the measured or anticipated frequency, magnitude, and duration of exposure can be used to define corresponding temporal scales in the ecosystem model. The spatial extent of an agent can be used to determine a relevant or necessary scale of the ecosystem model in order to effectively characterize risk. Clearly, there must be some overlap in ecological and agent scales for the model to be useful.

The *scales of measurement*, defined by the number, location, and frequency of samples in a monitoring program, determine the quality and quantity of data used to implement and subsequently evaluate the model. Simply stated, as the variability of the measured entity increases in space or time, more and more frequent samples will be required to accurately and precisely quantify it. Scales of measurement also pertain to the agents. As the scales of measurement become increasingly congruent with the inherent ecological or agent scales, the statistical variance estimated from the measured values ought to decrease to a minimum, whereupon additional sampling will not further reduce the variance.

The *scales of risk management* define the spatial–temporal characteristics of actions that risk managers have at their disposal for avoiding, minimizing, or mitigating risks. Management scales are also important in quantifying the possible significance of estimated risks.

In adapting or developing an ecosystem model for risk assessment, efforts should be made to obtain as much overlap as possible among these four scales.

28.4.2.4 Exposure–Response Functions

Given a model structured appropriately for an assessment, the next most important attribute to address is a functional relationship between the model structure representing the endpoints and the exposures central to the assessment. The model must be able to translate a quantitative description of the exposures to one or more agents to corresponding changes in the modeled values of the endpoint state variables so that it is useful in characterizing ecological risks. Ecological risk assessment can be fairly described as examining the implications of uncertain exposure–response functions.

The exposure–response function can assume different forms, depending in part on the nature of the exposure and the response. For chemicals, exposure–response functions are characteristically sigmoidal and monotonic (Figure 23.1). An additional consideration in developing these functions is the existence of a threshold value, the no observed effect concentration (NOEC), which can be incorporated into the overall formulation, shown conceptually in Figure 23.5. Probit functions (Figure 23.4) have proven useful in defining exposure–response functions used by ecosystem models for estimating risks posed by toxic chemicals (e.g., Bartell et al. 2000).

Regardless of the exact nature of an exposure–response function, the risk practitioner should attempt to quantify the uncertainties associated with the function. For example, an exposure–response function for a chemical might be more realistically described by a set of exposure–response functions that address the variability in response associated with organism size or age, depending on how such biological structure is represented in the ecosystem model.

28.4.2.5 Data

A perfect ecosystem model cannot inform the risk assessment process if the necessary supporting data are not available to perform the model calculations. While this is true for all ecological models, ecosystem models, being structurally complex by definition, commonly exhibit greater demands for data to perform the model calculations. The data needs of ecosystem models include values for initial conditions of the model state variables, values of the parameters in the governing equations, and values that quantify any necessary external forcing functions. The beginning values of all the model state variables define the initial conditions of the ecosystem model. The nature of the data required to quantify the initial conditions is largely a consequence of the units selected to describe the dynamics of the state variables. Initial conditions might include population sizes (numbers, biomass, or energy equivalents) of various biotic components. Initial values might also be required for environmental parameters (e.g., light, temperature, nutrient concentrations) that have been incorporated into the model.

The mathematical formulation of an ecosystem model will define the nature of the parameters that determine the dynamics of the state variables. Model parameters can range from simple linear constant coefficients to highly detailed values that are nonlinear functions of other biotic state variables and environmental forcing functions (e.g., temperature). Regardless of their nature, the values of model parameters (e.g., rates of growth, survival, and reproduction) are necessarily derived from site-specific or more generalized sources of data.

Site-specific monitoring programs can provide the physical–chemical data used by ecosystem models. Comprehensive databases maintained by various agencies (USEPA, USGS, USDA, etc.) can be used in the absence of monitoring or to augment sparse data.

Additional data needs include values of the state variables used to compare with the model calculations for purposes of model evaluation (validation). The nature of the calculations performed by the model delineates the data needed to assess model accuracy and precision. For example, time series of population sizes for key model components might be required to evaluate model performance. The criteria for assessing model performance (see below) will determine the level of effort necessary to acquire the needed data.

In practice, the often substantial data needs of ecosystem models are met through a collation of site-specific data, data from similar ecosystems, and data from the technical literature.

28.5   INNOVATIONS IN ECOSYSTEM MODELING

The preceding discussion emphasized more traditional approaches to ecosystem modeling for characterizing ecological risk. This modeling construct was developed by ecosystems ecologists and modelers primarily in the 1970s and has not changed dramatically since then. Nevertheless, there are opportunities for innovation in the development and application of ecosystem models used in risk assessment. Several possible opportunities are described in the following sections.

28.5.1 Structurally Dynamic Models

Traditional approaches to ecosystem modeling have relied on some initial description of system structure (e.g., Figure 28.1). Once implemented, the model structure typically does not change during the course of execution, except perhaps for some of the modeled state variables becoming zero, i.e., effectively removed from the system. Current ecosystem models seldom, if ever, permit the addition of new structure (state variables) while the model is running. Given observations that systems under stress might become increasingly susceptible to invasion by nonnative species (e.g., zebra mussel, golden mussel, round goby), risk assessors might desire a model construct that permits such dynamic addition or deletion of state variables to address this kind of assessment endpoint. It is entirely feasible to develop operating "rules" whereby modeled "novel species" can challenge the current model structure to become established and possibly persist throughout the course of simulation. In the case of invasive species, such rules would include, for example, ecological characteristics of the novel species, corresponding traits of the initial model components, and physical–chemical habitat requirements of the invaders.

28.5.2 Interactive Modeling Platforms

Risk assessors might desire the capability to interactively design and develop an ecosystem model for a particular risk application. User-friendly modeling platforms (e.g., STELLA) can provide this capability. These modeling platforms allow the user to (1) efficiently build and apply models and (2) explore the implications of alternative model formulations in relation to risk estimation. In the hands of a trained modeler, this interactive modeling capability can produce useful results with a minimum investment in time and resources. This same technology can, however, lead to fundamental mistakes in model development and application, if the user does not have the necessary training or experience in ecosystem modeling.

It is in the interest of risk assessors to advocate the continued development of such interactive modeling platforms to facilitate the development of ecosystem models for assessing risk. Commensurate with this continued innovation is the need to train risk assessors in the fundamentals of ecosystem modeling and analysis in order to make full and accurate use of this technology.

28.5.3 Network-Enabled Ecosystem Models

Ecosystem modelers or modeling centers can make ecosystem models accessible for use via the Internet. In addition to downloading existing models, network-enabled modeling capabilities would permit the assessor to actually execute the selected model on some remote server. One advantage of this service would be to make models that require substantial computational power (i.e., multiple, parallel processors; "super-computers") accessible to assessors who might lack access to these kinds of machines.

28.5.4 Ecosystem Animation

Continued advances in computational power and graphic software make it increasingly possible to present the results of ecosystem models as animated sequences of model results—either in space, time, or both. Inspection of such animated model output by the user can help identify interesting patterns in system response that are not obvious from looking at a number of tables or graphs. Mathematical techniques can be used to evaluate identified visual patterns to determine if they are numerically "real" or just perceptions.

Substantial amounts of information can be communicated efficiently through animation of ecosystem model results. Animation might more effectively enter the large volumes of results into risk management and decision making.

These opportunities for innovation in ecosystem risk assessment modeling are being realized to some extent through an integrated modeling and assessment effort involving 14 European countries (Brack et al. 2005). The MODELKEY project comprises an interdisciplinary approach to developing interactive, interlinked environmental fate and effects models (aquatic food web/ecosystem models) for characterizing risks posed by contaminants in freshwater and marine systems. The models are designed to be integrated with a user-friendly decision support system. The decision-support system will apply neural network and Geographical Information System (GIS)-based analysis of predicted effects and composite risk indices to evaluate risks, identify sources of contamination, and set priorities among contaminated sites. The developed models will be verified in case studies that focus on applications in the Mediterranean Sea, as well as selected river basins in western and central Europe.

As another example of future modeling approaches, Sydelko et al. (2001) describe plans for a dynamic information modeling architecture that permits efficient development of object-oriented (OO) simulations. This approach was used to develop an integrated dynamic landscape analysis and modeling system (OO-IDLAMS). The OO-IDLAMS was derived initially as a prototype resource conservation model to inform decision makers in natural resources planning and ecosystem management. Sydelko et al. (2001) emphasize the potential for integrating the OO-IDLAMS with ecological models of chemical uptake and effects in order to forecast the magnitude and extent of contamination and associated ecosystem risk.

28.6   ECOSYSTEM MODELS, RISK ASSESSMENT, AND DECISION MAKING

Ecosystem models have numerous possible roles in environmental decision making and management, as is briefly illustrated in this section. In the first case, an ecosystem model was used to both explore the implications of a test endpoint that is commonly used for screening benchmarks and quality criteria. In the second, the same model was used with the results of several microcosm and mesocosm tests for risks to ecosystems with different structures.

28.6.1   Model Results and NOECs

Recently, efforts have been made to understand the results of ecosystem models in the context of more traditional or familiar ecological benchmarks. Naito et al. (2002) used the CASM to estimate risks posed by seven quite different chemicals in Lake Suwa, Japan. The chemicals included insecticides, herbicides, organic contaminants, and one trace element. These investigators used the model to calculate changes in selected model populations compared to a reference simulation for various exposure scenarios. In addition, the model was used to estimate the probabilities of observing selected percentage decreases (or increases) in production for these same populations. However, a unique aspect of this risk assessment centered on "calibrating" the CASM results to chronic NOECs for zooplankton reported for these chemicals. The implicit hypothesis was that some constant degree of modeled effect on zooplankton would correspond to the zooplankton NOECs across this wide range of chemicals. Analysis of the model results for these chemicals demonstrated that the modeled endpoint of a 20% reduction in total annual zooplankton biomass (designated as the "BR20") correlated well with the NOECs (Figure 28.4).

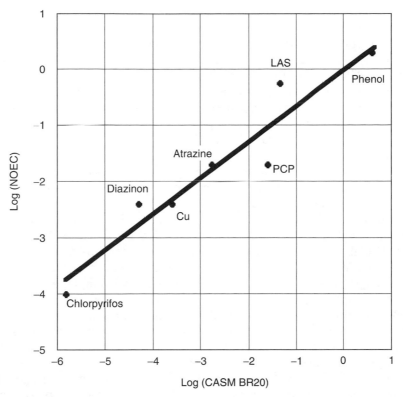

FIGURE 28.4 Correlations between Comprehensive Aquatic Systems Model (CASM)-modeled end-point of a 20% reduction in total annual zooplankton biomass (designated as the "BR20") and estimates of no observed effect concentrations (NOECs) for several toxic chemicals. (Redrawn from Naito et al. 2003. With permission.)

This result is important for several reasons. First, calibrating the model to more familiar toxicological endpoints provides a tool for estimating the zooplankton NOEC for chemicals that can be assessed using CASM, but for which NOECs have not yet been developed. Second, the result demonstrates the relevance and reliability of the CASM in assessing risks across a considerable range of chemicals with different patterns of exposure, modes of toxicity, and species sensitivities. Finally, the correlation of the zooplankton BR20 with the NOECs further supports previous claims (i.e., Bartell et al. 1992) that the underlying stress syndrome used in CASM (and SWACOM) is biased towards conservative (i.e., pessimistic) estimates of risk. That is, an exposure concentration corresponding to no observed effect in laboratory assays produces a 20% reduction in total annual production in the model. The model has a 20% bias towards overestimating risk compared to laboratory assay results. Furthermore, this bias appears consistent across a wide range of chemicals and NOECs. The correlation of the NOECs with the CASM BR20s also provides a decision maker with a better understanding and ability to interpret the ecosystem model results in relation to more traditional species-level endpoints.

28.6.2 Atrazine Levels of Concern

The previous examples and discussion hint at the use of ecosystem models in risk management and decision making. The following discussion describes the use of CASM in determining acceptable levels of atrazine in surface waters subject to agricultural runoff. CASM was implemented to represent, in a generalized manner (i.e., generic application), the food web structure and temporal patterns of production characteristic of second- and third-order Midwestern streams. A CASM reference simulation was developed using a collation of ecological and environmental data from Midwestern streams.

The novel application of CASM in this example lies in using the results of the generic stream ecosystem model to discriminate among the severity of measured effects in micro- and mesocosm studies of atrazine. In all, 25 separate studies were evaluated in relation to 77 reported effects (endpoints) on aquatic plants. Of these, 24 results were from tests on ponds or lakes, 20 on artificial streams, and 33 were microcosm tests. Typically, 1 to 3 concentrations of atrazine were tested in these studies, each with a single application to the test system at initiation. Atrazine concentrations were often kept constant for a variable duration period. Eight effects on plants were recorded on macrophytes, 29 on periphyton, and 40 on phytoplankton. Brock et al. (2000) analyzed a majority of the study results and quantified them as follows: 1 = no effect; 2 = slight effect; 3 = significant effect followed by return to control levels within 56 d; 4 = significant effect without return to control levels during an observation period of less than 56 d; 5 = significant effect without return to control levels for more than 56 d. Several studies not analyzed by Brock, but considered in this analysis, were scored with the same methods.

The 77 effect scores representing the results from the micro- and mesocosm studies were plotted against the study-specific test concentrations and exposure durations in Figure 28.5. The effects on plants observed in micro- and mesocosm studies generally became more severe with increasing exposure magnitude and duration.

The average daily percent difference in modeled plant community similarity between the CASM reference simulation and the simulated effects of atrazine was the principle model result used in evaluating the results of the micro- and mesocosm studies. Corresponding

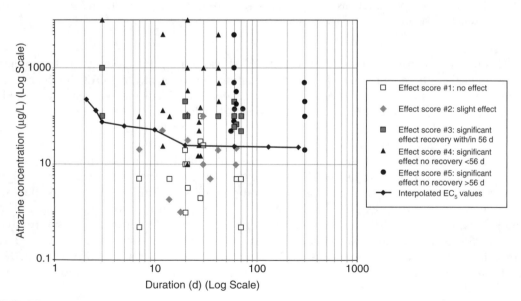

FIGURE 28.5 Concentration–duration interpolations of Comprehensive Aquatic Systems Model (CASM) simulations that produce a 5% change in average total producer community similarity.

relative changes in modeled biomass of phytoplankton, periphyton, and macrophytes were also examined. Based on the results of numerous CASM simulations that used exposure scenarios of varying atrazine concentrations and exposure durations, a decrease of 5% in average community similarity was found to discriminate the studies by Brock et al. with scores of 1 and 2 from the studies scored between 3 and 5 (Figure 28.5). The 5% average deviation in producer community similarity will be used to evaluate the results of field studies that monitor atrazine concentrations in surface waters. If the exposure profile developed from the monitoring results produces an average daily deviation in community similarity that is 5% or greater when analyzed using the generic CASM, the study site may become subject to additional monitoring or remediation. If the resulting modeled average deviation in producer community similarity is less than 5%, monitoring will simply continue.

Importantly, CASM was not used to forecast the site-specific effects anticipated for varying exposures to atrazine. Rather, the generic CASM was used as an ecosystem-modeling tool to assess the potential effects of atrazine in relation to observed responses in field and laboratory studies.

28.7 MODELS OR MODELERS

This chapter has discussed as much about the process of ecological modeling as it has about particular ecosystem models. Unarguably, the development of user-friendly, off-the-shelf, and readily applied ecosystem risk assessment models constitutes a worthwhile goal for model makers and produces desirable end products for model users. Progress in the previously outlined technical areas will generate increasingly sophisticated, readily accessible, and "user-seductive" ecosystem risk assessment models. Future assessment models will undoubtedly feature highly interactive and increasingly intelligent user interfaces. Model calculations will be completed in seconds on ever-increasingly fast computers and the rate-limiting step will become the user's ability to summarize, understand, interpret, and apply the voluminous model outputs in risk assessment, management, and decision making. Useful summarization and effective presentation of model results will be facilitated by advances in data visualization methods.

Despite a reasonable technological optimism, continuing challenges in improving quantitative ecosystem understanding (e.g., strange attractors, chaos) and the requirements of unique and novel assessments (e.g., genetically engineered organisms, invasive species, habitat degradation, landscape fragmentation) cannot overemphasize the importance of the ongoing scientific training of successive generations of ecosystem modelers. The technical skills, training, and experience of the "people behind the models" will continue to determine the success of ecosystem modeling in supporting ecological risk assessment and the intelligent management of valued natural resources.

For performing an evaluation of model realism and relevance, the categories of individual-based, population, ecosystem, and landscape provide a convenient (but somewhat arbitrary) grouping of ecological models. Models within each group reflect the ecological phenomena or topics of interest from different perspectives in quantitative ecology (Bartell et al. 2003). Associated with each perspective and resulting modeling approach are hypotheses concerning simplifying assumptions that facilitate the specification of model structure in relation to the ecological topic of interest. However, the ecologists, modelers, and observers are each exploring ecological complexity in the same natural world. Simplifying assumptions does not simplify nature. For example, the carrying capacity, K, in the logistic model simply represents all of the biotic and abiotic constraints on population size in a single aggregate parameter. In a real sense, the logistic model could be classified as a simple ecosystem model. The more structurally complex system models such as AQUATOX and CASM attempt to explicitly model many of the biotic and abiotic interactions that are believed to influence the production dynamics of aquatic populations included in these models. These structurally more complex models can justifiably be called population models. One important implication of this recognition is that a convenient categorization of ecological models does not imply that the measurable world can be similarly decomposed and categorized. Apart from evaluating the efficacy of existing models for assessing risks (e.g., Table 28.1), a major and continuing challenge lies in determining the necessary and sufficient structures (i.e., in terms of complexity and scale) of ecological models that provide risk estimates of known accuracy and precision.

The arbitrariness inherent in developing and classifying ecological models can lead to fatuous statements concerning the relative merit of different modeling approaches for assessing ecological risks. Models comparatively simpler in structure (e.g., demographic population models) might initially appear "better" for assessing risks than more structurally complex ecosystem models even though the demographic models are less realistic according to the criteria developed from a risk assessment viewpoint. Given that the model classifications derive as much from ecological shorthand and convenience as from ecological reality, assertions of this kind are analogous to positing that the "wave model" is better than the "particle model" for describing electromagnetic radiation. Such assertions are often framed in the context of model validation, i.e., simpler models might appear more easily validated than complex models. In other words, it would seem that accurately predicting the value of one or two state variables is a priori more likely than obtaining predictions of similar accuracy for 10 or 20 state variables. However, probability theory and modeling experience remind us that validating models, like proving hypotheses, cannot be done in any absolute or meaningfully relative sense. All possible future model/data comparisons cannot be made for any of the modeling approaches. No nonarbitrary baseline for comparing the relative validity of the different modeling approaches exists.

Selection and development of models for assessing ecological risks would benefit from focusing on the relative strengths and limitations of alternative modeling approaches. Evaluations of model realism, endpoint relevance, flexibility, ease of use, and other characteristics help to guide users in their choice of specific models for further development and for application to current risk assessment problems. Future efforts in ecological risk modeling should focus on identifying the necessary model complexity required to achieve sufficiently accurate and precise estimates of risk as determined by the needs of risk management and risk-based decision making. Working backwards from the perspective of risk management might provide additional insights concerning necessary and sufficient model structure for decision making. Working forwards from continuing advances in ecosystem understanding can help inform managers on scientifically defensible minimum model structures. Ecosystem models for risk assessment should be sufficiently complex, but no more so.

Part V

Risk Characterization

Statements about single events can't be decided by a calculator; they have to be hashed out by weighing the evidence, evaluating the persuasiveness of arguments, recasting the statements to make them easier to evaluate, and all the other fallible processes by which mortal beings make inductive guesses about an unknowable future.

Pinker (1997)

Risk characterization is the phase of ecological risk assessment that integrates the exposure and the exposure–response profiles to evaluate the likelihood of adverse ecological effects and uses those results to synthesize a useful conclusion. In other words, it is the process of estimating and interpreting the risks and associated uncertainties. There are two fundamentally different types of risk characterizations. Screening assessments are intended to quickly and easily divide risks into those that need more attention and those that can be ignored (Chapter 31). Definitive assessments are intended to inform a decision-making process by providing risk estimates for all assessment endpoints (Chapter 32).

Risk characterizations may be algorithmic in that they may use a standard procedure based on a standard set of input information using standard assumptions, scenarios, and models. Algorithmic approaches are used primarily in ecological risk assessments of pesticides and industrial chemicals (Luttik and van Raaij 2003; EPPO 2004). They are desirable in that context, because they are efficient and fair to all of the competitive products that come before a chemical regulator. They are popular with regulated parties, because the data requirements are clear, and the outcome of a regulatory assessment can be predicted. Algorithmic approaches are disadvantageous when chemicals have properties that are not considered in the algorithm. The obvious example is endocrine disruptors that are not addressed by standard test batteries or effects models.

Alternatively, risk characterization may be performed ad hoc. The advantage of ad hoc approaches is that they can be designed to provide the best estimate of risk and uncertainty given the types of information that are available and the particular circumstances of the assessment. Ad hoc approaches have been used for contaminated sites, because the conditions and information sets are highly variable. Ad hoc approaches are also employed when assessments are highly contentious or when unusual issues such as developmental deformities are involved.

Inference in risk characterization takes different forms depending on the type of assessment and the types of information that are available. They differ in how they use the available lines of evidence to reach a conclusion. In risk characterization, a line of evidence is an estimate of exposure and a corresponding exposure–response relationship.

Single line of evidence: The classic form of inference uses one line of evidence, which is either the only available evidence or the best evidence. For chemicals, the most common line of evidence is an exposure estimate from a mathematical model and a numerical endpoint from a toxicity test.

Weight of evidence: If multiple lines of evidence are available, they may be jointly considered. The multiple lines may be from a single type of evidence (e.g., exposure–response relationships from different tests) or from multiple types (e.g., chemical toxicity tests, tests of contaminated media, and biological surveys).

Risk characterizations may also be differentiated by the form of the inference.

Rule-based inference: Risk assessors may be provided with an inferential rule to determine whether a risk is acceptable. The simplest and most common is: if the exposure estimate exceeds the benchmark effects level (i.e., HQ > 1; Section 31.1), the risk is unacceptable. A more complex rule is: if the 90th percentile of the exposure distribution exceeds the 10th percentile of the effects distribution, the risk is unacceptable (Section 30.5). Rule-based inference is most common in algorithmic assessments of new chemicals. However, an inferential rule may be developed for an individual assessment during the problem formulation (Chapter 18). Rule-based inference may be applied to screening or definitive assessments. It is usually limited to a single line of evidence but, in its original form, the sediment quality triad is a rule-based inferential method for three lines of evidence (Chapter 32).

Ad hoc judgment: In many cases, risk characterizations include judgments concerning acceptability of a risk without a priori rules or guidance. This approach provides the greatest flexibility and influence to the assessors, but lacks transparency and diminishes the role of stakeholders and decision makers.

Structured judgment: Many risk characterizations are too complex and the evidence too ambiguous to allow rule-based inference, but ad hoc judgment gives too much latitude to assessors. In such cases, the assessor's judgment can be guided by an inferential structure including organization of the input data by type of evidence, the use of standard considerations to evaluate the evidence, and scoring systems. Examples of structures for judgment for causal analysis and risk characterization are presented in Chapter 4 and Chapter 32, respectively.

Risk estimation: One may estimate risks and uncertainties and report them to a risk manager who interprets the estimates and makes a decision. Risk estimation is used in definitive assessments and may be based on any number of lines of evidence. Risk estimates are essential if the results of risk characterization are to be used in an economic analysis, decision analysis, or other quantitative decision-support tool.

Comparison of alternatives: Rather than characterizing risks from an agent or activity to determine its acceptability, one may compare alternatives to determine which is preferable (Chapter 33). Examples include alternative chemicals with the same use, alternative remedial actions for a contaminated or disturbed site, and alternative management plans for a forest.

These approaches to inference are not mutually exclusive. For example, it is often appropriate to use structured judgment to determine whether significant effects are likely and then, if the results are positive, use risk estimation to inform a decision.

29 Criteria and Benchmarks

For various reasons, it is sometimes desirable to reduce the complexities of exposure–response relationships for various taxa, processes, and other ecological properties to a single number that is presumed to be a sufficiently protective level. Those that are used to separate acceptable from unacceptable concentrations for regulatory purposes are termed criteria or standards (henceforth, simply criteria). Those that are used for screening or prioritization are termed screening benchmarks or screening values.

29.1 CRITERIA

Criteria are concentrations of contaminants in water or other media that are intended to constitute the bounds of regulatory acceptability given prescribed conditions (Section 2.2). The only national ecological criteria in the United States are the acute and chronic National Ambient Water Quality Criteria (NAWQC). Criteria were proposed for sediments by the Environmental Protection Agency (EPA) but were converted to screening guidelines (Section 29.2.) The acute NAWQC are calculated by the EPA as half the final acute value, which is the 5th percentile of the distribution of 48 to 96 h LC_{50} values or equivalent median effective concentration (EC_{50}) values for each criterion chemical (Stephan et al. 1985). The acute NAWQC are intended to correspond to concentrations that would cause less than 50% mortality in 5% of exposed species in a relatively brief exposure. Because the criterion is not a no-effect level, the criterion is lowered if an important species is among the most sensitive 5% (Figure 29.1). The chronic NAWQC are final acute values divided by the final acute/chronic ratio, which is the geometric mean of quotients of at least three LC_{50}/CV ratios from tests of organisms belonging to different families of aquatic organisms (Stephan et al. 1985). Chronic NAWQC are intended to prevent significant toxic effects in most chronic exposures. Some, termed final residue values, are based on protection of humans or other piscivorous organisms rather than protection of aquatic organisms.

   Because criteria are applied to an entire state or nation, they should be derived in a way that accounts for variance among sites and uncertainty. Site-specific standards may incorporate site properties to reduce either variance or uncertainty. For example, the NAWQC for many metals are functions of hardness, so that important sources of variance can be eliminated in site-specific applications (Spehar and Carlson 1984; Stephan et al. 1985). Similarly, results from testing of local species may be used to modify national criteria in deriving site-specific standards. More broadly, standards may be derived for different classes of ecosystems (e.g., freshwater and saltwater standards in the United States), different uses (e.g., agricultural, residential, commercial, and industrial land uses in Canadian soil guidelines), or different levels of protection (e.g., the designation of National Parks as Class I under the US Clean Air Act).

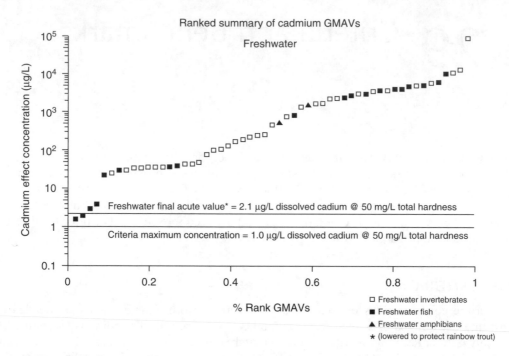

FIGURE 29.1 Acute and chronic ambient freshwater quality criteria for cadmium at 50 mg/L hardness (horizontal lines), and the acute species sensitivity distribution. The acute values (LC_{50}s and EC_{50}s) for species and genera are geometrically averaged so the points are genus mean acute values (GMAVs). (From EPA (U.S. Environmental Protection Agency), *2001 Update of Ambient Water Quality Criteria for Cadmium*, EPA-822-R-01-001, Office of Water, Washington, DC, 2006a. With permission.)

NAWQC are applicable regulatory criteria and are generally adequately protective, but they are often not good risk estimators for particular sites. If they are applied to a site, assessors should consider deriving site-specific criteria using the water effect ratio. This is a factor for adjusting criteria to site water that may be derived using an EPA procedure (EPA 1983; Office of Science and Technology 1994). It requires performing toxicity tests with the chemical in site waters, and, optionally, with site species (Figure 29.2). The time and expense

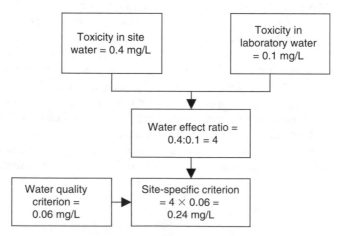

FIGURE 29.2 An illustration of the derivation and use of water effect ratios.

required to calculate site-specific criteria could be worthwhile if the water chemistry at a site differs significantly from conventional laboratory test waters. Otherwise, the effort is better expended on tests of ambient waters (Section 24.5).

Currently, in the United States, the methodology for deriving ambient water quality criteria is being reexamined, and the risk assessment framework is being applied. In particular, derivation of new criteria will begin with a problem formulation to determine the appropriate endpoints for the chemical, important exposure pathways, and the availability and utility of unconventional effects data. The more flexible approach is reflected in recent criteria and proposed criteria that use field data or novel modeling approaches (EPA 2000a, 2003a, 2004a, 2006a). For suspended and bedded sediments, a framework for deriving regional or watershed-specific values by multiple methods and weighing the results has been developed (EPA 2006b).

Many nations have criteria for water and other media, and comments about the utility of the US criteria may not apply to them. The utility of these criteria in risk assessments should be considered where they are potentially applicable. It is often appropriate to estimate the risk of exceeding a criterion in addition to estimating risks to ecological endpoints.

29.2 SCREENING BENCHMARKS

Screening benchmarks are concentrations of chemicals that are believed to constitute thresholds for potential toxic effects on some category of receptors exposed to the chemical in some medium. Since they are used for screening chemicals, they should be somewhat conservative so that chemicals that do in fact cause effects at a particular site are not screened out of the assessment (Chapter 31). It is more important to ensure that hazardous chemicals are retained than to avoid retention of chemicals that are not hazardous. However, excessive conservatism decreases the value of screening assessments, because effort is wasted on nonhazardous chemicals that might better be expended on the truly hazardous ones. Because of this deliberate conservatism, it is important to avoid adoption of screening benchmarks as remedial goals or other thresholds for action without some additional assessment to determine that they are appropriate.

There is little consensus about the best methods for deriving screening benchmarks. The following alternatives are based on US practices. Screening benchmarks used in Australia, Europe, and North America are reviewed by Barron and Wharton (2005).

29.2.1 Criteria as Screening Benchmarks

Criteria are commonly used as screening benchmarks because exceedence of one of these values constitutes cause for concern. The US NAWQC have been recommended for screening at contaminated sites by the EPA (Office of Emergency and Remedial Response 1996). However, it is not clear that they are sufficiently conservative, since they are assumed to be sufficiently close to the true threshold of effects to justify regulatory action and because of other concerns (Suter 1996c). These concerns are supported by the finding that nickel concentrations in a waste-contaminated stream on the Oak Ridge Reservation that were below chronic NAWQC were nonetheless toxic to daphnids (Kszos et al. 1992). When used for regulation of effluents— their intended purpose—these criteria achieve additional conservatism by being applied to relatively short exposure durations. That conservatism does not apply to contaminated sites.

29.2.2 Tier II Values

If NAWQC are not available for a chemical, the Tier II method described in the EPA Proposed Water Quality Guidance for the Great Lakes System or a slight variation used at

ORNL may be applied (EPA 1993e; Suter and Tsao 1996). Tier II values were developed so that aquatic life criteria could be conservatively estimated with fewer data than are required for the NAWQC. Tier II values are concentrations that would be expected to be higher than NAWQC in no more than 20% of cases, if sufficient test data were obtained to calculate NAWQC. For example, if there is only one acute value (LC_{50} or EC_{50}) for a chemical, that value is divided by 20.5 if it is a daphnid and 242 if it is not. Equivalent factors are available for other numbers of acute values in Appendix B of Suter and Tsao (1996). The sources of data for the Tier II values, and the procedure and factors used to calculate the SAVs and SCVs, are presented by EPA (1993e) and Suter and Tsao (1996).

29.2.3 BENCHMARKS BASED ON EXPOSURE–RESPONSE MODELS

Screening benchmarks might be based on low percentiles of exposure–response relationships. In particular, one can calculate an LC_0 or EC_0 for chemicals with apparent effects thresholds. Alternatively, the practice in human health risk assessment of using the lower 95% confidence limit on a benchmark dose (the EC_{10}) can be applied to nonhuman organisms (Linder et al. 2004). This value is considered by the US EPA to approximately correspond to a no observed adverse effect level (NOAEL) for human health effects, but is more consistent.

29.2.4 THRESHOLDS FOR STATISTICAL SIGNIFICANCE

Test endpoints based on statistical significance are commonly used as screening benchmarks. The endpoint used varies among media and receptors.

Lowest chronic values: Chronic values (CVs) are geometric means of no observed effect concentrations (NOECs) and lowest observed effect concentrations (LOECs). They are used to calculate the chronic NAWQC, and may be presented in place of chronic criteria by the EPA when chronic criteria cannot be calculated (EPA 1985). CVs are not conservative benchmark values.

Wildlife NOAELs: Screening benchmarks for wildlife are conventionally based on NOAELs from chronic or subchronic toxicity tests with mammals or birds. The major variables in derivation of wildlife benchmarks are the test endpoints used and whether allometric scaling or safety factors are used. Wildlife benchmarks use reproductive or other effects as endpoints, allometric equations for interspecies extrapolations, and factors to allow for shortcomings in the test design (Sample et al. 1996c; Office of Solid Waste and Emergency Response 2005). The resulting screening dose, termed the wildlife toxicity reference values (TRVs) must be converted to a concentration in soil or other medium to screen those media (Efroymson et al. 1997; Office of Solid Waste and Emergency Response 2005). That requires an exposure model (Chapter 22).

29.2.5 TEST ENDPOINTS WITH SAFETY FACTORS

Some states and EPA regions base screening benchmarks on test endpoints divided by safety factors. These factors do not have the scientific basis of the factors used to derive the Tier II values (above) or the factors proposed by Calabrese and Baldwin (Table 26.3). However, the use of factors of 10, 100, or 1000 have a long history in the EPA (Dourson and Stara 1983; Nabholz et al. 1997) (Table 26.1), and such factors can be easily applied to any test endpoint.

29.2.6 DISTRIBUTIONS OF EFFECTS LEVELS

Sets of screening benchmarks for sediments and soils have been derived from distributions of effects or no-effects levels. An estimate of the threshold effects concentration for a particular chemical is derived from a percentile of the distribution of reported effects or no-effects concentrations. These concentrations vary due to variance in the physical and chemical properties of soils or sediments, variance among the measured responses, and variance in the sensitivities of the species or communities. Therefore, the benchmarks derived in this way may be thought to protect some proportion of combinations of species, responses, and media. The following are examples of this approach.

Effects range-low and effects range-median for sediments: The National Oceanic and Atmospheric Administration (NOAA) uses three methods: (1) equilibrium partitioning; (2) spiked sediment toxicity tests; and (3) field surveys to develop exposure–response relationships (Long et al. 1995). Chemical concentrations observed or estimated to be associated with biological effects are ranked, and the lower 10th percentile (effects range-low, ERL) and the median (effects range-median, ERM) concentrations are identified. A variant of this approach is Florida's Threshold Effects Levels (MacDonald et al. 1996).

Screening level concentrations: These benchmarks are derived from synoptic data on sediment chemical concentrations and benthic invertebrate distributions. They are estimates of the highest concentration that can be tolerated by a specified percentage of benthic species. Examples include the Ontario Ministry of the Environment Lowest and Severe Effect Levels (Pesaud et al. 1993).

Oak Ridge National Laboratory benchmarks for soil: Benchmarks for toxicity to plants, soil invertebrates, and microbial processes have been developed from the 10th percentile distributions of toxicity test data (Efroymson et al. 1997a,b).

29.2.7 EQUILIBRIUM PARTITIONING BENCHMARKS

Equilibrium partitioning benchmarks are bulk sediment concentrations derived from aqueous criteria or benchmark concentrations based on the tendency of nonionic organic chemicals to partition between the sediment pore water and sediment organic carbon and for metals to be bound to sulfides (Section 22.3). The fundamental assumptions are that pore water is the principal exposure route for most benthic organisms and that the sensitivities of benthic species is similar to that of the species tested to derive the aqueous benchmarks, predominantly the water column species. Examples include the US EPA's equilibrium partitioning sediment guidelines (EPA 2000b, 2002c–f) and consensus sediment guidelines for PAHs (Swartz 1999).

29.2.8 AVERAGED VALUES AS BENCHMARKS

Sometimes the most sensitive response is thought to be too conservative, criteria for identifying the best value are not apparent, and there is no agreement concerning how to extrapolate to a safe level. In such cases, benchmarks may be derived by simply averaging test endpoints that are deemed to be relevant and of sufficient quality. This approach was used in the US EPA's soil screening values for plants and soil invertebrates (Office of Solid Waste and Emergency Response 2005).

29.2.9 ECOEPIDEMIOLOGICAL BENCHMARKS

When effects are observed in the field and the cause has been determined (Chapter 4), the effective exposure levels determined in those studies can be used as benchmarks at other sites. For example, tundra swans and other waterfowl were found dead or suffering toxicosis in the Coeur d'Alene Basin, Idaho, an area of lead mining. Field and laboratory studies were used to relate sediment lead to dietary lead to lead body burdens and effects. The result was an estimated toxic threshold of 530 μg lead per gram sediment dry weight and a lethal level of 1800 μg/g (Beyer et al. 2000; Henny 2003).

29.2.10 SUMMARY OF SCREENING BENCHMARKS

Currently the development of screening benchmarks is inconsistent across media. The large and relatively consistent body of data for aquatic animals has led to the development of more than a dozen alternative types of benchmarks. Similarly there are several alternative benchmarks for sediments, but they have been developed for fewer chemicals. Wildlife benchmarks are nearly always based on NOEC values, so usually only one type of benchmark is available. However, there is considerable variance in what effects are included and in the exposure models used to extrapolate back to soil concentrations. Finally, benchmarks for plants, invertebrates, and microbes in soil are inconsistent and are available only for few chemicals.

Given the lack of validation or even a common definition of validity, no single type of benchmark can be demonstrated to be consistently reliable. When there are multiple benchmarks for a chemical and none are clearly superior, "consensus" benchmark values may be simply derived by averaging. Swartz (1999) derived a threshold effects concentration for total PAHs (0.3 mg/g OC) as the arithmetic mean of five diverse benchmarks. He found that it was a reasonable threshold value for PAH effects in independent data sets from PAH-contaminated sites. Alternatively, the uncertainty concerning the most appropriate benchmark may be treated by choosing the lowest benchmark for each chemical.

Because the degree of conservatism of benchmarks is uncertain, concerns that truly toxic chemicals may be screened out may be relieved by using uncertainty factors. An example of the use of uncertainty factors for this purpose is the ecological risk assessment for the Rocky Mountain Arsenal, in which factors were applied to account for intrataxon variability, intertaxon variability, uncertainty of critical effect, exposure duration, endpoint extrapolation, and residual uncertainty (Banton et al. 1996). For each of these six issues, a factor of 1, 2, or 3 was applied signifying low, medium, or high uncertainty, respectively. Clearly, the magnitudes of these factors are not related to estimates of actual variance or uncertainty associated with each issue, and the multiplication of factors bears no relationship to any estimate of the total uncertainty in the benchmarks. However, uncertainty factors provide an assurance of conservatism without appearing completely arbitrary. An alternative is to derive uncertainty factors based on estimates of actual variance or uncertainty. An example is the prediction intervals on the intertaxon extrapolations and the uncertainty factors on the prediction intervals (PIs) for a given taxonomic level presented in Table 26.2 through Table 26.5.

30 Integrating Exposure and Exposure–Response

The primary task of risk characterization is to integrate the exposure estimates from the analysis of exposure with the exposure–response relationships from the analysis of effects to estimate the nature and magnitude of risks. In effect, response is estimated by solving the exposure–response function for the exposure estimate. In most assessments, this task has been performed by simple methods that require little thought. However, as more attention is paid to variability and uncertainty (Chapter 5), probabilistic methods are becoming more common.

30.1 QUOTIENT METHODS

If the analysis of exposure has generated a point estimate of exposure (e.g., the maximum measured concentration) and the analysis of effects has reduced the exposure–response relationship to a point (e.g., an LC_{50}), integration of the two reduces to the quotient method. The hazard quotient (HQ) is the quotient of an exposure concentration (C_e) divided by a toxicological benchmark concentration (C_b):

$$HQ = C_e/C_b \qquad (30.1)$$

Because this is a widely used assessment method, the terms have many representations. In Europe, C_e is usually termed the predicted environmental concentration (PEC) and C_b is termed the predicted no effect concentration (PNEC). If exposure and effect are expressed as doses, the HQ is equivalent $[D_e/D_b]$. The same simple model may be applied to a variety of agents such as temperature, percent fines, and radiation. Because of its simplicity, the quotient method is nearly always used in screening assessments, but it is also the most common method of risk characterization in definitive assessments.

Although some assessors have used Monte Carlo analysis (Chapter 5) to perform probabilistic analyses of HQs (as in the Hong Kong example, Section 30.7.3, and Zolezzi et al. 2005), they may be performed analytically (IAEA 1989; Hammonds et al. 1994). The quotient model can be expressed as:

$$\ln HQ = \ln C_e - \ln C_b \qquad (30.2)$$

HQ will be approximately log-normal even if the distributions assigned to C_e and C_b are not (IAEA 1989; Hammonds et al. 1994). Hence, the geometric mean of HQ is the antilog of the difference of the means of the logs of C_e and C_b, and the geometric variance is the antilog of the sum of the variances of the logs of C_e and C_b.

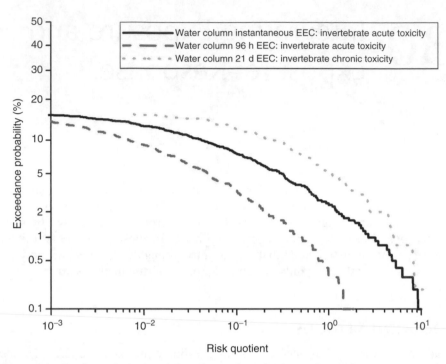

FIGURE 30.1 Distributions of acute and chronic quotients for invertebrates from an assessment of risks to pond communities from pyrethroid pesticides. EEC = estimated exposure concentration. (From Maund, S.J., Travis, K.R., Hendley, P., Giddings, J.M., and Solomon, K.R., *Environ. Toxicol. Chem.*, 20, 687, 2001. With permission.)

If the number of exposure values and effects values are finite, one may simply determine the distribution of all possible values of HQ. For example, in an assessment of risks to pond communities from pyrethroid pesticides, Maund et al. (2001) determined the distribution of quotients for 90th percentile concentrations in each of 72 pond categories with each acute and chronic toxicity datum (Figure 30.1).

While the HQ expresses how bad things are, a related concept, the margin of safety, expresses how good they are. The relative margin of safety is simply the inverse of the HQ. A relative margin of safety of 100 suggests that the exposure concentration must be increased by a factor of 100 to reach a toxic level. The absolute margin of safety is the difference between a toxic level and the exposure level. An absolute margin of safety of 100 mg/L suggests that the exposure concentration must be raised by that amount to reach a toxic level. An example of the use of margins of safety in ecological risk assessment is presented by Newsted et al. (2002).

30.2   EXPOSURE IS DISTRIBUTED AND RESPONSE IS FIXED

Frequently, the exposure–response relationship is reduced to a point, such as a criterion value, but the exposure estimate is distributed. The exposure distribution may come from the distribution of measured concentrations in the environment, from Monte Carlo analysis of a transport and fate model or from expert judgment. In such cases, the probability of exceeding the benchmark value (C_b) is the integral of the probability density function above C_b (i.e., 1—the cumulative probability at C_b). An example of this approach is the analyses of risks to herons and egrets in Hong Kong with determinate effects thresholds (Section 30.6).

30.3 BOTH EXPOSURE AND RESPONSE ARE DISTRIBUTED

Given distributions of exposure and response with respect to a common variable (e.g., concentration), one may calculate risk as the probability that a random draw from the exposure distribution exceeds a random draw from the response distribution (Suter et al. 1983). This concept of risk as the joint probability of exposure and effects distributions was applied to effects expressed as species sensitivity distributions (SSDs) by Van Straalen (1990) and Parkhurst et al. (1996a,b). Risk is the integral of the product of the probability density of the exposure concentration C_e and the cumulative distribution of the benchmark concentration C_b (Figure 30.2c). The derivation of this formula and alternatives, including a discrete approximation, are clearly presented by Van Straalen (2002b). This is conceptually equivalent to the probabilistic HQs (Section 30.1) but is both clearer and more elegant.

A variant of this approach is proposed by the ECOFRAM Aquatic Workgroup (1999) and applied to pesticide risk assessments (Giddings et al. 2005) as well as contaminated site assessments (Moore et al. 1999). From an exposure distribution (proportion of locations, times, or episodes, with respect to concentration) and an effects distribution (SSDs or other exposure–response distributions) one can derive a plot of exposure proportions vs. effects levels that is called a risk curve (Figure 30.3 and Figure 30.4). Since both proportions of exposures and responses are distributed with respect to concentrations, there are corresponding values of each. The area under the curve is called the mean risk. It is equivalent to risk estimated as a joint probability, discussed earlier.

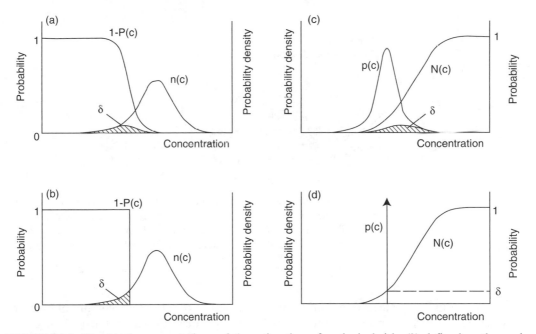

FIGURE 30.2 Graphical representations of the estimation of ecological risks (δ) defined as the probability that exposure concentrations are greater than no-effect concentrations (NECs). The probability density of exposure concentrations is denoted as $p(c)$, the distribution of NECs is denoted as $n(c)$. $P(C)$ and $N(C)$ are the corresponding cumulative distributions. In a and c, both variables are distributed. In b and d, the exposure concentration is assumed to be constant. (From Van Straalen, N.M., in L. Posthuma, G.W. Suter II, and T. Traas, eds., *Species Sensitivity Distributions in Ecotoxicology*, Lewis Publishers, Boca Raton, FL, 2002. With permission.)

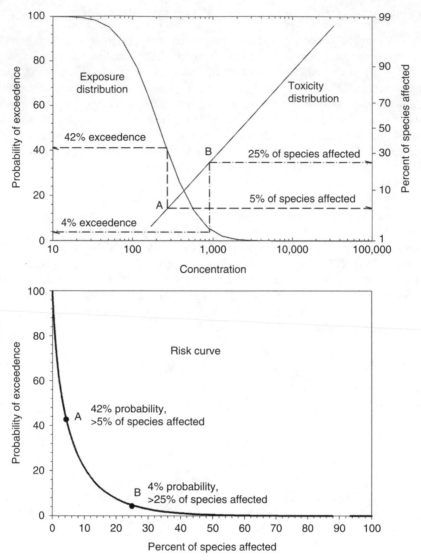

FIGURE 30.3 A demonstration of the derivation of a risk curve from distributions of exposure (probability of exceedence) and response (percent of species affected) with respect to concentration. (From Giddings, J.M., Anderson, T.A., Hall, L.W. Jr., et al., *Atrazine in North American Surface Waters: A Probabilistic Risk Assessment*, SETAC Press, Pensacola, FL, 2005. With permission.)

When exposure and effects distributions are used as part of a logical weighing of evidence (Chapter 32), it may be appropriate to logically interpret them rather than calculating joint probabilities. For example, the following interpretation occurs in the risk assessment for fish community of the Poplar Creek embayment of the Clinch River (Suter et al. 1999).

Copper. The distributions of ambient copper concentrations and aqueous test endpoints are shown in Figure 30.5. The ambient concentrations were dissolved phase concentrations in the subreaches (3.04 and 4.01) with potentially hazardous levels of Cu. The toxic concentrations were those from tests performed in waters with hardness approximately equal to the site

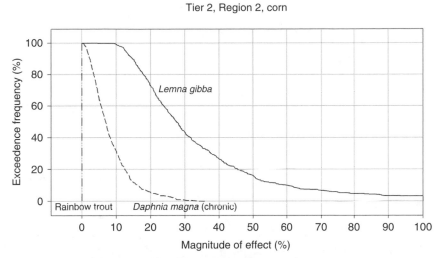

FIGURE 30.4 Risk curve for atrazine applied to corn based on estimated annual maximum instantaneous concentrations in pond water in a defined region, for three endpoints: duck weed (*Lemna gibba*) growth inhibition, cladoceran (*Daphnia magna*) reproduction inhibition, and rainbow trout mortality. Because of the relative insensitivity of acute lethality to trout, the corresponding line appears to be vertical in this plot. (From Giddings, J.M., Anderson, T.A., Hall, L.W. Jr., et al., *Atrazine in North American Surface Waters: A Probabilistic Risk Assessment*, SETAC Press, Pensacola, FL, 2005. With permission.)

water. The ambient concentrations fall into two phases. Concentrations below 0.01 mg/L display a fairly smooth increase suggestive of a log-normal distribution. The upper end of this phase of the distribution (above the 75th percentile of 4.01 and the 80th percentile of 3.04) exceed the lowest chronic value (CV) (a bluntnose minnow CV for reproductive effects). However, the distributions above the 90th percentile are not continuous with the other points. The break in the curve suggests that some episodic phenomenon causes exceptionally high concentrations. The two points in 4.01 and one in 3.04 that lie above this break exceed approximately 90% of the CVs, approximately 30% of the acute values, and both the acute and chronic National Ambient Water Quality Criteria (NAWQC). These results are suggestive of a small risk of chronic toxicity from routine exposures, but a high risk of short-term toxic effects of Cu during episodic exposures in lower Poplar Creek embayment and the Clinch River.

This sort of interpretation is a mixture of quantitative and qualitative analysis that can, as in this case, provide more information than a purely quantitative analysis. Had the Aquatic Risk Assessment and Mitigation Dialog Group criterion been applied or a joint probability been calculated, the results would have been less ad hoc but would have provided less basis for inference concerning the cause of observed effects and toxicity.

30.4   INTEGRATED SIMULATION MODELS

When a mathematical simulation model, such as a chemical transport and fate model (Chapter 21), is used to estimate exposure, one may simply add the exposure–response function to the exposure model so that the model output is an effects level. Similarly, when a population or ecosystem model is used to estimate responses to exposures, an exposure level

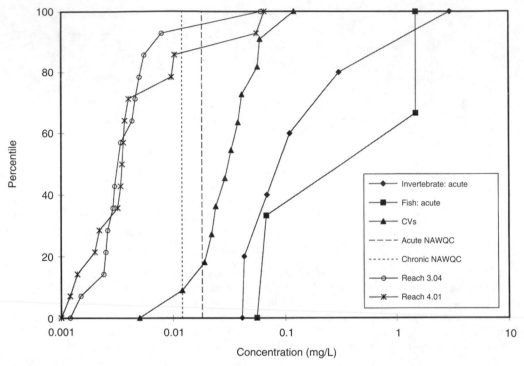

FIGURE 30.5 Empirical distribution functions (species sensitivity distributions, SSDs) for acute toxicity (LC_{50} and EC_{50} values) and chronic toxicity (chronic values) of copper to fish and aquatic invertebrates, and for individual measurements of copper in surface water from two stream reaches. Vertical lines are acute and chronic National Ambient Water Quality Criteria (NAWQC).

or an exposure model my be linked to produce an integrated model that estimates effects from loading rates or ambient levels (O'Neill et al. 1982). Monte Carlo analysis (Chapter 5) is used with such models to estimate risks as probabilities of effects.

30.5   INTEGRATION OF SENSE AND NONSENSE

When integrating exposure with exposure–response relationships, it is essential to ensure that they can be combined in a way that makes sense, i.e., they must be concordant. This requires first that the common units be consistent. This is not simply a matter of assuring that, for example, the exposure concentration and the concentration in the exposure–response relationship are both mg/L of copper. If the response concentration is a 96 h LC_{50} for dissolved copper, and the exposure concentration is an annual average of measured total copper concentrations, they are not concordant. Other measures of exposure or response must be used, or one of the measures must be adjusted to achieve concordance. For example, a metal speciation model may be used to estimate dissolved copper concentrations in the field and the peak 96 h concentration might be estimated from the time series of measurements.

Concordance becomes more complex when parameters are expressed as distributions and results are expressed as probabilities. For every distribution, it is essential to ask what is distributed and with respect to what it is distributed. Is a distribution of dose to mink (mg/kg/d) the distribution of the average dose across a mink population, the dose to the median mink, or the dose to an individual mink occupying a particular location? Is it distributed with respect to space (e.g., from sampling points on a site), to time (e.g., from year-to-year variation in diet), to individuals (e.g., from variance in size and dietary preference), or to degree of belief (e.g., an expression of an assessor's uncertainty concerning the dose estimate)? If it was generated by Monte Carlo analysis of an exposure model, the dose distribution might be a hodgepodge of variance of consumption rates across individuals, variance in drinking water contamination over time, variance in contaminant levels across individual fish in a pond, variance in contaminant levels in mice across space, and variance in dietary composition across different studies. The best that could be said of such a dose distribution is that the probabilities express the assessor's uncertainly concerning dose as degrees of belief.

A response distribution (e.g., one derived from a reproductive test) may also take different forms. In the simplest case, a dose–response distribution may be derived for the proportional reduction in the number of live births per female. That distribution might be used to estimate the average proportional reduction at a given dose, or the variance in the parameters of the fitted model might be used to estimate the distribution with respect to individual females of the dose causing a given proportional reduction (e.g., an ED_{10}). If the test was performed with rats rather than mink, an uncertainty factor may be applied resulting in a distribution of the ED_{10} with respect to degree of belief (Box 30.1).

Note that any paired exposure distribution and effects distribution from the previous two paragraphs superficially appear to be concordant, because they are all probabilities as functions of dose to mink. However, the probabilities express very different qualities.

The appropriate integration of exposure and response depends on the assessment endpoint, the preferences of the risk manager, and the available information. In the simplest case of the mink example, one might use an HQ. The point estimate of the annual average daily dose for the site might be divided by the ED_{10} and the risk could be declared significant if $HQ > 1$. To include uncertainty, one might subjectively estimate a lower confidence bound of $HQ/100$ for use in screening, or to be precautionary. One might estimate the distributions of both exposure and response doses and estimate the distribution analytically. However, to actually estimate the most likely effect or the risks of prescribed levels of effect, one would need to settle on a concordant set of distributions of exposure and response parameters. For example, to estimate the probability that a female mink on a contaminated site experiences a reproductive decrement, one could use the distribution of the ED_{10} (i.e., the proportion of females with fecundity at least 10% below controls) and a distribution of dose with respect to individual female mink estimated as described in Section 22.11. Since both the exposure dose and the effective dose would be distributed with respect to individual female mink, the joint probability would be the proportion of females in the site population estimated to experience a 10% or greater reduction in fecundity.

The problem of combining exposure and response distributions may be simplified by devising rules like the one provided by the Aquatic Risk Assessment and Mitigation Dialogue Group (1994) (see also Solomon et al. 1996). They reduced risk characterization for exposure and effects distributions to a dichotomous criterion; the risk is significant if the 90th percentile of the distribution of aqueous concentrations exceeds the 10th percentile of the SSD. Although this method has been recommended by a distinguished group of scientists, the criterion is not supported by legal or policy considerations. In addition, it does not interpret the distributions in terms of either variability or uncertainty. It is simply an easy and consistent rule, which may be adopted or adapted to an assessment if it makes sense.

The possible combinations of distributions to characterize ecological risks are effectively infinite, so this section can give only a sample of the possible problems in achieving concordance. It is incumbent on those who perform probabilistic risk assessments to carefully consider what is

---

BOX 30.1
Variability, Uncertainty, and Distributions of Effects

Conventionally, the relationship between an exposure variable (dose, concentration, duration, etc.) and a response variable (growth, death, etc.) is quantified using a distribution function (normal, logistic, etc.), the exposure–response function (Chapter 23). In ecological risk assessment, these functions are most commonly based on responses of exposed individual organisms (dose–response or concentration–response) or individual species (species sensitivity distributions, SSDs). These distribution functions are usually described as probability distributions without careful thought as to what mechanism generated the distribution.

Error: All individuals may be effectively the same or all species in a community may be effectively the same and the distribution is due to random effects in the tests (these random effects are termed error, but need not imply actual errors in conducting the tests). Hence, the output of the model is the probability of the prescribed response given the uncertainty due to experimental error (randomness).

Variability: There are actual differences among the individuals or species that are measured without effective error. Hence, the distribution describes that variability and the output of the model is the deterministic proportion of individuals or species responding at a given exposure level. This is the assumption underlying the estimation of risks to human populations, and the most common assumption in the case of species distributions, leading to the calculation of the potentially affected fraction (PAF) of species.

Identity: There are actual differences among the individuals or species that are measured without effective error. However, we are interested in the risk to an untested individual or species, rather than the proportion of individuals in a population or species in a community. This interpretation leads to an estimate of individual risks (probability of effects on an exposed organism other than a member of the test population) or species risks (probability of effects on an untested species). This is the assumption underlying the estimation of risks to human individuals. In the case of species distributions, this assumption implies that the endpoint species is a random draw from the same population as the set of test species used to define the distribution.

Extrapolation: The variability and uncertainty inherent in the test and the model fitted to the test data are negligible relative to the uncertainty associated with extrapolation to the endpoint species or community. In such cases, we may ignore the variance among organisms or other test units, estimate a midpoint of the observed responses (e.g., LC_{50} or EC_{50}) and assign a standard deviation, range, or other distribution parameter to express extrapolation uncertainty.

---

distributed, with respect to what it is distributed, and in what sense do probabilities from the distribution constitute risks to the assessment endpoint. If the nature of the probabilities is unclear in the assessor's mind, it will not be clear to the users and reviewers of the assessment, and there is a good chance that it is wrong. In such cases, assessors should consider seeking help or using a less complex analysis.

30.6   INTEGRATION IN SPACE

In regional assessments or other assessments at large spatial scales, the assessment must integrate risks over space. The most common approach is to divide the area into reasonably uniform units, estimate risks for each unit, and then generate a summary such as an area-weighted average effect or a distribution of effects across units to estimate risks on a site or in a region. The spatial units might be habitat types on a site, watersheds in a region, areas with distinct types of disturbance or contamination, or other relevant divisions of the area being assessed. In most cases, the exposure estimates vary among units, but the exposure–response relationships may vary as well due to differences in the biotic communities.

A sophisticated elaboration of this approach is found in the assessment of aquatic ecological risks from cotton pyrethroids in Yazoo County, Mississippi (Hendley et al. 2001; Travis and Hendley 2001). The units were ponds and their associated watersheds. A geographic information system (GIS) was used with transport and fate models to estimate 90th percentile concentrations in 597 ponds in the county and compare them to SSDs for acute and chronic effects of the pesticides. A further step would include spatial variance or even spatial dynamics of the endpoint organisms, populations, or communities along with the spatial variance in exposure.

Another approach estimates risks at points; usually, points at which soil or sediment has been sampled and analyzed. Kriging, Thiessen polygons, or some other geospatial statistical method is then used to define areas within which risks fall in defined ranges. This approach is appropriate for organisms with little mobility such as plants and benthic invertebrates. If toxicity tests are performed on soil or sediment samples, and if the tests are measures of effects on an assessment endpoint, this approach can be applied to those results as well (Figure 20.3).

Figure 30.6 illustrates a simple technique that is appropriate for organisms with territories or home ranges occurring on relatively simple contaminated sites. A contaminant has been dumped or spilled at a point, and the average soil concentration drops off approximately exponentially from that point as the area averaged increases. Equivalent curves could be plotted for other patterns of contamination around a point. Horizontal dashed lines indicate soil concentrations that are estimated to be thresholds for effects on small mammals and birds. The vertical dashed lines indicate the average home range or territory size for the endpoint species (shrew and woodcock in this case). If the average concentration falls below the effects concentrations before intersecting the home range size, not even one individual is

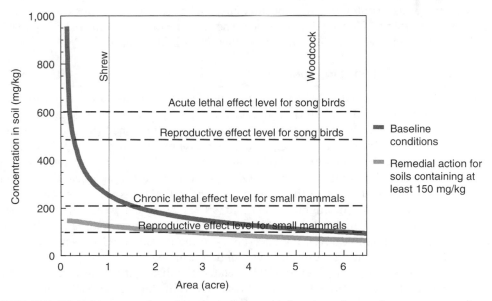

FIGURE 30.6 Mean concentration of a contaminant as a function of averaging area centered on the point of maximum concentration. Vertical lines indicate the home range area for a shrew and a woodcock. Horizontal dashed lines indicate soil concentrations estimated to produce toxic doses associated with specific effects. (Graphic redrawn from an unpublished graphic provided by C. Menzie).

expected to be affected. In Figure 30.6, no effects on woodcock are expected, but a shrew is estimated to experience reproductive effects and has a marginal risk of death.

To estimate risks to multiple organisms with territories or home ranges on a contaminated site or other defined area, a GIS can be used to cover the area with polygons having the area (on the map scale) of the territory or home range. Exposure levels in those areas can then be averaged to estimate risks to those individuals or reproducing pairs.

30.7 EXAMPLES

The following examples provide a taste of the range of approaches to integrating exposure and exposure–response information in ecological risk assessments.

30.7.1 SHREWS ON A MERCURY-CONTAMINATED SITE

Talmage and Walton (1993) collected shrews on the mercury-contaminated floodplain of East Fork Poplar Creek, Tennessee. They analyzed the mercury concentration in kidneys, the target organ, and compared them to the 20 μg/g threshold for mercury toxicity in rodents. They found that 75% of shrews exceeded that threshold.

30.7.2 EGRETS AND EAGLES IN SOUTH FLORIDA

A large stormwater treatment pond in South Florida was found to have high methyl mercury concentrations. To obtain a discharge permit, the State agreed to assess risks to great egrets and bald eagles foraging on the pond (Rumbold 2005). The exposure–response relationships were lowest observed adverse effect levels (LOAELs) divided by a factor of 3. Exposure was estimated from concentrations in fish collected from the pond and from dietary uptake models for pre-nesting females and nestlings. The nestling exposure model also included maternal mercury deposited in the eggs. Monte Carlo simulation of the uptake models was used to estimate the distributions of exposure, based on variance in the mercury concentrations in appropriate fish for each avian species. The assessment found that risks of exceeding the effects thresholds were low and similar to other areas in the region.

30.7.3 EGRETS AND HERONS IN HONG KONG

Connell et al. (2003) assessed risks from organochlorine compounds to the reproduction of black-crowned night herons and little egrets in the New Territories of Hong Kong. They estimated exposure by analyzing contaminants in eggs and established that DDE posed the greatest hazard. They developed a concentration–response relationship using published studies of the relationship between survival of young ardeids and DDE concentrations in eggs (Figure 30.7). They judged that 1000 ng/g DDE was a threshold for significant reduction in survival. By applying that value to the probability density functions for egg concentrations, they estimated that 12.4% of night herons and 40.9% of egrets were exposed at levels exceeding the threshold. Finally, they used Monte Carlo simulation to estimate the probability of exceeding the threshold given the uncertainty in the threshold. However, they considered only the possibility that the threshold was underestimated.

In a companion study, Connell et al. (2002) related the distributions of metal concentrations in heron and egret feathers to effects thresholds from the literature. In this case, the Monte Carlo analysis used the observed distribution of concentrations in feathers and a uniform distribution of the effects threshold between the highest level reported to have no effects (3 μg/g mercury) and the lowest level reported to reduce reproductive success in ardeids (5 μg/g).

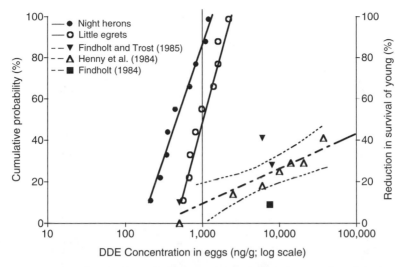

FIGURE 30.7 Cumulative distribution of concentrations of DDE in eggs for night herons and little egrets and a model with confidence interval fitted to data on percent reduction in the survival of young birds as a function of DDE in eggs. The vertical line is the estimated threshold effective concentration of 1000 ng/g. (From Connell, D.W., Fung, C.N., Minh, T.B., Tanabe, S., Lam, P.K.S., Wong, B.S.F., Lam, M.H.W., Wong, L.C., Wu, R.S.S., and Richardson, B.J., *Water Res.*, 37, 459, 2003. With permission.)

30.7.4 BIOACCUMULATIVE CONTAMINANTS IN A STREAM

After completion of the remedial investigation for East Fork Poplar Creek in Oak Ridge, Tennessee, the US Department of Energy commissioned a new ecological risk assessment to test new probabilistic techniques (Moore et al. 1999). Exposure of belted kingfishers and mink to mercury and polychlorinated biphenyls (PCBs) was estimated using Monte Carlo simulation of a multiroute model including inhalation, drinking, and feeding. The exposure–response relationships were estimated using generalized linear modeling of published test data. A single best study was used for kingfishers, but for mink data from multiple tests were combined to generate the dose–response functions. The exposure and exposure–response functions were combined to generate risk curves (Figure 30.8). These analyses showed significant risks to mink and kingfishers from mercury and to mink from PCBs. These results differed from the assessments for the remedial investigation, which found little risk to piscivorous wildlife. However, that assessment did not use measured concentrations in fish, but rather modeled uptake based on contaminants in the sediment, which was the medium that would be remediated under Superfund (Burns et al. 1997). The results by Moore et al. (1999) gave more similar results to a reservation-wide wildlife risk assessment that also used measured fish concentrations (Sample et al. 1996b). The exception was PCBs in mink, and the difference was that Sample et al. (1996b) used the LOAEL as a threshold. Consideration of the dose–response relationship showed that PCBs caused large reproductive effects at the LOAEL.

30.7.5 SECONDARY POISONING IN HAWAII

An ecological risk assessment of possible secondary poisoning in Hawaii provides an example of Monte Carlo simulation of an integrated exposure and effects model (Johnston et al. 2005). Broadcast baits containing rodenticides are consumed by invertebrates, which are in turn

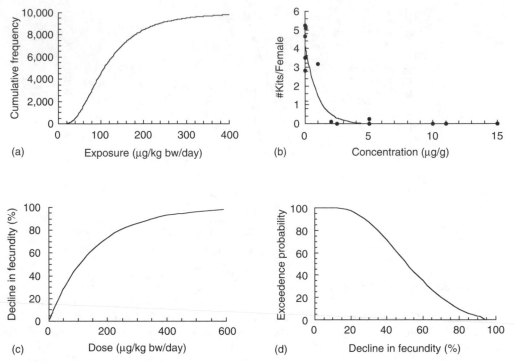

FIGURE 30.8 (a) Cumulative distribution function for female mink exposure to polychlorinated biphenyls (PCBs). (b) Dietary concentration–response curve for effects of PCBs on mink fecundity. (c) Dose–response distribution estimated from (b) and food intake rates. (d) Risk curve for fecundity of female mink exposed to PCBs. (From Moore, D.R.J., Sample, B.E., Suter, G.W., Parkhurst, B.R., and Teed, R.S., *Environ. Toxicol. Chem.*, 18, 2941, 1999. With permission.)

consumed by birds. As illustrated in Figure 30.9, the acute dose to Po'ouli (a honeycreeper) was estimated from the distribution of concentrations in snails and slugs from treated areas, and from estimated distributions of snail consumption rates derived from caloric requirements, fraction molluscs in diets, and energy content of molluscs. The exposure–response model was derived from distributions assigned to the avian LD_{50}, the slope of the dose–response relationship, and the sensitivity of honeycreepers relative to avian test species. Models were also developed for 5 and 14 d exposures. Median probabilities of acute mortality were 0.03% for adults and 0.57% for juveniles (Figure 30.10).

30.7.6 ATRAZINE

The ecological risk assessment for atrazine (Section 32.4.4) illustrates the use of risk curves to integrate exposure and exposure–response distributions (Section 30.3). The exposure–response distributions are SSDs. The exposure distributions include both distributions of measured concentrations and distributions of concentrations from Monte Carlo analyses of models of pond scenarios.

30.7.7 WARMING SUBALPINE FORESTS

Bolliger et al. (2000) estimated risks from climatic warming to the abundance and distribution of tree species in the Alps. The exposure–response relationships were logit regression models

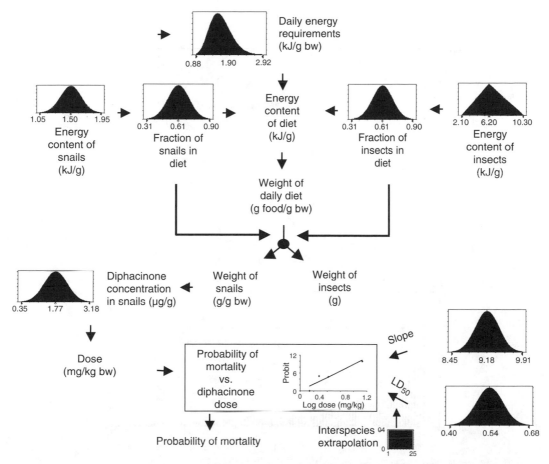

FIGURE 30.9 Diagram of a probabilistic ecological risk assessment for a single-day acute exposure of a Hawaiian honeycreeper to diphacinone. Dose is estimated from measured concentrations in snails and the consumption of snails estimated from energy requirements, dietary habits, and energy contents of dietary items. The exposure–response relationship is estimated from a reported avian LD_{50} and slope values and a distribution of safety factors (1–25) for interspecies extrapolation. (From Johnston, J.J., Pitt, W.C., Sugihara, R.T., Eisemann, J.D., Primus, T.M., Holmes, M.J., Crocker, J., and Hart, A., *Environ. Toxicol. Chem.*, 24, 1557, 2005. With permission.)

of the presence or absence of each major tree species at locations as a function of five biophysical habitat properties. This model was derived by fitting to forest inventory data, using a map of degree-days, radiation in July, summer frost frequency, July water budget, and slope. Exposure was expressed as current conditions and three scenarios: warming of 100, 200, and 400 degree-days, converted into the biophysical variables. The results showed little change in overall abundance, but distributions shifted, and species did not move together. In particular, spruce and beech, which currently occur together became segregated, leaving the subalpine belt dominated by spruce.

30.8 SUMMARY

The critical step in the characterization of ecological risks is the integration of exposure estimates with exposure–response relationships to estimate effects or probabilities of effects.

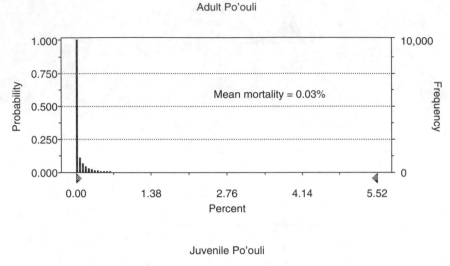

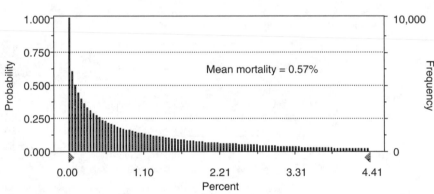

FIGURE 30.10 Results of the assessment methodology shown in Figure 30.9. Probabilities of mortality and equivalent frequencies are presented as functions of percent occurrence. (From Johnston, J.J., Pitt, W.C., Sugihara, R.T., Eisemann, J.D., Primus, T.M., Holmes, M.J., Crocker, J., and Hart, A., *Environ. Toxicol. Chem.*, 24, 1557, 2005. With permission.)

It may be deterministic or probabilistic based on variability or uncertainty in either component. The critical considerations are the relevance of any of the sources of variability and uncertainty to the decision and the concordance of the units of the two components and of the distributions.

31 Screening Characterization

Risk characterization in a screening ecological risk assessment consists of using exposure and effects information to screen the risks into categories. The most general categorization is:

- *De minimis*—risks that are clearly insignificant and can be ignored in subsequent assessments or decisions
- Indeterminate—risks that are not clearly significant or insignificant and must be resolved by further assessment or consideration of other issues such as costs, benefits, engineering feasibility, or public concerns
- *De manifestis*—risks that are clearly significant and need not be assessed further but should be referred to the risk manager for remediation or control

These Latinisms are derived from legal terminology. In particular, some risks or other legal issues are so small that they are considered trivial (*de minimis*) and therefore not worthy of the law's attention (Travis et al. 1987; Whipple 1987). Commonly, this three-part logic is reduced to two parts by replacing *de manifestis* and indeterminate risks with a single category, nontrivial risks. In that case, the nontrivial risks are carried forward to subsequent assessments.

Screening ecological risk assessments have a number of potential uses.

To prompt action: In some cases, a screening assessment will reveal that risks are manifestly significant, and remedial or preventative action should be taken without further data collection or assessment.

To determine the need for further assessment: A screening assessment may reveal that significant risks are highly unlikely or that the risks are of low priority relative to other risks or relative to the likely costs of assessment and management.

To define the scope of a definitive assessment: A screening assessment may screen out certain contaminants, receptors, media, routes of exposure, or portions of a site by demonstrating that they are associated with negligible risks.

To guide data collection: Data collection for subsequent tiers of assessment may be focused on the hazards that have not been screened out as clearly insignificant or to focus testing or modeling on routes of exposure or mechanisms of action that are likely to be associated with an agent.

In some assessment schemes, particularly rule-based assessments of industrial chemicals, there is no definitive assessment in the sense that risks are never estimated (Figure 3.8). Rather, data are generated and screening approaches are applied until the chemical falls into the acceptable or unacceptable category.

31.1 SCREENING CHEMICALS AND OTHER AGENTS

When characterizing risks in a screening assessment it is not necessary to estimate the nature, magnitude, or probability of effects, but it is necessary to assure that hazards fall into the correct category. It is especially important to avoid allowing a hazard to pass through the screens simply because few or no data are available or because the hazard is poorly characterized.

31.1.1 Quotients

The standard model for risk characterization in screening assessments is the hazard quotient (HQ) (Section 30.1). The benchmark and exposure levels may be derived in a variety of ways. The benchmark concentration or dose may be a regulatory standard or a value derived specifically for screening assessments or a standard test endpoint (Chapter 29). The simplest benchmark is the threshold for toxicological concern (TTC), "a level of exposure to chemicals below which no significant risk is expected to exist" (Kroes et al. 2000). For example, 0.01 μg/L has been proposed as a TTC for aquatic toxicity of organic chemicals (De Wolf et al. 2005). The exposure concentration for assessments of new pesticides or industrial chemicals is derived by employing standard use, release, or disposal scenarios to standard environmental models such as in the European Union system for the evaluation of substances (EUSES). For existing contamination, the derivation of an exposure concentration is complex, and is discussed with respect to contaminated site screening (Section 31.2).

Screening of mixtures of contaminants may employ tests of the mixture or models based on the toxicity of the constituents (Chapter 8). The only model that is routinely used for ecological screening assessments of mixtures is the hazard index. The assessor calculates:

$$HI = \sum (C_{ei}/C_{bi}),\qquad(31.1)$$

where HI is the hazard index, C_{ei}, the exposure concentration of chemical i, and C_{bi}, the corresponding benchmark concentration. If the sum is greater than 1, the mixture is potentially hazardous and must be retained for further assessment. If the individual constituents must be screened, one might retain those chemicals that contribute more than 10%, 1%, or some other percentile of the HI. Alternatively, one may include chemicals with individual quotients greater than some value. Parkhurst et al. (1996a) recommend a minimum quotient of 0.3.

If multiple chemicals occur at potentially toxic concentrations but the assumption of additive toxicity is not reasonable, it is useful to calculate the sum of toxic units (ΣTU) as an index of total toxicity (Section 8.1.2). This permits the assessor and reviewers to compare the relative contributions of chemicals to toxicity without necessarily assuming that the contaminants are concentration-additive. TUs are quotients of the concentration of a chemical in a medium divided by the standard test endpoint concentration for that chemical. A TU is similar to an HQ and a ΣTU is similar to an HI except that, because TUs are used for comparative purposes rather than to draw conclusions, a common test endpoint is used rather than conservative benchmarks or most relevant test endpoints. The expression of concentration and the test endpoint vary among media; for water they are typically the mean or upper 95% confidence limit exposure concentration and the 48 h EC_{50} for *Daphnia* sp. (the most common aquatic test endpoint). The chemicals that constitute a potentially significant component of toxicity (i.e., TU > 0.01) should be plotted for each reach or area for water, sediment, soil, and wildlife intake (e.g., Figure 20.2). The choice of a cutoff for inclusion is based on the fact that acute values are used in calculating the TUs, and chronic effects can occur at concentrations as much as two orders of magnitude below acute values. Other values may be used if specific circumstances warrant. The height of the plot at each subreach is the ΣTU for that medium and subreach (Figure 20.2). This value can be conservatively interpreted as the total toxicity-normalized concentration and therefore as a relative indication of the toxicity of the medium in that subreach.

31.1.2 SCORING SYSTEMS

When data are unavailable or there are no simple models of the hazard being assessed, expert judgment based on experience with the issues being assessed or similar issues may be used for screening (UK Department of the Environment 2000). These may be qualitative (e.g., high, medium, low) or semiquantitative. For example, for each hazard, scores (e.g., 1–5) may be applied to the source, transport pathways, receptor exposure, and response, and then summed. Because risk assessments should be transparent and subject to replication, it is important to clearly characterize the bases for judgments. This may be accomplished by developing a formal scoring system. Scoring systems have been used for decades to rank the risks from chemicals or from more diverse sets of agents (Harwell et al. 1992; Swanson and Socha 1997). However, to serve as screening tools, these systems should be calibrated to actual risks so that the total score is at least roughly linearly related to risk and cutoff scores can be defined for the screening categories. If scoring systems are subjective (i.e., not calibrated), it is important to avoid giving an impression of scientific accuracy to the numeric results.

31.1.3 SCREENING FOR PROPERTIES

Chemicals and other agents may be subject to particular assessment standards if they possess certain properties such as persistence, or particular modes of action such as mutagenicity or teratogenicity. For example, the US Food Quality Protection Act requires screening and subsequent testing of pesticides for endocrine-disrupting properties. This screening may be performed using in vitro tests (ER-CALUX test for estrogenicity), simple rapid whole organism tests (e.g., use of Japanese medaka embryos to screen waters for teratogenicity), or quantitative structure–activity relationships (QSARs) for particular modes of action (Section 26.1). If the chemical is not positive in the test or does not fit the model of the mode of action, it is screened out with respect to those standards.

31.1.4 LOGICAL CRITERIA

In some cases, particularly for agents other than chemicals, neither quantitative nor semi-quantitative methods are feasible, but simple logical criteria may be applied to determine whether a hazard exists. For example, to screen nonnative plant species to determine whether they should be assessed, Morse et al. (2004) asked: (1) Is the species established outside cultivation in the region of interest? (2) Is the species established in conservation areas or other native species habitats in that region? If the answer to either question is no, the plant species is screened out from further assessment.

31.2 SCREENING SITES[1]

While the screening of chemicals, materials, and other agents is largely constrained to comparisons of exposure and effects metrics, assessments of contaminated sites are more complex. The primary purpose of screening is to narrow the scope of subsequent assessment activities by focusing on those aspects of the site that constitute credible potential risks. Screening is performed by a process of elimination. Beginning with a site description and the full list of chemicals that are suspected to constitute site contaminants, one can potentially eliminate:

---

[1]This section is based on Chapter 5 of Suter et al. (2000).

- Particular chemicals or classes of chemicals as chemicals of potential ecological concern
- Particular media as sources of contaminant exposure
- Particular ecological receptors as credible assessment endpoints
- Ecological risks as a consideration in the remedial action

A secondary purpose of screening risk assessments is to identify situations that call for emergency responses. A screening assessment may identify ongoing exposures that are causing severe and clearly unacceptable ecological effects or potential sources of exposure that are likely to cause severe and clearly unacceptable ecological effects in the immediate future. In such cases, the usual remedial schedule is bypassed to perform a removal action or other appropriate response. No guidance is provided for such decisions because there are no generally applicable rules for defining an ecological emergency. They must be identified ad hoc.

Finally, screening assessments serve to identify data gaps. Media or classes of chemicals that have not been analyzed, for which analyses are of unacceptably low quality or quantity or for which the spatial or temporal distribution has been inadequately characterized, should be identified during screening assessments. This information serves as input to the assessment planning process for any subsequent assessments.

Screening assessments of sites are performed at three stages:

1. When a site is initially investigated, existing information is collected, and a screening assessment is performed to guide the development of an analysis plan (Chapter 18). The screening assessment is used to help focus the analysis plan on those elements of the site that require investigation and assessment.
2. In a phased assessment process, a screening assessment is performed after the preliminary phase to guide the development of the subsequent phase by focusing investigations on remaining uncertainties concerning credible potential risks.
3. Finally, as a preliminary stage to the definitive assessment, a screening assessment is performed to narrow the focus of the assessment on those contaminants, media, and receptors that require detailed assessment.

Site screening assessments are final assessments only when they indicate that no potential hazards to ecological receptors exist. Otherwise, they should prompt the parties to the remedial process to consider the need for additional data. Whether or not additional data are collected, a screening assessment that indicates that a site is potentially hazardous must be followed by a more definitive baseline ecological risk assessment that provides estimates of the risks and suggests whether remedial actions are needed.

31.2.1 SCREENING CHEMICALS AT SITES

At many sites, concentrations in environmental media will be reported for more than 100 chemicals, most of which are reported as undetected at some defined limit of detection. The assessor must decide which of these constitute chemicals of potential ecological concern (COPECs): which detected chemicals constitute a potential ecological hazard and which of the undetected chemicals may pose a hazard at concentrations below the reported detection limits. The concern about undetected chemicals results from the possibility that the detection limit may be higher than concentrations that cause toxic effects. This screening is done for each medium by applying one or more of the following criteria:

1. If the chemical is not detected and the analytical method is acceptable, the chemical may be excluded.

2. If the wastes deposited at the site are well specified, chemicals that are not constituents of the waste may be excluded.
3. If the concentration of a chemical in a medium is not greater than background concentrations, the chemical may be excluded.
4. If the application of physicochemical principles indicates that a chemical cannot be present in a medium in significant concentrations, the chemical may be excluded.
5. If the chemical concentration is below levels that constitute a potential toxicological hazard, the chemical may be excluded.

In the United States the list of chemicals to be screened is assumed to include the EPA Target Compound and Target Analyte Lists. Chemicals known to be associated with the site contaminants but not on those lists, particularly radionuclides, should be included as well.

Specific methods for applying these criteria are presented in the following subsections. The order of presentation is logically arbitrary, i.e., the screening methods can be applied in any order, and the order used in any particular assessment can be based on convenience. In addition, it is not necessary to use all of the five screening criteria in a screening assessment. Some criteria may be inappropriate to a particular medium or unit, and others may not be applicable because of lack of information.

Other criteria may be used, particularly if they are mandated by regulators or other risk managers. For example, the California EPA specifies that chemicals should be retained if they have regulatory standards, if they are difficult to treat, or if they are highly toxic or highly bioaccumulative, even if they occur at very low concentrations (Polisini et al. 1998). In the *Risk Assessment Guidance for Superfund*, the EPA has specified that chemicals found in less than 5% of samples may be excluded. These criteria are not recommended here, because they are not risk-based.

31.2.1.1 Screening Against Background

Waste sites should not be remediated to achieve concentrations below background; therefore, baseline risk assessments should not normally estimate risks from chemicals that occur at background concentrations. Chemicals that occur at background concentrations may be naturally occurring, may be the result of regional contamination (e.g., atmospheric deposition of cesium-137 from nuclear weapons testing or mercury from incinerators and coal combustion), or may be local contaminants that have been added in such small amounts that their concentrations in a medium have not been raised above the range of background concentrations. Screening against background requires that two issues be addressed. First, what locations constitute background for a particular site? Second, given a set of measurements of chemical concentrations at background locations, what parameter of that distribution constitutes the upper limit of background concentrations?

It must be noted that it has been the policy of the US EPA not to screen against background (Office of Solid Waste and Emergency Response 2002). While the Agency does not promote cleaning a site to below background concentrations, it states that consideration of background in screening assessments is likely to yield misleading results. Misleading screens against background are certainly possible, but a well-conducted screen can be an important and useful component of the assessment process (LaGoy and Schulz 1993; Smith et al. 1996).

31.2.1.1.1 Selection of Background Data

Background sites should be devoid of contamination from wastes or any other local source. For example, water from a location upstream of a unit cannot be considered background if there are outfalls or waste sites upstream of that location. To ensure that there is no local contamination, a careful survey of watersheds for potential background water or sediment sites should be performed, and for terrestrial sites, the history of land use must be determined. For example, although Norris Reservoir, upstream of Oak Ridge, Tennessee, is quite clean relative to Watts Bar Reservoir, the reservoir receiving contaminants from Oak Ridge, the occurrence of a chloralkali plant in the Norris Reservoir watershed has eliminated it as a background site for mercury. In theory, if a local source releases a small and well-characterized set of contaminants, a location that is contaminated by it could be used as a background site for other chemicals. However, wastes and effluents are seldom sufficiently well defined. Background can be defined at multiple scales: regional, local, and unit-specific. Each scale has its advantages and disadvantages.

National or regional background concentrations may be available from existing sources such as the US Geological Survey or state publications (Shacklette and Boerngen 1984; 'Slayton and Montgomery 1991; Toxics Cleanup Program 1994). National or regional background concentrations are advantageous in that they provide a broad perspective. It is not sensible to remove or treat soils at a site for a metal concentration that is higher than local background but well within the range of concentrations of that metal at uncontaminated sites across the region (LaGoy and Schulz 1993). However, one must be careful when using national or regional background values to ensure that the concentrations were measured in a manner that is comparable to the measurements at the waste site. For example, concentrations of metals from aqueous soil extractions should not be compared to total soil concentrations. Because the use of national or regional background concentrations is often less conservative than the use of local or unit-specific concentrations, the latter is often favored by regulators.

Local background measurements are generally the most useful. Local backgrounds are concentrations of chemicals in environmental samples collected to represent an entire site or a geologically homogeneous portion thereof. Examples include the background soil data for the Oak Ridge Reservation (Watkins et al. 1993). That study systematically collected soils from all geological units on the site. It provided high-quality data that was agreed by all parties to represent local background and its variance. In most cases, background measurements are less systematic. Examples include water collected upstream of a site or soil collected beyond the perimeter of a site. Local background measurements present some major disadvantages. Because samples are typically collected in the vicinity of the unit, there is some danger of undetected contamination. In addition, because local background measurements are often poorly replicated in space or time, their variance is often poorly specified. However, because the natural variance in background concentrations is lower in the vicinity of an individual site than across a nation or region, use of unit-specific background values is more likely (compared to background estimates on a larger scale) to suggest that concentrations on the site are above background.

Local background should be used when possible, with care taken to ensure that background variance is specified and that samples from background and contaminated locations are comparable. For example, because aqueous concentrations of naturally occurring chemicals are sensitive to hydrologic conditions, background samples should be taken at the same time as contaminated samples. Regional and national background values can be used to check the reasonableness of local and unit-specific background values.

31.2.1.1.2 *Quantitative Methods for Comparison to Background*
Various methods may be used for comparison of site concentrations to background concentrations. Because concentrations of chemicals at uncontaminated reference sites are variable, a definition of the upper limit of background must be specified. Some regulators use simple rules such as specifying that chemicals should be retained unless the maximum concentration

of the chemical in a medium on the unit is less than twice the mean background concentration. Other possible limits include the maximum observed value at reference sites, a percentile of the distribution of reference concentrations, or tolerance limits on a percentile. If there are multiple reference sites, one might use the site with the highest concentrations as best representing the upper limits of background, or statistically combine the sites to generate a distribution of concentrations across uncontaminated sites.

31.2.1.1.3 Treatment of Background in Multimedia Exposures

Wildlife species are exposed to contaminants in food, water, and soil. If concentrations of a chemical in all media are at background levels, the chemical can be screened out. However, if concentrations in one or more of the media are above background, the chemical cannot be eliminated from consideration in any of the media with respect to that wildlife endpoint, because all sources of the chemical contribute to the total exposure.

31.2.1.1.4 When Is a Concentration Not Comparable to Background?

If there is reason to believe that a chemical occurs in a form that is more toxic or more bioavailable than at background sites, it may be a chemical of concern even at concentrations that are within the range of background values. An example from Oak Ridge is the acidic and metal-laden leachate from the S-3 waste ponds that entered Bear Creek. Because metals are more bioavailable in acidic than in neutral waters, metal concentrations in a stream downgradient of the ponds that were within the range of background waters were not screened out. Considerations should include the major physical–chemical properties of the waste, such as pH, hardness, and concentrations of chelating agents relative to properties of the ambient media, and the species of the chemical in the waste relative to the common ambient species.

31.2.1.1.5 Screening Biota Contamination Using Background Concentrations
for Abiotic Media

It is possible to use background values for abiotic media to screen biota as sources of exposure to herbivores and predators. For example, if all metals in soil are at background concentrations, it can be assumed that plant and earthworm metal concentrations are also at background. Similarly, if concentrations in both water and sediment are at background levels, concentrations in aquatic biota can be assumed to be at background.

31.2.1.1.6 Screening Future Exposure Concentrations Against Background

If exposure concentrations may increase in the future, current concentrations should not be used to exclude chemicals from the baseline assessment, because future exposure scenarios must also be addressed. If the increased future exposures might result from movement of a contaminated ambient medium such as soil or groundwater, concentrations measured in those media should be screened against background. For example, if a plume of contaminated groundwater might intersect the surface in the future, concentrations in the plume should be screened against background. If the increased future exposures might result from changes in a source such as the failure of a tank, the contaminant concentrations predicted to occur in ambient media might often not be screened against background concentrations. That is because regulators often argue that the modeled future concentrations are, by definition, additions to background. Therefore, even if the predicted concentrations are within the range of local concentrations, they are not "really background."

31.2.1.2 Screening Against Detection Limits

Chemicals that are not detected in any sample of a medium may be screened out if the detection limits are acceptable to the risk manager. For example, EPA Region IV indicated that the Contract Laboratory Program's Practical Quantification Limits could be used for this purpose (Akin 1991). It should be noted that this screening criterion is not risk-based. It is entirely possible that undetected chemicals pose a significant risk to some receptors and may account for effects seen in media toxicity tests or biological surveys. The use of this criterion is based entirely on a policy that responsible parties should not be required to achieve lower detection limits than are provided by standard US EPA methods. An alternative is to use the limit of detection as the exposure concentration in the screening assessment.

Care should be taken when eliminating undetected chemicals that are known to bioaccumulate to concentrations in biota that are higher than in inorganic media. In particular, organic mercury and persistent lipophilic organic compounds such as polychlorinated biphenyls (PCBs) and chlordane may occur in significant amounts in aquatic biota when they cannot be detected in water or even sediment. If there are known sources of these chemicals, they should not be screened out until biota have been analyzed.

31.2.1.3   Screening Against Waste Constituents

If the waste constituents are well specified either because they were well documented at the time of disposal or because they are still accessible to sampling and analysis (e.g., leaking tanks), chemicals that are not waste constituents should be screened out. However, this is not possible at many sites because of imperfect record keeping, loss of records, poor control of waste disposal, disposal of ill-defined wastes, or occurrence of wastes in forms that do not permit definitive sampling and analysis.

31.2.1.4   Screening Against Physical–Chemical Properties

Chemicals can be screened out if their presence in significant amounts in a medium can be excluded by physicochemical principles. For example, volatile organic compounds (VOCs) were excluded from the risk assessment for Lower Watts Bar Reservoir, Tennessee, because any VOCs in Oak Ridge emissions would be dissipated by the time the contaminated waters reached the mouth of the Clinch River. Similarly, atmospheric routes of exposure have been eliminated from ecological risk assessments at most units, because significant atmospheric concentrations are implausible given the nature and concentrations of the soil and water contaminants.

31.2.1.5   Screening Against Ecotoxicological Benchmarks

Chemicals that occur at concentrations that are safe for ecological receptors can be excluded as COPECs. This screen is performed using HQs (Section 31.1). If the benchmark concentration or dose for a chemical exceeds its conservatively defined exposure concentration or dose, the chemical may be screened out. Although this is the typical approach to screening for toxic hazards, some risk managers require that the benchmark exceed the exposure level by some factor (e.g., 2 or 10). These safety factors are used because the benchmark or exposure levels are not conservative or not sufficiently conservative. The screening method developed by Parkhurst et al. (1996a) uses a factor of 3. A much more elaborate set of factors can be found in Kester et al. (1998).

Chemicals at contaminated sites typically occur as mixtures. One common approach to this problem is to screen the individual chemicals and assume that this issue is covered by the conservatism of the screening process: any chemical that significantly contributes to toxicity on a site will be retained by the conservative screening process. An alternative is to explicitly model the combined toxicity, which is typically done using an HI based on additive toxicity (Section 31.1). That is, if the sum of the HQs for the chemicals in the mixture, based on normalized toxicity benchmarks, is greater than 1, the mixture is retained for further assessment.

Petroleum and other complex and poorly specified materials present a particular problem. Currently the most common approach for sites contaminated with petroleum and its products is to screen total petroleum hydrocarbon (TPH) concentrations against TPH benchmarks. However, a representative chemical approach could be used to screen against benchmark concentrations. For example, when screening a mixture of polycyclic aromatic hydrocarbons (PAHs) in soil, assessors cannot obtain benchmarks or good toxicity data from which to derive benchmarks for many constituents. In that case, it is appropriate for the sake of screening to use a PAH for which a benchmark is available and which the assessor is confident is more toxic than average to represent all constituents of the mixture.

The calculation of exposure concentrations to be compared to the benchmarks depends on the characteristics of the receptor. In general, a concentration should be used that represents a reasonable maximum exposure given the characteristics of the medium and receptor. The fundamental distinction that must be made is between receptors that average their exposure over space or time and those that have essentially constant exposure.

- Terrestrial wildlife, like humans, move across a site potentially consuming soil, vegetation, or animal foods from locations that vary in their degree of contamination. Therefore, mean concentrations over space provide reasonable estimates of average exposure levels. For the conservative estimate to be used in the screening assessment, the 95% upper confidence limit (UCL) on the mean is as appropriate as in human health assessments (Office of Emergency and Remedial Response 1991, Office of Solid Waste and Emergency Response 2003).
- Fish and other aquatic organisms in flowing waters average their exposures over time. Therefore, the 95% UCL on the temporal mean is a reasonably conservative estimate of chronic aqueous exposure concentrations. If aqueous concentrations are known to be highly variable in time and if periods of high concentration that persist for extended durations can be identified, the averaging period should correspond to those periods.
- Wildlife that feed on aquatic biota average their dietary exposure across their prey organisms. Therefore, they average their exposure over space and time (i.e., over their feeding range and over time as their prey respond to variance in water quality). The 95% UCL on that mean is a reasonably conservative estimate of exposure concentrations.
- Soil and sediment concentrations are typically relatively constant over time, and plants, invertebrates, and microbes are immobile or effectively immobile. Therefore, there is effectively no averaging of concentrations over space or time. The reasonable maximum exposure for those media and receptors is the maximum observed concentration, if a reasonable number of samples have been analyzed. Some organisms occupy that maximally contaminated soil or sediment or would occupy it if it were not toxic. Therefore, exceedence of ecotoxicological benchmarks at any location implies a potential risk to some receptors. Alternatively, an upper percentile of the distribution of concentrations (e.g., 90th percentile) could be used. Such percentiles would be more consistent than maxima because they are less dependent on sample size. The US EPA recommends use of the 95% UCLs on mean soil concentrations for soil invertebrates and plants if many data are available (Office of Solid Waste and Emergency Response 2003), but that implies that these organisms are averaging across the site.

Screening against wildlife benchmarks requires specification of individual wildlife species, so that the concentration in the contaminated medium corresponding to the screening benchmark dose can be estimated using an appropriate exposure model (Section 22.8). Even if endpoint species have not yet been selected for the site through the assessment planning process, species should be selected for screening. The chosen species should include potentially sensitive representatives of trophic groups and vertebrate classes that are potentially exposed to contaminants on the site. The US EPA's surrogate species for deriving soil screening levels are: meadow vole (mammalian herbivore), short-tailed shrew (mammalian ground insectivore), long-tailed weasel (mammalian carnivore), mourning dove (avian grainivore), American woodcock (avian ground insectivore), and red-tailed hawk (avian carnivore) (Office of Solid Waste and Emergency Response 2003). For screening assessments, these species are assumed to be monophagous. For example, the short-tailed shrew is assumed to eat only earthworms.

If no appropriate benchmark exists for a chemical that cannot be screened out by other criteria, an effort should be made to find or develop a benchmark for that chemical. However, in some cases, there are no appropriate toxicity data available for a chemical–receptor combination. In such cases, the chemical cannot be eliminated by toxicity-based screening. The media in which such chemicals occur should not be eliminated from further assessment.

31.2.1.6   Screening Species Against Area

Wildlife endpoint species that have large home ranges relative to areas contaminated are often retained by screening assessments that assume that they spend 100% of their time in the contaminated area. They are then found to have very low risks in definitive assessments that consider spatial distributions of exposure. This wasted effort can be avoided by identifying an allowable contaminated area for screening purposes. One approach uses a percentage (e.g., 2%) of an animal's home range (Tannenbaum 2005b). Hence, since the smallest home range for a red fox is 123.5 acres, if the contaminated area were less than 2.5 acres, that species would be screened out. An alternative approach is based on density. An allowable contaminated area might be the area expected to contain a certain number of organisms of an endpoint species (e.g., 4) (Tannenbaum 2005a).

These approaches, like other screening criteria, must be applied with care. If a contaminated area is particularly attractive, because it provides water, mineral nutrients, or other features that do not occur elsewhere, screening based on area should be avoided or at least applied carefully. Also, the acceptability of area screening criteria should be determined by consulting the decision maker in advance.

31.2.2   Exposure Concentrations for Sites

The following issues must also be considered when deriving exposure concentrations for screening assessments of contaminated sites.

- The screening methods described here presume that measured chemical concentrations are available to define exposure. Use of measured concentrations implies that concentrations are unlikely to increase in the future. Where concentrations may increase in the future due to movement of a contaminated groundwater plume, failure of waste containment, or other processes, future concentrations must be estimated and used in place of measured concentrations for the future scenarios. For screening assessments, simple models and assumptions such as exposure of aquatic biota to undiluted groundwater are appropriate.
- For large sites, it is appropriate to screen contaminants within subunits such as stream reaches rather than in the entire site to avoid diluting out a significant contaminant exposure. The division of a site into areas or reaches should be done during the development of the analysis plan and should take into consideration differences in contaminant sources and differences in habitat.

- Some benchmarks are defined in terms of specific forms or species of chemicals. When forms are not specified in the data available for the screening assessment, the most toxic form should be assumed unless there are compelling reasons to believe that other forms predominate. For example, it has been generally recognized and confirmed in studies at Oak Ridge that the hexavalent form of chromium is converted to trivalent chromium in humid soils, sediments, and waters; therefore, EPA Region IV recommended against assuming that hexavalent chromium is present in significant amounts on that site.
- Measurements of chemicals in ambient media often include a mixture of detected concentrations and nondetects with associated detection limits. For screening of soil and sediment, the maximum value is still available in such cases. However, 95% UCLs on the mean concentration cannot be derived directly. If time and resources permit, these values should be estimated using a maximum likelihood estimator or a product limit estimator (Box 20.1). Otherwise, the 95% UCL can be calculated using the detection limits as if they were observed values. If the chemical was not detected in any sample and the analytical techniques did not achieve detection limits that were agreed to be adequate by the risk manager, the reported limit of detection should be screened in place of the maximum or 95% UCL value.

31.2.3 SCREENING MEDIA

If the screening of chemicals does not reveal any COPECs in a particular medium, and if the data set is considered adequate, that medium may be eliminated from further consideration in the risk assessment. However, if toxicity has been found in appropriate tests of the medium or if biological surveys suggest that the biotic community inhabiting or using that medium appears to be altered, the assessor and risk manager must consider what inadequacies in the existing data are likely to account for the discrepancy between the lines of evidence, and must perform appropriate investigations or reanalysis of the data to resolve the discrepancy.

31.2.4 SCREENING RECEPTORS

If all media to which an endpoint receptor is exposed are eliminated from consideration, that receptor is eliminated as well. For wildlife species that are exposed to contaminants in water, food, and soil, this means that all three media must be eliminated. Aquatic biota can be eliminated if both water and sediment have been eliminated. Plants and soil heterotrophs can be eliminated if soil has been eliminated. Any evidence of significant exposure to contaminants or injury of the receptor would prevent its elimination from the assessment.

31.2.5 SCREENING SITES

A site can be eliminated from a risk assessment if all endpoint receptors for that type of unit have been eliminated. However, it must be noted that even when there are no significant risks due to contaminant exposures on the site, the risk assessment must address fluxes of contaminants that may cause ecological risks off site or incidental use of the site by wildlife, which may cause risks to wide-ranging wildlife populations.

31.2.6 DATA ADEQUACY AND UNCERTAINTIES

Screening ecological risk assessments performed at the intermediate stages of a phased assessment process or as the initial step in the definitive assessment should have adequate data quality and quantity because the data set used should be the result of a proper assessment plan. However, the initial screening assessment is likely to have few data for some media, and the data quality may be questionable because the data are not adequately quality-assured or adequately documented to perform a full data evaluation. Sets of encountered data should at least be evaluated as far as possible to eliminate multiple reports of the same measurement, units conversion errors, and other manifest flaws. Further, because it is important to avoid screening out any potentially hazardous chemicals, the available data should be evaluated to eliminate data with an anticonservative bias. Once data evaluation has been carried out as far as possible, the proper response to remaining questions of data adequacy is to perform the screening assessment with the available data and describe the inadequacies of the data and the resulting uncertainties concerning the results. Highly uncertain screening results for a medium should then constitute an argument for making a broad chemical analysis part of the plan for the next phase of the assessment.

Screening assessments also must consider the relevance of historical data to current conditions. Issues to consider in deciding whether data are too old to be useful include the following:

- If contamination is due to a persistent and reasonably stable source that has been operating since before the date of the historic data, the data are likely to be relevant.
- If the source is not persistent and stable and the chemical is not persistent (e.g., it degrades or volatilizes), the data are unlikely to be relevant.
- If the ambient medium is unstable or highly variable, historic data are less likely to be relevant. Examples include highly variable aqueous dilution volumes and scouring of sediments.
- Human actions, particularly those taken to stabilize the wastes or partially remediate the site, may make historic data irrelevant.

31.2.7 PRESENTATION OF A SITE SCREENING ASSESSMENT

Because the screening assessment is not a decision document, the documentation of a screening assessment should be brief and need not provide the level of documentation or explanation that is required of a definitive assessment (Chapter 35). The results can be presented in two tables. The first table lists the chemicals that were retained with reasons for retaining any chemicals despite their passage through the screening benchmarks and background. The second table lists those chemicals that were rejected with the reason for rejection of each. These results may be presented for each medium, for each spatial unit within a site, and for each endpoint entity exposed to a medium. The assessment that generated these results should be presented in the format of the framework for ecological risk assessment (Chapter 3). However, the description and narrative should be minimal. Although it is important to do the screening assessment correctly, it is also important to ensure that the production of a screening assessment does not become an impediment to completion of a definitive assessment and remediation.

The following information is important to support the screening results:

- Rationale for the list of chemicals that was screened
- Sources of the site, reference, and background contaminant concentrations
- Justification of the use of any preexisting concentration data
- Methods used to derive any site-specific background concentrations
- Criteria used to determine the adequacy of the concentration data set
- Sources of existing screening benchmark values
- Methods used to derive any new screening benchmark values

31.3 EXAMPLES

Because they are a preliminary procedure, screening ecological risk assessments are rarely published in the open literature. However, Region V (2005a,b) has published screening assessments that represent the state-of-practice as part of its guidance for ecological risk assessment of contaminated sites. These assessments use simple conservative assumptions such as 100% area use factor and 100% bioavailability and risk characterizations based on deterministic HQs. The conclusions represent the roles that screening assessments can play. Camp Perry was found to have low ecological risks given current conditions, so it was recommended that efforts be focused on isolating the source, so that conditions will not worsen. Ecological risks were potentially significant at the Elliot Ditch/Wea Creek site, so the assessors recommended additional studies to reduce uncertainties. The examples included evaluation of the presence of endangered species, site-specific toxicity testing, bioaccumulation studies, and residue analyses of ecological receptors.

32 Definitive Risk Characterization by Weighing the Evidence

When you follow two separate chains of thought, Watson, you will find some point of intersection which should approximate the truth.

Sherlock Holmes, in *The Disappearance of Lady Frances Carfax*

Risk characterization for definitive risk assessments consists of integrating the available information about exposure and effects, analyzing uncertainty, weighing the evidence, and presenting the conclusions in a form that is appropriate to the risk manager and stakeholders. The integration of exposure and effects information should be carried out for each line of evidence independently so that the implications of each are explicitly presented. This makes the logic of the assessment clear and allows independent weighing of the evidence. For each line of evidence, it is necessary to evaluate the relationship of the measures of effect to the assessment endpoint, the quality of the data, and the relationship of the exposure metrics in the exposure–response data to the exposure metrics for the site. The actual characterization for ecological risk assessment is then performed by weight of evidence (Suter 1993a; EPA 1998a). Rather than simply running a risk model, ecological risk assessors should examine all available data from chemical analyses, toxicity tests, biological surveys, and biomarkers, and apply appropriate models to each to estimate the likelihood that significant effects are occurring or will occur and to describe the nature, magnitude, and extent of effects on the designated assessment endpoints. "Because so many judgments must be based on limited information, it is critical that all reliable information be considered" (The Presidential/ Congressional Commission on Risk Assessment and Risk Management 1997).

32.1 WEIGHING EVIDENCE

All types of evidence have strengths and weaknesses as bases for inference. By comparing multiple lines of evidence with independent strengths and weaknesses it is possible to identify inferences that are strongly supported and avoid those that have little support. In addition, consideration of all available and relevant evidence provides assurance to stakeholders that evidence is not being ignored or covered up (National Research Council 1994). Finally, the use of multiple lines of evidence provides a sort of replication in situations, such as contaminated or disturbed ecosystems, that are not replicated (Cavalli-Sforza 2000). If we apply three independent techniques to an assessment and get the same answer, it is analogous to applying the same technique to three different ecosystems and getting the same answer. Ecological risk characterization by weight of evidence is widely practiced and numerous methods have been devised (Chapman et al. 2002). Analyzing the weight of evidence is also the best general technique for determining causation in ecological epidemiology (Chapter 4).

Three terms are critical to this discussion.

Type of evidence: A category of evidence used to characterize risk. Each type of evidence is qualitatively different from others used in the risk characterization. Types of evidence are typically characterized by the source of the exposure–response relationship. The most commonly used types of evidence in ecological risk assessments of contaminants are (1) biological surveys, (2) toxicity tests of contaminated media, and (3) toxicity tests of individual chemicals.

Line of evidence: An exposure–response relationship and a corresponding estimate of exposure. A line of evidence (e.g., a fathead minnow LC_{50} and a 24 h maximum concentration estimated using EXAMS) is an instance of a type of evidence (e.g., laboratory test endpoints and modeled exposure levels). There may be multiple lines of evidence for a type of evidence in a risk characterization.

Weight of evidence: A process of identifying the best-supported risk characterization given the existence of multiple lines of evidence. Evidence to be weighed may include multiple lines of evidence of a single type or multiple types of evidence, each represented by one or more lines of evidence.

There are at least four approaches to weighing evidence in ecological risk characterization.

Best line of evidence: The simplest approach is to assemble all available lines of evidence, evaluate their strengths and weaknesses, identify the strongest, and use it to characterize the risks. This approach is most applicable when one line of evidence is much stronger than the others because of its quality or its relevance to the case. In such cases, one would not wish to dilute or cast doubt upon a high-quality line of evidence by combining it with lines of evidence that are likely to be misleading. This approach also has an advantage in that the methods and results are relatively clear and simple to present. The selection of the best line of evidence should be based on attributes like those developed for the Massachusetts weighting and scoring system (Table 32.1).

Tiered assessment: In this approach, lines of evidence (data and associated models) are developed sequentially, beginning with the simplest and cheapest. More complex and expensive lines of evidence are added until one gives a sufficiently clear result. This is a variant on the best line of evidence approach in that only one line of evidence is used in each tier and the risk characterization is based on the results of the highest implemented tier. This approach is illustrated by pesticide ecological risk assessments (Section 32.4.4).

Numerical weighting and scoring: Numerical weights can be applied to results for multiple lines of evidence to generate scores indicating the degree of support. A system of this sort was generated for assessing ecological risks at contaminated sites in Massachusetts (Massachusetts Weight-of-Evidence Work Group 1995; Menzie et al. 1996). Each line of evidence is scored for the indicated response (presence or absence of harm and high or low response), and then given a 1 to 5 score for each of 10 attributes and finally each attribute is weighted 0 to 1 for its relative importance. The 10 attributes have to do with the strength of association of the evidence with the site, the quality of the evidence, and the study design and execution (Table 32.1). Finally, the concurrence among the lines of evidence is scored. This system, inevitably with some modification, has been successfully applied to real sites (Johnston et al. 2002). Like other rule-based methods of risk characterization, numerical weighting and scoring is efficient and consistent but may not be optimal for an individual case.

Best explanation: Abductive inference, as opposed to deductive and inductive inference, reasons from a collection of evidence to the best explanation (Josephson and Josephson 1996) (Section 4.3). The best explanation is one that accounts for the apparent discrepancies among lines of evidence as well as the concurrences. For example, differences in bioavailability often explain differences in apparent toxicity between laboratory toxicity tests and ambient water

TABLE 32.1
Attributes for Scoring Lines of Evidence

Attributes	Explanation
Relationship of the line of evidence and the assessment endpoint	
Degree of association	The extent to which the response in the exposure–response relationship is representative of, or correlated with, the assessment endpoint
Exposure–response	The extent to which the exposure–response relationship is quantified and the strength of the relationship
Utility	The certainty, scientific basis, and sensitivity of the methods used to generate the exposure estimate and the exposure–response relationship
Data quality	
Quality of data	The quality of the methods used to generate the exposure estimate and the exposure–response relationship, including measurements, tests, and models
Study design	
Site specificity	Representativeness of the media, species, environmental conditions, and habitat type relative to those at the site; for non-site-specific assessments, the scenario is considered instead
Sensitivity	The ability of the exposure–response relationship to define endpoint effects of concern and to discriminate them from background variance or confounding causes
Spatial representativeness	Spatial overlap or proximity of the area sampled to generate the exposure estimate or the exposure–response relationship and the area being assessed
Temporal representativeness	Temporal overlap of the time when samples or measurements were taken to generate the exposure estimate or the exposure–response relationship and the time when effects were being induced or the frequency or duration of sampling or measurement relative to the temporal pattern of exposure and response
Quantitative measure	The degree to which the magnitude of response can be quantified by the line of evidence
Standard method	The degree to which exposure estimates or exposure–response relationships were generated using data produced by relevant standard methods and standard models

Source: Adapted from Menzie, C., Henning, M.H., Cura, J., Finkelstein, K., Gentile, J., Maughan, J., Mitchell, D., et al., *Hum. Ecol. Risk Assess.*, 2, 277–304, 1996; Massachusetts Weight-of-Evidence Work Group, *Draft Report: A Weight-of-Evidence Approach for Evaluating Ecological Risks*, 1995. (the Attributes are Somewhat Reworded, and the Explanations are Entirely Different).

toxicity tests. The best developed example of abductive inference in ecological risk characterization is the sediment quality triad, discussed below.

This chapter is devoted to weighing evidence to determine the best characterization of an ecological risk. Risk characterizations based on a best line of evidence or numerical weighting and scoring systems are program-specific and largely self-explanatory. In addition, they implicitly assume that each line of evidence is independent. Reasoning to the best explanation allows assessors to use knowledge of the relationships among lines of evidence to characterize the risks.

32.2 SEDIMENT QUALITY TRIAD: A SIMPLE AND CLEAR INFERENCE METHOD

Inference to the best explanation as an approach to weighing multiple lines of evidence is best illustrated by the sediment quality triad (Long and Chapman 1985; Chapman 1990). The three types of evidence from a site that form the triad are sediment chemistry, sediment toxicity, and sediment invertebrate community structure. Measures used for each line of evidence must be chosen and evaluated to be appropriately sensitive, so as not to cause frequent false negative or false positive results. For example, analyses of sediment chemicals must be sufficiently sensitive to detect potentially toxic concentrations, but they should also be compared to effects benchmarks or background concentrations to assure that trivial levels are not counted as positive results.

Assuming that all three components can be assigned a dichotomous score ($+/-$) and are determined with appropriate sensitivity and data quality, the rules in Table 32.2 can be used to reach a conclusion concerning the induction of effects by contaminants. The assumptions are critical. For example, in situation 5 the conclusion that the community alteration is not due to toxic chemicals depends on the assumptions that the chemical analyses include all potentially significant toxicants and are sufficiently sensitive, and that the sediment toxicity test was well conducted and was representative of the field exposure conditions and of the sensitivity of the more sensitive members of the community. If any of the data are not sufficiently sensitive or of high quality, one must weigh the evidence taking those issues into consideration. The sediment quality triad was developed for estuarine sediments and has been widely applied in those systems (Chapman et al. 1997). It has been adapted to the soft sediments of streams, rivers, and reservoirs (Canfield et al. 1994) and is applicable in principle to contaminated soils.

An alternative inferential triad, the exposure–dose–response triad, has been proposed for assessment of risks from contaminated water or sediments by Salazar and Salazar (1998). Exposure is estimated by analysis of the ambient medium, dose by analysis of tissue chemistry, and response by surveys of community properties or toxicity tests of the contaminated

TABLE 32.2
Inference Based on the Sediment Quality Triad

Situation	Chemicals Present	Toxicity	Community Alteration	Possible Conclusions
1	+	+	+	Strong evidence for pollution-induced degradation
2	−	−	−	Strong evidence that there is no pollution-induced degradation
3	+	−	−	Contaminants are not bioavailable, or are present at nontoxic levels
4	−	+	−	Unmeasured chemicals or conditions exist with the potential to cause degradation
5	−	−	+	Alteration is not due to toxic chemicals
6	+	+	−	Toxic chemicals are stressing the system but are not sufficient to significantly modify the community
7	−	+	+	Unmeasured toxic chemicals are causing degradation
8	+	−	+	Chemicals are not bioavailable or alteration is not due to toxic chemicals

Source: Chapman, P.M., *Sci. Total Environ.*, 97/98, 815–825, 1990. With permission.

Responses are shown as either positive ($+$) or negative ($-$), indicating whether or not measurable and potentially significant differences from control/reference conditions are determined.

media. Significant contaminant effects may be assumed to be occurring if effects are detected in the tests or surveys and if they are linked to the ambient contamination by body burdens, which are sufficient to indicate a toxic dose. In the examples provided by Salazar and Salazar (1998), the effects tests use growth responses of bivalves exposed in cages to the field. The tissue concentrations were from the same caged bivalves. Toxic tissue concentrations for the contaminants of potential concern were determined from controlled exposures. As the authors acknowledge, if the lines of evidence are not consistent, one must weigh the evidence, taking into consideration data quality and factors such as combined toxic effects, temperature, food availability, and variance in toxic tissue concentrations due to growth dilution. Subsequently, others have proposed adding bioaccumulation measurements to the sediment quality triad (Borgmann et al. 2001; Grapentine et al. 2002).

The original insight of the sediment quality triad—that the pattern of results from multiple types of evidence can be used to infer causal processes at a site—has been obscured by its adaptation to status and trends monitoring. Some users have reduced the triad to an index, simply counted the number of pluses and minuses, or integrated the area within triaxial plots (Long et al. 2004). The triad is presented here both for its utility as an inferential method when its assumptions are met and as a model of inference to the best explanation.

32.3   INFERENCE TO THE BEST CONCLUSION AT CONTAMINATED SITES

In most risk assessments, inference is not as simple as in the sediment quality triad. Data quality is mixed, the relevance of some data to the site may be questionable, the various types of evidence may not come from co-located samples, not all of the types of evidence may be available for all sampling locations, and results may not be unambiguously positive or negative. This section, which is condensed and adapted from Suter et al. (2000), presents an approach to organizing information from multiple types and lines of evidence to infer the existence, significance, and nature of effects on contaminated sites. This approach is based primarily on experience from the ecological risk assessments of the Oak Ridge Reservation (ORR) and Clinch River in Tennessee. Contaminated sites are emphasized, because they are the most common situation in which ecological risk assessors use diverse types of evidence.

32.3.1   SINGLE-CHEMICAL TOXICITY

This type of evidence uses concentrations of individual chemicals in environmental media (either measured or modeled) to estimate exposure, and uses results of toxicity tests for those individual chemicals to estimate effects (Figure 32.1). They are combined in two steps. First, the chemicals are screened against ecotoxicological benchmarks, against background exposures, and, where possible, against characteristics of the source to determine which are chemicals of potential ecological concern (COPECs) (Chapter 31). The results of the screening assessment should be presented in the definitive assessment as a table listing all of the chemicals that exceeded benchmarks, indicating which are COPECs, and stating the reasons for acceptance or rejection.

The integration of exposure with single-chemical toxicity data is minimally expressed as a hazard quotient (HQ) (Section 30.1). The effective concentration may be a test endpoint, a test endpoint corrected by a factor or other extrapolation model, or a regulatory criterion or other benchmark value. Compared to screening assessments, more realistic exposure estimates are used, and effects are expressed as test endpoints that are closely related to the assessment endpoint, or the effects threshold is estimated using an extrapolation model. In addition, in the definitive assessment one must be concerned about the magnitude of the quotient and not simply whether it exceeds 1. Large quotients suggest large effects or at least

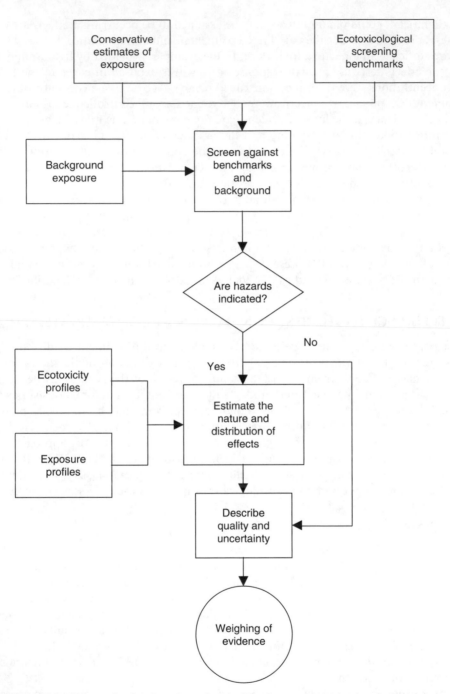

FIGURE 32.1 Risk characterization based on chemical analyses and single-chemical toxicity.

indicate that the uncertainty concerning the occurrence of the endpoint effect is likely to be low. Formal analyses of variance or uncertainty in HQs are also used (Section 30.1).

If multiple chemicals appear to be significantly contributing to toxicity, it is highly desirable to perform toxicity tests of the contaminated media to determine the nature of interactions (Section 24.5), or, if that is not possible, to perform tests of a synthetic mixture with laboratory media using conventional test methods. If even that is not possible, one must consider how to estimate the combined toxic effects (Chapter 8). In practice, estimates based on individual chemical exposures and responses may significantly underestimate toxicity, simply because analytical schemes do not include all chemicals that occur in complex mixtures (Barron and Holder 2003).

For all contaminants for each endpoint, exposures must be compared to the full toxicity profile of the chemical to characterize risk. For example, the distribution of concentrations in water would be compared to the distribution of concentrations of thresholds for chronic toxicity across endpoint species and across species upon which they depend (e.g., prey and habitat-forming species), the nature of the chronic effects would be described, and the exposure durations needed to achieve effects in the laboratory would be compared to temporal dynamics of concentrations in the field. Characteristics of the chemicals that are relevant to risks, such as the influence of metal speciation on toxicity and tendency of the chemical to accumulate in prey species, are also examined.

Inferences about the risk posed by the contaminants should be based on the probability distribution of exposure estimates relative to the exposure–response distribution. Distributions provide a better basis for inference than point estimates because they allow the consideration of variance in exposure over space or time and of sensitivity across species, measures of effects, media properties, or chemical forms. In all cases, risk is a function of the overlap between the exposure and exposure–response distributions, but the interpretation depends on the data that are used.

For all endpoints the risk characterization ultimately depends on weighing of all types of evidence. To facilitate the weighing of evidence and to make the bases clear to the reader, it is useful to summarize the results of the integration of single-chemical exposure and response information for each endpoint in each reach or area in which potentially toxic concentrations were found. Table 32.3 presents an approach for summarization in terms of a table of issues to be considered in the risk characterization and the type of results that are relevant. The information should be based on analyses of all lines of evidence within the single-chemical toxicology type of evidence.

32.3.1.1 Aquatic Organisms

Fish, aquatic invertebrates, and aquatic plants are exposed primarily to contaminants in water. Because water in a reach is likely to be more variable in time than space, due to the

---

TABLE 32.3
Summary Table for Integration of Single-Chemical Toxicity

Issue	Result of Risk Characterization for Single Chemicals
Taxa affected	List specific species or higher taxa, life stages, and proportion of tested species affected at estimated exposures
Estimated effect	List types and magnitudes of estimated effects at estimated exposures and credible upper bounds of effects
Spatial extent	Define the meters of stream, square meters of land, etc. estimated to experience specified effects
Frequency	Define the proportion of time or number of distinct episodes of prescribed effects
Association with source	Describe the spatial and temporal relationships of effects to sources
Confidence in results	Provide rating and supporting comments regarding confidence

rapid replacement of water in flowing systems, the mean chemical concentration in water within a reach or subreach is often an appropriate estimate of the chronic exposure experienced by fishes. The upper 95% confidence bound on the mean is commonly used as a conservative estimate for screening assessments but the full distribution of observed concentrations is used to estimate risks. Other logic may be applied to characterize exposures in lakes, estuaries, or wetlands.

Some fish and invertebrates spend most of their lives near the sediment, and the eggs and larvae of some species (e.g., most centrarchid fishes) develop at the sediment–water interface. These epibenthic species and life stages may be more highly exposed to contaminants than is suggested by analysis of samples from the water column. If available, water samples collected just above the sediments provide an estimate of this exposure. Alternately, the estimated or measured sediment pore water concentrations may be used as a conservative estimate of this exposure.

The aqueous toxicity data from the toxicity profiles and the aqueous chemical concentrations should be used to present distributions of exposure and effects. For exposure of fish and other aquatic organisms to chemicals in water, the exposure distributions are distributions of aqueous concentrations over time, and the effects distributions are usually distributions of sensitivities of species to acutely lethal effects (e.g., LC_{50}s) and chronically lethal or sublethal effects (e.g., chronic values, CVs). If the water samples were collected in a temporally unbiased design, overlap of these two distributions indicates the approximate proportion of the time when aqueous concentrations of the chemical are acutely or chronically toxic to a particular proportion of aquatic species. For example, 10% of the time copper concentrations in Reach 4.01 of the Clinch River, Tennessee, are at levels chronically toxic to approximately half of aquatic animals (Figure 30.5). Interpretation of this result depends on knowledge of the actual temporal dynamics of the exposures and effects. For example, 10% of a year is 36 d, which would cover the entire life cycle of a planktonic crustacean or the entire embryo-larval stage of a fish, so significant chronic effects are clearly possible. However, if the 36 d of high concentrations is associated with a number of episodes, the exposure durations are reduced. The 7 d duration of the standard EPA subchronic aqueous toxicity tests could be taken as an approximate lower limit for chronic exposures, so the proportion of the year with high copper concentrations could be divided into five equal episodes and still induce significant chronic effects on a large proportion of species. More precise interpretations would require knowledge of the actual duration of episodes of high concentrations and of the rate of induction of effects of copper on sensitive life stages.

Although the exposure and effects distributions described above are the most common in aquatic ecological risk assessments, numerous others are possible. Exposures may be distributed over space rather than time or may be distributed due to uncertainty rather than either spatial or temporal variance. Instead of distributions of test endpoints across species, effects may be distributed with respect to variance in a concentration–response model or uncertainties in extrapolation models. The risk estimates generated from these joint exposure and effects models must be carefully interpreted (Section 30.5).

32.3.1.2 Benthic Invertebrates

An appropriate estimate of risks to a benthic community is the percentage of samples exceeding particular effects levels. For each contaminant, the distributions of observed concentrations in whole sediment and pore water are compared with the distributions of effective concentrations in sediment and water. In the case of exposure of benthic invertebrates to sediment or pore water, the exposure distributions are interpreted as distributions over space, since sediment composition varies little over the period in which samples were collected, but samples were distributed in space within reaches or areas. For pore water, the effects distributions are the same as for surface water, i.e., distributions of species sensitivities in acute and chronic aqueous toxicity tests. Therefore, overlap of the distributions indicates the proportion of locations in the reach where concentrations of the chemical in pore water are acutely or chronically toxic to a particular proportion of species. If the samples are collected by an unbiased method, these proportions can be interpreted as proportions of the area of the reach. Therefore, an alternate expression of the result is that less than 10% of the area is estimated to be toxic to as much as 10% of benthic species.

In the case of exposure of benthic invertebrates to chemicals in whole sediment, the exposure distributions are, as with pore water, distributions in space within areas. Effects distributions may be distributions of thresholds for reductions in benthic invertebrate community parameters in various locations or distributions of concentrations reported to be thresholds for lethal effects in toxicity tests of various sediments. If a field effects data set is drawn from studies of a random sample of sediments so that the site sediments can be assumed to be a random draw from the same distribution, and if we assume that the reported community effects correspond to the community effects defined in the assessment endpoint, the effects distributions can be treated as distributions of the probability that the chemical causes significant toxic effects on the endpoint at a given concentration. Overlap of the exposure and effects distributions represents the probability of significant alteration in the benthic communities at a given proportion of locations in a reach or area.

32.3.1.3 Soil Exposure of Plants, Invertebrates, and Microbial Communities

Exposures to organisms rooted in, or inhabiting, soil are typically expressed as whole soil concentrations, though concentrations in soil pore water and concentrations normalized for soil characteristics are also potential measures of exposure (Section 22.4). For definitive assessments, the spatial distribution of observed concentrations of each chemical is the most generally useful estimate of exposure. The distribution should be defined for an assessment unit—an area which is expected to be treated as a single unit in remedial decisions.

The distribution of measured concentrations may be compared with the distributions of effective concentrations for plants, invertebrates, or microbial processes, depending on the assessment endpoints. The relevant level of effects should be defined in the problem formulation, though the paucity of data may prevent consistency in the distribution. For plants, the effects distributions are distributions of species–soil combinations in published, single-chemical soil toxicity tests. If it is assumed that characteristics of site soils are drawn from the same distribution of soil properties as the tested soils and that test plants have the same sensitivity distribution as plants at the contaminated site, the threshold concentrations for effects on site plants can be assumed to be drawn from the distributions of threshold concentrations for effects on plants in single-chemical toxicity tests. Therefore, an overlap of concentrations in the distributions indicates the proportion of locations in an area where concentrations of the chemical are expected to be toxic to a fraction of species in the site community (concentrations at half of the locations are apparently toxic to some of the plant species; concentrations at 20% of the locations are apparently toxic to more than 10% of the plant species; etc.). Assumptions for soil invertebrates are similar except that in ecosystems where earthworms are present, they are typically assumed to be representative of soil invertebrates. The distributions of concentrations toxic to microbial processes are not quite equivalent to the other two endpoints because some processes are carried out by individual microbial strains and others are carried out by entire microbial communities. Therefore, if processes comprise an endpoint, the distributions of effects data should be distributions of process–soil combinations rather than species–soil combinations. The overlap of the distributions indicates the proportion of locations in an area where concentrations of the chemical are expected to be toxic to a fraction of the microbial processes carried out in the area.

The definitive risk characterizations for soils differ from screening assessments in several ways. First, the role of chemical form and speciation must be considered. If a polycyclic aromatic hydrocarbon (PAH) is a constituent of petroleum, it may be less bioavailable than the chemical in laboratory toxicity tests. If the selenium concentration measured in a soil is mostly selenite, tests of selenate or organoselenium may not be as relevant. Even if these chemicals are equally toxic to plants, the typical measure of exposure—concentration in soil—requires that the assessor consider differences in uptake of the chemical species or forms. It is common for laboratory test chemicals to be more bioavailable than chemicals aged in contaminated soils. Second, the role of soil type must be considered. Tests in sandy or clay loams may have limited applicability to the estimation of toxicity in a muck soil. Third, differences in sensitivity between tested organisms and those at the site should be considered. The toxicity of chemicals to earthworms may differ substantially from the toxicity to arthropods. In addition to considering these factors in the summary of this single-chemical line of evidence, the assessor should incorporate them in decisions about how much to weigh this line of evidence, relative to media toxicity tests, biological surveys, and others.

32.3.1.4 Multimedia Exposure of Wildlife

In contrast to other ecological endpoints, wildlife exposure is typically estimated for multiple media (e.g., food, soil, and water), with exposure being reported as the sum of intake from each source in units of mg/kg/d (Section 22.8). For definitive assessments, more realistic estimates of exposure than those in the screening assessment should be generated using distributions of exposure parameters and uncertainty analysis (e.g., Monte Carlo simulation—see Chapter 7). Depending on the nature of the site and the goals of the assessment, exposure simulations are performed for each discrete area at the site of interest or for the range of individuals or populations of an endpoint species. Examples of this approach to wildlife risk characterization are presented in Sample and Suter (1999), Baron et al. (1999), and Sample et al. (1996a).

32.3.1.4.1 Integration of Exposure with Exposure–Response
Comparisons of exposure estimates may be performed in several ways. For screening assessments and even most definitive assessments, HQs are calculated. The exposure pathway that is driving risk may be identified by comparing HQs calculated for each pathway to the HQ for total exposure.

If exposure distributions are generated from Monte Carlo simulations, superposition of point estimates of toxicity values on these distributions provides an estimate of the likelihood of individuals experiencing adverse effects. The portion of the exposure distribution that exceeds the given toxicity value is an estimate of the likelihood of effects described by the toxicity value. For example, Figure 32.3 displays distributions for the dose to river otter of mercury within two subreaches of the Clinch River/Poplar Creek system. The lowest observed adverse effect level (LOAEL) for mercury to otters crosses the exposure distributions for Subreaches 3.01 and 3.02 at approximately the 85th and 15th percentiles, respectively (Figure 32.2). From this figure, there is a 85% likelihood of effects to individual otters from mercury within Subreach 3.01 and an 15% likelihood of effects within Subreach 3.02. Because the no observed adverse effect level (NOAEL) for mercury to otters was exceeded by the entire exposure distributions at both subreaches, the potential for effects at both locations is suggested. By superimposing toxicity values for different types and severity of effects (e.g., LD_{50}, growth, hepatotoxicity), additional information concerning the nature of risks associated with the estimated exposure may be produced.

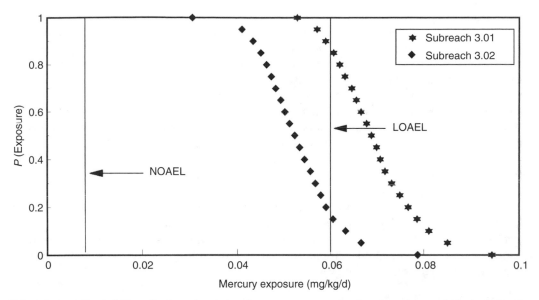

FIGURE 32.2 Probability density of estimated mercury exposures for otters in a subreach of Poplar Creek in relation to the allometrically scaled lowest observed adverse effect level (LOAEL). NOAEL = no observed adverse effect level.

Exposure distributions may also be compared to response distributions, such as from a population model (Chapter 27). Methods for comparison of exposure and effects distributions are discussed in Section 30.3.

For small sites or units, the chief problem is to determine whether even one organism will be exposed to a hazardous dose. Figure 30.6 illustrates this point. The vertical dashed lines indicate the home range or territory sizes of endpoint species (shrews and woodcock in this case). If the average concentration drops below the horizontal effects level lines before intersecting the home range size, not even one individual is expected to be affected. In Figure 30.6, no effects are expected on woodcock, but one shrew is estimated to experience reproductive effects and has a small risk of death.

32.3.1.4.2   Individual vs. Population Effects

Toxicity data largely represent organism-level responses with population-level implications (e.g., mortality, development, and reproduction). With the exception of endangered species that are protected as individuals, most ecological risk management decisions are based on populations. Therefore, unless population modeling is employed (Chapter 27), the results of the risk characterization must be expressed as the frequency or magnitude of organismal responses within populations. This may be done in several ways.

First, it may be assumed that there is a distinct population on the site, so that the exposure of the population is represented by the exposure of all of the individuals. All individuals at the site may be assumed to experience equivalent exposure. The exposure distributions represent distributions of total intake rate of the chemical across individuals in the population based on the distributions of observed concentrations in water, soil, and various food items within each modeled area. If it is assumed that the members of a population occurring in each modeled area independently sample the water, soil, and food items over the entire modeled area, proportions from the resulting exposure distribution represent estimates of the proportion of a population receiving a particular intake rate. This assumption is appropriate for nonterritorial organisms with limited home ranges, on large sites, particularly if the site constitutes a distinct habitat that is surrounded by inappropriate habitat. For example, a grassy site surrounded by forest or industrial development might support a distinct population of voles. The risks to that population can be estimated directly from the exposures of the individual organisms. Baron et al. (1999) used this approach to estimate population-level risks to a rough-winged swallow nesting colony on the Clinch River, Tennessee.

Another approach is to assume that a certain number of individuals out of a larger population are exposed to contaminants. The proportion of the local population exposed at levels that exceed toxic thresholds represents the proportion of the population potentially at risk. This was the logic underlying the preliminary assessment for wide-ranging wildlife on the ORR (Sample et al. 1997). On the ORR, most habitat for wide-ranging wildlife species exists outside of source units (directly contaminated areas), but some suitable habitat is present within source units. The proportion of the ORR-wide population potentially at risk from a source unit is estimated by the number of individuals that may use habitat on that unit. The degree to which a source unit is used (and therefore the risk that it may present) is dependent upon the availability of suitable habitat on the unit. An estimate of risks to reservation-wide populations was generated as follows:

1. Individual-based contaminant exposure estimates were generated for each unit using the generalized exposure model (Section 22.8). Contaminant concentrations were averaged over the entire unit.
2. Contaminant exposure estimates for each unit were compared to LOAELs to determine the magnitude and nature of effects that may result from exposure on the unit. If the exposure estimate was greater than the LOAEL, individuals on the unit were determined to potentially experience adverse effects.
3. The availability and distribution of habitat on the ORR and within each unit were determined using a satellite-generated land cover map for the ORR.
4. Habitat requirements for the endpoint species of interest were compared to the ORR habitat map to determine the area of suitable habitat on the ORR and within units.
5. The area of suitable habitat on the ORR and within units was multiplied by species-specific population density values (ORR-specific or obtained from the literature) to generate estimates of the ORR-wide population and the number of individuals expected to reside within each unit.
6. The number of individuals of a given endpoint species expected to be receiving exposures greater than LOAELs for each measured contaminant was totaled using the unit-specific population estimate from step 5 and the results from step 2. This sum was compared with the ORR-wide population to determine the proportion of the ORR-wide population that was receiving hazardous exposures.

This approach provides a very simple estimate of population-level effects. It is biased because it does not take wildlife movement into account. Wide-ranging species may travel among, and use, multiple units, and therefore receive exposures different from that estimated for a single unit. If this issue is important, a spatially dynamic approach must be used. In addition, application of this approach requires knowledge of the types, distribution, and quality of habitats present on the site.

A third approach is to combine the results of Monte Carlo simulation of exposure with literature-derived population density data to evaluate the likelihood and magnitude of population-level effects on wildlife. The number of individuals within a given area likely to experience exposures greater than LOAELs or other benchmarks can be estimated using cumulative binomial probability functions:

$$b(y; n; p) = \left[\frac{n}{y}\right] p^y (1 - p)^{ny} \tag{32.1}$$

where y = the number of individuals experiencing exposures > LOAEL; n = total number of individuals within the area; p = probability of experiencing an exposure in excess of the LOAEL; $b(y; n; p)$ = probability of y individuals, out of a total of n, experiencing an exposure > LOAEL, given the probability that exceeding the LOAEL = p.

By solving Equation 32.1 for $y = 0$ to $y = n$, a cumulative binomial probability distribution may be generated that can be used to estimate the number of individuals within an area who are likely to experience adverse effects.

This approach was used to estimate the risks that polychlorinated biphenyls (PCBs) and mercury in fish presented to the population of piscivores in watersheds on the ORR (Sample et al. 1997). Monte Carlo simulations were performed to estimate watershed-wide exposures. It was assumed that wildlife species were more likely to forage in areas where food is most abundant. The density or biomass of fish at or near locations where fish bioaccumulation data were collected were assumed to represent measures of food abundance. (Biomass data were preferred but, where unavailable, density data were used.) The relative proportion that each location contributed to overall watershed density or biomass was used to weight the contribution to the watershed-level exposure. The watershed-level exposure was estimated to be the weighted average of the exposure at each location sampled within the watershed. In this way, locations with high fish densities or greater fish biomass contributed more to exposure than locations with lower density or biomass. Because the watersheds were large enough to support multiple individuals, the weighted average exposure estimate was assumed to represent the exposure of all individuals in each watershed. While simplistic, this approach is believed to provide a better estimate of population-level effects than the previously described method. The use of this method, however, requires exposure data from multiple, spatially disjunct areas and data suitable to weight the potential exposure at each area. Sample and Suter (1999) and Hope and Peterson (2000) provide additional application of this method.

Freshman and Menzie (1996) present an additional approach for extrapolating to population-level effects. Their Population Effects Foraging (PEF) model estimates the number of individuals within a local population that may be adversely affected. The PEF model is an individual-based model that allows animals to move randomly over a contaminated site. Movements are limited by species-specific foraging areas and habitat requirements. The model estimates exposures for a series of individuals and sums the number of individuals that receive exposures in excess of toxic thresholds.

32.3.1.5 Body Burdens of Endpoint Organisms

Body burdens of chemicals are not a common line of evidence in risk assessments, and, when they are used, it is most often in assessments of aquatic toxicity. Although single-chemical internal exposure–response is not conceptually distinct from external exposure–response, it may be sufficiently independent in practice to be considered a distinct type of evidence. In that case, estimates of internal exposure from measurements of uptake models (Section 22.9) and exposure–response relationships based on internal concentrations (Section 23.2.7) should be integrated to estimate risks, and a summary table like Table 32.3 should be prepared.

Although nearly all toxicity data for fish are expressed in terms of aqueous concentrations, fish body burdens potentially provide an exposure metric that is more strongly correlated with effects. The correlation is most likely to be evident for chemicals that bioaccumulate in fish and other biota to concentrations greater than in water. For such chemicals dietary exposures may be more important than direct aqueous exposures, and concentrations that are not detectable in water may result in high body burdens in fish. Three common contaminants that accumulate in that manner are mercury, PCBs, and selenium. Since the individual body burden measurements correspond to an exposure level for an individual fish, the maximum value observed in an individual fish is used for screening purposes, and the risk estimate is based on the distribution of individual observations for each measured species. Measurements may be performed on muscle (fillet), carcass (residue after filleting), or whole fish. Since measurements in whole fish are most commonly used in the literature, concentrations of chemicals in whole fish should be either measured directly or reconstructed from fillet and carcass data to estimate exposure.

Body burdens may also be used to estimate risks to terrestrial wildlife, plants, and, less commonly, soil or sediment invertebrates. The justification of the use of this line of evidence for wildlife is that they, like fish, may bioaccumulate certain chemicals. Body burdens are not often measured in wildlife risk assessments because of the difficulty and ethical considerations inherent in sampling these vertebrates. Moreover, if the target organ is not known, it is unclear whether the chemical analysis of the whole animal, muscle tissue, or a particular organ is correlated with the concentration at the site of toxicity. For example, measurements of a particular metal in liver or kidney of deer are relevant to the risk assessment only if one of these organs is associated with the mechanism of toxicity or if the chemical is at equilibrium in all organs. In addition, for wide-ranging wildlife, the history of the animal (i.e., how much time it has spent on the site) is generally unknown. If concentrations of chemicals are measured in small mammals in order to estimate risks to predators, these concentrations may also be used to estimate risks to the small mammals. Otherwise, uptake models must be used.

Concentrations measured in plant tissues may also be used as estimates of exposure in risk assessments. As with wildlife, the concentrations are useful only if the organs (roots, leaves, reproductive structures) in which they were measured are either integral to the mechanism of toxicity or in equilibrium with organs involved in toxicity. Tissue concentration is a more direct estimate of exposure than soil concentrations, but far fewer studies relate these tissue concentrations to effects. Cases in which plant tissue concentrations would be more reliable (and therefore the associated evidence would be weighed more heavily) than soil concentrations would be instances where published single-chemical toxicity data for soils were irrelevant, because of differences in contaminant availability for uptake. Such an example is the estimate of toxicity to plants in sewage sludge-amended agricultural soils (Chang et al. 1992).

The estimation of body burdens from concentrations in abiotic media and the subsequent comparison to toxicity thresholds in published studies is rarely advisable, if site-specific measurements are possible. Because bioaccumulation models add considerable uncertainty to risk estimates, the measurement of tissue concentrations is generally well worth the expense.

32.3.2 AMBIENT MEDIA TOXICITY TESTS

Risk characterization for ambient media toxicity tests begins by determining whether the tests show significant toxicity (Figure 32.3). Although statistical significance has been used for this purpose, it is preferable to establish a significant level of toxicity during problem formulation. If no significant toxicity is found, the risk characterization consists of determining the likelihood that the result constitutes a false negative. False negatives could result from not collecting samples from the most contaminated sites or at the times when contaminant levels were highest, from handling the samples in a way that reduced toxicity, or from using tests that are not sufficiently sensitive to detect effects that would cause significant injuries to populations or communities in the field.

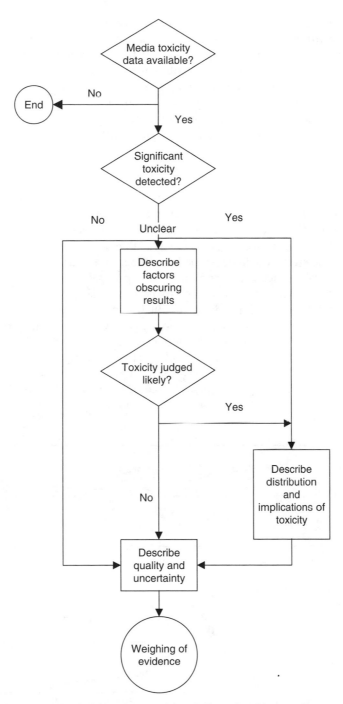

FIGURE 32.3 Risk characterization based on toxicity testing of ambient media.

If significant toxicity occurs in the tests, the risk characterization should describe the nature and magnitude of the effects and the consistency of effects among tests conducted with different species in the same medium.

Toxicity tests may produce ambiguous results in some cases due to poor performance of organisms in reference or control media (e.g., due to diseases, background contamination, inappropriate reference or control media, or poor execution of the test protocol). In such cases, expert judgment by the assessor in consultation with the individuals who performed the test should be used to arrive at an interpretation of the test results. One particular problem with reference and control media is their nutritional properties. This is particularly apparent with soil toxicity tests in which site soils may have very low levels of organic matter and nutrient elements due to site activities that scrape and mix the soil horizons. In contrast, reference soils may have higher nutrient and organic matter levels because they have been vegetated and have not been scraped or mixed, and control soils (e.g., potting mixes) may have much higher levels than all but the best site or reference soils. Such differences should be minimized as far as possible during test design but must still be considered and accounted for in the risk characterization.

If significant toxicity is found at any site, the relationship of toxicity to exposure must be characterized. The first way to do this is to examine the relationship of toxicity to concentrations of chemicals in the media. The manner in which this is done depends on the amount of data available. If numerous toxicity tests are available, the level of effects or the frequency of tests showing toxic effects could be defined as a function of concentrations of one or more chemicals. For example, concentrations of arsenic, copper, zinc, and, to a lesser degree, lead and cadmium in soils collected from the Anaconda mine site in Montana were positively correlated with phytotoxicity (Kapustka et al. 1995). Similarly, metal concentrations in soil were correlated with earthworm growth and mortality in Concord, California (Jenkins et al. 1995). If multiple chemicals with the same mode of action may be jointly contributing to toxicity, the aggregate concentration (e.g., total PAHs or toxicity equivalency factor (TEF)-normalized chlorinated dicyclic compounds—Section 8.1.2) could be used as the exposure variable. Alternatively, toxic responses might be plotted against the hazard index (HI) or sum of toxic units (ΣTU) (Figure 32.4). Finally, each chemical of concern may be treated as an

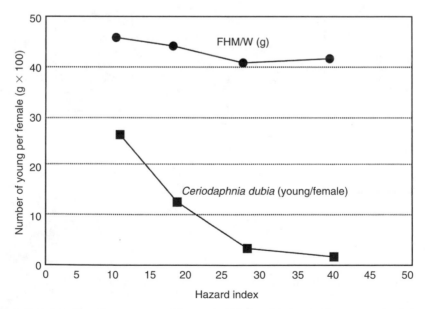

FIGURE 32.4 Results of ambient water toxicity tests (weight of fathead minnows and number of *Ceriodaphnia dubia* young per female) as a function of the chronic hazard index from a risk assessment of the Trinity River. (Redrawn from Parkhurst, B.R., Warren-Hicks, W., Cardwell, R.D., Volosin, J., Etchison, T., Butcher, J.B., and Covington, S.M., *Aquatic Ecological Risk Assessment: A Multi-Tiered Approach*, Project 91-AER-1, Water Environment Research Foundation, Alexandria, VA, 1996b. With permission.)

independent variable in a multiple regression or correlation. In general, if toxicity is occurring, it should be possible to identify exposure–response relationships. However, there are a number of reasons why a true causal relationship between a chemical and a toxic response may not be apparent, even if numerous samples are taken (Box 32.1). Therefore, the lack of an exposure–response relationship does not disprove that one or more chemicals caused an apparent toxic effect.

These results of relating contaminant levels in media to test results for those media may be used to examine the reliability of the single-chemical toxicity line of evidence (Section 32.3.1). For example, zinc was at least 10 times more toxic to *Eisenia foetida* in artificial soil than in contaminated soils collected from the field (Spurgeon and Hopkin 1996). The considerations in Box 32.1 can be used to help understand any discrepancies in the two sources of exposure–response relationships.

An alternative and potentially complementary approach to relating toxicity to exposure is to determine the relationship between the occurrence of toxicity and sources of contaminants (e.g., springs, seeps, tributaries, spills) or of diluents (i.e., relatively clean water or sediments). This may be done by simply creating a table of potential sources of contamination or dilution and indicating for each test whether toxicity increases, decreases, or remains unchanged below that source. The same information may be conveyed graphically. For a stream or river, toxicity may be plotted as a function of reach (if reach boundaries are defined by source locations) or distance downstream (with locations of sources marked on the graph) (Figure 32.5).

BOX 32.1
Why Contaminant Concentrations in Ambient Media may not be Correlated with Toxicity of those Media

Variation in bioavailability
  Due to variance in medium characteristics
  Due to variance in contaminant age among locations (contaminants added to soil and sediments may become less bioavailable over time due to sequestration)
  Due to variance in transformation or sequestration rates among locations
Variation in the form of the chemical (e.g., ionization state)
Variation in concentration over time or space (i.e., samples for analysis may not be the same as those tested)
  Spatial heterogeneity
  Temporal variability (e.g., aqueous toxicity tests last for several days, but typically water from only 1 d is analyzed)
Variation in composition of the waste
  Due to interactive toxicity and variance in the relative proportions of the contaminants
  Due to variance in concentrations of waste components that are not contaminants but that influence the toxicity of the contaminants
Variation in co-occurring chemicals
  Due to variance in upstream or other off-site contaminants
  Due to variance in background concentrations of chemicals
Inadequate detection limits (Correlations will be obscured if chemicals are not detected when they contribute to toxicity)
Variation in toxicity tests due to variation in test performance
Variation in toxicity test due to variance in medium characteristics (e.g., hardness, organic matter content, and pH)

TABLE 32.4
Summary Table for Integration of Results from Ambient Media Toxicity Tests

Issue	Result
Species affected	List species and life stages affected in the tests
Severity of effects	List types and magnitudes of effects in the tests
Spatial extent of effects	Delineate meters of stream, square meters of land, etc. for which media samples were toxic
Frequency of effects	If toxicity is episodic, calculate the proportion of time or the number of distinct toxic episodes per unit of time
Association with sources	Define spatial and temporal associations of toxic media to potential sources
Association with exposure	Define relationships to contaminant concentrations or other measures of exposure
Estimated effect	Summarize the nature and extent of estimated effects on the assessment endpoint and credible upper bounds
Confidence in results	Provide a rating and supporting comments

When sources of toxic water have been identified, and tests have been performed on dilution series of those waters, the transport and fate of toxicity can be modeled like that of individual chemicals (DiToro et al. 1991a). If toxicity can be assumed to be concentration-additive (Chapter 8), such models of toxicity can be used to explain ecological degradation observed in streams with multiple sources and to apportion causation among the sources.

Ambient soil tests may also be used to prioritize sites for remediation. For example, Kapustka et al. (1995) reported phytotoxicity scores for locations at the Anaconda site in Montana. Scores were calculated to combine results for three test species and six response parameters for soils from each location.

To facilitate the weight-of-evidence analysis and to make the bases clear to the reader, it may be useful to summarize the results of this integration for each reach or area where significant toxicity was found using Table 32.4.

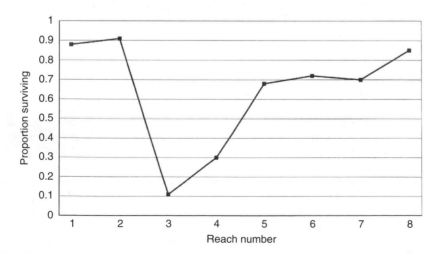

FIGURE 32.5 Mean proportional reduction in survival of fathead minnow larvae in ambient water relative to controls in numbered reaches. The source occurs in reach 2.

32.3.3 BIOLOGICAL SURVEYS

If biological survey data are available for an endpoint species or community, the first question is whether the data suggest that significant effects are occurring (Figure 32.6). For some groups, notably fish and benthic invertebrates, there is a good chance that reference streams have been identified and survey data are available for comparison. For most other endpoint groups, references must be established ad hoc, and the lack of temporal or spatial replication may make inference tenuous. For some taxa such as most birds, traditional survey data are not useful for estimating risks from contaminated sites because mobility, territoriality, or other factors obscure demographic effects. For example, attempts to use breeding bird survey data to characterize risks from the use of avicides were not sufficient to strengthen or weaken the risk characterization based on laboratory toxicology and modeled dose (Linder et al. 2004). Some exposed species populations were increasing and others were declining. However, survey results may be more reliable if efforts are made to control extraneous variance. For example, the reproductive success of birds can be estimated by setting out nest boxes on contaminated and reference sites.

Care must be taken to consider the sensitivity of field survey data to toxic effects relative to other lines of evidence. Some biological surveys are very sensitive (e.g., surveys of nesting success of colonial nesting birds or electrofishing surveys of wadeable streams), others are moderately sensitive (e.g., benthic macroinvertebrates), and still others are rather insensitive (e.g., fish community surveys in estuaries and small mammal surveys). Sensitivity is not just a matter of precision. For example, surveys of territorial breeding birds can be quite precise, but they are insensitive to toxic effects, because the number of breeding pairs on a site is often low, and because any mortality or reduction in reproductive success is unlikely to be reflected in subsequent densities of breeding pairs. However, even relatively insensitive surveys may be useful in assessments. For example, if the concentrations of chemicals suggest that a medium should be highly toxic, but toxicity tests of the medium find no toxicity, even a relatively insensitive survey, which found a community that was not highly modified, could indicate that the chemical analyses were misleading and the toxicity test data were probably correct. Conversely, a highly modified community in the absence of high levels of analyzed chemicals would suggest that combined toxic effects, toxic levels of unanalyzed contaminants, episodic contamination, or some other disturbance had occurred. However, field surveys interpreted in isolation without supporting data can be misleading, particularly when the absence of statistically significant differences are inappropriately interpreted as an absence of effects.

Biological surveys may also be insensitive, because organisms are adapted or resistant to contamination, so toxic effects are not apparent. Adaptation is most likely for short-lived species or long-contaminated sites. If it is suspected, it can be confirmed by testing organisms from the site along with organisms of the same species from uncontaminated sites. Depending on policies and attitudes, adaptation to pollution may be considered to have prevented an unacceptable effect, or to be an unacceptable effect itself. It is the basis for the use of the pollution-induced community tolerance (PICT) as an endpoint (Rutgers et al. 1998). The presence of a contaminant-adapted community is taken to be evidence of significant toxicity.

If biological survey data are consistent with significant reductions in abundance, production, or diversity, associations of apparent effects with causal factors must be examined. First, the distribution of apparent effects in space and time must be compared to the distribution of sources, contaminants, and habitat variables. Second, the distribution of apparent effects must be compared with the distribution of habitat factors that are likely to affect the organisms in question, such as stream structure and flow, to determine whether they account for the apparent effects (Kapustka 2003). For example, most of the variability in the benthic community of

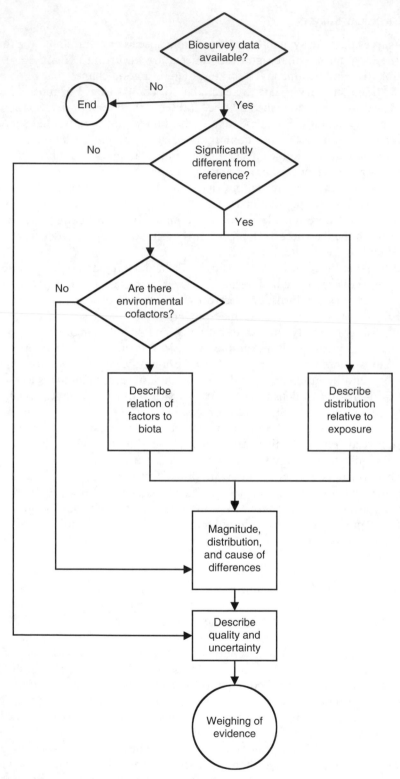

FIGURE 32.6 Risk characterization based on biological survey data.

Poplar Creek embayment was found to be associated with variance in sediment texture and organic matter (Jones et al. 1999). Only after that variance had been modeled by multiple regression could the residual variance be associated with contaminants. This process may be aided by the use of habitat models (available from the US Fish and Wildlife Service) or habitat indices (Rankin 1995). For example, when surveys of wildlife at the metal-contaminated Lehigh Gap, Pennsylvania, showed drastic reductions in abundance, the effects of vegetation loss were removed using habitat suitability models to determine the reduction that could be attributed to direct toxic effects (Beyer and Storm 1995). Even when available habitat models or indices do not characterize habitat effects at the site, they indicate which habitat parameters might be used to generate a site-specific model. If the important covariates are unknown, quantile regression can be used to account for their influence (Section 23.2.5). If any of these techniques reveal an apparent effect of contaminants on a survey metric, the relationship should be modeled or at least plotted (Section 23.2.5). As with ambient toxicity test results, the exposures may be expressed as concentrations of chemicals assessed individually, concentrations of individual chemicals used in a multiple regression, summed concentrations of related chemicals, ΣTU values, or HI values (Figure 32.7). Finally, if results are available for toxicity tests of ambient media, their relationships to the survey results should be determined. For example, soil from locations at the Naval Weapons Center in Concord, California, with reduced abundances of earthworm species or total macroinvertebrates, caused mortality and reduced growth in earthworm toxicity tests (Jenkins et al. 1995).

To facilitate the weight-of-evidence analysis and to make the bases clear to the reader, it may be useful to summarize the results of this integration for each endpoint in each reach or area using Table 32.5.

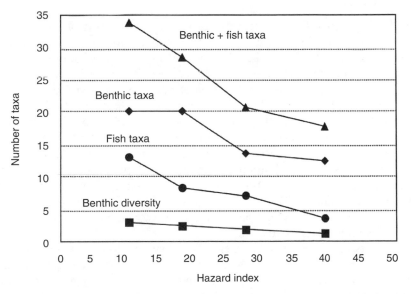

FIGURE 32.7 Results of biological surveys (number of fish taxa, number of benthic invertebrate taxa, and total taxa) as functions of the chronic hazard index from a risk assessment of the Trinity River. (Redrawn from Parkhurst, B.R., Warren-Hicks, W., Cardwell, R.D., Volosin, J., Etchison, T., Butcher, J.B., and Covington, S.M., *Aquatic Ecological Risk Assessment: A Multi-Tiered Approach*, Project 91-AER-1, Water Environment Research Foundation, Alexandria, VA, 1996b. With permission.)

TABLE 32.5
Summary Table for Integration of Biological Survey Results

Issue	Result
Taxa and properties surveyed	List species or communities and measures of effect
Nature and severity of effects	List types and magnitudes of apparent effects
Minimum detectable effects	For each measure of effect, define the smallest effect that could have been distinguished from the reference condition
Spatial extent of effects	Delineate meters of stream, square meters of land, etc., that are apparently affected
Number and nature of reference sites	List and describe reference sites including habitat differences from the contaminated site
Association with habitat characteristics	Describe any correlations or qualitative associations of apparent effects with habitat variables
Association with source	Describe any correlations or qualitative associations of apparent effects with sources
Association with exposure	Define relationships to ambient contaminant concentrations, body burdens, or other measures of exposure
Association with toxicity	Define relationships to toxicity of media
Most likely cause of apparent effects	Present the most likely cause of the apparent effects based on the associations described in previous items
Estimated effects	Summarize the estimated nature and extent of effects and credible upper bounds
Confidence in results	Provide rating and supporting comments

32.3.4   BIOMARKERS AND PATHOLOGIES

Biomarkers are physiological or biochemical measures, such as blood cholinesterase concentration, that may be indicative of exposure to, or effects of, contaminants. They are seldom useful for estimating risks by themselves, but they can be used to support other lines of inference. In particular, if the biota of a site are depauperate, biomarkers in the remaining resistant species may indicate what may have caused the loss of the missing species. The inference begins by asking if the levels of the biomarkers differ from those at reference sites (Figure 32.8). If they do, it is necessary to determine whether they are diagnostic or at least characteristic of any of the contaminants or of any of the habitat factors that are thought to affect the endpoint biota. If the biomarkers are characteristic of contaminant exposures, the distribution and frequency of elevated levels must be compared with the distributions and concentrations of contaminants. Finally, to the extent that the biomarkers are known to be related to overt effects such as reductions in growth, fecundity, or mortality, the implications of the observed biomarker levels for populations or communities should be estimated.

A potentially powerful use of biomarkers is as bioassays to estimate external or internal exposure to mixtures of chemicals. If chemicals in a mixture have a common mechanism of action, a biochemical measure of some step in that mechanism may be used as a measure of the total effective exposure. This approach has been applied to dioxin-like compounds. Generation of 7-ethoxyresorufin-o-deethylase (EROD) in rat hepatoma cell cultures has been used as a bioassay for 2,3,7,8-tetrachlorodibenzo-p-dioxin-equivalents (TCDD-EQs) in

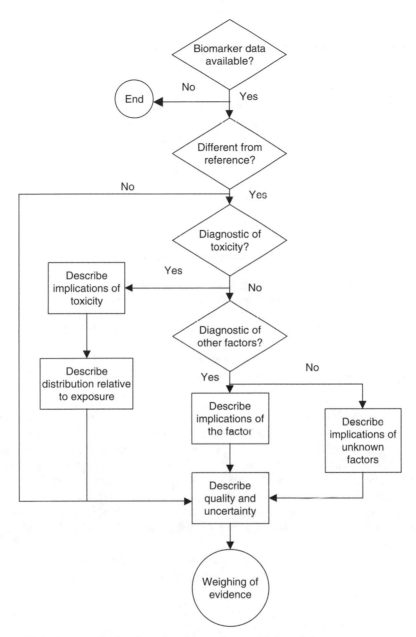

FIGURE 32.8 Risk characterization based on biomarker and injury data.

the food of endpoint organisms or in the organisms themselves (Tillitt et al. 1991). These can then be compared to TCDD levels that are known to cause effects in toxicity tests. The method has been used to determine the cause of deformities and reproductive failures in piscivorous birds of the Great Lakes (Ludwig et al. 1996). The technique is equivalent to the use of TCDD TEFs (Section 8.1.2) but eliminates the uncertainties associated with application of the factors to different species and conditions.

TABLE 32.6
Summary Table for Integration of Biomarker or Pathology Results

Issue	Result
Taxa and response	List the species and specific responses
Implications of responses for organisms and populations	Describe, as far as possible, the relationship between the biomarkers/ pathologies and population/community endpoints
Causes of the observed response	List chemicals, chemical classes, pathogens, or conditions (e.g., anoxia) that are known to induce the biomarker or pathology
Number and nature of reference sites	List and describe reference sites including habitat differences from the contaminated site
Association with habitat or seasonal variables	List habitat or life cycle variables that may affect the level of the biological response at the site
Association with sources	Describe any correlations or qualitative associations of the responses with sources
Association with exposure	Define relationships to contaminant concentrations or other measures of exposure
Most likely cause of response	Based on the associations described in previous items, present the most likely cause of the apparent responses
Estimated effects	Summarize the estimated nature and extent of effects associated with the biomarker or pathology and credible upper bounds if they can be identified
Confidence in results	Provide rating and associated comments

Pathologies include lesions, tumors, deformities, and other signs of disease. The occurrence of gross pathologies may be an endpoint itself because of public concern and loss of market value. However, they are more often used like biomarkers to help diagnose the causes of effects on organisms. Manuals are available for this purpose (e.g., Friend 1987; Meyer and Barclay 1990; Beyer et al. 1998). This type of evidence is particularly useful for identifying alternative potential causes of observed effects such as epizootics or anoxia. Greater diagnostic power can be obtained by combining pathologies with condition metrics and even population properties (Goede and Barton 1990; Gibbons and Munkittrick 1994; Beyer et al. 1998).

To facilitate the weight-of-evidence analysis and to make the bases clear to the reader, it may be useful to summarize the results of this integration for each relevant endpoint and each reach or area using Table 32.6.

32.3.5 WEIGHT OF EVIDENCE

Although the goal of a definitive risk characterization is to estimate the risks, risk characterizations by weight of evidence usually begin by determining whether the risk is significant or not. This reduction of risk to a dichotomous variable is usually desired by decision makers and stakeholders, and is usually presumed by regulatory risk assessment guidance such as the data quality objectives (DQO) process (Section 9.1.1). In addition, if a relatively simple categorical characterization process can determine that risks are clearly insignificant, the quantification of risks can be avoided. In other words, risk characterization by weight of evidence usually begins and often ends by classifying risks as in screening assessments (Chapter 31), but with all evidence considered and with uncertainty analysis rather than conservative assumptions. If risks to some endpoints are significant, the risk characterization can move on to estimate the nature and magnitude of those risks.

The weighing of evidence begins by summarizing the available types of evidence for each endpoint (Figure 32.9). For each type of evidence, one has to determine whether it is

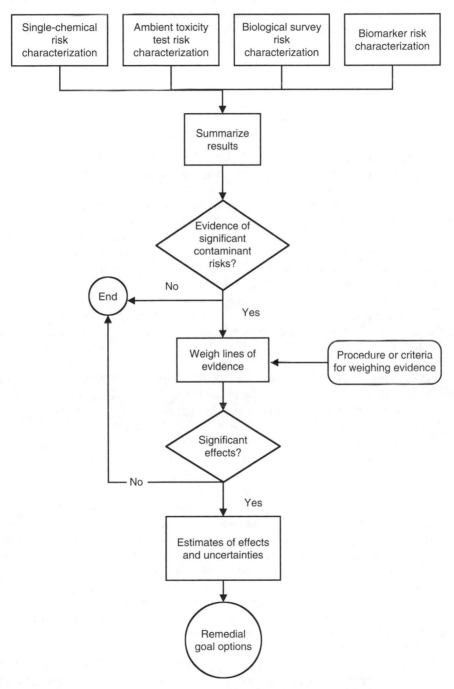

FIGURE 32.9 Risk characterization based on weighing of multiple lines of evidence.

consistent with exceedence of the threshold, inconsistent with exceedence, or ambiguous. If none of them are suggestive of significant risks, the risk characterization is done. If at least one type of evidence is indicative of significant risks, one must determine whether the results as a whole indicate that it is likely or unlikely that the threshold is exceeded. If there is no bias in the assessment that affects all lines of evidence, agreement among multiple lines of evidence is strong evidence supporting a significant risk. However, if there are inconsistencies, the true weighing of evidence must occur. The weights are determined based on the following considerations adapted from the Massachusetts attributes (Table 32.1) and from ecoepidemiological considerations (Suter 1998b).

32.3.5.1 Weighting Considerations

Relevance: Evidence is given more weight if the measure of effect is more directly related to (i.e., relevant to) the assessment endpoint.

- Effects are relevant if the measure of effect is a direct estimate of the assessment endpoint or if validation studies have demonstrated that the measurement endpoint is predictive of the assessment endpoint.
- The mode of exposure may not be relevant if the media used in a test are not similar to the site media. Normalization of media concentrations may increase the relevance of a test if the normalization method is applicable to the contaminant and site. Similarly, the relevance of tests in solution to sediment or soil exposures is low unless the models or extraction techniques used to estimate aqueous phase exposures at the site are reliable.
- Measures of effect derived from the literature rather than site-specific studies may have used a form of the chemicals that is not relevant to the chemical detected in the field. For example, is it the same ionization state, and has the weathering or sequestration of the field contaminant changed its composition or form in ways that are not reflected in the test?

In some cases, available information may not be sufficient to evaluate the relevance of a line of evidence. In such cases, relevance may be evaluated by listing the ways in which the results could be fundamentally inappropriate or so inaccurate as to nullify the results, and evaluating the likelihood that they are occurring in the case. For single-chemical toxicity tests, such a list could include the possibility that the test was (1) performed with the wrong form of the chemical, (2) performed in media differing from the site media in ways that significantly affect toxicity, or (3) insensitive due to short duration, a resistant species, or the lack of measures of relevant effects.

Exposure–Response: A line of evidence that demonstrates a relationship between the magnitude of exposure and the effects is more convincing than one that does not. For example, apparent effects in media toxicity tests may be attributed to the chemical of concern, but unless the tested medium is analyzed and an exposure–response relationship demonstrated, it may be suspected that effects are a result of other contaminants, nutrient levels, texture, or other properties. If an exposure–response relationship has not been demonstrated, consideration should be given to the magnitude of the observed differences. For example, if test data include only comparisons of contaminated and uncontaminated soils, the observed differences are less likely to be due to extraneous factors if they are large (e.g., 100% mortality rather than 25% less growth).

Temporal Scope: A line of evidence should be given more weight if the data encompass the relevant range of temporal variance in conditions. For example, if contaminated and reference soils are surveyed during a period of drought, few earthworms will be found at any site, so toxic effects will not be apparent. Temporal scope may also be inadequate if aqueous toxic effects occur when storm events flush contaminants into streams, but water for chemical analysis or toxicity testing is not collected during such events. This phenomenon was observed on the ORR and is probably a wide-spread problem. For example, studies of risks to fish from metals in the Clark Fork River, Montana, focused on chronic exposures, but fish kills occurred due to episodes of low pH and high metal concentrations following thunderstorms (Pascoe and Shazili 1986; Pascoe et al. 1994).

Spatial Scope: A line of evidence should be given more weight if the data adequately represent the area to be assessed, including directly contaminated areas, indirectly contaminated areas, and indirectly affected areas. In some cases the most contaminated or most susceptible areas are not sampled because of access problems or because of the sampling design (e.g., random sampling).

Quality: The quality of the data should be evaluated in terms of the protocols for sampling, analysis, and testing; the expertise of the individuals involved in the data collection; the adequacy of the quality control during sampling, sample processing, analysis, and recording of results; and any other issues that are known to affect the quality of the data for purposes of risk assessment (Chapter 9). Similarly, the quality of models and analytical methods used in the line of evidence need evaluation. Although the use of standard methods and models tends to increase the likelihood of high-quality results, they are no guarantee. Standard methods may be poorly implemented or may be inappropriate to a site. In contrast, a well-designed and performed site-specific measurement or testing protocol can give very high-quality results.

Quantity: The adequacy of the data needs to be evaluated in terms of the number of samples or observations taken. Results based on small sample sizes are given less weight than those based on large sample sizes. The adequacy of the number of observations must be evaluated relative to the variance as in any analysis of a sampling design, but it is also important in studies of this sort to consider their adequacy relative to potential biases in the sampling (see spatial and temporal scope, above).

Uncertainty: A line of evidence that estimates the assessment endpoint with low uncertainty should be given more weight. Uncertainty in a risk estimate is in part a function of the data quality and quantity, discussed above. In most cases, however, the major sources of uncertainty are the assumptions underlying the line of evidence and the extrapolations between the measures of effect and the assessment endpoint. In addition, the extrapolation from the measures of exposure to the exposure of the endpoint entities may be large due to considerations such as bioavailability and temporal dynamics.

These and other considerations can be used as points to consider in forming an expert judgment or consensus about which way the weight of evidence tips the balance. Table 32.7 presents an example of a simple summary of the results of weighing evidence based on this sort of process. The types of evidence are listed, and a symbol is assigned for each: (+) if the evidence is consistent with significant effects on the endpoint; (−) if it is inconsistent with significant effects; and (±) if it is too ambiguous to be assigned to either category. The last column presents a short summary of the results of the risk characterization for that type of evidence. If indirect effects are part of the conceptual model, they should be summarized in a separate line of the table. For example, effects on piscivorous wildlife could be due to direct toxic effects or inadequate food, which may be due to toxicity to fish. The last row of the table presents the weight-of-evidence-based conclusion concerning whether significant effects are occurring and a brief statement concerning the basis for the conclusion. This conclusion is not based simply on the relative number of (+) or (−) signs. The "weight" component of weight of evidence is the relative credibility and reliability of the conclusions of the various lines of evidence, based on the considerations discussed above. Additionally, the seven considerations

TABLE 32.7
A Summary of a Risk Characterization by Weight of Evidence for a Soil Invertebrate Community in Contaminated Soil

Evidence	Result[a]	Explanation
Biological surveys	−	Soil microarthropod taxonomic richness is within the range of reference soils of the same type, and is not correlated with concentrations of petroleum components
Ambient toxicity tests	−	Soil did not reduce survivorship of the earthworm *Eisenia foetida*; sublethal effects were not determined
Organism analyses	±	Concentrations of PAHs in depurated earthworms were elevated relative to worms from reference sites, but toxic body burdens are unknown
Soil analyses/single-chemical tests	+	If the total hydrocarbon content of the soil is assumed to be composed of benzene, deaths of earthworms would be expected; relevant toxicity data for other detected contaminants are unavailable
Weight of evidence	−	Although the earthworm tests may not be sensitive, they and the biological surveys are both negative, and are both more reliable than the single-chemical toxicity data used with the analytical results for soil

[a]Results of the risk characterization for each line of evidence and for the weight of evidence: + indicates that the evidence is consistent with the occurrence of a 20% reduction in species richness or abundance of the invertebrate community; − indicates that the evidence is inconsistent with the occurrence of a 20% reduction in species richness or abundance of the invertebrate community; ± indicates that the evidence is too ambiguous to interpret.

can be used to grade the weight assigned to each type of evidence (e.g., high, moderate, or low weight) (Table 32.8). This may be done informally by the assessors, or each type of evidence may be scored against each consideration as in the causal considerations (Table 4.3) or the Massachusetts approach (Section 32.1). This still leaves the inference to a process of expert judgment or consensus but makes the bases clearer to readers and reviewers.

Analogous systems have been proposed that include other types of evidence and other approaches to evaluating and combining them (Batley et al. 2002; Burton et al. 2002; Forbes and Calow 2002; Grapentine et al. 2002). They range from checklists to detailed procedures. Any of them might be formalized as a weighting and scoring system like the Massachusetts method (Section 32.1). Such systems have the advantage of being open, consistent, and less subject to hidden biases. However, as explained below, it is better to use the considerations, scores, and weights only to clarify the evidence before performing a logical analysis to the best explanation of the evidence.

The use of quantitative weighing of evidence or of an equivalent expert judgment about which lines of evidence are most reliable is based on an implicit assumption that the lines of evidence are logically independent. Another approach to weighing multiple lines of evidence is to determine whether there are logical relationships among the lines of evidence. Based on knowledge of site conditions and of environmental chemistry and toxicology, one may be able to explain why inconsistencies occur among the lines of evidence. For example, one may know that spiked soil tests tend to overestimate the availability and hence the toxicity of contaminants, and one may even be able to say whether the bias associated with this factor is sufficient to account for discrepancies with tests of site soils. Because it is mechanistic, this process of developing a logical explanation for differences among lines of evidence is

TABLE 32.8
Example of a Table Summarizing the Risk Characterization for the Species Richness and Abundance of a Fish Community in a Stream at a Waste Site

Evidence	Result[a]	Weight[b]	Explanation
Biological surveys	−	H	Fish community productivity and species richness are both high, relative to reference reaches; data are abundant and of high quality
Ambient toxicity tests	±	M	High lethality to fathead minnow larvae was observed in a single test, but variability is too high for statistical significance; no other aqueous toxicity was observed in 10 tests
Water analyses/single-chemical tests	+	M	Only zinc is believed to be potentially toxic in water and only to highly sensitive species
Weight of evidence	−		Reach 2 supports a clearly high-quality fish community. Other evidence, which suggests toxic risks, is much weaker (single-chemical toxicology) or inconsistent and weak (ambient toxicity tests)

[a]Results of the risk characterization for each line of evidence and for the weight of evidence: + indicates that the evidence is consistent with the occurrence of the endpoint effect; − indicates that the evidence is inconsistent with the occurrence of the endpoint effect; ± indicates that the evidence is too ambiguous to interpret.
[b]Weights assigned to individual lines of evidence: high (H), moderate (M), and low (L).

potentially more convincing than simple weighing of the evidence. However, it is important to remember that such explanations can degenerate into just-so stories if the relevance of the proposed mechanisms is not well supported. Therefore, studies should be designed and carried out specifically to support inference concerning reality and causality of inferred effects. For example, one might analyze aqueous extracts of site soils and spiked soils to support the inference that differences among types of evidence are due to relative bioavailability.

A real example is provided by assessments of risks to Hudson River striped bass from PCBs. High-quality species-specific toxicity tests and analyses indicated that PCB levels in the Hudson River were sufficient to cause toxic effects in striped bass larvae. However, Barnthouse et al. (2003) took advantage of a large long-term set of monitoring data to demonstrate that the striped bass population size showed no effects associated with changes in PCB exposures. No scoring system would reconcile the different results for these two high-quality lines of evidence. Barnthouse et al. (2003) made the case that compensatory processes had allowed the population to persist despite increased larval mortality. This inference shifts the question from "are PCBs are affecting Hudson River striped bass" to, "given that the adult population is not apparently affected, how important it is that some of the compensatory capacity of the population is being used to respond to PCB toxicity?"

In general, a logical analysis of the data should proceed from most realistic (i.e., site-specific) to most precise and controlled (e.g., single-chemical and species laboratory toxicity tests). Field surveys indicate the actual state of the receiving environment, so other lines of evidence that contradict the field surveys, after allowing for limitations of the field data, are clearly incorrect. For example, the presence of plants that are growing and not visibly injured indicates that lethal and gross pathological effects are not occurring, but does not preclude reductions in reproduction or growth rates or loss of sensitive species. These other effects could be addressed by more detailed field studies of growth rates, seed production and viability, and species composition. Similarly, the presence of individuals of highly mobile species such as birds indicates almost nothing about risks, because immigration replaces losses from mortality or reduced reproduction.

Ambient media toxicity tests indicate whether toxicity could be responsible for differences in the state of the receiving environment, including differences that may not be detectable in the field. However, effects in the field are usually more credible than negative test results, because field exposures are longer and otherwise more realistic, and species and life stages from the site may be more sensitive than test species and life stages.

Single-chemical toxicity tests indicate which components of the contaminated ambient media could be responsible for effects. Because they are less realistic than other lines of evidence, single-chemical toxicity tests are usually less credible than the other lines of evidence. They do not include combined toxic effects, the test medium may not represent the site media, the exposure may be unrealistic, and the chemicals may be in a different form from that at the site. However, because these studies are more controlled than those from other lines of evidence, they are more likely to detect sublethal effects. In addition, single-chemical toxicity tests may include longer exposures, more sensitive responses, and more sensitive species than tests of contaminated ambient media. These sorts of logical arguments concerning the interpretation of single-chemical toxicity test results must be generated ad hoc, because they depend on the characteristics of the data and the site.

It is interesting to note that this method of weighing evidence is similar to that used in ecological epidemiology (Chapter 4). In ecological epidemiology, the primary assessment problems are to determine whether a biotic population or community is impaired and then determine the cause. The goal is to identify and eliminate impairments. In the case of risk assessments of contaminated sites described in this chapter, contamination is known but the existence or future occurrence of effects has not been established. The goal of the assessment is to help determine what, if anything, should be done to remediate the contaminants.

However the weighing of evidence is performed, it is incumbent on the assessment scientist to make the basis for the judgment as clear as possible to readers and reviewers. Where multiple areas or reaches are assessed, it is helpful to provide a summary table for the weighing of evidence across the entire site as in Table 32.9 so that the consistency of judgment can be reviewed.

32.3.6   RISK ESTIMATION

After the lines of evidence have been weighed to reach a conclusion about the significance of risks to an assessment endpoint, it is usually appropriate to proceed to estimate the nature, magnitude, and distribution of any effects that were judged to be significant. A significant risk is sufficient to prompt consideration of remedial actions, but the nature, magnitude, and distribution of effects determine whether remediation is justified, given remedial costs and countervailing risks (Chapter 36). In general, it will be clear that one line of evidence provides the best estimate of effects. Some lines of evidence may be eliminated as inconsistent with the conclusion, and others may support the conclusion but not provide a basis for quantifying effects. If more than one line of evidence provides apparently reliable estimates of effects, their results should be presented, and, as far as possible, discrepancies should be explained. If one best estimate is identified, other lines of evidence may contribute by setting bounds on that estimate.

If a representative species has been chosen for one or more assessment endpoints (Box 16.5), it is important to estimate risks to the entire endpoint. That is, if night herons have been used to represent piscivorous birds, risks to all piscivorous birds on the site should be

TABLE 32.9
Summary of Weight-of-Evidence Analyses for Reaches Exposed to Contaminants in the Clinch River/Poplar Creek Operable Unit[a]

Reach	Biological Surveys	Bioindicators	Ambient Toxicity Tests	Fish Analyses	Water Analyses/ Single-Chemical Toxicity	Weight of Evidence
Upper Clinch River Arm	±	±		±	−	−
Poplar Creek Embayment	+	±	+	±	+	+
Lower Clinch River Arm	−	±	−	±	+	−
McCoy Branch Embayment			±			−

[a]Results of the risk characterization for each line of evidence and for the weight of evidence: + indicates that the evidence is consistent with the occurrence of the endpoint effect; − indicates that the evidence is inconsistent with the occurrence of the endpoint effect; ± indicates that the evidence is too ambiguous to interpret; blank cells indicate that data were not available for that line of evidence.

estimated. For example, one might estimate that complete reproductive failure is occurring in half of the nesting pairs in a night heron rookery, and therefore that reproductive failure is occurring in half the kingfisher territories that occur in the same area. If there is reason to believe that the kingfishers are less sensitive or less exposed, one might estimate that their reproduction is reduced by some lesser percentage. Alternatively, each species of the endpoint group may be independently assessed.

32.3.7 FUTURE RISKS

Baseline ecological risk assessments of contaminated sites typically focus on current risks as estimators of the risk that would occur in the near future in the absence of remediation. However, baseline risks in the distant future should also be characterized when:

- Contaminant exposures are expected to increase in the future (e.g., a contaminated groundwater plume will intersect a stream).
- Biological succession is expected to increase risks (e.g., a forest will replace a lawn).
- Natural attenuation of the contaminants is expected in the near term without remedial actions (i.e., the expense and potential ecological damage associated with remedial actions may not be justified).

Although these future baseline risks cannot be characterized by measuring effects or by testing future media, all lines of evidence that are useful for estimating current risks may be extended to them. As in human health risk assessments, risk models derived by epidemiological methods can be applied to future conditions and even applied to different sites. For example, if concentrations are expected to change in the future, the exposure–response relationship derived from biosurvey data (e.g., a relationship between contaminant concentration and invertebrate species richness) may provide a better estimate of future effects than a concentration–response relationship derived from laboratory test data. Results of toxicity tests of currently contaminated media may also be used to estimate future effects. For example, contaminated groundwater may be tested at full strength and diluted in stream water to generate an exposure–response relationship that may be used to estimate the nature and extent of future effects. The utility of the various risk models depends on their reliability, as suggested by the weight-of-evidence analysis, and their relevance to the future conditions.

32.4   EXAMPLES

Characterization of ecological risks by weighing multiple lines of evidence has become a common practice. However, most do not reach the open literature. The point of these examples is to highlight innovations in risk characterization, point to assessments that illustrate important features of risk characterization, and provide inspiration.

32.4.1   CHARACTERIZING CONTAMINATED SITE RISKS

Many ecological risk characterizations for contaminated sites have been produced in the United States because of the Superfund law and regulations. However, relatively few have been published, in part because of the long processes of decision making and litigation. The Clinch River assessment, referred to in this chapter, was published in *Environmental Toxicology and Chemistry*, vol. 18, no. 4. Some others are briefly described in this section.

The Baxter Springs/Treece Superfund Subsites in Kansas have contaminated streams with metals. The ecological risk assessment strategy involved calculating HQs, performing biological surveys, and explaining discrepancies by considering acclimation, adaptation, and metals speciation (Hattemer-Frey et al. 1995). The quotients suggested that the water could be toxic, but the condition factors for fish in the stream were similar to reference. Considering factors that could explain the discrepancy is a good practice, but the conclusion that the discrepancies could be explained would have been more convincing if site-specific evidence of acclimation, adaptation, or low bioavailability had been obtained.

The Elizabeth Mine is a former metal-sulfide mining site in South Strafford, Vermont. Mine waste leachate to the Copperas Brook watershed was acidic and had high metal content. An ecological risk assessment under Superfund included three types of evidence (Linkov et al. 2002b). Chemical analyses of water and sediment were used to calculate HQs and HIs. Standard EPA toxicity tests of water and sediment were performed. Surveys of fish and benthic invertebrate epifauna and infauna were conducted. All three types of evidence were consistent. Fish and invertebrate communities were clearly impaired, both water and sediment samples were toxic, and several metals in water and sediment had HQs well above 1. This consistency of the multiple types and lines of evidence led to a removal action, a type of early remedial action that is seldom prompted by ecological risks alone. One notable innovation is the use of maps to display the extent of impairment indicated by each type of evidence and highlighting the reaches within the watershed in which all evidence indicated significant risks (Figure 32.10).

The estuarine ecosystem of the Great Bay and Piscataqua River in Maine and New Hampshire is contaminated by various chemicals from the Portsmouth Naval Shipyard and other sources. An ecological risk assessment of that ecosystem employed a modified version of the Massachusetts weighting and scoring system (Section 32.1) (Johnston et al. 2002). The endpoints addressed the pelagic, epibenthic, and benthic community structures, eelgrass, saltmarsh cordgrass, and birds. Using qualitative weights, the assessors scored 54 measures of exposure and effects for data quality, strength of association, study design, and endpoint. For each endpoint, the results for the exposure and effects were qualitatively interpreted as evidence (e.g., evidence of elevated exposure or no evidence of effect) and combined weights

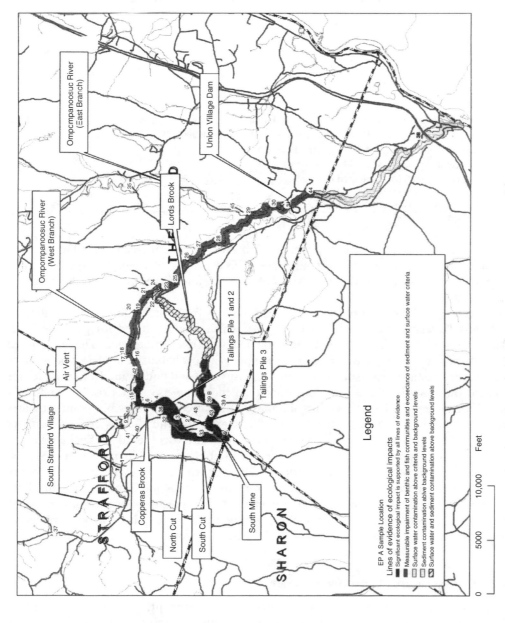

FIGURE 32.10 Use of a map to display the results of risk characterization. Different tones display positive results from particular lines of evidence or combinations of lines of evidence. (From Linkov I., Burmistrov, D., Cura, J., and Bridges, T.S., *Environ. Sci. Technol.*, 36, 238–246, 2002a. With permission.)

TABLE 32.10
Summary of a Qualitative Weight-of-Evidence Analysis for Six Endpoints in an Estuarine Ecosystem

Assessment Endpoint	Evidence of Effect[a]	Evidence of Exposure[b]	Magnitude of Risk	Confidence in Conclusions
Pelagic	Potential/M	Low/M	Low	Medium
Epibenthic	No/M	Elevated/M	Low	Medium
Benthic	No/H	Elevated/M	Low	High[c]
Eelgrass	Potential[d]/M	Elevated/M	Intermediate	Medium
Salt marsh	No/M	Elevated/M	Low	Medium
Birds		Negligible/M	Negligible	Medium

Source: Johnston, R.K., Munns, W.R. Jr., Tyler, P.L., Marajh-Whittemore, P., Finkelstein, K., Munney, K., Short, F.T., Melville, A., and Hahn, S.P., *Environ. Toxicol. Chem.*, 21, 182–194, 2002. With permission.
[a]Entry is evidence of effect/weight (M = medium, H = high).
[b]Entry is evidence of exposure/weight (M = medium).
[c]High concordance among highly weighted measures.
[d]Risk of dietary exposure for Portsmouth Harbor.

were derived (Table 32.10). The exposure and effects results were then combined to generate an overall risk score and associated confidence. This scoring and weighting system helped the assessors to keep track of the many lines of evidence and served to reveal the bases for the results.

Zolezzi et al. (2005) adapted the terrestrial methodology from the Ecological Committee on FIFRA Risk Assessment Methods (ECOFRAM), which was developed for pesticide registration (Section 32.4.4), to assess a site contaminated with trichlorobenzene. They performed the assessment in four tiers, with increasingly complex and probabilistic models. The primary difference from pesticide assessments was that exposures of soil and groundwater organisms were based on analysis of site samples rather than modeling use scenarios. Tier 1 used a simple quotient method. Tier 2 used distributions of measured exposures but point estimates of effects thresholds. Tier 3 used distributions of both exposure and effects (species sensitivity distributions, SSDs). The authors estimated a margin of safety as the difference between the 90th percentile of exposure and the 10th percentile of effects as well as risk curves (Section 30.3). Tier 4 applied Monte Carlo simulation to HQs for which both exposure and response were distributed. Hence, the same data sources were used throughout, and the multiple lines of evidence came from different methods for integrating exposure with exposure–response information. As expected, risks from shallow groundwater (pore water) were more evenly distributed than soil risks, which were associated with hot spots.

32.4.2 CHARACTERIZING CONTAMINATED SEDIMENT RISKS

A distinct tradition has grown up around sediment risk assessment, most notably as a result of Chapman's sediment quality triad (Section 32.2). However, most published studies have augmented the basic triad in some way.

The sediment quality triad was applied to the Los Angeles and Long Beach harbors, California, to determine the ecological risks from sediment toxicity (Anderson et al. 2001). In addition to the usual sediment chemical analyses, sediment toxicity tests with amphipods and benthic invertebrate surveys, pore water toxicity tests were performed with abalone embryos and larvae, which proved to be highly sensitive, as well as bioaccumulation studies with clams and analyses of contaminants in fish. The concentrations in sediment were compared to National Oceanic and Atmospheric Administration's (NOAA) effects range-medium (ERM) values or to the 90th percentiles of concentrations in a statewide data set. Because of the large number of sites, it was possible to statistically associate benthic species richness with amphipod survival and abalone development in toxicity tests. The results of the study were used to identify toxic hot spots for potential remediation.

The sediment quality triad was applied to lake sediments near the metal smelting complex at Sudbury, Ontario (Borgmann et al. 2001). The results were clear with greatly elevated concentrations of four metals, decreased abundance of amphipods, pisidiid clams, and tanytarsid midges, and severe sediment toxicity to amphipods and mayflies. However, the investigators wanted to know which metals caused the ambient effects. This was resolved with amphipod bioaccumulation studies. Only nickel was sufficiently accumulated to cause toxicity.

32.4.3 Characterizing Wildlife Risks

Ecological risk characterization by the weighing of evidence has primarily been applied to aquatic systems, because media toxicity testing and biological survey methods have been well developed for water and sediment. Wildlife risk assessments have emphasized a single type of evidence, the modeling of exposure, primarily dietary dose, and conventional laboratory dosing studies. However, Fairbrother (2003) has proposed that wildlife risk assessments for contaminated sites be performed in a tiered manner with increasingly realistic exposure models in each tier, with case-specific toxicity testing in tier 2 and with biological surveys in tier 3. The recommended case-specific toxicity tests are conventional tests but with relevant dosing and species selection. An additional type of evidence is sometimes provided by contaminated media testing of wildlife. An example is the mink reproductive test performed with PCB- and mercury-contaminated fish from Poplar Creek embayment in Oak Ridge (Halbrook et al. 1999a). Wildlife surveys at contaminated sites are seldom capable of distinguishing effects of concern, but colonial nesting birds and other specific circumstances make useful surveys possible (Halbrook et al. 1999b). Fairbrother (2003) suggested that biological surveys may be best used in an adaptive management framework. That is, biological surveys would be performed before and after an interim remedial action (e.g., hot spot cleanup), and the results used in another round of assessment that would take advantage of the remedial experiment.

The weighing of evidence is more common in wildlife epidemiology (Chapter 4) than in wildlife risk assessments. In particular, lead in the form of lead shot and contaminated sediments has been shown to be the cause of avian mortality observed in numerous cases (Bull et al. 1983; Eisler 1988; Burger 1995; Beyer et al. 1997, 1998, 2000; Henny 2003). These studies have used association of the birds with lead contamination, lead body burdens, the presence of lead shot in the gastrointestinal tract, exposure models, and assessment-specific toxicity tests to determine that lead caused the observed mortalities. The information from these assessments can be used for risk assessments when lead is known to be present but effects are uncertain, as demonstrated by Kendall et al. (1996).

Similarly, Dykstra et al. (1998, 2005) determined the cause of low bald eagle reproduction on Lake Superior using DDE and PCB levels in addled eggs and nestling blood, toxicity test results from the literature, and exposure factors including food delivery rate. Comparison of nesting success on Lake Superior and at inland sites and correlation with potential causes suggested that food availability and PCBs may have both been involved after the effects of

DDE declined, but eventually contaminant levels dropped and reproductive success achieved reference levels.

A final example of weighing multiple types of evidence in wildlife epidemiology is the studies to determine the cause of avian embryo mortalities and deformities at the Kesterson National Wildlife Refuge, California (Ohlendorf et al. 1986a,b; Heinz et al. 1987; Presser and Ohlendorf 1987; Ohlendorf and Hothem 1995). Selenium in agricultural drain water was shown to be the cause, based on avian toxicity tests showing that Se could cause the effects, studies showing that Se was elevated in birds on the site, exposure analyses showing that Se levels in food items were sufficient to cause an effective exposure, and geochemical studies showing that drain water was the source. Data and models from Kesterson have been used to assess risks from selenium at other wildlife refuges.

These ecoepidemiological assessments were prompted by observations of serious effects in wildlife. However, the data, models, and inferential techniques employed can be used in wildlife risk assessments when contamination is clear but wildlife effects are not.

32.4.4 CHARACTERIZING PESTICIDE RISKS

Because pesticides are toxic by design and are deliberately released to the environment in toxic amounts, they have been more strictly regulated than other chemicals. As a result, more data are available for ecological risk assessments of pesticides than for other individual chemicals, and more complex methods have been proposed and employed. Methods to make pesticide assessments more probabilistic and realistic were developed by the Aquatic Risk Assessment and Mitigation Dialog Group (Baker et al. 1994) and the Avian Effects Dialog Group (1994). These methodologies were further developed in the US EPA ECOFRAM program (ECOFRAM Aquatic Workgroup 1999; ECOFRAM Terrestrial Workgroup 1999). They use the tiered assessment approach to weighing evidence. They have been applied to numerous ecological risk assessments (e.g., Klaine et al. 1996; Solomon et al. 1996; Giesy et al. 1999, 2000; Hall et al. 1999, 2000; Giddings et al. 2000, 2001; Hendley et al. 2001; Maund et al. 2001; Solomon et al. 2001a,b).

The ecological risk assessment of the herbicide atrazine prepared for the manufacturer is currently the most complete implementation of the ECOFRAM aquatic methodology (Giddings et al. 2005). It has four tiers. The first is a screening assessment that applies the quotient method to exposure estimates from a simple model of a conservative scenario and effects estimates from lowest $LC_{50}s$ and no observed effect concentrations (NOECs). The second tier used Monte Carlo simulation of a coupled series of more complex exposure models to estimate distributions of exposure for prescribed scenarios in 11 regions and distributions of effects in individual toxicity tests to derive single-species risk curves (Figure 30.4). The third tier used the same exposure modeling method and scenarios, but with more realistic parameterizations. The distributions of effects were SSDs for aquatic plants and aquatic animals, and again the exposure and exposure–response distributions were integrated using risk curves. The fourth tier used exposure distributions based on field measurements of atrazine concentrations as well as distributions from more complex Monte Carlo simulations of a pond scenario than prior tiers. Effects were again expressed as SSDs and results of microcosm and mesocosm tests were also summarized. Risk curves and mean risks were estimated using the SSDs with both the modeled distributions and the distributions of observed concentrations. Temporal patterns of exposure, mixtures of metabolites and other triazine herbicides, and other issues were also discussed and analyzed. This assessment reflects a large body of data for atrazine and the high level of resources devoted to data analysis and modeling.

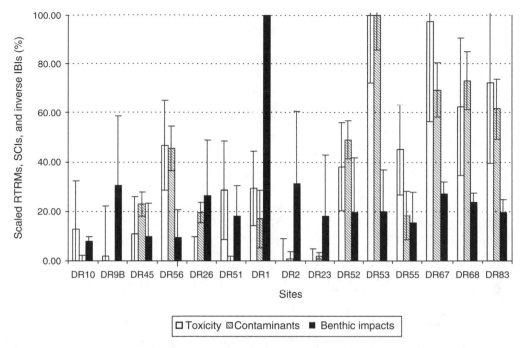

FIGURE 32.11 A summary of sediment quality triad analyses for 15 sites in the Delaware River estuary. The bars are indices of sediment toxicity, sediment contamination, and benthic invertebrate impairment that are scaled to the site means; 95% confidence limits are shown. (From Alden, R.W., Hall, L.W. Jr., Dauer, D., and Burton, D.T., *Hum. Ecol. Risk Assess.*, 11, 879, 2005. With permission.)

The ECOFRAM approach is based on tiered analyses of laboratory toxicity data and either measured or modeled exposure concentrations. Independent effects information such as microcosm and mesocosm results are used to determine the reasonableness of those results rather than as a line of evidence that generates its own risk estimate (Giddings et al. 2001, 2005). Ecosystem modeling results were not used (Section 28.6.2).

32.4.5 CHARACTERIZING EFFLUENT RISKS

Aqueous effluents in the United States are typically regulated by comparing their chemical concentrations and physical characteristics to limits in their permits or by effluent toxicity testing. However, in some cases they are subject to ecological risk assessments. Risks from the aqueous effluent of the Motiva refinery on the Delaware River were assessed under a court order (Hall and Burton 2005). The core of the assessment was a sediment quality triad analysis. Because the subject was the effluent rather than the contaminated sediment, the chemical concentration leg of the triad included analysis not only of the sediment but also of the effluent, transport modeling, PAH fingerprinting, and sediment cores. Together, these studies determined the levels of contamination at 15 locations in the river and the extent to which contamination at each was due to the effluent. Tests of survival, growth, and reproduction of two amphipod species were conducted. Benthic invertebrate community samples were used to derive diversity indices and an index of biotic integrity. Conventional triad analysis (Table 32.2) showed that some locations were clearly impaired by toxicity from PAHs and metals. However, much of the contamination was not attributable to the refinery. Benthic community effects were obscured by many factors operating in the estuary and particularly a drought that decreased flow and increased salinity in 2002. However, a clear community effect with moderate contamination and toxicity can be seen at site DR1, which is at the effluent canal (Figure 32.11). The authors concluded that the refinery's effluent does not pose a significant risk to benthic communities in this urban estuarine river.

To determine the effects of wastewater discharges on an urban stream, Winger et al. (2005) applied the rapid bioassessment procedure (Section 4.1), but found that it revealed only the effects of habitat degradation due to channelization. To assess toxic effects they used the sediment quality triad. Contamination was determined by analysis of whole sediment and pore water, toxicity was determined by amphipod tests with pore water and whole sediment, and the ambient effects metrics were benthic invertebrate taxa richness and Shannon-Weaver diversity. At two sites with impaired communities and toxic sediments, pore water concentrations of metals exceeded water quality criteria and PAHs were also at potentially toxic concentrations. At another degraded site the total simultaneously extracted metals greatly exceeded the acid volatile sulfides (SEM/AVS > 1—Section 22.3). However, the clearly toxic and degraded sites were upstream of the wastewater outfall, suggesting an unknown source near the headwaters.

32.5   INTERPRETATION

The results of ecological risk characterization require interpretation to aid the risk manager in making a remedial decision and to promote understanding by stakeholders and the public. The risk characterization should have determined, for each assessment endpoint, which risks exceed the threshold for significance and estimated the magnitude and probability of effects associated with the significant risks. Those significance thresholds should have been defined by the risk manager during the problem formulation (Chapter 10). That determination may have been made after considering the expressed values of stakeholders or the public. Therefore, this step should begin by reviewing the bases for having declared each of the assessment endpoints to be important and each threshold to be significant. If significance has not been defined, or if the significance criteria are no longer relevant because of changes in the decision makers, estimates of effects should simply be reported and explained.

One critical issue that requires interpretation in ecological risk assessment is the concept of adversity. The adversity of an effect depends in part on the nature of the value attached to it and the strength with which that value is held by the parties involved in the decision, but values with respect to ecological properties are complex and often not well formed. For example, an increase in phytoplankton production may be considered adverse if the concern is with the esthetics of swimming and boating or with blooms of noxious algae. However, at other sites, an equal percentage increase in phytoplankton production may be considered a benefit, because it causes an increase in fish production. In either case, the decision maker and the public are likely to be unaware of the relationship. Therefore, in some cases, the endpoint property may be affected, but once the implications have been explained, the effect may or may not be considered adverse.

Given that the assessment endpoint is estimated to change or has been changed in a way that is considered adverse, the significance of that change must be interpreted in terms of the nature and severity of the effects, its spatial and temporal scale, and the potential for recovery (EPA 1998a). These issues of significance should not be resolved by appeals to statistical significance, which has no relation to ecological or anthropocentric significance. Another oft-used criterion for determining whether the intensity of effects is significant is comparison to natural variation. When used, this criterion must be carefully defined. Over moderate time scales, sources of natural variation include droughts, floods, fires, late freezes, and other events that cause larger changes than would normally be acceptable for contaminant effects.

Risk characterizations must often be interpreted in terms of the potential for recovery. However, recovery is a more difficult issue than is generally recognized. The difficulty arises from the fact that ecosystems never recover to exactly the same state as existed prior to the contamination, and, even if they do, you cannot be sure of it because of limits on what can be measured. Therefore, it is necessary to define explicitly what would constitute sufficient recovery. For example, recovery of a forest might be defined as restoration of the precontamination canopy height and 80% of the precontamination diversity of vascular plants. Given such a definition, one may estimate the time to recovery based on successional studies in the literature such as Niemi et al. (1990), Yount and Niemi (1990), and Detenbeck et al. (1992). It is also possible to model the recovery processes of populations and ecosystems (Chapter 27 and Chapter 28). If the contaminants are persistent, the time required for degradation, removal, or dilution to nontoxic concentrations must be included in the recovery time. If recovery is an important component of the interpretation of risk for a site, that fact should be noted in the problem formulation. Because estimation of recovery is not simple, it may require site-specific studies or a significant modeling effort.

In general, the best interpretive strategies are comparative. The intensity, spatial and temporal extent, and the time to recovery can be compared to defined sources of natural variation, to other instances of contamination, or to relevant disturbances. The most relevant comparison is of the baseline effects of the contamination to the effects of the remediation (Chapter 33). Such comparisons provide a context for decision making.

The interpretation of ecological risk characterizations must include presentation of uncertainties (Chapter 34). The estimation and interpretation of uncertainty is discussed in Chapter 5. Here, it is necessary to emphasize the importance of correctly interpreting the results of uncertainty analyses. It is not sufficient to say that the probability is x.

33 Comparative Risk Characterization

To govern is to choose.

Duc de Le'vis

Comparative risk assessment often refers to the comparison of diverse hazards for the sake of prioritization (Section 1.3.1). However, it also refers, as here, to the assessment of risks associated with alternative actions in response to a particular hazard (Section 1.3.2). Examples include comparing the risks from alternative remedial actions for a contaminated site, alternative locations for a sewage treatment plant, alternative pest control strategies, or alternative detergent ingredients. What are the advantages of comparative risk characterization?

It can identify the best alternative: Critics of risk assessment have said that comparative assessment should replace assessment of acceptability (O'Brien 2000). They argue that it is inappropriate to consider an agent or action in isolation and determine whether it poses unacceptable risks. A proposed action or agent may not have been proven to be bad, but others may be better.

It can be less analytically demanding: In many cases, it is not possible to estimate the nature, magnitude, and probability of effects with reasonable confidence. However, relatively little information may be required to conclude that A is riskier than B. For example, the risks from an effluent are lower at a location with more dilution flow or with a less valued biota.

It provides context: An isolated ecological risk estimate may have little meaning for a decision maker or stakeholder. However, if they can see how the nature, magnitude, or probability of effects changes across alternatives, they have a better understanding of the significance of the results. For this purpose, the comparison need not be to real alternative decisions. Risks may be compared with those from other similar agents or situations that were previously assessed.

Comparative risk characterization also has potential disadvantages.

Little legal mandate: Most environmental laws do not give regulators a mandate for comparative risk characterization. That is, regulators are charged with blocking unacceptable products or actions, not with choosing the best. FIFRA, for example, does not mandate the Environmental Protection Agency (EPA) to pick the best pesticide for a particular use, or to reject a pesticide with acceptable risks if there are less risky ways to control the pest.

Difficulty in identifying alternatives: The proposed action is easily identified, but the universe of appropriate alternatives may be ill defined.

Alternatives may bias decisions: Any proposed action may be made to appear good by comparing it to sufficiently bad alternatives.

Alternatives may be irrelevant: In a case in the author's experience, the costs and benefits of dredging contaminated sediments in a polychlorinated biphenyl (PCB)-contaminated stream were qualitatively compared with those of remediating acid drainage from orphan mines in its headwaters. Although the mine reclamation would have been a better use of funds, it was not within the decision maker's power, so the comparison served only to reinforce his desire to avoid dredging. Hence, irrelevant alternatives can be a distraction or even a disincentive to possible action.

Risks may not be commensurable: Risk ranking is not easy when comparing risks to different endpoints. For example, allowing star thistle, an invasive weed, to remain uncontrolled on US public lands disrupts plant and animal communities and reduces wildlife habitat, livestock grazing, and recreational uses; control by herbicides results in nontarget toxicity and public concern; and biocontrol results in risks to native thistles including threatened and endangered species. These risks are not directly comparable because they are risks to different attributes of different entities, have qualitatively different consequences, and have different temporal and spacial scales.

It may not be easier: If, rather than simply ranking risks, the comparison must include benefits and costs, the relative simplicity is lost. In that case, the need to assess many alternatives becomes a burden.

It can accentuate ecological complexity: The comparison of risks may highlight ecological complexities that may result in ambiguous results and hence decisions based on other criteria such as cost or public preferences. For example, in a study comparing the effects of treating forests with *Bacillus thuringiensis* (Bt) to control gypsy moths with effects of allowing the gypsy moths to proliferate, Sample et al. (1996b) found that Bt reduced the abundance and species richness of native lepidoptera. However, they were also reduced, though not as much, on untreated plots, apparently by gypsy moth competition. Further, the relative long-term effects are unclear. Bt requires multiple years of application to be effective on many plots, which is likely to increase nontarget effects and inhibit recovery. However, allowing defoliation by gypsy moths can have long-term effects as well. Finally, year-to-year variation in weather during the study had equal or greater effects on lepidopteran population sizes. This not only obscured the treatment effects but also raised the issue of interactions between the effects of treatments and weather. The inclusion of pesticides in the comparative assessment would have further complicated the results due to their briefer but more severe direct effects on the wider range of species. Such complex results can cause decision makers to throw up their hands.

The balance between these advantages and disadvantages of comparative risk assessment will depend on the circumstances of the assessment. However, the advantage of identifying the best option is compelling, and in many cases ranking is the only feasible option for risk characterization.

33.1   METHODS OF COMPARATIVE RISK CHARACTERIZATION

Comparative risk characterizations can take various forms, ranging from the simple to the complex. Because few comparative risk assessments are limited to a single common endpoint, the methods must deal with complex comparisons. This may be done through qualitative and semiquantitative systems for ranking, scoring, or categorization, or by creating common quantitative scales such as net benefits (gain or loss in ecosystem services) or cost/benefit ratios.

33.1.1   RISK RANKING

The simplest approach to comparison is to rank the risks. This does not require estimating the nature of magnitude of risks, or even their relative magnitudes. It has been argued that risk ranking is often the only practical approach for regional ecological risk assessments (Landis 2005). Risk ranking is sufficient when considerations other than risk are not important. It may also serve for screening comparative assessments. That is, if there are many options to be compared, assessors might rank them and then choose the few highest ranking options for more detailed assessment. The chief advantage of ranking is that it requires relatively little information. We can determine that three alternative actions are ranked $A > B > C$ without knowing what sorts of effects might occur. The biggest disadvantage is that it supplies relatively little information, only an ordinal scale. If $A > B > C$ we do not know if all three are nearly equal, if A and B are similar but C is 1000 times as risky, or if there is some other underlying relationship.

33.1.2 RISK CLASSIFICATION

Risks may be compared by categorizing them. The most common is acceptable/unacceptable risks. For more resolution, one might use high/moderate/low risk. Categorical scales may be defined to suit the assessment. For example, risks of extirpation of a fishery from alternative management plans might be categorized as inevitable, likely, uncertain, or unlikely. If more than one option is classified as acceptable or some equivalently desirable category, the selection may be left to other criteria such as cost or esthetics, or a more detailed assessment of the options in the highest category may be performed.

33.1.3 RELATIVE RISK SCALING

Risk scaling is the assignment of a numerical scale to a set of risks that are not risk estimates but are related to, or in some sense indicative of, risks. For example, chemicals or sources of chemicals may be compared by scaling them in terms of exposure. One such comparative risk scale is the fraction of releases that is eventually taken up by one or more routes, the intake fraction (Bennett et al. 2002). When comparing actions that destroy or modify ecosystems or habitat for a valued species, the area affected may provide an adequate scale for comparison. In some other cases, the duration of an action may be an appropriate scale. The approaches used to scale risks for prioritization may also be employed to subjectively scale proposed actions (Section 1.3.1).

33.1.4 RELATIVE RISK ESTIMATION

If risks of each alternative are estimated, the relative risks, the ratios of risks or their quotients, may be calculated. For example, the risk of extinction from alternative A is estimated to be twice as great as from alternative B or alternative A is expected to extirpate twice as many species as alternative B. Alternatively, excess risk, the absolute difference in risk, may be calculated. For example, the risk of extinction from alternative A is estimated to be 0.25 greater than alternative B or alternative A is expected to extirpate four more species than alternative B. When background risks are important relative to the risks from the actions being compared, it may be important to first subtract out the background so as to calculate a relative excess risk (Suissa 1999). For example, while comparing the risks of mortality in an endangered species due to two pest control alternatives, it is important to know the total mortality under each alternative for population viability analysis, but the appropriate comparison of the risks from the actions under consideration is provided by the excess risk. It is preferable to the relative risk of death, which would tend to minimize the differences between the alternatives by adding them to the background.

33.1.5 NET ENVIRONMENTAL BENEFITS ANALYSIS

In some cases, one or more of the alternative management actions may pose a risk to the environment that may even exceed the benefits of the action. For example, a biocontrol agent for an invasive weed may damage native plants. Also, remedial actions at contaminated sites often involve processes such as dredging and capping that are themselves ecologically damaging. The relative benefits of alternatives are not routinely assessed, but must be considered in some contexts. For example, the US Guidelines for Specification of Disposal Sites for Dredged or Fill Material (Title 40, Part 230, Subpart G) state: "When a significant ecological change in the aquatic environment is proposed by a discharge of dredged or fill material, the permitting authority should consider the ecosystem that will be lost as well as the environmental benefits of the new system."

The need to analyze the benefits of remedial actions became obvious as a result of the damage to intertidal ecosystems caused by the cleaning of the shores of Prince William Sound following the EXXON Valdez oil spill. A result was the development of net environmental benefit analysis (NOAA Hazardous Materials Branch 1990; Efroymson et al. 2004). Net environmental benefits are the gains in environmental services or other ecological properties attained by remediation or ecological restoration minus the environmental injuries caused by those actions. The concept is illustrated by Figure 33.1. Following contamination of an ecosystem, damage occurs and then the system begins to recover, so that ideally, through natural attenuation of the contaminants and natural ecosystem recovery, the lost environmental services are restored. If the site is remediated to remove or reduce the contamination, the remedial activities will inevitably cause some additional damage, but recovery should be hastened. If, in addition, efforts are made to restore the ecosystem by planting native vegetation, restoring stream channel structure, etc., effects of remediation may be mitigated and recovery should be hastened, possibly resulting in higher-than-baseline levels of services. Note that qualitatively different trajectories may occur. The system may not recover through natural attenuation in any reasonable time, and remedial and restoration projects often fail to bring about recovery and may make things worse.

The net benefits of the alternatives can be estimated as the sums (or integrals) of the positive and negative areas between the curves and the reference line. Alternatively, net

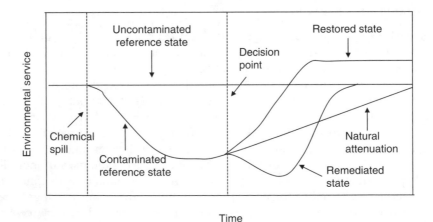

FIGURE 33.1 Hypothetical trajectory of an ecosystem service (or other endpoint attribute) with time following a chemical spill (contaminated reference state), the expected uncontaminated reference state, the expected trajectory of the remediated state, and the expected trajectory of the restored state. (From Efroymson, R.A., Nicollette, J.P., and Suter, G.W., II, *A Framework for Net Environmental Benefit Analysis for Remediation or Restoration of Contaminated Sites*, ORNL/TM-2003/17, Oak Ridge National Laboratory, Oak Ridge, TN, 2003. With permission.)

benefits might be the sums or integrals of the deviations of each remediation or restoration alternative from natural attenuation. In either case, net benefits may be positive or negative.

33.1.6 ECONOMIC UNITS

In the context of environmental management, cost–benefit analysis is usually a means of determining whether an action can be justified in terms of economic benefits (Chapter 37). However, economic analyses are also a way to provide a standard metric, money, for comparison of alternative actions. The same methods are used to monetize the environmental benefits of an action. The comparison may be based on the relative environmental benefits, the relative net benefits (i.e., net benefits analysis with monetary units), the relative cost effectiveness (i.e., which alternative meets the environmental goal at the least cost), the relative cost/benefit ratios, or the relative excess benefits (i.e., comparisons of benefits minus costs) of the alternatives.

33.1.7 REPORTING COMPARATIVE RISK

Conceptually, the simplest approach is to estimate and report to the decision maker and stakeholders the various risks to the appropriate endpoints for each alternatives. The responsible individuals may then balance those risks based on their own expressed or inchoate criteria. This approach requires that assessors be clear about the nature and extent of the effects and their probabilities of occurrence. It also requires that the assessors be fair to all alternatives. That is, the most sensitive endpoints for each alternative must be assessed as well as an equivalent range of endpoints. Equivalence in this case does not mean an equal number of endpoints. Rather, an alternative that potentially affects a wide range of entities and attributes would be assessed for more endpoints than one that potentially affects few. Hence, the lists should be cut off at similar levels of probability or severity. This alternative works best with experienced and knowledgeable decision makers.

33.2 COMPARISON AND UNCERTAINTY

All of the approaches described will provide a ranking of alternatives in terms of ecological risk, and some provide a much more thorough comparison. However, as with any environmental analysis these results are uncertain and may even change from place to place and year to year due to environmental variability (Chapter 4). As a result, the relative ranks of alternatives may be unclear. If two or more alternatives cannot be ranked with sufficient confidence, they may be treated as equally desirable and the choice may be made on bases other than ecological risk. However, it would be inappropriate in most cases to require that alternatives be discriminated with 95% confidence. In any case, the comparisons of ecological risks of alternatives are often subjective and qualitative, so judgments concerning confidence in rankings are even more subjective and qualitative.

33.3 SUMMARY

The chief advantage of comparative risk assessments is that it provides a better basis for decision making by revealing the implications of a choice among the range of feasible alternatives. However, because ecological risk assessment has been primarily concerned with supporting decisions concerning the acceptability of a proposed product or action, methods for comparison of ecological risks are poorly developed and relatively seldom demonstrated in the literature.

34 Characterizing Variability, Uncertainty, and Incomplete Knowledge

Statistical results must always be regarded as providing a minimum estimate of error and variability, and an extra margin of error—often large—must be added to account for uncertainty in the results not captured by the statistical analysis.

Bailar (2005)

The characterization of variability and uncertainty is an important component of risk characterization, and the Environmental Protection Agency (EPA) guidance states that both should be reported and distinguished (Science Policy Council 2000). As discussed in Chapter 5, variability refers to the inherent differences in nature while uncertainty is a result of lack of knowledge. In addition, potential inaccuracies may result from biases and limitations of time and resources. This step in the risk characterization process can be thought of as analysis of the components of data quality (Chapter 9).

34.1 CHARACTERIZING VARIABILITY

The variability of interest when characterizing ecological risks is the variance among instances of the endpoint entity or its constituents. For wildlife, this typically means variability among individuals within the assessment population. If, for example, we assess risks to mink on a contaminated lake, we are interested in the distribution among mink due to variance in size, diet, contaminant levels in food items, and other exposure parameters as well as variance in sensitivity as reflected in the dose–response distribution. However, if multiple populations are exposed, we may estimate the variance distribution among exposed populations such as different populations of sage grouse exposed to organophosphate pesticides applied to alfalfa fields in different irrigated valleys. Probabilistic models of exposure or exposure–response may be used to express variability as probabilities of effects or proportions of entities or constituents affected. See, for example, the case of the honeycreepers exposed to rodenticides (Section 30.6). These probabilities or proportions derived from variance distributions are expressions of risk, and, ideally, they should be reported as risk levels. However, when information is insufficient to define distributions, variability may be characterized by estimating effects for a typical individual, population, or ecosystem and for one that is highly exposed or highly susceptible.

34.2 CHARACTERIZING UNCERTAINTY

Characterization of uncertainty begins, and too often ends, with a listing of sources of uncertainty. Such ad hoc lists are of relatively little use without further analysis. It is desirable to have a systematic taxonomy or organization of uncertainties from which the lists are derived. The list may be presented as a table with an evaluation of whether it is a large, moderate, or small source of uncertainty and whether the assumptions associated with that uncertainty in the risk assessment had an identifiable bias. For example, in an assessment of risks to mink from mercury in water, the assumption that mink eat no terrestrial foods might be judged to be a moderate source of uncertainty that biases the risk estimate upward. If additional tiers of assessment are possible, these evaluations should be used to prioritize additional testing, measurement, or modeling. There are many taxonomies of uncertainty. The following types are of practical importance in ecological risk assessment:

Measurement uncertainty: This is the uncertainty due to error and inaccuracy in sampling and analysis, the domain of conventional sample statistics.

Model fitting uncertainty: Empirical models, such as a log probit model fitted to dose–response data, inevitably do not hit all of the data points. The lack of fit is in part due to measurement uncertainty and in part due to the model itself.

Extrapolation uncertainty: The available data inevitably are not from the entities and conditions that are the subject of the risk assessment. Extrapolation models may be used (Chapter 26 through Chapter 28), but they only reduce the uncertainty. Even when data are from the site and taxa of concern, they must be extrapolated to the future when risk management will be applied and conditions will be different.

Model selection uncertainty: This uncertainty results from the fact that models are always abstractions. Their results are uncertain because of choices that are made during their design such as the aggregation of taxa or life stages, the inclusion of processes, the choice of linear or nonlinear functional forms, etc. Model uncertainty is reduced by careful design or selection of models and is estimated by comparison of models (Chapter 9).

Subjective uncertainty: Because the statistics of uncertainty estimation and propagation are limited, they underestimate the degree of unpredictability of effects. In particular, they do not estimate the magnitudes of most errors or of confusion and ignorance. In ecological assessments, the potential for ignorance is large and often unappreciated (Box 34.1). Hence, assessors may use their expert judgment to subjectively define bounds on risk estimates. Various methods such as the Delphi technique are available to elicit and combine expert judgments, but experts tend to underestimate their true uncertainty (Fischoff et al. 1981; Morgan and Henrion 1990; Cooke 1991).

Identified uncertainties should be quantified as far as is reasonably possible. Uncertainties in the parameters due to measurement, extrapolation, and the fitting of any empirical models used in parameter derivation can be estimated by conventional statistics or expert judgment (Chapter 5) and propagated to estimate total uncertainty in the risk estimate, usually by Monte Carlo analysis (Chapter 30). Model uncertainty and subjective uncertainty are best estimated and reported separately.

The most generally useful method for reporting uncertainties is confidence intervals (CIs) (Section 5.5.2). However, they should not be limited to the conventional but arbitrary 95% CIs (Figure 5.2). In particular, the 50% CI, which may be thought of as the as-likely-as-not interval, should be considered the base. In addition to providing more information, a set of CIs avoids the impression imparted by wide 95% CIs that the assessment conveys no useful information.

BOX 34.1
Ecosystem Idiosyncracies and Uncertainty: The Case of Carbaryl and Larval Anurans

While it is commonplace that ecological risk assessments are not sufficiently ecological, and the influence of ecosystem context on the toxicity of chemicals should be considered, it is less clear how that can be done. This problem is illustrated by recent studies of the effects of three environmental factors on the effects of carbaryl on larval anurans (tadpoles). First, the pesticide is more lethal at higher temperatures. Second, it may be more or less lethal when larvae are exposed in the presence of competitors, depending on the species (Boone and Bridges 1999). In particular, Woodhouse's toad larvae experienced increased survival and growth in experimental ponds with carbaryl, apparently due to reduced competition with invertebrates for algae, but in one pond study carbaryl reduced survival (Boone and Semlitsch 2001, 2002; Boone et al. 2004). Different patterns of effect were seen in the same studies with larvae of leopard frogs, green frogs, and grey tree frogs (Boone and Semlitsch 2001, 2002; Mills and Semlitsch 2004). Third, carbaryl affects responses to predators. Sublethal levels of carbaryl affect the ability of larval anurans to avoid predators (Bridges 1999a,b). Exposure of larvae to chemical cues of the presence of a predator, the red-spotted newt, increased the lethality of carbaryl by factors of 8 and 46 in two species, caused no interactions in two other species, and increased toxicity in the early days of the experiment for a third pair (Relyea 2003).

These results are fascinating, but they suggest that there are severe limits on predictive assessment. The complexity of ecological systems and their responses to the various agents that must be assessed are beyond the limits of routine regulatory science. The effects of temperature are conceptually and methodologically straightforward, but tests for effects of temperature on toxicity are not routine. The possibility that by reducing competition from more sensitive species, a chemical could increase survival and growth of an endpoint species is also conceptually straightforward and could be predicted by generic ecosystem models (Chapter 28) or observed in mesocosm tests (Chapter 24). However, the inconsistency of this response among species and even among different sets of mesocosms suggests that generalizations are elusive. The synergistic effects of carbaryl and chemical signals of the presence of a predator are not expected or predicted by any ecosystem model and are not consistent among anuran species. What about predators other than newts, other competitors, other conditions, other cholinesterase-inhibiting pesticides? Without a lot of carefully designed and conducted fieldwork, one can only guess how these factors interact in real populations exposed to real pesticide applications.

Examples such as this suggest that ecological risk assessments will always be characterized by unquantifiable uncertainties. While numerous studies demonstrate that increased exposures to toxic chemicals cause increasingly severe effects in the field and those effects occur at levels roughly equal to those that affect sensitive species in the laboratory, there is much uncertainty in particular cases and specific predictions.

34.3 UNCERTAINTY AND WEIGHT OF EVIDENCE

When risks are characterized by weighing multiple lines of evidence (Chapter 32), uncertainty in the result cannot be quantified in any objective way. The best strategy is to estimate variability and uncertainty for each line of evidence separately. Then, if there is a best line of inference, its objectively estimated uncertainty can be reported. Finally, subjective uncertainty can be subjectively estimated, but, in these cases, the overall uncertainty is reduced by the increased confidence provided by the other lines of evidence. In some cases, this can be done objectively. For example, another line of evidence may place a limit on the possible effects and that knowledge can be used to truncate the confidence interval. More commonly, the totality of evidence will increase the subjective confidence of assessors allowing them to assign narrower confidence bounds than those provided by the statistics for the best line of evidence.

34.4   BIASES

Biases in risk characterization may result from the personal biases of assessors, but more important biases arise from institutional practices, policies, and laws. In particular, regulatory agencies are commonly mandated to ensure protection of public health and the environment. This is a requirement for a precautionary bias. For example, if alternative assumptions are plausible, a precautionary bias would choose the one that results in the greatest risk estimate. Industries and other responsible parties have no such requirements. They may have their own precautionary policies or they may have antiprecautionary biases based on economic interests. The results of these policy differences can be differences in risk characterizations and charges of bias. It is important to remember that bias can be a result of legitimate policy differences rather than an attempt to deceive.

The disclosure of biases, to the extent that they significantly influence the results, provides for a more informed decision and a more transparent risk characterization. Ideally, biases that are a result of policy should be revealed in regulations, guidance, or other policy documents. Biased assumptions, data selection, or model parameterizations should be described in individual assessments, but even there they should be as generic as possible. That is, decisions about the manner of performance of an assessment that would result in a precautionary or other bias should be agreed upon during the problem formulation and described in the analysis plan.

An oft-advocated alternative is to perform two assessments, one with the policy bias and the other with no bias. In practice, this not only increases the time and effort required, but also creates new sources of conflict. It is often unclear what assumption, model, or data set is unbiased, and therefore which would result in the "best estimate" of risk.

34.5   LIMITATIONS

In addition to the potentially quantifiable variability and uncertainty, assessors must report limitations of the risk assessment. The reporting of limitations includes meeting the requirements for transparency and reasonableness in reporting results (Box 35.1). In addition, it requires reporting the consequences of limited time and resources. This is not simply an opportunity for assessors to complain. Rather, the point is to clarify how additional time and resources (i.e., another tier of assessment) might clarify the risks and provide a better basis for decision making. These may range from modeling additional species to performing additional measurements or tests and even to defining a research program such as developing a new type of test or to determining the significance of a novel route of exposure.

There are at least four bases for making such recommendations. First, time and resources simply ran out before all endpoints and pathways were addressed. For example, the Clinch Valley watershed case study used the conceptual model to highlight those components of the system that had not been addressed (Serveiss et al. 2000). Second, new endpoints or pathways of importance to stakeholders may have been identified. For example, when presenting preliminary results of the Watts Bar Reservoir Ecological Risk Assessment, the author learned that people in a downstream community were concerned about turtles, which they ate and which had not been considered. Third, value of information arguments could be used. That is, one might demonstrate that a critical uncertainty in the results could be resolved by additional effort and would be likely to affect the outcome of the decision (Dakins 1999).

Finally, controversies may have arisen that could be addressed by additional data or analyses. These may be concerns or allegations raised by the public, regulators, or regulated parties that were not recognized during the problem formulation. In any case, it is better to disclose the existence of limitations in the assessment and suggest possible solutions, than to be accused of false confidence or even cover-up.

34.6 CONCLUSIONS

The results of characterizing variability, uncertainty, and incomplete knowledge are often enlightening to risk assessors and may lead to modification of the conclusions from screening or weighing evidence. The data must also be included in the reporting of results (Chapter 35). One potential way to organize and present the results is to use the numeral, unit, spread, assessment, and pedigree (NUSAP) system for data presentation (Section 9.1.4). It can systematize the qualitative and quantitative aspects of data quality, but is not commonly used, and so is not familiar to decision makers. NUSAP, or an equivalent system, should be considered at least for presenting influential data.

Part VI

Risk Management

Risk characterization serves to inform risk management. At minimum, ecological risk assessors must report the results of their risk characterization to the risk manager and any stakeholders who are involved in the decision-making process (Chapter 35). In some contexts, ecological risk assessors are also involved in the decision-making process (Chapter 36). To the extent that a formal analytical process is involved, risk assessors must, at least, present results that support that decision analysis. Because decisions are based on human health, legal, economic, ethical, and political considerations as well as ecological considerations, ecological risk assessors should be prepared to help integrate all of those considerations to support the risk manager (Chapter 37 and Chapter 38). Once the management decision has been made, ecological risk assessors may be involved in monitoring the results (Chapter 39).

This part of the process of environmental assessment and management is the least comfortable for most ecological risk assessors. However, if they are to be successful in influencing environmental management, environmental scientists must be prepared to engage with the social sciences and deal with the politics of decision making.

35 Reporting and Communicating Ecological Risks

It may not seem very important, I know, but it is, and that's why I'm bothering telling you so.

Dr. Seuss

It is important to distinguish the reporting of risk assessment results from communicating risks. Reporting ecological risks involves creating a document that will support the decision-making process by providing information for cost–benefit analysts or decision analysts, by informing the decision makers, by informing stakeholders such as manufacturers and responsible parties, and by serving as a basis for defense of the decision should it be challenged in court or sent for review to the National Research Council or similar body. Communicating ecological risk is a process in which risk assessors personally convey their findings to decision makers and stakeholders. Risk communication may also include written products, but they should be short statements (e.g., a one-page fact sheet) to prepare the audience or for the audience members to take away as a reminder of the message. Risk reporting and communication are similar in the need to convey results clearly to the intended audiences. However, a report must meet the needs of all likely audiences, while oral communication should be tailored to specific audiences.

35.1 REPORTING ECOLOGICAL RISKS

The form in which ecological risks are reported is an oft-neglected aspect of the practice of ecological risk assessment. The EPA's guidance for risk characterization states that a report of risk assessment results must be clear, transparent, reasonable, and consistent (Science Policy Council 2000). Considerations for achieving these goals are listed in Box 35.1. However, the goals of being brief (for clarity) and transparent are conflicting. If sufficient detail is presented for the reader to fully understand how the results were derived and to replicate them, the resulting multivolume report will be thicker than anyone will care to read. As discussed in Chapter 5, simply justifying the assignment of distributions to parameters may result in a sizable report. However, some critics have advocated more complete risk characterizations including multiple alternative risk estimates (Gray 1994). For ecological risk assessments, this means reporting not only risk estimates for all lines of evidence in all types of evidence for each endpoint, but also results for alternative assumptions within a line of evidence.

BOX 35.1
Clear, Transparent, Reasonable, and Consistent Risk Characterizations

For clarity:

- Be brief.
- Avoid jargon.
- Make language and organization understandable to risk managers and informed lay people.
- Explain quantitative results.
- Fully discuss and explain unusual issues specific to a particular risk assessment.

For transparency:

- Identify the scientific conclusions separately from policy judgments.
- Clearly articulate major differing viewpoints or scientific judgments.
- Define and explain the risk assessment purpose (e.g., regulatory purpose, policy analysis, priority setting).
- Describe the approaches and methods used.
- Explain assumptions and biases (scientific and policy) and their influence on results.

For reasonableness:

- Integrate all components into an overall conclusion of risk that is complete, informative, and useful in decision making.
- Acknowledge uncertainties and assumptions in a forthright manner.
- Describe key data as experimental, state-of-the-art, or generally accepted scientific knowledge.
- Identify reasonable alternatives and conclusions that can be derived from the data.
- Define the level of effort (e.g., quick screen, extensive characterization) along with the reason(s) for selecting this level of effort.
- Explain the status of peer review.

For consistency:

- Follow statutory requirements, guidelines, and precedents.
- Describe how the risks posed by one set of stressors compare with the risks posed by a similar stressor(s) or similar environmental conditions.
- Indicate how the strengths and limitations of the assessment compare with past assessments.

Source: Adapted from EPA (U.S.Environmental Protection Agency), *Guidelines for Ecological Risk Assessment*, EPA/630/R-95/002F, Risk Assessment Forum, Washington, DC, 1998; and Science Policy Council, *Risk Characterization Handbook*, EPA 100-B-00-002, US Environmental Protection Agency, Washington, DC, 2000.

The usual solution to this conflict between brevity and transparency is the executive summary. Unfortunately, executive summaries attempt to summarize the entire assessment and are seldom sufficient to stand alone if the "executive" is the risk manager. A report of results that neglected methods but presented risks in adequate detail for decision making would probably be more useful in most cases. In addition to summarizing the results, a report to the risk manager should explain the major issues, any controversies, and relevant precedents. Ideally, the contents and level of detail would be worked out between the risk assessors

and risk manager. Routine assessments, such as those for new chemicals, may have a standard form or format for reporting to the risk manager.

The needs of users other than the decision maker constitute a more serious conflict with the call for brevity. Cost–benefit analysts or decision analysts need detailed results to support their analyses. Risk assessments prepared by a responsible party must present data and methods in detail so that regulators can review their acceptability. Risk assessments prepared by regulators must present data and methods in sufficient detail that the responsible party can review their acceptability. In either case, the report must be sufficiently detailed to withstand legal scrutiny. As a result, the report of a complex ecological risk assessment may fill a library shelf. Simply providing data, models, and analytical results on a CD or DVD can help, but creative solutions to the problem are needed. The use of hypertext is promising in that it would allow a person reading a brief summary of the risk assessment to bore into a topic to the depth that is appropriate to interests and needs. However, creating a large hypertext document is not quick or easy, and many people do not like to read from a computer screen.

35.2 COMMUNICATING ECOLOGICAL RISKS

Risk communication is the process of conveying the results of risk assessments to decision makers, stakeholders, or the public, and of receiving and responding to their comments. (Some documents define risk communication to include consultation during planning and problem formulation.) It goes beyond the issue of reporting the results of the risk assessment in a clear and useful manner to actually conveying the results to a skeptical audience. It is difficult for two reasons. First, like any quantitative and scientific subject, it is difficult to convey to those who do not have the requisite training or experience. Risk assessments may be particularly difficult to explain, because they combine biological and physical sciences with mathematics and statistics. Second, situations that require risk assessments are often emotionally charged. People's health, livelihood, and property values are typically at stake, and distrust is typically high. Most of the literature of risk communication is directed at issues of managing emotions and gaining trust with respect to health risks, independent of the nature or quality of the technical message (NRC 1989; Fisher et al. 1995; Lundgren and McMakin 1998). Those issues will not be treated here, because emotional investment in ecological issues is usually lower and is likely to be qualitatively different. In many if not most ecological risk assessment cases with high levels of emotional investment, such as restrictions on harvesting, water withdrawals, or land use to protect resource species or endangered species, the issues are largely economic and the strong emotions are largely held by those who generate the risk. Fishermen, loggers, ranchers, and farmers are reluctant to believe that their activities damage the environment in ways that would justify restriction of their activities. Research is needed to guide risk communication in such situations.

The technical communication problems are more severe for ecological risk assessors than for health risk assessors. Ecological risk assessors not only deal with unfamiliar scientific and mathematical concepts, but also often estimate risks to unfamiliar entities and attributes. Decision makers and stakeholders know full well what a human is and have both knowledge of, and empathy for, the various fates that befall humans. However, many will not know what a spectacled eider or an Atlantic white cedar bog is, much less the implications of changes in their nesting success or in hydro-period. Humans have a sense of their own inherent value and their value to their family and community, but have little knowledge or appreciation of equivalent values of ecological entities. Hence, much of ecological risk communication is a matter of education. This education may require a little honest salesmanship as well as the basic description of an entity and attribute. This may involve the use of attractive photographs and explanations of values associated with the endpoint. Failure to protect the environment should not occur because the decision makers lack a vivid understanding of what may be lost or gained.

A related communication problem is the greater familiarity of most decision makers with the relatively simple methods and results of human health risk assessments. The numerous endpoints and multiple types of evidence employed in ecological risk assessments make them seem complex and ambiguous. This problem may be alleviated as health risk assessments begin to use multiple lines of evidence and to estimate the range of public health outcomes. However, in the meantime, decision makers tend to focus on aspects of ecological risk assessment that seem familiar. As a result of that tendency and the natural affinity of people for mammals and birds, risks expressed in the familiar terms of effects on survival or reproduction of such species tend to be inordinately influential. For other end-points, it is important to explain not only what they are and how they respond but also why they are assessed using unfamiliar methods and models. As far as possible, use analogies to health risk assessment. For example, analyses of biological survey data in risk assessments can be described as ecological epidemiology, and species sensitivity distributions can be described as dose–response models for ecological communities. Finally, the unfamiliarity of ecological methods and models often leads decision makers to ask whether they are official methods or have been used previously in decision making. Therefore, it is important to be prepared to cite guidance and precedents. If a genuinely novel method or model is used, be prepared to compare its results to those of more familiar methods or models and to explain the advantages of the innovation.

A more pervasive problem is the inherent difficulty of conveying scientific and mathematical concepts to people who are not trained in those fields. As Cromer (1993) explains, science is difficult to do and to convey, because it constitutes uncommon sense. It has been suggested that this is because the human mind evolved to extrapolate directly from experience, which works routinely but results in flawed logic in complex or unfamiliar inferences (Pinker 1997; Dawes 2001). Further, even when reasoning carefully, the mind deals more easily with some sorts of information and problems than others (Pinker 1997; Anderson 1998, 2001). From these generalizations, some advice can be derived.

Avoid probabilities: People, including scientifically trained experts, have difficulty with probabilities, but understand and manipulate frequencies relatively easily (Gigerenzer and Hoffrage 1995; Gigerenzer 2002). Whenever possible without distorting the results, translate probabilities to frequencies when communicating risks. This has the ancillary advantage of forcing you to determine exactly what you mean by a probability (Chapter 5).

Use discrete units: Although most properties of nature are continuous, the mind divides the continua of time and space into events and objects. For example, Bunnell and Huggard (1999) reduced the spatial continuum of forests to a hierarchy of units (patches, stands, landscapes, and regions), which could be more easily described to forest managers than sets or areas of forest.

Use categories: We not only discretize continuous variables, but also lump the units into like categories to which we assign names. Hence, "folk biology is essentialistic" (Pinker 1997). Long after ecology has revealed the variance in species composition over space, we continue to name vegetation and ecosystem types (e.g., mixed mesophytic forest). Similarly, it is often easier to communicate the frequencies of categories (e.g., high, moderate, or low) associated with ranges of a continuous variable (e.g., flow) than to communicate the meaning of the variable's probability density.

Use few categories: People tend to divide entities or events into only two or three categories such as drought, normal, and flood for flow regimes.

Tell stories: Information imbedded in a narrative is more convincing and remembered longer. Every conceptual model is a potential story.

Use multiple modes: Individual audience members will respond differently to verbal, diagrammatic, pictorial, or other modes of representation. By using multiple modes of presentation you are more likely to achieve comprehension with one. In particular, photographs of the effects being discussed such as photos comparing plants grown in site and reference soils or depictions of the diversity of fish species in a disturbed and reference stream can make unfamiliar effects vivid. In addition, repetition increases comprehension and retention and repetition using multiple modes avoids boredom. Similarly, in presentations it is advantageous to include speakers with different styles to engage the diverse members of the audience.

Use case studies: Even if your assessment is generic (e.g., national risks from mercury in coal combustion emissions), illustrate it with a particular case (e.g., loons in the boundary waters). People are more willing to extrapolate from one real case to many than from an abstraction to any real case.

Simplify carefully: Scientific concepts can be complex and difficult. It is often advantageous to use a simple analogy to convey a complex system. However, it is important to prepare qualifiers (e.g., of course it is not really that simple), because knowledgeable members of the audience, particularly those who are opponents, will pounce on an "oversimplification" (Schneider 2002).

Remember the human analogy: While you are talking about risks to otters from eating fish, many in the audience will be thinking about what it implies for people eating fish. Be prepared for the questions that such analogies imply and avoid statements that would alarm those who are focused on the human analogy or that contradict the human health risk assessment.

Avoid personalizing: Although personalizing a situation makes the message more vivid (e.g., "I would not let my daughter eat fish from that river"), it is a bad strategy for risk assessors. Present the results of your analysis and let the risk managers and stakeholders infer the personal implications.

Of course, this advice is superceded by the advice to know your audience. For example, former US EPA administrator Ruckleshaus (1984) preferred that risk assessment results be presented as cumulative distribution functions, contradicting some of the previous advice. This example also serves to remind assessors not to talk down to risk managers, stakeholders, or the public. An individual who is not an environmental scientist and does not know what Bayesian means or what a *Pimephales promelas* is should not be treated as unintelligent. A condescending or contemptuous attitude will be detected and will result in dismissal of your message. On the contrary, people can understand and appreciate descriptions of complex natural systems if they are well presented. The renowned conservation biologist Daniel Janzen has said of ecological complexity: "Audiences soak it up. I was told that it's too complicated and a lot of people won't understand it. Bullshit. They understand it perfectly well" (Allen 2001).

Risk communication is an opportunity to ensure that your efforts do some good. The audience is less likely to be hostile than in human health risk communication. Because ecological risks are not personal threats, audiences are open to learning a bit about nature and how plants and animals interact with contamination or disturbance. You have an opportunity not only to explain your results but also to educate and even entertain. In that way, you can help to create a constituency for good ecological management.

36 Decision Making and Ecological Risks

It is not industry, development or the nation's growing population that poses the greatest threat to the environment, it is shortcomings in the political process that perpetuate environmental degradation.

Howard R. Ernst (2003)

In the conventional environmental and ecological risk assessment frameworks, risk assessors communicate their results to risk managers and then leave the room. Decision making is viewed as a policy-based and science-informed political process that is best left to those who have political authority. This description is accurate in many cases. However, to clarify technical issues and avoid misunderstanding, risk assessors may be, and should be, involved in the decision-making process. In some cases, distinct analyses are performed after the risk assessment and before the decision by economists or experts on decision analysis. In such cases, risk assessors are likely to be involved in supporting or helping to perform those analyses. In addition, even when risk assessors are not involved, they should have some understanding of the decision-making process so that their results may be as useful as possible and to avoid unrealistic expectations about their ability to determine the outcome.

The bases for decision making discussed in this chapter are limited to those that are risk-related. As discussed in Chapter 2, some environmental decisions are made based on best available technology or some other criterion that does not include risk. It must be borne in mind that politics is the ultimate criterion. Any of the decision criteria discussed below may be overridden by political ideology or political expediency.

36.1 PREVENTING EXCEEDENCE OF STANDARDS

The simplest and least ambiguous decision criterion is that if an environmental standard is violated, action must be taken. To the extent that standards are based on ecological risks (Chapter 29), these are risk-based decisions.

36.2 PREVENTING ADVERSE EFFECTS

The national policy of the Netherlands states that exposure to substances should not result in adverse effects on humans or ecosystems (VROM 1994). Similarly, the US Clean Water Act prohibits "impairment of the physical, chemical, or biological integrity of the Nation's waters." Such policies provide the most clearly risk-based criteria for decision making. Given a definition of adverse effects, one can determine whether they are expected to occur or whether the probability of occurrence is excessive, and that is sufficient grounds for action.

36.3 MINIMIZING RISKS

Comparative risk assessments (Chapter 33) provide a basis for decision makers to choose the action that presents the least risk to human health or the environment. Comparative risk estimates can be based on a technical risk characterization as discussed in Chapter 33. Alternatively, it can be viewed more broadly as a type of policy analysis that includes relative risks and psychosocial preferences concerning different types of risks in a process of stakeholder consensus building. That approach can be thought of as a microscale application to a specific decision of the macroscale process of using comparative risk assessment for priority setting (Section 1.3.1) (Andrews et al. 2004).

36.4 ASSURING ENVIRONMENTAL BENEFITS

Often, either explicitly or implicitly, actions to protect the environment are judged to be defensible if they are expected to result in adequate benefits to the environment. Benefits are the complement of adverse effects and their risks. In the context of risk assessment, benefits are avoidance of a risk of adverse effects. In the context of remediation and restoration, risks are the probabilities that benefits will not be realized or that damage will occur due to poor planning or execution, or chance events. Benefits may be judged in terms of absolute benefits of an action, relative benefits of alternative actions (Section 33.1.5), or benefits relative to costs (Section 36.6).

Concerns for the benefits of actions may be expressed when a program has been ongoing for some time, costs have begun to mount, and questions arise about the cost-effectiveness of the expenditures. An example is the expenditure of $580 million on 38 projects to remediate contaminated sediments in the Laurentian Great Lakes without a clear linkage to increases in beneficial uses (Zarull et al. 1999). At some point, faith in such large expenditures fails without evidence of benefits. As a result, the Great Lakes Water Quality Board recommended the development of better methods to quantify the relationship between sediment contamination and use impairments, and to monitor the ecological benefits and beneficial uses of remediated sites.

36.5 MAXIMIZING COST-EFFECTIVENESS

Cost effectiveness analysis identifies the relative monetary costs of different means of achieving a standard, an acceptable risk level, or other goal. The decision maker could then choose the least-cost method. A variant of this idea that has had important environmental implications in the United States is the principles and guidelines framework of the Army Corps of Engineers which maximizes the net national economic benefits of project alternatives, as long as they do not cause significant environmental degradation (USACE 1983).

36.6 BALANCING COSTS AND BENEFITS

Cost–benefit analysis is increasingly applied to environmental regulatory and remedial actions. The decision model is that the public benefits of regulations should exceed the costs of compliance to the regulated parties. In some cases, it is sufficient to show that monetary and nonmonetary benefits qualitatively balance the costs. However, strict interpretations of cost–benefit requirements allow only monetary benefits. As discussed in Section 38.3, this requirement to monetize the benefits of ecological entities and processes can be a serious impediment to environmental protection.

36.7 DECISION ANALYSIS

Decision analysis comprises a diverse set of concepts and methods for informing a decision. Classic formal methods define the decisions to be made, the alternatives, goals, possible outcomes, their values to the decision maker (utility metrics), and their probabilities so as to calculate the expected value or utility of each alternative (Clemen 1996). An early example of an environmental application of formal decision analysis is an analysis of further research vs. remedial alternatives for polychlorinated biphenyl (PCB)-contaminated Salt Creek, Indiana (Parkhurst 1984). Other methods are less quantitative, less focused on quantifying the expected consequences, and more focused on clarifying the decision for the decision maker or stakeholders. In general, they include more considerations than the other decision-making approaches discussed in this chapter, and, because they do not require monetizing the decision criteria, decision analyses can include noneconomic criteria (Stahl et al. 2002). Utilities may be scaled in standard categories (e.g., low, medium, or high) or in terms of an environmental goal (e.g., hectares of wetland or abundance of game fish).

Although decision analysis has a large literature and commercial software for its implementation, it is rarely used in environmental regulation or management. In part, this is because the explicit inclusion and quantification of considerations other than risk is likely to anger some stakeholders (Hattis and Goble 2003). Also, decision makers have little incentive to go through the trouble of formal decision analysis when they have been successful using precedents and their own judgment.

The US Army Corps of Engineers has begun to use multicriteria decision analysis to support decisions concerning the management of contaminated sediments (Linkov et al. 2006). In an example from the Cocheco River, New Hamshire, the alternatives for sediment disposal were cement manufacture, flowable fill, wetland restoration, and upland disposal cell. The multiple criteria were cost, environmental quality, ecological habitat, and human habitat. A mail survey of stakeholders provided weights for the criteria that could be combined to produce a multicriteria score for each alternative. This sort of decision analysis serves primarily to inform the decision maker about preferences of stakeholder groups based on a structured elicitation process rather than the usual stakeholder meetings.

36.8 MISCELLANEOUS AND AD HOC CONSIDERATIONS

Many, if not most, environmental management decisions are made without any explicit criteria or any formal analysis of how the alternatives relate to criteria. Decision makers consider the information concerning risks, benefits, and costs, they may consult with stakeholders, they pay particular attention to legal and regulatory constraints and precedents, then they test the political winds, consult their gut, and make a decision.

37 Integration of Human Health Risk Assessment

People are the problem, but any solution which does not serve people will fail.

Marty Matlock (unpublished presentation)

Inevitably, when both human health and the environment are threatened by a common hazard, human health concerns dominate assessment and decision-making processes. Ecological risk assessors can use the concern for human health to their advantage by using wildlife as sentinels for health effects, by integrating the ecological assessment with the health assessment so as to combine resources, and by showing how ecological effects influence human health and welfare.

37.1 WILDLIFE AS SENTINELS

Wildlife may serve as sentinels, thereby strengthening the case for current or future risks to humans (NRC 1991; Burkhart and Gardner 1997; Peter 1998; Sheffield et al. 1998; van der Schalie et al. 1999; Colborn and Thayer 2000; Fox 2001). An example of wildlife as sentinels is the observation of thyroid pathology in wildlife due to halogenated organic chemicals that led to studies in humans (Fox 2001; Karmaus 2001). However, most reports of the use of animal sentinels of health effects have not provided the types and levels of evidence needed before health decisions can be based on sentinel responses (Rabinowitz et al. 2005). Although there are difficulties in extrapolating from wildlife to humans (Stahl 1997), they are conceptually no more severe than those associated with extrapolating from laboratory rats to humans.

Wildlife species are likely to be effective sentinels if they have a common source and route of exposure with humans but are more exposed, more sensitive, or more readily monitored. They are, in general, likely to be more exposed and therefore likely to respond more quickly and severely than humans (Box 2.1). In general, the use of wildlife as sentinels may be justified by the following factors:

Common routes: Wildlife may feed on the same organisms as humans, particularly in the case of piscivorous wildlife and subsistence or recreational fishermen. Similarly, wildlife may consume contaminated soil, as do children.

Stenophagy: Wildlife species are generally less omnivorous than humans and therefore species that consume a contaminated food item are likely to be more exposed.

Local exposure: Wildlife species do not obtain food or water from outside their home range, so they are more exposed to contaminated locations.

Same mixture: Wildlife species are exposed to the same mixture of contaminants as humans living in, or consuming foods from, the same contaminated system.

Nonenvironmental sources: Wildlife species do not have the occupational or lifestyle exposures that confound epidemiological studies of humans.

Variance: Most human populations are genetically diverse due to immigration and vary in their diets, religious practices, occupations, use of recreational and medicinal drugs, etc. These sources of variance, which do not occur in wildlife, tend to hide effects of environmental exposures.

Availability: Wildlife may be readily sampled, analyzed, and necropsied.

While wildlife are more commonly sentinels for humans, in some cases humans may be sentinels for wildlife. Birth, death, and disease records, which may be used to detect environmental effects in humans, are not available for wildlife. In addition, in some cases, humans may be exposed and affected where equivalent wildlife are rare or absent (Fox 2001). As a result, a field termed conservation medicine or conservation health has developed to jointly study effects on human and nonhuman organisms of environmental pollutants and pathogens (Weinhold 2003). This approach may result in the development of integrated sentinels. For example, harbor seals have been proposed as sentinels for other marine mammals and humans (Ross 2000).

A database of studies from the biomedical literature of the use of animal surrogates can be found at http://www.canarydatabase.org.

37.2   INTEGRATED ANALYSIS OF HUMAN AND ECOLOGICAL RISKS

For practical reasons, the methodologies for human health and ecological risk assessment were developed independently. However, for several reasons, the need for a more integrated practice of risk assessment has become clear. These issues led the World Health Organization to develop an integrated framework for health and ecological risk assessment (Section 3.2.1) (WHO 2001; Suter et al. 2003).

37.2.1   COHERENT EXPRESSION OF ASSESSMENT RESULTS

Decision makers must make a single decision with respect to an environmental hazard that is beneficial to both human health and the environment, but this goal is impeded by the presentation of incoherent results of health and ecological risk assessments. The results of independent health and ecological risk assessments may be inconsistent and the bases for the inconsistency may be unclear because the results of the health and ecological risk assessments are based on different spatial and temporal scales, different degrees of conservatism, or different assumptions, such as assumed parameter values or assumed land use scenarios. As a result, decision makers may find it difficult to decide whether, for example, the reported risks to humans are sufficient to justify taking a remedial action that will destroy an ecosystem. As another example, consider a decision to license a new pesticide that poses an increased risk to humans and a decreased risk to aquatic communities relative to a current pesticide. If the ecological risk estimates are based on expected effects on a spatially distributed community while the health risks are based on provision of a margin of safety on an effect level for a hypothetical maximally exposed individual, the two estimates of risk cannot be compared. Finally, if variance and uncertainty are not estimated and expressed equivalently for health and ecological risks, a decision maker cannot determine the relative need for additional research to support future assessments. For example, variance in aqueous dilution should be either included or excluded in both assessments, and, if it is included, the same estimates should be used. Integration of health and ecological assessments can avoid these impediments to defensible decisions.

An integrated comparative risk assessment of contaminated sediment management illustrates this issue (Driscoll et al. 2002). The assessment used a common set of alternatives and assumptions in a common health and ecological risk assessment that highlighted common results (the no-action alternative had the highest risk for both endpoints) and differences (island disposal had the highest ecological risk of the remedial alternatives but had relatively low health risk).

37.2.2 INTERDEPENDENCE

Ecological and human health risks are interdependent (Lubchenco 1998; Wilson 1998b). Humans depend on nature for food, water purification, hydrologic regulation, and other products and services, which are diminished by the effects of toxic chemicals or other disturbances. In addition, ecological injuries may result in increased human exposures to contaminants or other stressors. For example, addition of nutrients to aquatic ecosystems and the resulting changes in algal community structure may influence the occurrence of waterborne diseases such as cholera as well as toxic algae such as red tides. The need to assess ecological risks in order to estimate indirect effects on human health is particularly apparent in the assessment of climate change (Bernard and Ebi 2001).

37.2.3 QUALITY

The scientific quality of assessments is improved through sharing of information and techniques between assessment scientists in different fields. For example, in assessments of contaminated sites, human health assessors may use default uptake factors to estimate plant uptake, unaware that ecological assessors are measuring contaminant concentrations in plants from the site. The data sets available for the safety evaluation of chemicals in human food and drinking water are relatively large and are used to support intensive assessments. In contrast, ecological risk assessments for chemicals have relatively small data sets and few resources to perform assessments, even though the receptors include thousands of species such as plants, invertebrates, and vertebrates. Integration of efforts may help to alleviate these imbalances in quality.

37.2.4 EFFICIENCY

Integration of human health and ecological risk assessments offers significant increases in efficiency. In fact, isolated assessments are inherently incomplete when both humans and ecological systems are potentially at risk. For example, the processes of contaminant release, transport, and transformation are common to all receptors. Although only humans shower in water and only aquatic organisms respire water, the processes that introduce the contaminants to water, degrade or transform them, and partition them among phases are common to both. Therefore, there are clear advantages in an integrated exposure model. The development of risk assessment methods, which takes into account insights from both human and ecological risk assessments, will lead to improvements that can benefit both disciplines.

Integrated analysis of toxic risks will be facilitated by the increasingly mechanistic character of toxicology. Because the structure and function of vertebrate cells are highly conserved, the mechanism of action of a chemical is likely to be the same in all vertebrate species and even the effective concentration at the site of action is likely to be effectively constant (Section 23.1.6) (Escher and Hermens 2002). Hence, when toxicology is sufficiently mechanistic, the need for toxicity testing should decline, and a common effects analysis approach should serve for both humans and other vertebrates. However, this approach would increase the demand for toxicokinetic modeling to estimate site of action concentrations.

37.3   ENVIRONMENTAL CONDITION AND HUMAN WELFARE

One view of risk assessment and management is that we protect human health as well as the nonhuman environment, and the only connection between them is through environmentally mediated or transmitted threats to human health (Section 37.2.2). An alternative view is that we must consider not just human health but also human welfare, and that human welfare is influenced by environmental conditions.

Environmental quality influences human welfare through the provision of ecosystem services. Humans obtain a wide range of benefits from the environment, often termed services of nature, which make human life possible and enhance its quality (Daily et al. 1997, 2002). At the most basic level, plants capture the energy of the sun, produce food for humans and their livestock, and convert our CO_2 into O_2. Other life support functions of ecosystems include purification of water, air, and soil, and cycling of water and nutrients. In addition, ecosystems produce a variety of goods, such as timber, fisheries, biomass fuels, and medicinal herbs, roots, and spices. These services are estimated to be worth more than the entire monetary economy (Costanza et al. 1997; Naeem et al. 1999). The concept of restoring services is already contained in the Natural Resource Damage Assessment provisions of some US laws (Section 1.3.9).

A more subtle aspect of this issue is improvement in the quality of human life provided by nature (Keach 1998). An obvious example is the regenerative effect of a vacation spent fishing, hunting, bird-watching, hiking, photographing nature, or simply visiting natural places. However, day-to-day contact with nature such as walking in parks, feeding birds, and observing trees, flowers, and butterflies may be more important to the quality of life. The importance of this relationship to people's quality of life is reflected in various behavioral and economic measures including the billions spent on feeding birds and the value of houses on large lots in the outer suburbs or on bodies of water. The inverse of this relationship is the consternation people feel when they observe trees dying along highways, dead fish along a river, or even images of oiled birds on the television.

Finally, cultural values are often related to aspects of the environment. This is particularly true of indigenous peoples whose cultures require the use of certain natural resources and the presence of certain environmental features (Harris and Harper 2000). The same is true of nonindigenous cultures, although in less obvious ways. Examples from the United States include the bald eagle as a national emblem and unspoiled wide-open spaces as a backdrop for the national mythos of cowboys and hearty pioneers. Forests have played a similar role in German culture (Schama 1995).

It is important to distinguish risks to nonhuman organisms, populations, and communities from risks to human welfare through loss of ecosystem services. Most environmental legislation protects ecological entities irrespective of any requirement to demonstrate benefits to human welfare (EPA 2003c). However, when they can be demonstrated, risks to human welfare could significantly supplement ecological and health risks as justifications for environmental protection. At least, they provide a basis for estimating economic benefits of protective actions (Section 38.2).

37.4   SUMMARY

To become more influential in environmental decision making, ecological risk assessors must collaborate with, and supplement, health risk assessors. The performance of risk assessments for reductions in human welfare and indirect effects on human health is a large task that cannot be readily performed by ecological risk assessors. Health risk assessors must be encouraged to look beyond direct toxic effects to a broader view of risks to humans. Integration will require movement from both directions.

38 Integration of Risk, Law, Ethics, Economics, and Preferences

To derive conclusions about action, we need some statements about how the world works and some statements about what we believe are good and right.

Randall (2006)

Risks are never the sole basis for decision making (Chapter 36). Risks are more acceptable when there are countervailing benefits, and actions to reduce risks are more acceptable if there is a clear legal mandate, legal precedent, or public support. Increasingly, formal analyses of costs and benefits are required. Therefore, integration of the risk assessment with the economic analysis should be planned during the problem formulation. However, it is also important to be aware of the limitations of economic decision criteria and of the existence of other decision criteria. Environmental law, environmental economics, and environmental ethics are large and complex fields in their own rights, which are barely touched on here. This chapter is intended to simply create an awareness among ecological risk assessors of these other considerations so that they have an idea of how their risk estimates must be combined with these considerations during the decision-making process.

38.1 ECOLOGICAL RISK AND LAW

Risk assessment may contribute to both criminal and civil law. It may be used to demonstrate compliance with environmental laws or failure to comply. It may also be a tool to establish injuries in civil legal actions. The language of the law lends itself to concepts of risk: "more likely than not," "the preponderance of evidence," or "beyond a reasonable doubt." When laws provide clear and specific legal mandates, legal compliance is a sufficient justification for action. For example, the US Endangered Species Act provides clear legal protection for species that are listed as threatened or endangered and for their critical habitat. Efforts to restore the bald eagle and peregrine falcon were not subject to cost–benefit analyses or public surveys. Other laws such as the US Clean Water Act do not include economic considerations in its mandate to restore the "physical, chemical, and biological integrity of the Nation's waters," but the interpretation of these vague terms leaves room for balancing of interests in defining criteria and standards. However, once legal standards are established, risk assessment may simply estimate the probability that the standard will be exceeded. Other phrases in laws such as "reasonably achievable" create a mandate for balancing the environment against economic costs. Hence, the legal context of an assessment determines the degree of confidence required for an action and the extent to which risks must be balanced against other considerations.

38.2 ECOLOGICAL RISK AND ECONOMICS

In the United States and many other nations, efforts to improve environmental quality and protect nonhuman populations and ecosystems are increasingly subject to cost–benefit tests. This practice is based on welfare economics, which is based on the premise that when resources are limited, the general welfare is best achieved if individuals are free to maximize their individual welfare (also termed utility) through free markets for goods and services. In an ideal market, the decisions of rational individuals (including corporations as legal individuals) will lead to an efficient outcome. This invisible hand of the market often fails, particularly with respect to the environment. The main reason for these market failures is the absence of a market for environmental goods such as clean air or wild songbirds. Similarly, the services of nature such as water purification and soil formation are performed without anyone receiving a utility bill from nature. Polluters may destroy these goods and services without buying the right in an efficient market, replacing the lost goods and services, or compensating the users. Finally, for commonly held resources such as fisheries, there is an incentive for overexploitation. The economic gain goes to the irresponsible individual, but the resource loss is shared by all, leading to "the tragedy of the commons" (Hardin 1968). Regulation is needed to compensate for these market failures. To ensure that regulation is not excessive, welfare economists devised the cost–benefit analysis, which creates a pseudo market. Costs of regulation are assumed to be justified if they buy an equivalent benefit of environmental goods and services.

An obvious difficulty in this concept, even for welfare economists, is that environmental benefits are difficult to define and enumerate, much less quantify in monetary terms. Alternative approaches to estimating monetary benefits of environmental protection are listed in Table 38.1. All have severe limitations. The revealed preference methods require that the value of a resource be quantified in terms of some monetary expenditure to use the resource, such as using the cost to recreationists of visiting an ecosystem as an estimate of the value of the ecosystem. Clearly, such methods address a small fraction of the value of nature. Stated preference methods, particularly contingent valuation, are more commonly used to value the environment, because they use surveys to create an entirely hypothetical market. In particular, contingent valuation is used to estimate the monetary damages that polluters must pay to replace lost ecosystem services under Natural Resource Damage Assessment (Kopp and Smith 1993). These survey-based methods may be applied to any use or nonuse value, but they suffer from a number of problems including:

- The public has little understanding of most environmental resources and services.
- Even if they understand the resource or service, they may not have relevant well-defined values to be elicited by a survey.
- Even if respondents have well-defined values, the values may not encompass the full utility of the resource or service.
- Any attempt to educate the survey participants is likely to bias their responses.
- The respondents have no experience in pricing or paying for valued ecological resources or services.
- It is not known whether respondents would actually pay the stated amount if a market could be created.
- The number of resources or services that can be addressed is limited by the patience of the survey participants.
- People may not be willing to pay anything, because they believe the responsible party should pay.
- People may opt out because they object to putting a price on nature.

TABLE 38.1
Methods for Estimating Monetary Values of Environmental Goods and Services

Method	Description	Examples
Revealed preference methods (can estimate use values only)		
Market	When environmental goods are traded in markets, their value can be estimated from transactions	The benefits of an oil spill cleanup that would result in restoration of a commercial fishery can be projected from changes in markets for fish, before and after the spill, and their effects on fishermen and consumers
Production function	The value of an environmental good or service can be estimated when it is needed to produce a market good	If an improvement in air quality would lead to healthier crops, the value of the improvement includes, e.g., the reduction in fertilizer costs to produce the same amount of agricultural crops
Hedonic price method	The value of environmental characteristics can be indirectly estimated from the market, when market goods are affected by the characteristics	If an improvement in air quality improves a regional housing market, its value includes increases in housing value, which can be measured by statistically estimating the relationship between house prices and air quality
Travel cost method	The value of recreational sites can be estimated by examining travel costs and time	The value of a recreational fishing site to those who use it can be estimated by surveying visitors, to determine the relationship between the number of visits and the costs of time and travel
Stated preference methods (can estimate both use and nonuse values)		
Contingent valuation method	Individuals are surveyed regarding their willingness to pay for a specifically described nonmarket good	In a telephone survey, respondents are directly asked their willingness to pay, via a hypothetical tax increase, for a project that would reduce runoff, improving the health of a particular stream
Conjoint analysis	Survey respondents evaluate alternative descriptions of goods as a function of their characteristics, so the characteristics can be valued	In a mail survey, hypothetical alternative recreational fishing sites are described by type of fish, expected catch rate, expected crowding and round-trip distance; respondents' preferences are used to calculate value for changes in each of the characteristics

Source: Bruins, R.J.F., Heberling, M.T., eds., *Integrating Ecological Risk Assessment and Economic Analysis in Watersheds: A Conceptual Approach and Three Case Studies*, EPA/600/R-03/140R, Environmental Protection Agency, Cincinnati, OH, 2004.

These problems can introduce significant biases as well as uncertainty. For example, the last two problems described in the list cause people who are strongly pro-environment to opt out, thereby biasing the sample.

If one of these cost–benefit techniques is applied to an environmental decision, it is incumbent on ecological risk assessors to support that analysis. If a revealed preference method is used, the assessment endpoints must be identified and quantified in terms of the good or service that is provided. For example, if the market value of fish is used by the economists, reductions in harvestable mass of fish should be estimated. If a stated preference method is used, endpoints must be defined that are potentially understood and valued by the public. This may require converting a primary effect such as death of ugly forest lepidopteran larvae into valued secondary effects such as reduced abundance of beautiful birds, moths, and butterflies.

To ecologists, assigning monetary values to the natural environment is likely to be somewhat repugnant, and the techniques employed often seem scientifically suspect. However, in a decision context that requires cost–benefit analysis, nonparticipation of ecological risk assessors is likely to result in minimizing or even ignoring environmental benefits other than improved human health. This requires a different approach to risk assessment and decision making than the purely legal approach. It is not enough to establish that no unacceptable effects will occur. Rather, one must be able to estimate the risks of specific effects and the benefits of avoiding or remediating those risks. When working in an economic decision context, it is important to engage the economists in the problem formulation process and to understand their needs and methods (Bruins and Heberling 2004). An extensive summary of environmental economics can be found in van den Bergh (1999), a review of economics for watershed ecological risk assessment is provided by Bruins and Heberling (2005), and relevant guidance from a US regulatory perspective can be found in National Center for Environmental Economics (2000) and Science Policy Council (2002).

Analysis of economic and other benefits of ecological risk reduction can lead to a broadening of the scope of assessments. For example, the restoration of riparian vegetation in agricultural areas is justified under the Clean Water Act in terms of improved water quality. However, riparian communities have other benefits such as provision of habitat for birds (Deschenes et al. 2003). While the problem formulation for a risk assessment focuses on the endpoints that relate to the legal mandate, the accounting of benefits to justify the cost of riparian restoration is not constrained in that way. Just as all of the costs of requiring waste treatment, remediation, or restoration are identified and summed, so too should all of the benefits be identified and summed, not just those that were endpoints for the assessment. In fact, it is hard to know where to stop. The protection of an ecosystem results in a nearly infinite list of benefits. All of the species have value, all of the functions have value, and every visible or audible detail of an ecosystem has at least some esthetic value. The fact that the costs of regulation are relatively easily and completely tabulated (e.g., the costs of building and operating a waste treatment plant), while the benefits to human health and the environment are incompletely identified and estimated, has led to the idea that what is really practiced is "complete cost–incomplete benefits analysis" (Ackerman 2003). In fact, the costs to industry are routinely overestimated, largely because regulation leads to development of lower cost treatment technologies or waste minimization and reuse (Ruth Ruttenberg and Associates 2004).

In the face of highly uncertain benefits, it may be desirable to abandon cost–benefit analysis in favor of an insurance-based approach. For example, if the benefits of a policy such as greenhouse gas control are highly uncertain, a risk management approach can justify the cost of some control to avoid the risk of catastrophic losses. This approach is equivalent to the precautionary principle, but is based on economic rather than ethical principles.

A failure of contingent valuation and the other techniques in Table 38.1 is that they treat people strictly as consumers. People are also citizens who contribute to the formation of laws as discussed in Section 38.1 (Sagoff 1988). A person as a purely self-interested economic entity may be willing to pay very little to protect streams in other states, but that same person as a citizen of a democracy may agree that the political process should protect streams nationally, leading to legal criteria and standards. In addition, welfare economics ignores the fact that people are ethical and act in ways that are rational but do not maximize their economic welfare.

38.3 ECOLOGICAL RISK AND ETHICS

Welfare economics is not the only intellectual system for supporting environmental protection. Formal ethics may be applied. The fundamental difference between considering the environment in an ethical manner rather than an economic manner, no matter how broad, is illustrated by Routley's last man argument (Schmidtz and Willott 2002). Consider that you are the last person left in the world and you have little time to live. If you decided that it would be fun to cut down the last surviving redwoods, would there be anything wrong with that? Resource value, services of nature, and even esthetics would be irrelevant. Yet, I hope that most readers would agree that the act would be wrong. What would make it wrong is some ethical principle. While the public has been responsive to appeals for environmental protection that are effectively ethical, ethics has had little influence as a formal decision support tool relative to law or economics. This is in part because there is no generally accepted system of ethics analogous to classic welfare economics.

Ethicists have no standard classification system, but ethical principles and systems generally fall in the following categories:

Motivist: The acceptability of an act depends on the intent of the individual. This is the basis of the distinction between murder and manslaughter and for the requirement of "intent to deceive" in financial fraud. It has little relevance to modern environmental ethics, but may have been important in traditional cultures that allowed use and even waste of resources if it was done with the proper intent as displayed by rituals and incantations (Krech 1999).

Consequentialist: The acceptability of an act depends on its effects. This is the basis for utilitarian ethics and its offspring, welfare economics. However, consequentialist and even utilitarian interests may extend well beyond economic welfare. People may be willing to give up economic resources out of sympathy for nonhuman organisms or in order to avoid repugnant experiences.

Deontological: The acceptability of an act depends on its nature. This concept of ethics is related to concepts of duty, obligation, and rights. It is associated with Kant, but its best-known expression in the environmental literature is Aldo Leopold's land ethic, which expresses a human obligation to protect the land, by which he meant ecosystems. Those who ascribe moral standing to nonhuman organisms, populations or ecosystems are taking a deontological stance.

Environmental ethics are well summarized in Blamey and Comon (1999), and the range of positions in environmental ethics is represented in Schmidtz and Willott (2002).

The implications for risk assessment of ethical standards are less clear than those for legal or economic standards. Deontological ethics would tend to promote protective standards rather than the balancing of interests suggested by consequentialist ethics. However, some consideration of consequences is nearly inevitable. For example, if we have a duty to both aquatic and terrestrial ecosystems, the decision to sacrifice a terrestrial ecosystem for sewage sludge disposal must be based on a determination that the consequences of not generating the sludge or of disposing of it in the ocean would be more severe or at least less acceptable. Also, since different versions of the concept of duty toward the environment are conflicting, consequentialist ethics may be required to decide between them. For example, animal rights advocates condemn Leopold's land ethic because it implies management of animals for the sake of ecosystems. Therefore, a decision to protect an ecosystem by harvesting herbivores that are not naturally controlled would be based on a judgment about the consequences of not harvesting.

38.4   ECOLOGICAL RISK, STAKEHOLDER PREFERENCES, AND PUBLIC OPINION

Traditions and public opinions inform politics, which in turn drives the legal process and influences how individual management decisions are made. In addition, stakeholders may be involved in decision making as well as participating in planning and problem formulation. This trend has been encouraged in recent publications of advisory panels on risk assessment (National Research Council 1994; The Presidential/Congressional Commission on Risk Assessment and Risk Management 1997). This advice is based on concerns that decisions that are not accepted by the affected parties will be politically unacceptable and may be blocked or delayed. While these processes are usually informal, decision analytic methods that utilize multiple utilities can make this process as rigorous as cost–benefit analysis without requiring the conversion of preferences to monetary units (Brauers 2003).

From an ecological perspective, increasing stakeholder influence is unfortunate in that it tends to accentuate human health concerns and diminish ecological issues. The responsible parties do not push for ecological protection, the members of the public who are sufficiently motivated to participate are primarily those with health concerns, and environmental advocacy groups tend to adopt the public's health concerns or are simply absent. The stakeholders who do care passionately about ecological issues tend to be those who use or harvest natural resources (e.g., loggers, ranchers, fishermen, farmers) and therefore are opposed to protection. Hence, in most stakeholder processes, nobody speaks for the trees. Ecological risk assessors who participate in stakeholder-informed decision processes should be prepared to make the case for the environment by clearly presenting the results of their assessments in a way that the stakeholders can relate to (Section 34.2). If a management decision is of sufficient importance, national surveys of public opinion may serve to balance the interests of stakeholders. An example is the current conflict in the United States over oil development in the Arctic National Wildlife Refuges. Such ecological issues of national importance, for which environmental advocacy organizations can mobilize public opinion, are rare. In routine assessments, risk managers who are public officials must represent the public's interests, based on law and policy.

38.5   CONCLUSIONS

Most of this book has been concerned with how to estimate risks in an accurate and unbiased manner, addressing the endpoints of concern at the right spatial and temporal scales, and presenting results clearly and appropriately. The critical final step is ensuring that the results are influential in the full context of the decision. This chapter has described other considerations that may influence the interpretation of ecological risks. Ecological risk assessors who simply drop their results on the risk manager's desk are almost guaranteed to be frustrated by how little their work is reflected in the decision. Ecological risk assessors must learn to work with lawyers, economists, policy analysts, and others who have the ear of risk managers. For most of us, introverted ecologists who prefer dealing with the complexities of nature than with the complexities of human emotions or institutions, this is more of a challenge than calculating integrals of exposure and effects distributions.

39 Monitoring the Results of Risk Management

Many impacts cannot be foreseen and planning must therefore provide for monitoring and adaptation.

Holling (1978)

Results of ecological risk assessments are uncertain and regulatory and remedial actions may have unexpected effects. Therefore, risk-based decisions are not guaranteed to have the desired outcome. Environmental monitoring can reveal the actual outcome and guide further assessments, decisions, and actions. Unfortunately, the effects of environmental management actions on ecosystems are seldom monitored. Rather, it is typically assumed that the problem is resolved, and assessors and managers move on to the next problem. This is ostensibly a reasonable approach. Managers choose actions that they believe will be efficacious, and it seems reasonable to assume, at least, that unassessed and unremediated sites are worse. However, the efficacy of remediation and restoration techniques is often uncertain, they may be poorly carried out, and, in any case, the law of unintended consequences applies.

A fundamental question to be answered by monitoring is whether exposure and effects have been reduced by a remedial action? While it might seem self-evident that treatment of effluents or removal of contaminated media would reduce exposures, studies do not always bear that out. In particular, dredging of polychlorinated biphenyl (PCB)-contaminated sediments has not, in some cases, reduced exposure as measured by accumulation in bivalve molluscs and fish (Rice and White 1987; Voie et al. 2002). In effect, the mass of material decreased but availability to the biota did not. Further, remediation may reduce exposure but not eliminate toxicity. Following dredging and capping of the Lauritzen Channel of San Francisco Bay, California, concentrations of chlorinated pesticides declined in sediment and decreased in transplanted mussels at one site but increased at another (Anderson et al. 2000). One year after remediation, the benthic invertebrate community contained few species and individuals, and the sediment was more toxic to amphipods than prior to remediation.

The registration of new pesticides and new industrial chemicals lends itself to post-management monitoring and assessment. This practice is particularly common for pesticides. In the United States surveillance studies are sometimes required of the manufacturer. Because applications are replicate treatments, such studies are relatively easy to design. In fact, the availability of this option was a reason that requirements for preregistration field testing of pesticides have been largely eliminated in the United States (Tuart and Maciorowski 1997).

Because effluent permitting based on concentrations of individual chemicals does not account for combined toxic effects, effluent toxicity testing (Section 24.4) was developed as a means of monitoring the success of effluent permits. However, because effluent tests are periodic, they are likely to miss episodes of high toxicity due to treatment failures, temporary

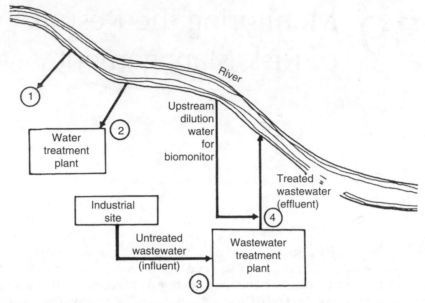

1. Surface water biomonitor or upstream biomonitor for a water treatment plant
2. Water treatment plant biomonitor
3. Influent wastewater biomonitor
4. Effluent wastewater biomonitor

FIGURE 39.1 Diagram of a system to biologically monitor the toxicity of waste water. (From van der Schalie, W.H., Gardner, H.S., Bantle, J.A., De Rosa, C.T., Finch, R.A., Reif, J.S., Reuter, R.H., et al., *Environ. Health Persp.*, 107, 309–315, 1989. With permission.)

changes in a process, or other incidents. Continuous biological monitoring of treated effluents has been proposed, but has not yet been adopted (Figure 39.1).

Effluent permitting may also fail to protect the environment because of the combined effects of multiple effluents and of nonpoint sources. In the United States, biological monitoring and bioassessment programs by the states and tribes are intended to detect those combined effects and provide a basis for regulating total pollutant loading, i.e., total maximum daily loads (TMDL findings) (Houck 2002). The primary limitation on this ecoepidemiological approach (Chapter 4) is the high cost of performing adequate monitoring at enough points at sufficient frequencies for the many water bodies of a state or nation.

The results of removing an exotic species should be monitored, because removal does not necessarily resolve ecological problems. For example, removal of livestock from Santa Cruz Island, California, and from one of the Mariana Islands resulted, in each case, in explosive growth of a formerly uncommon exotic weed and suppression of native plants (Simberloff 2003). Hence, monitoring is necessary to assure that ecosystem management goals are achieved as well as to confirm that the species has been removed.

Similarly, ecological restoration projects are often physically but not ecologically successful. In particular, wetland restoration projects often succeed in creating a wetland but fail to establish diverse wetland communities that support target species such as clapper rails or that adequately perform wetland functions such as nutrient retention.

When designing post-management monitoring programs, it is important to carefully consider the goals. The first goal will be to determine whether the endpoint attributes that prompted the management action were restored. For example, the assessment of the Lauritzen Channel remediation monitored changes in contamination of sediment and mussels, in toxicity to invertebrates, and in invertebrate community structure, but could not answer the question of efficacy because the endpoint assemblage, benthic fish, was not monitored (Anderson et al. 2000). The second common goal is to determine the cause of any management failure, which requires that the contaminant levels and other intermediate causal parameters be monitored along with the endpoints. This can be difficult, because failure often occurs due to some process that assessors did not consider. For example, the Lauritzen Channel assessment did not include possible sources of sediment contamination other than the United Heckathorn Superfund site (Anderson et al. 2000). Therefore, it is important to develop conceptual models of alternative candidate causes and identify variables that distinguish them and could be monitored.

Media toxicity tests (Section 24.5) can play an important role in monitoring to determine efficacy. By testing treated or remediated media, it is possible to distinguish ongoing effects that are due to residual toxicity from effects of other causes. Such tests can even precede remediation, if treated media are available from pilot-scale studies. For example, solvent extraction of soils contaminated primarily with PCBs removed 99% of the PCBs, but the toxicity of the soil to earthworms and plants was unchanged or even increased, depending on the species and test endpoint (Meier et al. 1997).

If monitoring reveals that the ecological system is not recovering or is still impaired after sufficient time has been allowed for recovery, an ecoepidemiological study should be conducted to determine the cause (Chapter 4). Residual impairment may be due to failure of the remedial action to sufficiently reduce exposure to the agent of concern or to the effects of other hazardous agents. This analysis requires that the residual impairment be clearly defined, that plausible candidate causes be listed, and that analyses be performed based on spatial and temporal associations of candidate causes with the impairment, experimental results, and mechanistic understanding. Successful causal analysis requires that the candidate causes and the residual impairment be monitored concurrently at the impaired site and at reference sites so that spatial and temporal associations can be determined. Once the most likely cause has been determined, an ecological risk assessment of the new remedial alternatives can inform a decision concerning appropriate management actions.

Another goal of monitoring management results is to improve understanding of what assessment practices are successful and to determine how and why other assessment practices fail, so that the practice of ecological risk assessment can advance. It is also valuable to determine whether failures tend to be overprotective or underprotective. There is evidence for both. For example, studies of toxic effects on striped bass embryos contaminated from maternal PCB burdens led the EPA Region II (2000) to conclude that effects on striped bass populations were likely in the Hudson River. However, a careful analysis of long-term striped bass population data showed no effect of PCBs on year-class strength, apparently due to density-dependent compensatory processes (Barnthouse et al. 2003). In contrast, continuous monitoring of Prince William Sound has shown long-term ecological effects that were not predicted by assessments performed shortly after the Exxon Valdez spill (Peterson et al. 2003). In particular, short-term toxicity studies of the water-soluble fraction of the oil—thought to be the toxicologically active fraction—resulted in predictions of no significant risk to fish. However, exposures of salmon eggs to persistent 3- to 5-ring hydrocarbons in field sediments and in the laboratory caused increased mortality for years after the spill. Hence, limitations in the state of the science and the need to employ simplifying assumptions in ecological risk assessments have caused errors in both directions.

In sum, monitoring of the results of management actions is essential to ensuring that the goals of those actions are achieved. In addition, real progress in ecological risk assessment depends on the feedback provided by well-designed and conducted monitoring programs. Neglected areas of environmental chemistry and toxicology such as the toxicology of reptiles will remain neglected until they are shown to be important in real-world decisions. Similarly, relatively neglected assessment techniques such as population and ecosystem simulation modeling are likely to remain neglected until it is demonstrated that management decisions have failed and would have succeeded if those techniques had been employed.

Part VII

The Future of Ecological Risk Assessment

Pursuing the whole spectrum of scientific activities essential for conservation requires taking creative advantage of all potential resources.

Gell-Mann (1994)

Since the publication of the first edition of this text, ecological risk assessment has become established as the dominant approach to informing decisions concerning the management of chemicals in the nonhuman environment. Now pressures from scientific advances and from changes in policy and public expectations are pushing the practice in different directions. The following are prognostications based on those pressures and a little wishful thinking.

Both the advance of science and the pressure of policy are pushing ecological risk assessment to be clearer and more specific in its predictions. Division of an LC_{50} by a factor of 1000 to estimate a threshold for unspecified ecological effects is justifiable if the science does not support anything more. However, the sciences that support ecological risk assessment now provide the bases for more sophisticated methods. Perhaps more important, public policymakers increasingly desire assurance that regulations or other management actions are justified. In particular, cost–benefit criteria are increasingly applied to environmental regulations. Similarly, demands for stakeholder involvement in decision making result in increasing requirements that risk assessors explain what is at risk and what is likely to happen under alternative actions. Hence, both capabilities and demands are increasing and assessors must respond.

If ecological risk assessors must apply more sophisticated tools and a wider range of data to estimate specific risks, those tools and data must be made more available and usable. Otherwise, ecological risk assessment will fail to meet the expectations of decision makers and stakeholders. This will require the development of better means of placing information and tools in the hands of assessors in forms that allow easy implementation. Publication in journals does not meet that need, and texts such as this one are a little more useful but can only skim the surface. Assessors need better access to information, better access to models to analyze the information, and help with organizing assessments and making proper inferences.

Information: Databases of secondary data, such as the EPA's ECOTOX, are useful but are difficult to sustain. Better platforms for sharing primary data are needed.

Models and modeling tools: As discussed in Chapter 28, both standard ecosystem models and easy-to-use systems for simulation modeling have become available. However, they still require more expertise than is possessed by most ecological risk assessors.

Assessment support systems: Computer-based systems like CADDIS (http:/cfpub.epa.gov/caddis/) should be the future of ecological risk assessment practice. CADDIS combines a framework and methodology of determining the most likely cause of ecological impairment, with worksheets, case studies, useful information, links to other useful information, and quantitative tools.

Uncertainty is poorly handled in ecological risk assessment, and one can hope that this will change in the next decade. Currently, the state of practice is to list sources of uncertainty without even ranking them or estimating their approximate magnitudes. Many techniques are available, but guidance is needed for their application.

The ongoing revolution in biology based on genomics, proteomics, and metabolomics will inevitably transform toxicology. In the coming years, organisms will be treated as systems responding to toxicological challenges rather than as black boxes. The resulting computational toxicology will facilitate extrapolation among species, life stages, and exposure regimes as well as allowing greatly enhanced prediction of effects of new chemicals.

In contrast, ecological risk assessment will become more attentive to the actual effects occurring in the field and more adaptively respond to those results. The increasing use of biological surveys by the US Environmental Protection Agency (US EPA) and other environmental regulatory agencies amounts to a tacit recognition that not all effects are predictable. This unpredictability results from both the complexity of ecological responses (e.g., Box 34.1) and the importance of unregulated agents such as agricultural runoff that affect ecosystems but are not subject to risk assessments. Increasingly in the coming years, predictive risk assessment and ecoepidemiological assessments based on field surveys must be linked by a common analysis of direct and indirect causal relationships.

Glossary[1,2]

abduction—*Inference* to the best explanation. An alternative logic to *deduction* and *induction*.

accuracy—Closeness of a measured or computed value to its true value.

acute—Occurring within a short period of time relative to the life span of an organism (conventionally <10%). Acute is also used to refer to severe effects, usually death, but that usage causes confusion.

adaptive management—The use of management actions as experimental treatments to test management *models* and thereby provide a better basis for future management actions.

advocacy science—Scientific studies performed for the purpose of supporting a particular position or sponsor.

aged chemical—A chemical that has resided in contaminated soil or sediment for a long period (e.g., years). Generally it is less bioavailable than a chemical freshly added to soil. Also termed a weathered chemical.

agent—Any physical, chemical, or biological entity or process that can potentially cause a response. It is synonymous with *stressor* but more general, because it includes nutrients, water flow, and other agents that may be beneficial or neutral rather than stressful. A synonym sometimes used in US Forest Service documents is affector.

ambient media toxicity test—A toxicity test conducted with environmental media (soil, sediment, water) from a contaminated *site*. Usually the media contain multiple chemicals.

analysis of effects—A phase in an *ecological risk assessment* in which the relationship between *exposure* to contaminants and effects on endpoint entities and properties and associated uncertainties are estimated.

analysis of exposure—A phase in an *ecological risk assessment* in which the spatial and temporal distribution of the intensity of the contact of endpoint entities with contaminants and associated uncertainties are estimated.

analysis plan—A plan for performing a risk assessment, including the data to be collected and the modeling and other analyses to be performed in order to provide the needed input to the environmental management decision.

antagonism—The process by which two or more chemicals cause joint effects that are less than additive (either exposure-additive or response-additive).

application niche—The range of conditions to which a *model* may be defensibly applied.

assessment endpoint—An explicit expression of the environmental value to be protected. An assessment endpoint must include an entity and specific attribute of that entity.

assessor—An individual engaged in the performance of a risk assessment or other assessment.

asymptotic LC_{50}—The minimum *median lethal concentration*, which occurs when *exposure* is extended until no more organisms die of toxicity. It is associated with the equilibrium *receptor* concentration in reversibly binding chemicals.

background concentration—The concentration of a substance in an environmental medium that is not contaminated by the *sources* being assessed or any other local *sources*. Background concentrations are due to natural occurrence or regional contamination.

baseline assessment—A risk assessment that determines the risks associated with current conditions so as to determine whether *remediation* is required.

---

[1]Some of these definitions are taken directly from, or are modified from, the glossary in EPA (1998).

[2]Terms within a definition that are also defined in this glossary are in italic font.

Bayesian—A branch of statistics characterized by the updating of prior knowledge and estimation of conditional probabilities using Bayes' theorem and by the treatment of probabilities as subjective degrees of belief.

benchmark dose (BMD)—A *dose* of a substance associated with a specified low level of an effect (usually 10%). The term is used in the US EPA's human health risk assessments and sometimes risk assessments for mammalian *wildlife*, equivalent to an ECp or ICp.

benchmark dose limit (BMDL)—A lower confidence limit on a *benchmark dose*.

bias—A systematic deviation of measured or computed values from true values.

bioassay—A procedure in which measures of biological responses are used to estimate the concentration or to determine the presence of some chemical or material. See *toxicity test*.

bioaccumulation—The net accumulation of a substance by an organism due to *uptake* from all environmental media.

bioaccumulation factor—The quotient of the concentration of element or compound in an organism divided by the concentration in an environmental medium, when the concentrations are near steady state, and when multiple *uptake* routes may contribute.

bioavailability—The extent to which a form of a chemical is susceptible to being taken up by an organism. A chemical is said to be bioavailable if it is in a form that is readily taken up (e.g., dissolved) rather than a less available form (e.g., sorbed to solids or to dissolved organic matter).

bioconcentration—The net accumulation of a substance by an organism due to *uptake* directly from aqueous solution.

bioconcentration factor—The quotient of the concentration of element or compound in an organism divided by the concentration in water, when the concentrations are near steady state and when only direct *uptake* from solution contributes.

bioindicator—A species or group of species that, by their presence or abundance, are indicative of a property of the *ecosystem* in which they are found. Enchytraid worms are bioindicators of low dissolved oxygen.

biomagnification—The increase in concentration of a chemical in a consumer species (or set of trophically similar species) relative to concentration in food species in a food web.

biomagnification factor—The ratio of the concentration of a chemical in organisms at a particular trophic level to the concentration at the next lower level. The factor may be defined for a particular consumer species and its food species or may be averaged across species at defined trophic levels.

biomarker—A measurable change in a biochemical, cellular, or physiological characteristic that may be used as a *measure of exposure* or effect.

biosurvey—A process of counting or measuring some property of biological *populations* or communities in the field. An abbreviation of biological survey.

biota/sediment accumulation factor—The ratio of the concentration of a chemical in a benthic organism to the concentration in sediment.

canopy cover—A measure of the degree to which the surface is covered by aboveground vegetation. It is related to the interception of solar radiation.

carbon mineralization—The process of conversion of the carbon in organic compounds to the inorganic state (usually carbon dioxide).

cation exchange capacity—A measure of the capacity of clay and organic colloids to remove positive ions from soil solution.

chlorosis—An abnormally yellow color of plant tissues resulting from partial failure to develop chlorophyll.

chronic—Occurring after a long period of time relative to the life span of an organism or effectively infinite in duration relative to the response rate of the exposed system. Chronic is also used to refer to nonlethal effects or effects on early life stages, but that usage causes confusion.

cleanup criterion—A concentration of a chemical in an environmental medium or other goal that is determined to be sufficiently protective of human health and ecological *assessment endpoints*.

community—A biotic community consists of all plants, animals, and microbes occupying the same area at the same time. However, the term is also commonly used to refer to a subset of the community such as the fish community or the benthic macroinvertebrate community. The latter is more properly termed an assemblage.

comparative risk assessment—Risk assessment used to rank or otherwise compare alternative actions to address a particular risk or to prioritize risks for remedial or regulatory action.

compensation—In *population* ecology, compensation is the increase in growth of a *population* at low densities due to decreased mortality, more rapid growth and maturation, and increased fecundity. In *ecosystem* ecology, compensation is the increased rate of performance of a process by one or more species as the abundance or activity of other species decline. For example, increased growth and mass production by chestnut oak compensated for the loss of American chestnut trees in southern Appalachian forests.

concentration additivity—A mode of combined toxicity in which each chemical behaves as a concentration or dilution of the other, based on their relative toxicities.

conceptual model—A representation of the hypothesized causal relationship between the *source* of a pollutant or other *agent* and the response of the endpoint entities. It typically includes a diagram and explanatory text.

confounding—A situation in which the effects of multiple *agents* or processes cannot be separated. In ecological field studies, an apparently causal relationship between an *agent* and an effect may be confounded by an unrecognized *agent* that is spatially or temporally correlated with the *agent* being studied.

contaminant—A substance that is present in the environment due to *release* from an anthropogenic *source* and is believed to be potentially harmful.

corrective action goal—A concentration of a chemical in an environmental medium or other goal that is determined to be protective of human health and ecological *assessment endpoints* (cleanup criterion).

cost–benefit analysis—Method for balancing the costs and benefits associated with an action or technology.

credibility—The estimated *probability* of a unique event given the *variability* of the system and the *assessor*'s *uncertainty*. The credibilities of a series of events should equal their frequencies in the long term.

cumulative distribution function (CDF)—A function expressing the *probability* that a random variable is less than, or equal to, a certain value. A CDF is obtained by integrating a *probability density function* (PDF) for a continuous random variable or summing the PDF for a discrete random variable.

deduction—*Inference* from a theorem or set of axioms to a particular conclusion. For example, if *bioconcentration factor* (BCF) $= 0.89 \log K_{ow} + 0.61$, then one may deduce that for a chemical with K_{ow} of 10, the BCF is 1.5. Deductive arguments are valid if the conclusions are always true when the premises are true. See *abduction, induction*.

definitive assessment—An assessment that is intended to support a remedial decision by estimating the *likelihood* of endpoint effects and risks, and to provide the basis for management decisions. See *scoping assessment* and *screening assessment*.

de manifestis—Sufficiently large to be obviously significant (i.e., risks so severe that actions are nearly always taken to prevent or remediate them).

de minimis—Sufficiently small to be ignored (i.e., risks low enough not to require actions to prevent or remediate them).

depensation—Depensation is the accelerated decline in a *population* at low densities due to reduced ability to find mates, increased predation, or decreased ability to condition the environment. It is the opposite of *compensation*.

detection limit—The concentration of a chemical in a medium that can be reliably detected by an analytical method. It is defined statistically (e.g., as the concentration that has a prescribed *probability* of being greater than zero, given *variability* in the analytical method).

deterministic—Having only one possible outcome.

direct effect—An effect resulting from an *agent* acting on the *assessment endpoint* or other ecological component of interest itself, not through effects on other components of the *ecosystem*. Synonymous with primary effect. See also *indirect effect* and *secondary effect*.

dose—The amount of a chemical, chemical mixture, pathogen, or radiation delivered to an organism. For example, mg of Cd per kg of mallard duck (mg/kg) administered by oral gavage.

dose additivity—A mode of combined toxicity in which each chemical behaves as a concentration or dilution of the other, based on their relative toxicities.

dose rate—The *dose* per unit time (e.g., mg/kg/d).

dredge spoil—Sediments dredged from a water body and deposited as waste to land or another aquatic location.

ecoepidemiology—The analysis of the causes and consequences of observed effects on ecological entities in the environment.

ecological entity—An *ecosystem*, functional group, *community*, *population*, or type of organism that may be exposed to a hazardous *agent* or may itself be a hazardous *agent*.

ecological risk assessment—A process that evaluates the *likelihood* that adverse ecological effects may occur or are occurring as a result of *exposure* to one or more *agents*.

ecosystem—The functional system consisting of the biotic community and abiotic environment occupying a specified location in space and time.

effects range-low for sediments—The lower 10th percentile of effects concentrations in coastal marine and estuarine environments (NOAA).

effects range-median for sediments—The median effects concentrations in coastal marine and estuarine environments (NOAA).

efficacy assessment—Analysis of the effectiveness of remedial actions.

empirical model—A mathematical *model* that is derived by fitting a function to data using statistical techniques or judgment. Purely empirical models summarize relationships in data sets and have no mechanistic interpretation.

endpoint entity—An organism, *population*, species, community, or *ecosystem* that has been chosen for protection. The endpoint entity is one component of the definition of an *assessment endpoint*.

environmental risk—A risk to humans or other entities due to hazardous *agents* in the environment. This definition applies to the United States, United Kingdom, and some other nations. However, some nations use environmental risk equivalently to ecological risk, as defined here.

equilibrium partitioning—The transfer of chemical among environmental media so that the relative concentrations of any two media are constant.

evidence—A summarization of data in the light of a hypothesis (a *model*).

excess risk—The difference between the risk given an *exposure* and the risk without the *exposure* or with an alternative *exposure*.

exotic species—A biological species that has been introduced from elsewhere, including species produced by biological engineering, selective breeding, or natural selection.

exposure—The contact or co-occurrence of a contaminant or other *agent* with a biological *receptor*.

exposure pathway—The physical route by which a contaminant moves from a *source* to a biological *receptor*. A pathway may involve exchange among multiple media and may include transformation of the contaminant.

exposure profile—The product of characterization of *exposure* in the analysis phase of *ecological risk assessment*. The *exposure profile* summarizes the magnitude and spatial and temporal patterns of *exposure* for the *scenarios* described in the conceptual *model*.

exposure–response—The functional relationship between the degree of *exposure* to an *agent* and the nature or magnitude of response of organisms, *populations*, or *ecosystems*.

exposure–response profile—The product of the characterization of ecological effects in the analysis phase of *ecological risk assessment*. The exposure–response profile summarizes the data on the effects of a contaminant, the relationship of the measures of effect to the *assessment endpoint*, and the relationship of the estimates of effects on the *assessment endpoint* to the measures of *exposure*.

exposure–response relationship—A quantitative relationship between the measures of *exposure* to an *agent* and a *measure of effect*. Exposure–response relationships may take various forms including thresholds (e.g., effects occur at concentrations greater than x mg/L), statistical *models* (e.g., the *probability* of death as a probit function of concentration), or mathematical process *models* (e.g., dissolved oxygen concentration as a function of phosphorous *loading* and other variables). Dose–response, concentration–response, and time-to-death *models* are specific examples of exposure–response relationships.

exposure route—The means by which a contaminant enters an organism (e.g., inhalation, stomatal *uptake*, ingestion).

exposure scenario—A set of assumptions concerning how an *exposure* may take place, including assumptions about the setting of the *exposure*, characteristics of the *agent*, activities that may lead to *exposure*, conditions modifying *exposure*, and temporal pattern of *exposure*.

extirpation—Effective elimination of a species from an *ecosystem*, watershed, or region. A synonym is functional extinction.

extrapolation—(1) The use of related data to estimate an unobserved or unmeasured value. Examples include use of data for fathead minnows to estimate effects on yellow perch, for individual organisms to estimate effects on communities, or for oxidation rates in 10°C water to estimate rates at 5°C, and (2) estimation of the value of an empirical function at a point outside the range of data used to derive the function.

feasibility study—The component of the CERCLA (*Superfund*) remedial investigation/feasibility study that is conducted to analyze the benefits, costs, and risks associated with *remedial alternatives*.

frequentist—A branch of statistics characterized by the analysis of a data set as one of a potentially infinite number of samples drawn from *population* with a particular distribution and by the treatment of probabilities as frequencies.

geographic information systems (GIS)—Software that uses spatial data to generate maps or to model processes in space.

geophagous—Eating soil. Usually refers to deliberate or at least not incidental ingestion.

habitat—An area that provides the needs of a particular species or set of species.

hazard—A situation that may lead to harm. In risk assessment, a hazard is a hypothesized association between an *agent* and a potentially susceptible *endpoint entity*. Identification of a hazard leads to assessment of the risk that the harm will occur.

hazard quotient—The quotient of the ratio of the estimated level of an *agent* divided by a level that is estimated to have no effect or to cause a prescribed effect. For example, the concentration of a chemical in water divided by its LC_{50}.

hyperaccumulator—An organism (usually plant) that accumulates high concentrations of an element or compound, relative to concentrations in soil or another medium.

indicator—A simple observation that indicates something about the *ecosystem* that is important, but not easy to observe.

indirect effect—An effect resulting from the action of an *agent* on components of the *ecosystem*, which in turn affect the *assessment endpoint* or other ecological component of interest. See *direct effect*. Indirect effects of chemical contaminants include reduced abundance due to toxic effects on food species or on plants that provide *habitat* structure. Equivalent to *secondary effects* but also includes tertiary and quaternary effects, etc.

induction—In logic, induction is the derivation of general principles from observations. For example, a series of observations of bioconcentration of different chemicals may allow us to induce that the *bioconcentration factor* (BCF) is a function of *octanol/water partitioning coefficients* (K_{ow}); in particular, $BCF = 0.89 \log K_{ow} + 0.61$. Inductive arguments are valid if the conclusions are usually true when the premises are true. See *abduction, deduction*.

inference—The act of reasoning from *evidence*.

interested party—See *stakeholder*.

intervention value—A screening criterion (the Netherlands) based on risks to human health and ecological *receptors* and processes. The ecotoxicological component of the intervention value is the hazardous concentration 50 (HC_{50}), the concentration at which 50% of species are assumed to be protected.

junk science—Scientific results that are said to be false because of perceived political, financial, or other motives other than a desire for truth. The term is itself political, having been developed by industry groups to discredit environmental and public health concerns. The antonym is *sound science*.

kinetic—Referring to movement. In particular, in toxicology and pharmacology, kinetic refers to the movement and transformation of a chemical in an organism (i.e., *toxicokinetic* or pharmacokinetic).

land farm—An area where organic wastes are tilled into the soil for disposal.

life-cycle assessment—A method for determining the relative environmental impacts of alternative products and technologies based on the consequences of their life cycle, from extraction of raw materials to disposal of the product following use.

likelihood—The hypothetical *probability* that events had a prescribed outcome. It may be thought of as the *probability* of *evidence* given a hypothesis $[P(E|H_x)]$ or as the *probability* of a sample $(x_1, x_2, \ldots, x_n)$ given a *probability density function*. Likelihoods are termed hypothetical probabilities, because the sum of likelihoods across a set of alternative hypotheses may be greater than 1. In ordinary English, it is synonymous with *probability*.

line of evidence—A set of data and associated analyses that can be used, alone or in combination with other lines of evidence, to estimate risks or determine causes. A line of evidence (e.g., a fathead minnow LC_{50} and a 24 h maximum concentration estimated using EXAMS) is an instance of a *type of evidence* (e.g., laboratory test endpoints and modeled *exposure* levels).

loading—The rate of input of a pollutant or other *agent* to a particular receiving system (e.g., nitrogen loading to the Chesapeake Bay).

lowest observed adverse effect level (LOAEL)—The lowest level of *exposure* to a chemical in a test that causes statistically significant differences from the controls in any measured response.

measure of effect—A measurable or estimable ecological characteristic that is related to the valued characteristic chosen as the *assessment endpoint* (equivalent to the earlier term "measurement endpoint").

measure of exposure—A measurable or estimable characteristic of a contaminant or other *agent* that is used to quantify *exposure*.

mechanism of action—The specific process by which an effect is induced. It is often used interchangeably with *mode of action* but is usually used to describe events at a lower level of organization than the effect of interest. For example, if the effect of interest is a reduction in survival rates, the *mode of action* of an *agent* may be *acute* lethality and its mechanism of action may be crushing, *acute* narcosis, cholinesterase inhibition, or burning.

mechanistic model—A mathematical *model* that estimates properties of a system by simulating its component processes rather than using empirical relationships.

media toxicity test—A *toxicity test* of water, soil, sediment, or biotic medium that is intended to determine the toxic effects of *exposure* to that medium. It includes *ambient media toxicity tests* plus tests of site media that have been spiked or otherwise treated.

median lethal concentration (LC_{50})—A statistically or graphically estimated concentration that is expected to be lethal to 50% of a group of organisms under specified conditions.

mesofauna—Animals that are barely visible such as nematodes and rotifers, which are larger than microfauna such as protozoans but smaller than macrofauna such as earthworms. The term is usually applied to soil or sediment communities.

mode of action—A phenomenological description of how an effect is induced. See *mechanism of action*. For example, if the effect of interest is local extinction of a species, the mode of action might be *habitat* loss and the *mechanism of action* might be fire, paving, or agricultural tillage.

model—A mathematical, physical, or conceptual representation of a system.

model uncertainty—The component of *uncertainty* concerning an estimated value that is due to possible misspecification of a *model* used for the estimation. It may be due to the choice of the form of the *model*, its component parameters, or its bounds.

Monte Carlo simulation—A resampling technique frequently used in *uncertainty* analysis in risk assessments to estimate the distribution of a *model*'s output parameter.

mycorrhiza—A symbiotic association of specialized mycorrhizal fungi with the roots of higher plants. The association often facilitates the *uptake* of inorganic nutrients by plants.

natural attenuation—Degradation or dilution of chemical contaminants by unenhanced biological and physicochemical processes.

net environmental benefits—The gains in environmental services or other ecological properties attained by *remediation* or ecological *restoration*, minus the environmental injuries caused by those actions. (Net benefits are also used in cost–benefit analysis as the difference between monetized benefits and costs.)

nitrification—The oxidation of ammonium to nitrate.

nitrogen fixation—The transformation of N_2 to ammonia by biological processes.

no observed adverse effect level (NOAEL)—The highest level of *exposure* to a chemical in a test that does not cause statistically significant differences from the controls in any measured response.

nonaqueous-phase liquid (NAPL)—A chemical or material present in the form of an oil phase.

normalization—Alteration of a chemical concentration or other property (usually by dividing by a factor) to reduce variance due to some characteristic of an organism or its environment (e.g., division of the body burden of a chemical by the organism's lipid content to generate a lipid-normalized concentration).

octanol/water partitioning coefficient (K_{ow})—The quotient of the concentration of an organic chemical dissolved in octanol divided by the concentration dissolved in water if the chemical is in equilibrium between the two solvents.

parties—The organizations that participate in making a decision. The representatives of all the parties are *risk managers*.

phytoremediation—*Remediation* of contaminated soil via the accumulation of the chemicals by plants or the promotion of degradation by plants.

phytotoxicity—Toxicity to plants.

population—An aggregate of interbreeding individuals of a species occupying a specific location in space and time.

precision—The exactitude with which a measurement or estimate can be specified or reproduced, usually determined by the similarity of independent determinations. The number of significant figures in a result is an expression of its precision.

preliminary remedial goal (PRG)—A concentration of a contaminant in a medium that serves as a default estimate of a *remedial goal* for *receptors* exposed to the contaminated medium.

primary data—Data obtained for the risk assessment and therefore designed to meet the *assessor*'s quality requirements and need to estimate a particular parameter or function.

probability—Two definitions (at least) are commonly used. (1) Objectivist and *frequentist*: The relative frequency of occurrence of an event in repeated trials, and (2) subjectivist and Bayesian: The degree of belief assigned to a hypothesis. Probability is scaled 0 to 1, with 0 indicating impossibility and 1 indicating inevitability.

probability density function (PDF)—For a continuous random variable, the PDF expresses the *probability* that the variable will occur in some very small interval. For a discrete random variable, the PDF expresses the *probability* that the variable assumes a prescribed value.

probable effects level for sediments—The geometric mean of the 50th percentile of effects concentrations and the 85th percentile of no-effects concentrations in coastal and estuarine sediment (Florida Department of Environmental Protection).

problem formulation—The phase in an *ecological risk assessment* in which the goals of the assessment are defined and the methods for achieving those goals are specified.

pseudoreplication—The treatment of multiple samples from a single treated location or system as if they were samples from multiple independently treated locations or systems. For example, multiple samples of benthic invertebrates from a stream reach below a wastewater outfall are pseudoreplicates.

quantal—Denoting an all-or-none response.

quantile—Any of the values that divide the range of a *probability* distribution into a given number of equal, ordered parts; examples are the median, quartiles, and percentiles. Each value divides the range into two parts: the part below the value corresponding to a prescribed fraction p and the part above to $1-p$.

quantitation limit—The concentration of a chemical in a medium that can be reliably quantified by an analytical method. Statistical definitions differ and are contentious, but are generally based on concentrations that can be estimated with prescribed *precision* (e.g., the true concentration that produces estimates having a relative standard deviation of 10%).

receptor—An organism, *population*, or community that is exposed to contaminants. Receptors may or may not be *assessment endpoint* entities.

record of decision—The document presenting the final decision resulting from the CERCLA remedial investigation/*feasibility study* process regarding selected alternative action(s).

recovery—The extent of return of a *population*, community, or *ecosystem* process to a condition with valued properties of a previous state. Due to the complex and dynamic nature of ecological systems, the attributes of a "recovered" system must be carefully defined.

reference, negative—A *site* or the information obtained from that *site* used to estimate the state of a receiving system in the absence of contamination or disturbance.

reference, positive—A *site* or the information obtained from that *site* used to estimate the state of a system exposed to contaminants other than the system that is being assessed.

reference value—A chemical concentration or *dose* that is a threshold for toxicity or significant contamination.

relative risk—The ratio of the risk given an *exposure* to the risk without the *exposure* or with an alternative *exposure*.

release—The movement of a contaminant from a *source* to an environmental medium.

remedial action objective—A specification of contaminants and media of concern, potential *exposure pathways*, and cleanup criteria (*remedial goal*).

remedial alternative—A potentially applicable remedial technology or action proposed in the *feasibility study* that is considered for *remediation* of a contaminated *site*. It may include controls on land use and the no action alternative (*natural attenuation*), as well as the usual engineered actions such as capping or thermal desorption.

remedial goal—A contaminant concentration, toxic response, or other criterion that is selected by the *risk manager* to define the condition to be achieved by remedial actions.

remedial goal option—A contaminant concentration, toxic response, or other criterion that is recommended by the *risk assessors* as likely to achieve conditions protective of the *assessment endpoints*.

remedial unit—An area of land or water to which a single *remedial alternative* applies.

remediation—Actions taken to reduce risks from contaminants including removal or treatment of contaminants and restrictions on land use. Note that, in contrast to *restoration*, remediation focuses strictly on reducing risks from contaminants and may actually reduce environmental quality.

removal action—An interim remedy for an immediate threat from *release* of hazardous substances.

restoration—Actions taken to make the environment whole, including restoring the capability of natural resources to provide services to humans. Restoration goes beyond *remediation* to include restocking, *habitat* rehabilitation, and reduced harvesting during a *recovery* period.

rhizosphere—The portion of a soil that is in the vicinity of, and influenced by, plant roots; includes enhanced microbial activity, nutrient mobilization, and other processes.

riparian—Occurring in, or by the edge of, a stream or in its floodplain.

risk assessor—An individual engaged in the performance of the technical components of risk assessments. Risk assessors may have expertise in the analysis of risk or specific expertise in an area of science or engineering relevant to the assessment.

risk characterization—A phase of *ecological risk assessment* that integrates the *exposure* and the *exposure–response profiles* to evaluate the *likelihood* of adverse ecological effects associated with *exposure* to the contaminants.

risk management—The processes of deciding whether to accept a risk or to take actions to reduce the risk, justifying the decision, and implementing the decision.

risk manager—An individual with the authority to decide what actions will be taken in response to a risk. Examples of risk managers include representatives of regulatory agencies, land managers, and investment managers.

rooting profile—The vertical spatial distribution of plant roots.

scenario—A possible future condition, given certain assumed actions and environmental conditions. In risk assessment, a scenario is a set of hypothetical or actual conditions under which *exposure* may occur and for which risks will be characterized.

scoping assessment—A qualitative assessment that determines whether a *hazard* exists that is appropriate for a risk assessment. For contaminated *sites*, it determines whether contaminants are present and whether there are potential *exposure pathways* and *receptors*.

screening assessment—A simple quantitative assessment performed to guide the planning of a subsequent assessment by eliminating *agents*, *receptors*, or areas from further consideration. That is, they are intended to screen out certain issues rather than to guide a management decision. See *scoping assessment* and *definitive assessment*.

screening benchmark—A concentration or *dose* that is considered a threshold for concern in the screening of contaminants.

screening level—An adjectival phrase applied to *models*, tests, or other sources of information that are adequate for use in *screening assessments* to sort risks into broad categories but not for risk estimation in a definitive assessment.

secondary data—Data obtained from the literature. Secondary data are not designed to meet the assessor's quality requirements or to estimate a particular assessment parameter or function.

secondary effect—An effect of an *agent* caused by effects on an entity that influences the *endpoint entity* rather than by *direct effects* on the *endpoint entity*. For example, herbicides kill plants (a primary or *direct effect*), which may cause loss of *habitat* structure and food, resulting in reduced herbivore abundance (the secondary effect). See also *indirect effect*, *direct effect*, and *primary effect*.

sensitivity—(1) In modeling, the degree to which model outputs are changed by changes in selected input parameters, and (2) in biology, the degree to which an organism or other entity responds to a specified change in *exposure* to an *agent*.

sentinel species—A species that displays a particularly sensitive response to a chemical or other *agent*. This property makes them useful *indicators* of the presence of hazardous levels of the *agent* to which they are sensitive.

single-chemical toxicity test—A *toxicity test* of an individual chemical administered to an organism or added to soil, sediment, or water to which an organism is exposed.

site—An area that has been identified as contaminated or disturbed and potentially in need of *remediation* or *restoration*.

sound science—Scientific results that are said to be credible. The term is usually used in a political context to describe results that support the speaker's positions. The antonym is *junk science*.

source—An entity or action that releases contaminants or other *agents* into the environment (primary source) or a contaminated medium that releases the contaminants into other media (secondary source). Examples of primary sources include spills, leaking tanks, dumps, and waste lagoons. An example of a secondary source is contaminated sediments that release contaminants by diffusion, bioaccumulation, and exchange. The term source is also used more generally to indicate the activities or drivers that are the sources of development, physical disturbance, or use.

species sensitivity distribution (SSD)—A distribution function, i.e., a *probability density function* (PDF) or cumulative distribution function (CDF), of the toxicity of a chemical or mixture to a set of species that may represent a taxon, assemblage, or community. In practice, SSDs are estimated from a sample of toxicity data for the specified species set. An SSD is equivalent to a conventional exposure–response *model*, but the points are effects levels for species rather than organisms.

stakeholder—An individual or organization that has an interest in the outcome of a regulatory or remedial action but is not an official party to the decision making. Examples include natural resource agencies and citizens groups. The synonym *interested party* is clearer but less commonly used.

stochasticity—Apparently random changes in a state or process that are attributed to inherent randomness of the system.

stressor—Stressor is commonly used in the United States in place of *agent*. It implies a prejudgment that the *agent* being assessed will have adverse effects. Just as the *dose* makes the poison, the level of *exposure*, the *receptor*, and the environmental conditions make an *agent* a stressor.

stressor–response—Synonymous with *exposure–response*, but (a) it incorporates the prejudgment implied by *stressor*, (b) it fails to recognize that it is *exposure* to an *agent* that causes response, not the existence of the *agent* per se, (c) it is nonparallel in that it pairs an entity (*stressor*) with a process (*response*), and (d) it obscures the relationship between *exposure* and the *exposure–response* relationship.

Superfund—The common name for the Comprehensive Environmental Response and Liability Act (CERCLA). It is the law in the United States that mandates the assessment and, as appropriate, the *remediation* of contaminated *sites*. The name comes from a fund that was created by taxing the chemical industry.

synergism—The process by which two or more chemicals or other *agents* cause joint effects that are more than additive (either exposure-additive or response-additive).

tertiary data—Data obtained from a published literature review or an electronic database derived from the literature. Like *secondary data*, tertiary data are not designed to meet the assessor's quality requirements or to estimate a particular assessment parameter or function. In addition, tertiary data may contain errors due to transcription or data entry and may not contain supporting information that is critical to interpretation.

threshold effects concentration—A concentration derived from various *toxicity test* endpoints, on which Canadian guidelines for soil contact are based (Canadian Council of Ministers of the Environment, CCME).

toxicity identification and evaluation (TIE)—A process whereby the toxic components of mixtures (usually aqueous effluents) are identified by removing components of a mixture and testing the residue, fractionating the mixture and testing the fractions, or adding components of the mixture to background medium and testing the artificially contaminated medium.

toxicity test—A procedure in which organisms or communities are exposed to defined levels of a chemical or material to determine the nature and magnitude of responses. See *bioassay*.

toxicodynamics—The study of the processes by which *exposure* to a chemical or mixture induces a toxic effect or a description of the results of such studies. In particular, toxicodynamics usually focuses on the biochemical processes by which an internal *exposure* induces injuries.

toxicokinetics—The study of the processes by which an external *exposure* to a potentially toxic chemical or mixture (e.g., a concentration in an ambient medium or a *dose*) results in an internal *exposure* (e.g., concentration at a *site* of action) or a description of results of such studies.

treatment endpoint—A concentration of a chemical in an environmental medium or other goal that is determined to be protective of human health and ecological *assessment endpoints* (a cleanup criterion).

type of evidence—A category of *evidence* used to characterize risk or identify a cause. Each type of evidence is qualitatively different from any others used in the *risk characterization* or causal analysis. The most commonly used types of *evidence* in *ecological risk assessments* of contaminated *sites* are (1) biological surveys, (2) *toxicity tests* of contaminated media, and (3) *toxicity tests* of individual chemicals. An individual instance of a type of evidence is termed a *line of evidence*.

uncertainty—Lack of knowledge concerning an event, state, *model*, or parameter. Uncertainty may be reduced by research or observation.

uncertainty factor—A factor applied to an *exposure* or effect estimate to correct for sources of *uncertainty*.

unit—An area that is the object of a risk assessment. A contaminated *site* may be assessed as a single unit, or there may be multiple units in a *site*. Common variants are "operable unit," "*remedial unit*," and "spatial unit."

uptake—Movement of a chemical from the environment into an organism as a result of any process.

uptake factor—The quotient of the concentration of element or compound in an organism divided by the concentration in an environmental medium. It is used interchangeably with *bioconcentration factor* and *bioaccumulation factor*, but is most often applied to *uptake* from food or ingested water by terrestrial species.

variability—Differences among entities or states of an entity attributable to heterogeneity. Variability is an inherent property of nature and may not be reduced by measurement. Examples include the differences in the weights of adult fathead minnows or differences among years in the minimum flow of a stream.

water effect ratio—A factor by which a water quality criterion or standard is multiplied to adjust for site-specific water chemistry.

watershed—An area of land from which water drains to a common surface water body.

weight of evidence—A process of identifying the best-supported *risk characterization* given the existence of multiple lines of *evidence* or the results of such a process.

wildlife—Nondomestic terrestrial or semiaquatic vertebrates. Wildlife includes mammals, birds, reptiles, and amphibians.

References

Ackerman, F. 2003. What's wrong with cost–benefit analysis? Risk Policy Report, 10:36–38.

Adams, D.F. 1963. Recognition of the effects of fluorides on vegetation. J. Air Pollut. Control Assoc., 13:360–362.

Adams, W.J. 1986. Toxicity and bioconcentration of 2,3,7,8-TCDD to fathead minnows (*Pimephales promelas*). Chemosphere, 15:1503–1511.

Adams, W.J. 1987. Bioavailability of neutral lipophilic organic chemicals contained on sediments: A review. Pages 219–244 in K.L. Dickson, A.W. Maki, and W.A. Brungs (eds.) Fate and Effects of Sediment-Bound Chemicals in Aquatic Systems. Pergamon Press, New York.

Adams, S.M. and DeAngelis, D.L. 1987. Indirect effects of early bass-shad interactions on predator population structure and food web dynamics. Pages 103–117 in W.C. Kerfoot and A. Sigh (eds.) Predation in Aquatic Ecosystems. University Press of New Hampshire, Hanover, NH.

ADEC (Alaska Department of Environmental Conservation). 2000. User's Guide for Selection and Application of Default Assessment Endpoints and Indicator Species in Alaska Ecoregions. Fairbanks, AK.

Adriaanse, P.I. 1996. Fate of pesticides in field ditches: The TOXSWA simulation model. DLO Winand Staring Centre, Report 90, Wageningen, The Netherlands.

Adriaanse, P.I. 1997. Exposure assessment of pesticides in field ditches: The TOXSWA model. Pestic. Sci., 49:210–212.

Akaike, H. 1973. Information theory as an extension of the maximum likelihood principle. Pages 267–281 in B.N. Petrov and F. Csaki (eds.) Second International Symposium on Information Theory. Akademia Kiado, Budapest.

Akin, E.W. 1991. Supplemental Region IV risk assessment guidance. US Environmental Protection Agency, Atlanta, GA.

Alabaster, J.S. and Lloyd, R. 1982. Water Quality Criteria for Freshwater Fish, Second Edition. Butterworth Scientific, London.

Alden, R.W., Hall, L.W. Jr., Dauer, D., and Burton, D.T. 2005. An integrated case study for evaluating the impacts of an oil refinery effluent on aquatic biota in the Delaware River: Integration and analysis of study components. Hum. Ecol. Risk Assess., 11:879–936.

Aldenberg, T. 1993. $E_T X$ 1.3a, a program to calculate confidence limits for hazardous concentrations based on small samples of toxicity data. RIVM, Bilthoven, The Netherlands.

Aldenberg, T. and Luttik, R. 2002. Extrapolation factors for tiny toxicity data sets from species sensitivity distributions with known standard deviation. Pages 103–118 in L. Posthuma, G.W. Suter II, and T. Traas (eds.) Species Sensitivity Distributions in Ecotoxicology. Lewis Publishers, Boca Raton, FL.

Aldenberg, T. and Slob, W. 1993. Confidence limits for hazardous concentrations based on logistically distributed NOEC toxicity data. Ecotoxicol. Environ. Saf., 25:48–63.

Alexander, M., Goldstein, L., Pauwels, S., Edwards, D., Zaborsky, O., Menzie, C., Heiger-Bernays, W., et al. 1995. Environmentally acceptable endpoints in soil: Risk-based approach to contaminated site management based on availability of chemicals in soil. GRI-95/0000. Gas Research Institute, Chicago, IL.

Allen, M. 2003. Initial sample preparation. Pages 34–63 in K.C. Thompson and C.P. Nathanail (eds.) Chemical Analysis of Contaminated Land. Blackwell, Oxford, UK.

Allen, T.F.H. and Starr, T.B. 1982. Hierarchy—perspectives for ecological complexity. University of Chicago Press, Chicago, IL.

Allen, W. 2001. Green Phoenix, Restoring the Tropical Forests of Guanacaste, Costa Rica. Oxford University Press, Oxford, UK.

Alsop, W.R., Hawkins, E.T., Stelljes, M.E., and Collins, W. 1996. Comparison of measured and modeled tissue concentrations for ecological receptors. Hum. Ecol. Risk Assess., 2:539–557.

Altenburger, R., Walter, H., and Grote, M. 2004. What contributes to the combined effects of a complex mixture? Environ. Toxicol. Chem., 38:6353–6362.

Ambrose, R.B. 1988. WASP4, a hydrodynamic and water quality model—model theory user's manual and programmers guide, EPA-600-3-87-039. USEPA, ERL, Athens, GA.

Anderson, B., Nicely, P., Gilbert, K., Kosaka, R., Hunt, J., and Phillips, B. 2003. Overview of freshwater and marine toxicity tests: A technical tool for ecological risk assessment. Environmental Protection Agency, Sacramento, CA.

Anderson, B.S., Hunt, J.W., Phillips, B.M., Stoelting, M., Becker, J., Fairey, R., Puckett, H.M., Stephenson, M., Tjeerdema, R.S., and Martin, M. 2000. Ecotoxicological change at a remediated superfund site in San Francisco, California, USA. Environ. Toxicol. Chem., 19:879–887.

Anderson, B.S., Hunt, J.W., Phillips, B.M., Fairey, R., Roberts, C.A., Oakden, J.M., Puckett, H.M., et al. 2001. Sediment quality in Los Angeles harbor, USA: A triad approach. Environ. Toxicol. Chem., 20:359–370.

Anderson, D.R., Burnham, K.P., and Thompson, W.L. 2000. Null hypothesis testing: Problems, prevalence, and an alternative. J. Wildl. Manag., 64:912–923.

Anderson, J.L. 1998. Embracing uncertainty: The interface of Bayesian statistics and cognitive psychology. Conserv. Ecol., 2(2):2.

Anderson, J.L. 2001. Stone-age minds at work on 21st century science. Conserv. Biol. Pract., 2:18–25.

Andrewartha, H.G. and Birch, L.C. 1954. The Distribution and Abundance of Animals. University of Chicago Press, Chicago, IL.

Andrewartha, H.G. and Birch, L.C. 1984. The Ecological Web. University of Chicago Press, Chicago, IL.

Andrews, C.J., Apul, D.S., and Linkov, I. 2004. Comparative risk assessment: Past experience, current trends and future directions. Pages 1–14 in I. Linkov and A.B. Ramadan (eds.) Comparative Risk Assessment and Environmental Decision Making. Kluwer Academic, Dordrecht, The Netherlands.

Ankley, G.T. and Schubauer-Berigan, M.K. 1995. Background and overview of current sediment toxicity identification evaluation procedures. J. Aquat. Eco. Health, 4:133–149.

Ankley, G.T., DiToro, D., and Hansen, D.J. 1996. Technical basis and proposal for deriving sediment quality criteria for metals. Environ. Toxicol. Chem., 15:2056–2066.

Ankley, G.T., Jensen, K.M., Kahl, M., and Korte, J.J.M.E.A. 2001. Description and evaluation of a short-term reproduction test with fathead minnow (*Pimephales promelas*). Environ. Toxicol. Chem., 20:1276–1290.

ANZ. 1995. Risk Management. AS/NZS 430:1995. Standards Australia and Standards New Zealand, Homebush, New South Wales, Australia, and Wellington, New Zealand.

ANZECC. 2000. Australian and New Zealand Guidelines for Fresh and Marine Water Quality. Volume 2. Aquatic Ecosystems—Rationale and Background Information. Paper No.4. 2000. Australia and New Zealand Environment and Conservation Council and Agriculture and Resource Management Council of Australia and New Zealand, Canberra, Australia.

APHA (American Public Health Association). 1999. Standard Methods for the Examination of Water and Waste Water. American Public Health Association, Washington, DC.

Aquatic Risk Assessment and Mitigation Dialogue Group. 1994. Final Report. Society for Environmental Toxicology and Chemistry, Pensacola, FL.

Arenal, C.A. and Halbrook, R.S. 1997. PCB and heavy metal contamination and effects in European starlings (*Sternus vulgaris*) at a Superfund site. Bull. Environ. Contam. Toxicol., 58:254–262.

Arnot, J.A., Mackay, D. and Webster, E. 2006. A screening level risk assessment model for chemical fate and effects in the environment. Environ. Sci. Technol. 40:2316–2323.

Arthur, J.W. and Aldredge, A.W. 1979. Soil ingestion by mule deer in North Central Colorado. J. Range Manag., 32:67–70.

Arthur, J.W. and Gates, R.J. 1988. Trace element intake via soil ingestion in pronghorns and in black-tailed jackrabbits. J. Range Manag., 41:162–166.

Ascher, W. 2006. Forecasting for environmental decision making: Research priorities. Pages 230–245 in National Research Council (ed.) Decision Making for the Environment: Social and Behavioral Science Research Priorities. National Academy Press, Washington, DC.

Asfaw, A., Ellersieck, M.R., and Mayer, F.L. 2003. Interspecies correlation estimations (ICE) for acute toxicity to aquatic organisms and wildlife. II. User manual and software. EPA/600/R-03/106. US Environmental Protection Agency, Washington, DC.

ASTM. 1994. Emergency standard guide for risk-based corrective action applied to petroleum release sites. ES 38-94. ASTM, Philadelphia.

ASTM. 1996. Standard practice for statistical analysis of toxicity tests conducted under ASTM guidelines. E 1847-96. ASTM, West Conshohocken, PA.

ASTM. 2002. Annual Book of ASTM Standards, Sec. 11, Water and Environmental Technology. ASTM, West Conshohocken, PA.

Avian Effects Dialog Group. 1994. Assessing pesticide impacts on birds: Final report of the Avian Effects Dialog Group, 1988–1993. RESOLVE, Washington, DC.

Baes, C.F., Sharp, R.D., Sjoreen, A.L., and Shor, R.W. 1984. A review and analysis of parameters for assessing transport of environmentally released radionuclides through agriculture. ORNL-5786. Oak Ridge National Laboratory, Oak Ridge, TN.

Bailer, A.J. and Oris, J.T. 1997. Estimating inhibition concentrations for different response scales using generalized linear models. Environ. Toxicol. Chem., 16:1554–1559.

Bailar, J.C. 2005. Redefining the confidence interval. Hum. Ecol. Risk Assess., 11:169–177.

Bailey, R.C., Kennedy, M.C., Dervish, M.C., and Taylor, R.M. 1998. Biological assessment of freshwater ecosystems using a reference condition approach. Freshw. Biol., 39:774.

Bailey, R.G. 1976. Ecoregions of the United States, US Forest Service, Ogden, UT.

Baird, D.J. and Burton, G.A. Jr. (eds.). 2001. Ecological Variability: Separating Natural from Anthropogenic Causes of Ecosystem Impairment. SETAC Press, Pensacola, FL.

Baker, J.L., Barefoot, A.C., Beasley, L.E., Burns, L., Caulkins, P., Clark, J., Feulner, R.L., et al. 1994. Final Report: Aquatic Risk Assessment and Mitigation Dialog Group. SETAC Press, Pensacola, FL.

Baker, J.P. and Harvey, T.B. 1984. Critique of acid lakes and fish population status in the Adirondack Region of New York State NAPAP Project E3-25. US Environmental Protection Agency, Corvallis, OR.

Balcomb, R., Bowen, C.A., II, Wright, D., and Law, M. 1984. Effects on wildlife of at-planting corn applications of granular carbofuran. J. Wildl. Manag., 48:1353–1359.

Banton, M.I., Klingensmith, J.S., Barchers, D.E., Clifford, P.A., Ludwig, D.F., Macrander, A.M., Sielken, R.L., and Valdez-Flores, C. 1996. An approach for estimating ecological risks from organochlorine pesticides to terrestrial organisms at Rocky Mountain Arsenal. Hum. Ecol. Risk Assess., 2:499–526.

Banuelos, G.S., Mead, R., Wu, L., Beuselinck, P., and Akohoe, S. 1992. Differential selenium accumulation among forage plant species grown in soils ammended with selenium-enriched plant tissue. J. Soil Water Cons., 47:338–342.

Barber, M.X., Suarez, L.A., and Lassiter, R.R. (1991) Modelling bioaccumulation of organic pollutants in fish with an application to PCBs in Lake Ontario salmonids. Can. J. Fish. Aquat. Sci., 48:318–337.

Barber, S.A. 1995. Soil Nutrient Bioavailability: A Mechanistic Approach. John Wiley, New York.

Barbour, M.T., Gerritsen, J., Griffith, G.O., Freydenborg, R., McCarron, E., White, J.S., and Bastian, M.L. 1996. A framework for biological criteria for Florida streams using benthic macroinvertebrates. J. N. Am. Benthol. Soc., 15:185–211.

Barbour, M.T., Gerritsen, J., Snyder, B.D., and Stribling, J.B. 1999. Rapid Bioassessment Protocols for Use in Streams and Wadeable Rivers: Periphyton, Benthic Macroinvertebrates and Fish, Second Edition. EPA 841-B-99-002. US Environmental Protection Agency, Washington, DC.

Baril, A., Jobin, B., Mineau, P., and Collins, B.T. 1994. A consideration of inter-species variability in the use of the median lethal dose (LD_{50}) in avian risk assessment. No. 216. Canadian Wildlife Service, Hull, PQ, Canada.

Barnthouse, L.W. 1996. Guide for developing data quality objectives for ecological risk assessment at DOE Oak Ridge Operations facilities. ES/ER/TM-815/R1. Environmental Restoration Risk Assessment Program, Lockheed Martin Energy Systems, Inc., Oak Ridge, TN.

Barnthouse, L.W. 2004. Quantifying population recovery rates for ecological risk assessment. Environ. Toxicol. Chem., 23:500–508.

Barnthouse, L.W. and Brown, J. 1994. Conceptual model development. Chapter 3 in Ecological Risk Assessment Issue Papers. EPA/630/R-94/009. US Environmental Protection Agency, Washington, DC.

Barnthouse, L.W. and Stahl, R.G. Jr. 2002. Quantifying natural resource injuries and ecological service reductions: Challenges and opportunities. Environ. Manag., 30:1–12.

Barnthouse, L.W. and Suter, G.W., II. 1986. User's manual for ecological risk assessment. ORNL-6251. Oak Ridge National Laboratory, Oak Ridge, TN.

Barnthouse, L.W., DeAngelis, D.L., Gardner, R.H., O'Neill, R.V., Suter, G.W., II, and Vaughan, D.S. 1982. Methodology for Environmental Risk Analysis. ORNL/TM-8167. Oak Ridge National Laboratory, Oak Ridge, TN.

Barnthouse, L.W., O'Neill, R.V., Bartell, S.M., and Suter, G.W., II. 1986. Population and ecosystem theory in ecological risk assessment. Pages 82–96 in T.M. Poston and R. Purdy (eds.) Aquatic Toxicology and Environmental Fate, 9th Symposium, ASTM STP 921. ASTM, Philadelphia, PA.

Barnthouse, L.W., Suter, G.W., II, Rosen, A.E., and Beauchamp, J.J. 1987. Estimating responses of fish populations to toxic contaminants. Environ. Toxicol. Chem., 6:811–824.

Barnthouse, L.W., Suter, G.W., II, and Rosen, A.E. 1988. Inferring population-level significance from individual-level effects: An extrapolation from fisheries science to ecotoxicology. Pages 289–300 in G.W. Suter II and M.A. Lewis (eds.) Aquatic Toxicology and Environmental Fate: 11th Symposium, ASTM STP-1007. ASTM, Philadelphia, PA.

Barnthouse, L.W., Suter, G.W., II, and Rosen, A.E. 1990. Risks of toxic contaminants to exploited fish populations: Influence of life history, data uncertainty, and exploitation intensity. Environ. Toxicol. Chem., 9:297–311.

Barnthouse, L.W., Glaser, D., and Young, J. 2003. Effects of historic PCB exposures on the reproductive success of the Hudson River striped bass population. Environ. Sci. Technol., 37:223–228.

Barnthouse, L.W., Munns, W.R. Jr., and Sorensen, M.T. (eds.). 2006. Population-Level Ecological Risk Assessment. SETAC Press, Pensacola, FL.

Baron, L.A., Sample, B.E., and Suter, G.W., II. 1999. Ecological risk assessment of a large river-reservoir: 5. Aerial insectivorous wildlife. Environ. Toxicol. Chem., 18:621–627.

Barrett, G.W. 1968. The effect of an acute insecticide stress on a semi-enclosed grassland ecosystem. Ecology, 49:1019–1035.

Barron, M.G. and Holder, E. 2003. Are exposure and ecological risks of PAHs underestimated at petroleum contaminted sites? Hum. Ecol. Risk Assess., 9:1533–1546.

Barron, M.G. and Wharton, S.R. 2005. Survey of methodologies for developing media screening values for ecological risk assessment. Integr. Environ. Assess. Manag., 1:320–332.

Barron, M.G., Mayes, M.A., Murphy, P.G., and Nolan, R.J. 1990. Pharmacokinetics and metabolism of triclopyr butoxyethyl ester in coho salmon. Aq. Toxicol., 16:9–32.

Bartell, S.M. 1978. Size-selective planktivory and phosphorus cycling in pelagic systems. Ph.D. Dissertation, University of Wisconsin, Madison.

Bartell, S.M. 2003. A framework for estimating ecological risks posed by nutrients and trace elements in the Patuxent River. Estuaries 26:385–397.

Bartell, S.M. and Brenkert, A.L. 1991. A spatial–temporal model of nitrogen dynamics in a deciduous forest watershed. Pages 379–398 in M.G. Turner and R.H. Gardner (eds.) Quantitative Methods in Landscape Ecology. Springer-Verlag, New York.

Bartell, S.M., Landrum, P.F., Giesy, J.P., and Leversee, G.J. 1981. Simulated transport of polycyclic aromatic hydrocarbons in artificial streams. Pages 133–144 in W.J. Mitsch, R.W. Bosserman, and J.M. Klopatek (eds.) Energy and Ecological Modelling. Elsevier, Amsterdam.

Bartell, S.M., Breck, J.E., Gardner, R.H., and Brenkert, A.L. 1986. Individual parameter perturbation and error analysis of fish bioenergetics models. Can. J. Fish. Aquatic Sci., 43:160–168.

Bartell, S.M., Gardner, R.H., and O'Neill, R.V. 1988. An integrated fates and effects model for estimation of risk in aquatic systems. Pages 261–274 in Aquatic Toxicology and Hazard Assessment: Volume 10. ASTM STP 971. ASTM, Philadelphia, PA.

Bartell, S.M., Brenkert, A.L., O'Neill, R.V., and Gardner, R.H. 1989. Temporal variation in the regulation of production in a pelagic food web model. Pages 101–118 in S.R. Carpenter (ed.) Complex Interactions in Lake Communities. Springer-Verlag, New York.

Bartell, S.M., Gardner, R.H., and O'Neill, R.V. 1992. Ecological Risk Estimation. Lewis Publishers, Chelsea, MI.

Bartell, S.M., Lefebvre, G., Kaminski, G., Carreau, M., and Campbell, K.R. 1999. An ecosystem model for assessing ecological risks in Québec rivers, lakes, and reservoirs. Ecol. Model., 124:43–67.

Bartell, S.M., Campbell, K.R., Lovelock, C.M., Nair, S.K., and Shaw, J.L. 2000. Characterizing aquatic ecological risks from pesticides using a diquat dibromide case study III: Ecological process models. Environ. Toxicol. Chem., 19:1441–1453.

Bartell, S.M., Pastorok, R.A., Akcakaya, H.R., Regan, H., Ferson, S., and Mackay, C. 2003. Realism and relevance of ecological models used in chemical risk assessment. Hum. Ecol. Risk Assess., 9:907–938.

Batley, G.E., Burton, G.A., Chapman, P.M., and Forbes, V.E. 2002. Uncertainties in sediment quality weight-of-evidence (WOE) assessments. Hum. Ecol. Risk Assess., 8:1517–1547.

Baum, E.J. 1997. Chemical Property Estimation: Theory and Application. Lewis Publishers, Boca Raton, FL.

Baumann, P.C., Smith, I.R., and Metcalfe, C.D. 1996. Linkage between chemical contaminants and tumors in benthic great lakes fish. J. Great Lakes Res., 22:131–152.

Bechtel-Jacobs. 1998. Empirical models for the uptake of inorganic echemicals from soil by plants. BJC/OR-133. Oak Ridge National Laboratory, Oak Ridge, TN.

Beck, L.W., Maki, A.W., Artman, N.R., and Wilson, E.R. 1981. Outline and criteria for evaluating the safety of new chemicals. Regul. Toxicol. Pharmacol., 1:19–58.

Begon, M., Townsend, C.R., and Harper, J.L. 1999. Ecology: Individuals, Populations, and Communities, Third Edition. Blackwell Science, London.

Beissinger, S.R. and McCollough, D.R. 2002. Population Viability Analysis. University of Chicago Press, Chicago, IL.

Belfroid, A., van den Berg, M., Seinen, W., Hermens, J., and van Gestel, K. 1995. Uptake, bioavailability, and elimination of hyrophobic compounds in earthworms (*Eisenis andrei*) in field-contaminated soil. Environ. Toxicol. Chem., 14:605–612.

Belfroid, A.C., Sijm, D.T.H.M., and van Gestel, C.A.M. 1996. Bioavailability and toxicokinetics of hydrophobic aromatic compounds in benthic and terrestrial invertebrates. Environ. Rev., 4:276–299.

Bellwood, D.R. and Hughes, T.P. 2001. Regional-scale assembly rules and biodiversity of coral reefs. Science, 292:1532–1534.

Beltman, W.H.J. and Adriaanse, P.I. 1999a. User's manual TOXSWA 1.2: Simulation of pesticide fate in small surface waters. DLO Winand Staring Centre, Technical Document 54, Wageningen, The Netherlands.

Beltman, W.H.J. and Adriaanse, P.I. 1999b. Proposed standard scenarios for an aquatic fate model in the Dutch authorization procedure of pesticides. Method to define standard scenarios determining exposure concentrations simulated by the TOXSWA model. DLO Winand Staring Centre, Report 161, Wageningen, The Netherlands.

Bence, A.E. and Burns, W.A. 1995. Fingerprinting hydrocarbons in the biological resources of the *Exxon Valdez* spill area. Pages 84–140 in P.G. Wells, J.N. Butler, and J.S. Hughes (eds.) Exxon Valdez Oil Spill: Fate and Effects in Alaskan Waters. ASTM, Philadelphia.

Benjamin, S.L. and Belluck, D.A. 2002. A Practical Guide to Understanding, Managing, and Reviewing Environmental Risk Assessment Reports. Lewis Publishers, Boca Raton, FL.

Bennett, D.H., Margni, M.D., McKone, T.P., and Jolliet, O. 2002. Intake fraction for multimedia pollutants: A tool for life-cycle analysis and comparative risk assessment. Risk Anal., 22:905–918.

Benoit, G. 1994. Clean technique measurement of Pb, Ag, and Cd in freshwater: A redefinition of metal pollution. Environ. Sci. Technol., 28:1987–1991.

Bergman, H.L. and Dorward-King, E.J. (eds.). 1997. Reassessment of Metals Criteria for Aquatic Life Assessment. SETAC Press, Pensacola, FL.

Berish, C.W.D.B.R., Harrison, W.A., Jackson, W.A., and Ritters, K.H. 1999. Conducting regional environmental assessments: The southern Appalachian experience. Pages 117–166 in J.D. Piene (ed.) Ecosystem Management for Sustainability: Principles and Practices. Lewis Publishers, Boca Raton, FL.

Bernard, S.B. and Ebi, K.L. 2001. Comments on the process and product of the health impact assessment component of the national assessment of the potential consequences of climate variability and change for the United States. Environ. Health Persp., 109:177–233.

Bernstein, P.L. 1996. Against the Gods: The Remarkable Story of Risk. John Wiley, New York.

Berry, D.A., Mueller, P., Grieve, A.P., Smith, M., Park, T., Blazek, R., Mitchard, N., and Krams, M. 2002. Adaptive Bayesian designs for dose-ranging drug trials. Pages 99–156 in C. Gastonis, B. Carlin, A. Carriquiry, A. Gelman, R.E. Kass, I. Verdinelli, and M. West (eds.) Case Studies in Bayesian Statistics, Volume V. Springer-Verlag, New York.

Bervoets, L., Baillieul, M., Blust, R., and Verheyen, R. 1996. Evaluations of effluent toxicity and ambient toxicity in a polluted lowland river. Environ. Pollut., 91:333–341.

Bevelhimer, M.S., Sample, B.E., Southworth, G.R., Beauchamp, J.J., and Peterson, M.J. 1996. Estimation of whole-fish contaminant concentrations from fish fillet data. ES/ER/TM-202. Oak Ridge National Laboratory, Oak Ridge, TN.

Beyers, D.W., Keefe, T.J., and Carlson, C.A. 1994. Toxicity of carbaryl and malathion to two federally endangered fishes, as estimated by regression and ANOVA. Environ. Toxicol. Chem., 13:101–107.

Beyer, W.N. and Storm, G. 1995. Ecotoxicological damage from zinc smelting at Palmerton, Pennsylvania. Pages 596–608 in D.J. Hoffman, B. Rattner, G.A. Burton, and J. Cairns (eds.) Handbook of Ecotoxicology. Lewis Publishers, Boca Raton, FL.

Beyer, W.N., Pattee, O.H., Siteo, L., Hoffman, D.J., and Mulhern, B.M. 1985. Metal contamination in wildlife living near two zinc smelters. Environ. Pollut. (Ser. A), 38:63–86.

Beyer, W.N., Connor, E.E., and Gerould, S. 1994. Estimates of soil ingeston by wildlife. J. Wildl. Manag., 58:375–382.

Beyer, W.N., Blus, L.J., Henny, C.J., and Audet, D.J. 1997. The role of sediment ingestion in exposing wood ducks to lead. Ecotoxicology, 6:181–186.

Beyer, W.N., Franson, J.C., Locke, L.N., Stroud, R.K., and Sileo, L. 1998. Retrospective study of the diagnostic criteria in a lead-poisoning survey of waterfowl. Environ. Contam. Toxicol., 35:506–512.

Beyer, W.N., Audet, D.J., Heinz, G.H., Hoffman, D.J., and Day, D. 2000. Relation of waterfowl poisoning to sediment lead concentrations in the Coeur d'Alene basin. Ecotoxicology, 9:207–218.

Beyers, D.W. 1998. Causal inference in environmental impact studies. J. N. Am. Benthol. Soc. 17:367–373.

Bilyard, G.R., Beckert, H., Bascietto, J.J., Abrams, C.W., Dyer, S.A., and Haselow, L.A. 1997. Using the data quality objectives process during the design and conduct of ecological risk assessment. DOE/EH-0544. Pacific Northwest National Laboratory, Richland, WA.

Birge, W.J., Black, J.A., and Ramey, B.A. 1986. Evaluation of effluent biomonitoring systems. Pages 66–80 in H.L. Bergman, R.A. Kimerle, and A.W. Maki (eds.) Environmental Hazard Assessment of Effluents. Pergamon Press, New York.

Blamey, R.K. and Comon, M.S. 1999. Valuation and ethics in environmental economics. Pages 809–823 in J.C.J.M. van den Bergh (ed.) Handbook of Environmental and Resource Economics. Edward Elgar, Cheltenham, UK.

Blus, L.J. and Henny, C.J. 1997. Field studies on pesticides and birds: Unexpected and unique relations. Ecol. Appl., 7:1125–1132.

Bobek, C., Embleton, K., Gorsky, L., and Knoop, K. 1995. Comparative Risk Assessment, Washington, DC.

Boeije, G. 1999. GREAT-ER Technical Documentation—Chemical Fate Models. Available at: http://www.great-er.org/files/techdoc_model.pdf.

Boelsterli, U.A. 2003. Mechanistic toxicology. Taylor & Francis, London.

Boersma, L., McFarlane, C., and Lindstrom, T. 1991. Mathematical model of plant uptake and translocation of organic chemicals: Application to experiments. J. Environ. Qual., 20:137–146.

Boethling R.S. and Mackay, D. 2000. Handbook of Property Estimation Methods for Chemicals: Environmental and Health Sciences. Lewis Publishers, Boca Raton, FL.

Bolen, E.G. and Robinson, W.L. 2002. Wildlife Ecology and Management, Fifth Edition. Prentice-Hall, Englewood Cliffs, NJ.

Bolliger, J., Kienast, F., and Zimmerman, N.E. 2000. Risks of global warming on montane and subalpine forests in Switzerland—a modeling study. Reg. Environ. Change, 1:99–111.

Bookhout, T.A. 1994. Research and Management Techniques for Wildlife and Habitats, Fifth Edition. The Wildlife Society, Bethesda, MD.

Boone, M.D. and Bridges, C.M. 1999. The effect of temperature on the toxicity of carbaryl for survival of tadpoles of the green frog (*Rana clamitans*). Environ. Toxicol. Chem., 18:1482–1484.

Boone, M.D. and Semlitsch, R.D. 2001. Interactions of an insecticide with larval density and predation in experimental amphibian communities. Conserv. Biol., 15:228–238.

Boone, M.D. and Semlitsch, R.D. 2002. Interactions of an insecticide with competition and pond drying in amphibian communities. Ecol. Appl., 12:307–316.

Boone, M.D., Semlitsch, R.D., Fairchild, J., and Rothermel, B.B. 2004. Effects of an insecticide on amphibians in large-scale experimental ponds. Ecol. Appl., 14:685–691.

Borgmann, A.I., Moody, M.J., and Scroggins, R.P. 2004. The lab-to-field (LTF) rating scheme: A new method of investigating the relationships between laboratory sublethal toxicity tests and field measurements in environmental effects monitoring studies. Hum. Ecol. Risk Assess., 10:683–707.

Borgmann, U., Norwood, W.P., Reynoldson, T.B., and Rosa, F. 2001. Identifying cause in sediment assessments: Bioavailability and the sediment quality triad. Can. J. Fish. Aquat. Sci., 58:950–960.

Bossert, I. and Bartha, R. 1984. The fate of petroleum in soil ecosystems. Pages 435–473 in R. Atlas (ed.) Petroleum Microbiology. Macmillan, New York.

Bovee, K.D. and Zuboy, J.R. 1988. Proceedings of a workshop on the development and evaluation of habitat suitability criteria. Biological Report 88(11). US Fish and Wildlife Service.

Brack, W., Bakker, J., De Deckere, E., Deerenberg, C., Van Gils, J., Hein, M., Jurajda, P., et al. 2005. MODELKEY. Models for assessing and forecasting the impact of environmental key pollutants on freshwater and marine ecosystems and biodiversity. Environ. Sci. Pollut. Res. Int. 12:252–256.

Bradbury, S.P., Henry, T.R., Niemi, G.J., Carlson, R.W., and Snarski, V.M. 1989. Use of respiratory-cardiovascular responses of rainbow trout (*Salmo gairdneri*) in identifying acute toxicity syndromes in fish. Part 3: Polar narcotics. Environ. Toxicol. Chem., 8:247–261.

Bradbury, S.P., Feijtel, T.C.J., and Van Leeuwen, C.J. 2004. Meeting the Scientific Needs of Ecological Risk Assessment in a Regulatory Context. Environ. Sci. Technol., 2004:463A–470A.

Brauers, W.K. 2003. Characterization Methods for a Stakeholder Society: A Revolution in Economic Thinking by Multi-Objective Optimization. Kluwer Academic, Dordrecht, The Netherlands.

Brazner, J.C., Heinis, L.J., and Jensen, D.A. 1989. A littoral enclosure for replicated field experiments. Environ. Toxicol. Chem., 8:1209–1216.

Breitburg, D.L., Sanders, J.G., Gilmour, C.G., Hatfield, C.A., Osman, R.W., Riedel, G.F., Seitzinger, S.P., and Sellner, K.G. 1999. Variability in responses to nutrients and trace elements, and transmission of stressor effects through an estuarine food web. Limnology and Oceanography. 44:837–863.

Brenkert, A.L., Gradner, R.H., Bartell, S.M., and Hoffman, F.O. 1988. Uncertainties associated with estimates of radium accumulation in lake sediments and biota. Pages 185–192 in G. Desmet (ed.) Reliability of Radioactive Transfer Models. Elsevier Applied Science, London.

Breton, R., Schurmann, G., and Purdy, R. 2003. Proceedings of QSAR 2002, QSAR and Combinatorial Science 22, Nos 1, 2 and 3, pp. 1–409.

Bridges, C.M. 1999a. Predator–prey interactions between two amphibian species: Effects of insecticide exposure. Environ. Toxicol. Chem., 33:205–211.

Bridges, C.M. 1999b. Effect of a pesticide on tadpole activity and predator avoidance. Environ. Toxicol. Chem., 33:303–306.

Briggs, G.G., Bromilow, R.H., and Evans, A.A. 1982. Relationship between lipophilicity and root uptake and translocation of nonionized chemicals by barley. Pesticide Sci., 13:495–504.

Brock, T.C.M. and Ratte, H.T. 2002. Ecological risk assessment of pesticides. Pages 33–41 in J.M. Giddings, T.C.M. Brock, W. Heger, F. Heimbach, S.J. Maund, S.M. Norman, H.T. Ratte, C. Schafers, and M. Streloke (eds.) Community-Level Aquatic System Studies— Interpretation Criteria. SETAC Press, Pensacola, FL.

Brock, T.C.M., Lahr, J., and Van den Brink, P.J., 2000. Ecological Risks of Pesticides in Freshwater Ecosystems. Part 1: Herbicides. Alterra-Rapport 088. Alterra, Green World Research, Wageningen, The Netherlands.

Brockwell, P.J., and Davis, R.A. 2003. Introduction to Time Series and Forecasting. Springer-Verlag, New York.

Broderius, S. and Kahl, M. 1985. Acute toxicity of organic chemical mixtures to the fathead minnow. Aquatic Toxicol., 6:307–322.

Bromilow, R.H. and Chamberlain, K. 1995. Principles governing uptake and transport of chemicals. Pages 37–68 in F. Trapp and J.C. McFarlane (eds.) Plant Contamination: Modeling and Simulation of Organic Chemical Processes. Lewis Publishers, Boca Raton, FL.

Brooks, A.S. and Seegert, G.L. 1977. The effect of intermittent chlorination on rainbow trout and yellow perch. Trans. Am. Fish. Soc., 106:278–286.

Bro-Rasmussen, F. and Lokke, H. 1984. Ecoepidemiology—a casuistic discipline describing ecological disturbances and damages in relation to their specific causes; exemplified by chlorinated phenols and chlorophenoxy acids. Reg. Toxicol. Pharmacol., 4:391–399.

Brownie, C., Glashow, H.B., Burkholder, J.M., Reed, R., and Tang, Y. 2002. Re-evaluation of the relationship between *Pfiesteria* and estuarine fish kills. Ecosystems, 6:1–10.

Brueske, C.C. and Barrett, G.W. 1991. Dietary heavy metal uptake by the least shrews, *Cryptotis parva*. Bull. Environ. Contam. Toxicol., 47:845–849.

Bruins, R.J.F. and Heberling, M.T. (eds). 2004. Integrating Ecological Risk Assessment and Economic Analysis in Watersheds: A Conceptual Approach and Three Case Studies. EPA/600/R-03/140R. US Environmental Protection Agency, Cincinnati, OH.

Bruins, R.J.F. and Heberling, M.T. (eds.). 2005. Economics and Ecological Risk Assessment: Applications to Watershed Management, CRC Press, Boca Raton, FL.

Brumbaugh, W.G., Krabbenhoft, D.P., Helsel, D.R., Wiener, J.G., and Echols, K.R. 2001. A national pilot study of mercury contamination in aquatic ecosystems along multiple gradients: Bioaccumulation in fish. USGS/BRD/BSR-2001-0009. Washington, DC.

Buchwalter, S.B. and Luoma, S.N. 2005. Differences in dissolved cadmium and zinc uptake among stream insects: Mechanistic explanations. Environ. Sci. Technol., 39:498–504.

Bull, K.R., Avery, W.J., and Freestone, P. 1983. Alkyl lead pollution and bird mortalities in the Mersey estuary, UK, 1979–1981. Environ. Pollut., 31A:239–254.

Bunce, N.J. and Remillard, R.B.J. 2003. Haber's rule: The search for quantitative relationships in toxicology. Hum. Ecol. Risk Assess., 9:1547–1559.

Bunnell, F.L. and Huggard, D.J. 1999. Biodiversity across spatial and temporal scales: Problems and opportunities. J. Forest Ecol. Manag., 115:113–126.

Burger, J. 1995. A risk assessment for lead in birds. J. Toxicol. Environ. Health, 45:369–396.

Burger, J. and Gochfeld, M. 1997. Risk, mercury levels, and birds: Relating adverse laboratory effects to field biomonitoring. Environ. Res., 75:160–172.

Burgess, R.M. and Lohmann, R. 2004. Role of black carbon in the partitioning and bioavailability of organic pollutants. Environ. Toxicol. Chem., 23:2531–2533.

Burgess, R.M., Cantwell, M.G., Pelletier, M.C., Ho, K.T., Serbst, J.R., Cook, H.F., and Kuhn, A. 2000. Development of toxicity identification evaluation procedures for characterizing metal toxicity in marine sediments. Environ. Toxicol. Chem., 19:981–991.

Burgman, M. 2005. Risks and Decisions for Conservation and Environmental Management. Cambridge University Press, Cambridge, UK.

Burgman, M.A., Ferson, S., and Akcakaya, H.R. 1993. Risk Assessment in Conservation Biology. Chapman & Hall, London.

Burkhard, L.P. 2000. Estimating dissoved organic carbon partition coefficients for nonionic organic chemicals. Environ. Sci. Technol., 34:4663–4668.

Burkhard, L.P., Endicott, D.D., Cook, P.M., Sappington, K.G., and Winchester, E.L. 2003. Evaluation of two methods for prediction of bioaccumulation factors. Environ. Toxicol. Chem., 22:351–360.

Burkhart, J.G. et al. 2000. Strategies for assessing the implications of malformed frogs for environmental health. Environ. Health Persp., 108:83–90.

Burkhart, J.G. and Gardner, H.S. 1997. Non-mammalian and environmental sentinels in human health: "Back to the future?" Hum. Ecol. Risk Assess., 3:309–328.

Burkholder, J.M., Glasgow, H.B., and Hobbs, C.W. 1995. Fish kills linked to a toxic ambush-predator dinoflagelate: Distribution and environmental conditions. Mar. Ecol. Prog. Ser., 124:43–61.

Burmaster, D.E. and Anderson, P.D. 1994. Principles of good practice for the use of Monte Carlo techniques in human health and ecological risk assessment. Risk Anal., 14:477–481.

Burmaster, D.E. and Hull, D.A. 1997. Using lognormal distributions and log-normal probability plots in probabilistic risk assessments. Hum. Ecol. Risk Assess., 3:235–255.

Burnham, K.P. and Anderson, D.R. 1998. Model Selection and Inference: A Practical Information Theoretic Approach. Springer-Verlag, New York.

Burnham, K.P. and Anderson, D.P. 2001. Kullback–Lieber information as a basis for strong inference in ecological studies. Wildl. Res., 28:111–119.

Burns, L. 2002. Exposure Analysis Modeling System (EXAMS): User manual and system documentation, EPA/600/R-00/81—revision F (June 2002). US Environmental Protection Agency, Research Triangle Park, NC.

Burns, T.P., Hadden, C.T., Cornaby, B.W., and Mitz, S.V. 1997. A food web model of mercury transfer from stream sediment to predators of fish for ecological risk based clean-up goals. Pages 7–27 in F.J. Dwyer, T.R. Doane, and M.L. Hinman (eds.) Environmental Toxicology and Risk Assessment: Modeling and Risk Assessment. ASTM, Philadelphia, PA.

Burton, G.A., Batley, G.E., Chapman, P.M., Forbes, V.E., Smith, E.P., Reynoldson, T., Schlekat, D.E., den Besten, P.J., Bailer, A.J., Green, A.S., and Dwyer, R.L. 2002. A weight-of-evidence framework for assessing sediment (or other) contamination: Improving certainty in the decision-making process. Hum. Ecol. Risk Assess., 8:1675–1696.

Bysshe, S.E. 1988. Uptake by biota. Pages 4.1–4.7.1 in I. Bodek, W.J. Luman, W.F. Reehl, and D.H. Rosenblatt (eds.) Environmental Inorganic Chemistry: Properties, Processes and Estimation Methods. Pergamon Press, New York.

Cade, B.S. and Noon, B.R. 2003. A gentle introduction to quantile regression for ecologists. Frontiers in Ecology, 1:412–420.

Cade, T.J. and Fyfe, R. 1970. The North American peregrine survey, 1970. Can. Fld. Nat., 84:231–245.

Cade, T.J., Lincer, J.L., White, C.M., Roseneau, D.G., and Swartz, L.G. 1971. DDE residues and eggshell changes in Alaskan falcons and hawks. Science, 172:955–957.

Cairns, J.J. and Pratt, J.R. 1993. A history of biological monitoring using macroinvertebrates. Pages 10–27 in D.M. Rosenberg and V.H. Resh (eds.) Freshwater Biomonitoring and Benthic Macroinvertebrates. Chapman & Hall, New York.

Cairns, J., Jr. 1983. Are single species toxicity tests alone adequate for estimating environmental hazards? Hydrobiologia, 100:47–57.

Cairns, J. Jr. 1986. The myth of the most sensitive species. Bioscience, 36:670–672.

Cairns, J. Jr., Dickson, K.L., and Maki, A.W. 1979. Estimating the hazard of chemical substances to aquatic life. Hydrobiologia, 64:157–166.

Calabrese, E.J. 1998. Toxicological and societal implications of hormesis—Part 2. Introduction. BELLE Newsletter, 7:1.

Calabrese, E.J. and Baldwin, L.A. 1993. Performing Ecological Risk Assessments. Lewis Press, Boca Raton, FL.

Calabrese, E.J. and Baldwin, L.A. 1994. A toxicological basis to derive a generic interspecies uncertainty factor. Environ. Health Persp., 102:14–17.

Calabrese, E.J. and Baldwin, L.A. 2000. Chemical hormesis: Its historical foundation as a biological hypothesis. Human Exper. Toxicol., 19:2–31.

Calabrese, E.J. and Baldwin, L.A. 2001. The frequency of U-shaped dose–response in the toxicological literature. Toxicol. Sci., 62:330–338.

Calder, C., Lavine, M., Muller, P., and Clark, J.S. 2003. Incorporating multiple sources of stochasticity into dynamic population models. Ecology, 84:1395–1402.

Calder, W.A.I. and Braun, E.J. 1983. Scaling of osmotic regulation in mammals and birds. Regulatory Integrative Comp. Physiol., 13:R601–R606.

Callahan, B.G. 1996. Special issue: Commemoration of the 50th anniversary of Monte Carlo. Hum. Ecol. Risk Assess., 2:627–1037.

Calow, P. (ed.). 1993. Handbook of Ecotoxicology. Blackwell Scientific, Oxford, UK.

Calow, P. and Sibley, R.M. 1990. A physiological basis of population processes: Ecotoxicological implications. Function. Ecol., 4:283–288.

Calow, P., Sibly, R.M., and Forbes, V. 1997. Risk assessment on the basis of simplified life-history scenarios. Environ. Toxicol. Chem., 16:1983–1989.

Campbell, P.J., Arnold, D., Brock, T., et al. 2003. Guidance Document on Higher-Tier Aquatic Risk Assessment for Pesticides. SETAC, Brussels, Belgium.

Campbell, P.G.C. (1995) Interactions between trace metals and aquatic organisms: A critique of the free-ion activity model. Pages 45–102 in A. Tessier and D.R. Turner (eds.) Metal Speciation and Bioavailability in Aquatic Systems. John Wiley, New York.

Campfens, J. and Mackay, D. (1997) Fugacity-based model of PCB bioaccumulation in complex aquatic food webs. Environ. Sci. Technol., 31:577–583.

Canfield, T.J., Kemble, N.E., Brumbaugh, W.G., Dwyer, F.J., Ingersoll, C.G., and Fairchild, J.F. 1994. Use of benthic community structure and the sediment quality triad to evaluate metal-contaminated sediment in the upper Clark Fork River, Montana. Environ. Toxicol. Chem., 13:1999–2012.

Carlisle, D.M. and Clements, W.H. 1999. Sensitivity and variability of metrics used in biological assessments of running waters. Environ. Toxicol. Chem., 18:285–291.

Carlson, R.W. and Bazzaz, F.A. 1977. Growth reduction in American sycamore (*Platanus occidentalis L.*) caused by Pb-Cd interaction. Environ. Pollut., 12:243–253.

Carlson, R.W. and Rolfe, G.L. 1979. Growth of rye grass and fescue as affected by lead–cadmium–fertilizer interactions. J. Environ. Qual., 8:348–352.

Carpenter, S.R. 1996. Microcosm experiments have limited relevance for community and ecosystem ecology. Ecology, 77:677–680.

Carsel, R.F., Imhoff, J.C., Hummel, P.R., Cheplick, J.M., Donigan, A.S. 2003. PRZM-3, A Model for Predicting Pesticide and Nitrogen Fate in the Crop Root and Unsaturated Soil Zones, User's Manual for Release 3.12, US Environmental Protection Agency, Athens, GA.

Carson, R. 1962. Silent Spring. Houghton Mifflin, Boston, MA.

Caswell, H. 2001. Matrix Population Models: Construction, Analysis, and Interpretation, Second Edition. Sinauer Associates, Sunderland, MA.

Caswell, H. and John, A.M. 1992. From the individual to the population in demographic models. Pages 36–61 in D.L. DeAngelis and L.J. Gross (eds.) Individual-Based Models and Approaches in Ecology. Chapman & Hall, New York.

Cataldo, D.A. and Wildung, R.E. 1978. Soil and plant factors influencing the accumulation of heavy metals by plants. Environ. Health Persp., 27:149–159.

Cavalli-Sforza, L. 2000. Genes, Peoples, and Languages. North Point Press, New York.

CCME (Canadian Council of Ministers of the Environment). 1996. A Framework for Ecological Risk Assessment: General Guidance. 108-4/10-1996e. National Contaminated Sites Remediation Program, Winnipeg, MB.

CCME (Canadian Council of Ministers of the Environment). 1999. Canadian Environmental Quality Guidelines. Canadian Council of Ministers of the Environment, Winnnipeg, MB.

CEC (Commission of the European Community). 1996. Technical Guidance Document in Support of Commission Directive 93/67/EEC on Risk Assessment for New Notified Substances and Commission Regulation (EC) 1488/94 on Risk Assessment of Existing Substances. EC Catalog

Numbers CR-48-96-001, 002, 003, 004-EN-C. Office of Official Publications of the European Community, Luxemburg.

Cestti, R., Srivastava, J., and Jung, S. 2003. Agricultural Non-Point Source Pollution Control: Good Management Practices—The Chesapeake Bay Experience, World Bank Publications, Washington, DC.

Chang, A.C., Granato, T.C., and Page, A.L. 1992. A methodology for establishing phytotoxicity criteria for chromium, copper, nickel, and zinc in agricultural land application of sewage sludge. J. Environ. Qual., 21:521–536.

Chapman, P.M. 1990. The sediment quality triad approach to determining pollution-induced degradation. Sci. Total Environ., 97/98:815–825.

Chapman, P.M. 1995. Extrapolating laboratory toxicity results to the field. Environ. Toxicol. Chem., 14:927–930.

Chapman, P.M. et al. 1997. Workgroup summary report on contaminated site cleanup decisions. Pages 83–114 in C.G. Ingersoll, T. Dillon, and G.R. Biddinger (eds.) Ecological Risk Assessment of Contaminated Sediments. SETAC Press, Pensacola, FL.

Chapman, P.M., Fairbrother, A., and Brown, D. 1998. A critical evaluation of safety (uncertainty) factors for ecological risk assessment. Environ. Toxicol. Chem., 17:99–108.

Chapman, P.M., McDonald, B.G., and Lawrence, G.S. 2002. Weight-of-evidence issues and frameworks for sediment quality (and other) assessments. Hum. Ecol. Risk Assess., 8:1489–1515.

Chappie, D.J. and Burton, G.A. Jr. 1997. Optimization of in situ bioassays with *Hyalella azteca* and *Chironomus tentans*. Environ. Toxicol. Chem., 16:559–564.

Chappie, D.J. and Burton, G.A. Jr. 2000. Applications of aquatic and sediment toxicity testing *in situ*. Soil Sediment Contam., 9:219–245.

Charbonneau, P. and Hare, L. 1998. Burrowing behavior and biogenic structures of mud-dwelling insects. J. N. Am. Benthol. Soc., 17:239–249.

Chaumot, A., Charles, S., Flammarion, P., Garric, J., and Auger, P. 2002. Using aggregation methods to assess toxicant effects on population dynamics in spatial systems. Ecol. Appl., 12:1771–1784.

Chaumot, A., Charles, S., Flammarion, P., and Auger, P. 2003. Ecotoxicology and spatial modeling in population dynamics: An illustration with brown trout. Environ. Toxicol. Chem., 22:958–969.

Christensen, N.L.C. et al. 1996. The report of the Ecological Society of America Committee on the Scientific Basis for Ecosystem Management. Ecol. Appl., 6:665–691.

Christensen, S.W. and Goodyear, C.P. 1988. Testing the validity of stock-recruitment curve fits. Am. Fish. Soc. Monogr., 4:219–231.

Christensen, V. and Pauly, D. 1992. ECOPATH II—a system for balancing steady-state ecosystem models and calculating network characteristics. Ecol. Model., 61:169–185.

Christian, J.J. 1983. Love Canal's unhealthy voles. Natl. His., 10:8–16.

Chung, N. and Alexander, M. 1998. Differences in sequestration and bioavailability of organic compounds aged in dissimilar soils. Environ. Sci. Technol., 32:855–860.

Claassen, M., Strydom, W.F., Murray, K., and Jooste, S. 2001. Ecological Risk Assessment Guidelines. WRC Report Number TT 151/01. Water Research Commission, Pretoria, South Africa.

Clark, B., Henry, J.G., and Mackay, D. 1995. Fugacity analysis and model of organic chemical fate in a sewage treatment plant. Environ. Sci. Technol., 29(6):1488–1494.

Clark, J.R., Goodman, L.R., Borthwick, P.W., et al. 1986. Field and laboratory toxicity tests with shrimp, mysids, and sheepshead minnows exposed to fenthion. Pages 161–176 in T.M. Posten and R. Purdy (eds.) Aquatic Toxicology and Environmental Fate: Volume 9. ASTM STP 921. ASTM, Philadelphia.

Clark, J.S. 2005. Why environmental scientists are becoming Bayesians. Ecol. Lett., 8:2–14.

Clark, M.M. 1996. Transport Modeling for Environmental Engineers and Scientists. John Wiley, New York.

Clemen, R. 1996. Making Hard Decisions: An Introduction to Decision Analysis, Second Edition. Duxbury Press, Belmont, CA.

Clements, W.H. 1997. Ecological significance of endpoints used to assess sediment quality. Pages 123–134 in C.G. Ingersoll, T. Dillon, and G.R. Biddinger (eds.) Ecological Risk Assessment of Contaminated Sediments. SETAC Press, Pensacola, FL.

Clements, W.H. and Newman, M.C. 2002. Community Ecotoxicology. John Wiley, Chichester, UK.

Clements, W.H., Cherry, D.S., and Van Hassel, J.H. 1992. Assessment of the impact of heavy metals on benthic communities at the Clinch River, Virginia. Can. J. Fish. Aquat. Sci., 49:1686–1694.

Clewell, H.J.I., Anderson, M.E., and Barton, H.A. 2002. A consistent approach for the application of pharmacokinetic modeling in cancer and noncancer risk assessment. Environ. Health Persp., 110:85–93.

Clifford, P.A., Barchers, D.E., Ludwig, D.F., Sielken, R.L., Klingensmith, J.S., Graham, R.V., and Banton, M.I. 1995. An approach to quantifying spatial components of exposure for ecological risk assessment. Environ. Toxicol. Chem., 14:895–906.

Codex. 1997. Codex Alimantarius Commission Procedural Manual, Tenth Edition. Joint FAO/WHO Food Standards Programme, FAO, Rome.

Cogliano, V.J. 1997. Plausible upper bounds: Are their sums plausible? Risk Anal., 17:77–84.

Colborn, T. and Thayer, K. 2000. Aquatic ecosystems: Harbingers of endocrine disruption. Ecol. Appl., 10:949–957.

Cole, L.C. 1954. The population consequences of life history phenomena. Q. Rev. Biol., 19:103–137.

Committee on Environment and Natural Resources. 1999. Ecological Risk Assessment in the Federal Government. CENR/5-99/001. National Science and Technology Council, Washington, DC.

Connell, D.W. and Markwell, R.D. 1990. Bioaccumulation in the soil to earthworm system. Chemosphere, 20:91–100.

Connell, D.W., Wong, B.S.F., Lam, P.K.S., Poon, K.F., Lam, M.H.W., Wu, R.S.S., Richardson, B.J., and Yen, Y.F. 2002. Risk to breeding success of ardeids by contaminants in Hong Kong: Evidence from trace metals in feathers. Ecotoxicology, 11:49–59.

Connell, D.W., Fung, C.N., Minh, T.B., Tanabe, S., Lam, P.K.S., Wong, B.S.F., Lam, M.H.W., Wong, L.C., Wu, R.S.S., and Richardson, B.J. 2003. Risk to breeding success of fish-eating ardeids due to persistent organic contaminants in Hong Kong: Evidence from organochlorine compounds in eggs. Water Res., 37:459–467.

Connolly, J.P. and Winfield, R.P. 1984. WASTOX: A framework for modeling toxic chemicals in aquatic systems. Part 1: Exposure concentration. EPA-600-3-84-077. US Environmental Protection Agency, Gulf Breeze, FL.

Cooke, R.M. 1991. Experts in Uncertainty: Opinion and Subjective Probability in Science. Oxford University Press, New York.

Copp, G.H., Garthwaite, R., and Gozlan, R.E. 2005. Risk identification and assessment of non-native freshwater fishes: Concepts and perspectives on protocols for the UK. Tech. Report No. 129. CEFAS, Lowestoft, UK.

Cormier, S.M., Smith, M., Norton, S., and Neiheisel, T. 2000. Assessing ecological risk in watersheds: A case study of problem formulation in the Big Darby Creek watershed, Ohio. Environ. Toxicol. Chem., 19:1082–1096.

Corp, N. and Morgan, A.J. 1991. Accumulation of heavy metals from soils by the earthworm *Lumbricus rubellus*: Can laboratory exposure of "control" worms reduce biomonitoring problems? Environ. Pollut., 74:39–52.

Costanza, R., d'Arge, R., deGroot, R., Farber, S., Grasso, M., Hannon, B., Limburg, K., et al. 1997. The value of the world's ecosystem services and natural capital. Nature, 387:253–260.

Costanza, R., Sklar, F., and White, M. 1990. Modeling coastal landscape dynamics. Bioscience, 40:91–107.

Cowan, C.E., Mackay, D., Feijtel, T.C.J., van de Meent, D., Di Guardo, A., Davies, J., and Mackay, N. 1995a. The multi-media fate model: A vital tool for predicting the fate of chemicals. Proceedings of a workshop organized by the Society of Environmental Toxicology and Chemistry (SETAC). SETAC Press, Pensacola, FL.

Cowan, C.E., Versteeg, D.J., Larson, R.J., and Kloepper-Sams, P.J. 1995b. Integrated approach for environmental assessment of new and existing substances. Reg. Toxicol. Pharmacol., 21:3–31.

Cowgill, U.M. 1988. Paleoecology and environmental analysis. Pages 53–62 in W.J. Adams, G.A. Chapmen, and W.G. Landis (eds.) Aquatic Toxicology and Hazard Assessment: Volume 10. ASTM, Philadelphia, PA.

Craft, R.A. and Craft, K.P. 1996. Use of free ranging American kestrels and nest boxes for contaminant risk assessment sampling: A field application. J. Raptor Res., 30:207–212.

Crane, M. and Godolphin, E. 2000. Statistical analysis of effluent bioassays. R&D Tech. Report E19. Environment Agency, Bristol, UK.

Crane, M. and Newman, M.C. 2000. What level of effect is a no observed effect? Environ. Toxicol. Chem., 19:516–519.

Crane, M., Newman, M.C., Chapman, P.F., and Fenlon, J. 2002. Risk Assessment with Time to Event Models. Lewis Publishers, Boca Raton, FL.

Crawford-Brown, D. 1999. Risk-Based Environmental Decisions. Kluwer Academic, Boston, MA.

Cromer, A. 1993. Uncommon Sense: The Heretical Nature of Science. Oxford University Press, Oxford, UK.

Cromey, C.J., Nickell, T.D., and Black, K.D. 2002. DEPOMOD–modelling the deposition and biological effects of waste solids from marine cage farms. Aquaculture 214:211–239.

Crommentuijn, T., Sijm, D., de Bruijn, J., van den Hoop, M., van Leeuwen, K., and van de Plassche, E. 2000. Maximum permissible and negligible concentrations for metals and metalloids in the Netherlands, taking into account background concentrations. J. Environ. Manag., 60:122–143.

Cronin, M.T.D., Walker, J.D., Jaworska, J., Comber, M.H.I., Watts, C.D., and Worth, A.P. 2003. Use of QSARs in international decision-making frameworks to predict ecological effects and environmental fate of chemical substances. Environ. Health Persp., 111:1376–1390.

Crossen, C. 1994. Tainted Truth: The Manipulation of Fact in America. Simon & Schuster, New York.

Crump, K.S. 1984. A new method for determining allowable daily intakes. Fundam. Appl. Toxicol., 4:854–871.

CSTE/EEC. 1994. EEC water quality objectives for chemicals dangerous to aquatic environments. Rev. Environ. Contam. Toxicol., 137:83–112.

Cullen, A.C. and Frey, H.C. 1999. Probabilistic Techniques in Exposure Assessment: A Handbook for Dealing with Variability and Uncertainty in Models and Inputs. Plenum Press, New York.

Currie, R.S., Fairchild, W.L., and Muir, D.C.G. 1997. Remobilization and export of cadmium from lake sediments by emerging insects. Environ. Toxicol. Chem., 16:2333–2338.

Custer, C.M., Custer, T.W., Archuleta, A.S., et al. 2003. A mining impacted stream: Exposure and effects of lead and other trace elements on tree swallows (*Tachycineta bicolor*) nesting in the upper Arkansas River Basin, Colorado. Pages 787–812 in D.J. Hoffman, B. Rattner, G.A. Burton, Jr., and J. Cairns, Jr. (eds.) Handbook of Ecotoxicology. Lewis Publishers, Boca Raton, FL.

Cuypers, C., Grotenhuis, T., Joziasse, J., and Rulkens, W. 2000. Rapid persulfate oxidation predicts PAH bioavailability in soils and sediments. Environ. Sci. Technol., 34:2057–2063.

Dai, J., Becquer, T., Rouiller, J.H., Reversat, G., Bernhardt-Reversat, F., Nahmani, J., and Lavelle, P. 2004. Heavy metal accumulation by two earthworm species and its relationship to total and DTPA-extractable metals in soils. Soil Biol. Biochem., 36:91–98.

Daily, G.C., Alexander, S., Ehrlich, P., Goulder, L., Lubchenco, J., Matson, P.A., Mooney, H.A., et al. 1997. Ecosystem services: Benefits supplied to human societies by natural ecosystems. Ecological Society of America, Washington, DC.

Daily G.C. and Ellison, K. 2002. The New Economy of Nature: The Quest to Make Conservation Profitable. Island Press, Washington, DC.

Dakins, M.E. 1999. The value of the value of information. Hum. Ecol. Risk Assess., 5:281–290.

Dale, V.H. and Gardner, R.H. 1987. Assessing regional impacts of growth declines using a forest succession model. J. Environ. Manag., 24:83–93.

Daniels, R.E. and Allan, J.D. 1981. Life table evaluation of chronic exposure to a pesticide. Can. J. Fish. Aquat. Sci., 38:485–494.

Danielson, T.J. 1998. Wetland Bioassessment Fact Sheets. US Environmental Protection Agency, Office of Water, Washington, DC.

Davidson, I.W.F., Parker, J.C., and Beliles, R.P. 1986. Biological basis for extrapolation across mammalian species. Regul. Toxicol. Pharmacol., 6:211–237.

Davies, J.C. 1996. Comparing Environmental Risks: Tools for Setting Government Priorities. Resources for the Future, Washington, DC.

Davis, B.M.K. and French, N.C. 1969. The accumulation of organochlorine insecticide residues by beetles, worms, and slugs in sprayed fields. Soil Biol. Biochem., 1:45–55.

Davis, L.S., Johnson, K.N., Bettinger, P., and Howard, T. 2000. Forest Management, Fourth Edition. McGraw-Hill, New York.

Davis, W.S. and Simon, T.P. (eds.). 1995. Biological Assessment and Criteria: Tools for Water Resource Planning and Decision Making. Lewis Publishers, Boca Raton, FL.

Dawes, R.M. 1993. Prediction of the future versus an understanding of the past: A basic asymmetry. Am. J. Psychol., 106:1–24.

Dawes, R.M. 2001. Everyday Irrationality. Westview Press, Boulder, CO.

Dawson, W.R., Ligon, J.D., Murphy, J.R., Myers, J.P., Simberloff, D., and Verner, J. 1987. Report of the advisory panel on the spotted owl. The Condor, 89:205–229.

DeAngelis, D.L. 1992. Dynamics of Nutrient Cycling and Food Webs. Chapman & Hall, London, UK.

DeAngelis, D.L. and Rose, K.A. 1992. Which individual-based approach is most appropriate for a given problem? Pages 367–387 in D.L. DeAngelis and L.J. Gross (eds.) Individual-Based Models and Approaches in Ecology. Chapman & Hall, New York.

DeAngelis, D.L., Bartell, S.M., and Brenkert, A.L. 1989. Effects of nutrient cycling and food chain length on resilience. Am. Natl., 134:788–805.

DeAngelis, D.L., Barnthouse, L.W., Van Winkle, W., and Otto, R.G. 1990. A critical appraisal of population approaches in assessing fish community health. J. Great Lakes Res., 16:576–590.

DeAngelis, D.L., Goudbout, L., and Shuter, B.J. 1991. An individual-based approach to predicting density-dependent compensation in smallmouth bass populations. Ecol. Model., 57:91–115.

DeBruyn, A.M.H., Marcogliese, D.J., and Rasmussen, J.B. 2003. The role of sewage in a large river food web. Can. J. Fish. Aquat. Sci., 60:1332–1344.

Deichmann, W.B., Henschler, D., Holmstedt, B., and Keil, G. 1986. What is there that is not a poison? A study of the Third Defense by Paracelsus. Arch. Toxicol., 58:207–213.

Deis, D.R. and French, D.P. 1998. The use of methods for injury determination and quantification from natural resource damage assessment in ecological risk assessment. Hum. Ecol. Risk Assess., 4:887–903.

deKroon, H.A., Plaisier, A., Van Groenendael, J., and Caswell, H. 1986. Elasticity: The relative contribution of demographic parameters to population growth rate. Ecology, 67:1427–1431.

Delorme, P., Francois, D., Hart, C., Hodge, V., Kaminski, G., Kriz, C., Mulye, H., Sebastien, R., Takacs, P., and Wandelmaier, F. 2005. Final report for the PMRA Workshop: Assessment Endpoints for Environmental Protection. Health Canada, Ottawa, Canada.

den Boer, P.J. 1968. Spreading of risk and stabilization of animal numbers. Acta Biotheoretica, 18:165–194.

Deneer, J.W., Sinnige, T.L. Seinen, W., and Hermens, J.L.M. 2005. The joint acute toxicity to Daphnia magna of industrial organic chemicals at low concentrations. Aquat. Toxicol., 12:33–38.

Dennis, B. 2004. Rejoinder. Pages 367–378 in M.L. Taper and S.R. Lele (eds.) The Nature of Scientific Evidence. University of Chicago Press, Chicago, IL.

Depledge, M.H. and Fossi, M.C. 1994. The role of biomarkers in environmental assessment: (2) invertebrates. Ecotoxicology, 3:173–179.

Depledge, M.H. and Galloway, T.S. 2005. Healthy animals, healthy ecosystems. Front. Ecol., 3:251–258.

De Roos, A.M., Diekmann, O., and Metz, J.A.J. 1992. Studying the dynamics of structured population models: A versatile technique and its application to Daphnia. Am. Natl., 139:123–147.

Deschenes, M., Belanger, L., and Giroux, J.-L. 2003. Use of farmland riparian strips by declining and crop damaging birds. Agr. Ecosys. Environ., 95:567–577.

DeShon, J.E. 1995. Development and application of the Invertebrate Community Index (ICI). Pages 217–243 in W.S. Davis and T.P. Simon (eds.) Biological Assessment and Criteria. Lewis Publishers, Boca Raton, FL.

Detenbeck, N.E., DeVore, P.W., Niemi, G.J., and Lima, A. 1992. Recovery of temperate-stream fish communities from disturbance: A review of case studies and synthesis of theory. Environ. Manag., 16:33–53.

Devillers, J. and Bintein, S. 1995. ChemFrance: A regional level III fugacity model applied to France. Chemosphere 30(3):457–476.

deVlaming, V. and Norberg-King, T. 1999. A review of single species toxicity tests: Are the tests reliable predictors of aquatic ecosystem community responses? EPA/600/R-97/114. US Environmental Protection Agency, Duluth, MN.

De Wolf, W., Seibel-Sauer, A., Lecloux, A., Koch, V., Holt, M., Feijtel, T., Comber, M., and Boeije, G. 2005. Mode of action and aquatic exposure thresholds of no concern. Environ. Toxicol. Chem., 24:479–485.

DeZwart, D. 2002. Observed regularities in species sensitivity distributions for aquatic species. Pages 133–154 in L. Posthuma, G.W. Suter II, and T. Traas (eds.) Species Sensitivity Distributions in Ecotoxicology. Lewis Publishers, Boca Raton, FL.

DeZwart, D. and Posthuma, L. 2005. Complex mixture toxicity for single and multiple species: Proposed methodologics. Environ. Toxicol. Chem., 24:2665–2676.

Diamond, J.M. and Serveiss, V.B. 2001. Identifying sources of stress to native aquatic fauna using a watershed ecological risk assessment framework. Environ. Sci. Technol., 35:4711–4718.

Diamond, J.M., Bressler, D.W., and Serveiss, V.B. 2002. Assessing relationships between human land uses and the decline of native mussels, fish, and macroinvertebrates in the Clinch and Powell River watershed, USA. Environ. Toxicol. Chem., 21:1147–1155.

Dickson, K.L., Maki, A.W., and Cairns, J. Jr. (eds.). 1979. Analyzing the Hazard Evaluation Process. American Fisheries Society, Washington, DC.

Dickson, K.L., Waller, W.T., Kennedy, J.H., and Ammann, L.P. 1992. Assessing the relationship between ambient toxicity and instream biological response. Environ. Toxicol. Chem., 11:1307–1322.

Dickson, K.L., Waller, W.T., Kennedy, J.H., et al. 1996. Relationship between effluent toxicity, ambient toxicity, and receiving system impacts: Trinity River dechlorination case study. Pages 287–308 in D.R. Grothe, K.L. Dickson, and D.K. Reed-Judkins (eds.) Whole Effluent Toxicity Testing: An Evaluation of Methods and Prediction of Receiving System Impacts. SETAC Press, Pensacola, FL.

Dieter, C.D., Flake, L.D., and Duffy, W.G. 1995. Effects of phorate on ducklings in northern prairie wetlands. J. Wildl. Manag., 59:498–505.

Di Guardo, A., Calamari, D., Zanin, G., Consalter, A., and Mackay, D. 1994. A fugacity model of pesticide runoff to surface water: Development and validation. Chemosphere, 28(3):511–531.

DiToro, D.M. 2001. Sediment Flux Modeling. John Wiley, New York.

DiToro, D.M. and McGrath, J.A. 2000. Technical basis for narcotic chemicals and polycyclic aromatic hydrocarbon criteria. I. Mixtures and sediments. Environ. Toxicol. Chem., 19:1971–1982.

DiToro, D.M., Halden, J.A., and Plafkin, J.L. 1991a. Modeling Ceriodaphnia toxicity in the Naugatuck River II. Copper, hardness, and effluent interactions. Environ. Toxicol. Chem., 10:261–274.

DiToro, D.M., Zarba, C.S., Hansen, D.H., Berry, W.J., Swartz, R.C., Cowan, C.E., Pavlou, S.P., Allen, H.E., Thomas, N.A., and Paquin, A.P.R. 1991b. Technical basis for establishing sediment quality criteria for nonionic organic chemicals using equilibrium partitioning. Environ. Toxicol. Chem., 10:1541–1583.

DiToro, D.M., Mahony, J.D., Hansen, D.J., Scott, K.J., Carlson, A.R., and Ankley, G.T. 1992. Acid volatile sulfide predicts the acute toxicity of cadmium and nickel in sediments. Environ. Sci. Technol., 26:96–101.

DiToro, D.M., McGrath, J.A., and Hansen, D.J. 2000. Technical basis for narcotic chemicals and polycyclic aromatic hydrocarbon criteria. I. Water and tissue. Environ. Toxicol. Chem., 19:1951–1970.

DiToro, D.M., Allen, H.E., Bergman, H.A., Meyer, J.S., Paquin, P.R., and Santore, R.C. 2001. A biotic ligand model of the acute toxicity of metals. I. Technical basis. Environ. Toxicol. Chem., 20:2383–2396.

Dixit, S.S., Smol, J.P., Kingston, J.C., and Charles, D.F. 1992. Diatoms: Powerful indicators of environmental change. Environ. Sci. Technol., 26:23–33.

Dobson, S. and Shore, R.F. 2002. Extrapolation for terrestrial vertebrates. Hum. Ecol. Risk Assess., 8:45–54.

DOE (Department of Energy). 1995. Remedial Investigation Report on Waste Area Grouping 5 at Oak Ridge National Laboratory, Oak Ridge, Tennessee. ORNL/ER-284. US Department of Energy, Office of Environmental Restoration and Waste Management, Washington, DC.

DOI (US Department of the Interior). 1986. Natural Resource Damage Assessments: Final Rule. Code of Federal Regulations, 43 CFR 11.

DOI (US Department of the Interior). 1987. Natural resource damage assessments: Final rule. Fed. Regist., 52:9042–9100.

Donker, M.H., Eijsackers, H., and Heimbach, F. 1994. Ecotoxicology of Soil Organisms. Lewis Publishers, Boca Raton, FL.

Donkin, P., Smith, E.L., and Rowland, S.J. 2003. Toxic effects of unresolved complex mixtures of aromatic hydrocarbons accumulated in mussels, *Mytilus edulis*, from contaminated field sites. Environ. Sci. Technol., 37:4825–4830.

Donkin, S.G. and Dusenbery, D.B. 1993. A soil toxicity test using the nematode *Caenorhabditis elegans* and an effective method of recovery. Environ. Contam. Toxicol., 25:145–151.

Doull, J. 1984. The past, present, and future of toxicology. Pharmacol. Rev., 36:15S–18S.

Dourson, M.L. 1986. New approaches in the derivation of acceptable daily intake (ADI). Comments Toxicol., 1:35–48.

Dourson, M.L. and Stara, J.F. 1983. Regulatory history and experimental support of uncertainty (safety) factors. Reg. Toxicol. Pharmacol., 3:224–238.

Dourson, M.L., Teuschler, L.K., Durkin, P.R., and Stiteler, W.M. 1997. Categorical regression of toxicity data: A case study using aldicarb. Reg. Toxicol. Pharmacol., 25:121–129.

Dowdy, D.L. and McKone, T.E. 1997. Predicting plant uptake of organic chemicals from soil or air using octanol/water and octanol/air partition rations and a molecular connectivity index. Environ. Toxicol. Chem., 16:2448–2456.

Driscoll, S.B.K., Wickwire, W.T., Cura, J., Vorhees, D.J., Butler, C.L., Moore, D.W., and Bridges, T.S. 2002. A comparative screening-level ecological risk assessment for dredged material management alternatives in New York/New Jersey Harbor. Hum. Ecol. Risk Assess., 8:603–626.

Driver, C.J., Ligotke, M.W., Van Voris, P., McVeety, B.D., and Brown, D.B. 1991. Routes of uptake and their relative contribution to the toxicologic response of northern bobwhile (*Colinus virginianus*) to an organophosphate pesticide. Environ. Toxicol. Chem., 10:21–33.

Drummond, D.B. and Russom, C.L. 1990. Behavioral toxicity syndromes: A promising tool for assessing toxicity metchanisms in juvenile fathead minnows. Environ. Toxicol. Chem., 9:37–46.

Drummond, R.A., Russom, C.L., Geiger, D.L., and DeFoe, D.L. 1986. Behavioral and morphological changes in fathead minnow (*Pimephales promelas*) as diagnostic endpoints for screening chemicals according to mode of action. Pages 415–435 in T.M. Poston and R. Purdy (eds.) Aquatic Toxicology and Environmental Fate: Volume 9. ASTM, Philadelphia, PA.

Duke, C.S. and Briede, J.W. 2001. Ecological risk assessment review. Pages 257–263 in S.L. Benjamin and D.A. Belluck (eds.) A Practical Guide to Understanding, Managing, and Reviewing Environmental Risk Assessment Reports. Lewis Publishers, Boca Raton, FL.

Dunning, J.B. 1993. CRC Handbook of Avian Body Masses. CRC Press, Boca Raton, FL.

Dykstra, C.R., Meyer, M.W., Warnke, D.K., Karasov, W.H., Andersen, D.E., Bowerman, W.W.I., and Giesy, J.P. 1998. Low reproductive rates of Lake Superior bald eagles: Low food delivery rates or environmental contaminants. J. Great Lakes Res., 24:32–44.

Dykstra, C.R., Meyer, M.W., Rasmussen, P.W., and Warnke, D.K. 2005. Contaminant concentrations and reproductive rate of Lake Superior bald eagles. J. Great Lakes Res., 31:227–235.

Echols, K.R., Tillitt, D.E., Nichols, J.W., Secord, A.L., and McCarty, J.P. 2004. Bioaccumulation of PCB congeners in nestling tree swallows (*Tachycineta bicolor*) from two contaminated sites on the upper Hudson River, New York. Environ. Sci. Technol., 38:6240–6246.

ECOFRAM. 1999. ECOFRAM Aquatic Report, Ecological Committee on FIFRA Risk Assessment Methods (ECOFRAM). US Environmental Protection Agency, Washington, DC.

ECOFRAM Aquatic Workgroup. 1999.ECOFRAM Aquatic Report. Available at: http://www.epa.gov/oppefed1/ecorisk/index.htm.

ECOFRAM Terrestrial Workgroup. 1999. ECOFRAM Terrestrial Draft Report. Available at: http://www.epa.gov/oppefed1/ecorisk/index.htm.

Ecological Committee on FIFRA Risk Assessment Methods. 1999a. ECOFRAM Aquatic Report. Available at: http://www.epa.gov/oppefed1/ecorisk/index.htm.

Ecological Committee on FIFRA Risk Assessment Methods. 1999b. ECOFRAM Terrestrial Draft Report. Available at: http://www.epa.gov/oppefed1/ecorisk/index.htm.

Efron, B. and Tibshirani, R. 1993. An Introduction to the Bootstrap. Chapman & Hall, New York.

Efroymson, R.E. and Suter, G.W., II. 1999. Finding a niche for soil microbial toxicity tests in ecological risk assessment. Hum. Ecol. Risk Assess., 5:715–727.

Efroymson, R.A. and Suter, G.W., II. 2001a. Ecological risk assessment framework for low-altitude aircraft overflights. I. Planning the analysis and estimating exposure. Risk Anal., 21:251–262.

Efroymson, R.A. and Suter, G.W., II. 2001b. Ecological risk assessment framework for low-altitude aircraft overflights. II. Estimating effects on wildlife. Risk Anal., 21:263–274.

Efroymson, R.A., Suter, G.W., II, Sample, B.E., and Jones, D.S. 1997. Preliminary remediation goals for ecological endpoints. ES/ER/TM-126/R2. Oak Ridge National Laboratory, Oak Ridge, TN.

Efroymson, R.E., Will, M.E., and Suter, G.W., II. 1997a. Toxicological benchmarks for contaminants of potential concern for effects on soil and litter invertebrates and heterotrophic processes: 1997 revision. ES/ER/TM-126/R2. Oak Ridge National Laboratory, Oak Ridge, TN.

Efroymson, R.E., Will, M.E., and Suter, G.W., II. 1997b. Toxicological benchmarks for screening contaminants of potential concern for effects on terrestrial plants. ES/ER/TM-85/R3. Oak Ridge National Laboratory, Oak Ridge, TN.

Efroymson, R.A., Sample, B.E., and Suter, G.W., II. 2001. Uptake of inorganic chemicals from soil by plant leaves: Regressions of field data. Environ. Toxicol. Chem., 20:2561–2571.

Efroymson, R.A., Nicollette, J.P., and Suter, G.W., II. 2003. A framework for net environmental benefit analysis for remediation or restoration of contaminated sites. ORNL/TM-2003/17. Oak Ridge National Laboratory, Oak Ridge, TN.

Efroymson, R.A., Nicollette, J.P., and Suter, G.W., II. 2004. A framework for net environmental benefit analysis for remediation or restoration of contaminated sites. Environ. Manag., 34:315–331.

Eganhouse, R.P. and Calder, J.A. 1976. The solubility of medium molecular weight aromatic hydrocarbons and the effects of hydrocarbon cosolvents and salinity. Geochem. Cosmochim. Acta, 40:555–561.

Eisler, R. 1995. Electroplating wastes in marine environments: A case history at Quonset Point, Rhode Island. Pages 539–548 in D.J. Hoffman, B. Rattner, G.A. Burton, and J. Cairns (eds.) Handbook of Ecotoxicology. Lewis Publishers, Boca Raton, FL.

Eisler, R. 1988. Lead hazards to fish, wildlife, and invertebrates: A synoptic review. Biological Report 85(1.14). US Fish and Wildlife Service, Laurel, MD.

Ellersieck, M.R., Asfaw, A., Mayer, F.L., Krause, G.F., Sun, K., and Lee, G. Acute to chronic estimation (ACE v 2.0) with time-concentration-effect models. User manual and software. EPA/600/R-03/107. 2003. US Environmental Protection Agency, Washington, DC.

Emans, H.J.B., Plassche, E.J. v.d., Canton, J.H., Okkerman, P.C., and Sparenburg, P.M. 1993. Validation of some extraplation methods used for effects assessment. Environ. Toxicol. Chem., 12:2139–2154.

Emlen, J.M. and Pikitch, E.K. 1989. Animal population dynamics: Identification of critical components. Ecol. Model., 44:253–274.

Environment Agency. 1996. LandSim: Landfill performance simulation by Monte Carlo method. CWM 094/96. Environment Agency, Bristol, UK.

Environment Canada. 1999. Guidance document on application and interpretation of single-species tests in environmental toxicology. EPS 1/RM/34. Method Development and Application Section, Ottawa, Ontario.

Environment Canada. 2005. Guidance document on statistical methods for environmental toxicity tests. EPS 1/RM/46. Method Development and Application Section, Ottawa, Ontario.

Environmental Response Team. 1994a. Plant biomass determination. SOP#: 2034. US Environmental Protection Agency, Edison, NJ.

Environmental Response Team. 1994b. Tree coring and interpretation. SOP#: 2036. US Environmental
    Protection Agency, Edison, NJ.
Environmental Response Team. 1994c. Terrestrial plant community sampling. SOP#: 2037. US Envir-
    onmental Protection Agency, Edison, NJ.
Environmental Response Team. 1995. Superfund program representative sampling guidance, Volume 1:
    Soil, interim final. EPA 540/R-95/141. US Environmental Protection Agency, Washington, DC.
Environmental Response Team. 1996. Vegetation assessment field protocol. SOP#: 2038. US Environ-
    mental Protection Agency, Washington, DC.
EPA (US Environmental Protection Agency). 1982. Air quality criteria for oxides of nitrogen. EPA-
    600/8-84-026f. Office of Air Quality Planning and Standards, Research Triangle Park, NC.
EPA (US Environmental Protection Agency) 1983. Water Quality Standards Handbook. Office of
    Water, Washington, DC.
EPA (US Environmental Protection Agency). 1985. Water quality criteria; availability of documents.
    Fed. Regist., 50:30784–30796.
EPA (US Environmental Protection Agency). 1989. Use of starling nest boxes for field reproductive
    studies. EPA 600/8-89/056. Office of Research and Development, Corvallis, OR.
EPA (US Environmental Protection Agency). 1990. National oil and hazardous substances pollution
    contingency plan: Final rule. Red. Reg., 55:8666–8873.
EPA (US Environmental Protection Agency). 1991a. Methods for aquatic toxicity identification evalu-
    ations: Phase I toxicity characterization procedures, Second Edition. EPA-600/6-91-003. US
    Environmental Protection Agency, Duluth, MN.
EPA (US Environmental Protection Agency). 1991b. Technical support document for water quality-
    based toxics control. EPA/505/2-90-001. Office of Water, Washington, DC.
EPA (US Environmental Protection Agency). 1992a. Framework for ecological risk assessment.
    EPA/630/R-92/001. Risk Assessment Forum, Washington, DC.
EPA (US Environmental Protection Agency). 1992b. Dermal exposure assessment: Principles
    and applications. EPA/6008-91/011B. Office of Health and Environmental Assessment,
    Washington, DC.
EPA (US Environmental Protection Agency). 1992c. Draft report: A cross-species scaling factor for
    carcinogen risk assessment based on equivalence of mg/kg3/4/day: Notice. Fed. Regist.,
    57:24152–24173.
EPA (US Environmental Protection Agency). 1993a. Methods for aquatic toxicity identification evalu-
    ations: Phase I toxicity characterization procedures. EPA-600/6-91-005F. Office of Research
    and Development, Duluth, MN.
EPA (US Environmental Protection Agency). 1993b. Methods for aquatic toxicity identification evalu-
    ations: Phase II toxicity identification procedures for samples exhibiting acute and chronic
    toxicity. EPA-600/6-92-080. Office of Research and Development, Duluth, MN.
EPA (US Environmental Protection Agency). 1993d. Technical basis for deriving sediment quality
    criteria for nonionic organic contaminants for the protection of benthic organisms by using
    equilibrium partitioning. EPA-822-R-93-001. Office of Water, Washington, DC.
EPA (US Environmental Protection Agency). 1993e. Water quality guidance for the Great Lakes system
    and correction: Proposed rules. Fed. Regist, 58:20802–21047.
EPA (US Environmental Protection Agency). 1993f. Wildlife criteria portions of the proposed water
    quality criteria for the Great Lakes System. EPA/822/R-93/006. Office of Science and Tech-
    nology, Washington, DC.
EPA (US Environmental Protection Agency). 1993g. Wildlife exposure factors handbook. EPA/600/R-
    93/187. Office of Health and Environmental Assessment, Washington, DC.
EPA (US Environmental Protection Agency). 1995. Mercury Study Report to Congress. EPA-452/R-
    96-011. Office of Air Planning and Standards and Office of Research and Development,
    Washington, DC.
EPA (US Environmental Protection Agency). 1996a. Proposed testing guidelines. Fed. Regist,
    61:16486–16488.
EPA (US Environmental Protection Agency). 1996b. Biological criteria: Technical guidance for streams
    and small rivers. EPA-822/B-96-001. Office of Water, Washington, DC.

EPA (US Environmental Protection Agency). 1997a. EPA's comparative risk projects: 1–3. Washington, DC.

EPA (US Environmental Protection Agency). 1997b. Estuarine and marine waters bioassessment and biocriteria technical guidance. EPA 822-B-97-002A. Office of Water, Washington, DC.

EPA (US Environmental Protection Agency). 1998a. Guidelines for ecological risk assessment. EPA/630/R-95/002F. Risk Assessment Forum, Washington, DC.

EPA (US Environmental Protection Agency). 1998b. Lake and reservoir bioassessment and biocriteria technical guidance document. EPA-841-B-98-007. Office of Water, Washington, DC.

EPA (US Environmental Protection Agency). 1999. Protocol for Developing Nutrient TMDLs. EPA 841-B-99-007. Office of Water, Washington, DC.

EPA (US Environmental Protection Agency). 2000a. Ambient aquatic life criteria for dissolved oxygen (saltwater): Caper Cod to Cape Hateras. EPA 822-R-00-012. Office of Water, Washington, DC.

EPA (US Environmental Protection Agency). 2000b. Technical basis for the derivation of equilibrium partitioning sediment guidelines (ESGs) for the protection of benthic organisms: Nonionic organics. EPA-822-R-00-001. Office of Water, Washington, DC.

EPA (US Environmental Protection Agency). 2000c. Stressor identification guidance document. EPA/822/B-00/025. Office of Water, Washington, DC.

EPA (US Environmental Protection Agency). 2002a. ECOTOX User Guide: EcoTOXicology database version 3.0. Office of Water, Washington, DC.

EPA (US Environmental Protection Agency). 2002b. Methods for measuring the acute toxicity of effluents to freshwater and marine organisms, Fifth edition. EPA/821/R-02/012. Office of Water, Washington, DC.

EPA (US Environmental Protection Agency). 2002c. Procedures for the derivation of equilibrium partitioning sediment benchmarks (ESBs) for the protection of benthic organisms: Dieldrin. EPA-600-R-02-010. Office of Water, Washington, DC.

EPA (US Environmental Protection Agency). 2002d. Procedures for the derivation of equilibrium partitioning sediment benchmarks (ESBs) for the protection of benthic organisms: Endrin. EPA-600-R-02-009. Office of Water, Washington, DC.

EPA (US Environmental Protection Agency). 2002e. Procedures for the derivation of equilibrium partitioning sediment benchmarks (ESBs) for the protection of benthic organisms: Metal mixtures. EPA-600-R-02-011. Office of Water, Washington, DC.

EPA (US Environmental Protection Agency). 2002f. Procedures for the derivation of equilibrium partitioning sediment benchmarks (ESBs) for the protection of benthic organisms: PAH mixtures. EPA-600-R-02-013. Office of Water, Washington, DC.

EPA (US Environmental Protection Agency). 2002g. Quality assurance project plans for modeling. QA/G-5M. US Environmental Protection Agency, Washington, DC.

EPA (US Environmental Protection Agency). 2002h. Short-term methods for estimating the chronic toxicity of effluents and receiving waters to freshwater organisms, Fourth Edition. EPA-821-R-02-013. Office of Research and Development, Washington, DC.

EPA (US Environmental Protection Agency). 2002i. Short-term methods for estimating the chronic toxicity of effluents and receiving waters to marine and estuarine organisms, Third Edition. EPA-821-R-02-013. Office of Research and Development, Washington, DC.

EPA (US Environmental Protection Agency). 2003a. 2003 draft update of ambient water quality criteria for copper. EPA 822-R-03-026. Office of Watter, Washington, DC.

EPA (US Environmental Protection Agency). 2003b. Developing relative potency factors for pesticide mixtures: Biostatistical analysis of joint dose-response. EPA/600/R-03/052. Office of Research and Development, Cincinnati, OH.

EPA (US Environmental Protection Agency). 2003c. Generic Ecological Assessment Endpoints (GEAEs) for Ecological Risk Assessment. EPA/630/P-02/004B. Risk Assessment Forum, Washington, DC.

EPA (US Environmental Protection Agency). 2003d. Methodology for deriving ambient water quality criteria for the protection of human health (2000). Technical support document, Volume 2: Development of national bioaccumulation factors. EPA-822-R-03-030. Office of Water, Washington, DC.

EPA (US Environmental Protection Agency). 2003e. Organophosphate pesticides: Revised OP cummulative risk assessment. Available at: http://www.epa.gov/pesticides/cumulative/rra-op/.

EPA (US Environmental Protection Agency). 2004a. Draft ambient aquatic life criteria for selenium—2004. EPA 822-R-04-001. Office of Water, Washington, DC.

EPA (US Environmental Protection Agency). 2005a. CatReg software documentation. EPA/600/R-98/052. National Center for Environmental Assessment, Research Triangle Park, NC.

EPA (US Environmental Protection Agency). 2005b. Microbial source tracking guide document. EPA/600-R-05-064. Office of Research and Development, Washington, DC.

EPA (US Environmental Protection Agency). 2006a. 2001 Update of ambient water quality criteria for cadmium. EPA-822-R-01-001. Office of Water, Washington, DC.

EPA (US Environmental Protection Agency). 2006b. Framework for developing suspended and bedded sediments water quality criteria. EPA-822-R-06-001. Office of Water, Washington, DC.

EPA (US Environmental Protection Agency). 2006c. Handbook for developing watershed plans to restore and protect our waters. Draft. Office of Water, Washington, DC.

EPA Region II. 2000. Further site characterization and analysis. Volume 2E—Revised baseline ecological risk assessment. Hudson River PCBs reassessment RI/FS, New York.

EPPO (European and Mediteranean Plant Protection Organization). 2004. Environmental Risk Assessment of Plant Protection Products. EPPO Bulletin 33. European and Mediteranean Plant Protection Organization, Paris.

Erickson, R.J. and Stephan, C.E. 1990. A model for exchange of organic chemicals at fish gills: Flow and diffusion limitation. Aquat. Toxicol., 18:175–197.

Eriksson, L., Jaworska, J., Worth, A.P., Cronin, M.T.D., McDowell, R.M., and Gramatica, P. 2003. Methods for reliability and uncertainty assessment and for applicability evaluations of classification- and regression-based QSARs. Environ. Health Persp., 111:1361–1375.

Ernst, H.R. 2003. Chesapeake Bay Blues. Rowman & Littlefield, Lanham, MD.

Escher, B.I. and Hermens, J. 2002. Modes of action in ecotoxicology: Their role in body burdens, species sensitivity, QSARs, and mixture effects. Environ. Sci. Technol., 36:4201–4217.

Escher, B.I. and Hermens, J. 2004. Internal exposure: Linking bioavailability to effects. Environ. Sci. Technol., December:455A–462A.

ESCORT (2001) Guidance document on regulatory testing and risk assessment procedures for plant protection products with non-target arthropods, Proceedings of ESCORT 2 workshop (European Standard Characteristics Of non-target arthropod Regulatory Testing), Wageningen, The Netherlands, 21–23 March 2000.

EUSES (1997) European Uniform system for the evaluation of substances (EUSES), version 1.0. European Chemical Bureau, Ispra, Italy.

Fairbrother, A. 2003. Lines of evidence in wildlife risk assessment. Hum. Ecol. Risk Assess., 9:1475–1491.

Fairbrother, A. and Kapustka, L.A. 1996. Toxicity extrapolation in terrestrial systems. California Environmental Protection Agency, Sacramento, CA.

Fairbrother, A., Kapustka, L.A., Williams, B.A., and Bennett, R.S. 1997. Effects-initiated assessments are not risk assessments. Hum. Ecol. Risk Assess., 3:119–124.

Fairbrother, A., Gentile, J., Menzie, C., and Munns, W. 1999. Report on the shrimp virus peer review and risk assessment workshop: Developing a qualitative ecological risk assessment. EPA/600/R-99/027. US Environmental Protection Agency, Washington, DC.

Farag, A., Woodward, D.F., Brumbach, W., Goldstein, J.N., and MacConell, E. 1999. Dietary effects of metals-contaminated invertebrates from the Coeur d'Alene River, Idaho, on cutthroat trout. Trans. Am. Fish. Soc, 128:578–592.

Feijtel, T., Boeije, G., Matthies, M., Young, A., Morris, G., Gandolfi, C., Hansen, B., et al. (1997) Development of a Geography-Referenced Regional Exposure Assessment Tool for European Rivers—GREAT-ER: Contribution to GREAT-ER #1. Chemosphere, 34:2351–2373.

Feldman, D.L., Hanahan, R.A., and Perhac, R. 1999. Environmental priority setting through comparative risk assessment. Environ. Manag., 23:483–493.

Ferson, S. 1996. Automated quality assurance checks on model structure in ecological risk assessment. Hum. Environ. Risk Assess., 2:558–569.

Field, L.J., MacDonald, D.D., Norton, S.B., Ingersoll, C.G., Severn, C.G., Smorong, D., and Lindskoog, R. 2002. Predicting amphipod toxicity from sediment chemistry using logistic regression models. Environ. Toxicol. Chem., 21:1993–2005.

Field, L.J., Norton, S.B., McDonald, D., Severn, C.G., and Ingersoll, C.G. 2005. Predicting toxicity to amphipods from sediment chemistry. EPA/600/R-04/030. US Environmental Protection Agency, Washington, DC.

Finkel, A. and Golding, D. 1995. Worst Things First? The Debate Over Risk-Based National Environmental Priorities. RFF Press, Washington, DC.

Finkelstein, M.E., Gwiazda, R.H., and Smith, D.R. 2003. Lead poisoning of seabirds: Environmental risk from leaded paint at a decommissioned military base. Environ. Sci. Technol., 37:3256–3260.

Finn, J.T. 1976. Measures of ecosystem structure and function derived from analysis of flows. J. Theoret. Biol., 56:363–380.

Fischoff, B., Lichtenstein, S., Slovic, P., Derby, A.S.L., and Keeney, R.L. 1981. Acceptable Risk. Cambridge University Press, Cambridge, UK.

Fisher, A., Emani, S., and Zint, M. 1995. Risk communication for industry practitioners: An annotated bibliography. Society for Risk Analysis, McLean, VA.

Fisher, R.A. 1930. The Genetical Theory of Natural Selection. Clarendon Press, Oxford, UK. (Reprinted in 1958 by Dover Publications, New York.)

Fletcher, J.S., Johnson, F.L., and McFarlane, J.C. 1990. Influence of greenhouse versus field testing and taxonomic differences on plant sensitivity to chemical treatment. Environ. Toxicol. Chem., 9:769–776.

Fogg, P. and Sangster, J. (2003) Chemicals in the Atmosphere Solubility, Sources and Reactivity. John Wiley, New York.

Foran, J.A. and Ferenc, S.A. 1999. Multiple Stressors in Ecological Risk and Impact Assessment. SETAC Press, Pensacola, FL.

Forbes, T.L. and Forbes, V.E. 1993. A critique of the use of distribution-based extrapolation models in ecotoxicology. Funct. Ecol., 7:249–254.

Forbes, V.E. 1999. Genetics and Ecotoxicology. Taylor & Francis, Philadelphia, PA.

Forbes, V.E. and Calow, P. 1999. Is the per capita rate of increase a good measure of population-level effects in ecotoxicology? Environ. Toxicol. Chem., 18:1544–1556.

Forbes, V.E. and Calow, P. 2002. Applying weight-of-evidence to retrospective ecological risk assessment when quantitative data are limited. Hum. Ecol. Risk Assess., 8:1625–1639.

Forbes, V.E., Calow, P., and Sibly, R.M. 2001. Are current species extrapolation models a good basis for ecological risk assessment? Environ. Toxicol. Chem., 20:442–447.

Foster, K.R., Vecchia, P., and Repacholi, M.H. 2000. Science and the precautionary principle. Science, 288:979–981.

Fox, G. 2001. Wildlife as sentinels of human health effects in the Great Lakes, St. Lawrence Basin. Environ. Health Persp., 109:853–861.

Fox, G.A. 1991. Practical causal inference for ecoepidemiologists. J. Toxicol. Environ. Health, 33:359–373.

Foxx, T.S., Tierney, G.D., and Williams, J.M. 1984. Rooting depths of plants relative to biological and environmental factors. Los Alamos National Laboratory, Los Alamos, NM.

Foy, C.D., Chaney, R.L., and White, M.C. 1978. The physiology of metal toxicity in plants. Annu. Rev. Plant Physiol., 29:511–566.

Francis, R.I.C.C. and Shotton, R. 1997. "Risk" in fisheries management: A review. Can. J. Fish. Aquat. Sci., 54:1699–1715.

French-McCay, D.P. 2002. Development and application of an oil toxicity and exposure model, OilToxEx. Environ. Toxicol. Chem., 21:2080–2094.

Freshman, J.S. and Menzie, C.A. 1996. Two wildlife exposure models to assess impacts at the individual and population levels and the efficacy of remediation. Hum. Ecol. Risk Assess., 2:481–498.

Friend, M. 1987. Field guide to wildlife diseases. Resource Pub. 167. US Fish and Wildlife Service, Washington, DC.

Froese, K.L., Verbrugge, D.A., Ankley, G.T., Niemi, G.J., Larsen, C.P., and Giesy, J.P. 1998. Bioaccumulation of polychlorinated biphenyls from sediments to aquatic insects and tree swallow eggs and nestlings in Saginaw Bay, Michigan, USA. Environ. Toxicol. Chem., 17:484–492.

Funtowicz, S.O. and Ravetz, J.R. 1990. Uncertainty and Quality in Science for Policy. Kluwer Academic, Dordrecht, The Netherlands.

FWS (US Fish and Wildlife Service). 1980. Habitat evaluation procedures (HEP). 870 FW-1. Division of Ecological Services, Washington, DC.

Galbraith, H., LeJeune, K., and Lipton, J. 1995. Metal and arsenic impacts to soils, vegetation communities and wildlife habitat in southwestern Montana uplands contaminated by smelter emissions. I. Field evaluations. Environ. Toxicol. Chem., 14:1895–1903.

Ganzelmeier, H., Rautmann, D., Spagenberg, R., Streloke, M., Hermann, M., Wenzelburger, H.J., and Walter, H.F. 1995. Studies on the Spray Drift of Plant Protection Products. Mitteilungen Aus Der Biologischen Bundesanstalt Fur Land-Und Fortwirtschaft, Berlin.

Gardner, R.H., O'Neill, R.V., Mankin, J.B., and Kumar, K.D. 1980. Comparative error analysis of six predator–prey models. Ecology, 61:323–332.

Gardner, R.H., O'Neill, R.V., Mankin, J.B., and Carney, J.H. 1981. A comparison of sensitivity and error analysis based on a stream ecosystem model. Ecol. Model., 12:173–190.

Gardner, R.H., Kemp, W.M., Kennedy, V.S., and Petersen, J.E. (eds.). 2001. Scaling Relations in Experimental Ecology. Columbia University Press, New York.

Garg, P., Tripathi, R.D., Rai, U.N., Sinha, S., and Chandra, P. 1997. Cadmium accumulation and toxicity in submerged plant Hydrilla verticillata (L.F.) Royle. Environ. Monitor. Assess., 47:167–173.

Garten, C.T. Jr. 1980. Ingestion of soil by hispid cotton rats, white-footed mice, and eastern chipmunks. J. Mammal., 6:136–137.

Garten, C.T. Jr. and Trabalka, J.R. 1983. Evaluation of models for predicting terrestrial food chain behavior of xenobiotics. Environ. Sci. Technol., 17:590–595.

Gaylor, D.W. 2000. The use of Haber's law in standard setting and risk assessment. Toxicology, 149:21–34.

Gell-Mann, M. 1994. The Quark and the Jaguar: Adventures in the Simple and the Complex. W.H. Freeman, New York.

Gentile, J.H., Gentile, S.M., and Hoffman, G. 1983. The effects of a chronic mercury exposure on survival, reproduction, and population dynamics of Mysidopsis bahia. Environ. Toxicol. Chem., 2:61–68.

Gentile, J.H., Harwell, M.A., Cropper, W., Harwell, C.C., DeAngelis, D., Davis, S., Ogden, J.C., and Lirman, D. 2001. Ecological conceptual models: A framework and case study on ecosystem management for South Florida sustainability. Sci. Tot. Environ., 274:231–253.

Gerard, P.D., Smith, D.R., and Weerakkody, G. 1998. Limits of retrospective power analysis. J. Wildl. Manag., 62:801–807.

Germano, J.D. 1999. Ecology, statistics, and the art of misdiagnosis: The need for a paradigm shift. Environ. Rev., 7:167–190.

Gersich, F.M., Blanchard, F.A., Applegath, S.L., and Park, C.N. 1986. The precision of daphnid (Daphnia magna Straus, 1820) static acute toxicity tests. Arch. Environ. Contam. Toxicol., 15:741–749.

Gezondheidsraad. 2003. Environmental health: Research for policy. Nr. 2003/20E. The Haag, The Netherlands.

Gibbons, W.N. and Munkittrick, K.R. 1994. A sentinal monitoring framework for identifying fish population responses to industrial discharges. J. Aquat. Eco. Health, 3:327–337.

Gibson, G.R., Barbour, M.T., Stribling, J.B., Gerritsen, J., and Karr, J.R. 1996. Biological criteria: Technical guidance for streams and small rivers (revised). EPA-B-96-001. US Environmental Protection Agency, Office of Water, Washington, DC.

Giddings, J.M. 1986. A microcosm procedure for determining safe levels of chemical exposure in shallow-water communities. Pages 121–134 in J. Cairns Jr. (ed.) Community Toxicity Testing. ASTM, Philadelphia, PA.

Giddings, J.M., Hall, L.W. Jr., and Solomon, K.R. 2000. Ecological risks of diazinin from agricultural use in the Sacramento–San Joaquin River basins, California. Risk Anal., 20:545–572.

Giddings, J.M., Solomon, K.R., and Maund, S.J. 2001. Probabilistic risk assessment of cotton pyrethroids: II. Aquatic mesocosm and field studies. Environ. Toxicol. Chem., 20:660–668.

Giddings, J.M., Brock, T.C.M., Heger, W., et al. 2002. Community-Level Aquatic System Studies—Interpretation Criteria. SETAC Press, Pensacola, FL.

Giddings, J.M., Anderson, T.A., Hall, L.W. Jr., et al. 2005. Atrazine in North American Surface Waters: A Probabilistic Risk Assessment. SETAC Press, Pensacola, FL.

Giesy, J. and Kannan, K. 1998. Dioxin-like and non-dioxin-like toxic effects of polychlorinated biphenyls (PCBs): Implications for risk assessment. Crit. Rev. Toxicol., 28:511–569.

Giesy, J.P., Ludwig, J.P., and Tillitt, D.E. 1994a. Deformities in birds of the Great Lakes region: Assigning causality. Environ. Sci. Technol., 28:128A–135A.

Giesy, J.P., Ludwig, J.P., and Tillitt, D.E. 1994b. Dioxins, dibenzofurans, PCBs and colonial fish-eating water birds. Pages 249–307 in A. Schecter (ed.) Dioxins and Health. Plenum Press, New York.

Giesy, J.P., Solomon, K.R., Coats, J.R., Dixon, K.R., Giddings, J.M., and Kenaga, E.E. 1999. Chlorpyrifos: Ecological risk assessment in North American aquatic environments. Rev. Environ. Contam. Toxicol., 160:1–129.

Giesy, J.P., Dobson, S., and Solomon, K.R. 2000. Ecotoxicological risk assessment for roundup herbicide. Rev. Environ. Contam. Toxicol., 167:35–120.

Gigerenzer, G. 2002. Calculated Risks. Simon & Schuster, New York.

Gigerenzer, G. and Hoffrage, U. 1995. How to improve Bayesian reasoning without instruction: Frequency formats. Psychol. Rev., 102:684–704.

Gigerenzer, G., Swijtink, Z., Porter, T., Beatty, J., and Kruger, L. 1989. The Empire of Chance: How Probability Changed Science and Everyday Life. Cambridge University Press, Cambridge, UK.

Giles, R.H. Jr. 1970. The ecology of a small-forested watershed treated with the insecticide Malathion-S35. Wildl. Monogr., 24.

Ginn, T.C. and Pastorok, R.A. 1992. Assessment and management of contaminated sediments in Puget Sound in G.A. Burton (ed.) Sediment Toxicity Assessment. Lewis Publishers, Boca Raton, FL.

Gobas, F.A.P.C. 1993. A model for predicting the bioaccumulation of hydrophobic organic chemicals in aquatic food webs: Application to Lake Ontario. Ecol. Model, 69:1–17.

Gobas, F.A.P.C. 2003. Mathematical Models of Bioaccumulation and Eco Fate. Simon Fraser University, Burnaby, British Columbia, Canada, Available at: www.rem.sfu.ca/toxicology/models

Gobas, F.A.P.C. and Morrison, H.A. 2000. Bioconcentration and Biomagnification in the Aquatic Environment. Chapter 9 in R.S. Boethling and D. Mackay (eds.) Handbook of Property Estimation Methods for Chemicals. Lewis Publishers, Boca Raton, FL.

Goede, R.W. and Barton, B.A. 1990. Organismic indices and an autopsy-based assessment as indicators of health and condition of fish. Am. Fish. Soc. Symp., 8:93–108.

Golley, F.B. 1993. A history of the ecosystem concept in ecology. Yale University Press, New Haven, CT.

Good, I.J. 1983. Good Thinking: The Foundations of Probability and Its Applications. University of Minnesota Press, Minneapolis, MN.

Goodman, D. 1976. Ecological expertise. Pages 317–360 in H.A. Feiveson, F.W. Sinden, and R.H. Socolow (eds.) Boundaries of Analysis: An Enquiry into the Tocks Island Dam Controversy. Ballinger, Cambridge, MA.

Goodman, D. 1987. The demography of chance extinction. Pages 11–34 in M.E. Soule (ed.) Viable Populations for Conservation. Cambridge University Press, Cambridge, UK.

Goodman, D. 2005. Taking the prior seriously: Bayesian analysis without subjective probability. Pages 379–410 in M.L. Taper and S.R. Lele (eds.) The Nature of Scientific Evidence. University of Chicago Press, Chicago, IL.

Goodman, S.N. and Berlin, J.A. 1994. The use of predicted confidence intervals when planning experiments and the misuse of power when interpreting results. Ann. Internal Med., 121:200–206.

Goodyear, C.P. 1993. Spawning stock biomass per recruit in fisheries management: Foundation and current use. Pages 67–81 in S.J. Smith, J.J. Hunt, and D. Rivard (eds.) Risk Evaluation and Biological Reference Points for Fisheries Management. Canadian Special Publications in Fisheries and Aquatic Sciences 120. National Research Council and Department of Fisheries and Oceans, Ottawa, Canada.

Goovarts, P. 1997. Geostatistics for Natural Resources Evaluation. Oxford University Press, New York.

Gordon, G.E. 1988. Receptor models. Environ. Sci. Technol, 22:1132–1142.

Graney, R.L., Giesy, J.P. Jr., and DiToro, D. 1989. Mesocosm experimental design strategies: Advantages and disadvantages in ecological risk assessment. Pages 74–88 in J.R. Voshell (ed.) Using Mesocosms to Assess the Aquatic Ecological Risk of Pesticides: Theory and Practice. Entomological Society of America, Lanham, MD.

Graney, R.L., Kennedy, J.H., and Rodgers, J.H. Jr. (eds.). 1994. Aquatic Mesocosm Studies in Ecological Risk Assessment. CRC Press, Boca Raton, FL.

Graney, R.L., Giesy, J.P., and Clark, J.R. 1995. Field studies. Pages 257–305 in G. Rand (ed.) Fundamentals of Aquatic Toxicology. Taylor & Francis, Washington, DC.

Grapentine, L., Anderson, J., Boyd, D., Burton, G.A., DeBarros, C., Johnson, G., Marvin, C., et al. 2002. A decision making framework for sediment assessment developed for the Great Lakes. Hum. Ecol. Risk Assess., 8:1655.

Gray, G.M. 1994. Complete risk characterization. Risk Persp., 2:1–2.

GREAT-ER Task Force (1997) GREAT-ER: Geography-referenced Regional Exposure Assessment Tool for European Rivers. European Centre for Ecotoxicology and Toxicology of Chemicals, Brussels.

Greene, J.C., Bartels, C.L., Warren-Hicks, W.J., et al. 1988. Protocols for short-term toxicity screening of hazardous waste sites. US Environmental Protection Agency, Corvallis, OR.

Greenland, S. 1988. Probability versus Popper: An elaboration of the insufficiency of current Popperian approaches for epidemiological analysis. Pages 95–104 in K.J. Rothman (ed.) Causal Inference. Epidemiology Resources, Chestnut Hill, MA.

Greger, M., Kautsky, L., and Sandberg, T. 1995. A tentative model of Df uptake in *Potamogeton pectinatus* in relation to salinity. Environ. Exp. Biology, 35:215–225.

Griffith, M.B., Lazorchak, J.M., and Herlihy, A.T. 2004. Relationships among exceedences of metals criteria, the results of ambient bioassays, and community metrics in mining-impacted streams. Environ. Toxicol. Chem., 23:1786–1795.

Grothe, D.R., Dickson, K.L., and Reed-Judkins, D.K. (eds.). 1996. Whole Effluent Toxicity Testing: An Evaluation of Methods and Prediction of Receiving System Impacts. SETAC Press, Pensacola, FL.

Grue, C.E., Hoffman, D.J., Beyer, W.N., and Franson, L.P. 1986. Lead concentrations and reproductive success in European starlings nesting within highway roadside verges. Environ. Pollut. Ser. A, 42:157–182.

Guinee, J.B. 2003. Handbook of Life Cycle Assessment. Kluwer Academic, Dordrecht, The Netherlands.

Gupta, M. and Chandra, P. 1998. Bioaccumulation and toxicity of mercury in rooted submerged macrophyte *Vallisneria spiralis*. Environ. Pollut., 103:327–332.

Gurney, W.S.C., McCauley, E., Nisbet, R.M., and Murdoch, W.W. 1990. The physiological ecology of *Daphina*: A dynamic model of growth and reproduction. Ecology, 71:716–732.

H. John Heinz III Center for Science Economics and the Environment. 2002. The State of the Nation's Ecosystems. Cambridge University Press, New York.

Haber, L., Strickland, J.A., and Guth, D.J. 2001. Categorical regression analysis of toxicity data. Comments Toxicol., 7:437–452.

Hacking, I. 1975. The Emergence of Probability. Cambridge University Press, Cambridge, UK.

Hacking, I. 2001. An Introduction to Probability and Inductive Logic. Cambridge University Press, Cambridge, UK.

Hackney, J.D. and Linn, W.S. 1979. Koch's postulates updated: A potentially useful application to laboratory research and policy analysis in environmental toxicology. Am. Rev. Respir. Dis., 1119:849–852.

Haddad, S. and Krishnan, K. 1998. Physiological modeling of toxicokinetic interactions: Implications for mixtures risk assessment. Environ. Health Persp., 106:1377–1384.

Haimes, Y.Y. 1998. Risk Modeling, Assessment, and Management. John Wiley, New York.

Hakoyama, H. and Iwasa, Y. 2000. Extinction risk of a density-dependent population estimated from a time series of population size. J. Theoret. Biol., 204:337–359.

Halbrook, R.S., Brewer, R.L. Jr., and Buehler, D.A. 1999a. Ecological risk assessment of a large river-reservoir: 8. Experimental study of the effects of polychlorinated biphenyls on reproductive success of mink. Environ. Toxicol. Chem., 18:649–654.

Halbrook, R.S., Brewer, R.L. Jr., and Buehler, D.A. 1999b. Ecological risk assessment of a large river-reservoir: 7. Environmental contaminant accumulation and effects in great blue herons. Environ. Toxicol. Chem., 18:641–648.

Halfon, E. (ed.). 1979. Theoretical Systems Ecology. Academic Press, New York.

Hall, C.A.S. and Day, J.W. Jr. (eds.). 1977. Ecosystem Modeling in Theory and Practice. John Wiley, New York.

Hall, L.W. Jr. and Burton, D.T. 2005. An Integrated Case Study for Evaluating the Impact of an Oil Refinery Effluent on Aquatic Biota in the Deleware River. Hum. Ecol. Risk Assess., 11:647–936.

Hall, L.W. Jr., Pinkney, A.E., and Horseman, L.O. 1985. Mortality of striped bass larvae in relation to contaminants and water quality in a Chesapeake Bay estuary. Trans. Am. Fish. Soc., 114:861–868.

Hall, L.W. Jr., Bushong, S.J., Ziegenfuss, M.C., Hall, W.S., and Herman, R.L. 1988. Concurrent mobile on-site and *in situ* striped bass environmental contaminant and water quality studies in the Choptank River and upper Chesapeake Bay. Environ. Toxicol. Chem., 7:815–830.

Hall, L.W. Jr., Giddings, J.M., Solomon, K.R., and Balcomb, R. 1999. An ecological risk assessment for the use of Irgarol 1051 as an algaecide for antifoulant paints. Crit. Rev. Toxicol., 29:367–437.

Hall, L.W. Jr., Scott, M.C., Killen, W.D., and Unger, M.A. 2000. A probabilistic ecological risk assessment of tributyltin in surface waters of the Chesapeake Bay watershed. Hum. Ecol. Risk Assess., 6:141–179.

Hallam, T.G. and Clark, C.E. 1983. Effects of toxicants on populations: A qualitative approach 1. Equilibrium environmental exposure. Ecol. Model., 18:291–304.

Hallam, T.G., Lassiter, R.R., Li, J., and Suarez, L.A. 1990. Modelling individuals employing an integrated energy response: Application to *Daphnia*. Ecology, 71:938–954.

Hammonds, J.S., Hoffman, F.O., and Bartell, S.M. 1994. An introductory guide to uncertainty analysis in environmental and health risk assessment. ES/ER/TM-35/R1. Oak Ridge National Laboratory, Oak Ridge, TN.

Hampton, N.L., Morris, R.C., and VanHorn, R.L. 1998. Methodology for conducting screening-level ecological risk assessments for hazardous waste sites. Part II: Grouping ecological components. Int. J. Environ. Pollut., 9:47–61.

Hannon, B. 1973. The structure of ecosystems. J. Theoret. Biol. 41:535–546.

Hanratty, M.P. and Stay, F.S. 1994. Field evaluation of the littoral ecosystem risk assessment model's predictions of the effects of chlorpyrifos. J. Appl. Ecol., 31:439–453.

Hansch, C. and Fujita, T. 1964 $\rho-\sigma-\pi$ analysis: A method for the correlation of biological activity and chemical structure. J. Am. Chem. Soc., 86:1616–1626.

Hansen, F. 1997. Policy for use of probabilistic analysis in risk assessment at the US Environmental Protection Agency. Environmental Protection Agency, Washington, DC, US. Available at: http://www.epa.gov/ncea/mcpolicy.htm.

Hanski, I. 1999. Metapopulation Ecology. Oxford University Press, Oxford, U.K.

Hanski, I. and Gilpin, M.E. 1996. Metapopulation Biology: Ecology, Genetics, and Evolution. Academic Press, San Diego, CA.

Hardin, G. 1968. The tragedy of the commons. Science, 162:1243–1248.

Hare, L., Carignan, R., and Huerta-Diaz, M.A. 1994. A field study of metal toxicity and accumulation by benthic invertebrates: Implications for the acid-volatile sulfide (AVS) model. Limnol. Oceanogr., 39:1653–1668.

Harrass, M.C. and Taub, F.B. 1985. Comparisons of laboratory microcosms and field responses to copper. Pages 57–74 in T.P. Boyle (ed.) Validation and Predictability of Laboratory Methods for Assessing the Fate and Effects of Contaminants in Aquatic Ecosystems. ASTM, Philadelphia, PA.

Harremoes, P., Gee, D., MacGarvin, M., Stirling, A., Keys, J., Wynne, B., and Guedes Vaz, S. 2001. Late lessons from early warnings: The precautionary principle 1896–2000. Environmental Issue Report No. 22. European Environmental Agency, Copenhagen, Denmark.

Harris, S.G. and Harper, B.L. 2000. Using eco-cultural dependency webs in risk assessment and characterization of risks to tribal health and culture. Environ. Sci. Pollut. Res., 2:91–100.

Hart, B., Burgman, M., Webb, A., Allison, G., Chapman, M., Duivenvoorden, L., Feehan, P., et al. 2005. Ecological Risk Management Framework for the Irrigation Industry. Water Studies Centre, Monash University, Clayton, Australia.

Hartwell, S.I., Dawson, C.E., Jordahl, D.H., and Durell, E.Q. 1995. Demonstrating a method to correlate measures of ambient toxicity and fish community diversity. CBRM-TX-95-1. Maryland Department of Natural Resources, Chesapeake Bay Research and Monitoring Division, Annapolis, MD.

Harwell, M.A. 1998. Science and environmental decision making in South Florida. Ecol. Appl., 8:580–590.

Harwell, M.A. and Gentile, J.H. 2000. Environmental decision-making for multiple stressors: Framework, tools, case studies and prospects. Pages 169–236 in S.A. Ferenc and J.A. Foran (eds.) Multiple Stressors in Ecological Risk Assessment: Approaches to Risk Estimation. SETAC Press, Pensacola, FL.

Harwell, M.A., Cooper, W., and Flaak, R. 1992. Prioritizing ecological and human welfare risks from environmental stresses. Environ. Manag., 16:451–464.

Hatfield, A.J. and Hipel, K.W. 2002. Risk and systems theory. Risk Anal., 22:1043–1057.

Hattemer-Frey, H.A., Quinlan, R.E., and Krieger, G.R. 1995. Ecological risk assessment case study: Impacts to aquatic receptors at a former metals mining Superfund site. Risk Anal., 15:253–265.

Hattis, D. and Goble, R. 2003. The red book, risk assessment, and policy analysis: The road not taken. Hum. Ecol. Risk Assess., 9:1297–1306.

Hatzinger, P.B. and Alexander, M. 1995. Effect of aging of chemicals in soil on their biodegradability and extractability. Environ. Sci. Technol., 29:537–545.

He, Q.B. and Singh, B.R. 1994. Crop uptake of cadmium from phosphorus fertilizers. I. Yield and cadmium content. Water Air Soil Pollut., 74:297–303.

Heaton, S.N., Bursian, S., Giesy, J.P., Tillitt, D.E., Render, R.A., Jones, P.D., Verbrugge, D.A., Kubiak, T., and Aulerich, R.J. 1995a. Dietary exposure of mink to carp from Saginaw Bay, Michigan. 2. Hematology and liver pathology. Arch. Environ. Contam. Toxicol., 28:411–417.

Heaton, S.N., Bursian, S., Giesy, J.P., Tillitt, D.E., Render, R.A., Jones, P.D., Verbrugge, D.A., Kubiak, T., and Aulerich, R.J. 1995b. Dietary exposure of mink to carp from Saginaw Bay, Michigan. 1. Effects on reproduction and survival and the potential risks to wild mink populations. Arch. Environ. Contam. Toxicol., 28:334–343.

Heiger-Bernays, W., Menzie, C., Montgomery, C., Edwards, D., and Panwels, S. 1997. A framework for biological and chemical testing of soil. Pages 388–420 in D.G. Linz and D. Nakles (eds.) Environmentally Acceptable Endpoints in Soil: Risk-Based Approach to Contaminated Site Management Based on Availability of Chemicals in Soil. American Academy of Environmental Engineers, Annapolis, MD.

Heikens, A., Peijnenburg, W., and Hendriks, A.J. 2001. Bioaccumulation of heavy metals in terrestrial invertebrates. Environ. Pollut., 113:385–403.

Heimbach, U., Leonard, P., Miyakawa, R., and Able, C. 1994. Assessment of pesticide safety to the carabid beetle, *Poecilus cupreus*, using two different semifield enclosures. Pages 205–240 in M.H. Donker, H. Eijsackers, and F. Heimbackers (eds.) Ecotoxicology of Soil Organisms. Lewis Publishers, Boca Raton, FL.

Heinz, G.H., Hoffman, D.J., Krynitsky, A.J., and Weller, D.M.G. 1987. Reproduction in mallards fed selenium. Environ. Toxicol. Chem., 6:423–433.

Henderson, J.D., Yamamoto, D.M., Fry, D.M., Seiber, J.N., and Wilson, B.W. 1994. Oral and dermal toxicity of organophosphate pesticides in the domestic pigeon (*Columba livia*). Bull. Environ. Contam. Toxicol., 52:633–640.

Hendley, P., Holmes, C., Kay, S., Maund, S.J., Travis, K.Z., and Zhang, M. 2001. Probabilistic risk assessment of cotton pyrethroids: III. A spatial analysis of the Mississippi, USA, cotton landscape. Environ. Toxicol. Chem., 20:669–678.

Hendriks, A.J., Ma, W.-C., Brouns, J.J., de Ruiter-Dijkman, E.M., and Gast, R. 1995. Modelling and monitoring organochlorine and heavy metal accumulation in soils, earthworms and shrews in Rhine-Delta floodplains. Arch. Environ. Contam. Toxicol., 29:115–144.

Henning, M.H., Shear Weinberg, N.M., Wilson, N.D., and Iannuzzi, T.J. 1999. Distributions of key exposure factors controling the uptake of xenobiotic chemicals by great blue herons (*Ardea herodius*) through ingestion of fish. Hum. Ecol. Risk Assess. 5:125–144.

Henny, C.J. 2003. Effects of mining lead on birds: A case history at Coeur d'Alene Basin, Idaho. Pages 755–766 in D.J. Hoffman, B. Rattner, G.A. Burton, Jr., and J. Cairns, Jr. (eds.) Handbook of Ecotoxicology. Lewis Publishers, Boca Raton, FL.

Henriques, W.D. and Dixon, K.R. 1996. Estimating spatial distribution of exposure by integrating radiotelemetry, computer simulation, and geographic information systems (GIS) techniques. Hum. Ecol. Risk Assess., 2:527–538.

Henshel, D.S., Martin, J.W., Norstrom, R., Whitehead, P., Steeves, J.D., and Cheng, K.M. 1995. Morphometric abnormalities in brains of great blue heron hatchlings exposed in the wild to PCDDs. Environ. Health Persp., 103(Suppl 4):61–66.

Herbes, S.E., Greist, W.H., and Southworth, G.R. 1978. Field site evaluation of aquatic transport of polycyclic aromatic hydrocarbons. Pages 221–230 in Proceedings of the Symposium on Potential Health and Environmental Effects of Synthetic Fossil Fuel Technologies, CONF-78093. Oak Ridge National Laboratory, Oak Ridge, TN.

Herkovits, J., Perez-Coll, C.S., and Herkovits, F.D. 1996. Ecotoxicity in the Reconqusta River, Province of Buenos Aires, Argentina: A preliminary study. Environ. Health Persp., 104:186–189.

Hertzberg, R.C. and MacDonald, M.M. 2002. Synergy and other ineffective mixture risk definitions. Sci. Total Environ., 288:31–42.

Hertzberg, R.C. and Teuschler, L.K. 2002. Evaluating quantitative formulas for dose–response assessment of chemical mixtures. Environ. Health Persp., 110:965–970.

Hettelingh, J.P. and Downing, R.J. 1991. Mapping critical loads for Europe. CCE Tech. Report No.1. Coordination Center for Effects, National Institute of Public Health and Environment Protection, Bilthoven, The Netherlands.

Hewlett, P.S. and Plackett, R.L. 1979. The Interpretation of Quantal Responses in Biology. Edward Arnold, London.

Heyer, W.R. (ed.). 1994. Measuring and Monitoring Biodiversity: Standard Methods for Amphibians. Smithsonian Press, Washington, DC.

Hickey, J.J. 1969. Peregrine Falcon Populations: Their Biology and Decline. University of Wisconsin Press, Madison, WI.

Hickey, J.J. and Anderson, D.W. 1968. Chlorinated hydrocarbons and eggshell changes in raptorial and fish-eating birds. Science, 162:271–273.

Hilborn, R. 1996. Do principles for conservation help managers? Ecol. Appl., 6:364–365.

Hilborn, R. and Mangel, M. 1997. The Ecological Detective: Confronting Models with Data. Princeton University Press, Princeton, NJ.

Hilborn, R. and Walters, C.J. 1992. Quantitative Fish Stock Assessment: Choice, Dynamics, and Uncertainty. Chapman & Hall, New York.

Hill, A.B. 1965. The environment and disease: Association or causation. Proc. Royal Soc. Med., 58:295–300.

Hill, B.H. 1997. The use of periphyton assemblage data in an index of biotic integrity. Bull. N. Am. Benthol. Soc., 14:158.

Hiraoka, Y. and Okuda, H. 1984. A tentative assessment of water pollution by medaka egg stationing method: Aerial application of fenitrothion emulsion. Environ. Res., 34:262–267.

Hobbs, E.A., Warne, M. St.J., and Markich, S.J. 2005. Evaluation of the criteria used to assess the quality of aquatic toxicity data. Integr. Environ. Assess. Manag., 1:174–180.

Hoekstra, J.A. and van Ewijk, P.H. 1993. Alternatives for the no-observed-effect level. Environ. Toxicol. Chem., 12:187–194.

Hoeting, J.A., Madigan, D., Raferty, A.E., and Volinski, C.T. 1999. Bayesian model averaging: A tutorial. Stat. Sci., 14:382–417.

Hoff, D.J. and Henningsen, G.M. 1998. Extrapolating toxicity reference values in terrestrial and semi-aquatic wildlife species using uncertainty factors. Abstract, 37th Annual Meeting, Society of Toxicology, Reston, VA.

Hoffman, D.J. et al. 1998. Comparative developmental toxicity of planar polychlorinated biphenyl congeners in chickens, American kestrels, and common terns. Environ. Toxicol. Chem., 17:747–757.

Hoffman, D.J., Rattner, B.A., Burton, G.A. Jr., and Cairns, J. Jr. 2003. Handbook of Ecotoxicology, Second Edition. Lewis Publishers, Boca Raton, FL.

Holchek, J.L., Pieper, R.D., and Herbal, C.H. 2003. Range Management: Principles and Practices, Fifth Edition. Prentice-Hall, New York.

Holcombe, G.W., Phipps, G.L., and Veith, G.D. 1988. Use of aquatic lethality tests to estimate safe chronic concentrations of chemicals in initial ecological risk assessments. Pages 442–467 in G.W. Suter II (ed.) Aquatic Toxicity and Hazard Assessment: Eleventh Symposium. ASTM, Philadelphia, PA.

Holdren, G.R. Jr., Strickland, M.D., Cosby, B.J., Marmorek, D., Bernard, D., Santore, R.C., Driscoll, C.T., Pardo, L., Hunsaker, C.T., and Turner, R.S. 1993. A national critical loads framework for atmospheric deposition effects assessment: V. Model selection, applications, and critical loads mapping. Environ. Manag., 17:355–363.

Holling, C.S. 1978. Adaptive Environmental Assessment and Management. John Wiley, Chichester, UK.

Hooda, P.A. and Alloway, B.J. 1993. Effects of time and temperature on the bioavaiabilty of Cd and Pb from sludge-amended soils. J. Soil Sci., 44:97–110.

Hope, B.K. 1995. A review of models for estimating terrestrial ecological receptor exposure to chemical contaminants. Chemosphere, 30:2267–2287.

Hope, B.K. 2004. The area use factor—is it right? SETAC Globe, 5:44.

Hope, B.K. and Peterson, J.A. 2000. A procedure for performing population-level ecological risk assessments. Environ. Manag., 25:281–289.

Hopkin, S.P. 1989. Ecophysiology of Metals in Soil Invertebrates. Elsevier Applied Sciences, London, UK.

Horness, B.H., Lomax, D.P., Johnson, L.L., Myers, M.S., Pierce, S.M., and Collier, T.K. 1998. Sediment quality thresholds: Estimates from hockey-stick regression of liver lesion prevalence in English sole (Pleuronectes vetulus). Environ. Toxicol. Chem., 17:872–882.

Hornshaw, T., Aulerich, R., and Johnson, H. 1983. Feeding Great Lakes fish to mink: Effects on accumulation and elimination of PCBs by mink. J. Toxicol. Environ. Health, 11:933–946.

Host, G.E., Regal, R.R., and Stephan, C.E. 1991. Analysis of acute and chronic data for aquatic life. PB93-154748. US Environmental Protection Agency, Duluth, MN.

Houck, O. 2002. The Clean Water Act TMDL Program: Law, Policy and Implementation, Second Edition. Environmental Law Institute, Washington, DC.

Houck, O. 2004. Tales from a troubled marriage: Science and law in environmental policy. Science, 302:1926–1929.

Howard, P.H. and Meylan, W.M. 1997. Handbook of Physical Properties of Organic Chemicals, Volume 1 to 5. CRC Press, Boca Raton, FL.

Huckabee, J.W., Carton, F.O., and Kennington, G.S. 1972. Environmental influence on trace elements in hair of 15 species of mammals. ORNL/TM-3747. Oak Ridge National Laboratory, Oak Ridge, TN.

Huggett, R.J., Kinerle, R.A., Mehrle, P.M., and Bergman, H.L. (eds.). 1992. Biochemical, Physiological, and Histological Markers of Anthropogenic Stress. Lewis Publishers, Boca Raton, FL.

Hughes, R.M., Larsen, D.P., and Omernik, J.M. 1986. Regional reference sites: A method for assessing stream potentials. Environ. Manag., 10:629–635.

Huijbegts, M.A.J., Gilijamse, W., Ragas, A.M.J., and Reijnders, L. 2003. Evaluating uncertainty in environmental life-cycle assessment: A case study comparing two insulation options for a Dutch one-family dwelling. Environ. Sci. Technol., 37:2600–2608.

Hunsaker, C.T. and Graham, R.L. 1991. Regional ecological assessment for air pollution. Page 312–334 in S.K. Majumdar, E.W. Miller, and J. Cahir (eds.) Air Pollution: Environmental Issues and Health Effects. Pennsylvania Academy of Sciences, Easton, PA.

Hunsaker, C.T., Graham, R.L., Suter, G.W., II, O'Neill, R.V., Barnthouse, L.W., and Gardner, R.H. 1990. Assessing ecological risk on a regional scale. Environ. Manag., 14:325–332.

Hunsaker, C.T., Graham, R.L., Ringold, P.L., Holdren, G.R. Jr., Turner, R.S., and Strickland, T.C. 1993. A national critical loads framework for atmospheric deposition effects assessment.

I. Defining assessment endpoints, indicators, and functional ecoregions. Environ. Manag., 17:335–341.

Hurlbert, S.H. 1984. Pseudoreplication and the design of ecological field experiments. Ecol. Monogr., 54:187–211.

Huston, M.A. 1997. Hidden treatments in ecological experiments: Re-evaluating the ecosystem function and biodiversity. Oecologia, 110:449–460.

Huston, M.A., and Smith, T.M. 1987. Plant succession: Life history and competition. Am. Natl., 130:168–198.

Hutchinson, T.H., Scholz, N., and Guhl, W. 1998. Analysis of the ECETOC aquatic toxicity (EAT) database: IV. Comparative toxicity of chemical substances to freshwater versus saltwater organisms. Chemosphere, 36:143–153.

Hydroqual, Inc. 2003. Biotic Ligand Model: Windows interface, version 2.0. US Environmental Protection Agency, Washington, DC.

IAEA (International Atomic Energy Agency). 1989. Evaluating the reliability of predictions made using environmental transfer models. IAEA Safety Series 100. Vienna, Austria.

IAEA (International Atomic Energy Agency). 1994. Handbook of Parameter Values for the Prediction of Radionuclide Transfer in Temperate Environments. Tech. Rep. Ser. No. 364. Vienna, Austria.

Ikonomou, M.G. 2002. PCBs in Dungeness crabs reflect distinct source fingerprints among harbor/industrial sites in British Columbia. Environ. Sci. Technol., 36:2545–2551.

Iman, R.L. and Helton, J.C. 1988. An investigation of uncertainty and sensitivity analysis techniques for computer models. Risk Anal., 8:71–90.

Ingersoll, C.G., Dillon, T., and Biddinger, G.R. (eds.). 1997. Ecological Risk Assessment of Contaminated Sediments. SETAC Special Publications Series. SETAC Press, Pensacola, FL.

International Joint Commission. 1989. Great Lakes Water Quality Agreement of 1978 as amended by Protocol signed November 19, 1987.

IPCS (International Programme on Chemical Safety). 2002. Principles and methods for assessment of risk from essential trace elements. Environmental Health Criteria 228. International Programme on Chemical Safety, WHO, Geneva.

IPCS (International Programme on Chemical Safety). 2004. Draft principles for modelling dose–response for the risk assessment of chemicals. Environmental Health Criteria XXX. International Program for Chemical Safety, WHO, Geneva.

ISO (International Organization for Standardization). 1997. Soil quality effects on earthworms (*Eisenia fetida*). ISO/DIS 11268-2. ISO, Geneva.

Jackson, D.R. and Watson, A.P. 1977. Disruption of nutrient pools and transport of heavy metals in a forested watershed near a lead smelter. J. Environ. Qual., 6:331–338.

Jackson, L., Kurtz, J., and Fisher, W. 2000. Evaluation guide for ecological indicators. EPA/620/R-99/005. US Environmental Protection Agency, Gulf Breeze, FL.

Jackson, R.B., Canadell, J.R., Ehleringer, J.R., Mooney, H.A., Sala, O.E., and Schulze, E.D. 1996. A global analysis of root distributions for terrestrial biomes. Oecologia, 108:489–511.

Jager, T., Fleuren, R.H.L.J., Hogendoorn, E.A., and de Korte, G. 2003. Elucidating the routes of exposure for organic chemicals in the earthworm, *Eisenia andrei* (Oligochaeta). Environ. Sci. Technol., 37:3399–3404.

Janssen, M.P.M., Bruins, A., DeVries, T.H., and Van Straalen, N.M. 1991. Comparison of cadmium kinetics in four soil arthropod species. Environ. Contam. Toxicol., 20:305–312.

Janssen, R.P.T., Posthuma, L., Baerselman, R., Den Hollander, H.A., Van Veen, R.P.M., and Peijenburg, W.J.G.M. 1997. Equilibrium partitioning of heavy metals in Dutch field soils. II. Prediction of metal accumulation in earthworms. Environ. Toxicol. Chem., 16:2479–2488.

Jarvinen, A.W. and Ankley, G.T. 1999. Linkage of Effects to Tissue Residues: Development of a Comprehensive Database for Aquatic Organisms. SETAC Press, Pensacola, FL.

Jarvis, N.J. 1994. The MACRO model (version 3.1). Technical description and sample simulations. Reports and Dissert. 19, Department of Soil Sciences, Swedish University of Agricultural Science, Uppsala, Sweden.

Jarvis, N.J. 1995. Simulation of soil-water dynamics and herbicide persistence in a silt loam soil using the MACRO model. Ecol. Model., 81:97–109.

Jarvis, N.J. 1998. Modelling the impact of preferential flow on nonpoint source pollution. Pages 195–221 in H.M. Selim and L. Wa (eds.) Physical Nonequilibrium in Soils: Modelling and Application. Ann Arbor Press, Chelsea, MI.

Jarvis, N.J., Nicholls, P., Hollis, J.M., Mayr, T., and Evans, S.P. 1996. Pesticide exposure assessment for surface waters and groundwater using the decision-support tool MACRO_DB. Pages 381–388 in A.A.M. Del Re, E. Capri, S.P. Evans, and M. Trevisan (eds.) Proceedings of the X Symposium of Pesticide Chemistry: The Environmental Fate of Xenobiotics. Piacenza, Italy.

Jaworska, J.S., Rose, K.A., and Brenkert, A.L. 1997a. Individual-based modeling of PCBs effects on young-of-the-year largemouth bass in southeastern U.S. reservoirs. Ecol. Model., 99:113–135.

Jaworska, J.S., Rose, K.A., and Barnthouse, L.W. 1997b. General Response Patterns of Fish Populations to Stress: An Evaluation Using an Individual-Based Simulation Model. J. Aqua. Ecosys. Stress Recovery, 6:15–31.

Jeffrey, K.A., Beamish, F.W.H., Ferguson, S.C., Kolton, R.J., and McMahon, P.D. 1986. Effects of the lampricide, 3-trifluoromethyl-4-nitrophenol (TFM) on the macroinvertebrates within the hyporheic region of a small stream. Hydrobiologia, 134:43–51.

Jenkins, D.W. 1979. Trace elements in mammalian hair and nails. EPA-600/4-79-049. US Environmental Protection Agency, Las Vegas, NV.

Jenkins, K.D., Lee, C.R., and Hobson, J.F. 1995. A hazardous waste site at the Naval Weapons Station, Concord, CA. Pages 883–901 in G. Rand (ed.) Fundamentals of Aquatic Toxicology: Effects, Environmental Fate, and Risk Assessment. Taylor & Francis, Washington, DC.

Jetz, W., Carbone, C., Fulford, J., and Brown, J.H. 2004. The scaling of animal space use. Science, 306:266–268.

Jiang, Q.Q. and Singh, B.R. 1994. Effect of different forms and sources of arsenic on crop yield and arsenic concentration. Water Air Soil Pollut., 74:321–343.

Johnson, D.H. 1999. The insignificance of statistical significance testing. J. Wildl. Manag., 63:763–772.

Johnson, G.D., Audet, D.J., Kern, J.W., LeCaptain, L.J., Strickland, M.D., Hoffman, D.J., and McDonald, L.L. 1999. Lead exposure in passerines inhabiting lead-contaminated floodplains in the Coeur d'Alene River Basin, Idaho, USA. Environ. Toxicol. Chem., 18:1190–1194.

Johnson, I., Whitehouse, P., and Crane, M. 2005. Effective montitoring of the environment. Pages 33–60 in K.C. Thompson, K. Wadhia, and A.P. Loibner (eds.) Environmental Toxicity Testing. Blackwell, Oxford, UK.

Johnston, J.J., Pitt, W.C., Sugihara, R.T., Eisemann, J.D., Primus, T.M., Holmes, M.J., Crocker, J., and Hart, A. 2005. Probabilistic risk assessment for snails, slugs, and endangered honeycreepers in diphacinone rodenticide baited areas on Hawaii, USA. Environ. Toxicol. Chem., 24, 1557–1567.

Johnston, R.K., Munns, W.R. Jr., Tyler, P.L., Marajh-Whittemore, P., Finkelstein, K., Munney, K., Short, F.T., Melville, A., and Hahn, S.P. 2002. Weighing the evidence of ecological risk from chemical contamination in the estuarine environment adjacent to the Portsmouth Naval Shipyard, Kittery, Maine. Environ. Toxicol. Chem., 21:182–194.

Jones, D.S., Barnthouse, L.W., Suter, G.W., II, Efroymson, R.E., Field, J.M., and Beauchamp, J.J. 1999. Ecological risk assessment of a large river-reservoir: 3. Benthic invertebrates. Environ. Toxicol. Chem., 18:599–609.

Jones, K.B., Ritters, K.H., Wickham, J.D., Tankersley, R.D., O'Neill, R.V., Chaloud, D.J., Smith, E.R., and Neale, A.C. 1997. An ecological assessment of the mid-Atlantic region. EPQA/600/R-97/130. US Environmentl Protection Agency, Washington, DC.

Jorgensen, S.E. 1994. Fundamentals of Ecological Modelling, Second Edition. Elsevier, Amsterdam.

Jorgensen, S.E. and Bendoriccio, G. 2001. Fundamentals of Ecological Modelling, Third Edition. Elsevier, Amsterdam.

Jorgensen, S.E., Halling-Sorensen, B. and Nielsen, S.N. (1996) Handbook of Environmental and Ecological Modeling. Lewis Publishers, Boca Raton, FL.

Jorgensen, S.E., Halling-Sorensen, B., and Mahler, H. (1998) Handbook of Estimation Method in Ecotoxicology and Environmental Chemistry. Lewis Publishers, Boca Raton, FL.

Josephson, J.R. and Josephson, S.G. 1996. Abductive Inference. Cambridge University Press, Cambridge, UK.

Kahkonen, M.A. and Manninen, P.K.G. 1998. The uptake of nickel and chromium from water by *Elodea canadensis* at different nickel and chromium exposure levels. Chemosphere, 36:1381–1390.

Kahkonen, M.A., Pantsar-Kallio, M., and Manninen, P.K.G. 1998. Analyzing heavy metal concentrations in the different parts of *Elodea canadensis* and surface sediment with PCA in two boreal lakes in southern Finland. Chemosphere, 36:2645–2656.

Kammenga, J. and Laskowski, R. 2000. Demography in Ecotoxicology. John Wiley, Chichester, UK.

Kammenga, J.E., Van Koert, P.H.G., Riksen, J.A.G., Korthals, G.W., and Bakker, J. 1996. A toxicity test in artificial soil based on the life history strategy of the nematode *Plectus acuminatus*. Environ. Toxicol. Chem., 15:722–727.

Kangas, M. 1996. Probabilistic risk assessment: Understanding uncertainties in estimates of risks for contaminated sites. ASTM Standardization News, June:28–33.

Kaplan, E.L. and Meier, P. 1958. Nonparametric estimation from incomplete observations. J. Am. Stat. Assoc., 53:457–481.

Kaplan, I., Lu, S.-T., Lee, R.-P., and Warrick, G. 1996. Polycyclic hydrocarbon biomarkers confirm selective incorporation of petroleum in soil and kangaroo rat liver samples near an oil well blowout site in the western San Joaquin Valley, California. Environ. Toxicol. Chem., 15:696–707.

Kapustka, L.A. 1997. Selection of phytotoxicity tests for ecological risk assessment. Pages 515–548 in W. Wang, J.W. Gorsuch, and J.S. Hughes (eds.) Plants for Environmental Studies. Lewis Publishers, Boca Raton, FL.

Kapustka, L.A. 2003. Rationale for use of wildlife habitat characterization to improve relevance of ecological risk assessment. Hum. Ecol. Risk Assess., 9:1425–1431.

Kapustka, L.A., Lipton, J., Galbraith, H., Cacela, D., and LeJeune, K. 1995. Metal and arsenic impacts to soils, vegetation communities and wildlife habitat in southwestern Montana uplands contaminated by smelter emissions: II. Laboratory phytotoxicity studies. Environ. Toxicol. Chem., 14:1905–1912.

Karabunarliev, S.H., Dimitrov, S., Nikolova, N., and Mekenyan, O. 2002. Prediction of acute aquatic toxicity of noncongeneric chemicals: Rule-based and quantitative structure–activity relationships. in J.D. Walker (ed.) Handbook on Quantitative Structure–Activity Relationships (QSARs) for Predicting Ecological Effects of Chemicals. SETAC Press, Pensacola, FL.

Karickhoff, W.W. 1981. Semi-empirical estimation of sorption of hydrophobic pollutants on natural sediments and soils. Chemosphere, 10:833–846.

Karmaus, W. 2001. Of jugglers, mechanics, communities and the thyroid gland: How do we achieve good quality data to improve public health? Environ. Health Persp., 109:863–869.

Karr, J.R. and Chu, E.W. 1997. Biological monitoring: Essential foundation for ecological risk assessment. Hum. Ecol. Risk Assess., 3:993–1004.

Karr, J.R. and Chu, E.W. 1999. Restoring Life in Running Waters: Better Biological Monitoring. Island Press, Washington, DC.

Karr, J.R., Fausch, K.D., Angermeier, P.L., Yant, P.R., and Schlosser, I.J. 1986. Assessing biological integrity in running waters: A method and its rationale. Illinois Natural History Survey Special Pub. 5. Champaigne, IL.

Keach, S. 1998. Assessing quality of life issues in EPA sponsored state and local comparative risk projects. US Environmental Protection Agency, Washington, DC. Available at: http://www.epa.gov/comp_risk/cr/qol3.htm.

Keddy, C.J., Greene, J.C., and Bonnell, M.A. 1995. Review of whole-organism bioassays: Soil, freshwater sediment, and freshwater assessment in Canada. Ecotoxicol. Environ. Chem., 30:221–251.

Keedwell, R.J. 2004. Use of population viability analysis in conservation management in New Zealand. Science for Conservation 243. Department of Conservation, Wellington, NZ.

Keith, L.H. 1994. Throwaway data. Environ. Sci. Technol., 28:389A–390A.

Kelly, M.E., Brauning, S.E., Schoof, R.A., and Ruby, M.V. 2002. Assessing Oral Bioavailability of Metals in Soil. Battelle Press, Columbus, OH.

Kelsey, J.W. and Alexander, M. 1997. Declining bioavailability and inappropriate extimates of risk of persistent compounds. Environ. Toxicol. Chem., 16:582–585.

Kendall, R.J., Lacher, T.E., Bunck, C., Daniel, B., Driver, C., Grue, C.E., Leighton, F., Stansley, W., Watanabe, P.G., and Whitworth, M. 1996. An ecological risk assessment of lead shot exposure in non-waterfowl avian species: Upland game birds and raptors. Environ. Toxicol. Chem., 15(1):4–20.

Kennedy, J.H., LaPoint, T., Balci, P., Stanley, J.K., and Johnson, Z.B. 2003. Model aquatic ecosystems in ecotoxicological research: Considerations of design, implementation and analysis. Pages 45–74 in D.J. Hoffman, B. Rattner, G.A. Burton Jr., and J. Cairns Jr. (eds.) Handbook of Ecotoxicology. Lewis Publishers, Boca Raton, FL.

Kerans, B.L. and Karr, J.R. 1992. A benthic index of biotic integrity (B-IBI) for rivers in the Tennessee Valley. Ecol. Appl., 4:785.

Kerans, B.L. and Karr, J.R. 1994. A benthic index of biotic integrity (B-IBI) for rivers in the Tennessee Valley. Ecol. Appl., 4:785.

Kerans, B.L., Karr, J.R., and Ahlstedt, S.A. 1992. Aquatic invertebrate assemblages: Spatial and temporal differences among sampling protocols. J. N. Am. Benthol. Soc., 11:377–390.

Kerr, D.R. and Meador, J.P. 1996. Modeling dose response using generalized linear models. Environ. Toxicol. Chem., 15:395–401.

Kester, J.E., VanHorn, R.L., and Hampton, N.L. 1998. Methodology for conducting screening-level ecological risk assessments for hazardous waste sites. Part III: Exposure and effects assessment. Int. J. Environ. Pollut., 9:62–89.

Ketcheson, G.L., Megahan, W.F., and King, J.G. 1999. "R1–R4" and "Boised" sediment prediction model tests using forest roads in granitics. J. Am. Water Resources Assoc., 35:83–98.

Kimball, K.D. and Levin, S.A. 1985. Limitations of laboratory bioassays: The need for ecosystem-level testing. Bioscience, 35:165–171.

Klaine, S.J., Cobb, G.P., Dickerson, R.L., Dixon, K.R., Kendal, R.J., Smith, E.E., and Solomon, K.R. 1996. An ecological risk assessment for the use of the biocide dibromonitrilopropionamide (DNBPA) in industrial cooling systems. Environ. Toxicol. Chem., 15:21–30.

Klapow, L.A. and Lewis, R.H. 1979. Analysis of toxicity data for California marine water quality standards. J. Water Pollut. Control Fed., 51:2054–2070.

Klein, M., Hosang, J., Schafer, H., Erzgraber, B., and Resseler, H., 2000. Comparing and evaluating pesticide leaching models. Results of simulations with PELMO. Agr. Water Manag., 44:263–282.

Klemm, D.J., Morrison, G.E., Norberg-King, T., Peltier, W., and Heber, M.A. 1994. Short-term methods for estimating the chronic toxicity of effluents and receiving waters to marine and estuarine organisms, Second Edition. EPA/600/4-91/003. US Environmental Protection Agency, Cincinnati, OH.

Klijn, F., DeWall, R., and Voshaar, J.H.O. 1995. Ecoregions and ecodistricts: Ecological regionalizations for the Netherlands environmental policy. Environ. Manag., 19:797–813.

Klimisch, H.-J., Andreae, M., and Tillmann, U. 1997. A systematic approach to evaluating the quality of experimental toxicological and ecotoxicological data. Reg. Tox. Pharmacol., 25:1–5.

Klopman, G., Saiakhov, R., and Rosenkranz, H. 2000. Multiple computer-automated structure evaluation study of aquatic toxicity II. Fathead minnow. Environ. Toxicol. Chem., 19:441–447.

Koch, A.L. 1966. The logarithm in biology 1. Mechanisms generating the log-normal distribution exactely. J. Theoret. Biol., 12:276–290.

Koenker, R. 2005. Quantile Regression. Cambridge University Press, New York.

Konemann, H. 1981a. Fish toxicity tests with mixtures of more than two chemicals: A proposal for a quantitative approach and experimental results. Toxicology, 19:229–238.

Konemann, H. 1981b. Quantitative structure–activity relationships in fish toxicity studies. Part 1: Relationship for 50 industrial chemicals. Toxicology, 19:209–221.

Kooijman, S.A.L.M. 1981. Parametric analysis of mortality rates in bioassays. Water Res., 15:107–119.

Kooijman, S.A.L.M. 1987. A safety factor for LC_{50} values allowing for differences in sensitivity among species. Water Res., 21:269–276.

Kooijman, S.A.L.M. 2000. Dynamic Energy and Mass Budgets in Biological Systems. Cambridge University Press, Cambridge, UK.

Kooijman, S.A.L.M. and Metz, J.A.J. 1984. On the dynamics of chemically stressed populations: The deduction of population consequences from effects on individuals. Ecotoxicol. Environ. Saf., 8:254–274

Kooijman, S.A.L.M. and Bedaux, J.J.M. 1996. The Analysis of Aquatic Toxicity Data. Free University Press, Amsterdam.

Kopp, R. and Smith, V. 1993. Valuing Natural Assets: The Economics of Natural Resource Damage Assessment. Resources for the Future, Washington, DC.

Korsloot, A., van Gestel, C.A.M., and Van Straalen, N.M. 2004. Environmental Stress and Cellular Response in Arthropods. CRC Press, Boca Raton, FL.

Kovacs, T.G., Martel, P.H., Voss, R.H., Wrist, P.E., and Willes, R.F. 1993. Aquatic toxicity equivalency factors for chlorinated phenolic compounds present in pulp mill effluents. Environ. Toxicol Chem., 12:684–691.

Kowal, N.E. 1971. Models of elemental assimilation by invertebrates. J. Theoret. Biol., 31:469–474.

Kraaij, R., Seinen, W., Tolls, J., Cornelissen, G., and Belfroid, A. 2002. Direct evidence of sequestration in sediments affecting the bioavailability of hydrophobic organic chemicals to benthic deposit feeders. Environ. Sci. Technol., 36:3525–3529.

Krebs, C.J. 2002. Ecology: The Experimental Analysis of Distribution and Abundance, Fifth Edition. Prentice-Hall, Upper Saddle River, NJ.

Krech, S., III. 1999. The Ecological Indian, W.W. Norton, New York.

Kreibel, D. et al. 2001. The precautionary principle in environmental science. Environ. Health Persp., 109:871–876.

Krishnan, K., Haddad, S., Beliveau, M., and Tardiff, R.G. 2002. Physiological modeling and extrapolation of pharmacokinetic interactions from binary to more complex interactions. Environ. Health Persp., 110:989–994.

Kroes, R., Galli, C., Schilter, B., Tran, L.-A., Walzer, R., and Wuertzen, G. 2000. Threshold of toxicological concern for chemical substances present in the diet: A practical tool for assessing the need for toxicity testing. Food Chem. Toxicol., 34:867.

Kszos, L.A., Stewart, A.J., and Taylor, P.A. 1992. An evaluation of nickel toxicity to *Ceriodaphnia dubia* and *Daphnia magna* in a contaminated stream and in laboratory toxicity tests. Environ. Toxicol. Chem., 11:1001–1012.

Kuhn, A., Munns, W.R. Jr., Champlin, D., McKinney, R., Tagliabue, M., Serbst, J., and Gleason, T. 2001. Evaluation of the efficacy of extrapolation population modeling to predict the dynamics of *Americamysis bahia* populations in the laboratory. Environ. Toxicol. Chem., 20:213–221.

Lackey, R.T. 1994. Ecological risk assessment. Fisheries, 19:14–18.

Lackey, R.T. 1998. Seven pillars of ecosystem management. Landsc. Urban Plan., 40:21–30.

Lacy, R.C. 1993. VORTEX: A computer simulation model for use in population viability analysis. Wildl. Res., 20:40–65.

LaGoy, P.K. and Schulz, C.O. 1993. Background sampling: An example of the need for reasonableness in risk assessment. Risk Anal., 13:483–484.

Lamberson, J.O., DeWitt, T.H., and Swartz, R.C. 1992. Assessment of sediment toxicity to marine benthos. Page 457 in G.A. Burton Jr. (ed.) Sediment Toxicity Assessment. Lewis Publishers, Boca Raton, FL.

Lande, R. 1988. Demographic models of the northern spotted owl (*Stix occidentalis caurina*). Oecologia 75:601–607.

Lande, R., Engen, S., Sæther, B.-E., Filli, F., Matthysen, E., and Weimerskirch, H. 2002. Estimating density-dependence from population time series using demographic theory and life-history data. Am. Natl., 159:321–337.

Landis, W.G. 2005. Regional Scale Ecological Risk Assessment Using the Relative Risk Model. CRC Press, Boca Raton, FL.

Landrum, P.F. 1988. Toxicokinetics of organic xenobiotics in the amphipod *Potoporeia hoyi*: Role of physiological and environmental variables. Aquat. Toxicol., 12:245–271.

Lane, P. and Collins, T. 1985. Food web models of a marine plankton community network: An experimental mesocosm approach. J. Exp. Mar. Biol. Ecol. 84:41–70.

Lanes, S.E. 1988. The logic of causal inference. Pages 59–76 in K.J. Rothman (ed.) Causal Inference. Epidemiology Resources, Chestnut Hill, MA.

Lange, R., Hutchinson, T.H., Scholz, N., and Solbe, J.F. 1998. Analysis of ECETOC aquatic toxicity (EAT) database II: Comparisons of acute to chronic ratios for various aquatic organisms and chemical substances. Chemosphere, 36:115–127.

LaPoint, T.W. 1995. Signs and measurements of ecotoxicology in the aquatic environment. Pages 13–24 in B.A. Hoffman, B.A. Rattner, G.A. Burton Jr., and J. Cairns Jr. (eds.) Handbook of Ecotoxicology. Lewis Press, Boca Raton, FL.

Larsson, P. 1984. Transport of PCBs from aquatic to terrestrial environments by emerging chironomids. Environ. Pollut. Ser. A, 34:283–289.

Laskowski, D. and Calder, W.A. Jr. 1971. A preliminary allometric analysis of respiratory variables in resting birds. Resp. Physiol., 11:152–166.

Laskowski, R. 1995. Some good reasons to ban the use of NOEC, LOEC, and related concepts in ecotoxicology. Oikos, 73:140–144.

Laskowski, R., Kramarz, P., and Jepson, P. 1998a. Selection of species for soil ecotoxicity testing. Pages 21–32 in H. Lokke and C.A.M. van Gestel (eds.) Handbook of Soil Invertebrate Toxicity Testing. John Wiley, Chichester, UK.

Laskowski, R., Kramarz, P., and Jepson, P. 1998b. Selection of species for soil ecotoxicity testing. Pages 21–32 in H. Lokke and C.A.M. van Gestel (eds.) Handbook of Soil Invertebrate Toxicity Testing. John Wiley, Chichester, UK.

Lassiter, R.R. and Hallam, T.G. 1990. Survival of the fattest: Implications for acute effects of lippophilic chemicals on aquatic populations. Environ. Toxicol. Chem., 9:585–595.

Law, R. 1983. A model for the dynamics of a plant population containing individuals classified by age and size. Ecology, 64:224–230.

Law, R. and Edley, M.T. 1990. Transient dynamics of populations with age- and size-dependent vital rates. Ecology, 71:1863–1870.

Lawrence, D.P. 2003. Environmental Impact Assessment: Practical Solutions to Recurrent Problems. John Wiley, Hoboken, NJ.

Lazim, M.N., Learner, M.A., and Cooper, S. 1989. The importance of worm identity and life-history in determining the vertical-distribution of tubificids (Oligochaeta) in a riverine mud. Hydrobiologia, 178:81–92.

LeBlanc, G.A. 1984. Interspecies relationships in acute toxicity of chemicals to aquatic organisms. Environ. Toxicol. Chem., 3:47–60.

Lee, J.-H., Landrum, P.F., and Koh, C.-H. 2002. Prediction of time-dependent PAH toxicity in *Hyalella azteca* using a damage assessment model. Environ. Sci. Technol., 36:3131–3138.

Lee, K.E. 1985. Earthworms: Their Ecology and Relationships with Soils and Land Use. Academic Press, Sydney.

Legierse, K.C.H.M., Verhaar, H.J.M., and Vaes, W.H.J. 1999. Analysis of time-dependent acute aquatic toxicity of organophosphate pesticides: The critical target occupation model. Environ. Sci. Technol., 33:917–925.

Leistra, M., van der Linden, J.J.T.I., Boesten, A., Tiktak, A., and van den Berg, F. 2001. PEARL model for pesticide behavior and emissions in soil-plant systems, description of the process. Alterra Report 13, RIVM Report 711401009. Alterra, Wageningen, The Netherlands.

LeJeune, K., Galbraith, H., Lipton, J., and Kapustka, L.A. 1996. Effects of metals and arsenic on riparian communities in southwest Montana. Ecotoxicology, 5:297–312.

Leslie, P.H. 1945. On the use of matrices in certain population mathematics. Biometrika, 33:183–212.

Levin, S. (ed.). 1974. Ecosystem analysis and prediction. Proceedings of a SIAM-SIMS Conference, Alta, Utah. Society for Industrial and Applied Mathematics, Philadelphia, PA.

Levins, R. 1969. Some demographic and genetic consequences of environmental heterogeneity for biological control. Bull. Entomol. Soc. Am., 15:237–240.

Levins, R. 1974. The qualitative analysis of partially specified systems. Annals NY Acad. Sci., 231:123–138.

Liao, K.H., Dobrev, I.D., Dennison, J.E. Jr., Anderson, M.E., Reisfeld, B., Reardon, K.F., Campain, J.A., Wei, W., Klein, M.T., Quann, R.J., and Yang, R.S.H. 2002. Application of biologically based computer modeling to simple or complex mixtures. Environ. Health Persp., 110:957–963.

Lincer, J.L. 1975. DDE-induced eggshell-thinning in the American kestrel: A comparison of the field situation and laboratory results. J. Appl. Ecol., 12:781–793.

Linder, G., Ingham, E., Brandt, C.J., and Henderson, G. 1992. Evaluation of terrestrial indicators for use in ecological assessments at hazardous waste sites. EPA/600/R-92/183. US Environmental Protection Agency, Corvallis, OR.

Linder, G., Krest, S.K., and Sparling, D.W. 2003. Amphibian Decline: An Integrated Analysis of Multiple Stressor Effects. SETAC Press, Pensacola, FL.

Linder, G., Harrahy, E., Johnson, L., et al. 2004. Sunflower depredation and avicide use: A case study focused on DRC-1339 and risks to non-target birds in North Dakota and South Dakota. Pages 202–220 in L.A. Kapustka, G.R. Biddinger, M. Luxon, and H. Galbraith (eds.) Landscaper Ecology and Wildlife Habitat Evaluation. ASTM, West Conshohocken, PA.

Linders, J.B.H.J. (ed.). 2001. Modelling of Environmental Chemical Exposure and Risk. Kluwer Academic, Dordrecht, The Netherlands.

Linkov, I., von Stackelberg, K.E., Burmistrov, D., and Bridges, T. 2001. Uncertainty and variability in risk from trophic transfer of contaminants in dredged sediments. Sci. Total Environ., 274:255–269.

Linkov, I., Burmistrov, D., Cura, J., and Bridges, T.S. 2002a. Risk-based management of contaminated sediments: Consideration of spatial and temporal patterns in exposure modeling. Environ. Sci. Technol., 36:238–246.

Linkov, I., Foster, S., Hathaway, E., and Suggat, R. 2002b. Preliminary ecological risk assessment for the Elizabeth Mine site, South Strafford, Vermont. Tailings and Mine Waste '02. A.A. Balkema, Leiden, The Netherlands.

Linkov, I., Satterstrom, F.K., Kiker, G., Seager, T.P., Bridges, T., Gardner, K.H., Rogers, S.H., Belluck, D.A., and Meyer, A. 2006. Multicriteria decision analysis: A comprehensive decision approach for management of contaminated sediments. Risk Anal., 26:61–78.

Liste, W.-H. and Alexander, M. 2002. Butanol extraction to predict bioavailability of PAHs in soil. Chemosphere, 46:1011–1017.

Liu, J. 1993. An introduction to ECOLECON: A spatially explicit model for ECOLogical ECONomics of species conservation in complex forest landscapes. Ecol. Model., 70:63–87.

Logan, D.T. and Wilson, H.T. 1995. An ecological risk assessment method for species exposed to contaminant mixtures. Environ. Toxicol. Chem., 14:351–359.

Loibner, A.P., Holzer, M., Gartner, O., Szolar, O.H.J., and Braun, R. 2000. The use of sequential supercritical fluid extraction for bioavailability investigations of PAH in soil. Die Bodenkultur, 225–233.

Loibner, A.P., Szolar, O.H.J., Braun, R., and Hirmann, D. 2003. Ecological assessment and toxicity screening in contaminated land analysis. Pages 229–267 in K.C. Thompson and C.P. Nathanail (eds.) Chemical Analysis of Contaminated Land. Blackwell, Oxford, UK.

Lokke, H. 1994. Ecotoxicological extrapolation: Tool or toy? Pages 411–426 in M.H. Donker, H. Eijsackers, and F. Heimbach (eds.) Ecotoxicology of Soil Organisms. Lewis Publishers, Boca Raton, FL.

Lokke, H. and van Gestel, C.A.M. 1998. Handbook of Soil Invertebrate Toxicity Testing. John Wiley, Chichester, UK.

Long, E.B. 2000. Degraded sediment quality in U.S. estuaries: A review of magnitude and ecological implications. Ecol. Appl., 10:338–349.

Long, E.R. and Chapman, P.M. 1985. A sediment quality triad: Measures of sediment contamination, toxicity and infaunal community composition in Puget Sound. Mar. Pollut. Bull., 16:405–415.

Long, E.R., MacDonald, D.D., Smith, S.L., and Calder, F.D. 1995. Incidence of adverse biological effects within ranges of chemical concentrations in marine and estuarine sediments. Environ. Manag., 19:81–97.

Long, E.R. and Morgan, L.G. 1991. The Potential for Biological Effects of Sediment-Sorbed Conta-minants Tested in the National Status and Trends Program. NOAA Technical Memorandum NOS OMA 52. National Oceanic and Atmospheric Administration, Seattle, Washington, DC, USA.

Long, E.R., Dutch, M., Aasen, S., and Welch, K. 2004. Sediment quality triad index in Puget Sound. Pub. No. 04-03-008. Washington State Department Ecology, Olympia, WA.

Longcore, T. and Rich, C. 2004. Ecological light pollution. Front. Ecol. Environ., 2:191–198.

Lotka, A.J. 1924. Elements of Physical Biology. Williams and Wilkins, Baltimore, MD. (Reprinted in 1956 by Dover Publications, New York, as Elements of Mathematical Biology.)

Loucks, D.P. 2003. Managing America's rivers: Who's doing it? Int. J. River Basin Manag., 1:21–31.

Loucks, O.L. 1972. Systems methods in environmental court actions. Pages 419–475 in B.C. Patten (ed.). Systems Analysis and Simulation in Ecology: Volume 2. Academic Press, New York.

Lozano, S.J., O'Halloran, S.L., Sargent, K.W., and Brazner, J.C. 2003. Effects of esfenvalerate on aquatic organisms in littoral enclosures. Environ. Toxicol. Chem, 11:35–47.

Lubchenco, J. 1998. Entering the century of the environment: A new social contract for science. Science, 279:491–495.

Ludwig, J.P., Kurita-Matsuba, H., Auman, H.J., Ludwig, M.E., Summer, C.L., Giesy, J.P., Tillitt, D.E., and Jones, P.D. 1996. Deformities, PCBs, and TCDD-equivalents in double-crested cormorants (*Phalacrocorax auritus*) and Caspian terns (*Hydroprogne caspia*) of the upper Great lakes 1986–1991: Testing a cause–effect hypothesis. J. Great Lakes Res., 22:172–197.

Lundgren, R. and McMakin, A. 1998. Risk Communication: A Handbook for Communicating Envir-onmental, Safety, and Health Risks. Battelle Press, Columbus, OH.

Luoma, S.N. 1995. Prediction of metal toxicity in nature from toxicity tests: Limitations and research needs. Pages 610–659 in A. Tessier and D. Turner (eds.) Metal Speciation and Bioavailability in Aquatic Systems. John Wiley, New York.

Luoma, S.N. and Fisher, N. 1997. Uncertainties in assessing contaminant exposure from sediments. Pages 211–238 in C.G. Ingersoll, T, Dillon, and G.R. Biddinger (eds.) Ecological Risk Assess-ment of Contaminated Sediments. SETAC Press, Pensacola, FL.

Luttik, R. and de Snoo, G.R. 2004. Characterization of grit in arable birds to improve pesticide risk assessment. Ecotoxicol. Environ. Saf., 57:319–329.

Lyman, W.J., Reehl, W.F., and Rosenblatt, D.H. (eds.). 1982. Handbook of Chemical Property Estima-tion Methods: Environmental Behavior of Organic Compounds. McGraw-Hill, New York.

Ma, W.-C. 1994. Methodological principles of using small mammals for ecological hazard assessment of chemical soil pollution, with examples of cadmium and lead. Pages 357–371 in M.H. Donker, H. Eijsackers, and F. Heimbach (eds.) Ecotoxicology of Soil Organisms. Lewis Publishers, Boca Raton, FL.

Ma, W.-C., van Kleunen, A., Immerzeel, J., and Gert-Jan de Maagd, P. 1998. Bioaccumulation of polycyclic aromatic hydrocarbons by earthworms: Assessment of equilibrium partitioning the-ory in in situ studies and water experiments. Environ. Toxicol. Chem., 17:1730–1737.

McCarty, L.S. and Mackay, D. 1993. Enhancing ecotoxicological modeling and assessment. Environ. Sci. Technol., 27:1719–1728.

McCarty, L.S. and Power, M. 2001. Approaches to developing risk management objectives: An analysis of international strategies. Environ. Sci. Policy, 3:311–319.

McCauley, E., Murdoch, W.W., Nisbet, R.M., and Gurney, W.S.C. 1990. The physiological ecology of *Daphnia*: A dynamic model of growth and reproduction. Ecology, 71:716–732.

McClung, G. and Sayre, P.G. 1994. Risk assessment for the release of recombinant *Rhizobia* at a small-scale agricultural field site. Pages 2–1–2–F3 in A Review of Ecological Assessment Case Studies from a Risk Assessment Perspective. EPA/630/R-94-003. US Environmental Protection Agency, Washington, DC.

MacDonald, D.D., Carr, R.S., Calder, F.D., Long, E.R., and Ingersoll, C.G. 1996. Development and evaluation of sediment quality guidelines for Florida coastal waters. Ecotoxicology, 5:253–278.

MacDonald, R., Mackay, D., and Hickie, B. 2002. Contaminant amplification in the environment. Environ. Sci. Technol., 36:456A–462A.

McGeer, J.C., Szebedinsky, C., McDonald, D.G., and Wood, C.M. 2002. The role of dissolved organic carbon in moderating the bioavailability and toxicity of Cu in rainbow trout during chronic waterborne exposure. Comp. Biochem. Physiol., 133C:147–160.

MacIntosh, D.L., Suter, G.W., II., and Hoffman, F.O. 1994. Uses of probabilistic exposure models in ecological risk assessments of contaminated sites. Risk Anal., 14:405–419.

Mackay, D. 1989. Modelling the Long-Term Behaviour of an Organic Contaminant in a Large Lake: Application to PCBs in Lake Ontario. J. Great Lakes Res., 15:283–297.

Mackay, D. 2001. Multimedia Environmental Models—The Fugacity Approach. Lewis Publishers, Boca Raton, FL.

Mackay, D. and Fraser, A. 2000. Bioaccumulation of persistent organic chemicals: Mechanisms and models. Environ. Pollut., 110:375–391.

Mackay, D., Paterson, S., Kicsi, G., Di Guardo, A., and Cowan, C.E. 1996a. Assessing the fate of new and existing chemicals: A five-stage process. Environ. Toxicol. Chem., 15(9):1618–1626.

Mackay, D., Paterson, S., Di Guardo, A., and Cowan, C.E. 1996b. Evaluating the environmental fate of a variety of types of chemicals using the EQC model. Environ. Toxicol. Chem., 15(9):1627–1637.

Mackay, D., Paterson, S., Kicsi, G., Cowan, C.E., Di Guardo, A., and Kane, D.M. 1996c. Assessment of chemical fate in the environment using evaluative, regional and local-scale models: Illustrative application to chlorobenzene and linear alkylbenzene sulfonates. Environ. Toxicol. Chem., 15(9):1638–1648.

Mackay, D., Shiu, W.Y., and Ma, K.C. 2006. Physical–Chemical Properties and Environmental Fate for Organic Chemicals. CRC Press, Boca Raton, FL.

McKay, D. and Singleton, P.C. 1974. Time required to reclaim land contaminated with crude oil: An approximation. B-612. Laramie, Agricultural Extension Service, University of Wyoming, Laramie, WY.

McKim, J.M. 1985. Early life stage toxicity tests. Pages 58–95 in G.M. Rand and S.R. Petrocelli (eds.) Fundamentals of Aquatic Toxicology. Hemisphere Publishing, Washington, DC.

McKim, J.M., Olson, G.F., Holcombe, G.H., and Hunt, E.P. 1976. Long-term effects of methylmercuric chloride on three generations of brook trout (*Salvelinus fontinalis*): Toxicity, accumulation, distribution, and elimination. J. Fisheries Res. Board Canada, 33:2726–2739.

McKone, T.E. 1993a. CalTOX, a Multi-media Total-Exposure Model for Hazardous Wastes Sites. Part II: The Dynamic Multi-media Transport and Transformation Model. A report prepared for the State of California, Department of Toxic Substances Control by the Lawrence Livermore National Laboratory, No. UCRL-CR-111456PtII, Livermore, CA.

McKone, T.E. 1993b. The precision of QSAR methods for estimating intermedia transfer factors in exposure assessments. SAR QSAR Environ. Res., 1:41–51.

McKone, T.E. 1994. Uncertainty and variability in human exposure to soil contaminants through home-grown food: A Monte Carlo assessment. Risk Anal., 14:449–463.

McLaughlin, M.J. 2001. Bioavailability of metals to terrestrial plants. Pages 39–68 in H.E. Allen (ed.) Bioavailability of Metals in Terrestrial Ecosystems. SETAC Press, Pensacola, FL.

McLaughlin, M.J., Smolders, E., and Merckx, R. 1998. Soil-root interface: Physicochemical processes. Pages 233–277 in P.M. Huang (ed.) Soil Chemistry and Ecosystem Health. Soil Science Society of America, Madison, WI.

McLaughlin, S.B. Jr. 1983. Effects of acid rain and gaseous pollutants on forest productivity: A regional scale approach. APCA J., 33:1042–1049.

McLaughlin, S.B. Jr. and Taylor, G.E. Jr. 1985. SO2 effects on dicot crops: Some issues, mechanisms, and indicators. Pages 227–249 in W.E. Winner, H.A. Mooney, and R.A. Goldstein (eds.) Sulfur Dioxide and Vegetation. Stanford University Press, Stanford, CA.

McLeay, D., Genthner, R., James, R., Lazarovits, G., and Percy, D. 2004. Guidance document for testing the pathogenicity and toxicity of new microbial substances to aquatic and terrestrial organisms. EPS 1/RM/44. Environment Canada, Ottawa, Ontario.

MacLeod, M., Woodfine, D., Mackay, D., McKone, T., Bennett, D., and Maddelena, R. 2001. BETR North America: A regionally segmented contaminant fate model of North America. Environ. Sci. Pollut. Res., 8(3):156–163.

Maclure, M. 1998. Inventing the AIDS virus hypothesis: An illustration of scientific vs unscientific induction. Epidemiology, 9:467–476.

Madenjian, C.P. and Carpenter, S.R. 1991. Individual-based model for growth of young-of-the-year walleye: A piece of the recruitment puzzle. Ecol. Appl., 1:267–278.

MADEP (Massachusetts Department of Environmental Protection). 2002. Characterizing risks posed by petroleum-contaminated sites: Implementation of the MADEP VPH/EPH approach. Department of Environmental Protection, Boston, MA.

Mahler, B.J., van Metre, P.C., Bashara, T.J., Wilson, J.T., and Johns, D.A. 2005. Parking lot sealcoat: An unrecognized source of urban polycyclic aromatic hydrocarbons. Environ. Sci. Technol., 39:5560–5566.

Marcot, B.G. and Holthausen, R. 1987. Analyzing population viability of the spotted owl in the Pacific Northwest. Trans. N. Am. Wildl. Nat. Res. Conf. 52:333–347.

Marine Protection Branch. 1991. Evaluation of dredged material proposed for ocean disposal, testing manual. EPA-503/8-91/001. US Environmental Protection Agency, Washington, DC.

Markwiese, J.T., Ryti, R.T., Nooten, M.M., Michael, D.I., and Hlohowskyj, I. 2001. Toxicity bioassays for ecological risk assessment in arid and semiarid ecosystems. Rev. Environ. Contam. Toxicol., 168:43–98.

Martinson, B.C., Anderson, M.S., and deVries, R. 2005. Scientists behaving badly. Nature, 435:737–738.

Massachusetts Weight-of-Evidence Work Group. 1995. Draft report: A weight-of-evidence approach for evaluating ecological risks, November 2, 1995.

Mathews, R.A., Mathews, G.B., and Landis, W.G. 2003. Application of community level toxicity testing to environmental risk assessment. Pages 225–253 in M.C. Newman and C.L. Strojan (eds.) Risk Assessment: Logic and Measurement. Lewis Publishers, Boca Raton, FL.

Maund, S.J., Travis, K.R., Hendley, P., Giddings, J.M., and Solomon, K.R. 2001. Probabilistic risk assessment of cotton pyrethroids: V. Combining landscape-level exposures and ecotoxicological effects data to characterize risks. Environ. Toxicol. Chem., 20:687–692.

Maurer, B.A. and Holt, R.D. 1996. Effects of chronic pesticide stress on wildlife populations in complex landscapes: Processes at multiple scales. Environ. Toxicol. Chem., 15:420–426.

Mauriello, D.A. and Park, R.A. 2002. An adaptive framework for ecological assessment and management. Pages 509–514 in A.E. Rizzoli and A.J. Jakeman (eds.) Integrated Assessment and Decision Support. International Environmental Modeling and Software Society, Manno, Switzerland.

Mayer, F.L., Krause, G.F., Buckler, D.R., Ellersieck, M.R., and Lee, G. 1994. Predicting chronic lethality of chemicals to fishes from acute toxicity test data: Concepts and linear regression analysis. Environ. Toxicol. Chem., 13:671–678.

Mayer, F.L., Ellersieck, M.R., and Asfaw, A. 2004. Interspecies correlation estimations (ICE) for acute toxicity to aquatic organisms and wildlife. I. Technical basis. EPA/600/R-03/106. US Environmental Protection Agency, Washington, DC.

Mayo, D.C. 2004. An error statistical philosophy of evidence. Pages 79–118 in M.L. Taper and S.R. Lele (eds.) The Nature of Scientific Evidence. University of Chicago Press, Chicago, IL.

Mazurek, M.A. 2002. Molecular identification of organic compounds in atmospheric complex mixtures and relationship to atmospheric chemistry and sources. Environ. Health Persp., 110:995–1003.

Meier, J.R., Chang, L., Jacobs, S., Torsella, J., Meckes, M.C., and Smith, M.K. 1997. Use of plant and earthworm bioassays to evaluate remediation of soil from a site contaminated with polychlorinated biphenys. Environ. Toxicol. Chem., 16:928–938.

Mekenyan, O.G., Veith, G.D., Call, D.J., and Ankley, G.T. 1996. A QSAR evaluation of Ah receptor binding of halogenated aromatic xenobiotics. Environ. Health Persp., 104:1302–1310.

Melnikov, A. 2003. Risk Analysis in Finance and Insurance. Chapman & Hall/CRC Press, Boca Raton, FL.

Menzel, D.B. 1987. Physiological pharmacokinetic modeling. Environ. Sci. Technol., 21:944–950.

Menzie, C.A. and Freshman, J.S. 1997. An assessment of the risk assessment paradigm for ecological risk assessment. Hum. Ecol. Risk Assess., 3:853–892.

Menzie, C.A., Burmaster, D.E., Freshman, D.S., and Callahan, C. 1992. Assessment of methods for estimating ecological risk in the terrestrial component: A case study at the Baird and McGuire Superfund Site in Holbrook, Massachusetts. Environ. Toxicol. Chem., 11:245–260.

Menzie, C., Henning, M.H., Cura, J., Finkelstein, K., Gentile, J., Maughan, J., Mitchell, D., et al. 1996. A weight-of-evidence approach for evaluating ecological risks: Report of the Massachusetts Weight-of-Evidence Work Group. Hum. Ecol. Risk Assess., 2:277–304.

Menzie, C.A., Hoeppner, S.S., Cura, J., Freshman, J.S., and LaFrey, E.N. 2002. Urban and suburban stormwater runoff as a source of polycyclic aromatic hydrocarbons (PAHs) in Massachusetts estuarine and coastal environments. Estuaries, 25:165–176.

Mertz, D.B. 1971.The mathematical demography of the California Condor. Am. Natl., 105:437–453.

Meyer, F.P. and Barclay, L.A. 1990. Field manual for the investigation of fish kills. Resource Pub. 177. US Fish and Wildlife Service, Washington, DC.

Meyer, J.S., Ingersoll, C.G., McDonald, L.L., and Boyce, M.S. 1986. Estimating uncertainty in population growth rates: Jacknife vs. Bootstrap techniques. Ecology, 67:1156–1166.

Meyer, J.S., Santore, R.C., Bobbitt, J.P., DeBrey, L.D., Boese, C.J., Paquin, P.R., Allen, H.E., Bergman, H.A., and DiToro, D. 1999. Binding of nickel and copper to fish gills predicts toxicity when water hardness varies but free ion activity does not. Environ. Sci. Technol, 33:913–916.

Meyer, J.S., Adams, W.J., Brix, K.V., et al. 2005. Toxicity of Dietborne Metals to Aquatic Organisms. SETAC Press, Pensacola, FL.

Meylan, W.M., Howard, P.H., Boethling, R.S., Aronson, D., Pruntup, H., and Gouchie, S. 1999. Improved methods for estimating bioconcentration/bioaccumulation factor from octanol/ partition coefficient. Environ. Toxicol. Chem., 18:664–672.

Michaels, D. and Wagner, W. 2003. Disclosure in regulatory science. Science, 302:2073.

Miles, L.J. and Parker, G.R. 1979. Heavy metal interaction with *Andropogon scoparious* and *Rudbeckia hirta* grown on soil from urban and rural sites with heavy metal additions. J. Environ. Qual., 8:443–449.

Millennium Ecosystem Assessment. 2005. Ecosystems and Human Well-Being Scenarios: Volume 2: Findings of the Scenarios Working Group. Island Press, Washington, DC.

Miller, F.J., Schlosser, P.M., and Janszen, D.B. 2000. Haber's rule: A special case in a family of curves relating concentration and duration of exposure to a fixed level of response for a given endpoint. Toxicology, 149:21–34.

Miller, J.E., Hassett, J.J., and Koeppe, D.E. 1977. Interactions of lead and cadmium on metal uptake and growth of corn plants. J. Environ. Qual., 6:18–20.

Mills, N.E. and Semlitsch, R.D. 2004. Competition and predation mediate the indirect effects of an insecticide on southern leopard frogs. Ecol. Appl., 14:1041–1054.

Mineau, P. 2002. Estimating the probability of bird mortality from pesticide sprays on the basis of the field study record. Environ. Toxicol. Chem., 21:1497–1506.

Mineau, P., Jobin, B., and Baril, A. 1994. A critique of the avian 5-day dietary test (LC_{50}) as the basis of avian risk assessment. Canadian Wildlife Service, Hull, PQ.

Mineau, P., Collins, B.T., and Baril, A. 1996. On the use of scaling factors to improve interspecies extrapolation of acute toxicity in birds. Regul. Toxicol. Pharmacol., 24:24–29.

Mitra, S. and Dickhut, R.M. 1999. Three-phase modeling of polycyclic aromatic hydrocarbon association with pore-water-dissolved organic carbon. Environ. Toxicol. Chem., 18:1144–1148.

Monaco, M.E. and Ulanowicz, R.E. 1997. Comparative ecosystem trophic structure of three U.S. mid-Atlantic estuaries. Mar. Ecol. Prog. Ser. 161:239–254.

Moore, D.R.J. and Caux, P.-Y. 1997. Estimating low toxic effects. Environ. Toxicol. Chem., 16:794–801.

Moore, D.R.J. and Bartell, S.G. 2000. Estimating ecological risks of multiple stressors: Advanced methods and difficult issues. Pages 117–168 in S.A. Ferenc and J.A. Foran (eds.) Multiple Stressors in Ecological Risk and Impact Assessment: Approches to Risk Estimation. SETAC Press, Pensacola, FL.

Moore, D.R.J., Sample, B.E., Suter, G.W., Parkhurst, B.R., and Teed, R.S. 1999. A probabilistic risk assessment of the effects of methylmercury and PCBs on mink and kingfishers along East Fork Poplar Creek, Oak Ridge, Tennessee, USA. Environ. Toxicol. Chem., 18:2941–2953.

Morel, F.M. 1983. Principles of Aquatic Chemistry. Wiley-Interscience, New York.

Morgan, M.G. and Henrion, M. 1990. Uncertainty: A Guide to Dealing with Uncertainty in Quantitative Risk and Policy Analysis. Cambridge University Press, Cambridge, UK.

Moriarty, F. 1988. Ecotoxicology: The Study of Pollutants in Ecosystems, Second Edition. Academic Press, New York.

Morse, L.E., Randall, J.M., Benton, N., Hiebert, R., and Lu, S. 2004. An invasive species assessment protocol: Evaluating non-native plants for their impact on biodiversity. Nature Serve, Arlington, VA.

Mount, D.I., Thomas, N.A., Norberg, T.J., Barbour, M.T., Roush, T.H., and Brandes, W.F. 1984. Effluent and ambient toxicity testing and instream community response on the Ottawa River, Lima, Ohio. EPA-600/3-84-080. US Environmental Protection Agency, Duluth, MN.

Mount, D.R., Dawson, T.D., and Burkhard, L.P. 1999. Implications of gut purging for tissue residues determined in bioaccumulation testing of sediment with *Lumbriculus variegatus*. Environ. Toxicol. Chem., 18:1244–1249.

Muirhead, E., Skillman, A.D., Hook, S.E., and Schultz, I.R. 2006. Oral exposure of PBDE-47 in fish: Toxicokinetics and reproductive effects in Japanese medaka (*Oryzias latipes*) and fathead minnows (*Pimephales promelas*). Environ. Sci. Technol., 40:523–528.

Mullins, J.A., Carsel, R.F., Scarborough, J.E., and Ivery, A.M. 1993. PRZM-2, a model for predicting pesticide fate in the crop root and unsaturated soil zones: User's manual for release 2.0. EPA/600/R93/046, Environmental Research Laboratory, Office of Research and Development, US Environmental Protection Agency, Athens, GA.

Munger, C., Hare, L., and Tessier, A. 1999. Cadmium sources and exchange rates for *Chaoborus* larvae in nature. Limnol. Oceanogr., 44:1763–1771.

Munkittrick, K.R. and Dixon, D.G. 1989. Use of white sucker (*Catastomus commersoni*) populations to assess the health of aquatic ecosystems exposed to low-level contaminant stress. Can. J. Fish. Aquat. Sci., 46:1455–1462.

Munkittrick, K.R., McMaster, M.E., Van Der Kraak, G., et al. 2000. Development of Methods for Effects-Driven Cumulative Effects Assessment Using Fish Populations: Moose River Project. SETAC Press, Pensacola, FL.

Munns, W.R. Jr., Black, D.E., Gleason, T., Salomon, K., Bengtson, D., and Gutjhar-Gobell, R. 1997. Evaluation of the effects of dioxin and PCBs on *Fundulus heteroclitus* populations using a modeling approach. Environ. Toxicol. Chem., 16:1074–1081.

Murdoch, W.W. 1994. Population regulation in theory and practice. Ecology, 75: 271–287.

Murphy, B.L. and Morrison, R.D. 2002. Introduction to Environmental Forensics. Academic Press, San Diego, CA.

Murray, N. 2002. Import Risk Analysis: Animals and Animal Products. New Zealand Ministry of Agriculture and Forestry, Wellington, NZ.

Musick, J.A. 1999. Criteria to define extinction risk in marine fishes. Fisheries, 24:6–12.

Myers, O.B. 1999. On aggregating species for risk assessment. Hum. Ecol. Risk Assess., 5:559–574.

Myers, R.A., Barrowman, N.J., Hilborn, R., and Kehler, D.G. 2001. Inferring Bayesian priors with limited direct data: Application to risk analysis. N. Am. J. Fish. Manag., 22:351–364.

Nabholz, J.V., Clements, R.G., and Zeeman, M.G. 1997. Information needs for risk assessment in EPA's Office of Pollution Prevention and Toxics. Ecol. Appl., 7:1094–1102.

Naeem, S., Chapin, F.S., III, Costanza, R., et al. 1999. Biodiversity and Ecosystem Functioning: Maintaining Natural Life Support Processes. Ecological Society of America, Washington, DC.

Nagy, K.A. 1987. Field metabolic rate and food requirement scaling in mammals and birds. Ecol. Monogr., 57:111–128.

Naito, W., Miyamoto, K., Nakanishi, J., Masunaga, S., and Bartell, S.M. 2002. Application of an ecosystem model for ecological risk assessment of chemicals for a Japanese lake. Water Res., 36:1–14.

Naito, W., Miyamoto, K., Nakanishi, J., and Bartell, S.M. 2003. Evaluation of an ecosystem model in ecological risk assessment of chemicals. Chemosphere, 53:363–375.

Nakamaru, M., Iwasa, Y., and Nakanishi, J. 2002. Extinction risk to herring gull populations from DDT exposure. Environ. Toxicol. Chem., 21:195–202.

Nash, C.E., Burbridge, P.R., and Volkman, J.K. 2005. Guidelines for ecological risk assessment of marine fish aquaculture. NMFS-NWFSC-71. National Oceanic and Atmospheric Administration, Washington, DC.

National Center for Environmental Economics. 2000. Guidelines for preparing economic analyses. EPA-240-R-00-003. US Environmental Protection Agency, Washington, DC.

NEPC. 1999. Schedule B (5) Guideline on Ecological Risk Assessment. National Environmental Protection Council, Australia.

Ness, E. 2005. Four futures. Conserv. Pract., 6:20–27.

Neuhauser, E.F., Cukie, Z.V., Malecki, M.R., Loehr, R.C., and Durkin, P.R. 1995. Bioconcentration and biokinetics of heavy metals in the earthworm. Environ. Pollut., 89:293–301.

Neuhold, J.M. 1986. Toward a meaningful interaction between ecology and aquatic toxicology. Pages 11–21 in T.M. Poston and R. Purdy (eds.) Aquatic Toxicology and Environmental Fate. ASTM, Philadelphia, PA.

Newbry, B.W. and Lee, G.F. 1984. A simple apparatus for conducting in-stream toxicity tests. J. Test. Eval., 12:51–53.

Newcombe, C.P. and Jensen, J.O.T. 1996. Channel suspended sediment and fisheries: A synthesis for quantitative assessment of risk and impact. N. Am. J. Fish. Manag., 16:693–727.

Newcombe, C.P. and MacDonald, D.D. 1991. Effects of suspended sediments on aquatic ecosystems. N. Am. J. Fish. Manag., 11:72–82.

Newman, M.C. 2001. Population Ecotoxicology. John Wiley, New York.

Newman, M.C. and Aplin, M.S. 1992. Enhancing toxicity data interpretation and prediction of ecological risk with survival time modeling: An illustration using sodium chloride toxicity to mosquitofish (*Gambusia holbrooki*). Aquat. Toxicol., 23:85–96.

Newman, M.C. and Dixon, P.M. 1990. UNCENSOR: A program to estimate means and standard deviations for data sets with below detection limit observations. Am. Environ. Lab., 2(2):26–30.

Newman, M.C. and Evans, D.A. 2002. Enhancing belief during causality assessments: Cognitive idols or Bayes's theorem? Pages 73–96 in M.C. Newman, M.H. Roberts Jr., and R.C. Hale (eds.) Coastal and Estuarine Risk Assessment. Lewis Publishers, Boca Raton, FL.

Newman, M.C. and Heagler, M.G. 1991. Allometry of metal bioaccumulation and toxicity. Pages 91–130 in M.C. Newman and A.W. McIntosh (eds.) Metal Ecotoxicology, Concepts and Applications. Lewis Publishers, Chelsea, MI.

Newman, M.C., Keklak, M.M., and Doggett, M.S. 1994. Quantifying animal size effects in toxicity: A general approach. Aquat. Toxicol., 28:1–12.

Newman, M.C., Greene, K.D., and Dixon, P.M. 1995. UNCENSOR v.4.0. SREL-44. Savannah River Ecology Laboratory, Aiken, SC.

Newman, M.C., Ownby, D.R., Mezin, L.C.A., Powell, D.C., Christensen, T.R.L., Lerberg, S.B., and Anderson, S.-A. 2000. Applying species sensitivity distributions in ecological risk assessment: Assumptions of distribution type and sufficient number of species. Environ. Toxicol. Chem., 19:508–515.

Newman, M.C., Ownby, D.R., Mezin, L.C.A., et al. 2002. Species sensitivity distributions in ecological risk assessment: Distributional assumptions, alternate bootstrap techniques, and estimation of adequate number of species. Pages 119–132 in L. Posthuma, G.W. Suter II, and T.P. Traas (eds.) Species Sensitivity Distributions in Ecotoxicology. Lewis Publishers, Boca Raton, FL.

Newsted, J.L., Nakanishi, J., Cousins, I., Werner, K., and Giesy, J. 2002. Predicted distribution and ecological risk assessment of a "segregated" hydrofluoroether in the Japanese environment. Environ. Sci. Technol., 36:4761–4769.

Nichols, J.W. 1999. Recent advances in the development and use of physiologically based toxicokinetic models for fish. Pages 87–103 in D.J. Smith, W.H. Gingerich, and M.G. Beconi-Barker (eds.) Xenobiotics in Fish. Kluwer Academic/Plenum Publishers, New York.

Nichols, J.W., Larson, C.P., McDonald, M.E., Niemi, G.J., and Ankley, G.T. 1995. Bioenergetics-based model for accumulation of polychlorinated biphenyls by nesting tree swallows, *Tachycineta bicolor*. Stud. Avian Biol., 6:121–136.

Nichols, J.W., Echols, K.R., Tillitt, D.E., Secord, A.L., and McCarty, J.P. 2004. Bioenergetics-based modeling of individual PCB congeners in nestling tree swallows from two contaminated sites on the upper Hudson River, New York. Environ. Sci. Technol., 38:6234–6239.

Niemi, G.J., DeVore, P., Detenbeck, N., Taylor, D., Lima, A., Pastor, J., Yount, J.D., and Naiman, R.J. 1990. Overview of case studies on recovery of aquatic systems from disturbance. Environ. Manag., 14:571–587.

Nirmalakhandan, N. 2002. Modeling Tools for Environmental Engineers and Scientists. CRC Press, Boca Raton, FL.

Nisbet, R.M. and Gurney, W.S.C. 1982. Modelling Fluctuating Populations. John Wiley, New York.

Niyogi, S. and Wood, C.M. 2004. Biotic ligand model, a flexible tool for developing site-specific water quality guidelines for metals. Environ. Sci. Technol., 38:6177–6192.

NOAA (National Oceanic and Atmospheric Administration). 1995. The utility of AVS/EqP in hazardous waste site evaluations. NOAA Technical Memorandum NOS ORCA 87. National Oceanic and Atmospheric Administration, Seattle, WA.

NOAA Hazardous Materials Branch. 1990. Excavation and rock washing treatment technology: Net environmental benefits analysis. National Oceanic and Atmospheric Administration, Seattle, WA.

Norberg, T.J. and Mount, D.I. 1985. A new fathead minnow (*Pimephales promelas*) subchronic toxicity test. Environ. Toxicol. Chem., 4:711–718.

Norberg-King, T. and Mount, D.I. 1986. Validity of effluent and ambient toxicity tests for predicting biological impact, Skeleton Creek, Enid, Oklahoma. EPA/600/30-85/044. Environmental Research Laboratory, Duluth, MN.

Norberg-King, T.J. 1993. A linear interpolation method for sublethal toxicity: The ICp approach (Version 2.0). Tech. Report 03-93. US Environmental Protection Agency, Duluth, MN.

Norberg-King, T.J., Ausley, L.W., Burton, D.T., et al. 2005. Toxicity Reduction and Toxicity Identification Evaluations for Effluents, Ambient Waters, and Other Aqueous Media. SETAC Press, Pensacola, FL.

Norris, R.H. and Georges, A. 1993. Analysis and interpretation of benthic macroinvertebrate surveys. Pages 234–286 in D.M. Rosenberg and V.H. Resh (eds.) Freshwater Biomonitoring and Benthic Macroinvertebrates. Chapman & Hall, New York.

Norton, S.B., Rodier, D.J., Gentile, J.H., van der Schalie, W.H., Wood, W.P., and Slimak, M.W. 1992. A framework for ecological risk assessment at the EPA. Environ. Toxicol. Chem., 11:1663–1672.

Norton, S.B., Cormier, S.M., Smith, M., and Jones, R.C. 2000. Can biological assessment discriminate among types of stress? A case study from the eastern cornbelt plains ecoregion. Environ. Toxicol. Chem., 19:1113–1119.

Norton, S.B., Cormier, S.M., Suter, G.W., II, Subramanian, B., Lin, E., Altfater, D., and Counts, B. 2002. Determining probable causes of ecological impairment in the Little Scioto River, Ohio, USA. Part I: Listing candidate causes and analyzing evidence. Environ. Toxicol. Chem., 21:1112–1124.

NRC (National Research Council). 1981. Testing for effects of chemicals on ecosystems. National Academy Press, Washington, DC.

NRC (National Research Council). 1983. Risk Assessment in the Federal Government: Managing the Process. National Academy Press, Washington, DC.

NRC (National Research Council). 1989. Improving Risk Communication. National Academy Press, Washington, DC.

NRC (National Research Council). 1991. Animals as Sentinels of Environmental Health Hazards. National Academy Press, Washington, DC.

NRC (National Research Council). 1993. Issues in Risk Assessment. National Academy Press, Washington, DC.

NRC (National Research Council). 1994. Science and Judgment in Risk Assessment. National Academy Press, Washington, DC.

NRC (National Research Council). 1999a. Downstream: Adaptive Management of Glen Canyon Dam and the Colorado River Ecosystem. National Academy Press, Washington, DC.

NRC (National Research Council). 1999b. Ecological Indicators for the Nation. National Academy Press, Washington, DC.

Nuutinen, V., Pitkanen, J., Kuusela, E., Widbom, T., and Lohilahti, H. 1998. Spatial variation of an earthworm community related to soil properties and yield in a grass-clover field. Appl. Soil Ecol., 8:85–94.

O'Brien, M. 2000. Making Better Environmental Decisions: An Alternative to Risk Assessment. MIT Press, Cambridge, MA.

O'Connor, D.J. 1988a. Models of sorptive toxic substances in freshwater systems. I: Basic equations. J. Environ. Eng., ASCE, 114(3):507–532.

O'Connor, D.J. 1988b. Models of sorptive toxic substances in freshwater systems. II: Lakes and reservoirs. J. Environ. Eng., ASCE, 114(3):533–551.

O'Connor, D.J. 1988c. Models of sorptive toxic substances in freshwater systems. III: Streams and rivers. J. Environ. Eng., ASCE, 114(3):552–574.

Odum, E.P. 1971. Fundamentals of ecology. WB Saunders, Philadelphia, PA.

Odum, H.T. and Odum, E.C. 2000. Modeling at All Scales. Academic Press, New York.

OECD. 1992. Report of the OECD workshop on the extrapolation of laboratory aquatic toxicity data to the real environment. OCDE/GD(92)169. Organization for Economic Cooperation and Development, Paris.

OECD. 1995. Guidance document for aquatic effects assessment. OCDE/GD(95)18. Organization for Economic Cooperation and Development, Paris.

OECD. 1998. Report of the OECD workshop on statistical analysis of aquatic toxicity data. ENV/MC/CHEM(98)18. Organization for Economic Cooperation and Development, Paris.

OECD. 2000. OECD guidelines for the testing of chemicals. Organization for Economic Cooperation and Development, Paris.

OECD. 2004. Draft guidance document on the statistical analysis of ecotoxicity data. Organization for Economic Cooperation and Development, Paris.

Office of Emergency and Remedial Response. 1991. Risk assessment guidance for Superfund: Volume 1—Human Health Evaluation Manual (Part C, Risk Evaluation of Remedial Alternatives) Publication 9285.7-01C. US Environmental Protection Agency, Washington, DC.

Office of Emergency and Remedial Response. 1992. Guidance for data useability in risk assessment. 9285.7-09A&B. US Environmental Protection Agency, Washington, DC.

Office of Emergency and Remedial Response. 1994a. Catalog of standard toxicity tests for ecological risk assessment. EPA 540-F-94-013. US Environmental Protection Agency, Washington, DC.

Office of Emergency and Remedial Response. 1994b. Field studies for ecological risk assessment. EPA 540-F-94-014. US Environmental Protection Agency, Washington, DC.

Office of Emergency and Remedial Response. 1994c. Field studies for ecological risk assessment. EPA 540-F-94-014. US Environmental Protection Agency, Washington, DC.

Office of Emergency and Remedial Response. 1994d. Selecting and using reference information in Superfund ecological risk assessments. EPA 540-F-94-015. US Environmental Protection Agency, Washington, DC.

Office of Emergency and Remedial Response. 1994e. Using toxicity tests in ecological risk assessment. EPA 540-F-94-012. US Environmental Protection Agency, Washington, DC.

Office of Emergency and Remedial Response. 1996. Ecotox Thresholds. EPA 540/F-95/038. US Environmental Protection Agency, Washington, DC.

Office of Environmental Information. 2002. Guidelines for ensuring and maximizing the quality, objectivity, utility, and integrity, of information disseminated by the Environmental Protection Agency. EPA/260R-02-008. US Environmental Protection Agency, Washington, DC.

Office of Environmental Policy and Assistance. 1996. Characterization of uncertainties in risk assessment with special reference to probabilistic uncertainty analysis. EH-413-068/0296. US Department of Energy, Washington, DC.

Office of Research and Development. 1998. Condition of the Mid-Atlantic Estuaries. EPA/600/R-98/147. US Environmental Protection Agency, Washington, DC.

Office of Science and Technology. 1994. Interim guidance on determination and use of water effect ratios for metals. EPA/823/B-94/001. US Environmental Protection Agency, Washington, DC.

Office of Science and Technology. 2001. Streamlined water-effect ratio procedure for discharges of copper. EPA-822-R001-005. US Environmental Protection Agency, Washington, DC.

Office of Science and Technology. 2003. Technical support document for the assessment of detection and quantitation approaches. EPA-821-R-03-005. US Environmental Protection Agency, Washington, DC.

Office of Solid Waste and Emergency Response. 1999. Farm food chain module: Backgound and implementation for the multimedia, multipathways, and multiple receptor risk assessment (3MRA) module for HWIR 99. US Environmental Protection Agency, Washington, DC.

Office of Solid Waste and Emergency Response. 2002. Role of background in the CERCLA cleanup program. OSWER 9285.6-07P. US Environmental Protection Agency, Washington, DC.

Office of Solid Waste and Emergency Response. 2003. Guidance for developing ecological soil screening levels. OSWER Directive 9285.7-55. US Environmental Protection Agency, Washington, DC.

Office of Solid Waste and Emergency Response. 2005. Guidance for developing ecological soil screening levels, revised. OSWER Directive 9285.7-55. US Environmental Protection Agency, Washington, DC.

Office of Water. 1999. Protocol for developing nutrient TMDLs. EPA 841-B-99-007. US Environmental Protection Agency, Washington, DC.

Office of Water. 2001. Methods for collection, storage, and manipulation of sediments for chemical and toxicological analysis, technical manual. EPA-828-F-01-023. US Environmental Protection Agency, Washington, DC.

Ohio EPA. 1998. Biological Criteria for the Protection of Aquatic Life: Volume 2: Users Manual for Biological Assessment of Ohio Surface Waters. WQMA-SWS-6. State of Ohio Environmental Protection Agency, Columbus, OH.

Ohlendorf, H.M. 1998. Evaluating bioaccumulation in wildlife food chains. Pages 65–109 in A. de Peyster and K.E. Day (eds.) Ecological Risk Assessment: A Meeting of Policy and Science. SETAC Press, Pensacola, FL.

Ohlendorf, H.M. and Hothem, R.L. 1995. Agricultural drainwater effects on wildlife in central California. Pages 577–595 in D.J. Hoffman, B. Rattner, and G.A. Burton (eds.) Handbook of Ecotoxicology. Lewis Publishers, Boca Raton, FL.

Ohlendorf, H.M., Hoffman, D.J., Saiki, M.K., and Aldrich, T.W. 1986a. Embryonic mortality and abnormalities of aquatic birds: Apparent impacts of selenium from irrigation drain waters. Sci. Total Environ., 52:49–53.

Ohlendorf, H.M., Hothem, R.L., Bunck, C.M., Aldrich, T.W., and Moore, J.F. 1986b. Relationships between selenium concentrations and avian reproduction. Trans. 51st N. Am. Wildl. Nat. Resour. Conf., 51:330–342.

OIE (Office International des Epizooties). 2001. International Animal Health Code. Office International des Epizooties, Paris.

Oliver, B.G. and Niimi, A.J. 1985. Bioconcentration factors for some halogenated organics for rainbow trout: Limitations in their use for prediction of environmental residues. Environ. Sci. Technol., 19:842–849.

Oliver, G.R. and Laskowski, D.A. 1986. Development of environmental scenarios for modeling fate of agricultural chemicals in soil. Environ. Toxicol. Chem. 5:225–232.

Oliver, B.G. and Niimi, A.J. 1988. Trophodynamic analysis of polychlorinated biphenyl congeners and other chlorinated hydrocarbons in the Lake Ontario ecosystem. Environ. Sci. Technol., 22:388–397.

Omernik, J.M. 1987. Ecoregions of the conterminous United States. Annals Assoc. Am. Geo., 77:118–125.

O'Neill, R.H., Gardner, R.V., Barnthouse, L.W., Suter, G.W., II, Hildebrand, S.G., and Gehrs, C.W. 1982. Ecosystem risk analysis: A new methodology. Environ. Toxicol. Chem., 1:167–177.

O'Neill, R.V. 2001. Is it time to bury the ecosystem concept? (With full military honors, of course!). Ecology, 82:3275–3284.

O'Neill, R.V. and Waide, J.B. 1981. Ecosystem theory and the unexpected: Implications for environmental toxicology. Pages 43–73 in B.W. Cornaby (ed.), Management of Toxic Substances in Our Ecosystems. Ann Arbor Science, Ann Arbor, MI.

O'Neill, R.V., DeAngelis, D.L., Waide, J.B., and Allen, T.F.H. 1986. A Hierarchical Concept of Ecosystems. Princeton University Press, Princeton, NJ.

OPP (Office of Pesticide Programs). 1989. Carbofuran: Special review technical support document. NTIS: PB 89168884. Environmental Protection Agency, Washington, DC.

OPP (Office of Pesticide Programs). 2004. A discussion with the FIFRA Scientific Advisory Panel regarding the Terrestrial and Aquatic Level II Refined Risk Assessment Models (Version 2.0). Available at: www.epa.gov/scipoly/sap/index.htm

OPPTS (Office of Pollution Prevention and Toxic Substances). 1996a. Ecological effects test guidelines: OPPTS 850.1900. Generic freshwater microcosm test guidelines. EPA 712-C-96-134. US Environmental Protection Agency, Washington, DC.

OPPTS (Office of Pollution Prevention and Toxic Substances). 1996b. Ecological effects test guidelines: OPPTS 850.2450. Terrestrial (soil-core) microcosm test. EPA 712-C-96-143. US Environmental Protection Agency, Washington, DC.

OPPTS (Office of Pollution Prevention and Toxic Substances). 1996c. Ecological effects test guidelines: OPPTS 850.2500. Field-testing for terrestrial wildlife. EPA 712-C-96-144. US Environmental Protection Agency, Washington, DC.

OPPTS (Office of Pollution Prevention and Toxic Substances). 1996d. Ecological effects test guidelines: OPPTS 850.3040. Field testing for pollinators. EPA 712-C-96-150. US Environmental Protection Agency, Washington, DC.

OPPTS (Office of Pollution Prevention and Toxic Substances). 1996e. Ecological effects test guidelines: OPPTS 850.4300. Terrestrial plants field study, Tier III. EPA 712-C-96-155. US Environmental Protection Agency, Washington, DC.

OPPTS (Office of Pollution Prevention and Toxic Substances). 1996f. Ecological effects test guidelines: OPPTS 850.5100. Soil microbial community toxicity test. EPA 712-C-96-161. US Environmental Protection Agency, Washington, DC.

OPPTS (Office of Prevention, Pesticides and Toxic Substances). 1996g. Microbial pesticide test guidelines: OPPTS 885.4000. Background for nontarget organism testing of microbial pest control agents. EPA 712-C-96-328. US Environmental Protection Agency, Washington, DC.

Oreskes, N. 1998. Evaluation (not validation) of quantitative models. Environ. Health Persp., 106:1453–1460.

Orr, R. 2003. Generic nonindigenous aquatic organisms risk analysis. Pages 415–438 in G.M. Ruiz and J.T. Carlton (eds.) Invasive Species: Vectors and Management Practices. Island Press, Washington, DC.

Ortiz, M. and Wolff, M. 2002. Application of loop analysis to benthic systems in northern Chile for the elaboration of sustainable management practices. Marine Ecol. Progress Ser., 242:15–27.

O'Shea, T.J. 2000a. Cause of seal die-off in 1988 is still under debate. Science, 290:1097.

O'Shea, T.J. 2000b. PCBs not to blame. Science, 288:1965–1966.

Ott, W.R. 1978. Environmental Indices—Theory and Practice. Ann Arbor Science Publishers, Ann Arbor, MI.

Owen, B.A. 1990. Literature-derived absorption coefficients for 39 chemicals via oral and inhalation routes of exposure. Regulat. Toxicol. Pharmicol. 11:237–252.

Paine, J.M., McKee, M., and Ryan, M.E. 1993. Toxicity and bioaccumulation of PCBs in crickets: Comparison of laboratory and field studies. Environ. Toxicol. Chem., 12:2097–2103.

Paquin, P.R., Gorsuch, J.W., Apte, S., Batley, G.E., Bowles, K.C., Campbell, P.G.C., Delos, C.G., et al. 2002. The biotic ligand model: A historical overview. Comp. Biochem. Physiol., Part C, 133:3–35.

Paquin, P.R., Farley, K., Santore, R.C., Kavvadas, C.D., Mooney, K.G., Winfield, R.P., Wu, K.-B., and Di Toro, D.M. 2003. Metals in Aquatic Systems: A Review of Exposure, Bioaccumulation, and Toxicity Models. SETAC Press, Pensacola, FL.

Park, R.A. et al. 1974. A generalized model for simulating lake ecosystems. Simulation 21:33–50.

Park, D., Hempleman, S.C., and Propper, C.R. 2001. Endosulfan exposure disrupts pheromone systems in the red-spotted newt: A mechanism for subtle effects of environmental chemicals. Environ. Health Persp., 109:669–673.

Park, R.A. 1998. AQUATOX for windows: A modular toxic effects model for aquatic ecosystems. US Environmental Protection Agency, Washington DC.

Park, R.A. and Clough, J.S. 2004. AQUATOX (Release 2). Volume 2: Technical documentation. Available at: www.epa.gov/waterscience/models/aquatox.

Parker, M.M. and van Lear, D.H. 1996. Soil heterogeneity and root distribution of mature loblolly pine stands in Piedmont soils. Soil Sci. Soc. Am. J., 60:1920–1925.

Parker, R.D., Nelson, H., and Jones, R.D. 1995. GENEEC: A screening model for pesticide environmental exposure assessment. Pages 485–490 in Water Quality Modelling Proceedings of the International Symposium, April 2–5, 1995, Orlando, FL.

Parkhurst, D.F. 1984. Decision analysis for toxic waste releases. J. Environ. Manag., 18:105–130.

Parkhurst, D.F. 1985. Interpreting failure to reject a null hypothesis. Bull. Ecol. Soc. Am., 66:301–302.

Parkhurst, D.F. 1998. Arithmetic versus geometric means for environmental concentration data. Environ. Sci. Technol. 32:92–98A.

Parkhurst, D.F. 1990. Statistical hypothesis tests and statistical power in pure and applied science. Pages 181–201 in G.M. von Furstenberg (ed.) Acting Under Uncertainty: Multidisciplinary Conceptions. Kluwer Academic, Boston, MA.

Parkhurst, B.R., Warren-Hicks, W., Cardwell, R.D., Volosin, J., Etchison, T., Butcher, J.B., and Covington, S.M. 1996a. Methodology for Aquatic Ecological Risk Assessment. No. RP91-AER-1. Water Environment Research Foundation, Alexandria, VA.

Parkhurst, B.R., Warren-Hicks, W., Cardwell, R.D., Volosin, J., Etchison, T., Butcher, J.B., and Covington, S.M. 1996b. Aquatic Ecological Risk Assessment: A Multi-Tiered Approach. Project 91-AER-1. Water Environment Research Foundation, Alexandria, VA.

Parmelee, R.W., Phillips, C.T., Checkai, R.T., and Bohlen, P.J. 1997. Determining the effects of pollutants on soil faunal communities and trophic structure using a refined microcosm system. Environ. Toxicol. Chem., 16:1212–1217.

Pascoe, D. and Shazili, N.A.M. 1986. Episodic pollution: A comparison of brief and continuous exposure of rainbow trout to cadmium. Ecotoxicol. Environ. Saf., 12:189–198.

Pascoe, D., Evans, S.A., and Woodworth, J. 1986. Heavy metal toxicity to fish and the influence of water hardness. Arch. Environ. Contam. Toxicol., 15:481–487.

Pascoe, G.A. and DalSoglio, J.A. 1994. Planning and implementation of a comprehensive ecological risk assessment at the Milltown Reservoir–Clark Fork River Superfund site, Montana. Environ. Toxicol. Chem., 13:1943–1956.

Pascoe, G.A., Blanchet, R.L., Linder, G., Palawski, D., Brumbaugh, W.G., Canfield, T.J., Kemble, N.E., Ingersoll, C.G., Farag, A., and DalSoglio, J.A. 1994. Characterization of the ecological risks at the Milltown Reservoir–Clark Fork River Superfund site, Montana. Environ. Toxicol. Chem., 13:2043–2058.

Pascual, J.A. 1994. No effects of a forest spraying of malathion on breeding blue tits (Parus caeruleus). Environ. Toxicol. Chem., 13:1127–1131.

Pascual, P., Stiber, N., and Sunderland, E. 2003. Draft guidance on the development, evaluation, and application of regulatory environmental models. The Council for Regulatory Environmental Modeling, US Environmental Protection Agency, Washington, DC.

Pastorok, R.A. and Akçakaya, H.R. 2002. Chapter 2—Methods. Pages 23–34 in R.A. Pastorok, S.M. Bartell, S. Ferson, and L.R. Ginzburg (eds.) Ecological Modeling in Risk Assessment: Chemical Effects on Populations, Ecosystems, and Landscapes. Lewis Publishers, Boca Raton, FL.

Pastorok, R.A., Butcher, M.K., and Nelson, R.D. 1996. Modeling wildlife exposure to toxic chemicals: Trends and recent advances. Hum. Ecol. Risk Assess., 2:444–480.

Pastorok, R.A., Bartell, S.M., Ferson, S., and Ginzburg, L.R. 2002. Ecological Modeling in Risk Assessment: Chemical Effects on Populations, Ecosystems, and Landscapes. CRC Press, Boca Raton, FL.

Paterson, S., Mackay, D., Tam, D., and Shiu, W.Y. 1990. Uptake of organic chemicals by plants: A review of processes, correlations and models. Chemosphere 21:297–331.

Patin, S.A. 1982. Pollution and the Biological Resources of the Oceans. Butterworth Scientific, London.

Patten, B.C. (ed.). 1971. Systems analysis and simulation in ecology, Volume 1. Academic Press, New York.

Patten, B.C. (ed.). 1972. Systems analysis and simulation in ecology, Volume 2. Academic Press, New York.

Patten, B.C., Bosserman, R.W., Finn, J.T., and Cale, W.G. 1976. Propagation of cause in ecosystems. Pages 458–579 in Patten, B.C. (ed.) Systems Analysis and Simulation in Ecology: Volume 4. Academic Press, New York.

Pauly, D., Christensen, V., and Walters, C. 2000. Ecopath, Ecosim, and Ecospace as tools for evaluating ecosystem impacts of fisheries. ICES J. Marine Sci., 57:697–706.

Pearson, W.H., Moksness, E., and Skalski, J.R. 1995. A field and laboratory assessment of oil spill effects on survival and reproduction of pacific herring following the Exxon Valdez spill. Pages 626–661 in P.G. Wells, J.N. Butler, and J.S. Hughes (eds.) Exxon Valdez Oil Spill: Fate and Effects in Alaskan Waters. ASTM, Philadelphia, PA.

Peijnenburg, W. 2001. Bioavailability of metals to soil invertebrates. Pages 87–109 in H.E. Allen (ed.) Bioavailability of Metals in Terrestrial Ecosystems. SETAC Press, Pensacola, FL.

Pesaud, D.R., Jaagumagi, R., and Hayton, A. 1993. Guidelines for the protection and management of aquatic sediment quality in Ontario. Ontario Ministry of Environment and Energy, Ontario, Canada.

Peter, S.R. 1998. Marine mammals as sentinels in ecological risk assessment: Approaches to Integrated Risk Assessment Meeting, (November 19, 1998). WHO, Charlotte, NC.

Peters, R.H. 1983. The Ecological Implications of Body Size. Cambridge University Press, Cambridge, UK.

Peters, R.H. 1991. A Critique for Ecology. Cambridge University Press, Cambridge, UK.

Peterson, C.H., Rice, S.D., Short, J.W., Esler, D., Bodkin, J.L., Ballachey, B.E., and Irons, D.B. 2003. Long-term ecosystem response to the Exxon Valdez oil spill. Science, 302:2082–2086.

Pinker, S. 1997. How the Mind Works. W.W. Norton, New York.

Pizl, V. and Josens, G. 1995. Earthworm communities along a gradient of urbanization. Environ. Pollut., 90:7–14.

Plackett, R.L. and Hewlett, P.S. 1952. Quantal responses to mixtures of poisons. J. Royal Stat. Soc., B14:141–163.

Platt, J.R. 1964. Strong inference. Science, 146:347–353.

Playle, R.C., Dixon, D.G., and Burnison, K. 1993. Copper and cadmium binding to fish gills: Estimates of metal-gill stability constants and modeling of metal accumulation. Can. J. Fish. Aquat. Sci., 50:2678–2687.

Pokras, M.A., Karas, A.M., Kirkwood, J.K., and Sedgewick, C.J. 1993. An introduction to allometric scaling and its uses in raptor medecinc. Pages 211–224 in P.T. Redig, J.E. Cooper, J.D. Remple, and D.B. Hunter (eds.) Raptor Biomedecine. University of Minnesota Press, Minneapolis, MN.

Polisini, J.M., Carlisle, J.C., and Valoppi, L.M. 1998. Guidance for performing ecological risk assessments at hazardous waste sites and permitted facilities in California. Pages 23–54 in A. de Peyster and K.E. Day (eds.) Ecological Risk Assessment: A Meeting of Policy and Science. SETAC Press, Pensacola, FL.

Popper, K.R. 1968. The Logic of Scientific Discovery. Harper & Row, New York.

Posthuma, L., van Gestel, C.A.M., Smit, C.E., Bakker, D.J., and Vonk, J.W. 1998. Validation of toxicity data and risk limits for soils: Final report. Report No. 607505. Bilthoven, RIVM, The Netherlands.

Posthuma, L., Suter, G.W., II, and Traas, T.P. (eds.). 2002. Species Sensitivity Distributions for Ecotoxicology. Lewis Publishers, Boca Raton, FL.

Power, M. and McCarty, L.S. 1998. A comparative analysis of environmental risk assessment/risk management frameworks. Environ. Sci. Technol., 32:224A–231A.

Power, M. and McCarty, L.S. 2002. Trends in the development of ecological risk assessment and management frameworks. Hum. Ecol. Risk Assess., 8:7–18.

Presser, T.A. and Ohlendorf, H.A. 1987. Biogeochemical cycling of selenium in the San Joaquin Valley, California, USA. Environ. Manag., 11:805–821.

Price, P.D., Pardi, R., Fthenakis, V.M., Holtzman, S., Sun, L.C., and Irla, B. 1996. Uncertainty and variation in indirect exposure assessments: An analysis of exposure to tetrachlorodibenzo-*p*-dioxin from a beef consumption pathway. Risk Anal., 16:263–277.

Prothro, M.G. 1993. Office of Water technical guidance on interpretation and implementation of aquatic life metals criteria. US Environmental Protection Agency, Washington, DC.

Quality Assurance Management Staff. 1994. Guidance for the data quality objectives process. EPA QA/G-4. US Environmental Protection Agency, Washington, DC.

Quality Assurance Management Staff. 1998. Guidance for data quality assessment. EPA/600/R-96/084. US Environmental Protection Agency, Washington, DC.

Quality Assurance Management Staff. 2000. Guidance for data quality assessment: Practical methods for data analysis. EPA/600/R-96/084 and EPA QA/G-9. 2000. US Environmental Protection Agency, Washington, DC.

Quinn, T.J. and Deriso, R.B. 1999. Quantitative Fish Dynamics. Oxford University Press, New York.

Rabinowitz, P.M., Gordon, Z., Taylor, B., Chudnov, W.M., Nadkarni, P., and Dein, F. 2005. Animals as sentinels of human environmental health hazards: An experience-based analysis. EcoHealth, 2:26–37.

Raffa, R.B. 2001. Drug-Receptor Thermodynamics: Introduction and Application. John Wiley, New York.

Raffensperger, C. and Tickner, J. (eds.). 1999. Protecting Public Health and the Environment: Implementing the Precautionary Principle. Island Press, Washington, DC.

Ram, R.N. and Gillett, J.W. 1993. An aquatic/terrestrial foodweb model for polychlorinated biphenyls (PCBs). Pages 192–212 in J. Hughes, W. Landis, and M. Lewis (eds.) Environmental Toxicology and Risk Assessment. ASTM, Philadelphia, PA.

Ramsey, J.C. and Gehring, P.J. 1980. Application of pharmacokinetic principles in practice. Federation Proc., 39:60–65.

Rand, G.M. 1995. Fundamentals of Aquatic Toxicology, Second Edition. Taylor & Francis, Washington, DC.

Randall, A. 2006. Risk and action—figuring out the right thing to do. in R.J.F. Bruins and M.T. Heberling (eds.) Economics and Ecological Risk Assessment. CRC Press, Boca Raton, FL.

Randall, R.C., Lee, H.H., Ozretich, R.J., Lake, J.L., and Pruell, R.J. 1998. Evaluation of selected lipid methods for normalizing pollutant bioaccumulation. Environ. Toxicol. Chem., 10:1431–1436.

Rankin, E.T. 1995. Habitat indices in water resource quality assessment. Pages 181–208 in W.S. Davis and T.P. Simon (eds.) Biological Assessment and Criteria: Tools for Water Resource Planning and Decision Making. Lewis Publishers, Boca Raton, FL.

Rasmussen, N.C. 1981. The application of probabilistic risk assessment techniques to energy technologies. Ann. Rev. Energy, 6: 123–138.

Rautmann, D., Streloke, M., and Winkler, R. 2001. New basic drift values in the authorization procedure for plant protection products. In R. Foster and M. Streloke (eds.) Workshop on Risk Assessment and Risk Mitigation Measures in the context of the Authorisation of Plant Protection Products (WORMM), September 27–29, 1999, Heft 383, Biologischen Bundesantalt für Land-und Fortwirtschaft, Berlin and Braunschweig, Germany.

Reagan, D. 2002. Determining values: A critical step in assessing ecological risk. Pages 1069–1098 in D.J. Paustenbach (ed.) Human and Ecological Risk Assessment: Theory and Practice. John Wiley, New York.

Reckhow, K.H. 1999. Lessons from risk assessment. Hum. Ecol. Risk Assess., 5:245–254.

Reddy, M.B., Yang, R.S.H., Clewell, H.J.I., and Anderson, M.E. 2005. Physiologically Based Pharmacokinetic (PBPK) Modeling. John Wiley, New York.

Redeker, E.S., Bervoets, L., and Blust, R. 2004. Dynamic model for the accumulation of cadmium and zinc from water and sediment by the aquatic oligochaete, *Tubifex tubifex*. Environ. Sci. Technol., 38:6193–6197.

Redman, C.L. 1999. Human Impact on Ancient Environments. University of Arizona Press, Tucson, AZ.

Reed, D.W. 2001. Natural hazards and the analysis of extremes. Pages 57–62 in S. Pollard and J. Guy (eds.) Risk Assessment for Environmental Professionals. Chartered Institute of Water and Environmental Management, London.

Region V. 2005a. Elliot Ditch/Wea Creek Ecological Risk Assessment. US Environmental Protection Agency, Chicago, IL. Available at: http://www.epa.gov/region5superfund/ecology/html/casestudies/elliotditch.html.

Region V. 2005b. Screening Level Ecological Risk Assessment of the Camp Perry Landfill, Port Clinton, Ohio. US Environmental Protection Agency, Chicago, IL. Available at: http://www.epa.gov/region5superfund/ecology/html/casestudies/campperry/html.

Reid, R.C., Prausnitz, J.M., and Poling, B.E. 1987. The Properties of Gases and Liquids, Fourth Edition. McGraw-Hill, New York.

Reilley, K.A., Banks, M.K., and Schwab, A. 1996. Dissipation of polycyclic aromatic hydrocarbons in the rhizosphere. J. Environ. Qual., 25:212–219.

Reinfelder, J.R., Fisher, N.S., Luoma, S.N., and Wang, W.-X. 1998. Trace element trophic transfer factor in aquatic organisms: A critique of the kinetic model approach. Sci. Total Environ., 219:117–135.

Reinhard, M. and Drefahl, A. 1999. Handbook for Estimating Physicochemical Properties of Organic Compounds. John Wiley, New York.

Relyea, R.A. 2003. Predator cues and pesticides: A double dose of danger for amphibians. Ecol. Appl., 13:1515–1521.

Restum, J., Bursian, S., Giesy, J., Render, J., Helferich, W., Shipp, E., Verbrugge, D., and Aulerich, R. 1998. Multigenerational study of the effects of consumption of PCB-contaminated carp from Saginaw Bay, Lake Huron, on mink. J. Toxicol. Environ. Health, Part A, 54:343–375.

Reynoldson, T.B., Norris, R.H., Resh, V.H., Day, K.E., and Rosenberg, D.M. 1997. The reference condition: A comparison of multimetric and multivariate approaches to assess water quality impairment using benthic macroinvertebrates. J. N. Am. Benthol. Soc., 16:833–852.

Rice, C.D. 2001. Fish immunotoxicology: Understanding mechanisms of action. Pages 96–138 in D. Schlenk and W. Benson (eds.) Target Organ Toxicity in Marine and Freshwater Teleosts, Volume 2—Systems. Taylor & Francis, New York.

Rice, C.P. and White, D.S. 1987. PCB availability assessment of river dredging using caged clams and fish. Environ. Toxicol. Chem., 6:259–274.

Richter, E.C. and Laster, R. 2005. The precautionary principle, epidemiology and the ethics of delay. Hum. Ecol. Risk Assess., 11:17–27.

Riedel, G.F., Sanders, J.G., and Breitburg, D.L. 2003. Seasonal variability in response of estuarine phytoplankton communities to stress: Linkages between toxic trace elements and nutrient enrichment. Estuaries, 26: 323–338.

Risk Assessment Forum. 1996. Summary report for the workshop on Monte Carlo analysis. EPA/630/R-96/010. US Environmental Protection Agency, Washington, DC.

Risk Assessment Forum. 1997. Guiding principles for Monte Carlo analysis. EPA/630/R-97/001. US Environmental Protection Agency, Washington, DC.

Risk Assessment Forum. 1999. Report of the workshop on selecting input distributions for probabilistic assessments. EPA/630/R-98/004. US Environmental Protection Agency, Washington, DC.

Risk Assessment Forum. 2000. Supplementary guidance for conducting health risk assessment of chemical mixtures. EPA/630/R-00/002. US Environmental Protection Agency, Washington, DC.

RIVM (National Institute for Public Health and the Environment). 1996. EUSES: European Union System for the Evaluation of Substances. European Chemicals Bureau, JRC Environmental Institute, Ispra, Italy.

Robbins, C.T. 1993. Wildlife Feeding and Nutrition. Academic Press, San Diego, CA.

Robinson, S.C., Kendall, R.J., Robinson, R., Driver, C.J., and Lacher, T.E. 1988. Effects of agricultural spraying of methyl parathion on cholinesterase activity and reproductive success in wild starlings. Environ. Toxicol. Chem., 7:343–349.

Roffe, T.J., Friend, M., and Lock, L.N. 1994. Evaluation of causes of wildlife mortality. Pages 324–348 in T.A. Brookhout (ed.) Research and Management Techniques for Wildlife and Habitats. The Wildlife Society, Bethesda, MD.

Rombke, J., Heimbach, F., Hoy, S., et al. 2003. Effects of Plant Protection Products on Functional Endpoints in Soil. SETAC Press, Pensacola, FL.

Roosenburg, W.M. 2000. Hypothesis testing, decision theory, and common sense in resource management. Conserv. Biol., 14:1208–1210.

Rose, K.A. and Cowan, J.H. Jr. 1993. Individual-based model of young-of-the-year striped bass population dynamics. I. Model description and baseline simulations. Transact. Am. Fish. Soc., 122:415–438.

Rose, K.A., Cowan, J.H., Houde, E.D., and Coutant, C.C. 1993. Individual-based modelling of environmental quality effects on early life-stages of fishes: A case study using striped bass. Am. Fish. Soc. Symp., 14:125–145.

Rose, K.A., Rutherford, E.S., McDermott, D., Forney, J.L., and Mills, E.L. 1999. An individual-based model of walleye and yellow pearch in Oneida Lake, New York. Ecol. Monogr. 69:127–154.

Rose, K.A., Cowan, J.H., Winemiller, J.O., Myers, R.A., and Hilborn, R. 2001. Compensatory density-dependence in fish populations: Importance, controversy, understanding, and prognosis. Fish Fish., 2:293–327.

Rose, K.A., Murphy, C.A., Diamond, S.L., Fuiman, L.A., and Thomas, P. 2003. Using nested models and laboratory data for predicting population effects of contaminants on fish: A step toward a bottom-up approach for establishing causality in field studies. Hum. Ecol. Risk Assess., 9:231–257.

Rose, K.R., Brenkert, A.L., Cook, R.B., and Gardner, R.H. 1991. Systematic comparison of ILWAS, MAGIC, and ETD watershed acidification models. 1. Monte Carlo under regional variability. Water Resources Res., 27:2577–2589.

Rosen, B.H. 1995. Use of periphyton in the development of biocriteria. Pages 209–215 in W.S. Davis and T.P. Simon (eds.) Biological Assessment and Criteria. Lewis Publishers, Boca Raton, FL.

Ross, P.S. 2000. Marine mammals as sentinels in ecological risk assessment. Hum. Ecol. Risk Assess., 6:29–46.

Ross, P.S., Vos, J.G., Birnbaum, L.S., and Osterhaus, A.D.M.E. 2000. PCBs are a health risk for humans and wildlife. Science, 289:1878–1879.

Rothman, K.J. 1986. Modern Epidemiology. Little, Brown, Boston, MA.

Roux, D.J., Jooste, S.H.J., and MacKay, H.M. 1996. Substance-specific water quality criteria for the protection of South African freshwater ecosystems: Methods for derivation and initial results for some inorganic toxic substances. S. Afr. J. Sci., 92:198–206.

Rowley, M.H., Christian, J.J., Basu, D.K., Pawlitowski, M.A., and Paigen, B. 1983. Use of small mammals (voles) to assess a hazardous waste site at Love Canal, Niagra Falls, New York. Arch. Environ. Contam. Toxicol., 12:383–399.

Royal Commission on Environmental Pollution. 2003. Chemicals in Products: Safeguarding the Environment and Human Health. 24th Report. The Stationary Office, London.

Royall, R. 1997. Statistical Evidence: A Likelihood Paradigm. Chapman & Hall, London.

Royall, R. 2004. The likelihood paradigm for statistical inference. Pages 119–152 in M.L. Taper and S.R. Lele (eds.) The Nature of Scientific Evidence. University of Chicago Press, Chicago, IL.

Rubinstein, R.Y. 1981. Simulation and Monte Carlo Method. John Wiley, New York.

Ruckleshaus, W.D. 1984. Risk in a free society. Risk Anal., 4:157–162.

Rumbold, D.G. 2005. A probabilistic risk assessment of effects of methylmercury on great egrets and bald eagles foraging at a constructed wetland in South Florida relative to the Everglades. Hum. Ecol. Risk Assess., 11:365–388.

Russell, B. 1948. Human Knowledge, Its Scope and Limits, Part V. Simon & Schuster, New York.

Russell, B. 1957. Mysticism and Logic. Doubleday, New York.

Russom, C.L., Bradbury, S.P., Broderius, S.J., Hammermeister, D.E., and Drummond, R.A. 1997. Predicting modes of toxic action from chemical structure: Acute toxicity in the fathead minnow (*Pimephales promelas*). Environ. Toxicol. Chem., 16:948–967.

Rutgers, M., van't Verlaat, I.M., Wind, B., Posthuma, L., and Breure, A.M. 1998. Rapid method for assessing pollution-induced community tolerance in contaminated soil. Environ. Toxicol. Chem., 17:2210–2213.

Ruth Ruttenberg and Associates. 2004. Not too costly after all: An examination of the inflated cost estimates of health, safety and environmental protections. Public Citizen Foundation, Washington, DC.

SAB (Science Advisory Board). 1990. Reducing Risk: Setting Priorities and Strategies of Environmental Protection. SAB-EC-90-021. US Environmental Protection Agency, Washington, DC.

SAB (Science Advisory Board). 1998. Summary of the US EPA workshop on the realtionship between exposure duration and toxicity. EPA 68-D5-0028. Science Advisory Board, US Environmental Protection Agency, Washington, DC.

SAB (Science Advisory Board). 2000. Toward integrated environmental decision-making. EPA-SAB-EC-00-011. Science Advisory Board, US Environmental Protection Agency, Washington, DC.

Sadana, U.S. and Singh, B. 1987. Yield and uptake of cadmium, lead and zinc by wheat grown in a soil polluted with heavy metals. J. Plant Sci. Res. 3:11–17.

Sadiq, M. 1985. Uptake of cadmium, lead and nickel by corn grown in contaminated soils. Water Air Soil Pollut., 26:185–190.

Sadiq, M. 1986. Solubility relationships of arsenic in calcareous soils and its uptake by corn. Plant Soil, 91:241–248.

Safe, S. 1998. Hazard and risk assessment of chemical mixtures using the toxic equivalency factor (TEF) approach. Environ. Health Persp., 106:1051–1058.

Safe S.H., Pallaroni, L., Yoon, K., Gaido, K., Ross, S., and McDonald, D. 2002. Problems for risk assessment of endocrine-active estrogenic compounds. Environ. Health Persp., 110:925–929.

Sagoff, M. 1988. The Economy of the Earth. Cambridge University Press, Cambridge, UK.

Salazar, M. and Salazar, S. 1998. Using caged bivalves as part of the exposure–dose–response triad to support an integrated assessment strategy. Pages 167–192 in A. de Peyster and K.E. Day (eds.) Ecological Risk Assessment: A Meeting of Policy and Science. SETAC, Pensacola, FL.

Salice, C.J. and Miller, T.J. 2003. Population-level responses to long-term cadmium exposure in two strains of the freshwater gastropod *Biomphalaria glabrata*: Results from a life-table response experiment. Environ. Toxicol. Chem., 22:678–688.

Salsburg, D.S. 2001. The Lady Tasting Tea: How Statistics Revolutionized Science in the Twentieth Century. Henry Holt, New York.

Salwasser, H. 1986. Conserving a regional spotted owl population. Pages 227–247 in N. Grossblatt (ed.) Ecological Knowledge and Environmental Problem Solving: Concepts and Case Studies. National Academy Press, Washington, DC.

Sample, B.E. and Arenal, C.A. 1999. Allometric models for interspecies extrapolation for wildlife toxicity data. Bull. Environ. Contam. Toxicol., 62:653–663.

Sample, B.E. and Suter, G.W. 1994. Estimating exposure of terrestrial wildlife to contaminants. ES/ER/TM-125. Oak Ridge National Laboratory, Oak Ridge, TN.

Sample, B.E. and Suter, G.W., II. 1999. Ecological risk assessment of a large river-reservoir: 4. Piscivorous wildlife. Environ. Toxicol. Chem., 18:610–620.

Sample, B.E. and Suter, G.W., II. 2002. Screening evaluation of the ecological risks to terrestrial wildlife associated with a coal ash disposal site. Hum. Ecol. Risk Assess., 8:637–656.

Sample, B.E., Butler, L., Zivkovich, C., Whitmore, R.C., and Reardon, R. 1996a. Effects of *Bacillus thuringiensis* Berliner var. *kurstaki* and defoliation by gypsy moth [*Lymantria dispar* (L.) (Lepidoptera: Lymantriidae)] on native arthropods in West Virginia. Can. Entomol., 128:573–592.

Sample, B.E., Hinzman, R.L., Jackson, B.L., and Baron, L.A. 1996b. Preliminary assessment of the ecological risks to wide-ranging wildlife species on the Oak Ridge Reservation. DOE/OR/01-1407&D2. US Department of Energy, Oak Ridge, TN.

Sample, B.E., Opresko, D.M., and Suter, G.W., II. 1996c. Toxicological benchmarks for wildlife. ES/ER/TM-86/R3. Oak Ridge National Laboratory, Oak Ridge, TN.

Sample, B.E., Aplin, M.S., Efroymson, R.E., Suter, G.W., II, and Welsh, C.J.E. 1997. Methods and Tools for Estimation of the Exposure of Terrestrial Wildlife to Contaminants. Oak Ridge National Laboratory, Oak Ridge, TN.

Sample, B.E., Beauchamp, J., Efroymson, R.A., Suter, G.W., II, and Ashwood, T.L. 1998. Development and validation of literature-based bioaccumultion models for small mammals. ES/ER/TM-219. Oak Ridge National Laboratory, Oak Ridge, TN.

Sample, B.E., Suter, G.W., II, Beauchamp, J., and Efroymson, R.A. 1999. Literature-derived bioaccumulation models for earthworms: Development and validation. Environ. Toxicol. Chem., 18:2110–2120.

Santore R.C. and Driscoll, C.T. 1995. The Chess model for calculating chemical equilibria in solids and solutions. Pages 357–375 in R.H. Loeppert, A. Schwab, and S. Goldberg (eds.) Chemical Equilibrium and Reaction Models. American Society of Agronomy, Madison, WI.

Sarda, N. and Burton, G.A. Jr. 1995. Ammonia variation in sediments: Spatial, temporal and method-related effects. Environ. Toxicol. Chem., 14:1499–1506.

SAS Institute, Inc. 1989. SAS/STAT User's Guide, Version 6, Fourth Edition, Volume 2. SAS Institute, Cary, NC.

Sauve, S. 2001. Speciation of metals in soil. Pages 7–38 in H.E. Allen (ed.) Bioavailability of Metals in Terrestrial Ecosystems. SETAC Press, Pensacola, FL.

Saxe, J.K., Impellitteri, C.A., and Allen, H.E. 2001. Novel model describing trace metal concentrations in the earthworm *Eisenia andrei*. Environ. Sci. Technol., 35:4522–4529.

Schama, S. 1995. Landscape and Memory. Vintage Books, New York.

Schecher, W.D. and McAvoy, D.C. 1994. MINEQL+: A chemical equilibrium program for personal computers. Version 3.01. Environmental Research Software, Hallowell, ME.

Scheff, P.A. and Wadden, R.A. 1993. Receptor modeling of volatile organic compounds. 1. Emission inventory and validation. Environ. Sci. Technol., 27:617–625.

Scheringer, M., Vogel, T., von Grote, J., Capaul, B., Schubert, R., and Hungerbuhler, K. 2001. Scenario-based risk assessment of multi-use chemicals: Application to solvents. Risk Anal., 21:481–497.

Scheuhammer, A.M. and Templeton, D.M. 1998. Use of stable isotope ratios to distinguish sources of lead exposure in wild birds. Ecotoxicology, 7:37–42.

Scheunert, I., Topp, E., Attar, A., and Korte, A. 1994. Uptake pathways for chlorobenzenes in plants and their correlation with N-octanol/water partition coefficients. Ecotoxicol. Environ. Saf., 27:90–104.

Schindler, D.W. 1974. Eutrophication and recovery in experimental lakes: Implications for management. Science, 184:897–899.

Schindler, D.W. 1987. Detecting ecosystem responses to anthropogenic stress. Can. J. Fish. Aquat. Sci., 44:6–25.

Schindler, D.W. 1998. Replication versus realism: The need for ecosystem-scale experiments. Ecosystems, 1:323–334.

Schindler, D.W., Mills, K.H., Malley, D.F., Findlay, D.L., Shearer, J.A., Davies, I.J., Turner, M.A., Linsey, G.A., and Cruikshank, D.R. 1985. Long-term ecosystem stress: The effects of years of experimental acidification on a small lake. Science, 228:1395–1401.

Schlenk, D. and Bensen, W.H. 2001. Target Organ Toxicity in Marine and Freshwater Teleosts (2 volumes). Taylor & Francis, New York.

Schmidtz, D. and Willott, E. 2002. Environmental Ethics. Oxford University Press, Oxford, UK.

Schmieder, P.K., Tapper, M.L.A., Denny, J., Kolanczyk, R., and Johnson, R. 2000. Optimization of a precision-cut trout liver tissue slice assay as a screen for vitellogenin induction. Aquat. Toxicol., 49:251–268.

Schmoyer, R.L., Beauchamp, J.J., Brandt, C.C., and Hoffman, F.O. 1996. Difficulties with the log-normal model in mean estimation and testing. Environ. Ecol. Stat., 3:81–97.

Schneider, S.H. 2002. Keeping out of the box. Am. Sci., 90:496–498.

Schoener, T.W. 1968. Sizes of feeding territories among birds. Ecology, 49:123–136.

Schwarz, R.C., Schults, D.W., Ozretich, R.W., Lamberson, J.O., Cole, F.A., DeWitt, T.H., Redmond, M.S., and Ferraro, S.P. 1995. Sigma PAH: A model to predict the toxicity of polynuclear aromatic hydrocarbon mixtures in field-collected sediments. Environ. Toxicol. Chem., 14:1977–1978.

Schwarzenbach, R.P., Gschwend, P.M., and Imboden, D.M. 1993. Environmental Organic Chemistry. John Wiley, New York.

Science Policy Council. 2000. Risk Characterization Handbook. EPA 100-B-00-002. US Environmental Protection Agency, Washington, DC.

Science Policy Council. 2002. A Framework for the Economic Assessment of Ecological Benefits. US Environmental Protection Agency, Washington, DC.

Seife, C. 2000. CERN's gamble shows the perils, rewards of playing the odds. Science, 289:2260–2262.

Serveiss, V., Cox, J.P., Moses, J., and Yeager, B.L. 2000. Workshop report on characterizing ecological risk at the watershed scale. EPA/600/R-99/111. US Environmental Protection Agency, Washington, DC.

Serveiss, V.B. 2002. Applying ecological risk principles to watershed assessment and management. Environ. Manag., 29:145–154.

Shacklette, H.T. and Boerngen, J.G. 1984. Elemental concentrations in soils and other surficial materials of the conterminous United States. Professional Paper 1270. US Geological Survey, Washington, DC.

Shaver, J.P. 1999. What statistical significance testing is, and what it is not. J. Exp. Edu., 61:293–316.

Shaw, B.P. and Panigrahi, A.K. 1986. Uptake and tissue distribution of mercury in some plant species collected from a contaminated area in India. Arch. Environ. Contam. Toxicol., 15:439–446.

Shaw-Allen, P. and Suter, G.W., II. 2005. Methods/indicators for determining when metals are the cause of biological impairments of rivers and streams. NCEA-C-1494. US Environmental Protection Agency, Washington, DC.

Sheffield, S.R., Matter, J.M., Rattner, B.A., and Guiney, P.D. 1998. Fish and wildlife species as sentinels of environmental endocrine disruptors. Pages 369–430 in R.J. Kendall (ed.) Principles and Processes for Evaluating Endocrine Disruption in Wildlife. SETAC Press, Pensacola, FL.

Sheppard, M.I., Sheppard, S.C., and Amiro, B.D. 1991. Mobility and plant uptake of inorganic ^{14}C and ^{14}C-labelled PCB in soils of high and low retention. Health Phys., 61:481–492.

Sheppard, S.C., Gaudet, C., Sheppard, M.I., Cureton, P.M., and Wong, M.P. 1992. The development of assessment and remediation guidelines for contaminated soils, a review of the science. Can. J. Soil Sci., 72:359–394.

Shugart, H.H. 1984. A Theory of Forest Dynamics. Springer, New York.

Shugart, H.H. and O'Neill, R.V. 1979. Systems ecology. Dowden, Hutchinson & Ross, Stroudsburg, PA.

Shugart, H.H. and West, D.C. 1977. Development and application of an Appalachian deciduous forest succession model and its application to assessment of the impact of the chestnut blight. J. Environ. Manag., 5:161–179.

Sigal, L.L. and Suter, G.W., II. 1989. Potential effects of chemical agents on terrestrial resources. Environ. Profess., 11:376–384.

Sijm, D., van Wezel, A.P., and Crommentuijn, T. 2002. Environmental risk limits in the Netherlands. Pages 221–253 in L. Posthuma, G.W. Suter II, and T. Traas (eds.) Species Sensitivity Distributions in Ecotoxicology. Lewis Publishers, Boca Raton, FL.

Silva, M. and Downing, J.A. 1995. CRC Handbook of Mammalian Body Masses. CRC Press, Boca Raton, FL.

Simberloff, D. 1998. Flagships, umbrellas, and keystones: Is single-species management passe in the landscape era? Biol. Conserv., 83:247–257.

Simberloff, D. 2003. Community and ecosystem impacts of single-species extinctions. Pages 221–233 in P. Kareiva and S.A. Levin (eds.) The Importance of Species. Princeton University Press, Princeton, NJ.

Simon, T.P. 2002. Biological Response Signatures. CRC Press, Boca Raton, FL.

Simon, T.P. and Lyons, J. 1995. Application of the index of biotic integrity to evaluate water resource integrity in freshwater ecosystems. Pages 245–262 in W. Davis and T.P. Simon (eds.) Biological Assessment and Criteria: Tools for Water Resource Planning and Decision Making. Lewis Publishers, Boca Raton, FL.

Simpson, J., Norris, R., Barmuta, L., and Blackman, P. 1996. Australian river assessment system: National river health program predictive model manual. Available at: http://ausrivas. canberra.au/ausrivas

Simpson, J.M., Santo Domingo, J.W., and Reasner, D.J. 2002. Microbial source tracking: State of the science. Environ. Sci. Technol., 36:5279–5288.

Sims, J.T. and Kline, J.S. 1991. Chemical fractionation and plant uptake of heavy metals in soils amended with co-composted sewage sludge. J. Environ. Qual., 20:387–395.

Sinko, J.W. and Streifer, W. 1967. A new model for age-structure of a population. Ecology, 48:910–918.

Sinko, J.W. and Streifer, W. 1969. Applying models incorporating age-size structure of a population of *Daphnia*. Ecology, 50:608–615.

Sjogren-Gulve, P. and Ebenhard, T. 2000. The Use of Population Viability Analysis in Conservation Planning. Blackwell, New York.

Skelly, J.M., Davis, D.D., Merrill, W., Cameron, E.A., Brown, H.D., Drummond, D.B., and Dochinger, L.S. 1990. Diagnosing Injury to Eastern Forest Trees. Pennsylvania State University, University Park, PA.

Slayton, D. and Montgomery, D. 1991. Michigan background soil survey. Michigan Department of Natural Resources, Ann Arbor, MI.

Slikker, W. Jr., Anderson, M.E., Bogdanffy, M.S., Bus, J.S., Cohen, S.D., Conolly, R.B., David, R.M., et al. 2004a. Dose-dependent transitions in mechanisms of toxicity. Toxicol. Appl. Pharmacol., 201:203–255.

Slikker, W. Jr., Anderson, M.E., Bogdanffy, M.S., Bus, J.S., Cohen, S.D., Conolly, R.B., David, R.M., et al. 2004b. Dose-dependent transitions in mechanisms of toxicity: Case studies. Toxicol. Appl. Pharmacol., 201:226–294.

Sloof, W., van Oers, J.A.M., and deZwart, D. 1986. Margins of uncertainty in ecotoxicological hazard assessment. Environ. Toxicol. Chem., 5:841–852.

Smith, F.E. 1969. Effects of enrichment in mathematical models. Pages 631–645 in Eutrophication: Causes, Consequences, Correctives. Proceedings of a Symposium, National Academy of Sciences, Washington, DC.

Smith, A.H., Sciortino, S., Goeden, H., and Wright, C.C. 1996. Consideration of background exposures in the management of hazardous waste sites: A new approach to risk assessment. Risk Anal., 16:619–625.

Smith, E.P. and Cairns, J. Jr. 1993. Extrapolation methods for setting ecological standards for water quality: Statistical and ecological concerns. Ecotoxicology, 2:203–219.

Smith, E.P., Robinson, T., Field, J.M., and Norton, S.B. 2003. Predicting sediment toxicity using logistic regression: A concentration-addition approach. Environ. Toxicol. Chem., 22:565–575.

Snell, T.W. and Serra, M. 2000. Using probability of extinction to evaluate the ecological significance of toxicant effects. Environ. Toxicol. Chem., 19:2357–2363.

Solomon, K.R., Baker, D.B., Richards, R.P., Dixon, K.R., Klaine, S.J., La Point, T.W., Kendall, R.J., et al. 1996. Ecological risk assessment for atrazine in North American surface waters. Environ. Toxicol. Chem., 15:31–76.

Solomon, K.R., Giddings, J.M., and Maund, S.J. 2001a. Probabilistic risk assessment of cotton pyrethroids. I. Distributional analysis of laboratory aquatic toxicity data. Environ. Toxicol. Chem., 20:652–659.

Solomon, K.R., Giesy, J.P., Kendall, J.R., Best, L.B., Coats, J.R., Dixon, K.R., Hooper, M.J., Kenaga, E.E., and McMurry, S.T. 2001b. Chlorpyrifos: Ecotoxicological risk assessment for birds and mammals in corn ecosystems. HERA, 7:497–632.

Sonnemann, G., Castells, F., and Schuhmacher, M. 2004. Integrated Life-Cycle and Risk Assessment for Industrial Processes. Lewis Publishers, Boca Raton, FL.

Sparling, D. 2003. White phosphorus at Eagle River Flats, Alaska: A case history of waterfowl mortality. Pages 767–785 in D.J. Hoffman, B. Rattner, G.A. Burton Jr., and J. Cairns Jr. (eds.) Handbook of Ecotoxicology. Lewis Publishers, Boca Raton, FL.

Spehar, R.L. and Carlson, A.R. 1984. Derivation of site-specific water quality criteria for cadmium and the St. Louis River basin, Duluth, Minnesota. EPA-600/3-84-029. US Environmental Protection Agency, Duluth, MN.

Sperber, O., From, J., and Sparre, P. 1977. A method to estimate the growth rate of fishes, as a function of temperature and feeding level, applied to rainbow trout. Meddelerser fra Danmarks Fiskeriog Havundersogelser, 7:275–317.

Sprague, J.B. 1970. Measurement of pollutant toxicity to fish. Water Res., 4:3–32.

Spray Drift Task Force. 1997. A summary of aerial application studies. Spray Drift Task Force, c/o Stewart Agricultural Research Services, Macon, MO.

Sprenger, M.D. and Charters, D.W. 1997. Ecological risk assessment guidance for Superfund: Process for designing and conducting ecological risk assessment, interim final. Environmental Response Team, US Environmental Protection Agency, Edison, NJ.

Spromberg, J.A. and Birge, W.J. 2005. Modeling the effects of chronic toxicity on fish populations: The influence of life-history strategies. Environ. Toxicol. Chem., 24:1532–1540.

Spromberg, J.A., John, B.M., and Landis, W.G. 1998. Metapopulation dynamics: Indirect effects and multiple distinct outcomes in ecological risk assessment. Environ. Toxicol. Chem., 17:1640–1649.

Spurgeon, D.J. and Hopkin, S.P. 1996. Risk assessment of the threat of secondary poisoning by metals to predators of earthworms in the vicinity of a primary smelting works. Sci. Total Environ., 187:167–183.

Stahl, C.H., Cimorelli, A.J., and Chow, A.H. 2002. A new approach to environmental decision analysis: Multi-criteria integrated resource assessment (MIRA). Bull. Sci. Technol. Soc., 22:443–459.

Stahl, R.G. Jr. 1997. Can mammalian and non-mammalian "sentinel species" data be used to evaluate the human health implications of environmental contaminants? Hum. Ecol. Risk Assess., 3:328–335.

Stahl, W.R. 1967. Scaling of respiratory variables in mammals. J. Appl. Physiol., 22:453–460.

Stanley-Horn, D.E., Matilla, H.R., Sears, M.K., Dively, G., Rose, R., Hellmich, R.L., and Lewis, L. 2001. Assessing impact of Cry 1Ab-expressing corn pollen on monarch butterfly larvae in field studies. Proc. Natl. Acad. Sci. USA 10.1073/pnas.211277798.

Stansley, W., Roscoe, E., Hawthorne, E., and Meyer, R. 2001. Food chain aspects of chlordane poisoning in birds and bats. Arch. Environ. Contam. Toxicol., 40:285–291.

Stavric, B. and Klassen, R. 1994. Dietary effects on the uptake of benzo(a)pyrene. Food Chem. Toxicol., 8:727–734.

Stearns, S.C. 1977. The evolution of life history traits: A critique of the theory and a review of the data. Annual Rev. Ecol. Systematics, 8:145–171.

Stein, B.A., Kutner, L.S., and Adams, J.S. (eds.). 2000. Precious Heritage: The Status of Biodiversity in the United States. Oxford University Press, Oxford, UK.

Stelfox, H., Chua, G., O'Rourke, K., and Detsky, A. 1998. Conflict of interest in the debate over calcium-channel antagonists. NEJM, 338:101–106.

Stephan, C.E., Mount, D.I., Hanson, D.J., Gentile, J.H., Chapman, G.A., and Brungs, W.A. 1985. Guidelines for deriving numeric National Water Quality Criteria for the protection of aquatic organisms and their uses. PB85-227049. US Environmental Protection Agency, Washington, DC.

Stevens, D., Linder, G., and Warren-Hicks, W., 1989. Data interpretation. Pages 9–11 in Warren-Hicks, W., Parkhurst, B.R., and Baker, S.S. Jr. (eds.) Ecological Assessment of Harardous Waste Sites: A Field and Laboratory Reference Document. EPA 600/3-89/013. Corvallis Environmental Research Laboratory, Corvallis, OR.

Stewart-Oaten, A. 1995. Rules and judgments in statistics: Three examples. Ecology, 76:2001–2009.

Stohlgren, T.J. and Schnase, J.L. 2006. Risk analysis for biological hazards: What we need to know about invasive species. Risk Anal., 26:163–173.

Stow, C.A. 1999. Assessing the relationship between *Pfiesteria* and estuarine fishkills. Ecosystems, 2:237.

Stow, C.A. and Borsuk, M.E. 2003. Enhancing causal assessment of estuarine fishkills using graphical models. Ecosystems, 6:11–19.

Strickland, T.C., Holdren, G.R. Jr., Ringold, P.L., Bernard, D., Snarski, V.M., and Fallon, W. 1993. A national critical loads framework for atmospheric deposition effects assessment: Method summary. Environ. Manag., 17:329–334.

Strojan, C.L. 1978. Forest litter decomposition in the vicinity of a zinc smelter. Oecologia (Berl.), 32.203–219.

Stubblefield, W.A., Hancock, G.A., Ford, W.H., Prince, H.H., and Ringer, R.K. 1995a. Evaluation of the toxic properties of naturally weathered *Exxon Valdez* crude oil to surrogate wildlife species. Pages 665–692 in P.G. Wells, J.N. Butler, and J.S. Hughes (eds.) Exxon Valdez Oil Spill: Fate and Effects in Alaskan Waters. ASTM, Philadelphia, PA.

Stubblefield, W.A., Hancock, G.A., Ford, W.H., and Ringer, R.K. 1995b. Acute and subchronic toxicity of naturally weathered *Exxon Valdez* crude oil in mallards and ferrets. Environ. Toxicol. Chem., 14:1941–1950.

Stubblefield, W.A., Hancock, G.A., Ford, W.H., and Ringer, R.K. 1995c. Effects of naturally weathered *Exxon Valdez* crude oil on mallard reproduction. Environ. Toxicol. Chem., 14:1951–1960.

Suissa, S. 1999. Relative excess risk: An alternative measure of comparative risk. Am. J. Epi., 150:279–282.

Sun, K., Krause, G.F., Mayer, F.L., Ellersieck, M.R., and Basu, A.P. 1995. Estimation of acute toxicity by fitting a dose–time–response surface. Risk Anal., 15:247–252.

Sun, M. 1983. Missouri's costly dioxin lesson. Science, 219:367–369.

Susser, M. 1988. Falsification, verification and causal inference in epidemiology: Reconsideration in light of Sir Karl Popper's philosophy. Pages 33–58 in K.J. Rothman (ed.) Causal Inference. Epidemiology Resources, Chestnut Hill, MA.

Suter, G.W., II. 1989. Ecological endpoints. Pages 2–1 in W. Warren-Hicks, B.R. Parkhurst, and S.S. Baker Jr. (eds.) Ecological Assessment of Hazardous Waste Sites: A Field and Laboratory Reference Document. EPA 600/3-89/013. Corvallis Environmental Research Laboratory, Corvallis, OR.

Suter, G.W., II. 1990. Use of biomarkers in ecological risk assessment. Pages 419–426 in J.F. McCarthy and L.L. Shugart (eds.) Biomarkers of Environmental Contamination. Lewis Publishers, Ann Arbor, MI.

Suter, G.W., II. 1993a. Ecological Risk Assessment. Lewis Publishers, Boca Raton, FL.

Suter, G.W., II. 1993b. A critique of ecosystem health concepts and indices. Environ. Toxicol. Chem., 12:1533–1539.

Suter, G.W., II. 1996a. Abuse of hypothesis testing statistics in ecological risk assessment. Hum. Ecol. Risk Assess., 2:331–349.

Suter, G.W., II. 1996b. Interpreting probability distributions as an expression of ecological risk. SETAC Annual Meeting Abstracts, 17:44.

Suter, G.W., II. 1996c. Toxicological benchmarks for screening contaminants of potential concern for effects on freshwater biota. Environ. Toxicol. Chem., 15:1232–1241.

Suter, G.W., II. 1997. Guidance for treatment of variability and uncertainty in ecological risk assessment. ES/ER/TM-228. Oak Ridge National Laboratory, Oak Ridge, TN.

Suter, G.W., II. 1998a. Comments on the interpretation of distributions in "Overview of recent developments in ecological risk assessment". Risk Anal., 18:3–4.

Suter, G.W., II. 1998b. Retrospective assessment, ecoepidemiology, and ecological monitoring. Pages 177–217 in P. Calow (ed.) Handbook of Environmental Risk Assessment and Management. Blackwell Scientific, Oxford, UK.

Suter, G.W., II. 1999a. A framework for assessment of ecological risks from multiple activities. Hum. Ecol. Risk Assess., 5:397–414.

Suter, G.W., II. 1999b. Developing conceptual models for complex ecological risk assessments. Hum. Ecol. Risk Assess., 5:375–396.

Suter, G.W., II. 2001. Applicability of indicator monitoring to ecological risk assessment. Ecological Indicators, 1:101–112.

Suter, G. W, II and Bartell, S.M. 1993. Ecosystem-level effects. Pages 275–310 in G.W. Suter II (ed.) Ecological Risk Assessment. Lewis Publishers, Chelsea, MI.

Suter, G.W., II and Rosen, A.E. 1988. Comparative toxicology for risk assessment of marine fishes and crustaceans. J. Environ. Sci. Technol., 22:548–556.

Suter, G.W., II and Sharples, F.E. 1984. Examination of a proposed test for effects of toxicants on soil microbial processes. Pages 327–344 in D. Liu and B.J. Dutka (eds.) Toxicity Screening Procedures Using Bacterial Systems. Marcel Dekker, New York.

Suter, G.W., II and Tsao, C.L. 1996. Toxicological benchmarks for screening potential contaminants of concern for effects on aquatic biota: 1996 revision. ES/ER/TM-96/R2. Oak Ridge National Laboratory, Oak Ridge, TN.

Suter, G.W., II, Vaughan, D.S., and Gardner, R.H. 1983. Risk assessment by analysis of extrapolation error, a demonstration for effects of pollutants on fish. Environ. Toxicol. Chem., 2:369–378.

Suter, G.W., II, Barnthouse, L.W., Baes, C.F., Bartell, S.G., Cavendish, R.H., Gardner, R.H., O'Neill, R., and Rosen, A.E. 1984. Environmental Risk Analysis for Direct Coal Liquefaction. ORNL/TM_9074. Oak Ridge National Laboratory, Oak Ridge, TN.

Suter, G.W., II, Rosen, A.E., Linder, E., and Parkhurst, D.F. 1987. Endpoints for responses of fish to chronic toxic exposures. Environ. Toxicol. Chem., 6:793–809.

Suter, G.W., II, Rosen, A.E., Beauchamp, J.J., and Kato, T.T. 1992. Results of analysis of fur samples from the San Joaquin kit fox and associated water and soil samples from the Naval Petroleum

Reserve No. 1, Tupman, California. ORNL/TM-12244. Oak Ridge National Laboratory, Oak Ridge, TN.

Suter, G.W., II, Luxmore, R.J., and Smith, E.D. 1993. Compacted soil barriers at abandoned landfill sites are likely to fail in the long term. J. Environ. Qual., 22:217–226.

Suter, G.W., II, Sample, B.E., Jones, D.S., and Ashwood, T.L. 1994. Approach and strategy for performing ecological risk assessments for the Department of Energy's Oak Ridge Reservation. ES/ER/TM-33/R1. Environmental Restoration Division, Oak Ridge National Laboratory, Oak Ridge, TN.

Suter, G.W., II, Barnthouse, L.W., Efroymson, R.E., and Jager, H. 1999. Ecological risk assessment of a large river-reservoir: 2. Fish community. Environ. Toxicol. Chem., 18:589–598.

Suter, G.W., II, Efroymson, R.A., Sample, B.E., and Jones, D.S. 2000. Ecological Risk Assessment for Contaminated Sites. Lewis Publishers, Boca Raton, FL.

Suter, G.W., II, Norton, S.B., and Cormier, S.M. 2002a. A methodology for inferring the causes of observed impairments in aquatic ecosystems. Environ. Toxicol. Chem., 21:1101–1111.

Suter, G.W., II., Traas, T., and Posthuma, L. 2002b. Issues and practices in the derivation and use of species sensitivity distributions. Pages 437–474 in L. Posthuma, G.W. Suter II, and T. Traas (eds.) Species Sensitivity Distributions in Ecotoxicology. Lewis Publishers, Boca Raton, FL.

Suter, G.W., II, Vermier, T., Munns, W.R. Jr., and Sekizawa, J. 2003. Framework for the integration of health and ecological risk assessment. Hum. Ecol. Risk Assess., 9:281–302.

Suter, G.W., II, Rodier, D.J., Schwenk, S., Troyer, M.W., Tyler, P.L., Urban, D.J., Wellman, M.C., and Wharton, S. 2004. The U.S. Environmental Protection Agency's generic ecological assessment endpoints. Hum. Ecol. Risk Assess., 10:1–15.

Suter, G.W., II., Norton, S.B., and Fairbrother, A. 2005. Individuals versus organisms versus populations in the definition of ecological assessment endpoints. Integr. Environ. Assess. Manag., 1:397–400.

Swanson, M.B. and Socha, A.C. 1997. Chemical Ranking and Scoring: Guidelines for Relative Assessment of Chemicals. SETAC Press, Pensacola, FL.

Swartz, R.C. 1999. Consensus sediment quality guidelines for polycyclic aromatic hydrocarbon mixtures. Environ. Toxicol. Chem., 18:780–787.

Swartz, R.C., Schults, D.W., Ozretich, R.J., Lamberson, J.O., Cole, F.A., DeWitt, T.H., Redmond, M.S., and Ferraro, S.P. 1995. Sigma PAH: A model to predict the toxicity of polynuclear aromatic hydrocarbon mixtures in field-collected sediments. Environ. Toxicol. Chem., 14:1977–1987.

Swartzman, G.L. and Kaluzny, S.P. 1987. Ecological Modeling Primer. Macmillan, New York.

Sydelko, P.J., Hlohowskyj, I., Majerus, K., Christiansen, J., and Dolph, J. 2001. An object-oriented framework for dynamic ecosystem modeling: Application for integrated risk assessment. Sci. Total Environ., 274:271–281.

Syracuse Research Corp. 2003. EPI Suite Estimation Methods. Available at: http://esc.syrres.com

Tal, A. 1997. Assessing the environmental movement's attitudes toward risk assessment. Environ. Sci. Technol., 31:470A–476A.

Talmage, S.S. and Walton, B.T. 1993. Food chain transfer and potential renal toxicity of mercury to small mammals at a contaminated terrestrial field site. Ecotoxicology, 2:243–256.

Tamura, H., Yoshikawa, H., Gaido, K.W., Ross, S., DeLisle, S., Welsh, W.J., and Richard, A.M. 2003. Interaction of organophosphate pesticides and related compounds with the androgen receptor. Environ. Health Persp., 111:545–552.

Tannenbaum, L.V. 2005a. A critical assessment of the ecological risk assessment process: A review of misapplied concepts. Integr. Environ. Assess. Manag., 1:66–72.

Tannenbaum, L.V. 2005b. Two simple algorithms for refining mammalian receptor selection in ecological risk assessment. Integr. Environ. Assess. Manag., 1:290–298.

Taper, M.L. and Lele, S.R. 2004. The Nature of Scientific Evidence: Statistical, Philosophical and Empirical Considerations. University of Chicago Press, Chicago, IL.

Taub, F.B. 1969. Gnotobiotic models of freshwater communities. Verh. Internat. Verein. Limnol., 17:485–496.

Taub, F.B. 1997. Unique information contributed by multispecies systems: Examples from the standardized aquatic microcosm. Ecol. Appl., 7:1103–1110.

Taub, F.B. and Read, P.L. 1982. Model Ecosystems: Standardized Aquatic Microcosm Protocol Food and Drug Administration Contract No. 223-80-2352. Food and Drug Administration, Washington, DC.

Taylor, B.R. 1997. Rapid assessment procedures: Radical re-invention or just sloppy science. Hum. Ecol. Risk Assess., 3:1005–1016.

Taylor, F. 1979. Convergence to the stable age distribution in populations of insects. Am. Natl., 113:511–530.

Taylor, L.N., Wood, C.M. and McDonald, D.G. 2003. An evaluation of sodium loss and gill metal binding properties in rainbow trout and yellow perch to explain species differences in copper tolerance. Environ. Toxicol. Chem., 23:2159–2166.

Tcuschler, L.K., Dourson, M., Stiteler, W.M., McCollough, D.R., and Tully, H. 1999. Health risk above the reference dose for multiple chemicals. Reg. Tox. Pharmacol., 30:S19–S26.

The Presidential/Congressional Commission on Risk Assessment and Risk Management. 1997. Risk Assessment and Risk Management in Regulatory Decision-Making. Government Printing Office, Washington, DC.

Thibodeaux, L.J. 1996. Environmental Chemodynamics: Movement of Chemicals in Air, Water and Soil, Second Edition. Wiley-Interscience, New York.

Thiessen, K.M., Hoffman, F.O., Rantavaara, A., and Hossain, S. 1997. Environmental models undergo international test. Environ. Sci. Technol., 31:358–363.

Thomann, R.V. 1989. Bioaccumulation model of organic chemical distribution in aquatic food chains. Environ. Sci. Technol. 23:699–707.

Thomas, J.M., Skalski, J.R., Cline, J.F., McShane, M.C., Simpson, J.C., Miller, W.E., Peterson, S.A., Callahan, C.A., and Greene, J.C. 1986. Characterization of chemical waste site contamination and determination of its extent using bioassays. Environ. Toxicol. Chem., 5:487–501.

Thomas, P. 2003. Metal analysis. Pages 64–98 in K.C. Thompson and C.P. Nathanail (eds.) Chemical Analysis of Contaminated Land. Blackwell, Oxford, UK.

Thomp, K.C. and Nathanail, C.P. 2003. Chemical Analysis of Contaminated Land, Blackwell, Oxford, UK.

Tiktak, A., de Nie, D.S., van der Linden, J.J.T.I., and Kruijne, R. 2002. Modelling the leaching and drainage of pesticides in the Netherlands. Agronomie 22:373–387.

Tiktak, A., van der Linden, J.J.T.I., and Boesten, A. 2003. The GeoPEARL model: Model description, applications and manual. RIVM Report 716601007/2003. RIVM, Bilthoven, The Netherlands.

Tiktak, A., de Nie, D.S., Pineros Garcet, J.D., Jones, A., and Vanclooster, M., 2004. Assessment of the pesticide leaching risk at the Pan-European level, the EuroPEARL approach. J. Hydrobiol. 289:222–238.

Tillitt, D.E., Ankley, G.T., and Giesy, J.P. 1989. Planar chlorinated hydrocarbons (PCHs) in colonial fish-eating waterbird eggs from the Great Lakes. Mar. Environ. Res., 28:505–508.

Tillitt, D.E., Giesy, J.P., and Ankley, G.T. 1991. Characterization of the H4IIE rat hepatoma cell bioassay as a tool for assessming toxic potency of planar halogenated hydrocarbons in environmental samples. Environ. Sci. Technol., 25:87–92.

Tipping, E. 1994. WHAM—a chemical equilibrium model and computer code for waters, sediments and soils incorporating a discrete site/electrostatic model of ion binding by humic substances. Comput. Geosci., 20(6):973–1023.

Tones, S.J., Ellis, S.A., Breeze, V.G., Fowbert, J., Miller, P.C.H., Oakley, J.N., Parkin, C.S., and Arnold, D.J. 2001. Review and evaluation of test species and methods for assessing exposure of non-target plants and invertebrates to crop pesticide sprays and spray drift. Report: DEFRA project number PN 0937.

Toose, L., Woodfine, D.G., MacLeod, M., Mackay, D., and Gouin, T. (2004) BETR-World—a geographically explicit model of chemical fate: Application to transport of α-HCH to the Arctic. Environ. Pollut., 128:223–240.

Topp, E., Schenert, I., Attar, A., Korte, A., and Korte, F. 1986. Factors affecting the uptake of C×14(-labelled organic chemicals by plants from soil. Ecotoxicol. Environ. Saf., 11:219–231.

Topping, C.J. and Odderskær, P. 2004. Modeling the influence of temporal and spatial factors on the assessment of impacts of pesticides on skylarks. Environ. Toxicol. Chem., 23:509–520.

Toxics Cleanup Program. 1994. Natural background soil metals concentrations in Washington State. Pub. #94–115. Washington State Department of Ecology, Olympia, WA.

Traas, T.P., van de Meent, D., Posthuma, L., et al. 2002. The potentially affected fraction as a measure of ecological risk. Pages 315–344 in L. Posthuma, G.W. Suter II, and T.P. Traas (eds.) Species Sensitivity Distributions in Ecotoxicology. Lewis Publishers, Boca Raton, FL.

Trapp, F. 1995. Model for uptake of xenobiotics into plants. Pages 107–151 in F. Trapp and J.C. McFarlane (eds.) Plant Contamination: Modeling and Simulation of Organic Chemical Processes. Lewis Publishers, Boca Raton, FL.

Trapp, S. and Matthies, M. 1997. Modeling volatilization of PCDD/F from soil and uptake into vegetation. Environ. Sci. Technol., 31:71–74.

Travis, C.C. and Arms, A.D. 1988. Bioconcentration of organics in beef, milk, and vegetation. Environ. Sci. Technol., 22.271–292.

Travis, C.C., Baes, C.F., III, Barnthouse, L.W., Etnier, E.L., Holton, G.A., Murphy, B.D., Thompson, G.P., Suter, G.W., II, and Watson, A.P. 1983. Exposure Assessment Methodology and Reference Environments for Synfuels Risk Analysis. ORNL/TM-8672. Oak Ridge National Laboratory, Oak Ridge, TN.

Travis, C.C., Richter, S.A., and Crouch, E.A.C. 1987. Cancer risk management. Environ. Sci. Technol., 21:415 420.

Travis, K.Z. and Hendley, P. 2001. Probabilistic risk assessment of cotton pyrethroids: IV. Landscape-level exposure characterization. Environ. Toxicol. Chem., 20:679–686.

Tuart, L.W. 1988. Hazard evaluation division technical guidance document: Aquatic mesocosm tests to support pesticide registration. EPA-540/09-88-035. US Environmental Protection Agency, Washington, DC.

Tuart, L.W. and Maciorowski, A.F. 1997. Information needs for pesticide registration in the United States. Ecol. Appl., 7:1086–1093.

Tucker, K.A. and Burton, G.A. Jr. 1999. Assessment of nonpoint-source runoff in a stream using in situ and laboratory approaches. Environ. Toxicol. Chem., 18:2797–2803.

Tufte, E.R. 1983. The Visual Display of Quantitative Information. Graphics Press, Cheshire, CT.

Tufte, E.R. 1990. Envisioning Information. Graphics Press, Cheshire, CT.

Tufte, E.R. 1997. Visual Explanations. Graphics Press, Cheshire, CT.

Tuljapurkar, S.D. 1990. Population dynamics in variable environments. Springer, New York.

Turner, D.B. 1994. Atmospheric Dispersion Estimates, Second Edition. Lewis Publishers, Boca Raton, FL.

Turner, M.G., 1993. A landscape simulation model of winter foraging by large ungulates. Ecol. Model., 69:163–184.

Turner, M.G., Wu, Y., Romme, W.H., Wallace, L.L., and Brenkert, A.L. 1994. Simulating winter interactions among ungulates, vegetation, and fire in northern Yellowstone Park. Ecol. Appl., 4:472–496.

Tyler, G. 1984. The impact of heavy metal pollution on forests: A case study of Gusum, Sweden. AMBIO, 13:18–26.

Uhler, A.D., Emsbo-Mattingly, S., Liu, B., and Hall, L.W. Jr. 2005. An integrated case study for evaluating impacts of an oil refinery effluent on aquatic biota in the Deleware River: Advanced chemical fingerprinting of PAHs. Hum. Ecol. Risk Assess., 11:771–836.

UK Department of the Environment, Food and Rural Affairs. 2000. Guidelines for Environmental Risk Assessment and Management. Stationary Office, London.

Underwood, A.J. 2000. Importance of experimental design in detecting and measuring stresses in marine populations. J. Aquat. Ecosys. Stress Recovery, 7:3–24.

Urban, D.J. and Cook, N.J. 1986. Hazard Evaluation, Standard Evaluation Procedure, Ecological Risk Assessment. EPA-540/9-85-001. US Environmental Protection Agency, Washington, DC.

USACE (US Army Corps of Engineers). 1983. Economic and environmental principles for water and related land resources implementation studies. Headquarters, Washington, DC.

US Army BTAG. 2002. Selection of assessment and measurement endpoints for ecological risk assessment. Biological Technical Assistance Group, Department of the Army, Washington, DC.

US Department of Health Education and Welfare. 1964. Smoking and Health: Report of the Advisory Committee to the Surgeon General. Public Health Service Publication 1103, US Department of Health, Education and Welfare, Public Health Service, Washington, DC.

US Environmental Protection Agency (EPA). 1998. Guidelines for Ecological Risk Assessment. EPA/630/R-95/002F. Risk Assessment Forum, US Environmental Protection Agency, Washington, DC.

US Environmental Protection Agency (EPA). 2003. Generic Ecological Assessment Endpoints (GEAEs) for Ecological Risk Assessment. EPA/630/P-002/004F. Risk Assessment Forum, US Environmental Protection Agency, Washington, DC.

USEPA. 1995. User's Guide for the Industrial Source Complex (ISC3) Dispersion Models, OAQPS. US Environmental Protection Agency, Research Triangle Park, NC. Report EPA-454/B-95-003a. Available at: http://www.epa.gov/ttn/scram.

US Fish and Wildlife Service. 1987. Type B technical information document: Guidance on use of habitat evaluation procedures and suitability index models for CERCLA applications. PB88-100151. US Department of the Interior, Washington, DC.

US Fish and Wildlife Service. 2001. National Wild Fish Health Survey. US Department of the Interior, Washington, DC.

US Geological Survey. 1999. Field Manual of Wildlife Diseases: General Filed Procedures and Diseases of Birds. Information and technology Report 1999-001. US Geological Survey National Wildlife Center, Washington, DC.

Vaal, M., van der Wal, J.T., Hoekstra, J., and Hermens, J. 1997. Variation in the sensitivity of aquatic species in relation to the classification of environmental pollutants. Chemosphere, 35:1311–1327.

Valoppi, L., Petreas, M., Donohoe, R.M., Sullivan, L., and Callahan, C.A. 1999. Use of PCB congener and homologue analysis in ecological risk assessment. Pages 147–161 in F.T. Price, K.V. Brix, and N.K. Lane (eds.) Environmental Toxicology and Risk Assessment: Recent Achievements in Environmental Fate and Transport. ASTM, West Conshohocken, PA.

Van Brummelen, T.C., Verweij, R.A., Wedzinga, S.A., and van Gestel, C.A.M. 1996. Polycyclic aromatic hydrocarbons in earthworms and isopods from contaminted forest soil. Chemosphere, 32:315–341.

van de Meent, D. and Toet, D. 1992. Dutch priority setting system for existing chemicals. Report No. 679120001. National Institute for Public Health and Environmental Protection, Bilthoven, The Netherlands.

van den Berg, M. et al. 1998. Toxic equivalency factors (TEFs) for PCBs, PCDDs, PCDFs for humans and wildlife. Environ. Health Persp., 106:775–792.

van den Bergh, J.C.J.M. (ed.). 1999. Handbook of Environmental and Resource Economics. Edward Elgar, Cheltenham, UK.

van den Brink, P.J., Brock, T.C.M., and Posthuma, L. 2002. The value of the species sensitivity distribution concept for predicting field effects: (Non-)confirmation of the concept using semi-field experiments. Pages 155–193 in L. Posthuma, G.W. Suter II, and T.P. Traas (eds.) Species Sensitivity Distributions in Ecotoxicology. Lewis Publishers, Boca Raton, FL.

van der Boesten, J.J.T.I. and Linden, A.M.A. 2001. Effect of long-term sorption kinetics on leaching as calculated with the PEARL model for FOCUS scenarios. Pages 27–32 in BCPC Symposium, Pesticide Behavior in Soils and Water, Symposium Proceedings no. 78. Brighton, UK.

van der Schalie, W.H., Gardner, H.S., Bantle, J.A., De Rosa, C.T., Finch, R.A., Reif, J.S., Reuter, R.H., et al. 1989. Animals as sentinels of human health hazards of environmental chemicals. Environ. Health Persp., 107:309–315.

van der Sluijs, J.P., Craye, M., Funtowicz, S., Kloprogge, P., Ravetz, J.R., and Risbey, J. 2005. Combining qualitative and quantitative measures of uncertainty in model-based environmental assessment: The NUSAP system. Risk Anal., 25:481–492.

van Gestel, C.A.M. and Ma, W.C. 1988. Toxicity and bioaccumulation of chlorophenols in earthworms in relation to bioavailability in soil. Ecotoxicol. Environ. Saf., 15:289–297.

van Gestel, C.A.M. and Van Straalen, N.M. 1994. Ecotoxicological test systems for terrestrial invertebrates. Pages 205–240 in M.H. Donker, H. Eijsackers, and F. Heimbackers (eds.) Ecotoxicology of Soil Organisms. Lewis Publishers, Boca Raton, FL.

van Gestel, C.A.M., Ma, W., and Smit, C.E. 1991. Development of QSARs in terrestrial ecotoxicology: Earthworm toxicity and soil sorption of chlorophenols, chlorobenzenes, and chloroaniline. Sci. Total Environ., 109/110:589–604.

Van Hook, R.I. and Yates, A.J. 1975. Transient behavior of cadmium in a grassland arthropod food chain. Environ. Res., 9:76–83.

Van Straalen, N.M. 1990. New methodologies for estimating the ecological risk of chemicals in the environment. Pages 165–173 in A.A. Balkema (ed.) Proceedings of the 6th Congress of the International Association of Engineering Geology, Rotterdam, The Netherlands.

Van Straalen, N.M. 2002a. Theory of ecological risk assessment based on species sensitivity distributions. Pages 37–48 in L. Posthuma, G.W. Suter II, and T. Traas (eds.) Species Sensitivity Distributions in Ecotoxicology. Lewis Publishers, Boca Raton, FL.

Van Straalen, N.M. 2002b. Threshold models for species sensitivity distributions applied to aquatic risk assessment for zinc. Environ. Toxicol. Pharmacol., 11:167–172.

Van Straalen, N.M. and Denneman, G.A.J. 1989. Ecological evaluation of soil quality criteria. Ecotoxicol. Environ. Saf., 18:241–245.

Vannote, R.L., Minshall, G.W., Cummins, K.W., Sedell, J.R., and Cushing, C.E. 1980. The river continuum concept. Can. J. Fish. Aquat. Sci., 37:130–137.

Van Voris, P., Tolle, D., and Arthur, M.F. 1985. Experimental terrestrial soil-core microcosm test protocol. PA 600/3-85-047. National Technical Information Service, Springfield, VA.

Veith, G.D. and Kosian, P. 1983. Estimating bioconcentration potential from octanol/water partitioning coefficients. Pages 269–282 in D.R. Mackay, S. Patterson, S. Eisenreich, and M. Simmons (eds.) Physical Behavior of PCBs in the Great Lakes. Ann Arbor Press, Ann Arbor, MI.

Veith, G.D., DeFoe, D.L., and Bergstedt, B.V. 1979. Measuring and estimating the bioconcentration factor for chemicals in fish. J. Fish. Res. Board Can., 36:1040–1048.

Veith, G.D., Call, D.J., and Brook, L.T. 1983. Structure–toxicity relationships for fathead minnow, *Pimephales promelas*: Narcotic industrial chemicals. Can. J. Fish. Aquat. Sci., 40:743–748.

Verhaar, H.J.M., Van Leeuwen, C., and Hermens, J. 1992. Classifying environmental pollutants. Chemosphere, 25:471–491.

Verhaar, H.J.M., De Wolf, W., Dyer, S.A., Legierse, K.C.H.M., Seinen, W., and Hermens, J. 1999. An LC_{50} vs time model for the aquatic toxicity of reactive and receptor mediated compounds. Consequences for bioconcentration kinetics and risk assessment. Environ. Sci. Technol., 33:758–763.

Verhaar, H.J.M., Solbe, J.F., Speksnijder, J., Van Leeuwen, C., and Hermens, J. 2000. Classifying environmental pollutants. Part 3: External validation of the classification system. Chemosphere, 40:875–883.

Verschueren, K. 1996. Handbook of Environmental Data on Organic Chemicals, Second Edition. Van Nostrand Reinhold, New York.

Vighi, M., Altenburger, T., Arrhenius, A., Backhaus, T., Bodeker, W., Blanck, H., Consolaro, F., et al. 2002. Water quality objectives for mixtures of toxic chemicals: Problems and perspectives. Ecotoxicol. Environ. Saf., 54:139–150.

Vogelbein, W.K., Shields, J.D., Haas, L.W., Reece, K.S., and Zwerner, D.E. 2001. Skin ulcers in estuarine fishes: A comparative pathological evaluation of wild and laboratory-exposed fish. Environ. Health Persp., 109:687–693.

Voie, O.A., Johnsen, A., and Fossland, H.K. 2002. Why biota still accumulate high levels of PCB after removal of PCB-contaminated sediments in a Norwegian fjord. Chemosphere, 46:1367–1372.

Voinov, A., Fritz, C., and Costanza, R. 1998. Surface water flow in landscape models: Everglades case study. Ecol. Model., 108:131–144.

Vollenweider, R.A. 1976. Advances in defining critical loading levels for phosphorus in lake eutrophication. Mem. 1st. Ital. Idrobiol., 33:53–83.

von Stackelberg, K.E. and Menzie, C.A. 2002. A cautionary note on the use of species presence and absence data in deriving sediment criteria. Environ. Toxicol. Chem., 21:466–472.

Vose, D. 2000. Risk Analysis: A Quantitative Guide. John Wiley, New York.

VROM. 1994. Environmental quality objectives in the Netherlands. Ministry of Housing, Spatial Planning, and the Environment, The Hague, The Netherlands.

Walker, J.D. 1993. The TSCA Interagency Testing Committee, 1977 to 1992: Creation, structure, functions and contributions. Pages 451–462 in J.W. Gorsuch, F.J. Dwyer, C.G. Ingersoll, and T.W. La Point (eds.) Environmental Toxicology and Risk Assessment: Volume 2. ASTM, Philadelphia, PA.

Walker, J.D. (ed.). 2003. Annual Review: Quantitative Structure–Activity Relationships. Environ. Toxicol. Chem., 22:1651–1935.

Walker, J.D. and Schultz, T.W. 2003. Structure–activity relationships for predicting ecological effects of chemicals. Pages 893–910 in D.J. Hoffman, B. Rattner, G.A. Burton Jr., and J. Cairns Jr. (eds.) Handbook of Ecotoxicology, Second Edition. Lewis Publishers, Boca Raton, FL.

Walters, C.J. 1986. Adaptive management of renewable resources. MacMillan, New York.

Walthall, W.K., and J.D. Stark. 1997. Comparison of two population-level ecotoxicological endpoints: The intrinsic (r_m) and instantaneous (r_i) rates of increase. Environ. Toxicol. Chem., 16:1068–1073.

Wang, J.X. and Roush, M.L. 2000. Risk Engineering and Management. Marcel Dekker, New York.

Wang, M.-J. and Jones, K.C. 1994. Behavior and fate of chlorobenzenes (CBs) introduced into soil–plant systems by sewage sludge application: A review. Chemosphere, 21:297–331.

Wang, X., White-Hull, C., Dyer, S. and Yang, Y. 2000. GIS-ROUT: A River Model for Watershed Planning. Environ. Plan. B: Plan. Des., 27, 231–246.

Wania, F. 2003. Environmental Models (Globo-POP, POPCYCLING-Baltic and CoZMo-POP). University of Toronto at Scarborough, Ontario, Canada. Available at: www.utsc.utoronto.ca/wania.

Warren-Hicks, W.J. and Moore, D.R.J. (eds.). 1998. Uncertainty Analysis in Ecological Risk Assessment. SETAC Press, Pensacola, FL.

Wasserman, L. 2000. Bayesian model selection and averaging. J. Math. Psych., 44:92–107.

Watkins, D.R. et al. 1993. Final report of the background soil characterization project at the Oak Ridge Reservation, Oak Ridge, Tennessee. DOE/OR/01-1175. Oak Ridge National Laboratory, Oak Ridge, TN.

Weaver, R.W., Melton, J.R., Wang, D., and Duble, R.L. 1984. Uptake of arsenic and mercury from soil by bermuda grass *Cynodon dactylon*. Environ. Pollut., 33:133–142.

Webb, D.A. 1992. Background metal concentrations in Wisconsin surface waters. Wisconsin Department of Natural Resources, Madison, WI.

Webster, E., Mackay, D., Di Guardo, A., Kane, D., and Woodfine, D. 2004. Regional Differences in Chemical Fate Model Outcome. Chemosphere, 55:1361–1376.

Webster, J.A. and Crossley, D.A. Jr. 1978. Evaluation of two models for predicting elemental accumulation by arthropods. Environ. Entomol., 7:411–417.

Weed, D.L. 1988. Causal criteria and Popperian refutation. Pages 15–32 in K. J. Rothman (ed.) Causal Inference. Epidemiology Resources, Chestnut Hill, MA.

Weed, D.L. 1997. On the use of causal criteria. Internat. J. Epidemiol., 26:1137–1141.

Weidema, B.P., Ekvall, T., Pesonen, H.-L., et al. 2004. Scenarios in Life-Cycle Assessment. SETAC Press, Pensacola, FL.

Weinhold, B. 2003. Conservation medicine, combining the best of all worlds. Environ. Health Persp., 111:A525–A529.

Weis, J.S. 1996. Scientific uncertainty and environmental policy: Four pollution case studies. Pages 160–187 in J. Lemons (ed.) Scientific Uncertainty and Environmental Problem Solving. Blackwell Science, Cambridge, MA.

Welshons, W.V., Thayer, K.A., Judy, B.M., Taylor, J.A., Curran, E.M., and vom Saal, F.S. 2003. Large effects from small exposures. I. Mechanisms for endocrine-disrupting chemicals with estrogenic activity. Environ. Health Persp., 111:994–1006.

Weng, L., Temminghoff, E.J.M., Tipping, E., and van Reimsdijk, W.H. 2002. Complexation with dissolved organic matter and solubility control of heavy metals in a sandy soil. Environ. Sci. Technol., 36:4804–4810.

Wentsel, R.S., Beyer, W.N., Edwards, C.A., Kapustka, L.A., and Kuperman, R.G. 2003. Effects of contaminants on ecosystem structure and function. Pages 117–159 in R.P. Lanno (ed.) Contaminated Soils: From Soil–Chemical Interactions to Ecosystem Management. SETAC Press, Pensacola, FL.

Wenzel, A., Nendza, M., and Kanne, R. 1997. Testbattery for the assessment of aquatic toxicity. Chemosphere, 35:307–322.

Westall, J. 1979. MICROQL: I. A chemical equilibrium program in BASIC. EAWAG CH-8600. Swiss Federal Institute of Technology, Duebendorf, Switzerland.

Whipple, C. 1987. De Minimis Risk. Plenum Press, New York.

WHO (World Health Organization). 2001. Report on Integrated Risk Assessment. WHO/IPCS/IRA/01/12. World Health Organization, Geneva, Switzerland.

Whyte, I.J. 2002. Headaches and heartaches: The elephant management dilemma. Pages 293–305 in D. Schmidtz and E. Willott (eds.) Environmental Ethics. Oxford University Press, Oxford, UK.

Wickwire, W.T., Menzie, C.A., Burmistrov, D., and Hope, B.K. 2004. Incorporating spatial data into ecological risk assessments: The spatially explicit exposure module for ARAMS. Pages 297–310 in L.A. Kapustka, H. Galbraith, M. Luxon, and G.R. Biddinger (eds.) Landscape Ecology and Wildlife Habitat Evaluation. ASTM International, West Conshohocken, PA.

Wiegers, J.K. and Landis, W.G. 2005. Application of the relative risk model to the fjord of Port Valdez, Alaska. Pages 53–90 in W.G. Landis (ed.) Regional Scale Ecological Risk Assessment Using the Relative Risk Model. CRC Press, Boca Raton, FL.

Wiemeyer, S.N. and Porter, R.D. 1970. DDE thins eggshells of captive American kestrels. Nature, 227:737–738.

Williams, R.B. 1971. Computer simulation of energy flow in Cedar Bog Lake, Minnesota, based on the classical studies of Lindeman. Pages 544–582 in B.C. Patten (ed.) Systems Analysis and Simulation in Ecology: Volume 1. Academic Press, New York.

Wilson, D.E. (ed.). 1996. Measuring and Monitoring Biodiversity: Standard Methods for Mammals. Smithsonian Institution Press, Washington, DC.

Wilson, E.O. 1998a. Consilience: The Unity of Knowledge. A.A. Knopf, New York.

Wilson, E.O. 1998b. Integrated science and the coming century of the environment. Science, 279:2048–2049.

Windom, II.L., Byrd, J.T., Smith, R.G. Jr., and Huan, F. 1991. Inadequacy of NASQAN data for assessing metal trends in the nation's rivers. Environ. Sci. Technol., 25:1137–1142.

Wing, S., Freedman, S., and Band, L. 2002. The potential impact of flooding on confined animal feeding operations in Eastern North Carolina. Environ. Health Persp., 110:387–391.

Winger, P.V., Lasier, P.J., and Bogenrieder, K.J. 2005. Combined use of rapid bioassessment protocols and sediment quality triad to assess stream quality. Environ. Monitor. Assess. 100:267–295.

Wipf, H.K. and Schmidt, S. 1981. Seveso: An environmental assessment. Pages 255–274 in R.E. Tucker, A.L. Young, and A.P. Gray (eds.) Human and Environmental Risks of Chlorinated Dioxins and Related Compounds. Plenum Press, New York.

Woodman, J.N. and Cowling, E.B. 1987. Airborne chemicals and forest health. Environ. Sci. Technol., 21:120–126.

Woodward, D.F., Brumbaugh, W.G., DeLonay, A.J., Little, E.E., and Smith, C.E. 1994a. Effects on rainbow trout fry of a metals-contaminated diet of benthic macroinvertebrates from the Clark Fork River, Montana. Trans. A. Fish. Soc., 123:51–62.

Woodward, D.F., Farag, A., Bergman, H.L., DeLonay, A.J., Little, E.E., Smith, C.E., and Barrows, F.T. 1994b. Metals-contaminated benthic invertebrates in the Clark Fork River, Montana: Effects on age-0 brown and rainbow trout. Can. J. Fish. Aquat. Sci., 52:1994–2004.

Wright, J.F., Armitage, P.D., Furse, M.T., and Moss, D. 1989. Prediction of invertebrate communities using stream measurements. Reg. Riv.: Res. Manag., 4:147–155.

Wright, J.F., Furse, M.T., and Armitage, P.D. 1993. RIVPACS: A technique for evaluating the biological quality of rivers in the UK. Eur. Water Pollut. Control, 3:15–25.

Wu, J. and Loucks, O.L. 1995. From balance of nature to hierarchical patch dynamics: A paradigm shift in ecology. Quart. Rev. Biol., 70:439–465.

Yang, R.S.H., Thomas, R.S., Gustafson, D.L., Campian, J., Benjamin, S.A., Verhaar, H.J.M., and Mumtaz, M.M. 1998. Approaches to developing alternative and predictive toxicology based on PBPK? PD and QSAR modeling. Environ. Health Persp., 106:1385–1393.

Yarie, J. and Van Cleve, K. 1996. Effects of carbon, fertilizer, and drought on foliar chemistry of tree species of interior Alaska. Ecol. Appl., 6:815–827.

Yeates, G.W., Orchard, V.A., Speir, T.W., Hunt, J.L., and Hermans, M.C.C. 1994. Impact of pasture contamination by copper, chromium, arsenic timber preservative on soil biological activity. Biol. Fertil. Soils, 18:200–208.

Yerushalmy, J. and Palmer, C.E. 1959. On the methodology of investigations of etiologic factors in chronic disease. J. Chronic Dis., 10:27–40.

Yoder, C.O. and Rankin, E.T. 1995a. Biological response signatures and the area of degradation value: New tools for interpreting multi-metric data. Pages 263–286 in W.S. Davis and T.P. Simon (eds.) Biological Assessment and Criteria: Tools for Water Resource Planning and Decision Making. Lewis Publishers, Boca Raton, FL.

Yoder, C.O. and Rankin, E.T. 1995b. Biological criteria program development and implementation. Pages 109–144 in W.S. Davis and T.P. Simon (eds.) Biological Assessment and Criteria: Tools for Water Resource Planning and Decision Making. Lewis Publishers, Boca Raton, FL.

Yoder, C.O. and Rankin, E.T. 1998. The role of biological indicators in a state water quality management process. Environ. Monitor. Assess., 51:61–88.

Yosioka, Y., Ose, Y., and Sato, T. 1986. Correlation of five test methods to assess chemical toxicity and relation to physical properties. Ecotoxicol. Environ. Saf., 12:15–21.

Yount, J.D. and Niemi, G.J. 1990. Recovery of lotic communities and ecosystems following disturbance: Theory and application. Environ. Manag., 14(5):515–516.

Zak, D.R., Holmes, W.E., Finzi, A.C., Norby, R.J., and Schlesinger, W.H. 2003. Soil nitrogen cycling under elevated CO_2: A synthesis of forest FACE experiments. Ecol. Appl., 13:1508–1514.

Zarull, M.A., Hartig, J.H., and Maynard, L. 1999. Ecological benefits of contaminated sediment remediation in the Great Lakes basin. Great Lakes Water Quality Board, International Joint Commission, Windsor, Ontario.

Zeeman, M.G. 1995. Ecotoxicity testing and estimation methods developed under Section 5 of the Toxic Substances Control Act (TSCA). Pages 703–715 in G. Rand (ed.) Fundamentals of Aquatic Toxicology: Effects, Environmental Fate, and Risk Assessment. Taylor & Francis, Washington, DC.

Zelles, L., Scheunert, I., and Korte, F. 1986. Comparison of methods to test chemicals for side effects on soil microorganisms. Ecotoxicol. Environ. Saf., 12:53–69.

Zheng, J. and Frey, H.C. 2005. Quantitative analysis of variability and uncertainty with known measurement error: Methods and case study. Risk Anal., 25:663–675.

Zolezzi, M., Cattaneo, C., and Tarazona, J.V. 2005. Probabilistic ecological risk assessment of 1,2,4-trichlorobenzene at a former industrial contaminated site. Environ. Sci. Technol., 39:2920–2926.

Index